Interaction of
Atoms and Molecules
with Solid Surfaces

PHYSICS OF SOLIDS AND LIQUIDS

AMORPHOUS SOLIDS AND THE LIQUID STATE
Edited by Norman H. March, Robert A. Street, and Mario P. Tosi

CHEMICAL BONDS OUTSIDE METAL SURFACES
Norman H. March

CRYSTALLINE SEMICONDUCTING MATERIALS AND DEVICES
Edited by Paul N. Butcher, Norman H. March, and Mario P. Tosi

ELECTRON SPECTROSCOPY OF CRYSTALS
V. V. Nemoshkalenko and V. G. Aleshin

FRACTALS
Jens Feder

**INTERACTION OF ATOMS AND MOLECULES
WITH SOLID SURFACES**
Edited by V. Bortolani, N. H. March, and M. P. Tosi

MANY-PARTICLE PHYSICS, Second Edition
Gerald D. Mahan

ORDER AND CHAOS IN NONLINEAR PHYSICAL SYSTEMS
Edited by Stig Lundqvist, Norman H. March, and Mario P. Tosi

THE PHYSICS OF ACTINIDE COMPOUNDS
Paul Erdös and John M. Robinson

**POLYMERS, LIQUID CRYSTALS, AND LOW-DIMENSIONAL
SOLIDS**
Edited by Norman H. March and Mario P. Tosi

THEORY OF THE INHOMOGENEOUS ELECTRON GAS
Edited by Stig Lundqvist and Norman H. March

Interaction of Atoms and Molecules with Solid Surfaces

Edited by

V. Bortolani

University of Modena
Modena, Italy

N. H. March

University of Oxford
Oxford, England

and

M. P. Tosi

International Center for Theoretical Physics
Trieste, Italy

Plenum Press • **New York and London**

0393245x

PHYSICS

Library of Congress Cataloging-in-Publication Data

Interaction of atoms and molecules with solid surfaces / edited by
V. Bortolani, N.H. March, and M.P. Tosi.
     p.   cm. -- (Physics of solids and liquids)
  Includes bibliographical references and index.
  ISBN 0-306-43424-5
  1. Surfaces (Physics) 2. Surface chemistry. 3. Atoms.
4. Molecules. I. Bortolani, V. II. March, Norman H. (Norman
Henry), 1927- . III. Tosi, M. P. IV. Series.
QC173.4.S94I56 1990
530.4'27--dc20   90-7272
                                          CIP

© 1990 Plenum Press, New York
A Division of Plenum Publishing Corporation
233 Spring Street, New York, N.Y. 10013

Printed in the United States of America

Contributors

C. M. Bertoni, Dipartimento di Fisica, Seconda Università di Roma "Tor Vergata," 00173 Rome, Italy

A. M. Bradshaw, Fritz-Haber-Institut der Max-Planck-Gesellschaft, D-1000 Berlin 33, Federal Republic of Germany

A. E. DePristo, Department of Chemistry, Iowa State University, Ames, Iowa 50011, USA

M. C. Desjonquères, IRF/DPhG/PAS, Centre d'Etudes Nucléaires de Saclay, 91191 Gif-sur-Yvette Cédex, France

M. Djafari Rouhani, Laboratoire Physique des Solides, Université P. Sabatier, 31062 Toulouse Cedex, France

D. Estève, LAAS–CNRS, 31077 Toulouse Cedex, France

R. M. Feenstra, IBM Research Division, T. J. Watson Research Center, Yorktown Heights, New York 10598, USA

F. García-Moliner, Instituto de Ciencia de Materiales, C.S.I.C., 28006 Madrid, Spain

T. B. Grimley, Donnan Laboratories, The University of Liverpool, Liverpool L69 3BX, England

A. Hamnett, Inorganic Chemistry Laboratory, University of Oxford, Oxford OX1 3QR, England

S. Holloway, Surface Science Research Centre, University of Liverpool, Liverpool L69 3BX, England

H. Ibach, Institut für Grenzflächenforschung und Vakuumphysik, Kernforschungs-anlage Jülich, D-5170 Jülich, Federal Republic of Germany

J. E. Inglesfield, SERC Daresbury Laboratory, Warrington, England

A. L. Johnson, National Institute of Standards and Technology, Gaithersburg, Maryland 20899, USA. Present address: Department of Chemistry, Cambridge University, Lensfield Road, Cambridge, England.

S. A. Joyce, National Institute of Standards and Technology, Gaithersburg, Maryland 20899, USA

W. Kohn, Department of Physics, University of California at Santa Barbara, Santa Barbara, California 93106, USA

J. Lapujoulade, Centre d'Etudes Nucléaires de Saclay, Service de Physique des Atomes et des Surfaces, 91191 Gif-sur-Yvette Cedex, France

B. I. Lundqvist, Institute of Theoretical Physics, Chalmers University of Technology, S-412 96 Göteborg, Sweden

T. E. Madey, Department of Physics and Astronomy, Rutgers, The State University, Piscataway, New Jersey 08855, USA

N. H. March, Theoretical Chemistry Department, University of Oxford, Oxford OX1 3UB, England

J. B. Pendry, Imperial College, London, London SW7 2BZ, England

D. Spanjaard, Laboratoire de Physique des Solides, Université Paris-Sud, 91405 Orsay Cédex, France

G. Wahnström, Institute of Theoretical Physics, Chalmers University of Technology, S-412 96 Göteborg, Sweden

Preface

There is considerable interest, both fundamental and technological, in the way atoms and molecules interact with solid surfaces. Thus the description of heterogeneous catalysis and other surface reactions requires a detailed understanding of molecule–surface interactions.

The primary aim of this volume is to provide fairly broad coverage of atoms and molecules in interaction with a variety of solid surfaces at a level suitable for graduate students and research workers in condensed matter physics, chemical physics, and materials science. The book is intended for experimental workers with interests in basic theory and concepts and had its origins in a Spring College held at the International Centre for Theoretical Physics, Miramare, Trieste.

Valuable background reading can be found in the graduate-level introduction to the physics of solid surfaces by Zangwill[1] and in the earlier works by Garcia-Moliner and Flores[2] and Somorjai.[3] For specifically molecule–surface interactions, additional background can be found in Rhodin and Ertl[4] and March.[5]

V. Bortolani
N. H. March
M. P. Tosi

References

1. A. Zangwill, *Physics at Surfaces*, Cambridge University Press, Cambridge (1988).
2. F. Garcia-Moliner and F. Flores, *Introduction to the Theory of Solid Surfaces*, Cambridge University Press, Cambridge (1979).
3. G. A. Somorjai, *Chemistry in Two Dimensions: Surfaces*, Cornell University Press, Ithaca, New York (1981).
4. T. N. Rhodin and G. Ertl, *The Nature of the Surface Chemical Bond*, North-Holland, Amsterdam (1979).
5. N. H. March, *Chemical Bonds outside Metal Surfaces*, Plenum Press, New York (1986).

Contents

4. Basic Vibrational Properties of Surfaces   71

F. García-Moliner

5. Electronic Structure of Metal Surfaces   117

J. E. Inglesfield

6. Basic Structural and Electronic Properties of Semiconductor Surfaces   155

C. M. Bertoni

9. Electronic Theory of Chemisorption **255**

D. Spanjaard and M. C. Desjonquères

19. Growth Processes at Surfaces: Modeling and Simulations **657**

M. Djafari Rouhani and D. Estève

Chemical Bonds outside Solid Surfaces

N. H. March

1.1. Introduction

This chapter deals with the changes induced in a chemical bond when it is brought up from infinite separation to a finite distance from a planar solid surface. Attention will focus mainly on the chemical bond parallel to the surface, though some discussion will be included of the perpendicular configuration and, indeed, the bond at an arbitrary orientation to the surface.

Clearly, the surface might be a whole variety of materials: ionic, covalently bonded networks such as Si or Ge, metallic such as Pt, or semimetallic such as graphite. Furthermore, the results will depend not only on the nature of the surface, whether insulating, semiconducting, or metallic, but on the question as to whether the electron cloud associated with the chemical bond overlaps appreciably with the electron distribution of the solid surface. This serves to give us a gross classification into two regimes:

1. Physisorption, where there is negligible overlap of electronic charge distributions.
2. Chemisorption, where the molecular electrons overlap with the valence or conduction electrons in the planar surface.

It is natural to begin with the physisorption regime and, to be definite, let us consider first a metal surface.

N. H. March • Theoretical Chemistry Department, University of Oxford, Oxford OX1 3UB, England.

1.2. Physisorption on a Planar Conducting Surface

We shall consider two cases below which, though the same general conclusions emerge from both, require different treatment. The first is when the bond is built from atoms with rather well-defined cores at the equilibrium separation. Then we shall deal with the H_2 molecule; for the present purposes the Heitler–London treatment proves amenable to generalization to account for the proximity of the solid surface.

1.2.1. Bond in a Molecule with Well-Defined Cores

The usual potential energy curve of the homonuclear molecule with which we shall specifically be concerned will be represented by a Lennard-Jones 6–12 potential. The interaction energy, $\Delta E(R)$ say, between the pair of like atoms at separation R is therefore given by

$$\Delta E(R) = \frac{B}{R^{12}} - \frac{A}{R^6} \qquad (1.2.1)$$

in free space.

The effect of the proximity of a metallic conducting surface on $\Delta E(R)$ is now examined, say when the bond is parallel to the surface. The first term on the rhs of equation (1.2.1) originates from core repulsion, which can be expected to be insensitive to the proximity of the surface. Thus, attention below will focus on the way the surface influences the magnitude of the dispersion force, characterized by the constant A in equation (1.2.1).

To see how changes in A affect properties of the bond, we calculate the equilibrium bond length R_e for which $d\Delta E/dR = 0$. It then follows almost immediately from equation (1.2.1) that

$$R_e = (2B/A)^{1/6} \qquad (1.2.2)$$

which shows, for assumed fixed B, that as A changes from its free-space value, R_e is pretty insensitive to modest changes. However, if the result (1.2.2) is now substituted back into equation (1.2.1) in order to calculate the well depth, then one readily finds that

$$\Delta E|_{min} = -A^2/4B \qquad (1.2.3)$$

which depends on the square of A. Hence the well depth is plainly quite sensitive to changes in A brought about by the presence of the solid surface.

The calculation of the dispersion force in the presence of a planar conducting surface will now be treated. The reason why the constant A will be modified when the bond is brought up to the metal surface is readily recognized. Charge fluctuations in the molecule will lead to a response of the itinerant electrons in the metal, which will in turn induce effects in the molecule.

Following pioneering work by McLachlan,[1] Mahanty and March[2] calculated A in the presence of the conductor using essentially the Lifshitz theory of dispersion forces. If the dimensionless variable

$$s = \frac{2(\text{Distance } z \text{ from surface})}{\text{Diatomic bond length}} \qquad (1.2.4)$$

is introduced, then one can write

$$A = A_{\text{London}} F(s) \qquad (1.2.5)$$

where the free-space constant A_{London} may be expressed in the form

$$A_{\text{London}} = \tfrac{3}{4} \hbar \alpha_1(0) \alpha_2(0) \qquad (1.2.6)$$

with $\alpha_j(\omega)$ the polarizability of the jth constituent atom at frequency ω.

By anticipating the result of the argument sketched in Appendix 1.1 for the bond parallel to a perfect planar conductor, the function $F(s)$ in equation (1.2.5) is given explicitly by Mahanty and March[2] as

$$F(s) = 1 + \frac{1}{(1+s^2)^3} - \frac{4}{3} \frac{(1+s^2/4)}{(1+s^2)^{5/2}} \qquad (1.2.7)$$

This function varies monotonically from $\tfrac{2}{3}$ at $s = 0$ to 1 as $s \to \infty$. Insertion of the maximum reduction of A by $\tfrac{2}{3}$ into the equilibrium bond length leads to a small increase of a few percent. But clearly, the same assumed reduction factor will make the well shallower than its free-space form, according to equation (1.2.3), by a factor of $\tfrac{4}{9}$, which is evidently the maximum reduction possible with the above simple model.

1.2.2. Application to Coverage Dependence of Desorption Energy of Xe on Pt

Evidence in favor of a reduction in the well depth discussed above comes from the work of Redondo et al.[3] They used free-space interatomic potentials to discuss desorption energy as a function of coverage for Xe on W. Their conclusion was that the desorption energy calculated from these free-space potentials varied more strongly as a function of coverage than did that extracted from these potentials.

Subsequently, Joyce et al.[4] used the Mahanty–March lateral interaction discussed above, and Figure 1.1 shows the marked reduction in the variation of desorption energy with coverage from the free-space potential prediction. This is clearly in the right direction to improve the agreement with experiment. Joyce et al. also calculated thermal desorption spectra, to compare with the measurements of Opila and Gomer.[5] Quite reasonable agreement with experiment resulted.

The conclusion from this work is that the data support the theory of lateral interactions outside a metal surface in the physisorption regime.

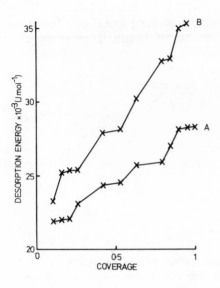

Figure 1.1. Variation of desorption energy with coverage. Upper curve B shows result with free-space Xe–Xe lateral interaction. Lower curve A is with image effects due to perfect conductor included in lateral interaction (after Joyce *et al.*[4]).

We note here that Mahanty and March[2] also considered the perpendicular configuration. The function F in equation (1.2.5) is now an enhancement factor, with a maximum value of $\frac{8}{3}$ and minimum value unity. However now, unlike the parallel configuration, interaction of individual atoms with the surface is very different and must be included. One anticipates in this perpendicular configuration that the bond length will be reduced, while the bond strength will be more dependent on detail for this orientation than for the parallel case. Finally, Mahanty and March considered a bond at arbitrary orientation; this work was extended subsequently by Mahanty *et al.*[6] to whose paper the interested reader is referred for full details. It may be noted here, however, that these workers also took into account surface plasmon excitations.

1.2.3. Case of a H_2 Molecule: Modification of Heitler–London Theory to Include Image Effects

As already mentioned, the previous argument is only useful when cores are well defined at the equilibrium separation. The interesting case of the H_2 molecule therefore requires separate treatment. This was given by Flores *et al.*,[7] their argument resting on two assumptions: (1) the Heitler–London theory affords a useful starting point, for a semiquantitative study; (2) the Coulombic interaction $1/|\mathbf{r}_a - \mathbf{r}_b|$ between two charges a and b is to be replaced, in the presence of a conducting surface, by

$$\frac{1}{|\mathbf{r}_a - \mathbf{r}_b|} - \frac{1}{|\mathbf{r}_a - \mathbf{r}_b^I|}$$

where I refers to an image distance.

The approach subsequently follows closely the traditional Heitler–London calculation. Thus the molecular energy E is written as

$$E = 2E_{1s} + \frac{J}{1 + S^2} + \frac{K}{1 + S^2} \qquad (1.2.8)$$

and one must then examine the behavior of the Coulomb piece J and the exchange term K in the presence of the metal surface, S being the overlap integral between hydrogen 1s wave functions.

Calculation shows J to play an unimportant role and the correction to the Heitler–London energy due to the metal surface may be expressed as

$$\delta E = \delta K / (1 + S^2) \qquad (1.2.9)$$

With the interaction between charges modified by the image term as in assumption (2) above, δK was calculated by Flores et al.[7] Again, in the parallel configuration, the bond is weakened and (slightly) lengthened.

1.2.4. Insulating and Semiconducting Surfaces as Well as Metals

More briefly, we shall now comment on lateral interactions in the physisorption regime for some insulating and semiconducting surfaces in addition to metals, the details being relegated to Appendix 1.3. The account follows fairly closely that given by Girard and Girardet.[8]

Considerable effort has been expended at identifying all the contributions of the interaction of adatoms with both dielectric and metallic substrates according to the adsorbate coverage rate. Bruch[9] has discussed the various dominant terms in the lateral interaction and has analyzed the dominance of the substrate-mediated energy for rare gas monolayers adsorbed on noble metals (cf. Section 1.2.2) or the semimetal graphite.

As Girard and Girardet[8] emphasize, a fully quantitative determination of the latter energy per adatom requires the calculation of (1) the corrugation energy of the surface, (2) all the substrate-mediated energies including the long-range dispersion contribution discussed in Appendix 1.3, the long- and short-range electrostatic terms, and the elastic distortion of the surface, (3) the proper energy of the adatom in the monolayer including many-body effects, and (4) the influence of finite temperature.

Hence, it is difficult to discuss the relative importance of the long-range substrate-mediated contribution if the monolayer configuration is not known. One can, however, assert that the corrugation decreases when the metal face becomes more dense and, in contrast, that the substrate-mediated long-range energy increases with the atomic density of the face and with the size of the adatom. This fact necessitates an accurate determination of the substrate-mediated term for graphite and noble or transition metals with the more dense (0001) and (111) faces, respectively. Indeed, it appears that this term, in this case, is dominant in determining the structure of the adsorbed monolayer.

For molecules adsorbed on ionic surfaces, the substrate-mediated dispersion energy remains weak when compared with the multipolar induction terms. These terms depend fundamentally on the relative orientations of the adsorbed molecules and on the orientation of the molecules in their adsorption sites. For ionic crystals, the corrugation energy is generally greater than for metals and the study of the relative magnitude of the contributions to the lateral energy is required to determine the monolayer structure.

A remark should be added concerning other interaction mechanisms which can be classified in the substrate-mediated energy (see, for example, Bruch[9]). For instance, the interaction between dipole moments induced by the proximity of the metal in the case of rare gas atoms needs a fuller study. In a similar way, the quadrupole and higher multipole moments would modify the form and magnitude of the multipolar induction energy between adsorbed molecules.

1.3. Chemisorption on Metallic Surfaces

Having discussed especially lateral interactions in the physisorption range at some length, we must turn now to the case when the metal–chemical bond interaction is strong, due to appreciable overlap of the molecular electron distribution with the itinerant electrons in the metal.

Then the question immediately arises: how does a "perturbation" embedded in an electron gas affect the metal electron distribution? We shall see below that the answer to this question has an important bearing on the nature of the lateral interactions between atomic or molecular species outside a conducting surface in the chemisorption regime.

1.3.1. Elementary Metal Models

Let us start from electrons in a one-dimensional box of length l, with origin $z = 0$ at one edge of this box. Evidently the electron density $\rho(z)$ for N singly filled levels is given by the sum of the squares of the normalized wave functions:

$$\rho(z) = \frac{2}{l} \sum_{n=1}^{N} \sin^2 \frac{n\pi z}{l} \qquad (1.3.1)$$

This can be summed exactly to yield, with $y = \pi z / l$,

$$\rho(z) = \frac{N + \frac{1}{2}}{l} - \frac{1}{2l} \frac{\sin (2N + 1)y}{\sin y} \qquad (1.3.2)$$

We shall be interested in the "perturbation" induced by the edge of the box at $z = 0$ deep in the Fermi gas: this can be obtained by taking the limit when the length of the box l tends to infinity. One can either take this limit, such that $N/l \to \rho_0$, the constant "bulk" density, in equation (1.3.2), or quite equivalently one can return to equation (1.3.1) and replace the summation by an integration. Either way, one readily obtains the desired result

$$\rho(z) = \rho_0 \left(1 - \frac{\sin 2k_f z}{2k_f z}\right) \qquad (1.3.3)$$

where we have introduced, via the usual phase-space argument, the Fermi wave number k_f related to ρ_0 for singly filled levels by $k_f = \pi \rho_0$.

Two points are to be noted from the elementary model result (1.3.3):

1. The electron density ρ rises from zero at $z = 0$, the position of the infinite barrier, to its asymptotic bulk value ρ_0, oscillating about its asymptote with wavelength π/k_f; i.e., the long-range oscillations are determined by the de Broglie wavelength of an electron at the Fermi surface.
2. The density ρ first reaches its bulk value ρ_0 after a distance given by $2k_f z = \pi$, and so one can say that the "thickness" of the surface "spill-out" of the electron density distribution in a simple metal is also about π/k_f which is approximately 1 Å.

In a semi-infinite three-dimensional electron gas with an infinite barrier, equation (1.3.3) is replaced by the result

$$\rho(z) = \rho_0 \left(1 + 3 \frac{\cos(2k_f z)}{(2k_f z)^2} - \frac{\sin(2k_f z)}{(2k_f z)^3}\right) \qquad (1.3.4)$$

which was first obtained by Bardeen.[10] In Appendix 1.2, his model is considered in more detail following the work of Moore and March.[11]

1.3.2. Relation of Long-Range Oscillations in Perturbed Electron Density to Lateral Interactions between Chemisorbed Species

Of course, we want to relate what has been learnt about long-range oscillations induced by perturbations in a Fermi gas of itinerant metal electrons to the (indirect) lateral interactions between chemisorbed species outside a planar metal surface. The detailed argument is given in Appendix 1.2, but we note here that for a weak perturbing potential $V(\mathbf{r})$ the charge $\rho(\mathbf{r}) - \rho_0$ displaced by the perturbation follows from linear response theory as

$$\Delta \rho \equiv \rho(\mathbf{r}) - \rho_0 = \int F(\mathbf{r r'}) V(\mathbf{r'}) \, d\mathbf{r'} \qquad (1.3.5)$$

The kernel F must be calculated for the appropriate situation. For a perturbation in a bulk electron gas, function F was obtained by March and Murray[12] in the form

$$F = \frac{-k_f^2}{2\pi^3} \frac{j_1(2k_f t)}{t^2} \qquad (1.3.6)$$

with $t \equiv |\mathbf{r} - \mathbf{r'}|$.

If $V(\mathbf{r}') = \lambda \delta(\mathbf{r}')$, it follows immediately from equation (1.3.5) that

$$\Delta\rho(\mathbf{r}) \sim \frac{j_1(2k_f r)}{r^2} \qquad \text{where } j_1(x) = \frac{\sin x - x \cos x}{x^2} \qquad (1.3.7)$$

or, at sufficiently large r,

$$\Delta\rho(r) \sim \frac{\text{constant} \cos (2k_f r)}{r^3} \qquad (1.3.8)$$

This form is to be compared with the form (1.3.4) or (1.3.3): all exhibit long-range oscillations, but one must be careful about the inverse powers of distance; this is the essential point we shall stress below. Actually, for a test charge perturbation, equation (1.3.5) can be combined with Poisson's equation and leads to a self-consistent field problem. When solved, precisely the form (1.3.8) is regained, with the same oscillatory decay with distance for the screened potential round the test charge. The so-called electrostatic model then leads to the interaction energy between test charges at separation R in the form

$$\Delta E(R) \sim \frac{A \cos 2k_f R}{R^3} \qquad (1.3.9)$$

As Corless and March[13] demonstrated, equation (1.3.9) is valid for charges interacting in a bulk metal. Near a surface, one can use the Bardeen model developed in Appendix 1.2 to recalculate the function F in equation (1.3.5). Then, for a pair of charges parallel to the surface, the argument of Appendix 1.2 reveals a shorter-range oscillatory interaction, still valid asymptotically, having the form

$$\Delta E(R)\big|^{\text{linear response}}_{\text{chemisorption}} \sim \frac{A \cos 2k_f R}{R^5} \qquad (1.3.10)$$

Following pioneering work by Grimley and by Einstein and Schrieffer, the linear response form (1.3.10) was derived independently by Flores et al.[7] and by Lau and Kohn.[14]

Just as Friedel and co-workers showed for the bulk metal result (1.3.9), when the "perturbation" scatters the itinerant electrons strongly, there is a phase shift ϕ introduced into the asymptotic form (1.3.10) to yield for the indirect lateral interaction

$$\Delta E(R)\big|_{\text{chemisorption}} \sim \frac{A \cos (2k_f R + \phi)}{R^5} \qquad (1.3.11)$$

Below, we shall mention two applications of the indirect long-range oscillatory interaction between chemisorbed species: the first for H on Pt and the second for CO molecules interacting also outside a Pt surface.

1.3.3. Hydrogen Chemisorbed on a Pt Surface

Flores et al.[15] considered the indirect lateral interaction (1.3.11) while attempting to interpret some experimental findings for H on Pt(111). By using such a pair potential description of lateral H–H interactions, these workers were able to interpret experimental information on surface stoichiometries and heat of adsorption and desorption. In order to apply the above result (1.3.11), one wished to establish that the H atoms occupy sites which enable H to lie in a region of high electron density. Flores et al. cited such evidence as coming from (1) magnetic resonance measurements, (2) infrared data, and (3) neutron inelastic scattering. Elaborating (1) to illustrate the point, the magnetic resonance data show a state of adsorbed H with a large chemical shift in the opposite direction to the Knight shift. This can be interpreted as due to a proton lying in the region of high d-electron density from the metal atoms.

Flores et al. employed equation (1.3.11) for $\Delta E(R)$ and reached the following conclusions:

1. One can derive only one surface structure consistent with the requirement that the minimum value of the pair potential must be associated with all possible near-neighbor distances and the Pt/H ratios found for the (111) plane of Pt.
2. The existence of two surface states of different bonding energies is required by the potential; however, they are structurally and energetically equivalent at complete coverage of both sites. (This is in agreement with the convergence of the thermal desorption energies noted by Christmann et al.[16])
3. The structure of the first H sites, those adsorbed on the clean surface, is consistent with the LEED studies of Weinberg and Merrill.[17]

A fuller discussion of the work of Flores et al.[15] can be found in the review by Burch;[18] see also March.[19] The work lends strong support to the importance of the asymptotic lateral interaction (1.3.11).

1.3.4. Lateral Interaction between Two CO Molecules Chemisorbed on Pt

Joyce et al.[20] have considered the same type of lateral interaction (1.3.11) in attempting to interpret thermal desorption measurements as a function of coverage for CO chemisorbed on Pt.

In particular, these workers set up a model based on the molecular $2\pi^*$ and 5σ levels lying within the metal bands (cf. Figure 1.2); the former above the Fermi level and the latter well below. Broadening of the former into a virtual bound state which can overlap the Fermi level can lead to a situation in which it is demonstrated that one can get strong indirect interactions between CO molecules. The amplitude β and phase ϕ of the long-range oscillatory form $\beta \cos(2k_f R + \phi)/R^5$ discussed in detail above can thereby be estimated. Table 1.1 presents the values of ϕ and β for an assumed coupling strength V between admolecule and substrate. The phase ϕ is seen to be large, testifying to strong scattering of metal electrons by

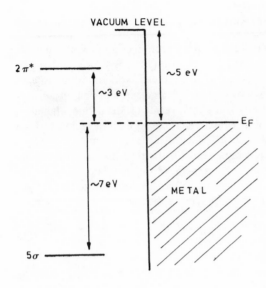

Figure 1.2. $2\pi^*$ and 5σ levels of CO molecules chemisorbed on a Pt metal surface, positioned relative to Fermi energy E_F of metal (after Joyce et al.[20]).

the adsorbate. We shall later briefly discuss the amplitudes β recorded also for various plausible values of the coupling strength in Table 1.1.

The results of the above model have been compared by Joyce et al.[20] with an analysis they made of thermal desorption as a function of coverage for CO on Pt. The strong scattering situation predicted by the above model is immediately revealed, with a phase shift between $\pi/4$ and $\pi/2$ and a large amplitude β. To press the comparison with the model, a least-squares fit of the experimental decrease in the desorption energy performed by Joyce et al. gave $\phi = 0.95$ and $\beta = 1.0 \times 10^7$ J mol^{-1} Å^5. Though the agreement for β is only semiquantitative, it seems remarkable that such a simple model gives the gist of the experimental findings.

1.4. Molecular versus Dissociative Adsorption on Transition Metal Surfaces

We conclude with a discussion of trends in chemisorption, of a more general kind. Specifically, one wants to analyze the behavior of small molecules like CO, N_2, NH_3, etc. on transition metal surfaces. A pattern emerges from an analysis of

Table 1.1. Values of Amplitude β and Phase ϕ in Lateral Long-Range Oscillatory Interaction between CO Molecules on Pt as a Function of Coupling Parameter V (After Joyce et al.[20])

V (eV)	3.5	4.0	4.5	5.0
β (J mol^{-1} Å^5)	0.26×10^7	0.41×10^7	0.58×10^7	0.79×10^7
ϕ	0.75	0.98	1.19	1.39

data on various adsorbate-substrate systems. For a given molecule, metals to the left of the line dividing the Periodic Table dissociate the molecule, while those to the right adsorb it with its chemical bond(s) intact (see Table 1.4 below).

Below, we shall present a chemical approach which is based on Pauling's treatment[21] of bond energies in free molecules. This treatment was adapted by Eley[22] and other workers to the chemisorbed state. Specifically, Eley obtained estimates of adsorption heats by employing Pauling's equation in calculating the metal atom-adsorbate atom bond energy denoted by $E(M\text{-}A)$. This quantity is thereby expressed in the form

$$E(M\text{-}A) = \tfrac{1}{2}[E(M\text{-}M) + E(A\text{-}A)] + (x_m - x_A)^2 \tag{1.4.1}$$

where $E(M - M)$ and $E(A\text{-}A)$ are the single bond energies of the metal and diatomic molecule, respectively, while x_m and x_A are the corresponding atomic electronegativities. Equation (1.4.1) is written in electron volts.

It is true that equation (1.4.1) has limitations as a route to the calculation of accurate bond energies, because of its sensitivity to the electronegativity difference. Nevertheless, the work of Flores et al.,[23] utilized in the ensuing discussion, demonstrates that equation (1.4.1) is adequate for the purpose of interpreting the general pattern of dissociative versus molecular adsorption, provided a term due to metal-metal bond breaking is added. To the writer's knowledge, a fully quantitative basis for such a term from wave mechanics has not as yet been given. Therefore Flores et al.[23] use as input data into this term, representing the breaking of metal bonds, a measure of valence as obtained from wave mechanical band-structure calculations. With this addition, Flores et al. demonstrate that equation (1.4.1) is indeed adequate to expose a pattern of behavior for the chemisorption of a given adsorbate on a variety of transition metals.

1.4.1. Reactivity of Metal toward a Molecular Adsorbate

If geometric factors are ignored, as well as metal bond breaking for the time being, the reactivity of a metal toward a molecular adsorbate is determined by:

1. The dissociation energy of the molecule in free space.
2. The strength of the bonds formed between its constituent atoms and the metal.
3. Molecular chemisorption energies.

For most of the systems of interest here, (3) is considerably smaller than (1) or (2), and provided one is focusing on the pattern of behavior rather than fully quantitative work, Flores et al.[23] argue that it may be neglected. Then, for instance in the case of a diatomic molecule, one must clearly compare the energies involved in (1) and (2) above. To study (2), it is important to consider next a tractable, if somewhat oversimplified, way in which bonding in transition metals is to be handled.

1.4.2. Bonding and Valence in Transition Metals

A useful description of bonding in transition metals can be given in terms of overlapping hybrids on neighboring atoms. These hybrids are formed from the s, p, and d orbitals in such a manner that they are directed toward nearest and next-nearest neighbors in the crystal.

1.4.2.1. Resonating Bonds

Let us consider as a specific example the case of a face-centered-cubic (fcc) transition metal for which the 12 nearest neighbors make the major contribution to the bonding. In principle, then, the number of hybrids is six or less. This calls for a description of the ground electronic state in terms of a number of resonating states so that, on average, there is a hybrid pointing toward each of the 12 near-neighbor atoms.

Of course, a direct consequence of this picture is that only a fraction of an electron can be shared in each bond between atoms; this is expressed in Pauling's concept of metallic valence.[21] One then wishes to associate the fractional occupation of the hybrids with some average valence, defined as the total number of electrons used for bonding to the neighboring atoms.

Though the above discussion referred specifically to a fcc crystal, in fact for a hexagonal closed packed (hcp) metal the local coordination number is also 12 while in a body-centered-cubic (bcc) structure there are eight near neighbors and six next-near neighbors. Thus, even for the bcc structure, the assumption of 12 hybrids around each atom should provide a useful average description.

1.4.2.2. Calculation of Average Valence

On returning to the question of the average valence, it is important to choose this consistent with the available knowledge on the electronic charge distribution in the energy bands of transition metals. For in spite of the strong electron–electron correlation that Pauling's ideas correctly incorporate, it is known from density functional theory that correlation can be subsumed into a one-body potential when calculating, in principle exactly, the electronic charge density in the ground state from energy band theory.

Therefore Flores *et al.*[23] made use of the augmented plane wave (APW) calculations of Moruzzi *et al.*[24] for the energy bands of transition metals in order to obtain the number of interstitial electrons per atom in the crystal. This number they defined from the APW calculation as the number of electrons outside the muffin-tin sphere. Their proposal was to correlate this quantity with the number of bonding electrons, the reason being that the localized electrons are not shared in bonds and contribute only to the charge in the immediate vicinity of one atom.

Then it was natural to assume the average valence proportional to the number of interstitial electrons. By choosing the proportionality constant to yield a valence of 1 for Na and 3 for Al with bcc and fcc structures, respectively, the average valences obtained were as recorded in Table 1.2.

Table 1.2. Average Valences Assigned to Transition Metals (After Flores et al.[23])

Ti	V	Cr	Mn	Fe	Co	Ni	Cu
3.4	3.3	3.4	3.5	3.5	3.2	2.9	2.6
Zr	Nb	Mo	Tc	Ru	Rh	Pd	Ag
4.0	4.0	4.2	4.3	4.1	3.6	3.1	2.7

At the time of writing, we are unaware of similar quantitative calculations for the number of interstitial electrons in the third transition series. Therefore, in order to extend their analysis to these elements, Flores et al.[23] assigned the valences of the second-row elements to them since, in general, properties of the third transition series resemble more closely the second series than the first. It is clear, however, that because of this oversimplified treatment of valence in the third series one ought, at most, only to consider the general trends in this case.

It is of some interest to comment briefly on the average valence recorded in Table 1.2 in relation to Pauling's maximum valence. Pauling assigned a valence of 6 to the elements Cr to Ni and $5\frac{1}{2}$ to Cu; these are 2–3 units larger than the proposals of Flores et al.[23] It had been suggested already by Hume-Rothery et al.[25] that the metallic valence is less for the elements of the first long period than it is in the later periods. On the basis of the physical properties of the elements, these workers[25] suggest that V and Cr do not involve more than four bonding electrons per atom and that Mn has a relatively low valence of about two, followed by a rise in passing to Fe. A decrease in valence through group VIII is more evident in the second and third periods. Flores et al. drew attention to the fact that their proposed scheme is supported by the relative importance of the valences exhibited in the chemistry of the transition elements.

Table 1.2 giving the average valence can now be employed to estimate metal–metal bond energies by dividing twice the cohesive energy by the average valence. These values are recorded in Table 1.3. These, it should be added, differ substantially from those given in the much earlier work of Eley et al.,[22,26] who obtained metal–metal bond energies by dividing twice the heat of sublimation by the number of nearest neighbors.

Table 1.3. Metal–Metal Bond Energies in Electron Volts[a]

Ti	V	Cr	Mn	Fe	Co	Ni	Cu
2.89	3.00	2.26	1.66	2.57	2.81	3.12	2.70
Zr	Nb	Mo	Tc	Ru	Rh	Pd	Ag
3.20	3.60	3.06	3.07	3.30	3.18	2.55	2.24
Hf	Ta	W	Re	Os	Ir	Pt	Au
(3.14)	(3.86)	(3.88)	(3.77)	(4.01)	(3.80)	(3.82)	(2.85)

[a] Numbers in parentheses were found using valences as for the second-row series (after Flores et al.[23]).

1.4.3. Admolecule Bonding to a Metal Surface

Given the concept of average valence as above, and the estimated metal–metal bond energies thereby obtained, as in Table 1.3, Flores et al.[23] discussed the bonding at the surface along the following lines.

They regarded each surface atom as having a number of dangling bonds, determined by the crystal structure and the particular crystal face, while the number of electrons occupying these bonds is governed by the average valence, v say, of Table 1.2. For instance, each surface atom on the (111) face of a fcc crystal has three dangling bonds with $v/12$ electrons in each. Due to the similarities of local coordination in fcc, hcp, and bcc crystals referred to above, Flores et al. assumed that this is true for a dangling bond on any face. This implies that in order to form one bond with the adsorbate, other $(1 - v/12)$ metal electrons are required for sharing. More generally, a number of metal bonds, n say, depending on the particular crystal face and on the adsorbate, will have to be broken in order to saturate the metal–adatom bond. Thus equation (1.4.1) can be modified to read

$$E(\text{M-A}) = \tfrac{1}{2}[E(\text{M-M}) + E(\text{A-A})] + (x_m - x_A)^2 - nE(\text{M-M}) \qquad (1.4.2)$$

which was the starting point for the calculations of Flores et al.[23]

Flores et al. consider, on the basis of equation (1.4.2), the energetics of N_2, O_2, CO, and NO chemisorption on transition metals. Without appeal to specific crystal structures, for the reasons given above, Flores et al. assume that, for oxygen and nitrogen, every hybridized atomic orbital can form a bond with the metal surface. For carbon, in contrast, it is not clear how many bonds can be formed with the surface, since the interaction with the surface influences the valence of carbon.

As one example, merely to illustrate trends, Flores et al. write on the basis of the arguments given above that the energy of chemisorption of one nitrogen atom is given by

$$E(\text{M-N}) = 3\{\tfrac{1}{2}[E(\text{M-M}) + E(\text{N-N})] + (x_m - x_N)^2 - (1 - v/12)E(\text{M-M})\}$$
$$(1.4.3)$$

In this equation the factor 3 accounts for the number of bonds formed between nitrogen and the metal. Clearly, equation (1.4.3) is sensitive to the difference $(x_M - x_N)$, as considered in the calculations now to be briefly summarized.

Equation (1.4.3) must only be applied when $E(\text{M-N})$ is greater than that energy obtained by assuming a fractional bond order given by v and no breaking of the metal bonds. Anticipating the results in Table 1.4 below, it is noteworthy that this latter energy is in fact higher for Pt and Ir, and in these two cases it is then the appropriate values that are recorded. For these two metal surfaces, instead of $E(\text{M-N})$ given by equation (1.4.3), the bond energy $\{\tfrac{1}{2}[E(\text{M-N}) + E(\text{N-N})] + (x_m - x_N)^2\}$ must be multiplied by $v/12$, the fractional bond order referred to above.

Table 1.4. Energy of Metal–Nitrogen Atom Bond in Electron Volts (After Flores et al.[23])[a]

Cr^D	Mn^D	Fe^D	Co	Ni^M	Cu^M
7.0	8.3	5.8	5.2	3.8	4.2
Mo^D	Tc	Ru	Rh^M	Pd^M	Ag
6.2	—	4.1	3.4	3.1	5.0
W	Re	Os^M	Ir^M	Pt^M	Au^M
(5.2)	(4.3)	(3.5)	(3.5)	(2.6)	(2.5)

[a] Estimated errors are given by Flores et al.[23] M denotes molecular adsorption while D refers to dissociation. Absence of labeling means no decisive conclusion. (Values in parentheses indicate that the same valences were used in series 3 as in series 2.)

The metal–nitrogen bond energies obtained from equation (1.4.3) combined with data from Tables 1.2 and 1.3 are collected in Table 1.4. These should be compared with 4.8 eV, which is half the dissociation energy of N_2, in order to determine whether dissociation or molecular adsorption is likely to occur.

The solid line in Table 1.4 shows the experimental boundary according to Brodén et al.[27] Agreement between theory and experiment seems satisfactory when one bears in mind the large errors originating from the theoretical estimates, due to the wide extremes of electronegativity quoted by Gordy and Thomas.[28]

1.5. Summary and Some Future Directions

It is clear from this discussion that for fully quantitative results there is obviously no substitute for fairly detailed computations on the specific system under discussion.

Thus, as Appendix 1.3 makes clear, the type of argument presented for physisorption in Section 1.2 needs refinement to take account of the discrete nature of the solid substrate.

There is evidently scope for much further work on indirect lateral interactions mediated by the itinerant metal electrons, along lines laid down for CO–CO interactions in Section 1.3.4. As stressed there, one probably needs to appeal to photoelectron spectroscopy to "place" the molecular energy levels relative to the metal Fermi level. Given such an empirical level diagram, we expect the method outlined in Section 1.3.4 to have wide applicability, even to substantially more complex molecules than considered here.

While conformational studies remain of considerable interest, it is evident for the future that reaction rates of chemical reactions outside solid surfaces remains a challenging area for theoretical chemical physics. Can simple arguments and/or concepts be found which will allow an understanding of the way the lowering of activation barriers can come about in catalyzed reactions? Can likewise concepts be isolated to permit deeper understanding of Sabatier's concept of activated complex? Of course, much computational study related to the above has already been recorded in the literature. What seems needed is a synthesis of this and, more importantly, experiment, using arguments and concepts that combine collective excitations possible for the solid surface and reactants in strong interaction with the above ideas of activation barrier lowering and intermediate complexes.

Appendixes

Appendix 1.1. Image Theory of Dispersion Force in the Presence of a Solid Surface for the Physisorption Regime

Here we sketch the main steps involved in setting up a calculation of function $F(s)$ in equation (1.2.5), which determines the change in the free-space van der Waals constant A_{London} due to the presence of a solid substrate.

The desired result is obtained most readily[2] in terms of the quantity G connecting the electric field $E(r)$ at r to a dipole source $\mu(\omega)$ at r' oscillating with frequency ω, namely

$$E(r) = G(rr'; \omega)\mu(\omega) \tag{A1.1.1}$$

In technical language, G is a "dyadic Green function." What is important in the present context is that, in the presence of a dielectric surface taken to be the (x, y) plane, when both r and r' are outside the dielectric (i.e. $z, z' > 0$), G can be expressed as the sum of two parts:

$$G(rr'; \omega) = G_D(r - r') - \Delta(\omega)G_I(rr') \tag{A1.1.2}$$

Here, the direct part G_D follows from electrostatics and is given by

$$G_D(r - r') = -(\nabla\nabla')\frac{1}{|r - r'|} \tag{A1.1.3}$$

while G_I in equation (A1.1.2) is the indirect contribution, resulting from the mediation of the semi-infinite dielectric medium with dielectric constant $\varepsilon(\omega)$. Application of image theory yields, with $\Delta(\omega) = [\varepsilon(\omega) - 1]/[\varepsilon(\omega) + 1]$,

$$G_I(rr') = -(\nabla\nabla')\frac{1}{|r - r'_{im}|} \tag{A1.1.4}$$

where r'_{im} is evidently the position of the image of the point r'. In the book by Mahanty and Ninham,[29] it is shown from what is essentially the Lifshitz theory of dispersion forces that the interaction energy $E_D(12)$ of the two atoms, to leading order in the polarizabilities $\alpha_1(\omega)$ and $\alpha_2(\omega)$, can be written as

$$E_D(12) = \frac{-\hbar}{4\pi} \int_{-\infty}^{\infty} d\xi\, \alpha_1(i\xi)\alpha_2(i\xi)\, \text{Tr}\, G(21)G(12)] \tag{A1.1.5}$$

where, in the trace of the Green function product, the notation $1 \equiv \mathbf{r}_1$, $2 \equiv \mathbf{r}_2$ has been used.

Mahanty and March[2] studied this method for the case of a diatomic molecule (or a dimer) parallel and perpendicular to a planar metal surface. Furthermore, the solid was taken to be an ideal conductor corresponding to the choice $\Delta = 1$. Mahanty et al.[6] have subsequently considered the effects of a dielectric function $\varepsilon(\omega)$ into which is subsumed electronic excitations in the form of surface plasmons. With $\Delta = 1$, equation (A1.1.5) reduces to the form

$$E_D(12) = \tfrac{1}{6} \operatorname{Tr}\left[\mathbf{G}(21)\mathbf{G}(12)\right]\left[\frac{3\hbar}{2\pi}\int_{-\infty}^{\infty} \alpha_1(i\xi)\alpha_2(i\xi)\, d\xi\right] \qquad (A1.1.6)$$

The last bracketed term on the rhs of this equation is the constant A of equation (1.2.1); the equivalence of the two is known from the work of Margenau and Kestner.[30]

The traces required to evaluate E_D from equation (A1.1.6) are given explicitly in equations (11) and (12) of Mahanty and March.[2] One can thus immediately determine E_D for parallel and perpendicular configurations. In the former case, the expression for $F(s)$ quoted in equation (1.2.7) is the desired result.

In the perpendicular configuration, the atoms are assumed to be situated at z_1 and z_2. Equation (A1.1.6) then yields

$$E_D(12) = E_{\text{London}}(z_{12})F_\perp(z_1/z_{12}) \qquad (A1.1.7)$$

where

$$F_\perp(t) = 1 + \frac{1}{(1+2t)^6} + \frac{2}{3(1+2t)^3} \qquad (A1.1.8)$$

Here $F_\perp(z_1/z_{12})$ is an enhancement factor, decreasing monotonically from $\tfrac{8}{3}$ at $z_1/z_{12} = 0$ to 1 as z_1/z_{12} tends to infinity.

Finally, for the case of arbitrary orientation of the chemical bond relative to the planar metal surface, the reader may consult the account of Mahanty et al.;[6] see also March.[19]

Appendix 1.2. Linear Response Function for Bardeen's Infinite Barrier Model of a Planar Metal Surface

Moore and March[11] calculated the linear response function of the Bardeen model[10] in which a semi-infinite electron gas is confined by an infinite barrier following earlier work by Brown et al. The electron density ρ in this model has already been given in equation (1.3.4). Specifically, Moore and March[11] determine the function $F(\mathbf{rr}')$ in equation (1.3.5), its bulk form corresponding to a homogeneous electron gas having the form (1.3.6).

The tool employed in the above work was the so-called canonical density matrix $C(\mathbf{rr}'\beta)$ defined by

$$C(\mathbf{rr}'\beta) = \sum_{\text{all } i} \psi_i^*(\mathbf{r})\psi_i(\mathbf{r}') \exp(-\beta E_i), \qquad \text{where } \beta = (k_B T)^{-1} \quad (A1.2.1)$$

where ψ_i and E_i are the wave functions and corresponding energies of the barrier model. The appropriate Schroedinger equation $H\psi = E\psi$ yields for $C(\mathbf{r}\mathbf{r}'\beta)$ the Bloch equation

$$H_r C(\mathbf{r}\mathbf{r}'\beta) = -\frac{\partial C}{\partial \beta} \qquad (A1.2.2)$$

It also follows from definition (A1.2.1) that

$$C(\mathbf{r}\mathbf{r}'0) = \delta(\mathbf{r} - \mathbf{r}') \qquad (A1.2.3)$$

which expresses the completeness relation for the eigenfunctions $\psi_i(\mathbf{r})$.

Moore and March[11] exploit the analogy between the classical equation of heat conduction[31] and the Bloch equation (A1.2.2) to calculate the canonical density matrix analytically for noninteracting electrons in both infinite and finite-step models of a metal surface.

Below, results will be given only for the infinite barrier limit, the z axis being taken perpendicular to the planar metal surface. One then finds that

$$C(\mathbf{r}\mathbf{r}'\beta) = (2\pi\beta)^{-1} \exp\left(-|\mathbf{x} - \mathbf{x}'|^2/2\beta\right) f(zz'\beta) \qquad (A1.2.4)$$

where $\mathbf{x}$ denotes a two-dimensional vector in the (x, y) plane taken to lie in the planar metal surface, while for free electrons

$$f_{\text{free}}(zz'\beta) = (2\pi\beta)^{-1/2} \exp\left(-|z - z'|^2/2\beta\right) \qquad (A1.2.5)$$

Substitution of equation (A1.2.4) into (A1.2.2) readily yields a differential equation for f in the case of general potential energy $V(z)$, equation (A1.2.5) then being the solution for $V = 0$. For the infinite barrier limit, when $z' < 0$ and $z \le 0$ the solution is readily verified to be

$$f(zz'\beta) = \frac{\exp\left[-|z - z'|^2/2\beta\right]}{(2\pi\beta)^{1/2}} - \frac{\exp\left[-(z + z')^2/2\beta\right]}{(2\pi\beta)^{1/2}} \qquad (A1.2.6)$$

In the other regions, $f = 0$ in this infinite barrier limit.

Moore and March also record the result for the linear response function $F(\mathbf{r}\mathbf{r}')$, their infinite barrier result being explicitly

$$F(\mathbf{r}\mathbf{r}'E_f) = \begin{cases} -\frac{k_f^2}{(2\pi)^3}\left(\frac{j_1(2k_f s)}{s^2} + \frac{j_1(2k_f s')}{s'^2} - \frac{2j_1[2k_f(s + s')]}{ss'}\right) \\ \qquad \text{for } z, z' < 0 \\ \\ \qquad 0 \quad \text{otherwise} \end{cases} \qquad (A1.2.7)$$

where $s = |\mathbf{r} - \mathbf{r}'|$ and $s' = [|\mathbf{x} - \mathbf{x}'|^2 + (z + z')^2]^{1/2}$. Here again $\mathbf{x}$ and $\mathbf{x}'$ are vectors in the plane parallel to the surface.

This result (A1.2.7), inserted into equation (1.3.5), leads after some calculation to an interaction energy $\Delta E(R)$ between a pair of test charges at separation R and parallel to the metal surface, with asymptotic form

$$\Delta E(R) \sim \frac{A \cos (2k_f R)}{R^5} \tag{A1.2.8}$$

For a treatment transcending the above linear response theory, nonlinear effects can be subsumed into the asymptotic form (A1.2.8) by introducing a phase shift ϕ.

Appendix 1.3. Physisorption Regime with a Discrete Solid Lattice

In the text, long-range interactions mediated by the substrate were treated, disregarding the discrete character of the solid surface.

The purpose of this Appendix is to assess the effects due to this granularity of the surface, following the treatment of Girard and Girardet.[8] Indeed, lattice models using reaction-field techniques had been employed earlier by Macrury and Linder[32] to introduce the discrete description of the solid surface.

With the advent of accurate probes for exploring cleaner surfaces and for finding the structure of adsorbed monolayers and multilayers,[33] much valuable experimental data have been obtained pertaining to the heat of adsorption (desorption) of atoms and molecules of various dielectric or metallic surfaces and the commensurate/incommensurate structure transitions of adsorbed layers.[34,35]

In this Appendix, the susceptibility formalism used by Girard and Girardet[36] and Girard et al.[37] will be employed to discuss further the determination of substrate-mediated interactions. One can then incorporate: (1) the discrete structure of the substrate, and hence face specific effects; (2) the influence of the nonadditive induction contributions in the substrate-mediated energy for adsorbed multipolar molecules; (3) the possibility for treating either dielectric or metallic surfaces from knowledge of the dynamical susceptibility of the medium.

We give below only a brief summary of the main results of Girard and Girardet. Then the influence of point (1) on the interaction between two atoms close to a metal surface will be reviewed. Finally, an assessment of the effect of points (1) and (2) on the interaction between two dipolar molecules adsorbed on a dielectric surface will be made.

Formula for Substrate-Mediated Interaction

Briefly, we shall expose the general structure of the theory for substrate-mediated interactions between two molecules labeled a and b via a solid S. Quantities $\mathbf{R}_a$ and $\mathbf{R}_b$ will be used to denote the locations of the centers of mass of the molecules with respect to an absolute frame (X, Y, Z) fixed to the solid [see Figure 1 of Girard and Girardet[8] for the specific example of two NH_3 molecules adsorbed on a diatomic (100) surface]. The solid is described by a set of parallel infinite planes containing the centers of mass of the atoms, the pth plane being located distance z_p from the surface plane.

It is then useful to separate the substrate-mediated energy for the system into multipolar inductive and dispersive terms.

Multipolar Inductive Part

This contribution can be expressed in terms of the linear susceptibilities $^\rho\chi^\rho_\alpha(\mathbf{r}, \mathbf{r}')$ ($\alpha = a, b,$ or S) of each partner and of the effective electrical potentials $\tilde{\phi}_\alpha(\mathbf{r})$ experienced by the molecules or the solid. The structure of this part is then of the form

$$V_1 = -\tfrac{1}{2} \sum_{\alpha = a,b,S} \iint d\mathbf{r}\, d\mathbf{r}'\, {}^\rho\chi^\rho_\alpha(\mathbf{r}, \mathbf{r}')\tilde{\phi}_\alpha(\mathbf{r})\tilde{\phi}_\alpha(\mathbf{r}') \qquad (A1.3.1)$$

In the linear approximation with respect to the susceptibilities $^\rho\chi^\rho_\alpha$, the effective potential $\tilde{\phi}_\alpha(\mathbf{r})$ is a single sum of the local electrostatic potentials $\phi^l(\mathbf{r})$ due to the other partners, e.g.,

$$\tilde{\phi}_a(\mathbf{r}) = \phi^l_b(\mathbf{r}) + \phi^l_S(\mathbf{r}) \qquad (A1.3.2)$$

$\phi^l_{a,b}(\mathbf{r})$ is a function of the molecular charge density $\rho_{a,b}$,

$$\phi^l_{a,b}(\mathbf{r}) = \int d\mathbf{r}'\frac{\rho_{a,b}(\mathbf{r}')}{|\mathbf{r} - \mathbf{r}'|} \qquad (A1.3.3)$$

and $\phi^l_S(\mathbf{r})$ is constant for a metal surface or depends on the charge distribution in a dielectric substrate. Then, substituting equations (A1.3.2) and (A1.3.3) into (A1.3.1) yields the substrate-mediated energy in the form

$$V_1^M = -(1 + P_{a,b}) \int d\mathbf{r}\, d\mathbf{r}'\, d\mathbf{r}''\frac{{}^\rho\chi^\rho_a(\mathbf{r}, \mathbf{r}')\phi^l_S(\mathbf{r})\rho_b(\mathbf{r}'')}{|\mathbf{r}' - \mathbf{r}''|}$$

$$+ \tfrac{1}{2} \int d\mathbf{r}\, d\mathbf{r}'\rho_a(\mathbf{r})K_S(\mathbf{r}, \mathbf{r}')\rho_b(\mathbf{r}') \qquad (A1.3.4)$$

Here $P_{a,b}$ is the permutation operator for subscripts a and b and the potential susceptibility is defined as

$$K_S(\mathbf{r}, \mathbf{r}') = \int dr_1\, dr_2 \frac{1}{|\mathbf{r} - \mathbf{r}_1|} {}^\rho\chi^\rho_S(\mathbf{r}_1, \mathbf{r}_2)\frac{1}{|\mathbf{r}_2 - \mathbf{r}'|} \qquad (A1.3.5)$$

Approximate expressions for K_S are given in Appendix A of Girard and Girardet[8] for a dielectric or a metal substrate.

Dispersive Contribution

From the book by Mahanty and Ninham[29] one can write [cf. equation (A1.1.5)] the dispersion energy of the total system in the form

$$V_D = -\frac{\hbar}{2\pi} \int_0^\infty d\xi \iint d\mathbf{r}\, d\mathbf{r}'\, {}^\rho\chi^\rho_a(\mathbf{r}, \mathbf{r}', i\xi)K_{S+b}(\mathbf{r}', \mathbf{r}, i\xi) \qquad (A1.3.6)$$

The susceptibility of the system formed by the molecule b and the substrate S, labeled K_{S+b}, is obtained from an iterative method (see Appendix A of Girard and Girardet[8]) as a series expansion in terms of the susceptibilities $^\rho\chi^\rho_\alpha$. When this expansion is taken to third order, one obtains for the substrate-mediated term

$$V^m_b = -\frac{\hbar}{2m} \int_0^\infty d\xi \int d\mathbf{r}\, d\mathbf{r}'\, d\mathbf{r}_1\, d\mathbf{r}_2$$

$$\times [(1 + P_{a,b})K_S(\mathbf{r}, \mathbf{r}', i\xi)^\rho\chi^\rho_b(\mathbf{r}', \mathbf{r}_1, i\xi)|\mathbf{r}_1 - \mathbf{r}_2|^{-1}\chi^\rho_a(\mathbf{r}_2, \mathbf{r}, i\xi)$$

$$+ K_S(\mathbf{r}, \mathbf{r}', i\xi)^\rho\chi^\rho_b(\mathbf{r}', \mathbf{r}_1, i\xi)K_S(\mathbf{r}_1, \mathbf{r}_2, i\xi)\chi^\rho_a(\mathbf{r}_2, \mathbf{r}, i\xi)] \qquad (A1.3.7)$$

where the presence of frequency-dependent quantities (i.e., ξ) characterizes the dynamic as opposed to the static quantities appearing in the inductive part.

Equations (A1.3.4) and (A1.3.7) therefore exhibit the general structure of the substrate-mediated energy between two interacting molecules adsorbed on a solid surface. We conclude the Appendix with two examples.[8]

Example A. Two Adatoms on a Metal Surface

The multipolar inductive contribution vanishes in this case as both the permanent multipole moments of atoms and the gradient of the local potential ϕ^l_S are zero.

The dispersive part is then calculated, in the dipolar approximation for the susceptibility, by introducing the polarizability tensor $\alpha_{a,b}$ of the atoms a and b:

$$^\rho\chi^\rho_{a,b}(\mathbf{r}, \mathbf{r}') \cong \alpha_{a,b} : \nabla_\mathbf{r}\nabla_{\mathbf{r}'}\delta(\mathbf{r} - \mathbf{R}_{a,b})\delta(\mathbf{r}' - \mathbf{R}_{a,b}) \qquad (A1.3.8)$$

The energy V^m_D is usefully separated into two parts (cf. McLachlan[1]):

$$V^m_{D_1} = -\frac{\hbar}{2\pi}(1 + P_{ab})\int_0^\infty d\xi \, \mathrm{Tr}\,[\alpha_b(i\xi)\cdot\mathbf{T}(\mathbf{R}_b, \mathbf{R}_a)\cdot\alpha_a(i\xi)\,{}^1\mathbf{S}^1(\mathbf{R}_a, \mathbf{R}_b, i\xi)] \qquad (A1.3.9)$$

and

$$V^m_{D_2} = -\frac{\hbar}{2\pi}\int_0^\infty d\xi \, \mathrm{Tr}\,[\alpha_b(i\xi)\cdot{}^1\mathbf{S}^1(\mathbf{R}_b, \mathbf{R}_a, i\xi)\cdot\alpha_a(i\xi)\cdot{}^1\mathbf{S}^1(\mathbf{R}_a, \mathbf{R}_b, i\xi)] \qquad (A1.3.10)$$

where ${}^1\mathbf{S}^1$ and $\mathbf{T}$ are second-rank tensors connected to the double gradient of K_S and $1/|\mathbf{r} - \mathbf{r}'|$, respectively.

Our object in writing equations (A1.3.9) and (A1.3.10) is to display the general structure of the theory.

Girard and Girardet[8] then show that for a metal, with a symmetric configuration of the two adatoms $\mathbf{R}_a(\frac{1}{2}D, 0, d)$ and $\mathbf{R}_b(-\frac{1}{2}D, 0, d)$ where D is the interatomic distance and d the substrate–atom distance, quantities $V_{D_1}^m$ and $V_{D_2}^m$ assume the forms

$$V_{D_1}^m = \frac{2LC_{S1}}{D^3}\left\{\sum_{p=0}^{\infty} \frac{Z(6D^2 + Z^2)}{(D^2 + Z^2)^{7/2}} - \frac{1}{3\pi}\sum_{g_1,g_2}^{*} \cos{(\mathbf{g}\cdot\boldsymbol{\tau}_p + g_xD/2)}I_1(\mathbf{g}, d, D)\right\}$$

(A1.3.11)

and

$$V_{D_2}^m = -6L^2C_{S2}\left\{\sum_{p=0,p'=0}^{\infty} [ZZ'(6Z^2Z'^2 - 9D^2(Z^2 + Z'^2) + 26D^4)\right.$$

$$+ 2D^2(D^2 - 4Z^2)(D^2 - 4Z'^2)][(D^2 + Z^2)^{7/2}(D^2 + Z'^2)^{7/2}]^{-1}$$

$$- \sum_{g_1,g_2}\sum_{g_1',g_2'} \cos{(\mathbf{g}\cdot\boldsymbol{\tau}_p + g_xD/2)}$$

$$\left. \times \cos{(\mathbf{g}'\cdot\boldsymbol{\tau}_p + g_x'D/2)}I_2(\mathbf{g}, \mathbf{g}', d, D)\right\}$$

(A1.3.12)

Here $Z = 2(d + pL)$ and $Z' = 2(d + p'L)$, where L defines the distance between consecutive atomic planes of the solid; C_{S1} and C_{S2} are the usual mediated dispersion coefficients defined by McLachlan, namely [cf. equation (A1.1.5)]

$$C_{Sj} = \frac{3\hbar}{\pi}\int_0^{\infty} d\xi \, \Delta^j(i\xi)\alpha_a(i\xi)\alpha_b(i\xi)$$

(A1.3.13)

The first sum in equations (A1.3.11) and (A1.3.12) is over all the planes p of the metal, and the second sum is over the values of the reciprocal translation vector $\mathbf{g}$ of the two-dimensional lattice parallel to the surface $(g \neq 0)$. The functions I_1 and I_2 describe the spatial dependence of the harmonic terms in the Fourier space and are given in Appendix B of Girard and Girardet.[8]

Equations (A1.3.11) and (A1.3.12) contain terms which do not depend on the discrete characteristics of the various planes, and other terms which account for the effect of the atomic corrugation of the planes. The second type of terms remain generally small, particularly for planes with large atomic density. For example, for the (111) face of Pd and Ag and for the (0001) face of graphite, these terms do not exceed 10% of the magnitude of the others.

Girard and Girardet[8] present some illuminating numerical plots to illustrate the effect of the discrete substrate; the interested reader is referred to their paper for further numerical details.

Example B. Two Dipolar Molecules on a Dielectric Surface

We conclude with the second example treated by Girard and Girardet.[8] The multipolar inductive contribution connected to the substrate-mediated interaction, expressed in equation (A1.3.4), can be elaborated in the dipolar approximation by using the relation for the charge density of the molecules a and b:

$$\rho_{a,b}(\mathbf{r}) = -\mathbf{\mu}_{a,b} \cdot \nabla \delta(\mathbf{r} - \mathbf{R}_{a,b}) \qquad (A1.3.14)$$

where $\mathbf{\mu}_a$ denotes the permanent dipole of the molecule a. Quantity V_1^m is then separated into two parts: V_{11}^m is connected with the induction energy of one molecule by the local field $E_S^l = \nabla \phi_S^l$ of the substrate and by the dipolar field of the other molecule while V_{12}^m characterizes the induction energy in the substrate due to the dipolar field of the two molecules. These two contributions take the form

$$V_{11}^m = -(1 + P_{a,b})[E_S^l(\mathbf{R}_a) \cdot \mathbf{\alpha}_a \cdot \mathbf{T}(\mathbf{R}_a, \mathbf{R}_b) \cdot \mathbf{\mu}_b) \qquad (A1.3.15)$$

and

$$V_{12}^m = -\tfrac{1}{2}(1 + P_{a,b})\mathbf{\mu}_a \cdot {}^1S^1(\mathbf{R}_a, \mathbf{R}_b) \cdot \mathbf{\mu}_b \qquad (A1.3.16)$$

These contributions have been calculated by Girard *et al.* for two NH_3 molecules adsorbed on a NaCl surface. Quantities E_S and ${}^1S^1$ for the ionic crystal can be characterized by effective ionic charges and atomic polarizabilities. Some numerical results are recorded by Girard and Girardet[8] for this system (see especially their Table 3).

Notes Added in Proof

Since this Chapter was completed, Gumhalter *et al.*[38] have proposed a unified interpretation of a variety of experimental spectroscopic properties of chemisorbed CO on transition-metal surfaces, which utilizes the CO $2\pi^*$ resonance as in Section 1.3.4 (see also March[39]). Further support for the model comes from the study of Brookes *et al.*[40] on the interaction of CO with Fe (001).

References

1. A. D. McLachlan, *Mol. Phys.* **7**, 381 (1964).
2. J. Mahanty and N. H. March, *J. Phys. C* **9**, 2905 (1976).
3. A. Redondo, Y. Zeiri, and W. A. Goddard, *Surf. Sci.* **136**, 41 (1984).
4. K. Joyce, P. J. Grout, and N. H. March, *Surf. Sci. Lett.* **181**, L141 (1987).
5. R. Opila and R. Gomer, *Surf. Sci.* **112**, 1 (1981).
6. J. Mahanty, N. H. March, and B. V. Paranjape, *Appl. Surf. Sci.* **33–34**, 309 (1988).
7. F. Flores, N. H. March, and I. D. Moore, *Surf. Sci.* **69**, 133 (1977).
8. C. Girard and C. Girardet, *Surf. Sci.* **195**, 173 (1988).
9. L. W. Bruch, *Surf. Sci.* **125**, 194 (1983); *J. Chem. Phys.* **79**, 3148 (1983).
10. J. Bardeen, *Phys. Rev.* **49**, 653 (1936).
11. I. D. Moore and N. H. March, *Annu. Phys.* **97**, 136 (1976).

12. N. H. March and A. M. Murray, *Phys. Rev.* **120**, 830 (1960).
13. G. K. Corless and N. H. March, *Philos. Mag.* **6**, 1285 (1961).
14. K. H. Lau and W. Kohn, *Surf. Sci.* **75**, 69 (1978).
15. F. Flores, N. H. March, and C. J. Wright, *Phys. Letts.* **64A**, 231 (1977).
16. K. Christmann, G. Ertl, and T. Pignet, *Surf. Sci.* **54**, 365 (1976).
17. W. H. Weinberg and R. P. Merrill, *Surf. Sci.* **33**, 493 (1972).
18. R. Burch, *Spec. Per. Rep. Chem. Soc.* **8**, 1 (1980).
19. N. H. March, *Prog. Surf. Sci.*, **25**, 229 (1988).
20. K. Joyce, A. Martin-Rodero, F. Flores, P. J. Grout, and N. H. March, *J. Phys. C* **20**, 3381 (1987).
21. L. Pauling, *Nature of the Chemical Bond*, Cornell University Press, Ithaca (1960).
22. D. D. Eley, *Discuss. Faraday Soc.* **8**, 34 (1950).
23. F. Flores, I. Gabbay, and N. H. March, *Chem. Phys.* **63**, 391 (1981).
24. V. L. Moruzzi, J. F. Janak, and A. R. Williams, in: *Calculated Electronic Properties of Metals*, Pergamon, Oxford (1978).
25. W. Hume-Rothery, H. M. Irving, and R. J. P. Williams, *Proc. R. Soc. London, Ser. A* **208**, 431 (1951).
26. I. Higuchi, T. Ree, and H. Eyring, *J. Am. Chem. Soc.* **77**, 4969 (1955); **79**, 1330 (1957).
27. G. Bróden, T. N. Rhodin, C. Brucker, R. Benbow, and Z. Hurych, *Surf. Sci.* **59**, 593 (1976).
28. W. Gordy and W. J. O. Thomas, *J. Chem. Phys.* **24**, 439 (1956).
29. J. Mahanty and B. W. Ninham, *Dispersion Forces*, Academic Press, London (1976).
30. H. Margenau and N. R. Kestner, *Theory of Intermolecular Forces*, 2nd edn., Ch. 2, Pergamon Press, Oxford (1971).
31. H. Carslaw and J. C. Jaeger, *Conduction of Heat in Solids*, Clarendon Press, Oxford (1959).
32. T. B. Macrury and B. Linder, *J. Chem. Phys.* **54**, 2056 (1971); **56**, 4368 (1972).
33. See, for instance, M. Bienfait, *Surf. Sci.* **162**, 411 (1985).
34. J. G. Dash and J. Ruvalds, eds., *Phase Transitions in Surface Films*, NATO Adv. Study Inst. Ser., Vol. B **51**, Plenum Press, New York (1980).
35. T. Meichel, J. Suzanne, and J. M. Gay, *C. R. Acad. Sci. Paris* **303**, 989 (1986).
36. C. Girard and C. Girardet, *J. Chem. Phys.* **86**, 6531 (1987).
37. C. Girard, J. M. Vigoureux, and C. Girardet, *J. Chim. Phys.* **84**, 809 (1987).
38. B. Gumhalter, K. Wandelt, and Ph. Avouris, *Phys. Rev.* **B37**, 8048 (1988).
39. N. H. March, *Phys. Rev.* **B39**, 1385 (1989).
40. N. B. Brookes, A. Clarke, and P. D. Johnson, *Phys. Rev. Lett.* **63**, 2764 (1989).

Gas–Surface Interactions
Basic Thermodynamics and Recent Work on Sticking

T. B. Grimley

2.1. Introduction

In this chapter an introduction is provided to two aspects of gas–solid interactions: classical and statistical thermodynamics of adsorption, and the quantum theory of sticking. The thermodynamics of adsorption finds applications principally, but not exclusively, to *physisorption* systems. Many *chemisorption* systems widely studied are irreversible; an equilibrium pressure is never measured. This does not mean that desorption of chemisorbed material is not studied. It is, but in an irreversible way, by temperature programmed desorption, for example.

Thermodynamics is concerned with certain macroscopic properties of matter in equilibrium, pressure, volume, entropy, free energy, enthalpy, for example. The laws of thermodynamics lead only to relations among them so that a certain experimental input is required to enable these quantities to be calculated. Electrochemical cells provide a well-known example. Measurements of the cell EMF and its temperature coefficient give the Gibbs free energy change and the entropy change, respectively, in the cell reaction. From these, the enthalpy change is calculated from $G = H - TS$, and calorimetry is avoided. In this chapter we consider how calorimetric determination of the enthalpy of (physical) adsorption is avoided in favor of measuring the temperature dependence of the equilibrium pressure. On the other hand, for chemisorption, calorimetric determination of the enthalpy change seems to be essential for most systems, and a detailed feasibility study of a single-crystal calorimeter has been made.[1]

T. B. Grimley • Donnan Laboratories, The University of Liverpool, Liverpool L69 3BX, England.

Statistical thermodynamics tells us how to *calculate* all thermodynamic quantities from first principles once we have settled on a microscopic theoretical model. Of course our models may not always be adequate for all aspects of the thermodynamic behavior, but this is a familiar enough situation in science; more refined models may be too difficult to compute.

Statistical thermodynamics has a breadth and importance not shared with classical thermodynamics, because the microscopic theoretical models we construct to calculate the thermodynamic quantities are also valid models for understanding and computing many other things, such as the dynamical behavior. A single instance will suffice. When we construct a model for the thermodynamics of the adsorption of a rare gas on a metal, we have to decide whether the gas is adsorbed on specific sites (localized adsorption), or whether a two-dimensional gas is a more appropriate model. Exactly the same decision has to be made when we contemplate the calculation of the angle-resolved thermal desorption flux, and in fact our theoretical model for thermal desorption is only more general than that for the equilibrium thermodynamics in that, for the former, we need to handle arbitrarily large gas–solid separations.

The sticking of an atom on a cold solid is perhaps the simplest of all phenomena in the general category of *Process Dynamics*, and there is no doubt that a full quantum mechanical computation of at least one model system ought to be given. Of course this will not be done here, but the quantum theory of sticking seems to contain some pitfalls, and even if these are well known in other branches of theoretical physics, it is useful to see them also in the context of our subject. At a more practical level, we need to think about the energy loss mechanism responsible for sticking, to obtain a formula from which the sticking coefficient could be computed, and to look at the validity of semiclassical methods in this field.

This chapter will now deal with the meaning of the terms "physisorption" and "chemisorption," introducing the concept of a "surface lifetime" and discussing its immediate significance. This concept is also relevant in sticking as we shall see later.

2.2. Physisorption and Chemisorption

2.2.1. General

Although there is no clear-cut distinction between physisorption and chemisorption, it is useful to have the notion that, in the latter, a chemical bond is formed to the surface and the adsorbate may, or may not, retain its identity, while in the former, no surface chemical bond is formed and the adsorbate always retains its identity. Simple examples of chemisorption systems are:

1. H_2 on (100)W. The H—H bond is broken, and the hydrogen is chemisorbed as atoms:

This is the situation in the p(1 × 1) structure.

2. CO on (100)Ni. The CO molecule retains its identity, but the C—O *bond order* is reduced because a C—Ni surface bond is formed:

This is the situation in the c(2 × 2) structure.

Physisorption occurs, for example, with rare gases on metals. The binding energy is typically less than about 20 kJ mol^{-1} (200 meV/atom), and although the system wave functions are only slightly affected by adsorption, some measurable properties, such as the work function of a metal adsorbent, *are* changed by rare gas adsorption.

Binding energies in chemisorption are typically in excess of 30 kJ mol^{-1} (300 meV/particle), and even where the adsorbate retains its identity on the surface, substantial changes in the electronic structure occur. As an example we again mention CO on (100)Ni. The three highest occupied molecular orbitals in the free CO molecule and their energies with respect to vacuum are 4σ (−16 eV), 1π (−13 eV), and 5σ (−11 eV), so that the highest occupied molecular orbital (HOMO) is the 5σ. The lowest unoccupied molecular orbital (LUMO) is the antibonding $2\pi^*$ (−4 eV). The Fermi level of Ni is about −4.5 eV. According to *Frontier Orbital* theory, interaction with the metal will depopulate the 5σ orbital and populate the $2\pi^*$ orbital, thereby reducing the C—O bond order. Because electrons repel one another, these population changes lead to changes in the orbital energies. In particular the 5σ orbital energy is pulled down to be near the 1π level; the 1π orbital interacts only weakly with the metal. The result of a rather elaborate calculation[2] is that the ordering of the 1π and 5σ levels is reversed by chemisorption. From this we learn that, although the OC-metal bond is rather weak (59 kJ mol^{-1}) so that an equilibrium pressure of CO can be measured in this chemisorption system,[3] significant changes in the electronic structure have indeed occurred.

2.2.2. Surface Residence Times

An important concept is the average time a single adsorbate spends on the surface. This is t_d, where t_d^{-1} is the thermal desorption rate. The value of t_d can be determined experimentally or estimated from the surface bond vibration frequency ν and the binding energy D (the surface bond dissociation energy) using

$$t_d = \nu^{-1} \exp(D/RT) \tag{2.2.1}$$

When t_d is determined experimentally, equation (2.2.1) is often used to fit the data; some values at 300 K are given in Table 2.1.

Table 2.1. Surface Lifetimes t_d(s) at 300 K and Parameters ν and D in Equation (1)

	H/(100)W	Hg/(100)Ni	CO/(111)Ni	N$_2$/(100)Ru	Xe/(111)W
ν (s^{-1})	3×10^{13}	10^{12}	8×10^{15}	10^{13}	10^{15}
D (kJ)	268	115	125	31	40
t_d (s)	10^{33}	10^8	7×10^5	3×10^{-8}	6×10^{-9}

The values of t_d span an enormous range. One significance of t_d is that it tells us whether equilibrium thermodynamics can be applied to the adsorption system. Roughly speaking, if t_d exceeds the time needed for an experimental measurement, thermodynamic equilibrium cannot be maintained in the experiment. Of course an experimentalist does not construct Table 2.1 to see whether or not thermodynamic equilibrium can be maintained in an adsorption system.

The significance of t_d in sticking on a finite-temperature solid will be mentioned later.

2.3. Thermodynamics of Adsorption

Classical thermodynamics is concerned with precise definitions of certain macroscopic quantities, with the relations between them, and with their determination from experimental data. Many books and reviews dealing with the classical thermodynamics of adsorption are available.[4-7] If one aims at total rigor, the equations are in the main useless to the experimentalist. Therefore, in this chapter reference will be made only to the approaches developed by Hill[7] and Everett[8] and attention confined to the enthalpy of adsorption (in fact, the isosteric heat of adsorption) and its determination from the adsorption isotherm.

The adsorption system is regarded as a two-phase one: a two-component condensed phase consisting of the adsorbent (S), the adsorbed particles (A), and a gas phase (G). Of course it is not necessary to include all the solid in S, but we must go far enough into the solid so that the disturbance caused by the adsorbate has "healed." The resemblance to a binary solution with one volatile component is intentional.

The condensed phase has volume V, temperature T, entropy S, n_A moles of adsorbed gas, n_S moles of adsorbent, with chemical potentials μ_A, and μ_S. Consequently, the change in internal energy dU due to changing S, V, n_S, and n_A is given by

$$dU = T\,dS - p\,dV + \mu_S\,dn_S + \mu_A\,dn_A \tag{2.3.1}$$

where p is the gas pressure. The chemical potentials depend on the composition $x = n_A/n_S$ of the condensed phase, and on the other intensive variables T and p. Thus $\mu_A = \mu_A(T, p, x)$, and therefore

$$d\mu_A = (\partial\mu_A/\partial T)_{p,x}\,dT + (\partial\mu_A/\partial p)_{T,x}\,dp + (\partial\mu_A/\partial x)_{T,p}\,dx \tag{2.3.2}$$

Since $\mu_A = (\partial G/\partial n_A)_{T,p,n_S}$ and $S = -(\partial G/\partial T)_{p,n_S,n_A}$, equation (2.3.2) can be written as

$$d\mu_A = -s_A dT + v_A dp + (\partial\mu_A/\partial x)_{T,p}\, dx \tag{2.3.3}$$

where

$$s_A = (\partial S/\partial n_A)_{T,p,n_S} \quad \text{and} \quad v_A = (\partial V/\partial n_A)_{T,p,n_S}$$

are so-called partial molar quantities.

For the gas phase

$$d\mu_G = -S_{G,m}\, dT + V_{G,m}\, dp \tag{2.3.4}$$

where

$$S_{G,m} = (\partial S_G/\partial n_G)_{T,p} = S_G/n_G \quad \text{and} \quad V_{G,m} = (\partial V_G/\partial n_G)_{T,p} = V_G/n_G$$

are, for a one-component system, the molar entropy and molar volume.

At equilibrium $d\mu_A = d\mu_G$ and $dx = 0$ ($\mu_A = \mu_G$ and there is no net adsorption or desorption, so x cannot change), and we have from equation (2.3.3)

$$(V_{G,m} - v_A)dp = (S_{G,m} - s_A)\, dT \tag{2.3.5}$$

If, as will usually be the case, $v_A \ll V_{G,m}$ (condensed matter has a much smaller molar volume than a gas at ordinary temperature and pressure), and if we regard the gas as ideal so that $V_{G,m} = RT/p$, then equation (2.3.5) becomes

$$(\partial \ln p/\partial T)_x = (S_{G,m} - s_A)/RT \tag{2.3.6}$$

By introducing the partial molar enthalpy h_A and the molar enthalpy $H_{G,m}$ so that

$$\mu_A = h_A - Ts_A \quad \text{and} \quad \mu_G = H_{G,m} - TS_{G,m}$$

we have at equilibrium ($\mu_A = \mu_G$)

$$H_{G,m} - h_A = T(S_{G,m} - s_A) \tag{2.3.7}$$

Thus equation (2.3.6) can be written in the form

$$(\partial \ln p/\partial T)_x = q_{st}/RT^2 \tag{2.3.8}$$

where $q_{st} = H_{G,m} - h_A$ is the *isosteric heat of adsorption* defined at constant (T, p, x); q_{st} is the heat evolved when one mole of gas is transferred to the adsorbed state under conditions such that the gas pressure, the temperature, and the ratio of the number of moles of the gas adsorbed to the number of moles of the adsorbent change by only negligible amounts.

Of course equation (2.3.8) looks like the Clausius–Clapeyron equation, and it is precisely the same as the equation for the vapor pressure of a two-component liquid with one volatile component. This illustrates the Hill–Everett contribution.

Equations (2.3.7) and (2.3.8) show how q_{st} and the partial molar entropy of adsorption are obtained by plotting $\ln p$ (p is the equilibrium pressure at temperature T) against $1/T$ at constant adsorbed amount n_A. The relation of q_{st} to the quantity measured in an isothermal calorimeter needs careful consideration, however.[6]

It is worth stating explicitly that, while the above discussion evidently covers cases where the adsorbed molecules and those in the gas phase are the same, it applies equally to the dissociative adsorption of a diatomic gas, for example. For such a system, there will be both atoms and undissociated molecules in the condensed phase, but n_A is the number of moles of adsorbed molecules, dissociated or not.

2.4. Statistical Thermodynamics of Adsorption

2.4.1. Basics

Classical thermodynamics ignores everything we know about the microscopic structure of matter. Statistical thermodynamics feeds this knowledge into the problem at the beginning via detailed molecular models of the condensed phase and of the gas phase, constructs partition functions for the two phases, and calculates the thermodynamic functions from these by well-known formulas.[4,5]

If $Q(T, V, n_S, n_A)$ is the canonical partition function for the condensed phase, then the thermodynamic functions are

$$F = -kT \ln Q, \qquad U = kT^2(\partial \ln Q/\partial T), \qquad S = k\,\partial(T \ln Q)/\partial T,$$

$$p = kT(\partial \ln Q/\partial V)$$

Here $F(T, V, n_S, n_A)$ is the Helmholtz free energy, and the variables held constant in the partial derivatives are obvious and need not be displayed. This is usually the case in statistical thermodynamics. Of course, the partition function Q can only be constructed either analytically or by computer simulation, for rather simple models. Here it is assumed that the adsorption system consists of an ideal gas phase and an adsorbed phase *on an inert adsorbent*. Thus asorbate-induced expansion, contraction, or reconstruction of the adsorbent are not treated. For such a model, the partition function $Q_A(T, V_A, n_A, \sigma)$ for the adsorbed phase depends on the volume V_A of the phase, and on the surface area σ of the adsorbent, thus allowing the *spreading pressure* ϕ to be defined as well as the hydrostatic pressure:

$$\phi = kT(\partial \ln Q_A/\partial \sigma) \qquad \text{and} \qquad p = kT(\partial \ln Q_A/\partial V_A) \qquad (2.4.1)$$

The chemical potential is defined in the usual way from the Helmholtz free energy $F_A(T, V_A, n_A, \sigma)$,

$$\mu_A = \partial F_A/\partial n_A = -kT(\partial \ln Q_A/\partial n_A)$$

and the corresponding quantity for an ideal gas is

$$\mu_G = \partial F_G(T, V_G, n_G)/\partial n_G = kT \ln (n_G/f_G) \qquad (2.4.2)$$

where f_G is the molecular partition function for the gas. If M is the molecular mass and h is Planck's constant,

$$f_G = (2\pi MkT/h^2)^{3/2} V_G f_G^{int}$$

where f_G^{int} is the partition function for the internal degrees of freedom (vibrations and rotations). The conditions for thermodynamic equilibrium are

$$\mu_A = \mu_G \qquad \text{and} \qquad \sigma = \text{constant} \qquad (2.4.3)$$

2.4.2. Langmuir's Model

This is the simplest possible model of localized adsorption. The n_A molecules are distributed at random over n_S adsorption sites with no more than one molecule per site and with no interactions between the adsorbed molecules. All adsorption sites are identical and adsorbed molecules do not dissociate. Thus

$$Q_A = \Omega(n_A, n_S)(f_A)^{n_A} \qquad \text{and} \qquad \Omega(n_A, n_S) = n_S!/n_A!(n_S - n_A)! \quad (2.4.4)$$

where Ω is the number of arrangements of n_A identical molecules on n_S sites. For monatomic molecules, f_A may be approximated by three simple harmonic oscillator terms, one for each vibrational mode against the rigid substrate. For polyatomics, there will be factors for internal vibrations and for (hindered) rotations.

From equation (2.4.4) after using Stirling's approximation on the factorials ($\ln N! \simeq N \ln N - N$), we obtain

$$\mu_A = kT[\ln (\theta/1 - \theta) - \ln f_A] \qquad (2.4.5)$$

where $\theta = n_A/n_S$ is the fraction of the surface covered. At thermodynamic equilibrium, equations (2.4.2)–(2.4.5) yield

$$\theta/(1 - \theta) = n_G f_A/f_G = pK \qquad (2.4.6)$$

where p is the gas pressure and K is the equilibrium constant. This is the famous isotherm obtained originally by Langmuir from a simple kinetic argument. Of course, partition functions for the molecules in the two phases must be constructed from the same energy zero which is usually chosen as the energy of a molecule at rest in the gas phase. Then we display the adsorption energy D explicitly and write $f_A \exp(D/kT)$ in place of f_A, so that the equilibrium constant is

$$K = (f_A V_G/f_G kT) \exp (D/kT)$$

The T-dependence of the equilibrium pressure p at constant surface coverage yields, according to equation (2.3.8), the isosteric heat of adsorption,

$$q_{st} = kT^2(\partial \ln p/\partial T) = -kT^2(\partial \ln K/\partial T)$$

$$= D - kT^2\partial \ln (f_A/f_G)/\partial T + kT \qquad (2.4.7)$$

For the dissociative adsorption of a diatomic gas X_2 we proceed as follows. Choose atoms at rest in the gas phase as the energy zero, then for adsorbed atoms [see equation (2.4.5)]

$$\theta = \lambda_{AX}(1 - \theta)f_{AX} \exp (E/kT)$$

where E is the adsorption energy for *atoms* and we have introduced the *absolute activity* $\lambda_{AX} = \exp(\mu_{AX}/kT)$. For gas-phase molecules at pressure p_{X_2} we have from equation (2.4.2), after introducing the absolute activity λ_{X_2},

$$p_{X_2} = \lambda_{X_2}(f_{X_2}kT/V_G) \exp [D(X_2)/kT]$$

where $D(X_2)$ is the dissociation energy of X_2. Since the equilibrium $X_2 \rightleftharpoons 2X$ exists in the gas phase, we have that $(\lambda_{AX})^2 = \lambda_{X_2}$, and so we derive the adsorption isotherm

$$\theta/(1 - \theta) = p_{X_2}^{1/2}(f_{AX}^2 V_G/f_{X_2}kT)^{1/2} \exp \{[E - \tfrac{1}{2}D(X_2)]/kT\} = p_{X_2}^{1/2}K'$$

and

$$q_{st} = 2E - D(X_2) - kT^2\partial \ln (f_{AX}^2/f_{X_2})/\partial T + kT$$

We remark that for Langmuir's model, q_{st} is always independent of the surface coverage.

2.4.3. Two-Dimensional Lattice Gas

In Langmuir's model (the ideal two-dimensional lattice gas), the lack of interactions between the adsorbed molecules means that surface phase transitions do not occur. The simplest improvement is to introduce interactions between *nearest neighbors*. This gives the two-dimensional Ising model. It is very easy to treat this model in the lowest-order approximation.

Let z be the coordination number (the number of nearest neighbors) of the two-dimensional lattice of adsorption sites. When the surface coverage is θ, the arrangements of the admolecules are assumed to be partitioned into sets characterized by the number N_p of nearest-neighbor pairs of occupied sites. For the Ising model, all arrangements in such a set have the same interaction energy N_pw where w is the pair interaction energy, and consequently

$$Q_A = (f_A)^{n_A} \sum_{N_p} \Omega(n_A, n_S, N_p) \exp (-N_pw/kT) \qquad (2.4.8)$$

where $\Omega(n_A, n_S, N_p)$ is the number of arrangements of the admolecules keeping N_p nearest-neighbor pairs; N_p goes from 0 to $\frac{1}{2}n_A^2$ [strictly $\frac{1}{2}n_A(n_A - 1)$, but n_A is large]. At the upper limit, the admolecules form an island in a sea of unoccupied sites.

The first approximation to the sum in equation (2.4.8) is obtained if N_p is replaced by its average value $\bar{N}_p$ in the exponential. Then

$$Q_A = (f_A)^{n_A} \exp\left(-\bar{N}_p w/kT\right) \sum_{N_p} \Omega(n_A, n_S, N_p)$$

$$= (f_A)^{n_A} \exp\left(-\bar{N}_p w/kT\right)\Omega(n_A, n_S)$$

$$= Q_A^{(L)} \exp\left(-\bar{N}_p w/kT\right) \tag{2.4.9}$$

where $Q_A^{(L)}$ is the partition function for Langmuir's model, equation (2.4.4). The value of $\bar{N}_p$ is easily calculated. Each site is occupied with probability θ, so $z\theta$ is the average number of admolecules on nearest-neighbor sites round a given site. The given site is occupied with probability θ, hence for n_S sites the average number of nearest-neighbor pairs of admolecules is therefore

$$\bar{N}_p = \frac{1}{2}n_S z\theta^2 = \frac{1}{2}z n_A^2/n_S$$

When this is employed in equation (2.4.9) we find an extra term in μ_A of $zw\theta$, so that

$$\mu_A = kT \ln\left(\theta/1 - \theta\right) - \ln f_A + zw\theta$$

and therefore in place of equation (2.4.6) we have *Fowler's* adsorption isotherm

$$pK = [\theta/1 - \theta)] \exp\left(zw\theta/kT\right) \tag{2.4.10}$$

The isosteric heat of adsorption now depends linearly on the surface coverage,

$$q_{st} = q_{st}^{(L)} + zw\theta \tag{2.4.11}$$

where $q_{st}^{(L)}$ is the value for Langmuir's model, equation (2.4.7).

2.4.4. Two-Dimensional van der Waals Gas

Although this adsorption isotherm is not derivable from a well-founded microscopic model, it is mentioned here because it is often used by experimentalists to try to fit their data (see Jones and Tong,[9] for example).

The canonical partition function is expressed for the adsorbed phase in the "ideal gas" form[7]

$$Q_A = (f_A)^{n_A}/n_A! \tag{2.4.12}$$

with

$$f_A = (2\pi MkT/h^2)(\sigma - n_A a)f_A^{int} \exp(D/kT) \exp(\alpha n_A/\sigma kT) \quad (2.4.13)$$

where σ is the surface area of the adsorbent, f_A^{int} is the molecular partition function for the surface vibrations and internal motions, a is the excluded area per molecule, and α is the correction to the spreading pressure for attractive forces, both of which appear in the two-dimensional van der Waals equation of state

$$(\phi + n_A^2 \alpha/\sigma^2)(\sigma - n_A a) = n_A kT \quad (2.4.14)$$

deduced from equations (2.4.12) and (2.4.13) using the first member of equation (2.4.1). If $\alpha = a = 0$, then equations (2.4.12)–(2.4.14) collapse to those of the ideal two-dimensional adsorbed gas. If $\alpha = 0$ but $a \neq 0$, we have the slightly imperfect two-dimensional adsorbed gas, and if $a = 0$ but $\alpha \neq 0$, we have the "smoothed potential" model of a two-dimensional fluid, where each molecule moves in a constant potential proportional to the surface density n_A/σ. All three are well-founded, but not particularly useful, theoretical models.

The adsorption isotherm for the present model is

$$pK'' = [\theta/(1 - \theta)] \exp[(\theta/1 - \theta) - (2\alpha\theta/akT)]$$

if the surface coverage is defined as an_A/σ. The definition of K'' is

$$K'' = [a/kT(2\pi MkT/h^2)^{1/2}](f_A^{int}/f_G^{int}) \exp(D/kT)$$

and, neglecting any T-dependence of a, the isosteric heat of adsorption is

$$q_{st} = D + 3kT/2 + 2\alpha\theta/a - kT^2 \partial \ln(f_A^{int}/f_G^{int})/\partial T$$

2.5. Rate of Adsorption

2.5.1. General

One of the simplest rate processes in surface science is the adsorption (sticking) of a single atom on a cold ($T = 0$) solid. In this process, the kinetic energy of the incident atom is lost to internal excitations of the stuck system, atom plus solid. This loss might be to phonons or to electrons (electron–hole pairs), and an important question is which mechanism operates in a specified adsorption system. Generally speaking, we should expect to have to examine the electron mechanism for nondissociative chemisorption systems, while the phonon mechanism should be important in physisorption. This statement is based on the simple notion that, if a new chemical bond is being made, as it is in chemisorption, then energy transfer to the electron system might be facilitated because profound changes in the electron distribution must occur during the collision if the electronic system is to remain unexcited. If a chemical bond is not formed, only slight changes in the electron distribution are needed to keep the electronic system in its ground state, and energy transfer to the electron system would seem to be less likely in this case. But no generalization can be universally valid; a light atom impinging on a heavy substrate cannot easily lose energy to the phonon system (think of the fairground coconut shy), so the electron loss mechanism might sometimes have to be investigated for physisorption systems.

It will be noted that nondissociative chemisorption was specified above. This is because dissociative chemisorption often proceeds via a weakly-bound precursor state, and sticking into this precursor is the rate-determining step. Consequently, although chemical bonds are ultimately made and broken, the critical energy loss process is probably to phonons, not electrons.

2.5.2. Theoretical Description of Sticking

2.5.2.1. General

We consider first sticking on a cold solid S so that S is in its ground state initially, but the stuck system $(AS)^*$ has internal excitation denoted by the asterisk $*$:

$$A + S \rightarrow (AS)^*$$

It should be noted at once that, for a finite solid, the stuck system will eventually (in a Poincaré time) decompose; the atom desorbs, and the solid reverts to its ground state. Thus, for a finite solid, the adsorption is not permanent. Only for an infinite solid is the process permanent and irreversible, and since we can easily build the essential feature of an infinite solid (a continuum of excitations) into the theoretical model, sticking on a cold solid is rather different from sticking on a finite-temperature one. But if we insist that in practice the solid *is* finite, then sticking on a cold solid only appears irreversible if the observation time t_{ob} satisfies $t_{ob} \ll \hbar/\Delta E$, where ΔE measures the spacing of the relevant internal excitations of the system. In theoretical models we simulate an infinite system by replacing discrete excitations by Gaussians, by averaging over an energy interval large compared with ΔE so as to produce a continuous function, or by some other device.

On a finite-temperature solid, the long-time behavior is that the atom desorbs (no sticking) *whether the solid is infinite or not.* This is because the stuck atom finds itself in the bath of thermal excitations already present in the solid, so that its residence time is essentially t_d of Section 2.2.2. But again, sticking *appears* irreversible if the observation time satisfies $t_{ob} \ll t_d$. For dissociative chemisorption where sticking into an undissociated precursor is the critical step, the thermal bath effects the rapid transition to the dissociated state and the relevant residence time is that for the latter, not for the precursor.

2.5.2.2. Sticking on a Cold Solid

The best general reference for the material of this section is still Chapter 3 of Goldberger and Watson[10]; see also Chapter 2.7 of Levine.[11]

Sticking is a reactive collision which, on a cold solid, can be written as the reaction

$$A + S \to AS + K$$

where A is the particle (atom), S the solid, and K the excitation. The Hamiltonian can be written in two ways:

$$H = H_A + H_S + V_{A-S} = H_{AS} + H_K + V_{AS-K} \qquad (2.5.1)$$

where the first form describes the initial channel in the sense that the whole interaction V_{A-S}, elastic and inelastic, between the particle and the solid is initially zero. The second form describes the final channels; H_{AS} describes the particle and the solid interacting via the elastic potential so that bound states of H_{AS} correspond to the stuck particle, while V_{AS-K} is the interaction of the excitation with the stuck system. Standard scattering theory requires V_{A-S} to vanish in the remote past ($t \to -\infty$) and V_{AS-K} in the distant future ($t \to +\infty$). The former is true because the particle is far from the solid in this limit, and correspondingly, the latter requires the excitation to be far from the stuck particle in the distant future. Thus, the final state must have the excitation in the form of an outgoing wave packet just as the initial state has an incoming wave packet describing the particle. It is well known (see, for example, relevant books[10-12]) that states of this sort are formed by linear superposition of stationary scattering states which are solutions of a Lippmann–Schwinger equation. For a structureless particle traveling in the direction specified by the unit momentum vector $\hat{p}$ with energy E and incident on a solid in its ground state $|0\rangle$, the Lippmann–Schwinger scattering state $|\hat{p}, E; 0 +\rangle$ is defined by

$$(E + i0 - H_A - H_S - V_{A-S})|\hat{p}, E; 0 +\rangle = (E + i0 - H_A - H_S)|\hat{p}, E; 0\rangle$$

or

$$|\hat{p}, E; 0 +\rangle = |\hat{p}, E; 0\rangle + G_0(E^+)V_{A-S}|\hat{p}, E; 0 +\rangle$$

$$G_0(E^+) = (E + i0 - H_A - H_S)^{-1} \qquad (2.5.2)$$

Here $|\hat{p}, E; 0\rangle = |\hat{p}, E\rangle|0\rangle$ describes the atom in a plane-wave state $|\hat{p}, E\rangle$ and the solid in its ground state $|0\rangle$ without interaction between them; $E + i0$, which we often write as E^+, stands for

$$\lim_{\eta \to 0} (E + i\eta)$$

The normalization of $|\hat{p}, E\rangle$ is

$$\langle E', \hat{p}'|\hat{p}, E\rangle = \delta[E(p') - E(p)]\delta(\hat{p}' - \hat{p})$$

The particle wave packet prepared by the experimentalist is a superposition of stationary states $|\hat{p}, E\rangle \exp{(-iEt/\hbar)}$

$$|\phi(t)\rangle = \int d\hat{p} \int dE \, A(\hat{p}, E) \exp{(-iEt/\hbar)}|\hat{p}, E\rangle \tag{2.5.3}$$

the function $A(\hat{p}, E)$ being defined by the experimental arrangement. Consequently, the initial state is

$$|\Phi_i(t)\rangle = \int d\hat{p} \int dE \, A(\hat{p}, E) \exp{(-iEt/\hbar)}|\hat{p}, E; 0\rangle \tag{2.5.4}$$

In the case of the analog of equation (2.5.3) for the excitation wave packet, the situation is different; in sticking, the excitation is not measured but is determined from the initial state by the equations of motion. What *is* measured is the probability to observe a stuck particle irrespective of the excitation state, and this is the quantity we wish to calculate. To do this, we evolve the initial state in time under the full Hamiltonian until the collision is over $(t \to +\infty)$, project this time-evolved state onto the set of stuck states of H_{AS} (these are products of a stuck state $|s\rangle$, and an excitation $|n\rangle$), and then sum on the excitation state. We note that the above projection, if summed on the stuck states, determines the excitation wave packet.

From Schroedinger's time equation

$$i\hbar(\partial/\partial t)|\Phi_i(t)\rangle = H|\Phi_i(t)\rangle$$

the state at time t which evolves from $|\Phi_i(t')\rangle$, $t' \to -\infty$, is

$$|\Phi_i^+(t)\rangle = \exp{[-iH(t - t')/\hbar]}|\Phi_i(t')\rangle \tag{2.5.5}$$

It will be convenient to relate $|\Phi_i^+(t)\rangle$ to $|\Phi_i(t)\rangle$, and therefore we evolve $|\Phi_i(t)\rangle$ *backward* under $H_0 = H_A + H_S$ to time t' $(t' \to -\infty)$ so that

$$|\Phi_i(t')\rangle = \exp{[-iH_0(t' - t)/\hbar]}|\Phi_i(t)\rangle \tag{2.5.6}$$

Consequently

$$|\Phi_i^+(t)\rangle = \lim_{t' \to -\infty} \exp{[-iH(t - t')/\hbar]} \exp{[-iH_0(t' - t)/\hbar]}|\Phi_i(t)\rangle$$

$$= \Omega^+|\Phi_i(t)\rangle \tag{2.5.7}$$

where Ω^+, which is defined by equation (2.5.7), is the Møller wave operator. It is independent of time, being equally well defined by

$$\Omega^+ = \lim_{\tau \to -\infty} \exp{(iH\tau/\hbar)} \exp{(-iH_0\tau/\hbar)}$$

Its important property is that it generates the stationary scattering state $|\hat{p}, E; 0+\rangle$ from $|\hat{p}, E; 0\rangle$,

$$|\hat{p}, E; 0+\rangle = \Omega^+|\hat{p}, E; 0\rangle \qquad (2.5.8)$$

This is no more than a special case of equation (2.5.7), the arbitrary state $|\Phi_i(t)\rangle$ being replaced by $|\hat{p}, E; 0\rangle$. Thus, if we apply the operator Ω^+ to both sides of equation (2.5.4), then we have with the aid of equations (2.5.7) and (2.5.8)

$$|\Phi_i^+(t)\rangle = \int d\hat{p} \int dE\, A(\hat{p}, E) \exp\left(-iEt/\hbar\right)|\hat{p}, E; 0+\rangle \qquad (2.5.9)$$

This shows how an arbitrary collision can be discussed in terms of Lippmann-Schwinger stationary scattering states.

The state $|\hat{p}, E; 0+\rangle$ can be expressed in terms of the complete set of eigenstates of $H_{AS} + H_K$. These are products of a particle state and an excitation $|n\rangle$, and for our problem the particle states describe the particle either stuck, scattering elastically off the solid, or desorbing, the latter two being needed for elastic and inelastic scattering. We denote the stationary states of $H_{AS} + H_K$ which have a stuck particle by $|s, n : t\rangle$,

$$|s, n : t\rangle = \exp\left(-iE_{sn}t/\hbar\right)|s, n\rangle$$

where $E_{sn} = E_s + E_n$ is the energy. The quantity we need is the scalar product

$$\langle n, s : t|\Phi_i^+(t)\rangle = \int d\hat{p} \int dE\, A(\hat{p}, E) \exp\left[i(E_{sn} - E)t/\hbar\right]$$

$$\times \langle n, s|\hat{p}, E; 0+\rangle \qquad (2.5.10)$$

specifically, the limit of this projection for $t \to \infty$ summed on the excitation n, since this is the amplitude to find a stuck particle after the collision is over.

Next we need the formula (see Appendix 2.1)

$$\langle n, s|\hat{p}, E; 0+\rangle = (E + i0 - E_{sn})^{-1}$$

$$\times \left[i0\langle n, s|\hat{p}, E; 0\rangle + \langle n, s|V_{AS-K}|\hat{p}, E; 0+\rangle\right] \qquad (2.5.11)$$

which with the aid of the formula (see Appendix 2.2)

$$\lim_{t \to +\infty} \int dz\, g(z)(z - i0)^{-1} \exp\left(izt\right) = 2\pi i g(0)$$

enables us to perform the integration on E in equation (2.5.10). The first term in equation (2.5.11) does not contribute because the presence of $i0$ in the numerator makes $g(0) = 0$ for this term. The second term yields

$$\lim_{t \to +\infty} \langle n, s : t | \Phi_i^+(t) \rangle = 2\pi i \int d\hat{p} \, A(\hat{p}, E_{sn}) \langle n, s | V_{AS-K} | \hat{p}, E_{sn}; 0+ \rangle$$

$$= B(s, n)$$

(2.5.12)

It should be noted that the matrix element in equation (2.5.12) is nothing more than the element of the T-matrix on the energy shell $E = E_{sn}$:

$$\langle n, s | V_{AS-K} | \hat{p}, E_{sn}; 0+ \rangle = \langle n, s | T(E_{sn}) | \hat{p}, E_{sn}; 0 \rangle$$

(2.5.13)

Equation (2.5.12) is a basic result, where $B(s, n)$ is the amplitude to find the stuck state $|s, n\rangle$ occupied after the collision is over. Consequently, the probability to observe the particle in the stuck state $|s\rangle$ after the collision is over, irrespective of the excitation state, is

$$P_s = \sum_n |B(s, n)|^2$$

(2.5.14)

and the sticking probability is

$$P_{\text{stick}} = \sum_s P_s$$

$$= 4\pi^2 \sum_{s,n} \left| \int d\hat{p} \, A(\hat{p}, E_{sn}) \langle n, s | V_{AS-K} | \hat{p}, E_{sn}; 0+ \rangle \right|^2$$

(2.5.15)

There is only one collision (only one incident wave packet), so P_{stick} is equal to the sticking coefficient s (the sticking probability per collision).

According to equation (2) the sticking coefficient depends on $A(\hat{p}, E)$, i.e., on the way the initial particle wave packet is prepared. This is contrary to experience. To resolve this contradiction we note that, for heavy particles (say $M \sim 4000$ au, which is H_2) as opposed to electrons ($M = 1$), the Uncertainty Principle imposes only negligible restrictions on our ability to prepare the particle with a well-defined momentum $\hbar p$. For example, such a particle with kinetic energy 27 meV (10^{-3} au) can be localized to within 10^{-6} m at a cost in energy uncertainty of $\Delta E / E = 4 \times 10^{-5}$ (momentum uncertainty $\Delta p / p = 2 \times 10^{-5}$). Consequently, the particle wave packet has an extremely narrow momentum distribution so that we are allowed to set

$$A(\hat{p}, E) = \delta(\hat{p} - \hat{p}_i)\delta(E - E_i)$$

(2.5.16)

and to refer to $\hat{p}_i$ and E_i as the momentum direction and the energy, respectively, of the incident particle. Substitution of equation (2.5.16) in equation (2.5.12) and (2.5.14) yields

$$P_s = (2\pi)^2 \sum_n |\langle n, s| V_{AS-K}|\hat{p}_i, E_s + E_n; 0+\rangle|^2 \delta(E_s + E_n - E_i) \qquad (2.5.17)$$

Since the spectrum of excitations is essentially continuous, we introduce the replacements

$$\sum_n \to \int dE_n \rho_K(E_n) \qquad \text{and} \qquad |n\rangle \to |E_n\rangle/[\rho_K(E_n)]^{1/2}$$

where $\rho_K(E)$ is the density of excitations of energy E. Consequently

$$P_s = (2\pi)^2 |\langle E_i - E_s, s| V_{AS-K}|\hat{p}_i, E_i; 0+\rangle|^2 \qquad (2.5.18)$$

This is the *yield* of transitions to the stuck state $|s\rangle$ from the initial wave packet state with a well-defined particle momentum. The expression for the sticking coefficient

$$s = \sum_s P_s$$

is now exactly the same as that obtained by calculating the *constant* rate of transitions out of an initial *plane wave* particle state $|\mathbf{p}; 0\rangle$, namely

$$R_i = (2\pi/\hbar) \sum_{s,n} |\langle n, s| V_{AS-K}|\mathbf{p}_i; 0+\rangle|^2 \delta(E_s + E_n - E_i) \qquad (2.5.19)$$

and dividing by the incoming flux in $|\mathbf{p}_i\rangle$ (see Appendix 2.3). The reason for this equivalence is of course contained in the discussion leading to equation (2.5.16); the particle exists inside the wave packet in what is, for practical purposes, a momentum eigenstate because the de Broglie wavelength is very small compared with the size of the wave packet.

2.5.2.3. The Hamiltonian

The Hamiltonian can be written in the representation afforded by the states $|\mathbf{p}, n\rangle$ which are product states of plane waves $|\mathbf{p}\rangle$ for the atom, and excitations $|n\rangle$ which need not refer to the isolated solid. The usual form is

$$H = \sum_{\mathbf{p}} |\mathbf{p}\rangle E_\mathbf{p} \langle \mathbf{p}| + \sum_n |n\rangle E_n \langle n| + \sum_{\mathbf{p}, \mathbf{p}'} |\mathbf{p}\rangle\langle\mathbf{p}| U |\mathbf{p}'\rangle\langle\mathbf{p}'|$$

$$+ \sum_{n,n'} \sum_{\mathbf{p},\mathbf{p}'} |\mathbf{p}, n\rangle\langle n, \mathbf{p}| W |\mathbf{p}', n'\rangle\langle n', \mathbf{p}'| \qquad (2.5.20)$$

$$= H_0 + U + W$$

where U is the elastic and W the inelastic part of the scattering potential. We note that in equation (2.5.1), $H_A + H_S$ is H_0, and H_{AS} is $H_0 + U$.

The replacement of the set of plane waves $\{|\mathbf{p}\rangle\}$ by the *contracted** set, $\{|C\rangle\}$ say, of elastic scattering states $\{|\mathbf{P}\rangle\}$ and stuck states $\{|s\rangle\}$, which are eigenstates of $H_0 + U$, so that

$$H \to \sum_C |C\rangle E_C \langle C| + \sum_n \langle n|E_n|n\rangle$$

$$+ \sum_{n,n'} \sum_{C,C'} |C, n\rangle\langle n, C|W|C', n'\rangle\langle n', C'| \qquad (2.5.21)$$

leads to problems. Thus, if $|\mathbf{P}_i; 0\rangle$ is the initial state and we calculate the T-matrix element

$$\langle n, s|W|\mathbf{P}_i; 0 + \rangle = \langle n, s|T(E_i)|\mathbf{P}_i; 0\rangle$$

which appears in equation (2.5.19) or (2.5.15) then using the Hamiltonian (2.5.21) the *result is exactly zero.*[13] If this T-matrix element is only calculated approximately, as $\langle n, s|W|\mathbf{P}_i; 0\rangle$, for example, the result is clearly not zero, and although such a result is, strictly speaking, spurious, it is validated by the gettering-theory approach which is mentioned below. This vanishing of the T-matrix element, and hence of the sticking coefficient, seems to arise because the Hamiltonian (2.5.21) cannnot describe a beam scattering experiment. In fact equation (2.5.21) has the form of a *gettering Hamiltonian*, which can provide an entirely different theoretical approach to sticking and inelastic scattering.[13,14] Gettering was the name given to the process of removing residual gas from a vacuum tube by adsorbing it on a reactive metal. The same term is applied to the decay of an atomic *standing wave* state in a vessel due to sticking collisions with the walls, and the Hamiltonian (2.5.21) can describe this situation exactly. However, the decay of a prepared standing wave state is an idealization of gettering not so far realized in practice. The importance of the gettering Hamiltonian (2.5.21) is that it is significantly easier to handle than the Hamiltonian (2.5.20), because the basis states used to write the Hamiltonian (2.5.21) appear again in the definition of the *gettering T*-matrix, and this is not the case for the scattering theory approach based on equations (2.5.13) and (2.5.20). Owing to this simplification, Brivio and Grimley[13,14] were able to compute sticking and inelastic scattering essentially exactly for a simple theoretical model of an H/metal system. It should be said, however, that the model only allowed energy loss to *single excitations* (in the case computed, to single electron-hole pairs); multiexcitation final states pose a computational problem so far unsolved.

---

* This terminology comes from quantum chemistry. The members of a contracted basis set are fixed linear combinations of the original (uncontracted) basis.

2.5.2.4. Sticking on a Solid at Finite Temperature

It was noted in Section 2.5.2.1 that in order to investigate sticking, we confine attention to times short compared with the residence time t_d, but much interesting chemical physics is contained in the time evolution of the initial state taken all the way through t_d to infinity. These aspects will not be considered here.

The situation now is that, at time t' ($t' \to -\infty$), the system is represented by a *density operator* $\rho(t')$ describing the gas atom wave packet $|\phi(t')\rangle$ and the solid in a state specified by the density operator $\rho_S(t')$:

$$\rho(t') = |\phi(t')\rangle\rho_S(t')\langle\phi(t')|$$

Function $\rho(t')$ evolves under the full Hamiltonian H, and at time $t > t'$

$$\rho(t) = \exp\left[-iH(t - t')\right]\rho(t')\exp\left[iH(t - t')\right]$$

It will be assumed that $\rho_S(t')$ describes the canonical ensemble at temperture T,

$$\rho_S(t') = \exp\left(-\beta H_S\right)/Z_S$$

where $\beta = 1/k_B T$ (k_B is Boltzmann's constant) and

$$Z_S = \text{Trace} \exp\left(-\beta H_S\right)$$

is the partition function. The eigenstates of H_S being $|n\rangle$, the probability $P_s(t)$ to observe the gas atom in a stuck state $|s\rangle$ irrespective of the excitation state is (note that $|s, n\rangle = |s\rangle|n\rangle$)

$$P_s(t) = \sum_n \langle n, s|\rho(t)|s, n\rangle$$

$$= \sum_n \exp\left(-\beta E_n\right)\sum_{n'} |\langle n', s|\Phi^+_{in}(t)\rangle|^2/Z_S$$

where $|\Phi^+_{in}(t)\rangle = \exp\left[-iH(t - t')\right]|\phi(t')\rangle|n\rangle$ is the state which evolves from $|\phi\rangle|n\rangle$ at time t' under the influence of H. The quantity $|\Phi^+_{in}\rangle$ is the obvious generalization of $|\Phi^+_i\rangle$ in Section 2.5.2.2, and we easily find that

$$P_s = \sum_n \exp\left(-\beta E_n\right)\sum_{n'} |B(s, n, n')|^2/Z_S$$

$$(2.5.22)$$

$$B(s, n, n') = 2\pi i \int d\hat{p}\, A(\hat{p}, E_{sn'})\langle n', s|V_{\text{AS-K}}|\hat{p}, E_{sn'}; n+\rangle$$

Compared with equation (2.5.12) the general excitation state $|n\rangle$ replaces the ground state $|0\rangle$, and the corresponding Boltzmann factor appears in the formula for P_s. As in Section 2.5.2.2, the excitations may, or may not, refer to the isolated solid; it depends how the inelastic operator W is defined (see Section 2.5.2.5).

2.5.2.5. Loss to Electrons

If $\mathbf{R}$ is the position of the gas atom and $\mathbf{r}$ denotes all coordinates necessary to describe the electrons in the system, then with all nuclei in the solid clamped

$$H = K_A(\mathbf{R}) + H_{EI}(\mathbf{r}, \mathbf{R}) \qquad (2.5.23)$$

Here K_A is the kinetic energy operator for the gas atom and H_{EI} is the Hamiltonian for the electrons in the field of the nuclei. The exact eigenfunctions are denoted by $\chi(\mathbf{r}, \mathbf{R})$, where

$$(H - E)\chi(\mathbf{r}, \mathbf{R}) = 0 \qquad (2.5.24)$$

If $\{\phi_m(\mathbf{r}; \mathbf{R})\}$ is a complete orthonormal set of functions of the electron coordinates with the gas atom clamped at $\mathbf{R}$, then we can write

$$\chi(\mathbf{r}, \mathbf{R}) = \sum_m \phi_m(\mathbf{r}; \mathbf{R})\psi_m(\mathbf{R}) \qquad (2.5.25)$$

When this expression is substituted into equation (2.5.24), the result multiplied by ϕ_n, integrated on $\mathbf{r}$, and Dirac brackets $\langle \ \ | \ \ \rangle$ used to denote the $\mathbf{r}$-integration, is

$$\sum_m (\langle \phi_n | K_A \phi_m \psi_m \rangle + \langle \phi_n | H_{EI} - E | \phi_m \rangle \psi_m) = 0 \qquad (2.5.26)$$

By noting that

$$\langle \phi_n | K_A \phi_m \psi_m \rangle = -(\hbar^2/2M)(\langle \phi_n | \phi_m \rangle \nabla_{\mathbf{R}}^2 \psi_m + 2\langle \phi_n | \nabla_{\mathbf{R}} | \phi_m \rangle \nabla_{\mathbf{R}} \psi_m$$

$$+ \langle \phi_n | \nabla_{\mathbf{R}}^2 | \phi_m \rangle \psi_m)$$

equation (2.5.26) can be expressed in the form

$$(K_A + E_n - E)\psi_n + \sum_m C_{nm}\psi_m + \sum_{m \neq n} \langle \phi_n | H_{EI} | \phi_m \rangle \psi_m = 0 \qquad (2.5.27)$$

where

$$E_n(\mathbf{R}) = \langle \phi_n | H_{EI} | \phi_n \rangle$$

and

$$C_{nm}(\mathbf{R}) = (-\hbar^2/M)(\langle \phi_n | \nabla_{\mathbf{R}} | \phi_m \rangle \nabla_{\mathbf{R}} + \tfrac{1}{2}\langle \phi_n | \nabla_{\mathbf{R}}^2 | \phi_m \rangle) \qquad (2.5.28)$$

If the second and third terms on the left in equation (2.5.27) could be neglected, the functions $\psi_n(\mathbf{R})$ would describe the gas atom in an energy eigenstate on the *potential energy surface* $E_n(\mathbf{R})$. If these eigenstates are labeled α, they satisfy the equation

$$(K_A + E_n(\mathbf{R}) - E_{n\alpha})\psi_{n\alpha} = 0 \qquad (2.5.29)$$

where $\psi_{n\alpha}$ is the αth state on the nth potential energy surface $E_n(\mathbf{R})$. From equation (2.5.25) the complete wave functions would be

$$\chi_{\alpha n}(\mathbf{r}, \mathbf{R}) = \phi_n(\mathbf{r}; \mathbf{R})\psi_{n\alpha}(\mathbf{R}) \qquad (2.5.30)$$

However, equation (2.5.27) tells us that the Hamiltonian (2.46) is not diagonal in the basis (2.5.30). In fact, if we denote the states corresponding to those in equation (2.5.30) by $|\alpha, n\rangle$, then

$$H = \sum_{\alpha,n} |\alpha, n\rangle E_{n\alpha}\langle n, \alpha| + \sum_{\alpha,\alpha'} \sum_{n,n'} |\alpha, n\rangle\langle n, \alpha|W|\alpha', n'\rangle\langle n', \alpha'| \qquad (2.5.31)$$

where

$$\langle n, \alpha|W|\alpha', n'\rangle = \int d\mathbf{R} \, \psi_{n\alpha}^*(\mathbf{R})C_{nn'}(\mathbf{R})\psi_{n'\alpha'}(\mathbf{R})$$

$$+ (1 - \delta_{nn'}) \int d\mathbf{R} \, \psi_{n\alpha}^*(\mathbf{R})\langle n|H_{EI}|n'\rangle\psi_{n'\alpha'}(\mathbf{R}) \qquad (2.5.32)$$

and the operator W causes transitions between the states (2.5.30) with both the electronic state and the state of motion of the gas atom changing. There are two contributions to W: the first (that involving $C_{nn'}$) depends on the velocity of the gas atom, the second depends on the choice of the basis set of electron functions $\phi(\mathbf{r}; \mathbf{R})$. Formal simplicity is achieved if we choose quantities ϕ_m to be the *adiabatic* electron wave functions satisfying

$$H_{EI}(\mathbf{r}, \mathbf{R})\phi(\mathbf{r}; \mathbf{R}) = E_m(\mathbf{R})\phi(\mathbf{r}, \mathbf{R}) \qquad (2.5.33)$$

because then the second term on the right in equation (2.5.32) vanishes, and the operator W is only nonzero because the velocity of the gas atom is nonzero. In this case, a gas atom approaching the solid infinitely slowly stays on the same *electronically adiabatic potential energy surface* $[E_0(\mathbf{R})$ for a cold solid], and no inelastic events occur. Inelastic processes, sticking and inelastic scattering, result from the finite velocity of the gas atom. We note that W is a *one-electron* operator in this case. If the choice (2.5.33) is not made, the basis is called *diabatic*, and because H_{EI} contains the Coulomb repulsions of the electrons, W now contains a two-electron operator.

In the gettering-theory approach, the initial state for a cold solid is the adiabatic state $|\alpha, 0\rangle$, so the Hamiltonian is used directly in the form of equation (2.5.31). The electronic excitations (electron–hole pairs) are defined in the coupled system, gas atom plus solid, so in this approach a state with one electron–hole pair corresponds in general to a superposition of states of the uncoupled system with $0, 1, 2, \ldots$ electron–hole pairs.

For the scattering-theory approach, the Hamiltonian (2.5.23) is expressed in terms of the basis $|\mathbf{p}, n\rangle$, which are products of plane waves $|\mathbf{p}\rangle$ for the atom and $\mathbf{R}$-dependent adiabatic electron states $|n\rangle$. By noting that

$$\langle n, \mathbf{p}|K_A|\mathbf{p}', n'\rangle = \delta_{nn'}\langle \mathbf{p}|K_A|\mathbf{p}'\rangle + \langle \mathbf{p}|C_{nn'}(\mathbf{R})|\mathbf{p}'\rangle$$

and

$$\langle n, \mathbf{p}|H_{EI}|\mathbf{p}', n'\rangle = \delta_{nn'}\langle \mathbf{p}|E_n(\mathbf{R})|\mathbf{p}'\rangle$$

we have, after using $\sum_n |n\rangle\langle n| = 1$,

$$H = \sum_{\mathbf{p}} |\mathbf{p}\rangle E_{\mathbf{p}}\langle \mathbf{p}| + \sum_{n,\mathbf{p},\mathbf{p}'} |n, \mathbf{p}\rangle\langle \mathbf{p}|E_n(\mathbf{R})|\mathbf{p}'\rangle\langle \mathbf{p}', n|$$

$$+ \sum_{n,n'}\sum_{\mathbf{p},\mathbf{p}'} |n, \mathbf{p}\rangle\langle \mathbf{p}|C_{nn'}(\mathbf{R})|\mathbf{p}'\rangle\langle \mathbf{p}', n'| \qquad (2.5.34)$$

where $E_{\mathbf{p}} = p^2/2M$. Comparison of this with equation (2.5.20) shows that the inelastic part of the scattering potential is given by

$$\langle n, \mathbf{p}|W|\mathbf{p}', n'\rangle = \langle \mathbf{p}|C_{nn'}(\mathbf{R})|\mathbf{p}'\rangle$$

but otherwise equation (2.5.34) does not have the form of equation (2.5.20). In order to obtain this form, it is necessary to make the assumption that the potential energy surfaces $E_n(\mathbf{R})$ all have the same shape, being simply shifted vertically by the electronic excitation energy E_n:

$$E_n(\mathbf{R}) = E_0(\mathbf{R}) + E_n \qquad (2.5.35)$$

Then the second term on the right in equation in (2.5.34) becomes

$$\sum_n |n\rangle E_n\langle n| + \sum_{\mathbf{p},\mathbf{p}'} |\mathbf{p}\rangle\langle \mathbf{p}|E_0(\mathbf{R})|\mathbf{p}'\rangle\langle \mathbf{p}'|$$

and we have exactly the form of equation (2.5.20). The assumption (2.5.35) will often be valid, but one can foresee[13] circumstances in which it will fail.

2.5.2.6. Loss to Phonons

The Hamiltonian is

$$H = K_A + K_L + V(\mathbf{R}, \mathbf{U}_L) \qquad (2.5.36)$$

where K_A and K_L are the kinetic energy operators for the gas atom A and the substrate lattice L, $\mathbf{R}$ is the gas atom's position, $\mathbf{U}_L$ is a many-dimensional vector specifying the displacements $\mathbf{u}_l$ of all the lattice atoms from their equilibrium positions, and $V(\mathbf{R}, \mathbf{U}_L)$ is the potential energy of the system.

The gas–solid potential, $V(\mathbf{R}, 0) = E_0(\mathbf{R})$ say, is separated for the rigid lattice [$E_0(\mathbf{R})$ is the ground-state electronically adiabatic potential energy surface] and $V(\mathbf{R}, \mathbf{U}_L)$ expanded in a Taylor series to obtain

$$
\left.
\begin{aligned}
H &= H_A + H_{ph} + W_{A-ph} \\[1em]
H_A &= K_A + E_0(\mathbf{R}) \\[1em]
H_{ph} &= K_L + V(\infty, \mathbf{U}_L) \\
&= K_L + \tfrac{1}{2} \sum_{l,l'} (\mathbf{u}_l \cdot \nabla_l)(\mathbf{u}_{l'} \cdot \nabla_{l'}) V(\infty, 0) + \cdots \\[1em]
W_{A-ph} &= \sum_l \mathbf{u}_l \cdot \nabla_l V(\mathbf{R}, 0) \\
&\quad + \tfrac{1}{2} \sum_{l,l'} (\mathbf{u}_l \cdot \nabla_l)(\mathbf{u}_{l'} \cdot \nabla_{l'})[V(\mathbf{R}, 0) - V(\infty, 0)] + \cdots
\end{aligned}
\right\}
\qquad (2.5.37)
$$

This model, with linear and quadratic couplings to the phonons, should be adequate for most investigations, and although the quadratic term is necessary for momentum conservation, the model with only the linear coupling term has a 50-year history and is still very useful. In connection with the Hamiltonian (2.5.37), one must beware of converting it into a gettering Hamiltonian by using a basis set representation with only the elastic scattering states of H_A in the basis, and then using scattering theory (see Section 2.5.2.3).

Another way of treating Hamiltonian (2.5.36) is to introduce $H_{d\,ph}$, the Hamiltonian for the phonons of the solid in the presence of the gas atom clamped at $\mathbf{R}$, i.e., the Hamiltonian for *displaced* or *adiabatic* phonons. Now

$$
H_{d\,ph} = K_L + V(\mathbf{R}, \mathbf{U}_L) - V(\mathbf{R}, 0) \qquad (2.5.38)
$$

so that

$$
H = H_A + H_{d\,ph} \qquad \text{and} \qquad H_A = K_A + V(\mathbf{R}, 0) \qquad (2.5.39)
$$

and subsequent developments follow exactly as in the electron case (Section 2.5.2.5) with adiabatic phonon states replacing the adiabatic electron states. Thus the coupling to adiabatic phonons has only linear and quadratic terms. There is no truncated Taylor series. This approach has recently been used[15] in the gettering-theory approach to examine some aspects of the sticking of H atoms into a Morse potential on copper.

Certainly the Hamiltonian has its simplest form if adiabatic phonon states are used in the basis, but the situation is slightly different from the electron case because the displaced phonon states having energies $E_n(\mathbf{R})$, a gas atom approaching a cold solid infinitely slowly, moves on the potential energy surface $V(\mathbf{R}, 0) + E_0(\mathbf{R})$, where $E_0(\mathbf{R})$, the displaced phonon ground-state energy, represents a certain polarization of the lattice by the gas atom at $\mathbf{R}$. This polarization energy is not difficult to compute for force-constant models of the substrate phonons when $V(\mathbf{R}, 0)$ is given. In the gettering-theory approach, we need the eigenstates of $H_A + E_0(\mathbf{R})$, and these are determined by computation after $E_0(\mathbf{R})$ has been found. Since the eigenstates of H_A would in general have to be found by computation, the situation regarding those of $H_A + E_0(\mathbf{R})$ is no different.

2.5.2.7. The Semiclassical Method (Trajectory Approximation)

In this approximation the gas atom is given its classical trajectory. The excitations of the system are treated quantum mechanically, and they are supposed simply to respond to the gas atom moving classically on the ground-state potential energy surface $E_0(\mathbf{R})$. The Hamiltonian for the excitations (the target) is

$$H = H_0 + V(t)$$

where H_0 is the Hamiltonian when the gas atom is far away. The initial state Φ_i of the target at $t \to -\infty$ evolves under H to Φ_f at $t \to +\infty$; Φ_i is an eigenstate of H_0. The state at time t, $\Phi(t)$, satisfies Schroedinger's time equation

$$i\hbar\, \partial\Phi/\partial t = H\Phi \qquad \text{with } \Phi(-\infty) = \Phi_i$$

The solution can be written as

$$\Phi(t) = U(t)\Phi_i$$

where $U(t)$ is the evolution operator satisfying

$$i\hbar\, \partial U/\partial t = HU \qquad \text{with } U(-\infty) = 1$$

We ask for the probability to find the target in the excited state Φ_α of H_0 as $t \to +\infty$ when the collision is over. Thus we need the amplitude

$$\langle \Phi_\alpha | \Phi_f \rangle = \langle \Phi_\alpha | \Phi(+\infty) \rangle = \langle \Phi_\alpha | U(t \to +\infty) | \Phi_i \rangle$$

For a cold target Φ_i is the ground state, and the above amplitude gives the probability of an energy loss E_α, the energy of Φ_α. The probability of energy loss E is therefore

$$P(E) = \sum_\alpha |\langle \Phi_\alpha | \Phi_f \rangle|^2 \delta(E - E_\alpha) \qquad (2.5.40)$$

and since

$$\sum_\alpha |\Phi_\alpha\rangle \delta(E - E_\alpha)\langle\Phi_\alpha| = \delta(E - H_0)$$

we can write equation (2.5.40)

$$P(E) = \langle\Phi_f|\delta(E - H_0)|\Phi_f\rangle \tag{2.5.41}$$

If E_A is the kinetic energy of the incoming atom, the sticking probability is given by

$$s = \int_{E_A}^\infty dE\, P(E) \tag{2.5.42}$$

because the atom cannot leave the potential well in $E_0(\mathbf{R})$ if it has lost energy greater than its initial energy E_A.

Once $V(t)$ is given, multiexcitation final states, which pose a serious computational problem in fully quantum mechanical models, can sometimes be included. But obtaining $V(t)$ is not easy because the response of the quantum system (the target) to the incoming particle reacts back on the particle to influence its trajectory, and so to modify $V(t)$. This reaction is expected to be large when the inelasticity is large ($s \sim 1$), which is just the situation where multiexcitation final states are important. In other words, when multiexcitation final states are populated, a self-consistent trajectory method is needed. A non-self-consistent trajectory approximation has recently been used[16] for Ne on (100)Cu with energy loss to phonons. The semiclassical method has been comprehensively dealt with in the literature.[17]

The Hamiltonian for which exact results can be obtained is that describing a system of forced Bosons,

$$H_0 = \hbar \sum_q \omega_q b_q^+ b_q \quad \text{and} \quad V(t) = \sum_q g_q(t)(b_q + b_q^+) \tag{2.5.43}$$

Here b_q^+ and b_q are the usual creation and destruction operators for Bosons of frequency ω_q, and $g_q(t)$ is determined from the gas atom's trajectory and the details of the Boson system. The index q contains a wave vector (whose perpendicular component will be complex for any surface modes) as well as another index, but these details do not concern us here. Function $V(t)$ is assumed to have the property that it vanishes for $t \to \pm\infty$, i.e., in the remote past and in the distant future, but this cannot be true for a self-consistent trajectory if the particle sticks, because then it remains coupled to the substrate Bosons forever. Overlooking this and other inconsistencies (nonconservation of energy, for example) the Hamiltonian (2.5.43) can obviously describe loss to phonons in the linear coupling approximation.[19] It can also describe loss to electrons because electron–hole pairs can be described in a Boson approximation.[17]

Our ability to treat the Hamiltonian (2.5.43) exactly stems from the fact that the equation of motion of the operator b_q can be integrated explicitly. Bearing in mind the equal-time commutation rule

$$[b_q(t), b_q^+(t)] = 1$$

and that the Boson operators commute otherwise, the equation of motion for $b_q(t)$ is

$$i\hbar\, \partial b_q^+(t)/\partial t = [b_q(t), H(t)] = \hbar\omega_q b_q(t) + g_q(t)$$

This first-order linear equation can be integrated at once. If we set

$$b_q(t) = a_q(t) \exp(-i\omega_q t)$$

and let $a_q(t \to -\infty) = a_{qi}$, the initial value, and $a_q(t \to +\infty) = a_{qf}$, the final value, then

$$a_{qf} - a_{qi} = -ig_q(\omega_q) \tag{2.5.44}$$

where

$$g_q(\omega) = \int_{-\infty}^{+\infty} dt \exp(i\omega t) g_q(t) \tag{2.5.45}$$

Owing to the simple result (2.5.44), we can find the unitary operator $\exp(S_q)$ which transforms a_{qi} into a_{qf} according to the relationship

$$a_{qf} = \exp(-S_q) a_{qi} \exp(S_q)$$

It is given by

$$S_q = -[ig_q(\omega_q)/\hbar](a_{qi} + a_{qi}^+) \tag{2.5.46}$$

Now, if the eigenvectors of $a_{qi}^+ a_{qi}$ and $a_{qf}^+ a_{qf}$ are denoted $|n_{qi}\rangle$ and $|n_{qf}\rangle$, respectively, the relation

$$|n_{qf}\rangle = \exp(-S_q)|n_{qi}\rangle \tag{2.5.47}$$

holds. If we start with a cold solid, then the mode q is in its ground state. In this case $|n_{qi}\rangle = |0_q\rangle$, and therefore after the collision is over its state is, according to equation (2.5.47)

$$|n_{qf}\rangle = \exp(-S_q)|0_q\rangle$$

Consequently, the amplitude to observe n_q Bosons in the mode q after the collision is over is

$$\langle n_q | n_{qf} \rangle = \langle n_q | \exp(-S_q) | 0_q \rangle$$

If we employ equation (2.5.46) and the identity

$$\exp(A + B) = \exp(A) \exp(B) \exp(-\tfrac{1}{2}[A, B])$$

for operators A and B which commute with their commutator, then the probability $P(n_q)$ to observe n_q Bosons in the mode q after the collision is given by the *Poisson distribution*

$$P(n_q) = \exp(-X_q)(X_q)^{n_q}/n_q!$$

$$X_q = [g_q(\omega_q)/\hbar]^2$$

(2.5.48)

This shows how the creation of many phonons in a single mode is taken into account. But the general multiexcitation final state has many phonons in many modes, and the loss function we need is

$$P(E) = \sum_{\substack{n_1, n_2, \ldots \\ (\sum n_q \hbar \omega_q = E)}} P(n_1) P(n_2) \cdots$$

(2.5.49)

If we work with the Fourier transform of $P(E)$, namely

$$P(t) = \int_{-\infty}^{+\infty} dE \exp(-iEt/\hbar) P(E)$$

then the E-integration removes the restriction on the n_q values in equation (2.5.49) and we have

$$P(t) = \exp(-2W) \prod_q \left\{ \sum_{n_q} [X_q \exp(-i\omega_q t)]^{n_q}/n_q! \right\}$$

$$= \exp(-2W) \exp\left[\sum_q X_q \exp(-i\omega_q t) \right]$$

(2.5.50)

where

$$2W = \sum_q X_q = \sum_q [g_q(\omega_q)/\hbar]^2$$

is the Debye–Waller factor. The required loss function is now obtained from the inverse transformation

$$P(E) = (2\pi\hbar)^{-1} \int_{-\infty}^{+\infty} dt \exp(iEt/\hbar) P(t)$$

It can be seen from equations (2.5.48) and (2.5.49) that the no-loss intensity is

$$P(0) = \exp(-2W)$$

while in a completely classical theory the no-loss intensity from a vibrating lattice is zero; elastic scattering from a vibrating lattice is a quantum phenomenon.

Appendixes

Appendix 2.1

The Lippmann–Schwinger equation (2.5.2) is obtained from

$$(E + i0 - H_A - H_S - V_{A-S})|\hat{p}, E; 0+\rangle = (E + i0 - H_A - H_S)|\hat{p}, E; 0\rangle$$

by premultiplying by $(E + i0 - H_A - H_S)^{-1}$. If we start from

$$(E + i0 - H_{AS} - H_K - V_{AS-K})|\hat{p}, E; 0+\rangle = (E + i0 - H_A - H_S)|\hat{p}, E; 0\rangle$$

and premultiply with $(E + i0 - H_{AS} - H_K)^{-1}$, then we obtain[18]

$$|\hat{p}, E; 0+\rangle = (E + i0 - H_{AS} - H_K)^{-1}$$
$$\times [(E + i0 - H_A - H_S)|\hat{p}, E; 0\rangle + V_{AS-K}|\hat{p}, E; 0+\rangle]$$

Consequently, because $|s, n\rangle$ is an eigenstate of $H_{AS} + H_K$ and $|\hat{p}, E; 0\rangle$ is an eigenstate of $H_A + H_S$, we have the result quoted in equation (2.5.11).

Appendix 2.2

To evaluate

$$\underset{t \to \infty}{\mathrm{Lim}} \int dz\, g(z)(z - i0)^{-1} \exp(izt)$$

we close the contour in the upper half-plane, where the exponential vanishes if $t \to \infty$. Then the integral is $2\pi i$ times the residue at the pole $z = i0$ inside the contour, namely $2\pi i\, g(0)$.

Appendix 2.3

We adopt the normalizations

$$\langle E', \hat{p}'|\hat{p}, E\rangle = \delta(E' - E)\delta(\hat{p}' - \hat{p})$$
$$= (M/\hbar^2 p)\delta(p' - p)\delta(\hat{p}' - \hat{p})$$

and

$$\langle p', \hat{p}' | \hat{p}, p \rangle = 2\pi\delta(p' - p)\delta(\hat{p}' - \hat{p})$$

whence

$$|\hat{p}, E\rangle = (M/2\hbar^2 p)^{1/2}|\hat{p}, p\rangle$$

Consequently equation (2.5.19) for the constant transition rate becomes, after passing to a continuous spectrum of excitations E_n,

$$R_i = (2\pi/\hbar)(2\pi\hbar^2 p_i/M) \int dE_n |\langle E_n, s| V_{\text{AS-K}}|\hat{p}_i, E_i; 0+\rangle|^2 \delta(E_s + E_n - E_i)$$

$$= (2\pi)^2(\hbar p_i/M)|\langle E_i - E_s, s| V_{\text{AS-K}}|\hat{p}_i, E_i; 0+\rangle|^2$$

Since the incoming flux is $\hbar p_i/M$, we see that

$$R_i/(\text{incoming flux}) = P_s$$

References

1. D. A. King, Private communication.
2. K. Hermann and P. Bagus, *Phys. Rev. B* **16**, 4195–4202 (1977).
3. J. C.Tracy, *J. Chem. Phys.* **56**, 2736–2742 (1972).
4. W. A. Steele, *The Interactions of Gases with Solid Surfaces*, Pergamon Press, Oxford (1974).
5. J. G. Dash, *Films on Solid Surfaces*, Academic Press, New York (1975).
6. S. Cerny, in: *The Chemical Physics of Solid Surfaces and Heterogeneous Catalysis* (D. A. King and D. P. Woodruff, eds.), Vol. 2, pp. 1–57, Elsevier, Amsterdam (1983).
7. T. L. Hill, *J. Chem. Phys.* **17**, 520–529 (1949); *Adv. Catal.* **4**, 212–263 (1952).
8. D. H. Everett, *Trans. Faraday Soc.* **46**, 453–463, 942–957, 957–965 (1950); *Proc. Chem. Soc. London*, 38–51 (1957).
9. R. G. Jones and A. W.-L. Tong, *Surf. Sci.* **188**, 87–106 (1987).
10. M. L. Goldberger and K. M. Watson, *Collision Theory*, Wiley, New York (1964).
11. R. D. Levine, *Quantum Mechanics of Molecular Rate Processes*, Oxford University Press, London (1969).
12. E. Merzbacher, *Quantum Mechanics*, Wiley, New York (1970).
13. G. P. Brivio and T. B. Grimley, *Phys. Rev. B* **35**, 5969–5974 (1987).
14. G. P. Brivio, *Phys. Rev. B* **35**, 5975–5984 (1987).
15. G. P. Brivio, T. B. Grimley, and A. Devescovi, *J. Electron Spectrosc. Relat. Phenom.* **45**, 391–402 (1987).
16. M. Persson and J. Harris, *Surf. Sci.* **187**, 67–85 (1987).
17. K. Schoenhammer and O. Gunnarsson, in: *Many Body Phenomena at Surfaces* (D. Langreth and H. Suhl, eds.), pp. 421–449, Academic Press, Orlando (1984).
18. B. A. Lippmann, *Phys. Rev.* **91**, 264–270 (1956).
19. J. I. Kaplan and E. Drauglis, *Surf. Sci.* **36**, 1–14 (1973).

3

Theory of Atom–Surface Collisions

W. Kohn

3.1. Introduction

The field of atom–surface collisions (ASC) has, in the last two decades, become an important and precise subfield of surface science. (Much of the material of this chapter applies also to collisions of ions and molecules with surfaces.) This progress has consisted of major experimental advances—preparation of better controlled, cleaner surfaces, preparation of very well-defined incident beams, development of high-resolution detectors—as well as significant theoretical progress. In the writer's opinion the experimental advances have outstripped theoretical progress, posing interesting challenges to the theoreticians.

What distinguishes ASC from other experimental techniques, such as radiation and electron (or positron) spectroscopies, is the comparatively *large mass* of the probe. This can lead to significant energy exchange with the *vibrations of the target atoms*, which are of great interest.

Related to the large mass is the comparatively *long time scale* of the interactions (typically 10^{-13} to 10^{-14} s for energies in the range of 0.02 to 0.5 eV), much longer than typical electronic time scales (10^{-15} s). As a result, in general, the electrons of the "incidon"–target system follow the instantaneous positions of the atoms (Born–Oppenheimer regime) and electronic excitations (or de-excitations) are unimportant. The subject of ASC is of interest for many reasons, including the following:

1. One of the central quantities in surface science is the incidon–target *interaction potential* $V(r)$ and its dependence on the displacements u_l of the target atoms; the phenomena of adsorption and desorption, surface diffusion, surface-mediated (or heterogeneous) catalysis, epitaxial growth are all governed by it. ASC provide rich information about this interaction and, in a few favorable cases, have allowed its accurate determination.

W. Kohn • Department of Physics, University of California at Santa Barbara, Santa Barbara, California 93106, USA.

2. Inasmuch as, roughly speaking, the repulsive part of the interaction is proportional to the unperturbed *electron density* $n(r)$ of the target, ASC provide some information about this quantity.
3. ASC are an excellent means* for studying those *vibrational modes of the target* (either clean or with adsorbates) which have substantial amplitudes in the last surface layer, especially, of course, localized surface modes.
4. ASC are the simplest case of *molecule–surface collisions* and provide information about the role of the translational degree of freedom of the incidon.
5. Finally, ASC represent a challenging case of a many-body problem with unique features: the target consists of N ($\to\infty$) harmonically coupled particles in thermal equilibrium, the incidon interacts strongly with only the outermost one or two layers, necessarily nonharmonically (since the interaction vanishes at large distances). It is the interesting task of the theorist to develop useful and manageable methodologies for this system.

In this chapter we shall limit ourselves to incidon energies in the approximate range of 0.01 to 1 eV. In this range incident beams with energy resolutions in the range of 1-5% can be obtained and the interatomic forces of the target can (except for quantum crystals, like solid He) be adequately described as harmonic.

We divide our material into two parts: Part 1 treats the target atoms as fixed in their equilibrium positions. This is a good approximation when the target atoms are at a low temperature and sufficiently heavy so that zero-point oscillations are small, and when the incident atom is of sufficiently low energy and mass to minimize the perturbation of the target atoms. In this regime the quantity of central interest is the incidon–target interaction potential $V(r)$, which requires a knowledge of the many-body ground state energy as a function of r. If $V(r)$ is considered as known, the motion of the incidon is described by a simple one-particle Schroedinger equation.

In Part 2, we include the dynamics of the target atoms. We shall regard the interaction potential $V(r; u_1, u_2, \ldots, u_N)$ as known and discuss various approaches to deal with the $(N + 1)$-particle dynamics.

3.2. Part 1: The Rigid Target Model

3.2.1. The Atomic–Surface Interaction Potential—*A Priori* Calculations

3.2.1.1. Long-Range Polarization (van der Waals) Attraction

The asymptotic interaction of a distant incidon with the target is of the form

$$V(r) \sim -Cz^{-3} \tag{3.2.1}$$

where z is the distance from the surface. It is a polarization potential of the same general nature as the well-known van der Waals r^{-6} potential between two atoms.

* The other, competitive, technique is electron energy loss spectroscopy (EELS).

The form (3.2.1) can be obtained by the following elementary, *rough* argument. For large z, the incidon-target interaction can be regarded as the sum of the r^{-6} van der Waals potentials between the incidon and the individual target atoms and the sum can be replaced by a three-dimensional integral. The elementary integration accounts for the difference in powers between r^{-6} and z^{-3}. However, in a classic paper,[1] Lifshitz derived an *exact* expression for the coefficient C of equation (3.2.1) in terms of the dielectric constant $\varepsilon(\omega)$ of the target material and the polarizability $\alpha(\omega)$ of the incidon:

$$C = \frac{1}{4\pi} \int_0^\infty dn\, \alpha(iu) \left(\frac{\varepsilon(iu) - 1}{\varepsilon(iu) + 1} \right) \qquad (3.2.2)$$

where $\varepsilon(\omega)$ and $\alpha(\omega)$ are independently measurable, optically. The constant C has been computed for various incidon-target combinations, such as rare gas incidons-noble metal targets[2,3] and He-LiF.[4,5]

Expression (3.2.1) is the leading term in an asymptotic expansion. If the origin of the z-axis is selected in some definite (but arbitrary) way, say at the plane passing through the last layer of nuclei, then, up to the next order in z^{-1}, we can write

$$V(r) \sim -\frac{C}{z^3} + \frac{D}{z^4} + \cdots = -\frac{C}{(z - \bar{z})^3} + O(z^{-5}) \qquad (3.2.1)$$

where $\bar{z} \equiv -D/3C$ and can be regarded as the physical reference plane for the Lifshitz potential. Since for ASC the relevant values of z are small (approximately 1-2 Å), an estimate of $\bar{z}$ is important. This was provided elsewhere[2,3] for rare gas-noble metal systems. Celli *et al.*[5] deal with the higher-order terms in equation (3.2.1') in a different, semiempirical manner for the LiF system.

3.2.1.2. The Potential at Shorter Distances

From the quantum theory of molecules one knows that the total interaction between two atoms has the qualitative shape shown in Figure 3.1, where r_0 is the

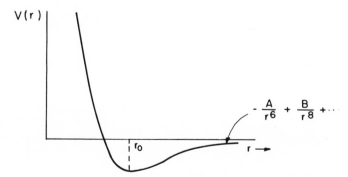

Figure 3.1. Interaction potential between two atoms (schematic).

equilibrium interatomic distance. For $r < r_0$ the potential rises rapidly due to Coulomb repulsion and electron exchange effects (correlation slows the rise); for $r < r_0$ the van der Waals attraction is dominant.

For a first orientation, the interaction potential between an incidon and a surface can be regarded as the sum of all incidon–surface atom potentials. This potential is obviously invariant under translations τ_1, τ_2 in the x-y plane which take the target into itself. Consequently, it can be written in the form

$$V(r) = V_0(z) + \sum_{g \neq 0} V_g(z) \, e^{ig*\rho} \tag{3.2.3}$$

where g are the two-dimensional reciprocal lattice vectors corresponding to τ_1 and τ_2 and $\rho \equiv (x, y)$. The function $V_0(z)$ is the interaction potential averaged over x and y and has the qualitative form shown in Figure 3.2, where we note the long-range z^{-3} attraction and the laterally averaged short-range repulsion of the atoms. The coefficients $V_g(z)$ of the "corrugation potential" [the sum in equation (3.2.3)] are all decreasing exponentially (see Figure 3.2).

The above features of the potential $V(r)$ derived from a sum of incidon–target atom interactions are, in fact, also features of the *exact* $V(r)$. This may be demonstrated by using the exact lateral periodicity of the Hamiltonian and the fact that the corrugated parts of the polarization interaction all decay exponentially.

A major *unsolved problem* is the formulation of a unified systematic theory which will yield simultaneously the long-range attraction in $V_0(z)$ as well as the short-range behavior of $V_0(z)$ and $V_g(z)$. In the absence of such a theory $V(r)$ is written in the form

$$V(r) = V^{\text{long}}(z) + V^{\text{short}}(r) \tag{3.2.4}$$

and, by a combination of theory and experiment, an attempt is made to determine the long- and short-range parts of $V(r)$. To the writer's knowledge, however, the separation into long- and short-range terms has not been sharply defined in a

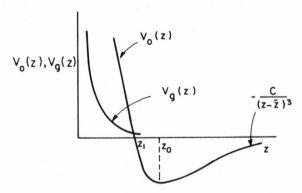

Figure 3.2. The laterally averaged potential $V_0(z)$ and a coefficient $V_g(z)$ of the corrugation potential (schematic); see equation (3.2.3).

physically and mathematically satisfactory way. The difficult region, where polarization and electron-overlap forces are comparable, lies typically near the minimum (z_0) of $V_0(z)$ and therefore is generally of great importance. For $z \gg z_0$, the accurately known Lifshitz force is dominant; for $z \ll z_0$, local density functional theory (and possibly the effective medium theory; see below) gives satisfactory descriptions. For more detailed discussion see a recent paper by Takada and Kohn.[6]

With these caveats, several approaches to the calculation of $V^{short}(r)$ will be described below.

a. Density Functional Theory. Lang and Williams[7] have obtained accurate energies for incidon–jellium systems, for H, Li, O, Na, Si, Cl, as a function of z. The jellium (extending over the interval $-\infty < z \le 0$) is a much-used model for a metal surface. Unfortunately, this paper does not include the rare gas atoms He and Ne, which are used extensively in ASC. Also, of course, in this model $V(r)$ is independent of x and y, i.e., the corrugation portion of $V(r)$ is absent. Finally, since the local density approximation is used, there is no long-range polarization.

b. Heitler–London Theory. In the spirit of the Heitler–London approach to molecular energies, one can in principle use trial functions of the form

$$\Psi = P\Psi_t \Psi_i \qquad (3.2.5)$$

where Ψ_t and Ψ_i are, respectively, trial functions for the target and the incidon, and P is the electron permutation operator. The simplest choice is to take for Ψ_t and Ψ_i trial functions for the bare target and the isolated incidon (centered at r); but one can also allow for *mean* mutual peturbations by allowing Ψ_t and Ψ_i to depend on r. Such functions can include intraincidon and intratarget correlation effects, but they do not include the dynamic correlations between incidon and target electrons which are responsible for the long-range van der Waals forces.

c. Hartree-Fock Theory. It is well known from studies of diatomic molecules that the Hatree-Fock equations may have solutions of two kinds: (1) the eigenfunctions are localized on one or another of the constituent atoms; or (2) they can be extended over both atoms.

The interaction of He with noble metals corresponds to regime (1) above. It was treated by Zaremba and Kohn[8] and by Harris and Liebsch[9] using an antisymmetrized product of a Slater determinant of functions to describe the metallic electrons (perturbed by the He atom), and a Slater determinant of two localized functions to describe the He $(1s)^2$-state. This resulted in an exponentially decreasing, repulsive interaction. To this was added the z^{-3} Lifshitz attraction. It is believed that, *in this particular case*, such a simple addition of short-range repulsion and long-range attraction should be quantitatively rather accurate, because the point z_1 (see Figure 3.2) separating the attractive and repulsive regions is situated far (3-4 Å) from the surface.

d. S-Matrix Theory.[6,10,11] There is an *exact* way of obtaining the repulsive interaction between an incidon and target for large z, *if* the Coulomb interactions are cut off at some finite, arbitrary distance a. In that case the interaction energy of the incidon with the evanescent target electrons, whose energy E relative to the vacuum level is *negative*, can be obtained by appropriate analytic continuation from the scattering phase shifts $\eta_l(E)$, describing collisions of *positive* energy electrons with the incidon.[6] This theory applies in the usual case when the ionization energy of the incidon is greater than the work function of the atom.

The results can be expressed as follows. We let $n^0(r)$ be the unperturbed density of the target electrons and Fourier expand it in the form

$$n^0(r) = n_0(z) + \sum_{g \neq 0} n_g(z) \, e^{ig \cdot \rho} \tag{3.2.6}$$

Similarly, we expand the interaction potential as

$$V(r) = V_0(z) + \sum_{g \neq o} V_g(z) \, e^{ig \cdot \rho} \tag{3.2.7}$$

Then, for $z \to \infty$, one obtains the proportionalities

$$V_0(z) = \alpha_0 n_0(z) \quad \text{and} \quad V_g(z) = \alpha_g n_g(z) \tag{3.2.8}$$

where the coefficients α_0 and α_g are independent of z but do depend on both the incidon and the target. The coefficient α_0, for example, is given by

$$\alpha_0 = \sum_{l=0}^{\infty} F_l \tag{3.2.9}$$

where

$$F_l = \pi(2l + 1)(-1)^l \{S_l[i(2W)^{1/2}] - 1\}/2\omega^{1/2} \tag{3.2.10}$$

W is the target work function, and

$$S_l(k) \equiv \exp[2i\eta_l(k) - 1 \tag{3.2.11}$$

is the S-matrix element corresponding to angular momentum l and wave number k. For real k this element can be determined from electron–incidon scattering experiments or calculations. The quantity $S_l(k)/k$ is an analytic function of $E = k^2$ and must be continued to negative energies. (If the Coulomb interactions are *not* cut off, this theory does not work. The long- and short-range forces become entangled and $S_l(k)/k$ is not analytic at $E = 0$.)

Specifically, for He incident on Cu one finds $\alpha_0 \approx 490 \, \text{eVa}_0^3$; the values of α_g (for the smallest value of g) were subject to controversy. In principle α_g can be very different from α_0 (Takada and Kohn[6] proposed a value of $270 \, \text{eVa}_0^3$). However, for this particular incidon–target pair and this particular g, the most likely value of α_g is about α_0.[10,11] We remark here that this value of α_g gives a corrugation potential about twice as large as that deduced empirically from diffraction scattering. As far as we know this major failure of the theory still lacks a satisfactory explanation.

e. Effective Medium Theory.[12-14] This may be regarded as a simplified version of the S-matrix theory just described. It is also conceptually very closely related to density functional theory.

Let us consider a given atom placed in *uniform* electron gas of density n_0 (we again ignore the troublesome issue of charge neutrality due to the infinite range of the Coulomb forces). The total energy is expressed in the form

$$E(n_0) = E_0(n_0) + E_i + \Delta E(n_0) \tag{3.2.12}$$

where $E_0(n_0)$ is the energy of the gas without the atom, E_a is the energy of the isolated atom, and ΔE, by definition, is the interaction energy.

For an *inhomogeneous* electron density, such as that outside the target surface, we can write, *if density gradients are ignored,*

$$E = E_0[n(r)] + E_i + \Delta E[n(r)] \tag{3.2.13}$$

where the first and second terms describe the energies of the infinitely separated target–incidon system, and the third term represents the interaction energy, depending only on the density at the position r of the incidon. For an incidon far from the surface, $n(r)$ is small. In this limit

$$\Delta E(n) = \alpha_{00} n \tag{3.2.14}$$

where α_{00} can be expressed in terms of the electron atom scattering length as

$$\alpha_{00} = \frac{2\pi\hbar^2}{m} a_s \tag{3.2.15}$$

For He, $\alpha_{00} = 330 \text{ eVa}_0^3$. This may be compared with the value of 490 eVa_0^3 from the complete S-matrix theory of the He–Cu system. Thus we see that for realistic incidon–target systems, the neglect of density gradients in the effective medium theory leads to significant errors for α_0. The effective medium theory also neglects the g-dependence of α_g (this is, of course, due to the neglect of density gradients). Various refinements of this theory have given uncertain results for α, in the range of about $200\text{--}1000 \text{ eVa}_0^3$.

f. Generalized Valence Bond Method.[15] This method is a generalization of the Heitler–London and Hartree–Fock methods. It is limited to finite systems, which are described by trial functions built up from atomic-type orbitals of well-defined angular momenta whose amplitudes and radial behavior are chosen so as to minimize the total energy. This highly computer-intensive variational method has been used for atom–surface systems by replacing the target by a suitable finite cluster of atoms. It is primarily geared for systems with chemical bond formation and less so for simple metal targets and rare gas incidons.

3.2.2. Calculations of Atom–Surface Scattering

In Section 3.2.1 we saw how *a priori* calculations can lead to more or less accurate approximations for the incidon–surface interaction potential $V(r)$. To predict or understand experimental results one must solve the one-particle Schroedinger equation for the incidon,

$$\left(-\frac{\hbar^2}{2M} \nabla^2 + V(r) - E \right) \psi(r) = 0 \tag{3.2.16}$$

As a result of the lateral periodicity of $V(r)$ given by equation (3.2.3) the solutions of equation (3.2.16) have lateral Bloch-wave character, i.e., they have the form

$$\psi_k(r) = \sum_g e^{i(k+g)\cdot\rho} \psi_g(z) \tag{3.2.17}$$

Substitution into equation (3.2.16) yields the coupled equations

$$-\frac{1}{2M} \frac{d^2\psi_g(z)}{dz^2} + \sum_g V_{g-g'}(z)\psi_{g'}(z) = k_{gz}^2 \psi_g(z) \tag{3.2.18}$$

where

$$V_{g-g'}(z) = \int e^{-i(g-g')\cdot\rho} V(r)\, d\rho \qquad \text{and} \qquad k_{gz}^2 = \frac{2ME}{\hbar^2} - (k+g)^2 \tag{3.2.19}$$

These equations must be solved subject to the asymptotic conditions

as $z \to -\infty$, $\quad \psi_g(z) \to 0$ $\hspace{5cm}$ (3.2.20a)

as $z \to +\infty$, $\quad \psi_0(z) \to \exp(-k_z z) + c_0 \exp ik_z z \quad$ and $\quad \psi_g(z) \to c_g \exp ik_{gz} z$
$$\hspace{11cm} \text{(3.2.20b)}$$

The outgoing waves represent specular ($g = 0$) and diffraction ($g \neq 0$) scattering. Some or all wave numbers k_{gz} are imaginary [see equation (3.2.19)] describing evanescent waves.

The solution of the scattering problem (3.2.18)–(3.2.20) is elementary. First, one truncates the sum over g' in equation (3.2.18), including say $\bar{N}$ two-dimensional g-vectors. Next, one chooses a point z_0 well behind the classical turning point (really, surface) and, instead of condition (3.2.20a), one imposes the boundary conditions

$$\psi_g^\mu(z_0) = 0 \qquad \text{and} \qquad d\psi_g^\mu(z)/dz = \delta_{gg}^\mu, \qquad \mu = 1, \ldots, \bar{N} \tag{3.2.21}$$

Now one integrates numerically the $\bar{N}$ coupled differential equations (so-called coupled channel equations) forward toward $z \to \infty$. Of course, each ψ_g^μ does *not* satisfy the asymptotic conditions (3.2.20b). However, the totality of $\bar{N}$ solutions ψ_g^μ represents a complete system and one can select the linear superposition which does satisfy them.

In general, the Schroedinger equation (3.2.16) has both scattering states $(E > 0)$ and bound states $(E < 0)$. The *a priori* calculated $V(r)$ may then be adjusted empirically to fit both binding energies and scattering amplitudes. For He scattering, a corrugated hard-wall (CHW) model often gives satisfactory results. It approximates $V(r)$ by

$$V_{\text{CHW}}(r) = \infty \qquad z \le \xi(\rho)$$

$$= 0 \qquad z > (\rho) \qquad (3.2.22)$$

where

$$\xi(\rho) = \sum_0^g \xi_g(E)\, e^{ig \cdot \rho} \qquad (3.2.23)$$

and usually one or two inequivalent gs are sufficient. The position of the CHW depends on the incident energy E; it is given, qualitatively, by the locus of classical turning points. When the attractive van der Waals force is important, such as near a resonance with an $E > 0$ "bound" state which is localized in the z-direction but traveling in the ρ-plane, the CHW model fails.

In this way, in some favorable cases an accurate and apparently unique interaction potential has been obtained which fits, with an error of about 1%, both bound state energies as well as diffraction intensities over a wide energy range.[19]

3.2.3. The Atom–Surface Interaction Potential from Empirical Data

It follows from the lateral periodicity of the incidon–target interaction $V(r)$, given by equation (3.2.3), that the incidon eigenfunctions have the lateral Bloch form

$$\psi_{n,k}(r) = u_{n,k}(r)\, e^{ik \cdot \rho} \qquad (3.2.24)$$

where $\rho \equiv (x, y)$, k is a two-dimensional wave vector in the fundamental lateral Brillouin zone, and $u_{n,k}$ is laterally periodic satisfying

$$u_{n,k}(r + \tau_j) = u_{n,k}(r), \qquad j = 1, 2 \qquad (3.2.25)$$

n denoting the remaining quantum numbers. When coupling between the lateral and perpendicular motions is negligible, n refers to the motion in the z-direction as well as to lateral band indices. The function $u_{n,k}(r)$ may always be laterally Fourier-analyzed as follows:

$$u_{n,k}(r) = \sum_g e^{i(k+g) \cdot \rho} f_{n,g+k}(z) \qquad (3.2.26)$$

There are now two types of solution: (1) bound states, in which *all* fs decay exponentially with z, and (2) scattering states in which one or more fs extend sinusoidally to $z = +\infty$.

3.2.3.1. Bound State Levels; $V_0(r)$

For the negative energy bound states, the classical turning points lie at relatively large values of z. Except for very highly corrugated surfaces, the eigenfunctions extend only slightly into the regions in which $V_g(z)$ is appreciable (see Figure 3.2) and therefore, to a good approximation, these functions have the simple form

$$\psi_{m,g,k}(r) = e^{i(k+g)\cdot\rho}f_m(z) \tag{3.2.27}$$

where $f_m(z)$ satisfies the one-dimensional Schroedinger equation

$$\left[-\frac{\hbar^2}{2m}\frac{d^2}{dz^2} + V_0(z) - E_m \right] f_m(z) = 0 \tag{3.2.28}$$

Since the shape of function $V_0(z)$ is known qualitatively (see Figure 3.2) one can make small adjustments, $\delta V_0(z)$, to fit observed bound states energies, ε_m. These energies can be determined experimentally by several methods; ε_0, the lowest eigenvalue, is the ionization energy of the adsorbed atom in its ground state and is reflected in adsorption isotherms.[16] Differences $E_m - E_0$ are obtainable from optical excitations of the adatom vibrations.[16] Finally, the quantities E_m show up as resonances in elastic specular and diffraction scattering.[17]

3.2.3.2. Elastic Diffraction Scattering; $V_g(z)$

Elastic diffraction scattering is caused by the corrugation potentials $V_g(z)$. Recent experiments with rare gas incidons show of the order of 5–10 diffraction peaks.[18] Their positions are, of course, determined by consideration of lateral momentum, modulo g, and total energy:

$$k' = k + g; \qquad k_z'^2 + k'^2 = k_z^2 + k^2 \tag{3.2.29}$$

However, for given $V_0(z)$, the *intensities* depend on the components of the corrugation potential $V_g(z)$. Since, experimentally, the diffraction intensities depend on the angle of incidence and on the incident energy, they provide extensive information about the functions $V_g(z)$. (In practice, only one or two independent functions $V_g(z)$ are significant.) For *fixed* incidon energy, the hard corrugated surface (HCS) model has been useful, in which the *total* potential $V(r)$ is represented by

$$V(r) = +\infty, \qquad z < \xi_0 + \xi(x, y)$$

$$= 0, \qquad z > \xi_0 + \xi(x, y) \tag{3.2.30}$$

i.e., the incidon is viewed as being scattered by an impenetrable, corrugated wall. However, to fit the data, the shape of the wall $\xi(x, y)$ must be taken as energy dependent. There is an important interplay between the bound states of equation (3.2.28) (with $V_g = 0$) and the scattering states. If conservation of lateral crystal momentum and of energy are satisfied, i.e.,

$$\frac{\hbar^2}{2m}(k^2 + k_z^2) = E_m + \frac{\hbar^2}{2m}(k + g)^2 \qquad (3.2.31)$$

then the scattering state (k, k_z) is coupled resonantly to the bound state $(E_m, k + g)$ by $V_g(z)\, e^{ig \cdot \rho}$. Since the HCS model has no bound states, it cannot describe these important phenomena and more realistic models of $V(r)$ are needed.

3.2.3.3. Debye–Waller Reduction of Elastic Scattering Processes

We have so far discussed only elastic collisions and bound states, and not included any real transitions of the lattice (phonon emission and absorption). However, as is well known, the existence of real, inelastic channels reduces the intensity of the elastic channels by the so-called Debye–Waller factors, conventionally written as e^{-W}. This factor depends strongly on the momentum transfer and the target temperature. Very qualitatively

$$W \sim \tfrac{1}{2}\langle (q \cdot u)^2 \rangle \qquad (3.2.32)$$

where q is the momentum transfer and u is the excursion of a surface atom from its equilibrium position. For incidons with atomic weight in excess of, say, 40 even at the lowest available incidon energies (about 0.01 eV) and target temperatures, this reduction is so considerable that elastic collisions can generally not be seen against the broad background of inelastic collisions. The most favorable atoms for the observation of elastic specular and diffraction scattering are low-energy He atoms, for which the Debye–Waller factor aproaches 1 at low temperatures.

The Debye–Waller factor provides a measure of the adequacy of a description of elastic collisions in terms of a static interaction potential $V(r)$. When this factor is substantially less than 1, the dynamics of the target atoms must be taken into account. This subject is addressed in Part 2 in the following section.

3.3. Part 2: Dynamics of Interacting Incidon and Target Atoms

3.3.1. Formulation

In Part 1, we have treated the target atoms as stationary while using the Born–Oppenheimer (B–O) approximation for target and incidon electrons. This resulted in a one-particle Schroedinger equation for the incidon in which the potential $V(r)$ was given by the total ground state energy of stationary target nuclei in their equilibrium positions, a stationary incidon in position r, and all electrons.

This model is correct semiquantitatively for *light, low-energy incidons and heavy target atoms*, e.g., low-energy He–Ag scattering. In line with the model, most of the observed scattering is elastic.

However, when the conditions just mentioned are not satisfied, the dynamics of the target atoms must be included. In most cases the B–O approximation for electrons is still very accurate and we shall ignore real excitations of electrons.*

The B–O Hamiltonian for the target-incidon system is then given by

$$H = H_{\text{target}} + H_{\text{inc}} + H' \tag{3.3.1}$$

where

$$H_{\text{target}} = \sum_{1}^{N} \frac{p_l^2}{2M_t} + U(u_1, \ldots, u_N) \tag{3.3.2}$$

Here, u_l and p_l are, respectively, displacements and momenta of the target atoms of mass M_t and U is their B–O potential energy in the absence of the incidon. We shall assume harmonic forces, i.e.,

$$U = \tfrac{1}{2} \sum A_{ll'} u_l u_l' \tag{3.3.3}$$

H_{inc} is given simply by

$$H_{\text{inc}} = p^2/2M \tag{3.3.4}$$

where p and M are, respectively, the momentum and mass of the incidon at position r, and

$$H' = V(r, u_1, \ldots, u_N) \tag{3.3.5}$$

describes the interaction Hamiltonian. For most practical purposes, only displacements u_l of atoms in one, or at most two, layers significantly enter V. If the target atoms move only slightly compared to the range of incidon–target forces, then V can be represented adequately as

$$V(r; u_1, \ldots, u_N) \approx V(r) + \sum_l K_l(r) u_l \tag{3.3.6}$$

where $K_l(r)$ has the periodicity of the target surface. The potential $V(r)$, corresponding to all $u_l = 0$, is the static incidon–target interaction of Part 1.

In Part 2 we shall regard as known: the lattice force constants $A_{ll'}$ of equation (3.3.3), including their modified values near the surface; the eigenmodes and frequencies of the target; and the incidon–target interaction H' of equations (3.3.5) and (3.3.6). We shall address the problem of how an incidon, striking the surface with initial momentum k, is scattered into a final state of momentum k', or possibly trapped at the surface. We shall confine ourselves to the case where the target is at temperature $T = 0$.

---

* For scattering of rare gas atoms by metals, the probability of exciting a real electron–hole pair has been estimated as 10^{-5}. For electronically reactive incidons, electronic excitations may occur in special situations.

3.3.2. Regimes

At first sight the $(N + 1)$ quantum-mechanical many-body collision problem seems to be extremely formidable. Fortunately, in the multidimensional parameter space describing the collisions there are a number of regions ("regimes") where major simplifications are possible. Relevant parameters describing an atom–surface collision include: the incidon mass M, the mass of the target atoms M_t (if we limit ourselves to elemental targets), parameters characterizing the interaction potential, the well depth $-V_{00}$, the range d (more accurately, attractive and repulsive ranges d_a and d_r), the 2D surface lattice parameter a, the amplitude V_c of the corrugation potential at the point $\bar{z}$ where the laterally averaged potential vanishes, the energy of incidence E and a characteristic energy loss E_l, the Debye frequency of the target ω_D, and Planck's constant $\hbar$. From these parameters* seven (!) dimensionless parameters can be constructed, indicating dramatically the wide variety of possible regimes. We shall not attempt an exhaustive discussion of these but limit ourselves to a few important ones. (We recall again that the target is assumed to have temperature $T \approx 0$ K, i.e., $kT \ll$ all relevant energies.)

3.3.3. Classical Regime

Quantum effects will be negligible if the De Broglie wavelength of the incidon is small compared to the range d of the potential,† and if the energy loss E_l is large compared to ω_D. These are sufficient conditions. (The classical solutions may well give satisfactory results even when they are not satisfied.)

Calculations in this regime are straightforward, though highly computer intensive. One replaces the semi-infinite target by a finite slab containing N atoms (typically several hundred). For a given direction and energy of incidence one chooses a sufficient sampling of points of incidence and, for each, solves the $3(N + 1)$ coupled Newton's equation with the given initial conditions. Reliable results are entirely within the reach of present-day computers. In this regime, the theoretical uncertainties come entirely from the potential energy $V(r; u_1, u_2, \ldots, u_N)$. It should be noted also that *every* classical collision results in an energy loss, strictly elastic collisions being a pure quantum phenomenon. Examples of classical calculations by Tully and comparisons with experiment are given elsewhere.[21]

3.3.4. Quantum Effects

Two regimes are especially accessible to theoretical calculations: the extreme quantum regime, and the nearly classical regime with small quantum corrections. We shall consider these in turn.

---

* Not counting E_l, which is a dynamical consequence of the others.
† We believe that the attractive range d_a is relevant since there is a classical limit even for a hard-core repulsion ($d_r = 0$).

3.3.4.1. The Extreme Quantum Regime

In this regime the wave nature of the incidon and/or target atoms plays a decisive role. In practice one encounters this regime mostly with He incidons of low energy, say approximately 0.01 eV. When incident on typical target atoms of heavier mass, say W, owing to the mismatch of masses the scattering is primarily elastic (specular and diffraction). Next in importance is the scattering with the production of one phonon, either a surface phonon or a bulk phonon with significant amplitude in the surface region. Surface phonons are completely characterized by their parallel wave vector q and polarization vector, denoted by j. Their dispersion relation is of the form $\omega_j(q)$ (q_z is not a good quantum number). Conservation of parallel crystal momentum and of energy give the following relationships for one phonon production:

$$k = k' - q + g$$

$$\frac{\hbar^2}{2M}(k'^2 + k_z'^2) = \frac{\hbar 2}{2M}[k^2 + k_z^2] - \hbar\omega_j(q) \qquad (3.3.7)$$

where M is the incidon mass; k and k_z refer respectively to initial, parallel and perpendicular wave numbers of the incidon; similarly for the final wave numbers k' and k_z'; g is a two-dimensional reciprocal lattice vector parallel to the surface. Similar relations hold for one-phonon annihilation. Thus, measurements of the initial and final wave vectors of the incidon allow the simultaneous determination of q (modulo g) and of ω_j, i.e., they provide a point on the dispersion curve $\omega_j(q)$. In this way, very precise measurements of surface phonon dispersion relations have been made.[22]

Bulk phonons in the presence of the surface can nevertheless be labeled not only by q and j but also by q_z, their perpendicular wave vector in the bulk. The conservation laws are still of the form (3.3.7) but with $\omega_j(q)$ replaced by $\omega_j^B(q, q_z)$. Clearly a one-phonon ASC *cannot* determine q_z of the bulk phonon.

In a one-phonon ASC, the surface phonons stand out as peaks (say as a function of k_z' for fixed k, k_z, k') while the bulk phonons provide a continuous background which allows at least the determination of the range as a function of q_z and of $\omega_j^B(q, q_z)$ for fixed q.

For several targets, the information gained from the ASC's absent surface and bulk phonon dispersion relations agrees very well with independent lattice dynamics calculations. We note that this comparison uses only conservation principles and does not require a knowledge of the interaction potential between the incidon and target, nor the details of the interaction dynamics.

The *intensities* of the one-phonon processes, however, require detailed dynamical calculations.[17] This is a difficult problem for the following reasons:

(1) It requires a knowledge of $V(r, u_i, \ldots, u_N)$ for finite displacements u_l of the target atoms; this is much less completely known than $V(r; 0, \ldots, 0)$.

(2) Even though quantities u_l are small compared with the lattice parameters a, owing to the nearly "hard wall" character of the repulsive potential between

incidon and target the derivatives $\partial V/\partial u_l$ for surface atoms are very large and a simple Taylor expansion of V in powers of u_l is questionable. So are simple Born approximations (so-called "distorted wave" Born approximations or DWBA) starting with the uncoupled, unperturbed system in which $V(r; u, \ldots, u_N)$ is replaced by $V(r; 0, \ldots, 0)$.

(3) Finally, the long-range van der Waals potential $-C/z^3$ causes problems (logarithmic divergences) with another possible approach, an expansion of the cross section in powers of the incident momentum (k, k_z).

We refer the reader to a recent review[23] for an account of theoretical work on the difficult problem of the intensities. This continues to be an active area of research.

3.3.4.2. Quantum Corrections in Nearly Classical Regimes

a. Trajectory Approximation (TA). A simple, partially quantum mechanical method is the so-called trajectory approximation,[24-26] which has been used for both $T = 0$ and $T > 0$. We limit our discussion again to $T = 0$. The procedure is, in principle, as follows:

(1) One first carries out a fully classical calculation, resulting in an incidon trajectory $r(t)$.

(2) Next, regarding $r(t)$ as known, one considers $V(r(t), u_1, \ldots, u_N)$ as a given time-dependent perturbation acting on the target.

(3) V is expanded up to first order in the u, resulting in a u-independent part and the effective perturbation

$$H' = \sum K_l(r(t)) \cdot u_l \tag{3.3.8}$$

(4) The function H', combined with the unperturbed Hamiltonian for the harmonic target lattice

$$H_t = \sum_1^N \frac{p_i^2}{2M_t} + \tfrac{1}{2}\sum A_{ll'} u_l u_{l'} \tag{3.3.9}$$

[see equations (3.3.2) and (3.3.3)], results in a problem of driven coupled harmonic oscillators, which can be solved analytically.

(5) The final target state is a coherent superposition of eigenstates,

$$\Psi'_{\text{target}} + \sum C(Q_n, E_n)\Phi_{Q_n, E_n} \exp\left(-iE_n t/\hbar\right) \tag{3.3.10}$$

where Φ_{Q_n, E_n} is a state with total lateral phonon momentum Q_n and energy E_n (Φ_{Q_n, E_n} generally has several excited phonons).

(6) One now recalls the total momentum and energy conservation of the incidon–target system and obtains

$$P(Q, E)\,dQ\,dE = \sum_{Q < Q_n < Q+dQ} \sum_{E < E_n < E+dE} |C(Q_n, E_n)|^2 \tag{3.3.11}$$

for the probability $P(Q, E)$ of a transfer of lateral momentum Q and energy E from the incidon to the target.

(7) Finally, one averages over incident impact parameters.

In practice calculations several simplifications have been introduced.

Although this quantum method may be conceptually appealing and is practically quite tractable, it has a number of logical difficulties.

1. The incidon dynamics is treated fully classically and not self-consistently.
2. The phonon lateral momentum is meaningful only to within a lateral reciprocal lattice vector g, and so intrinsically this method does not define Q unambiguously.

It is therefore not clear in what sense, if any, the TA provides a consistent quantum correction to classical calculations.

b. Consistent Quantum Correction to First Order in $\hbar$. Can one develop a practically useful theory which provides consistent lowest-order quantum corrections to classical ASC calculations? The answer appears to be yes.

J. Jensen and coauthors (to be published) have completed calculations for the following simple model. A one-dimensional incidon (M, x) interacts with a one-dimensional harmonic oscillator (M_t, y, ω) by a smooth repulsive potential $V(x - y)$ where

$$H = \frac{p_x^2}{2M} + \frac{p_y^2}{2M} + \frac{1}{2} M_t \omega^2 y^2 + V_0 e^{-(x-y)/d} \tag{3.3.12}$$

and is reflected back, leaving the oscillator in an eigenstate of quantum number n.

The Heisenberg equations of motion for the operators x, y, p_x, and p_y are

$$\dot{x} = p_x/M, \qquad \dot{p}_x = -V'(x - y)$$

$$\dot{y} = p_y/M_t, \qquad \dot{p}_y = -M_t\omega^2 y + V'(x - y) \tag{3.3.13}$$

We now make the Ansatz

$$x = X(t) + u, \qquad p_x = P_x(t) + p_u$$

$$y = Y(t) + v, \qquad p_y = P_y(t) + p_v \tag{3.3.14}$$

where $X(t)$, $Y(t)$, $P_x(t)$, and $P_y(t)$ are the classical trajectories which we regard as known while u, v, p_u, and p_v are treated as small. They will be found to be $O(\hbar^{1/2})$. By using the fact that $X(t), \ldots, P_y(t)$ satisfy the classical equations of motion, one obtains the following quantum equations for $u, \ldots, p_u$ up to the order $\hbar^{1/2}$:

$$\dot{u} = p_u/M, \qquad \dot{p}_u = -(u - v)V''(X(t) - Y(t))$$

$$\dot{v} = p_v/M, \qquad \dot{p}_v = -(v - u)V''(X(t) - Y(t)) - M_t\omega^2 v \qquad (3.3.15)$$

The variables u and v can thus be thought of as two coupled quantum oscillators with Hamiltonian

$$H(u, v) = \frac{p_u^2}{2M} + \left[\frac{p_v^2}{2M(t)} + \frac{1}{2M_t}\omega^2 v^2\right] + \tfrac{1}{2}(u - v)^2 V''(X(t) - Y(t)) \qquad (3.3.16)$$

The coupling is again harmonic with a known, time-dependent spring constant $V''(X(t) - Y(t))$. The quantum equations (3.3.15) are linear, so their solution can be obtained from that of the corresponding set of classical equations. If the variables u, v, p_u, and p_v are denoted by the vector Q_j ($j = 1, \ldots, 4$), then the solution of the classical *and* quantum equations (3.3.15) has the form

$$Q_j(t) = \sum_{j'} A_{jj'}(t)Q_j(0)$$

The matrix $A_{jj'}(t)$, which solves the quantum-mechanical initial-value problem (3.3.15), can thus be obtained conveniently by finding the *classical* trajectories of the time-dependent harmonic problem (3.3.15). Thus, interestingly, only *classical* trajectories need to be determined in order to obtain the leading *quantum* corrections.

From these calculations a *specific* quantum effect can be calculated. While, for given incident momentum, the classical energy loss is unique, in quantum mechanics the energy loss has a *width* of order $\hbar^{1/2}$. Our group is studying potential experiments to show this effect.

We can now also study the validity of the trajectory approximation (TA). The latter corresponds to replacing $(v - u)$ by v in equation (3.3.15). It is therefore not surprising that, escept for special limiting regimes $[p_x(0)/\omega d \ll M$ or $M_t]$, the TA does not give the correct leading quantum corrections to classical results.

The equations-of-motion method is now being extended to realistic three-dimensional incidon–surface collisions.

References

1. E. M. Lifshitz, *Zh. Teor. Fiz.* **29**, 94 (1955).
2. E. Zaremba and W. Kohn, *Phys. Rev. B* **12**, 2270 (1976).
3. P. Feibelman, *Surf. Sci.* **12**, 287 (1982).
4. N. R. Hill *et al.*, *J. Chem. Phys.* **73**, 363 (1982).
5. V. Celli *et al.*, *J. Chem. Phys.* **83**, 2504 (1985).
6. Y. Takada and W. Kohn, *Phys. Rev. Lett.* **54**, 470 (1985); *Phys. Rev. B* **37**, 826 (1988).
7. N. D. Lang and A. R. Williams, *Phys. Rev. B* **18**, 616 (1978).
8. E. Zaremba and W. Kohn, *Phys. Rev. B* **15**, 1769 (1977).
9. J. Harris and A. Liebsch, *J. Phys. C* **15**, 2275 (1982).
10. J. Tersoff, *Phys. Rev. Lett.* **55**, 140 (1985).
11. Y. Takada and W. Kohn, *Phys. Rev. Lett.* **55**, 141 (1985).
12. J. K. Nørskov and N. Lang, *Phys. Rev. B* **21**, 2131 (1980).

13. M. J. Stott and E. Zarembo, *Phys. Rev. B* **22**, 1564 (1980).
14. M. Manninen *et al.*, *Phys. Rev. B* **29**, 2314 (1984).
15. W. A. Goddard III and L. B. Harding, *Ann. Rev. Phys. Chem.* **29**, 363 (1978).
16. W. A. Steele, *The Interaction of Gases with Solid Surfaces*, Pergamon, New York (1974).
17. M. Garcia *et al.*, *Phys. Rev. B* **19**, 634 (1979).
18. D. Gorse *et al.*, *Surf. Sci.* **147**, 611 (1984).
19. M. G. Dondi *et al.*, *Phys. Rev. B* **34**, 5897 (1986); *Phys. Rev. B* (to appear).
20. O. Gunnarsson and K. Schönhammer, in: *Many Body Phenomena at Surfaces* (D. Length and H. Suhl, eds.), Academic Press, New York (1984).
21. J. C. Tully, in: *Many Body Phenomena at Surfaces* (D. Langreth and H. Suhl, eds.), Academic Press, New York (1984).
22. J. P. Toennies, J. Vac. Sci. Technol. **A2**, 1055 (1984).
23. V. Bartolani and A. C. Levy, *Atom Surface Scattering Theory*, in: *Rev. Nuovo Cimento* **9**, No. 11 (1986).
24. W. Brening *et al.*, *Z. Phys. B* **36**, 81 (1979); **36**, 245 (1980); **41**, 243 (1981).
25. B. Brako and D. M. Newns, *Surf. Sci.* **117**, 42 (1981).
26. D. M. Newns, *Surf. Sci.* **154**, 658 (1985).

Basic Vibrational Properties of Surfaces

F. García-Moliner

4.1. Generalities on Lattice Dynamics

A first principles formulation of the study of lattice vibrations in a crystal would start from the total Hamiltonian

$$H = T_e + T_i + \Phi_{ii}(R) + \Phi_{ee}(r) + \Phi_{ie}(r, R) \qquad (4.1.1)$$

where T denotes kinetic energy, e electrons, i ions, R the ion coordinates, and r denotes the electron coordinates. For an eigenstate of this system

$$H\Psi(r, R) = E\Psi(r, R) \qquad (4.1.2)$$

To solve this problem rigorously is a formidable task, so we would like to simplify it. Ions are of order 10^3 to 10^5 times heavier than electrons, so their motions may be expected to involve sufficiently different time scales that they may be decoupled to a good approximation. It does not seem unreasonable to expect that when the ions move, the electrons can follow their motions *adiabatically*, i.e., that their states are deformed without undergoing abrupt transitions, adapting themselves to the instantaneous configuration of ionic positions such that their interactions can be evaluated as if they "saw" the ions frozen in their instantaneous positions. This is the idea of the *Born–Oppenheimer* or *adiabatic approximation*, in which one imagines some frozen configuration R and writes down the R-dependent Schroedinger equation

$$[T_e + \Phi_{ee}(R) + \Phi_{ie}(r, R)]\chi_R(r) = E_e(R)\chi_R(r) \qquad (4.1.3)$$

F. García-Moliner • Instituto de Ciencia de Materiales, C.S.I.C., 28006 Madrid, Spain.

for the electronic system, for which the eigenfunction and the eigenvalue contain the ionic coordinates R as parameters. The task of lattice dynamics is to study the validity of equation (4.1.3).

Details can be found in standard textbooks. It is noteworthy that $E_e(R)$ includes not only the electron–electron and electron–ion interactions in the presence of the ions, but also the kinetic energy of the electrons, and this also depends on the ion coordinates R. In the spirit of the adiabatic approximation we set

$$\Psi(r, R) = \chi_R(r)\psi(R) \tag{4.1.4}$$

On substituting this in equation (4.1.2) and using equation (4.1.3) one finds that for this to be an eigenfunction of H, it must be possible (1) to neglect certain terms and (2) to have $\psi(R)$ satisfy

$$[T_i + \Phi(R)]\psi(R) = E\psi(R) \tag{4.1.5}$$

where

$$\Phi(R) = \Phi_{ii}(R) + E_e(R) \tag{4.1.6}$$

In the eigenvalue equation for the ionic wave function we have introduced the *effective potential energy function* (4.1.6), to which the electrons contribute through function $E_e(R)$. It turns out that, although due to different reasons, the adiabatic approximation can be usually justified to study lattice dynamics in insulators, semiconductors, and metals. The validity of this approximation does not depend on having a weak electron–ion coupling but on whether it is the *static* or the *dynamic* aspects of this interaction that mainly intervene in the phenomenon under study. If we wish to study simply lattice dynamics as such, i.e., we want to obtain a phonon band structure, then the static aspects of the electron–ion interaction are included explicitly in the function $\Phi_{ie}(r, R)$ and the issue is decided by the other considerations just outlined. However, this ceases to hold for problems where the electron dynamics is involved explicitly, such as superconductivity or ordinary resistivity due to electron–phonon scattering, in which case we must take explicit account of (virtual or real) abrupt transitions between electronic states, the absence of which constitutes the very notion of *adiabatic* approximation.

Now, in this approximation we are supposed to start from the effective potential $\Phi(R)$, given in equation (4.1.6), for the ionic motions. As noted above, an *ab initio* calculation of this quantity entails a previous calculation of $E_e(R)$, which is still a major computational task. However, the adiabatic argument provides a suggestive hint for the form of equation (4.1.5). If we know the type of ion–ion interactions to be expected for a given crystal, then we can start from some appropriate phenomenological model, i.e., we can attempt to design $\Phi(R)$ as a phenomenological potential which contains the appropriate type of interactions, e.g., central forces (two-body interactions), angular forces (three-body interactions), nearest-neighbor (NN) or next-nearest-neighbor (NNN) interactions. Let R_0 be the equilibrium positions and R the actual positions when the ions are displaced through a displacement $\mathbf{u} = \mathbf{R} - \mathbf{R}_0$. For small displacements we have the expansion

$$\Phi = \Phi_0 + \sum_r f_r u_r + \frac{1}{2} \sum_{rs} f_{rs} u_r u_s + \frac{1}{3!} \sum_{rsp} f_{rsp} u_r u_s u_p + \cdots \quad (4.1.7)$$

The labels can be regarded as spanning all ionic positions and spatial coordinates. It suffices to consider succintly the form of Φ. Retaining second-order terms yields the *harmonic approximation*. The *anharmonic terms*, beyond second order, are essential for the study of some phenomena such as thermal expansion or thermal conductivity, but the harmonic approximation suffices to study the phonon band structure and this will be the level of our discussion.

4.2. Equations of Motion and Force Constants

We let $\mathbf{a}_1$, $\mathbf{a}_2$, and $\mathbf{a}_3$ be three linearly independent *primitive translation vectors* defining a *primitive unit cell*. The equilibrium position of the lth unit cell is denoted as

$$\mathbf{r}(l) = l_1 \mathbf{a}_1 + l_2 \mathbf{a}_2 + l_3 \mathbf{a}_3 \quad (4.2.1)$$

In *primitive* or *Bravais lattices* there is just one atom per unit cell and equation (4.2.1) gives the positions of the atoms in the crystal. In general we have s (>1) atoms per unit cell. These form the *basis* of the crystal structure and we need an inner index k to label the atoms of the basis. The atomic positions in the crystal are then

$$\mathbf{r}(l, k) = \mathbf{r}(l) + \mathbf{r}(k) \quad (4.2.2)$$

In fact these—discrete—positions correspond to what we have termed R_0, the equilibrium positions. The actual ionic positions in the vibrating crystal are

$$\mathbf{R}(l, k) = \mathbf{r}(l, k) + \mathbf{u}(l, k) \quad (4.2.3)$$

The task of lattice dynamics is to study $\mathbf{u}(l, k)$, with components $u(l, k)$, $\alpha = x, y, z$. We cast Φ as a function of the instantaneous atomic positions

$$\Phi = \Phi\{\ldots, \mathbf{r}(l, k) + \mathbf{u}(l, k), \ldots\} \quad (4.2.4)$$

and expand for small displacements up to second order:

$$\Phi = \Phi_0 + \sum_{l\alpha} \Phi_\alpha(l, k) u_\alpha(l, k) + \frac{1}{2} \sum_{\substack{lk\alpha \\ l'k'\beta}} \Phi_{\alpha\beta}(l, k; l', k') u_\alpha(l, k) u_\beta(l', k') \quad (4.2.5)$$

where

$$\Phi_\alpha(l, k) = \left[\frac{\partial \Phi}{\partial u_\alpha(l, k)} \right]_0 \quad (4.2.6a)$$

$$\Phi_{\alpha\beta}(l, k; l', k') = \left[\frac{\partial^2\Phi}{\partial u_\alpha(l, k)\partial u_\beta(l, k)}\right]_0 \tag{4.2.6b}$$

The force acting in the direction α on the atom (l, k) for a given system of displacements of all the other atoms is given by

$$F_\alpha(l, k) = -\frac{\partial\Phi}{\partial u(l, k)} = -\Phi_\alpha(l, k) - \sum_{l'k'\beta} \phi_{\alpha\beta}(l, k; l', k')u_\beta(l', k') \tag{4.2.7}$$

We consider all atoms at their equilibrium positions. Then all displacements and forces vanish. Hence

$$\Phi_\alpha(l, k) = 0 \tag{4.2.8}$$

so that the harmonic potential is

$$\Phi - \Phi_0 = \frac{1}{2} \sum_{\substack{lk\alpha \\ l'k'\beta}} \Phi_{\alpha\beta}(l, k; l', k')u_\alpha(l, k)u_\beta(l', k') \tag{4.2.9}$$

the force is

$$F_\alpha(l, k) = -\sum_{l'k'\beta} \phi_{\alpha\beta}(l, k; l', k')u_\beta(l', k') \tag{4.2.10}$$

and the equation of motion for the atom (l, k) with mass m_k is

$$m_k\ddot{u}_\alpha(l, k) = -\sum_{l'k'\beta} \phi_{\alpha\beta}(l, k; l', k')u_\beta(l', k') \tag{4.2.11}$$

The coefficients $\Phi_\beta(l, k; l', k')$ are called the *force constants*, to be obtained either from empirical fitting, or from *ab initio* calculation, or from a combination of hard labor, approximations, and fitting.

The force constants satisfy some general relationships. In essence:

1. The order of differentiation in equation (4.2.6b) can be inverted. Hence

$$\Phi_{\alpha\beta}(l, k; l', k') = \phi_{\beta\alpha}(l', k'; l, k) \tag{4.2.12}$$

2. Let us consider a rigid translation of the entire crystal and note that there are three such independent translations assumed valid through a given distance v_β. All associated forces must vanish, i.e.,

$$\sum_\beta \left[\sum_{l'k'} \Phi_{\alpha\beta}(l, k; l', k')\right]v_\beta = 0 \tag{4.2.13}$$

for every independent value of v_β ($\beta = x, y, z$). Hence

$$\sum_{l'k'} \Phi_{\alpha\beta}(l, k; l', k') = 0 \tag{4.2.14}$$

It is useful to express this as

$$\Phi_{\alpha\beta}(l, k; l', k') = - \sum_{(l', k') \neq (l, k)} \Phi_{\alpha\beta}(l, k; l', k') \qquad (4.2.15)$$

3. Let us consider a primitive translation $t(m) = m_1 a_1 + m_2 a_2 + m_3 a_3$. This carries atom (l, k) to $(l + m, k)$ and (l', k') to $(l' + m, k')$. The interactions cannot be different. Thus

$$\Phi_{\alpha\beta}(l + m, k; l' + m, k') = \Phi_{\alpha\beta}(l, k; l', k') \qquad (4.2.16)$$

If $m = -l$, then

$$\Phi_{\alpha\beta}(l, k; l', k') = \Phi_{\alpha\beta}(0, k; l' - l, k') \qquad (4.2.17)$$

and if $m = -l'$, then

$$\Phi_{\alpha\beta}(l, k; l', k') = \Phi_{\alpha\beta}(l' - l, k; 0, k') \qquad (4.2.18)$$

In short, the force constants depend on l and l' only through their difference.

4.3. Dynamical Matrix and Eigenvectors

A crystal with N atoms and Born-von Karman periodic boundary conditions is considered. We seek solutions of the form

$$u_\alpha(l, k) = \frac{1}{2\sqrt{N m_k}} \{A(\mathbf{q}) e_\alpha(k|\mathbf{q}) \, e^{i[\mathbf{q} \cdot \mathbf{r}(l,k) - \omega(\mathbf{q})t]} + \text{c.c.}\} \qquad (4.3.1)$$

where $\mathbf{q}$ is the wave vector, $A(\mathbf{q})$ the amplitude, and $e(k|\mathbf{q})$ the polarization vector, A general vibrational state is a combination of waves like equation (4.3.1). Substitution of this in equation (4.2.11) yields the equation of motion in the Fourier transform

$$\omega^2(\mathbf{q}) e_\alpha(k|\mathbf{q}) = \sum_{k'\beta} D_{\alpha\beta}(\mathbf{q}; k, k') e_\beta((k'|\mathbf{q}) \qquad (4.3.2)$$

This introduces the *dynamical matrix* $\mathbf{D}$ of elements

$$D_{\alpha\beta}(\mathbf{q}; k, k') = \frac{1}{\sqrt{m_k m_{k'}}} \sum_{l'} \Phi_{\alpha\beta}(l, k; l', k) \, e^{i\{\mathbf{q} \cdot [\mathbf{r}(l, k) - \mathbf{r}(l',k')]\}} \qquad (4.3.3)$$

By using equations (4.2.1), (4.2.2), and (4.2.16)-(4.2.18) matrix $\mathbf{D}$ can be expressed in the form

$$D_{\alpha\beta}(\mathbf{q}; k, k') = \frac{e^{i\mathbf{q} \cdot [\mathbf{r}(k') - \mathbf{r}(k)]}}{\sqrt{m_k m_{k'}}} \sum_L \Phi_{\alpha\beta}(0, k; L, k') \, e^{i\mathbf{q} \cdot \mathbf{r}(L)} \qquad (4.3.4)$$

where $L = l' - l$. We recall that $\mathbf{r}(k)$ gives the position of the kth atom within one unit cell, while $\mathbf{r}(L)$ locates the different cells. Hence $\mathbf{D}$ is like a kind of Fourier transform of Φ, though not quite. Mass factors enter $\mathbf{D}$ due to the inertial term in the equation of motion, and there is only a summation over the primitive lattice, the geometry of the basis remaining displayed explicitly in the prefactor, which depends explicitly on the atomic positions within one unit cell.

The first task is to decide on the model for Φ. This is physics. Then matrix $\mathbf{D}$ must be obtained. This is geometry. Later on we shall see how the dynamical matrix of a given crystal is conveniently cast in a different form in which a three-dimensional (3D) crystal is described as a linear sequence of two-dimensional (2D) atomic layers, the choice being adapted to the crystal orientation of the surface to be studied. For the time being we retain the 3D crystal treated as a bulk and express equation (4.3.2) as

$$\sum_{k'\beta} [\omega^2 \delta_{\alpha\beta}\delta_{kk'} - D_{\alpha\beta}(\mathbf{q}; k, k')]e_\beta(k'|\mathbf{q}) = 0 \tag{4.3.5}$$

This is the $3s \times 3s$ matrix representation of an eigenvalue equation: $\mathbf{e}(k'|\mathbf{q})$ is the eigenvector and $\omega^2(\mathbf{q})$ the eigenvalue. The secular equation is

$$\det |\omega^2(\mathbf{q})\delta_{\alpha\beta}\delta_{kk'} - D_{\alpha\beta}(\mathbf{q}; k, k')| = 0 \tag{4.3.6}$$

an algebraic equation of degree $3s$. For a given vector $\mathbf{q}$ there are in general $3s$ (different or maybe some equal) eigenvalues or branches $\omega_j^2(\mathbf{q})$ with $j = 1, 2, \ldots, 3s$. It is easily seen that matrix $\mathbf{D}$ is Hermitian, whence the eigenvalues are real and all quantities $\omega_j(\mathbf{q})$ are either real or imaginary. The latter would blow up timewise. Thus only real values of $\omega_j(\mathbf{q})$ are physically admissible solutions. Each one gives a branch in the phonon band structure of the crystal.

Having found one eigenvalue we substitute it back into equation (4.3.5), which can be written in compact matrix form as

$$\mathbf{D}(\mathbf{q}) \cdot \mathbf{e}(\mathbf{q}|j) = \lambda_j(\mathbf{q})\mathbf{e}(\mathbf{q}|j), \qquad \lambda_j(\mathbf{q}) = \omega_j^2(\mathbf{q}) \tag{4.3.7}$$

This yields the eigenvectors up to an amplitude factor, which can be chosen to satisfy the *orthonormality* and *closure* relations

$$\sum_{k\alpha} e_\alpha^*(k|\mathbf{q}j)e_\alpha(k|\mathbf{q}j') = \delta_{jj'} \tag{4.3.8a}$$

$$\sum_j e_\alpha(k|\mathbf{q}j)e_\beta^*(k'|\mathbf{q}j) = \delta_{\alpha\beta}\delta_{kk'} \tag{4.3.8b}$$

It is further seen that

$$e_\alpha^*(k|-\mathbf{q}j) = e_\alpha(k|\mathbf{q}j) \tag{4.3.9}$$

Equations (4.3.7) and (4.2.14) enable one to prove that of the $3s$ branches $\omega_j^2(\mathbf{q})$ for each given value of $\mathbf{q}$, there are three which tend to zero as $q \to 0$. These are the *acoustic modes*. In the limit $q = 0$ they describe the three independent rigid translations of the entire crystal. The other branches (when $s > 1$) correspond to the so-called *optical modes*. Further details and alternative ways to define quantities $\mathbf{D}$ and $\mathbf{e}$ can be found in standard textbooks.

4.4. Some Practical Examples

It is instructive to work out the details of some force constant models. Appendix 4.1 gives an example: a simple cubic lattice with nearest-neighbor (NN) and next-nearest-neighbor (NNN) interactions and *central* and *angular* forces. Here we consider an fcc lattice with central forces and NN interactions only. This is a very idealized model, but it possesses the virtue of having a real crystalline lattice and serves very well to illustrate the main features of the results appearing in an actual surface phonon calculation. We shall see later that sometimes it can even have a reasonable, almost quantitative value.

In the fcc lattice there are atoms at the corners and at the face centers of a cube of side a. The rhombohedral unit cell of volume $a^3/4$ contains *one* atom. Geometrical details are given in Figure 4.1. We concentrate on one atom defined as the central atom. It has *twelve* NN. The unit vectors from the central atom to its NN are listed in Figure 4.1. Our central force model is

$$\Phi = \frac{K}{4} \sum_{ll'} [(\mathbf{u}_l - \mathbf{u}_{l'}) \cdot \mathbf{r}_{ll'}]^2 \tag{4.4.1}$$

where $\mathbf{r}_{ll'}$ is the unit vector of $\mathbf{r}(l') - \mathbf{r}(l)$. In this case there is no internal index k (one atom per unit cell).

The formula for the dynamical matrix (4.3.4) in this case can be simplified to

$$D_{\alpha\beta}(\mathbf{q}) = \frac{1}{m} \sum_{\mathbf{L}} \Phi_{\alpha\beta}(\mathbf{L}) \, e^{i\mathbf{q} \cdot \mathbf{L}} \tag{4.4.2}$$

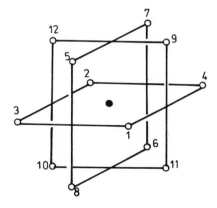

Figure 4.1. Fcc lattice. One atom (●), taken as the central atom, is surrounded by 12 NN (○) labeled 1 to 12. The unit vectors from the central atom to each one of the NN are $\mathbf{r}_{0j} = (1/\sqrt{2})$ $(L_{j1}^0, L_{j2}^0, L_{j3}^0)$. We list $(L_{j1}^0, L_{j2}^0, L_{j3}^0)$ for all 12 cases: (1) $(1, 1, 0)$, (2) $(\bar{1}, \bar{1}, 0)$, (3) $(1, \bar{1}, 0)$, (4) $(\bar{1}, 1, 0)$, (5) $(1, 0, 1)$, (6) $(\bar{1}, 0, \bar{1})$, (7) $(\bar{1}, 0, 1)$, (8) $(1, 0, \bar{1})$, (9) $(0, 1, 1)$, (10) $(0, \bar{1}, \bar{1})$, (11) $(0, 1, \bar{1})$, (12) $(0, \bar{1}, 1)$.

where **L** spans, from the origin at the central atom, the position vectors of all twelve NN *and also* the central atom itself ($\mathbf{L} = 0$). Expression (4.2.15) can be employed to express equation (4.4.2) in the form

$$D_{\alpha\beta}(\mathbf{q}) = \frac{1}{m} \sum_{\mathbf{L} \neq 0} \Phi_{\alpha\beta}(\mathbf{L})\, e^{i\mathbf{q}\cdot\mathbf{L}} \tag{4.4.3}$$

Equation (4.4.1) yields

$$\frac{\partial^2 \Phi}{\partial u_{l,\alpha}\, \partial u_{l',\beta}} = -\frac{K}{2} r_{ll',\beta} r_{ll',\alpha} \qquad (l' \neq l) \tag{4.4.4}$$

We use this relationship for $l = 0$ (central atom) and $\mathbf{r}_{ll'}$ denoting the quantities $\mathbf{r}_{0j}$ listed in Figure 4.1. This yields the following nonvanishing force constants:

$$\Phi_{xx}(1) = \Phi_{xx}(2) = \Phi_{xx}(3) = \Phi_{xx}(4) = \Phi_{xx}(5) = \Phi_{xx}(6)$$

$$= \Phi_{xx}(7) = \Phi_{xx}(8) = -K/4$$

$$\Phi_{xy}(1) = \Phi_{xy}(2) = -K/4 = -\Phi_{xy}(3) = -\Phi_{xy}(4)$$

$$\Phi_{xz}(5) = \Phi_{xz}(6) = -K/4 = -\Phi_{xz}(7) = -\Phi_{xz}(8)$$

$$\Phi_{yy}(1) = \Phi_{yy}(2) = \Phi_{yy}(3) = \Phi_{yy}(4) = \Phi_{yy}(9) = \Phi_{yy}(10)$$

$$= \Phi_{yy}(11) = \Phi_{yy}(12) = -K/4$$

$$\Phi_{yz}(9) = \Phi_{yz}(10) = -K/4 = -\Phi_{yz}(11) = -\Phi_{yz}(12)$$

$$\Phi_{zz}(5) = \Phi_{zz}(6) = \Phi_{zz}(7) = \Phi_{zz}(8) = \Phi_{zz}(9) = \Phi_{zz}(10)$$

$$= \Phi_{zz}(11) = \Phi_{zz}(12) = -K/4$$

$$\Phi_{yx}(l') = \Phi_{xy}(l'); \qquad \Phi_{zx}(l') = \Phi_{xz}(l'); \qquad \Phi_{yz}(l') = \Phi_{zy}(l')$$

Equation (4.4.2), with $\phi = aq/2$, yields

$$D_{xx}(\mathbf{q}) = \frac{K}{m}\left(\sin^2 \frac{\phi_x + \phi_y}{2} + \sin^2 \frac{\phi_x - \phi_y}{2} + \sin^2 \frac{\phi_x + \phi_z}{2} + \sin^2 \frac{\phi_x - \phi_z}{2} \right)$$

$$D_{xy}(\mathbf{q}) = \frac{K}{m}\left(\sin^2 \frac{\phi_x + \phi_y}{2} - \sin^2 \frac{\phi_x - \phi_y}{2} \right) = D_{yx}(\mathbf{q})$$

$$D_{xz}(\mathbf{q}) = \frac{K}{m}\left(\sin^2 \frac{\phi_x + \phi_z}{2} - \sin^2 \frac{\phi_x - \phi_z}{2} \right) = D_{zx}(\mathbf{q})$$

$$D_{yy}(\mathbf{q}) = \frac{K}{m}\left(\sin^2\frac{\phi_x + \phi_y}{2} + \sin^2\frac{\phi_x - \phi_y}{2} + \sin^2\frac{\phi_y + \phi_z}{2} + \sin^2\frac{\phi_y - \phi_z}{2}\right)$$

$$D_{yz}(\mathbf{q}) = \frac{K}{m}\left(\sin^2\frac{\phi_y + \phi_z}{2} - \sin^2\frac{\phi_y - \phi_z}{2}\right) = D_{zy}(\mathbf{q})$$

$$D_{zz}(\mathbf{q}) = \frac{K}{m}\left(\sin^2\frac{\phi_x + \phi_z}{2} + \sin^2\frac{\phi_x - \phi_z}{2} + \sin^2\frac{\phi_y + \phi_z}{2} + \sin^2\frac{\phi_y - \phi_z}{2}\right)$$

We can now study the eigenvalue problem (4.3.5), (4.3.6). We consider propagation along symmetry directions (Figure 4.2) and take $\mathbf{q} = (q, 0, 0)$. This yields one longitudinal mode

$$\omega_L^2 = D_{xx} = \frac{4K}{m}\sin^2\frac{aq}{4} \qquad (4.4.5)$$

and two degenerate transverse modes

$$\omega_{T_1}^2 = \omega_{T_2}^2 = D_{yy} = D_{zz} = \frac{2K}{m}\sin^2\frac{aq}{4} \qquad (4.4.6)$$

If $\mathbf{q} = 2^{-1/2}(q, q, 0)$, then $D_{xz} = D_{yz} = 0$ but $D_{xy} \neq 0$. Thus one decoupled transverse solution is

$$\omega_{T_2}^2(q) = D_{zz} = \frac{4K}{m}\sin^2\frac{aq}{4\sqrt{2}} \qquad (4.4.7)$$

corresponding to a vibration in the z-direction. For the rest we have a 2×2 dynamical matrix with

$$D_{xx} = D_{yy} = \frac{K}{m}\left(\sin^2\frac{aq}{2\sqrt{2}} + \sin^2\frac{aq}{4\sqrt{2}}\right)$$

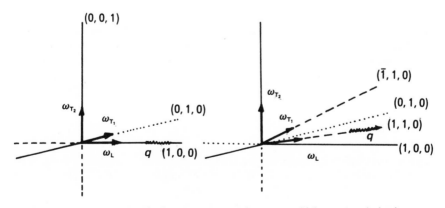

Figure 4.2. Cubic symmetry. Wiggly arrow: propagation vector; thick arrow: polarization vector.

$$D_{xy} = D_{yx} = \frac{K}{m} \sin^2 \frac{aq}{4\sqrt{2}} \tag{4.4.8}$$

This 2×2 secular problem has the two roots

$$\omega_{\pm}^2 = D_{xx} \pm D_{xy} \tag{4.4.9}$$

The $+$ sign corresponds to the normalized eigenvector $e_L = 2^{-1/2} (1, 1, 0)$, which describes a longitudinal mode with

$$\omega_L^2 = \omega_+^2 = \frac{2K}{m} \left(\sin^2 \frac{aq}{2\sqrt{2}} + \sin^2 \frac{aq}{4\sqrt{2}} \right) \tag{4.4.10}$$

The other root is a transverse mode (T_1) with normalized eigenvector $e_{T_1} = 2^{-1/2}(-1, 1, 0)$ and eigenvalue

$$\omega_{T_1}^2 = \omega_-^2 = \frac{2K}{m} \sin^2 \frac{aq}{4\sqrt{2}} \tag{4.4.11}$$

The normalized eigenvector of solution (4.4.7) is obviously $e_{T_2} = (0, 0, 1)$.

For the $(1, 1, 1)$ direction we also find three eigenvectors independent of q. However, for any other q not in a symmetry direction the eigenvectors depend on both the modulus and the direction of q.

4.5. The Layer Description of Lattice Dynamics

We now single out the plane orientation which the surface will eventually have and let $L = (l_1, l_2) = (l, m)$ label 2D atomic positions within an atomic layer with this orientation while n labels the successive layers, redefining the basis vectors so that a_1 and a_2 are in the plane $n = $ const and a_3 is perpendicular to this plane. This is the appropriate geometry for the eventual study of the surface. We then have a layer description of the bulk. for instance, instead of writing $\Phi(l, l')$ or $\Phi(l, k; l', k')$ as in Section 4.2, where l is an abbreviated notation for the full 3D (l, m, n), we write

$$\Phi_{\alpha\beta}(L, L'; n, k, n', k') = \frac{\partial^2 \Phi}{\partial u_\alpha(L, n, k) \, \partial u_\beta(L', n', k')} \tag{4.5.1}$$

We stress that the meaning of L here is *not* the same as in Section 4.4. As in the 3D bulk, equation (4.5.1) depends on L and L' only through their difference and

$$\Phi_{\alpha\beta}(L - L'; n, k, n', k') = \Phi_{\beta\alpha}(L' - L; n', k, n, k) \tag{4.5.2}$$

There are other important properties of the bulk which it is convenient to express in layer notation. The two basic ones are *translational invariance*

$$\sum_{L',n',k'} \Phi_{\alpha\beta}(L - L'; n, k, n', k') = 0 \qquad (4.5.3)$$

and *rotational invariance*

$$\sum_{L',n',k'} [\Phi_{\alpha\beta}(L - L'; n, k, n', k')r_\gamma(l', n', k')$$

$$- \Phi_{\alpha\gamma}(L - L'; n, k, n', k')r_\beta(l', n', k')] = 0 \qquad (4.5.4)$$

The proof of equation (4.5.4) is a bit more involved, but here we merely quote the results. These invariance conditions do not depend on any particular details of the symmetry group of the crystal. They hold even for any group of atoms, since no forces can be created when these are translated or rotated or displaced in any way as long as it is a rigid bodily motion of the entire group of atoms. These invariance conditions play an important role in some formulations of surface lattice dynamics (a significant case will be referred to in Section 4.11) and are of fundamental importance for a correct description of the long-wave (elastic) limit.

Now, the equation of motion in the layer description is

$$m_k(n)\ddot{u}_\alpha(L; n, k) = -\sum_{L',n',k',\beta} \Phi_{\alpha\beta}(L - L'; n, k, n', k')u_\beta(L'; n', k')$$

$$(4.5.5)$$

At this stage we Fourier transform in 2D, introducing the 2D wave vector κ conjugate to the 2D position $\mathbf{r}(L)$. The normal-mode analysis of equation (4.5.5) then rests on waves of the form

$$\mathbf{u}(\mathbf{r}; n, k) = \frac{1}{2\sqrt{N_2 m_k(n)}} \tilde{\mathbf{e}}(n, k | \kappa, j) e^{i[\kappa \cdot \mathbf{r} - \omega_j(\kappa)t]}, \qquad \mathbf{r} = \mathbf{r}(L) \qquad (4.5.6)$$

where N_2 is the number of unit cells in the 2D lattice, i.e., the number of values that L takes, and we use the 2D periodic Born–von Karman boundary conditions. It is noteworthy that here we are using the alternative definition of the polarization vectors

$$\tilde{\mathbf{e}}(n, k | \kappa, j) = e^{i\kappa \cdot \mathbf{e}(k)} \tilde{\mathbf{e}}(n, k | \kappa, j) \qquad (4.5.7)$$

The secular system corresponding to equation (5.6) is

$$\omega_j(\kappa)\tilde{e}_\alpha(n'k | \kappa, j) = \sum_{n'k'\beta} D_{\alpha\beta}(n, k, n', k'; \kappa)\tilde{e}_\beta(n'k' | \kappa, j) \qquad (4.5.8)$$

where

$$D_{\alpha\beta}(n, k, n', k'; \mathbf{\kappa}) = \frac{1}{\sqrt{m_k(n)m_k(n')}} \sum_{L'} \Phi_{\alpha\beta}(L - L'; n, k, n', k') \, e^{i\mathbf{\kappa}\cdot[\mathbf{r}(L)-\mathbf{r}(L')]}$$

$$(4.5.9)$$

For this dynamical matrix only the 2D wave vector κ is introduced and explicit layer indices are displayed.

So far this is a mere rewriting of the bulk lattice dynamics in layer notation. In spite of the 2D character emphasized by these equations the problem remains 3D, but equation (4.5.6), for instance, displays explicitly the harmonic form of the wave only in 2D. If the actual wave vector is equal to κ, then the n-dependent quantities appearing in equation (4.5.6) have the same value for all n. This would then amount to a standard description, with n omitted, of a 3D bulk wave which happens to travel with $\mathbf{q} = \mathbf{\kappa}$. If the wave travels at an angle with the planes $n = \text{const}$, then equation (4.5.6) has an n-dependence which would have to be determined. The layer description would then be very clumsy. The only natural thing to do would then be to carry on with the complete Fourier transform for the third spatial component, which would lead us back to the scheme of Section 4.4.

The layer description is set up to study the surface problem, for which the layer representation constitutes a natural language as there is an explicit reference to the atomic layers. We expect to find a different situation as we study first the surface and then the successive layers, moving down to the bulk. Before proceeding with this, it helps to have a quick look at the long-wave limit.

4.6. Surface Waves: A Summary of the Long-Wave Limit

Let us consider a bulk, isotropic, elastic medium with two degenerate transverse modes ω_T and one longitudinal mode ω_L such that

$$\omega_T = v_T q \qquad \text{and} \qquad \omega_L = v_L q \qquad (4.6.1)$$

We shall now examine the planes $z = \text{const}$ and define the vector $\mathbf{q}$ as (ω, q_z), where ω is the projection of $\mathbf{q}$ on the said planes. Let us consider a bulk branch $\omega = vq$ (Figure 4.3); we take a fixed value κ' of the κ projection of $\mathbf{q}$ and let q_z grow with this fixed κ'. For $q_z = 0$, the value of ω is $\omega' = v\kappa'$ (a) (in Figure 4.3). As q_z grows, so do mod $\mathbf{q} = q$ and ω, i.e.,

$$q = \sqrt{(\kappa^2 + q_z^2)} \qquad \text{and} \qquad \omega = vq \qquad (4.6.2)$$

For instance, when treating $\mathbf{q}_1'$ (b) we move in (a) to the right of κ', i.e., we move to *higher* frequencies. If this is plotted in (c) we obtain a point like (q_1'). If expressions (4.6.2) are now increased to the value corresponding to $\mathbf{q}_2'$ (b), an even higher frequency (q_2') is obtained in (c). If the tip of $\mathbf{q}$ varies continuously along the line from $\mathbf{q}_1'$ to $\mathbf{q}_2'$, then a vertical line is generated in diagram (c) starting from $\omega = \omega'$. The same process is repeated for another value κ''. In the diagram (c) another vertical line of points is generated starting from $\omega = \omega''$. All these points

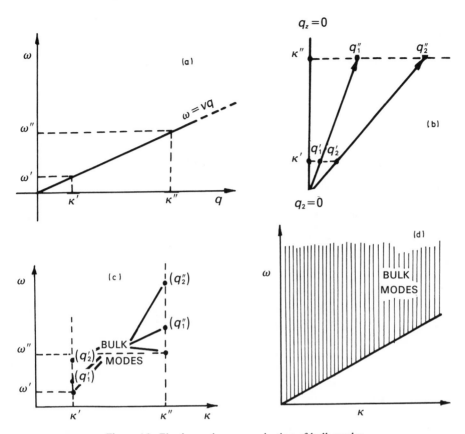

Figure 4.3. Elastic continuum: projection of bulk modes.

correspond to the bulk modes. As we vary κ continuously, the shaded area of diagram (d) is generated. This describes the same spectrum as (a) but presented in a different manner which emphasizes a preferred orientation, something like a continuum counterpart of the layer description of bulk lattice dynamics. If we do this for all the branches of the spectrum, i.e., just for $\omega_L(\mathbf{q})$ and $\omega_T(\mathbf{q})$ in this case, then we have a *projection of the "bulk band structure,"* this projection being referred to a given orientation.

Clearly we have prepared the description of the bulk for the study of surface waves. Figure 4.4 shows the well-known *Rayleigh mode* which appears below the lowest bulk band threshold. This is a new *surface mode*, made possible by the change in boundary conditions. Indeed, this is a usual way to find this type of solution. One writes down a trial form of the solution of the equations of motion for elastic waves that contains some parameters. These are determined by imposing the surface boundary conditions and this yields the Rayleigh mode.[1] But this is not all. If we think in terms of a change in the mode density, not only does a new (surface) mode appear, but also the bulk modes are affected, as *all* solutions must obey the new boundary conditions. If, in Figure 4.4, a vertical line is drawn

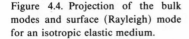

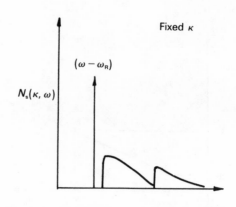

Figure 4.4. Projection of the bulk modes and surface (Rayleigh) mode for an isotropic elastic medium.

Figure 4.5. Qualitative form of the surface-projected, fixed-κ mode density for an elastic isotropic medium.

corresponding to a fixed value of κ, then the new local value $N_s(\kappa, \omega)$ of the fixed-κ mode density as a function of ω, projected at the surface, looks like the picture shown in Figure 4.5.

The two key features of this qualitative picture are (1) the presence of the δ-function peak $\delta(\omega - \omega_R)$, which signals the existence of the Rayleigh mode, and (2) the characteristic humps near the bulk thresholds, where $N_s(\kappa, \omega)$ looks very different from the bulk homogeneous values $N(\kappa, \omega)$. Frequencywise, the distortion of the bulk modes piles up near the band edges. Spacewise, it tends to accumulate near the surface, as is intuitively obvious. The surface spectrum can be determined by surface Brillouin scattering spectroscopy. Figure 4.6 shows experimental results for polycrystalline Al. The incoming probe (a photon) undergoes inelastic scattering during which it exchanges some energy (frequency) and momentum (wave vector) parallel to the surface. These are, respectively, the ω and κ of the surface mode absorbed or emitted in the inelastic scattering event. It is also clear intuitively that the scattering cross section must be essentially proportional to the density of modes to be emitted or absorbed. Figure 4.6 also shows a theoretical calculation[3] of $N_s(\kappa, \omega)$, in very good agreement with experiment.

This discussion bears out two important points: (1) When the surface is excited it is *not only the surface modes* but *the entire lattice dynamics* that is involved. (2) We have just seen the relevance of an important concept, namely, $N_s(\kappa, \omega)$. This is an example of a *spectral function,* a concept to which we shall return in Section 4.11.

In the example of Figure 4.6 the calculation was conducted for an isotropic average of Al, but crystal anisotropy is actually important,[4] In general the two transverse thresholds split apart. Let us plot this in the following manner. First, for instance, a continuous elastic but anisotropic medium with cubic symmetry is

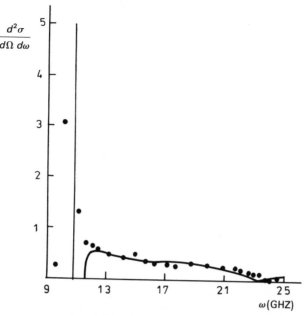

Figure 4.6. Dots refer to experimental[2] surface Brillouin scattering data for polycrystalline Al and continuous lines to surface Green function matching calculation[3] of the surface projection of the fixed-κ mode density. Calculated for an isotropic average of elastic (long-wave) Al with κ corresponding to the experiment.

considered. The $(0, 0, 1)$ surface is examined and an analysis similar to that leading to Figure 4.3d repeated, but this time the phase velocity is plotted instead of the frequency. Bulk waves with $\mathbf{q} = (\kappa, 0, 0)$ are treated. As seen in Section 4.4, the two transverse modes are degenerate for $\mathbf{q}$ in this high-symmetry direction. Thus the two transverse bulk thresholds are equal. Above this threshold we have bulk modes with the same surface projection $\boldsymbol{\kappa} = (\kappa, 0)$ of $\mathbf{q}$ and somewhere above there is a longitudinal threshold.

Now, starting from the $(1, 0, 0)$ direction, the direction of κ is tilted (Figure 4.7) so that it forms an angle θ with $(1, 0, 0)$. Then the two transverse thresholds

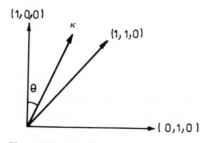

Figure 4.7. Definition of θ for Figure 4.8.

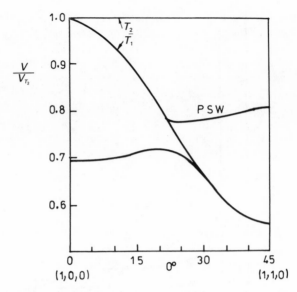

Figure 4.8. Cubic elastic medium: $(0, 0, 1)$ surface. Variation of bulk transverse threshold and surface wave phase velocities for different propagation directions. The actual results correspond to Cu.

split apart. If this is represented in terms of phase velocities (Figure 4.8), then $T_2 = T_1$ for $\theta = 0$ and, as θ increases, one of the two thresholds (T_2) remains constant while the other (T_1) varies. For some materials $T_1 > T_2$ and for others $T_2 > T_1$.[1,4] The latter are more interesting. We rotate the (x, y, z) axes so that x is always the propagation direction (that of κ), y is *transverse horizontal* (parallel to the surface), and z is *transverse vertical* (perpendicular to the surface). If κ is in the $(1, 0, 0)$ direction, $\theta = 0$, then T_1 is transverse horizontal with amplitude $(0, u_y, 0)$, T_2 is transverse vertical with amplitude $(0, 0, u_z)$, and the L (longitudinal) mode higher up in Figure 4.8 has amplitude $(u_x, 0, 0)$. Below T_1 we have the surface Rayleigh mode. This, as is well known, has *sagittal polarization* $(u_x, 0, u_z)$. As θ increases, R moves typically as shown in Figure 4.8 and, at the same time, its polarization changes gradually so that its sagittal strength *decreases* while it develops an *increasing* transverse horizontal amplitude u_y.[4] When $\theta = 45°$, the $(1, 1, 0)$ direction, the sagittal amplitude has decreased to zero while u_y has reached its full value and the surface mode has become degenerate with the T_1 threshold. For this reason it is usually termed a *generalized Rayleigh* (GR) mode. Thus when $\theta = 45°$ it has ceased to exist as a surface mode. However, *another* surface mode appears, between T_1 and T_2, which is purely sagittal, like R when $\theta = 0$. This is actually a distinct surface mode, but only for $\theta = 45°$. As θ *decreases*, the sagittal strength *decreases*, and this mode, usually termed a *pseudosurface wave* (PSW), develops an *increasing* horizontal transverse amplitude u_y of increasing strength until, for $\theta = 0$, all sagittal strength vanishes and it becomes degenerate with the lowest bulk threshold again.

The behavior of the PSW is the opposite to that of R, but this mode has a new feature. For any arbitrary θ it is coupled to the bulk modes, so there is some leakage, i.e., energy from this mode is radiated into the bulk. For real κ the frequency eigenvalue ω is complex, corresponding to a finite lifetime (decay) of the mode. For real ω it is κ that is complex, corresponding to a damping of the mode in its spatial propagation. Technically, the PSW is in fact a *resonance*, which in practice tends to be very sharp (typically, the imaginary part is of order 10^{-4} times the real part), i.e., sufficiently long-lived that it shows up in the experiments. Furthermore, in particular, the imaginary part strictly vanishes when $\theta = 45°$. The PSW is then a true surface state, or a *virtual surface state*. Finally, we have referred only to the surface mode solutions, but of course there are also threshold effects and distortions of the bulk continuum, as in the isotropic case, but they are more complicated.[5]

In conclusion, (1) surface dynamics means more than simply surface modes, (2) crystalline anisotropy is important, and (3) even when referring only to surface modes we must anticipate also the possibility of other solutions of resonant type. The elasticity limit, describing only the long-wave part, gives a limited picture. It is clear that the projection of the bulk phonon band structure and the different kinds of solutions and effects will show up in a more complicated pattern. However, the basic principles are the same; the details must be found by explicit calculation.

4.7. Surface Phonon Calculations

We shall now return to the dynamical matrix of a crystal expressed in the layer notation (4.5.9). For fixed n and n' this is a $3s \times 3s$ matrix. If a finite slab containing N_L atomic layers is now considered, then the n-dependence becomes an explicit feature of the problem, to begin with, for the force constants themselves. The forces acting on the surface atoms are obviously different. At the very least the outer neighbors are missing and we must account for this by "cutting the links." This is sometimes described as introducing a "cleavage perturbation." Besides, there may also be other changes in these force constants due, for instance, to surface *relaxation* and/or *reconstruction*. It is also conceivable that these perturbations may affect one or more atomic layers. After this we are left with a $3sN_L \times 3sN_L$ matrix. In the often practiced *slab calculations* one sets up the eigenvalue problem for all the N_L atomic layers contained in the slab, simultaneously, i.e., one uses precisely this inhomogeneous n- and n'-dependent $3sN_L \times 3sN_L$ dynamical matrix to form the corresponding secular system and then seeks the eigenvalues of its determinant and associated eigenvectors. The solutions thus obtained may also exhibit an n-dependence, but this can be very different for the different eigenmodes of the system. There are modes with amplitudes extending across the slab which form different branches that get closer together as N_L increases. For $N_L \to \infty$ these would span the projection of the continuum of the bulk phonon band structure. Others have amplitudes localized near the surface and remain as distinct branches outside the bulk bands. These are identified as surface modes. Frequencywise, they may appear within the projection of the bulk

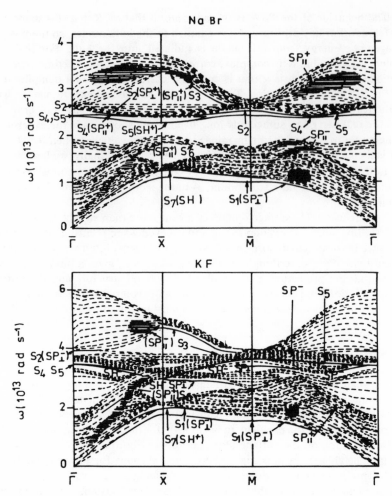

Figure 4.9. Slab calculation[6] of surface phonon dispersion relations with corresponding projection—(0, 0, 1) surface—of the bulk bands.

spectrum and couple to bulk modes, though this is not necessarily implied. Figure 4.9 shows an example of results of this type of calculation.[6] In this case the calculations include surface relaxation, but this is not too significant.

The purpose of this discussion is not to analyze all details but just to gain some idea of what real calculations look like and to discern some general features. For example, let us examine the surface modes for κ on the segment $\bar{X}\bar{M}$ of the 2D Brillouin zone. These are well identified and correspond to short-wave phonons. We now move from either $\bar{X}$ or $\bar{M}$ toward Γ, i.e., toward long waves. As we approach $\bar{\Gamma}$ ($\kappa = 0$), the Rayleigh mode ceases to be well obtained in this calculation: long waves penetrate far into the bulk and eventually the two surfaces are coupled. If the two surfaces are very far away, then distinct surface modes are doubly degenerate (assuming inversion symmetry), with only one well-defined eigenvalue. If the two surfaces become significantly coupled, then the degeneracy is lifted and the method ceases to be practical. One cannot increase indefinitely the number of atomic layers in the slab because the secular matrix becomes too large. (Ten to twenty layers are usually employed. Fifteen-layer slabs were used for the calculations of Figure 4.9.) It should also be noted that only in the long-wave regions, and *only for low frequencies* (acoustic modes), does the projection of the bulk bands look more or less like Figure 4.4. Elsewhere, the picture for a real crystal is much more complicated, as is to be expected, with optical modes at higher frequencies and with "windows" of forbidden bulk frequencies where more surface modes can appear (see also Figures 4.10 and 4.12 through 4.15 below). The discussion at the end of Sections 4.6 implies that one may also expect to find *resonances*. As a result of their coupling to bulk modes, the resonances also penetrate far into the bulk and again these tend to be elusive in this type of calculation, in spite of quite considerable efforts made to try and obtain them.

Thus slab calculations are in practice very useful when attempting to obtain (albeit at the expense of considerable computation) distinct surface phonon bands, but they have practical shortcomings when resonances or long-wave acoustic modes are involved.

Slab calculations are very frequent but, of course, there are many more methods.[7-13] However, the purpose of these notes is not to review them all but to introduce some general concepts in lattice dynamics for which it proves convenient to formulate the problem in term of Green functions.

4.8. Lattice Dynamics in Terms of Green Functions

If $R, R', \ldots$ denote 3D atomic positions in a lattice and the force constants matrix is expressed in concise form as $\langle R|\Phi|R'\rangle$, then the force acting on the atom at R is

$$\langle R|f\rangle = -\langle R|\Phi \cdot u\rangle = -\sum_{R'} \langle R|\phi|R'\rangle \cdot \langle R'|u\rangle \tag{4.8.1}$$

The various details like contracted products (such as $\Phi \cdot u$) will be understood implicitly everywhere. The equation of motion is then

$$\langle R|(\phi - \Omega)u|R'\rangle = 0 \qquad \text{where } \Omega = m\omega^2 \tag{4.8.2}$$

We note that (1) here ω^2 is employed as the natural variable because it appears naturally in the inertial term $\ddot{u}$, and (2) a diagonal mass matrix M has been introduced, meaning that different atoms may have different mass. This situation may arise in the bulk if there exists more than one atom per unit cell, and also in surface problems if there is an adsorbed layer, so it is convenient to stress the mass factor explicitly. The meaning of the mass matrix is that $\langle R|M|R\rangle$ is the mass

m_R of the atom at R and the corresponding eigenvalue is $m_R \omega^2$. We then define the Green function **G** (a matrix) associated with equation (4.8.2) by

$$\langle R|(\boldsymbol{\phi} - \boldsymbol{\Omega})\mathbf{G}|R'\rangle = \delta(R - R') \tag{4.8.3}$$

Thus $\langle R|\mathbf{G}|R'\rangle$ gives at **R** the vibration amplitude responding to a standard unit input at R'. To put it intuitively, if we wiggle an atom at R', then $\langle R|\mathbf{G}|R'\rangle$ describes the propagation of this vibration from R' to R. Thus G is a *propagator* or *dynamical response function* for lattice vibrations. Equation (4.8.3) can be expressed in the concise form

$$(\boldsymbol{\Phi} - \boldsymbol{\Omega}) \cdot \mathbf{G} = \mathbf{I} \tag{4.8.4}$$

where **I** is the unit of the complete space of all atoms linked by the force constants contained in **Φ**. The normal-mode eigenvalues are the zeros of $\boldsymbol{\Phi} - \boldsymbol{\Omega}$, i.e., *the poles of its inverse* **G**.

From now on all matrices **G**, etc., will be written as G, etc. In concise form G has the spectral representation

$$G(\Omega) = \sum_j \frac{|j\rangle\langle j|}{\Phi_j - \Omega} \tag{4.8.5}$$

If

$$\Omega = \Omega^+ = \lim_{\varepsilon \to 0} (\Omega + i\varepsilon)$$

for a causal description and the relationship

$$\frac{1}{x + i\varepsilon} \to P\frac{1}{x} - i\pi\delta(x) \tag{4.8.6}$$

is used, then

$$\text{Im } G(\Omega^+) = \pi \sum_j |j\rangle\langle j|\delta(\Phi_j - \Omega) \tag{4.8.7}$$

is obtained. With an orthogonal basis

$$\text{Im} \sum_i \langle i|G(\Omega^+)|i\rangle = \text{Im Tr } G(\Omega^+) = \pi \sum_j \sum_i \delta_{ij}\delta_{ji}\delta(\Phi_j - \Omega)$$

$$= \pi \sum_i \delta(\Phi_i - \Omega) \tag{4.8.8}$$

The mode density $N(\Omega)$ is now examined. By definition

$$\int_1^2 d\Omega \, N(\Omega) \tag{4.8.9}$$

is the number of eigenvalues in the interval $(1, 2)$. However, this can also be expressed as

$$\int_1^2 d\Omega \sum_i \delta(\Phi_i - \Omega) \tag{4.8.10}$$

whence

$$N(\Omega) = \frac{1}{\pi} \text{Im Tr } G(\Omega^+) \tag{4.8.11}$$

The derivation just given is very informal, but the result (4.8.11) is correct, quite general, and very important. It should be noted, incidentally, that expression (4.8.11) defines a mode density in the variable $m\omega^2$. This can immediately be changed into a density $N(\omega)$ by simply multiplying $M(\Omega)$ by $d\Omega/d\omega = 2m\omega$. Often, the natural variable Ω is defined as ω^2, without mass factor. All these optional definitions entail corresponding different definitions of the dynamical matrix and all these details are unessential. The point is that equation (4.8.11) makes G very useful: *The spectrum of the system is contained in the singularities of G.* Its poles yield the eigenvalues and equation (4.8.11) yields the mode density. Quantity $N(\Omega)$ is another example of a spectral function.

4.9. Surface Dynamics in Terms of Green Functions

A semi-infinite crystal with a free surface is now considered. It is assumed that an appropriate force constants model is embodied in Φ_s, which includes the interactions between the atoms in the half crystal, the effects of termination, possible surface changes in the force constants between existing atoms at or near the surface, and perhaps also other "*surface perturbations*" due to relaxation and/or reconstruction. If I_s is the unit of the half crystal, the Green function of the half crystal defined by

$$(\Phi_s - \Omega)G_s = I_s \tag{4.9.1}$$

is sought. There are many different ways to find G_s. We summarize here the basis of one method, explained in detail elsewhere,[14] and can also be found in the literature with a different presentation designed for a different purpose.[15]

In a bulk crystal the propagator $\langle R|G|R'\rangle$ can be pictured as a line going from R' to R. However, in the presence of a surface, R can also be reached by propagation to the surface, reflection, and then propagation again.

Thus the form of G_s can be expressed as

$$\langle R|G_s|R'\rangle = \langle R|G|R'\rangle + \langle R|GRG|R'\rangle \tag{4.9.2}$$

meaning that R is some surface object which describes reflection. This can be expressed in the form

$$G_s = G + GRG \tag{4.9.3}$$

on the understanding that *only atomic positions within the existing half crystal intervene in this and all subsequent formulas.* It is important to remember this point.

As seen in Section 4.5, a layer description is convenient to study surface problems. Everything is Fourier transformed in 2D and equation (4.9.2) then reads

$$\langle n|G_s|n'\rangle = \langle n|G|n'\rangle + \sum_{s,s'} \langle n|G|s\rangle\langle s|R|s'\rangle\langle s'|G|n'\rangle \tag{4.9.4}$$

Wherever not shown explicitly, the (κ, Ω)-dependence is to be understood everywhere. Here s denotes the atomic layers forming the *surface domain*. By definition this consists of the layers affected by the creation of the surface. Thus the surface domain may (and usually does) contain more than one atomic layer. Let $\mathcal{T}$ denote the projector of the surface domain. We use script characters for all the *surface objects* which result either from direct definition, like R, or from projection of a bulk object onto the domain $\mathcal{T}$. An object like R or $\mathcal{G}$ is identically $\mathcal{T}R\mathcal{T}$ or $\mathcal{T}G\mathcal{T}$, but $\mathcal{T}$ will not be displayed explicitly unless it is necessary for clarity. Two specially important surface objects are $\mathcal{G}_s$ and $\mathcal{G}$, the surface projection of the unknown G_s and of the known bulk G. If there are ν_s atomic layers in the surface domain, then $\mathcal{J}$ is the diagonal $3\nu_s \times 3\nu_s$ unit matrix and R, $\mathcal{G}_s$, $\mathcal{G}$, and similar objects are (κ, Ω)-dependent $3\nu_s \times 3\nu_s$ matrices. Their inverses, *defined within the subspace of $\mathcal{T}$*, satisfy

$$\mathcal{G}_s^{-1}\mathcal{G}_s = \mathcal{G}^{-1}\mathcal{G} = \mathcal{T} \tag{4.9.5}$$

The idea of a formulation of this kind is to project the problem in the space of $\mathcal{T}$ and then proceed there with all the relevant algebra, a substantially different philosophy from that of slab calculations. It can be expressed in the following intuitive manner. We know what happens in the bulk. We want to know what happens in the semi-infinite crystal and we hope that this can be achieved if only we know what happens in the surface domain. Indeed equation (4.9.3), or (4.9.4), says that if we can find R then we know the entire G_s. Thus it all hinges on finding R, and we aim at doing this by means of an algebra to be carried out in the domain having as unit

$$\mathcal{T} = \sum_s |s\rangle\langle s| \tag{4.9.6}$$

Now, equation (4.9.3) yields the surface projection

$$\mathcal{G}_s = \mathcal{G} + \mathcal{G}R\mathcal{G} \tag{4.9.7}$$

whence

$$R = \mathcal{G}^{-1}(\mathcal{G}_s - \mathcal{G})\mathcal{G}^{-1} \tag{4.9.8}$$

On substituting this back in equation (4.9.3), we obtain

$$G_s = G + G\mathcal{G}^{-1}(\mathcal{G}_s - \mathcal{G})\mathcal{G}^{-1}G \qquad (4.9.9)$$

which expresses the same idea more emphatically: *In order to know the complete G_s it suffices to know its surface projection $\mathcal{G}_s$.*

We note, in particular, that the new surface modes must be those singularities of equation (4.9.9) that are not bulk modes, and these can only come from $\mathcal{G}_s$, so that the zeros of its inverse are the surface mode eigenvalues, i.e., the secular equation is

$$\det|\mathcal{G}_s^{-1}(\boldsymbol{\kappa}, \Omega)| = 0 \qquad (4.9.10)$$

The argument is again very informal but the result (4.9.10) can be proved rigorously. The point is that here we have a secular equation and the size of the secular matrix is only that of the surface domain. Whether the surface modes penetrate more or less into the bulk is immaterial. A resonance appears as a complex eigenvalue but neither this nor the long waves present any essential difficulty. In fact the elastic limit can be explicitly recovered as the long-wave limit of this analysis.[14]

Now in order to find $\mathcal{G}_s$ we return to equation (4.9.1) and take the surface projection

$$\Omega\mathcal{G}_s - \mathcal{T}\Phi_s G_s\mathcal{T} = \mathcal{T} \qquad (4.9.11)$$

whence

$$\mathcal{G}_s^{-1} = \mathcal{T}(\Omega - \Phi_s G)\mathcal{T}\mathcal{G}_s^{-1} \qquad (4.9.12)$$

If the surface projection of equation (4.9.9) is taken *from the right only* then

$$G_s\mathcal{J} = G\mathcal{G}^{-1}\mathcal{G}_s \qquad (4.9.13)$$

the meaning of this is that the matrix element between any layer n and a surface layer s is

$$\langle n|G_s|s\rangle = \sum_{s',s''} \langle n|G|s'\rangle\langle s'|\mathcal{G}^{-1}|s''\rangle\langle s''|\mathcal{G}_s|s\rangle \qquad (4.9.14)$$

For expression (4.9.12) we need $G_s\mathcal{G}_s^{-1}$ which, in explicit layer representation, is

$$\sum_{s'} \langle n|G_s|s'\rangle\langle s'|\mathcal{G}_s^{-1}|s\rangle = \sum_{s'} \langle n|G|s'\rangle\langle s'|\mathcal{G}^{-1}|s\rangle \qquad (4.9.15)$$

i.e.,

$$G_s\mathcal{G}_s^{-1} = G\mathcal{G}^{-1} \qquad (4.9.16)$$

Thus

$$\mathcal{G}_s^{-1} = \mathcal{T}\Omega - \mathcal{T}\Phi_s G \mathcal{G}^{-1} \qquad (4.9.17)$$

The important difference between expressions (4.9.12) and (4.9.17) is that G_s has disappeared altogether: relationship (4.9.17) is a *formula*. From this, as has just been seen, we can obtain directly the surface-mode secular equation or evaluate the complete G_s, which contains *all* the physical information on the semi-infinite crystal.

Now, in Section 4.6 we saw the importance of another spectral function, namely, the surface projection of the fixed κ-mode density. In a crystal we may have more than one atomic layer, so that

$$N_s(\mathbf{\kappa}, \Omega) = \frac{1}{\pi} \operatorname{Im} \operatorname{Tr} \mathcal{G}_s(\mathbf{\kappa}, \Omega^+) = \frac{1}{\pi} \operatorname{Im} \sum_s \operatorname{tr} \langle s| \mathcal{G}_s(\mathbf{\kappa}, \Omega^+)|s\rangle \qquad (4.9.18)$$

Here tr means the summation over the three diagonal (x, x), (y, y), and (z, z) terms, i.e., the contributions from all possible polarizations, which can be identified separately if desired. In practice the most interesting local spectral function is just the projection on the terminal atomic layer with index $s = n = 0$. We shall indicate this by

$$\rho_s(\kappa, \Omega) = \frac{1}{\pi} \operatorname{Im} \operatorname{tr} \langle 0| \mathcal{G}_s(\mathbf{\kappa}, \Omega^+)|0\rangle \qquad (4.9.19)$$

which is the discrete crystal analog of $N_s(\mathbf{\kappa}, \Omega)$ (Section 4.6). The study of this spectral function provides, among other things (see Section 4.11 below), an alternative way to calculate surface phonon dispersion relations. For fixed κ, the values of Ω corresponding to surface modes are those at which $\rho_s(\mathbf{\kappa}, \Omega)$ has a δ-function singularity (in numerical terms, a conveniently characterized sharp peak) but, of course, $\rho_s(\mathbf{\kappa}, \Omega)$ contains also additional information of interest.

Thus we have a formalism ready for conducting practical calculations. The surface phonon band structure can be obtained either by looking for the roots of the secular equation, i.e., the determinant of expression (4.9.17), or else by looking for peaks in the spectral function (4.9.18). The latter, evaluated as a function of Ω for fixed κ, contains also spectral information of interest for all Ω, not only for the eigenvalues $\Omega(\mathbf{\kappa})$, as will be seen later in Section 4.11.

4.10. Examples of Application

It is instructive to start with some academic example in order to get acquainted with the practical use of the formal analysis. First we note the following point about expression (4.9.17): $\mathcal{T}\Phi_s G \mathcal{T}$ has layer matrix elements

$$\sum_{n \geq 0} \langle s|\Phi_s|n\rangle\langle n|G|s'\rangle = \langle s|\Phi_s G|s'\rangle \qquad (4.10.1)$$

if the crystal is contained in $n \geq 0$, so that $n = 0$ is the terminal surface layer. It should be remembered that s spans the (surface) domain of $\mathcal{T}$. If this consists of only one atomic layer, then the sum (4.10.1) includes $n_{at} = 0$ (which is s) and also $n_{at} = 1$. But $\mathcal{T}$ usually includes more than one atomic layer and, in practice, often just two. In this case s includes $n_{at} = 0$ and $n_{at} = 1$ and the sum (4.10.1) includes up to $n_{at} = 3$.

The elements $\langle s|\phi_s|n\rangle$ for $n \geq 0$ may or may not be equal to the same $\langle s|\Phi|n\rangle$ of the bulk crystal. They are equal for *ideal cleavage*. Otherwise we have some *surface perturbations*, $\Delta\Phi$, as stressed earlier. By definition $\Delta\Phi$ is all contained in $\mathcal{T}$, with matrix elements

$$(\mathcal{T}\Delta\Phi\mathcal{T})_{ss'} = \langle s|\Delta\Phi|s'\rangle \tag{4.10.2}$$

Then

$$\mathcal{T}\Phi_s G\mathcal{G}^{-1} = \mathcal{T}\Phi I_s G\mathcal{G}^{-1} + \mathcal{T}\Delta\Phi\mathcal{T}G\mathcal{G}^{-1} = \mathcal{T}\Phi I_s G\mathcal{G}^{-1} + \mathcal{T}\Delta\Phi\mathcal{T}\mathcal{G}\mathcal{G}^{-1}$$

$$= \mathcal{T}\Phi I_s G\mathcal{G}^{-1} + \mathcal{T}\Delta\Phi\mathcal{T} \tag{4.10.3}$$

and expression (4.9.17) reads

$$\mathcal{G}_s^{-1} = \mathcal{T}\Omega - \mathcal{T}\Phi I_s G\mathcal{G}^{-1} - \mathcal{T}\Delta\Phi\mathcal{T} \tag{4.10.4}$$

For ideal cleavage ($\Delta\Phi = 0$), $\mathcal{G}_s^{-1}$ has the value that we may call "unperturbed"

$$\mathcal{G}_{s0}^{-1} = \mathcal{T}\Omega - \mathcal{T}\Phi I_s G\mathcal{G}^{-1} \tag{4.10.5}$$

and equation (4.10.4) becomes

$$\mathcal{G}_s^{-1} = \mathcal{G}_{s0}^{-1} - \mathcal{T}\Delta\Phi\mathcal{T} \tag{4.10.6}$$

which is nothing but a Dyson equation *in the $\mathcal{T}$ domain*, where $\mathcal{T}\Delta\Phi\mathcal{T}$ is the perturbation.

It is clear that the hard work lies in determining $\mathcal{G}_{s0}^{-1}$ given by equation (4.10.5), which amounts to redoing the lattice dynamics for the half crystal. We shall consider some examples of ideally cleaved surfaces which suffice to acquire familiarity with the basic features involved in surface phonon calculations. Before doing this we note that the layer matrix elements of the term $\mathcal{T}\Phi_s^0 G\mathcal{T}$ involve sums of the form

$$\langle s|\Phi_s^0|s'\rangle = \sum_{n\geq 0}' \langle s|\Phi^0|n\rangle\langle n|G|s'\rangle \tag{4.10.7}$$

The prime in $\sum'$ indicates that *not all terms with $n = 0$ enter the definition* of this sum. The point is that $\sum'$ includes only the interactions in the existing half of the crystal and leaves out, by definition, those associated with the missing atoms.

We now imagine just a linear chain and a force term of the type

$$\sum_n \Phi_{0n} u_n = K(u_0 - u_{-1}) + K(u_0 - u_1) \tag{4.10.8}$$

corresponding simply to NN interactions.

If a surface is created at $n = 0$, so that the half "crystal" is in $n \geq 0$, then

$$\sideset{}{'}\sum_{n \geq 0} \Phi_{0n} u_n = \Phi_{00} u_0 + \Phi_{01} u_1 = K(u_0 - u_1) \tag{4.10.9}$$

i.e., Φ_{00} is actually $2K$, but Φ'_{00} which enters $\sum'_{n \geq 0}$ is only K. The other contribution to Φ_{00} enters the definition of the complementary sum

$$\sideset{}{'}\sum_{n \leq 0} \Phi_{0n} u_n = \bar{\Phi}_{00} u_0 + \Phi_{0,-1} u_{-1} = K(u_0 - u_{-1}) \tag{4.10.10}$$

In 3D the details are more complicated, but this example suffices to illustrate the idea. In 3D the two sums are less symmetrical because $\sum'_{n \geq 0}$ includes also by definition the interactions between atoms in the surface layers, which are all excluded from $\sum'_{n \leq 0}$. For instance, let us imagine a simple square lattice with NN interactions. The central atom under study is at $(0, 0)$ and is coupled to its neighbors at $(-1, 0)$, $(1, 0)$, $(0, 1)$, and $(0, -1)$. If the terminal surface layer is $(0, m)$, then $\sum'_{n \geq 0}$ includes only the term connecting with $(-1, 0)$. This can be used to simplify the secular equation in the following manner.

We consider the definition of G for the bulk crystal

$$(\Omega - \Phi)G = I \tag{4.10.11}$$

and take the projection in the domain $\mathscr{T}$, which will become the surface domain when the surface is created by (in this case) ideal cleavage. then

$$-\mathscr{T}\Phi G\mathscr{T} = \mathscr{T} - \Omega \mathscr{G} \tag{4.10.12}$$

i.e.,

$$-\sideset{}{'}\sum_{n \geq 0} \langle s|\Phi|n\rangle\langle n|G|s'\rangle = \langle s|(\mathscr{T} - \Omega\mathscr{G})|s'\rangle + \sum_{n \leq 0} \langle s|\Phi|n\rangle\langle n|G|s'\rangle \tag{4.10.13}$$

Equations (4.10.13) and (4.10.5) yield

$$\langle s|\mathscr{G}_{s0}^{-1}|s'\rangle = \langle s|\Omega + \mathscr{G}^{-1} - \Omega|s'\rangle$$

$$+ \sideset{}{'}\sum_{n \leq 0} \sum_{s''} \langle s|\Phi|n\rangle\langle n|G|s''\rangle\langle s''|\mathscr{G}^{-1}|s'\rangle \tag{4.10.14}$$

i.e.,

$$\mathscr{G}_{s0}^{-1} = (\mathscr{T} + \mathscr{T}\bar{\Phi}G\mathscr{T})\mathscr{G}^{-1} \tag{4.10.15}$$

where

$$\langle s|\mathscr{T}\bar{\Phi}G\mathscr{T}|s'\rangle = \sum_{n\leq 0}{}' \langle s|\Phi|n\rangle\langle n|G|s'\rangle \tag{4.10.16}$$

Thus the secular equation can be expressed simply as

$$\det |\mathscr{T} + \mathscr{T}\bar{\Phi}G\mathscr{T}| = 0 \tag{4.10.17}$$

instead of

$$\det |(4.10.5)| = \det |\mathscr{T}\Omega - \mathscr{T}\Phi I_s G\mathscr{G}^{-1}| = 0 \tag{4.10.18}$$

The algebra is simplified for two reasons: (1) $\mathscr{T}\Phi G\mathscr{T}$ includes fewer terms than $\mathscr{T}\Phi_s^0 G\mathscr{T}$, as stressed above. (2) We can omit the factor $\mathscr{G}^{-1}$. However, *this is necessary for the spectral function* (4.9.18).

We now take as an exercise Rosenszweig's model: a simple cubic lattice, central forces, NN interactions with force constant K and NNN interactions with force constant $\frac{1}{2}K$. This is obtained from the model worked out in Appendix 4.1 by setting $K' = \frac{1}{2}K$ and $K'' = 0$. This simple model is isotropic in the long-wave limit, which is a shortcoming, but it satisfies the condition of rotational invariance (4.5.5) and yields the Rayleigh mode in the long-wave limit, a property not shared by all simple models which have often been used. It also has just sufficient complexity to exhibit one of the characteristic features of a 3D crystal, namely, the appearance of "windows" in the continuum obtained from the projection of the bulk band structure. In this case the surface domain consists of one atomic layer ($n = 0$). We must now be a bit more precise concerning notation. In the above equations we have emphasized only the layer index, but in 3D we have the three labels (l_1, l_2, l_3), which we write as (l, m, n), as in Appendix 4.1. We set $L = (l, m)$ and remember that Φ and G are 3×3 matrices. Thus for $\mathscr{T}\Phi G\mathscr{T}$ we must evaluate the sums

$$\sum_{\alpha\beta} (L, L') = \sum_{n\leq 0}{}' \sum_L \sum_\gamma \langle L, 0|\Phi_{\alpha\gamma}|L + \Lambda, n\rangle\langle L + \Lambda, n|G_{\gamma\beta}|L', 0\rangle \tag{4.10.19}$$

where Λ is also a 2D discrete label $(\Delta l, \Delta m)$ giving the 2D position from the central atom $L = (l, m)$.

It remains to define the surface orientation. We shall study the $(0, 0, 1)$ surface, with (l, m, n) associated with (x, y, z). The Φ terms involved in equation (4.10.19) can be obtained from the results given in Appendix 4.1, remembering that (u_x, u_y, u_z) are written as (u, v, w). Quantities Λ entering equation (4.10.19) are obtained by inspection of Figure A4.1. For instance, when dealing with $\langle L, 0|(\bar{\Phi}G)|_{x\beta}|L', 0\rangle$ we start from $-F_x$ (with $K' = K/2$ and $K'' = 0$) and identify the coefficients of all u terms. The coefficients thus obtained are to be multiplied by $G_{x\beta}$ terms. Then the coefficients of v terms contribute to Φ_{xy} terms, to be multiplied by $G_{y\beta}$ terms (as a matter of fact, in this case this contribution is zero), and the coefficients of w terms contribute to Φ_{xz} terms, to be multiplied by $G_{z\beta}$ terms. The result is

$$\langle L, 0|(\bar{\Phi}G)_{x\beta}|L', 0\rangle = -\tfrac{1}{2}K[\langle L + (1,0), -1|G_{x\beta}|L', 0\rangle + \langle L + (\bar{1},0), -1|G_{x\beta}|L', 0\rangle$$
$$- 2\langle L, 0|G_{x\beta}|L', 0\rangle + \langle L + (\bar{1},0), -1|G_{z\beta}|L', 0\rangle$$
$$+ \langle L + (1,0), -1|G_{z\beta}|L', 0\rangle] \qquad (4.10.20)$$

The other terms are obtained in a similar manner.

Another technical point to note is that the first result which comes from the analysis depends on 2D (discrete) position variables and we still have to perform the 2D Fourier transform. The actual position vectors are indicated by $L = (l, m)a$, etc. A matrix element $\langle L, n|G|L', n'\rangle$ is selected with 2D Fourier transform (Ω-dependence understood)

$$\langle n|G(\boldsymbol{\kappa})|n'\rangle = \sum_{L,L'} e^{-i\boldsymbol{\kappa}\cdot L}\langle L, n|G|L', n'\rangle\, e^{i\boldsymbol{\kappa}\cdot L'} \qquad (4.10.21)$$

Then the 2D Fourier transform of $\langle L + \Lambda, n|G|L', n'\rangle$ is

$$e^{i\boldsymbol{\kappa}\cdot\Lambda}\langle n|G(\boldsymbol{\kappa})|n'\rangle \qquad (4.10.22)$$

This holds for functions like G_β, which depend only on the difference $L - L'$. Thus, for the evaluation of equation (4.10.20) it suffices to evaluate the transform of $\langle L, n|G|L', 0\rangle$ for $n = 0, -1$ and to use expression (4.10.22). For instance, as regards κ in the $(1, 0, 0)$ direction and putting $\phi = a\kappa$ we find the terms listed in Appendix 4.2. We note that the (xy), (yx), (yz), and (zy) elements of $\langle 0|(\bar{\Phi}G)_{\alpha\beta}|0\rangle$ vanish. Finally we obtain

$$\mathcal{G}_{s0}^{-1} = \begin{Vmatrix} \dfrac{D_{xx}}{D} & 0 & \dfrac{D_{xz}}{D} \\[2mm] 0 & \dfrac{D_{yy}}{\langle 0|G_{yy}|0\rangle} & 0 \\[2mm] \dfrac{D_{zx}}{D} & 0 & \dfrac{D_{zz}}{D} \end{Vmatrix} \qquad (4.10.23)$$

The various quantities D are listed in Appendix 4.2 and the quantities G for this model are easily evaluated. We have included the factor $\mathcal{G}^{-1}$, because we want to see (below) all effects contained in the spectral function (4.9.18).

The discussion in Section 4.6 is now recalled where we saw that in cubic materials, for propagation in the $(1, 0, 0)$ direction of the $(0, 0, 1)$ surface, there is a decoupling of a shear horizontal mode (degenerate with the bulk threshold) and a surface sagittal mode with amplitudes $(u_x, 0, u_z)$. This feature is not restricted to the long-wave limit. It is a consequence of the cubic symmetry, which shows up again in the form of equation (4.10.23). The (yy) term yields the transverse shear horizontal mode and the rest is a (2×2) determinant with sagittal polarization. Figure 4.10 shows the projection of the bulk bands for this model and surface orientation, as well as the surface modes resulting from equation (4.10.23). These include the Rayleigh mode, below the lowest bulk threshold, and another branch appearing in the window which opens up inside the projection of the bulk continuum. Let us now take a given value of κ corresponding to $\phi = a\kappa = 0.63\pi$ and cut vertically across the bands shown in Figure 4.10. We consider the spectral function $\rho_s(\kappa, \Omega)$ given by equation (4.9.19). For fixed κ this gives the surface projection as a function of the frequency variable Ω, which runs along the corresponding vertical line. The most significant values correspond to the points labeled (1) to (6) in Figure 4.10. Actually we change from $\Omega = m\omega^2$ to ω, so ρ_s becomes $\rho_s(\kappa, \Omega)$, as discussed in Section 4.8. Figure 4.11 gives the corresponding spectral function exhibiting the two surface modes and the bulk threshold effects.

This is a very simple model, but in essence the above exercise demonstrates the kind of analysis involved in real surface phonon calculations and the results exhibit many of the characteristic features of real crystals, including a resonance (ω_r). In Section 4.4 we saw the details of another simple example, namely, an fcc

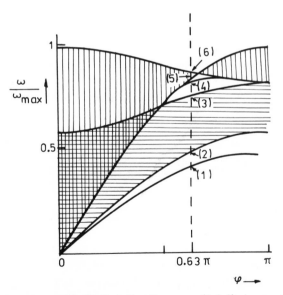

Figure 4.10. Rosenszweig's model with $(0, 0, 1)$ surface, $\kappa = \kappa(1, 0, 0)$, $\phi = a\kappa$; projection of bulk bands and surface modes.

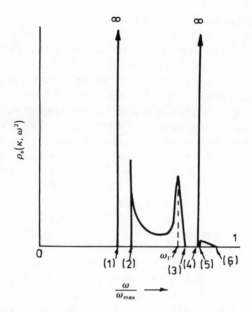

Figure 4.11. Local value, projected at the surface, of the fixed κ-mode density for the same situation as that of Figure 4.10. The points labeled (1) to (6) correspond to those shown in Figure 4.10. Note that δ-function peaks for surface states, bulk threshold effects, and a resonance (ω_r) near the bulk threshold (3).

lattice with central force NN interactions. This is not only an academic model but also one which, used with caution, may have some reasonable practical value. From a rigorous point of view the important role played by d electrons in a metal like, e.g., Ni would require, for a microscopic calculation, a nondiagonal dielectric matrix, a proper treatment of anisotropic ionic potentials, and a calculation altogether very complicated indeed. However, this simple force constants model actually yields a not unreasonable picture of bulk surface phonons in Ni.[16] We shall consider it here as a working tool to study surfaces of fcc metals. There has been considerable experimental activity producing metal surface phonon data from inelastic atom scattering[17-21] and electron energy loss spectroscopy,[22] although the interpretation of the latter type of experimental data tends to be rather more involved.

   A calculation of surface phonon dispersion relations for the Ni–(0, 0, 1) surface can be carried out by following the same pattern as expounded above for Rosensweig's model and using the results for the bulk dynamics seen in Section 4.4. For this surface the x, y, z axes are the $(1, 0, 0)$, $(0, 1, 0)$, and $(0, 0, 1)$ symmetry axes. The primitive translation vectors can be taken as

$$\mathbf{t}_1 = \frac{a}{2}(1, 1, 0), \qquad \mathbf{t}_2 = \frac{a}{2}(1, \bar{1}, 0), \qquad \mathbf{t}_3 = \frac{a}{2}(0, 0, 1) \qquad (4.10.24)$$

with associated 2D reciprocal lattice vectors

$$\mathbf{g_1} = 2\pi \frac{\mathbf{t_2} \wedge \mathbf{t_3}}{\mathbf{t_1} \cdot (\mathbf{t_2} \wedge \mathbf{t_3})} = \frac{2\pi}{a}(1, 1, 0), \qquad \mathbf{g_2} = 2\pi \frac{\mathbf{t_3} \wedge \mathbf{t_1}}{\mathbf{t_2} \cdot (\mathbf{t_3} \wedge \mathbf{t_1})} = \frac{2\pi}{a}(1, \bar{1}, 0)$$

$$(4.10.25)$$

Figure 4.12 shows the calculated[23] surface phonon dispersion relations for Ni-$(0, 0, 1)$ when the tip of κ spans the lines $\bar{\Gamma}\bar{M}$, $\bar{M}\bar{X}$, and $\bar{\Gamma}\bar{X}$ of the said 2D Brillouin zone. We note the usual Rayleigh mode (S_1), modes S_6 and S_7 appearing in the windows of the projection of the bulk bands (compare with Figure 4.10 for a simple cubic lattice), and the S_4 mode which starts out as a sharp resonance just above the lowest transverse bulk threshold and eventually (near the Brillouin zone boundary at $\bar{X}$) becomes a distinct surface mode again. This model contains one parameter, which was adjusted to fit the maximum bulk frequency. The results are very similar to those obtained in other calculations.[24,25] The S_4 mode has been found experimentally[25] and the overall agreement is quite reasonable considering the extreme simplicity of the model. A considerably better agreement is naturally obtained with a more elaborate model[26] involving central and angular forces with up to NNN interactions. On the other hand, it can be argued that the failure of the simple calculation to follow the downward bending of this phonon branch near the Brillouin zone edge is mainly due to the assumption of an ideally unrelaxed surface, a deficiency which is easily remedied. An inward relaxation of the surface relaxation resulting in a 20% increase in the force constant parameter between the first and second atomic layers appears to be capable of accounting for this bending.[25] Thus it is reasonable to say that between some elaboration of the

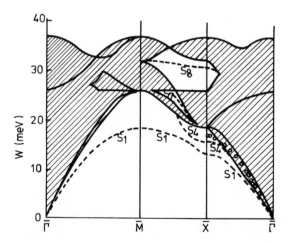

Figure 4.12. Projected bulk bands and calculated[23] and measured[25] surface phonon bands of Ni-$(0, 0, 1)$. The calculation was conducted for an ideal, unrelaxed surface.

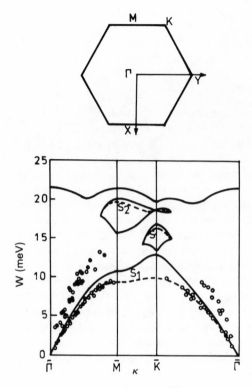

Figure 4.13. Same as Figure 4.12 for Ag-$(1, 1, 1)$. The calculation was set up only for distinct (real eigenvalue) surface modes. Shown above is the corresponding $2D$ Brillouin zone.

model—mainly inclusion of three body forces—and some relaxation of the surface layer one can have a rather reliable picture.

A little more complete are the experimental data[21] on Ag-$(1, 1, 1)$, another fcc metal. We can also try to study this in terms of the same simple model. The geometry for this surface is different. For this we rotate the axes x, y, z to x', y', z' in the directions $(1, 1, 2)$, $(1, 1, 0)$, $(1, 1, 1)$—always the z-axis perpendicular to the surface. The primitive lattice translations can then be chosen as

$$\mathbf{t}_1 = \left(\frac{a}{\sqrt{2}}, 0, 0\right), \qquad \mathbf{t}_2 = \left(\frac{-a}{\sqrt{8}}, \frac{3a}{\sqrt{24}}, 0\right), \qquad \mathbf{t}_3 = (0, 0, a) \qquad (4.10.26)$$

with 2D reciprocal lattice vectors

$$\mathbf{g}_1 = \frac{\sqrt{2}\pi}{a}\left(1, \frac{1}{\sqrt{3}}, 0\right), \qquad \mathbf{g}_2 = \frac{\sqrt{8}\pi}{a}\left(0, \frac{2}{\sqrt{3}}, 0\right) \qquad (4.10.27)$$

The 2D Brillouin zone is shown in Figure 4.13, which also shows the experimental data[21] and the results of a calculation[23] carried out by using again the same method and model as for Ni-$(0, 0, 1)$ and fitting the one force constant parameter to the maximum bulk frequency.[27] Only distinct surface states were sought in this calculation. The resulting dispersion curves for this Rayleigh-type mode are in close agreement with those obtained in other calculations for the same type of model.[24,28] Experimental evidence shows in this case another mode in the $\bar{\Gamma}\bar{M}$ and $\bar{\Gamma}\bar{K}$ directions. These are resonance branches, which one could also obtain by looking for complex eigenvalues of the secular equation (4.10.17) or (4.10.18) or by studying the spectral function $\rho_s(\kappa, \Omega)$ given by (4.9.19) for fixed κ as a function of Ω and looking for a second peak above the Rayleigh peak. But again the ideal surface calculation does not follow the downward bending of the experimental branches near the Brillouin zone boundaries. This is also well reproduced by combining a more elaborate model with some changes in the surface force constants.[12] It is not unlikely that just the latter, even with the simple model, could yield a reasonable fit to experimental data, but these are detailed technicalities which do not concern us here. The most relevant feature of this case is the appearance of distinctly identifiable resonance branches.

4.11. Surface Phonon Spectra and Inelastic Atom–Surface Scattering Data

Although it has been known for quite some time that inelastic atom scattering off solid surfaces could constitute a desirable spectroscopic method to study surface phonons and that the theoretical foundation existed already in the late sixties,[29,30] it was not until the eighties that sufficient resolution was achieved[21,31] so that this has become a very useful experimental-research tool capable of producing good surface phonon spectra. Basically, the method works better for low excitation energies because at higher energies (say, optical phonon frequencies in ionic crystals) multiphonon processes might be involved, rendering the interpretation of experimental data none too clear. Yet the first important experiments of this kind were conducted mainly with ionic crystals, as the acoustic-mode region of these involves lower frequencies than that of, say, metals. More recently the technique has been improved considerably and even optical modes can also be detected.[32]

In an elementary analysis of ionic crystal dynamics one assumes rigid-ion displacements, but recognizes that these entail (in the optical modes) dipole moments and associated electric fields. The rigid-ion *lattice polarizability* thus enters the lattice dynamics of ionic crystals in the usual manner explained in textbooks. The idea of the *shell models*[33] is to intrdoduce a new concept, namely, that the ions are not rigid. The positive ion core is assumed to be rigid but the electronic charge around it is deformable, i.e., the ions themselves are also polarizable. These deformations entail at least another contribution to the dipole moment, with associated electric field, and this is then included in the lattice dynamics so that the *ionic polarizability* is now involved. In the simplest approximation—within a shell model—one considers a bodily displacement of the electronic charge, undeformed, relative to the positive ion core. This is sufficient to produce a dipole. In the next step, called the *breathing shell model*, one assumes also radial deformations of the electronic charge such that only dipolar deformations are involved. This suffices to introduce sufficient anisotropy so that Cauchy's relations do not hold in the long-wave limit. Of course one could go on and allow also for quadrupole deformations. This would surely amount to some improvement, but in practice the breathing shell model involving only dipolar deformations appears to work so well that it is used profusely with good results for the lattice dynamics of ionic crystals.

Green functions can also be used to study (bulk or) surface lattice dynamics in different ways. There is one, called the Invariant Green Function (IGF) method,[34] which is specially useful when studying surface lattice dynamics of ionic crystals. In this method one starts formally from a slab, however eventually one allows its thickness to tend to infinity so that the two surfaces are completely decoupled, but this is a mere technicality. The idea is to describe the introduction of the surface as a perturbation of the bulk crystal (which leads to work in the restricted localized space of the surface domain) and impose the conditions that the force constants for both the complete and incomplete crystal must satisfy translational (and parallel to the surface) as well as rotational invariance. Thus the terms of the perturbation which creates the surface must also satisfy the same invariance conditions. This is expressed by means of a projection on the surface domain and this is the way to express the surface (boundary) conditions. The term *perturbation* has here only a formal meaning. It does not imply solving the problem by some approximation scheme. We have returned to the idea of studying the surface problem by using only the projection in the localized surface domain and doing there all the algebra. In this analysis[34] the layer description of the crystal is carried out in such a way that the long-range Coulomb interactions are summed by layers, and one is left with effective interlayer interactions which decay very rapidly with distance. This makes the IGF method very useful to study ionic crystal surfaces. Combined with the breathing shell model it has produced very useful results in practice.

Figure 4.14 shows the results of such a calculation[34] for LiF-$(0, 0, 1)$, in excellent overall agreement with experimental data except for one significant detail. The data near the Brillouin zone edge again bend downward below the theoretical (broken line) curve for an ideal surface. The same behavior as that of the broken line of Figure 4.14 was found earlier in slab calculations.[24]

The breathing shell model is very useful here in helping to understand the situation. An adjustment of the polarizability of the bulk F^- ions gives already a better result (heavy line), but a 17% increase in the polarizability of the surface F^- ions is still needed to obtain perfect agreement.[35] A detailed study shows that in LiF for the Rayleigh modes near the Brillouin zone edge it is mainly the anions that move, and thus their polarizability is important in determining the outcome. NaF is different in that, for the same type of Rayleigh modes (near $\bar{M}$), the anions are at rest, while they move in the modes of the optical branches (for long waves these are referred to as Lucas modes[36]). Owing to their larger size, the anions usually interact more strongly with the incident He atoms used in the experiment and so it has become possible to detect the optical surface modes in NaF $(0, 0, 1)$ by inelastic atom scattering, as shown in Figure 4.15. The results are also in very good agreement with IGF calculations based on the breathing shell model.[31]

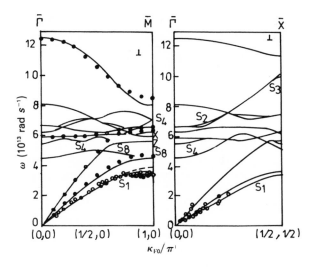

Figure 4.14. Li-(0, 0, 1): projected bulk bands and surface phonon bands. Discussion in the text.

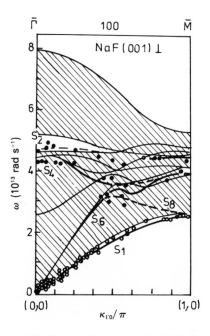

Figure 4.15. Same as Figure 4.14, for NaF-(0, 0, 1).

It is clear that the IGF method and the method described in Sections 4.9 and 4.10 have many features in common. Both use Green functions, both deal with the surface as a distinct problem, and in both one ends up by projecting in the localized surface domain, so that the information on the bulk crystal is embodied in its corresponding bulk G, but only the surface projection ($\mathcal{G}$ in Sections 4.9 and 4.10) is used to study surface phonons. In practical terms, the secular matrix is never larger than necessary and in both cases the size of its matrix is the same. Moreover, it can be proved[37] that although the two methods originated differently and have a different scope, in the part where they overlap they are identically equivalent. More precisely, the object called $\mathcal{G}_s$ in Sections 4.9 and 4.10, i.e., the surface projection of the Green function of the semi-infinite crystal, is identically the object called g_s in the IGF method. Moreover, in one method the worker is led naturally to $\mathcal{G}_s^{-1}$ and in the other he is led naturally to g_s^{-1}; the reason is that this is precisely the secular matrix. But the interesting thing is that this object, $\mathcal{G}_s$ or g_s, contains more information than just the surface phonon eigenvalues, as was stressed earlier. For fixed κ the secular equation yields the surface eigenfrequencies, but $\mathcal{G}_s(\kappa, \Omega)$, or $g_s(\kappa, \Omega)$, studied as a function of Ω yields the spectral function $\rho_s(\kappa, \Omega)$ given by equation (4.9.19) which, as we saw, is the local value for fixed κ of the projected mode density for the terminal atomic layer.

What makes the latter particularly interesting is that this is precisely what is involved in the interpretation of inelastic atom scattering spectra. Let us translate ρ_s into $\rho_s(\kappa, \Omega)$. In the experiments, $\hbar\kappa$ is the momentum parallel to the surface exchanged between incident atom (the probe) and semi-infinite crystal (the target) and one makes an energy analysis: $\hbar\omega$ is the energy exchanged and the experiment yields the inelastic scattering cross section $\sigma(\kappa, \omega)$ as a function of ω for fixed κ. The detailed analysis requires some care and some labor, as surface modes may be *created* or *annihilated*, and the scattering may be *backward* or *forward*, but such details lie outside our scope. What is relevant from our standpoint is the conspicuous role played by $\rho_s(\kappa, \omega)$. In any study of this problem there are three distinct ingredients. One is the model used for the interaction between incident atom and surface atoms. The second is the approximation made in formulating and solving the scattering problem itself. The third ingredient is the description of the spectrum of surface modes created or absorbed. Without going into the details of this analysis, it is clear that $\sigma(\kappa, \omega)$ is going to be in effect proportional to $\rho_s(\kappa, \omega)$, since σ essentially maps out the spectral density of the modes involved in the process. We can also look at it in this way: The cross section is ultimately proportional to the mean-square value of the vibrating amplitudes at the surface, $\langle u_s^2 \rangle$, which is classically the size of the obstacle with which the incoming atom collides. More technically, this is to be Fourier-analyzed so it is a function of (κ, ω). Now, $\langle u_s^2 \rangle$ is a (surface) autocorrelation function. In general fluctuation-dissipation theory, via thermal averages, etc., the autocorrelation function is in the end proportional to the imaginary part of the corresponding response function, which in this case is just $\mathcal{G}_s$. Thus we expect that σ is essentially proportional to Im Tr $\mathcal{G}_s$, i.e., to ρ_s. The trace is due to the fact that we must add up the contributions from all possible polarizations.

The interchangeability between the two Green function methods just mentioned provides a very convenient scheme, because formulas from either method

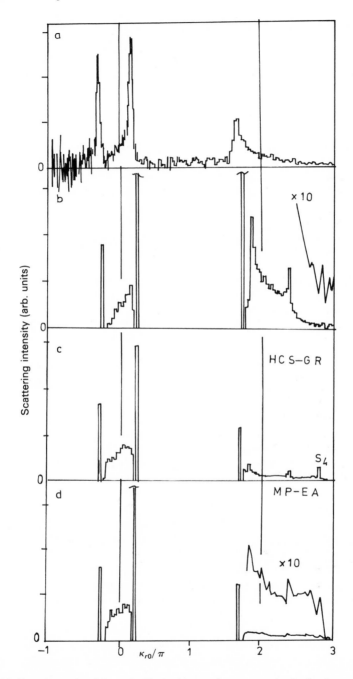

Figure 4.16. Experimental and theoretical inelastic scattering cross section for fixed κ as a function of energy for LiF-(0, 0, 1). κ_{r0}/π is a measure of the energy change of the reflected atoms. Discussion in the text.

can be combined in practice to obtain ρ_s (κ, ω). Figure 4.16 shows experimental results for LiF-(0, 0, 1) and theoretical results obtained in this way.[37] Three different calculations are presented. They differ in the atom–atom potentials and in the approximations used to solve the scattering problem, but all three use the same function ρ_s (κ, ω) obtained in the manner indicated above and based on a breathing shell model (actually, weighted by Bose–Einstein and Debye–Waller factors which a complete analysis requires). The point is that the details vary, of course, but the key features of the spectrum remain and are in good agreement with experiment. This demonstrates how important it is to have a reliable description of surface phonon dynamics for the analysis of inelastic atom scattering spectra. It also bears out the observation anticipated in Section 4.6 and evident also in Brillouin scattering data (Figure 4.6), that when the surface modes are excited (or absorbed) it is *the entire surface dynamics* that is excited (or absorbed), not only the surface modes.

Further details on atom–surface scattering lie outside the scope of this chapter and can be found elsewhere,[38] including plenty of information related to surface lattice dynamics.

Appendixes

Appendix 4.1. Force Constants and Equations of Motion for a Simple Cubic Lattice with Nearest-Neighbor and Next-Nearest-Neighbor Interactions and Central and Angular Forces

This was first employed for a model calculation by D. C. Gazis, R. Herman, and R. F. Wallis, *Phys. Rev.* **119**, 533 (1960). The geometrical details are shown in Figure A4.1. The notation is simplified and (l, m, n) used instead of (l_1, l_2, l_3), in this case associated with (x, y, z). We also write (u, v, w) instead of (u_x, u_y, u_z). As regards the various terms, atomic positions and displacements are viewed from the n (or z, or 0, 0, 1) axis.

Central forces. The contributions to Φ^{NN} which originate from central forces acting on the central atom under consideration, (l, m, n), are due to:

1. u displacements from atoms with the same (m, n)
2. v displacements from atoms with the same (l, n)
3. w displacements from atoms with the same (l, m)

Thus

$$\Phi^{NN}_{central} = K[(u_{l,m,n} - u_{l+1,m,n})^2 + (u_{l,m,n} - u_{l-1,m,n})^2 + (v_{l,m,n} - v_{l,m+1,n})^2$$

$$+ (v_{l,m,n} - v_{l,m-1,n})^2 + (w_{l,m,n} - w_{l,m,n+1})^2 + (w_{l,m,n} - w_{l,m,n-1})^2]$$

$$(A4.1.1)$$

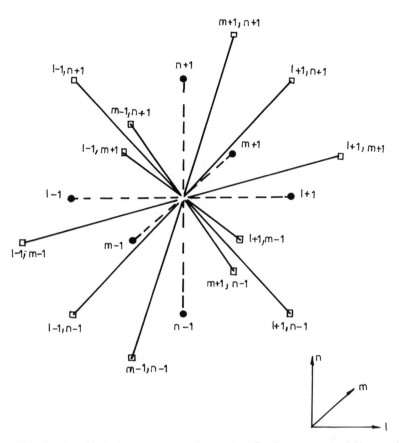

Figure A4.1. Simple cubic lattice, central atom l, m, n, the 6 ● refer to nearest neighbors and the 12 □ to next-nearest neighbors. Position labels not explicitly shown are l, m, or n; e.g., the central atom is (l, m, n), $l + 1$, $n + 1$ is $(l + 1, m, n + 1)$.

where K is one of the parameters of the model. We note that a displacement perpendicular to the line joining two atoms also contributes to central forces (the spring is also stretched as the distance varies). However, for small displacements

$$\frac{\delta a}{a_0} = \frac{(a_0^2 + v^2)^{1/2} - a_0}{a_0} = \frac{1}{2}\left(\frac{v}{a_0}\right)^2$$

which gives a contribution to Φ of order $(\delta a/a_0)^2 \propto (v/a_0)^4$, i.e., a higher-order term beyond the harmonic approximation. The force term derived from equation (A4.1.1) and acting in the x-direction is

$$-F_{x,\text{central}}^{\text{NNN}} = 2K[(u_{l,m,n} - u_{l+1,m,n}) + (u_{l,m,n} - u_{l-1,m,n}) \qquad (A4.1.2)$$

We now denote by ${}^l\Phi_{\text{central}}^{\text{NNN}}$, ${}^m\Phi_{\text{central}}^{\text{NNN}}$, ${}^n\Phi_{\text{central}}^{\text{NNN}}$ the contributions to Φ_{central} originating from NNN atoms contained in the planes l = const, m = const, respectively. Then

$$
{}^{n}\Phi_{\text{central}}^{\text{NNN}} = K'\{[(u_{l,m,n} + v_{l,m,n}) - (u_{l+1,m+1,n} + v_{l+1,m+1,n})]^2
$$

$$
+ [(u_{l,m,n} + v_{l,m,n}) - (u_{l-1,m-1,n} + v_{l-1,m-1,n})]^2
$$

$$
+ [(u_{l,m,n} - v_{l,m,n}) - (u_{l+1,m-1,n} - v_{l+1,m-1,n})]^2
$$

$$
+ [(u_{l,m,n} - v_{l,m,n}) - (u_{l-1,m+1,n} - v_{l-1,m+1,n})]^2\}
$$

$$(A4.1.3)$$

whence the contribution to F_x is given by

$$
- {}^{n}F_{x,\text{central}}^{\text{NN}} = 2K'\{[(u_{l,m,n} + v_{l,m,n}) - (u_{l+1,m+1,n} + v_{l+1,m+1,n})]
$$

$$
+ [(u_{l,m,n} + v_{l,m,n}) - (u_{l-1,m-1,n} + v_{l-1,m-1,n})]
$$

$$
+ [(u_{l,m,n} - v_{l,m,n}) - (u_{l+1,m-1,n} - v_{l+1,m-1,n})]
$$

$$
+ [(u_{l,m,n} - v_{l,m,n}) - (u_{l-1,m+1,n} - v_{l-1,m+1,n})]\}
$$

$$(A4.1.4)$$

where K' is another parameter of the model. Similarly, we find

$$
- {}^{m}F_{x,\text{central}}^{\text{NNN}} = 2K'\{[(u_{l,m,n} + w_{l,m,n}) - (u_{l+1,m,n+1} + w_{l+1,m,n+1})]
$$

$$
+ [(u_{l,m,n} + w_{l,m,n}) - (u_{l-1,m,n-1} + w_{l-1,m,n-1})]
$$

$$
+ [(u_{l,m,n} - w_{l,m,n}) - (u_{l+1,m,n-1} - w_{l+1,m,n-1})]
$$

$$
+ [(u_{l,m,n} - w_{l,m,n}) - (u_{l-1,m,n+1} - w_{l-1,m,n+1})]\}
$$

$$(A4.1.5)$$

It is easily seen that ${}^{l}F_{x,\text{central}}^{\text{NNN}} = 0$ because ${}^{l}\Phi_{\text{central}}^{\text{NNN}}$ involves only quantities v and w, but not u. We have obtained all contributions to $F_{x,\text{central}}$, whence the total force, including NN and NNN interactions. Knowing F_x we obtain F_y and F_z by appropriate cyclic permutations.

Angular bending forces. We use the same sort of notation, which should be self-explanatory. The different contributions are as follows.

$$
{}^{n}\Phi_{(i)} = K''[(u_{l,m+1,n} - u_{l,m,n}) + (v_{l+1,m,n} - v_{l,m,n})]^2 \qquad (A4.1.6)
$$

$$
- {}^{n}F_x^{(i)} = -2K''[(u_{l,m+1,n} - u_{l,m,n}) + (v_{l+1,m,n} - v_{l,m,n})] \qquad (A4.1.7)
$$

$$
- {}^{n}F_x^{(ii)} = -2K''[(u_{l,m+1,n} - u_{l,m,n}) - (v_{l-1,m,n} - v_{l,m,n})] \qquad (A4.1.8)
$$

$$- {}^{n}F_{x}^{(\text{iii})} = -2K''[(u_{l,m-1,n} - u_{l,m,n}) + (v_{l-1,m,n} - v_{l,m,n})] \tag{A4.1.9}$$

$$- {}^{n}F_{x}^{(\text{iv})} = -2K''[(u_{l,m-1,n} - u_{l,m,n}) - (v_{l,m+1,n} - v_{l,m,n})] \tag{A4.1.10}$$

$${}^{n}\Phi^{(\text{v})} = K''[(u_{l,m,n} - u_{l,m+1,n}) - (v_{l+1,m+1,n} - v_{l,m+1,n})]^{2} \tag{A4.1.11}$$

$$- {}^{n}F_{x}^{(\text{v})} = 2K''[(u_{l,m,n} - u_{l,m+1,n}) - (v_{l+1,m+1,n} - v_{l,m+1,n})] \tag{A4.1.12}$$

$${}^{n}\Phi^{(\text{vi})} = K''[(u_{l,m,n} - u_{l,m-1,n}) + (v_{l+1,m-1,n} - v_{l,m-1,n})]^{2} \tag{A4.1.13}$$

$$- {}^{n}F_{x}^{(\text{vi})} = 2K''[(u_{l,m,n} - u_{l,m-1,n}) + (v_{l+1,m-1,n} - v_{l,m-1,n})] \tag{A4.1.14}$$

$${}^{n}\Phi^{(\text{vii})} = K''[(u_{l+1,m-1,n} - u_{l+1,m,n}) + (v_{l,m,n} - v_{l+1,m,n})]^{2} \tag{A4.1.15}$$

$$- {}^{n}F_{x}^{(\text{vii})} = 0 \tag{A4.1.16}$$

$${}^{n}\Phi^{(\text{viii})} = K''[(u_{l-1,m-1,n} - u_{l,m-1,n}) - (v_{l,m,n} - v_{l-1,m,n})]^{2} \tag{A4.1.17}$$

$$- {}^{n}F_{x}^{(\text{viii})} = 0 \tag{A4.1.18}$$

$${}^{n}\Phi^{(\text{ix})} = K''[(u_{l-1,m+1,n} - u_{l-1,m,n}) - (v_{l,m,n} - v_{l-1,m,n})]^{2} \tag{A4.1.19}$$

$$- {}^{n}F_{x}^{(\text{ix})} = 0 \tag{A4.1.20}$$

$${}^{n}\Phi^{(\text{x})} = K''[(u_{l+1,m+1,n} - u_{l+1,m,n}) - (v_{l,m,n} - v_{l+1,m,n})]^{2} \tag{A4.1.21}$$

$$- {}^{n}F_{x}^{(\text{x})} = 0 \tag{A4.1.22}$$

$${}^{n}\Phi^{(\text{xi})} = K''[(u_{l,m,n} - u_{l,m+1,n}) - (v_{l-1,m,n} - v_{l,m+1,n})]^{2} \tag{A4.1.23}$$

$$- {}^{n}F_{x}^{(\text{xi})} = 2K''[(u_{l,m,n} - u_{l,m+1n}) - (v_{l-1,m,n} - v_{l,m+1,n})] \tag{A4.1.24}$$

$${}^{n}\Phi^{(\text{xii})} = K''[(u_{l,m,n} - u_{l,m-1,n}) - (v_{l-1,m-1,n} - v_{l,m-1,n})]^{2} \tag{A4.1.25}$$

$$- {}^{n}F_{x}^{(\text{xii})} = 2K''[(u_{l,m,n} - u_{l,m-1,n}) - (v_{l-1,m-1,n} - v_{l,m-1,n})] \tag{A4.1.26}$$

Then the total values are given by

$${}^{n}F_{x,\text{ang}} = \sum_{(\nu)=(\text{i}),\dots,(\text{xii})} {}^{n}F_{x}^{(\nu)} \tag{A4.1.27}$$

Quantity ${}^{m}F_{x,\text{ang}}$ is obtained from ${}^{n}F_{x,\text{ang}}$ by changing $v \to w$, $m + \varepsilon \to n + \varepsilon$ ($\varepsilon = \pm 1$), and $n \to m$. ${}^{l}F_{x,\text{ang}} = 0$ ($u_{l,m,n}$ does not enter ${}^{l}\Phi^{\text{ang}}$).

The preceding relationships enable the equation of motion for $u_{l,m,n}$ to be written. Some cancellations among the above contributions take place on summing.

Hence

$$M\ddot{u}_{l,m,n} = 2K[u_{l+1,m,n} + u_{l-1,m,n} - 2u_{l,m,n}]$$

$$+ 2K'[u_{l+1,m+1,n} + u_{l-1,m-1,n} + u_{l+1,m-1,n} + u_{l-1,m+1,n}$$

$$+ u_{l+1,m,n+1} + u_{l-1,m,n-1} + u_{l+1,m,n-1} + u_{l-1,m,n+1} - 8u_{l,m,n}]$$

$$\ast \qquad + 2K'[v_{l+1,m+1,n} + v_{l-1,m-1,n} - v_{l+1,m-1,n} - v_{l-1,m+1,n}$$

$$+ w_{l+1,m,n+1} + w_{l-1,m,n-1} - w_{l+1,m,n-1} - w_{l-1,m,n+1}]$$

$$\ast\ast \qquad + 2K''[v_{l+1,m+1,n} + v_{l-1,m-1,n} - v_{l+1,m-1,n} - v_{l-1,m+1,n}$$

$$+ w_{l+1,m,n+1} + w_{l-1,m,n-1} - w_{l+1,m,n-1} - w_{l-1,m,n+1}]$$

$$+ 8K''[u_{l,m+1,n} + u_{l,m-1,n} + u_{l,m,n+1} + u_{l,m,n-1} - u_{l,m,n}] \qquad (A4.1.28)$$

We note that * and ** possess the identical form but are of different physical origin.

Quantities $Mv_{l,m,n}$ and $Mw_{l,m,n}$ are obtained from equation (A4.1.28) by cyclic permutation of (u,v,w) and of the increments $\delta = \pm 1$, $\varepsilon = \pm 1$ on the indices l,m,n. For instance, in the case of $Mv_{l,m,n}$

$$\left.\begin{matrix}u\\v\\w\end{matrix}\right\} \to \left.\begin{matrix}v\\w\\u\end{matrix}\right\} \quad \text{and} \quad \left.\begin{matrix}l+\delta\\m\\n\end{matrix}\right\} \to \left.\begin{matrix}l\\m+\delta\\n\end{matrix}\right\}; \quad \left.\begin{matrix}l+\delta\\m+\varepsilon\\n\end{matrix}\right\} \to \left.\begin{matrix}l\\m+\delta\\n+\varepsilon\end{matrix}\right.$$

$$\left.\begin{matrix}l+\delta\\m\\n+\varepsilon\end{matrix}\right\} \to \left.\begin{matrix}l+\varepsilon\\m+\delta\\n\end{matrix}\right\}; \quad \left.\begin{matrix}1\\m+\delta\\n\end{matrix}\right\} \to \left.\begin{matrix}l\\m\\n+\delta\end{matrix}\right\} \to \left.\begin{matrix}l+\delta\\m\\n\end{matrix}\right\}$$

For this one can express equation (A4.1.28) in the equivalent condensed form

$$M\ddot{u}_{l,m,n} = 2K \sum_{\delta=\pm 1} (u_{l+\delta,m,n} - u_{l,m,n})$$

$$+ 2K' \sum_{\delta=\pm 1} \sum_{\varepsilon=\pm 1} (u_{l+\delta,m+\varepsilon,n} + u_{l+\delta,m,n+\varepsilon} - 2u_{l,m,n})$$

$$+ 2(K' + K'') \sum_{\delta=\pm 1} \sum_{\varepsilon=\pm 1} \varepsilon\delta(v_{l+\delta,m+\varepsilon,n} + w_{l+\delta,m,n+\varepsilon})$$

$$+ 8K'' \sum_{\delta=\pm 1} (u_{l,m+\delta,n} + u_{l,m,n+\delta} - 2u_{l,m,n}) \qquad (A4.1.29)$$

In the elastic limit

$$aC_{11} = 2(K + 4K'), \qquad aC_{12} = 4K', \qquad aC_{44} = 2(2K' + K'') \qquad (A4.1.30)$$

$C_{11} \neq C_{12} + 2C_{44}$ (anisotropic). Cauchy's relations are *not* obeyed.

For $K'' = 0$ (central forces only) the long-wave limit yields an elastic isotropic medium with $C_{11} = C_{12} + 2C_{44}$ (Cauchy's relations).

Appendix 4.2. Simple Cubic Lattice with Central Force Nearest-Neighbor and Next-Nearest-Neighbor Interactions. Rosenszweig's Model

We give here some results for κ in the $(1, 0, 0)$ direction in layer notation for $(0, 0, 1)$ layers, with layer index n and $\phi = a\kappa$. We put

$$\sum_{\alpha\beta} = \langle 0|(\bar{\Phi}G)_{\alpha\beta}|0\rangle \tag{A4.2.1}$$

This is the κ-dependent 2D Fourier transform of $\langle L, 01|(\bar{\Phi}G)_{\alpha\beta}|L', 0\rangle$ for $\kappa = (\kappa, 0, 0)$. Ω-dependence is understood throughout. The terms $\langle n|G_\beta|n'\rangle$ are likewise the 2D Fourier transforms of $\langle L, n|G_\beta|L, n'\rangle$. The property (4.10.22) has been used and the symbols $\sum'_s$ satisfy

$$\overline{\sum_{xy}} = \overline{\sum_{yx}} = \overline{\sum_{yz}} = \overline{\sum_{zy}} = 0 \tag{A4.2.2}$$

$$\overline{\sum_{xx}} = K[-\cos\phi\langle-1|G_{xx}|0\rangle + \langle0|G_{xx}|0\rangle + i\sin\phi\langle-1|G_{xz}|0\rangle] \tag{A4.2.3}$$

$$\overline{\sum_{xz}} = K[-\cos\phi\langle-1|G_{xz}|0\rangle + i\sin\phi\langle-1|G_{zz}|0\rangle] \tag{A4.2.4}$$

$$\overline{\sum_{zx}} = K[-(2+\cos\phi)\langle-1|G_{zx}|0\rangle + i\sin\phi\langle-1|G_{xx}|0\rangle] \tag{A4.2.5}$$

$$\overline{\sum_{zz}} = K[-(2+\cos\phi)\langle-1|G_{zz}|0\rangle + 3\langle0|G_{zz}|0\rangle + i\sin\phi\langle-1|G_{xz}|0\rangle] \tag{A4.2.6}$$

$$\overline{\sum_{yy}} = K[\langle0|G_{yy}|0\rangle - \langle-1|G_{yy}|0\rangle] \tag{A4.2.7}$$

The terms entering equation (4.10.23) are

$$D_{xx} = \{1 - K[-\cos\phi\langle-1|G_{xx}|0\rangle + \langle0|G_{xx}|0\rangle$$
$$+ i\sin\phi\langle-1|G_{zx}|0\rangle]\}\langle0|G_{zz}|0\rangle \tag{A4.2.8}$$

$$D_{xz} = -K[-\cos\phi\langle-1|G_{xz}|0\rangle + i\sin\phi\langle-1|G_{zz}|0\rangle]\langle0|G_{xx}|0\rangle \tag{A4.2.9}$$

$$D_{zx} = K[-(2+\cos\phi)\langle-1|G_{zx}|0\rangle + i\sin\phi\langle-1|G_{xx}|0\rangle]\langle0|G_{zz}|0\rangle \tag{A4.2.10}$$

$$D_{zz} = K\{1 - K[-(2+\cos\phi)\langle-1|G_{zz}|0\rangle + 3\langle0|G_{zz}|0\rangle$$
$$+ i\sin\phi\langle-1|G_{zz}|0\rangle]\}\langle0|G_{xx}|0\rangle \tag{A4.2.11}$$

$$D_{yy} = 1 - K[\, -\langle -1|G_{yy}|0\rangle + \langle 0|G_{yy}|0\rangle] \tag{A4.2.12}$$

$$D = \langle 0|G_{xx}|0\rangle\langle 0|G_{zz}|0\rangle - \langle 0|G_{zx}|0\rangle\langle 0|G_{xz}|0\rangle \tag{A4.2.13}$$

The matrix elements $G_{\alpha\beta}$ for this simple model are easily obtained by direct evaluation of the spectral representation

$$\langle n|G_{\alpha\beta}|n'\rangle = \frac{1}{2\pi}\sum_j \int_{-\pi}^{\pi} d\psi \lim_{\varepsilon \to 0} \frac{e_\alpha(\psi|j)e_\beta^*(\psi|j)\, e^{i\psi(n-n')}}{\Omega_j(\psi) - \Omega + i\varepsilon} \tag{A4.2.14}$$

Here $\psi = aqz$ and j labels the different eigenvalues and eigenvectors obtained from the dynamical matrix as explaiined in the text.

This is not always the most practical way to evaluate Green functions, but it is the most direct when the model is, as in this case, sufficiently simple than it can be worked out analytically.

Acknowledgment

The author is very grateful to V. R. Velasco for his generous help in the preparation of this chapter.

Bibliography

A. General Background References: Bulk

P. Brüesch, *Phonons: Theory and Experiments. I. Lattice Dynamics and Models of Interatomic Forces*, Springer-Verlag, Berlin (1982). Very good for physical models.

J. W. Cochran, *The Dynamics of Atoms in Crystals*, E. Arnold, Ltd., London (1973). Also very intuitive and good for physical models, although less up to date and more restricted in scope than Brüesch.

G. W. Leibfried, "Lattice dynamics," in: *Theory of Condensed Matter*, ICTP International College, 1967, p. 175, International Atomic Energy Agency, Vienna (1968). A good general introduction. Authoritative pedagogic presentation of general principles. Formal aspects made easy.

A. A. Maradudin, E. W. Montroll, G. H. Weiss, and I. P. Ipatova, "Theory of lattice dynamics in the harmonic approximation," in: Supplement 3 of the series *Solid State Physics* (H. Ehrenreich, F. Seitz, and D. Turnbull, eds.), Academic Press, New York (1971). Formal and advanced aspects. For professionals. Contains also a chapter on surfaces.

Also, many standard textbooks on physics of solids contain general treatments of bulk lattice dynamics at different levels to suit all tastes.

B. General Background References: Surface

G. W. Farnell, "Properties of elastic surface waves," in: *Physical Acoustics*, Vol. III (W. P. Mason, ed.), Academic Press, New York (1969). Rather extensive and dealing exclusively with the long-wave (elastic-wave) limit. Formulation based on traditional matching method. Contains a great deal of information.

F. García-Moliner, "Theory of surface waves," in *Crystalline Semiconducting Materials and Devices*, ICTP International College, 1984 (P. N. Butcher, N. H. March, and M. P. Tosi, eds.), p. 483, Plenum Press, New York (1986). Less extensive and more didactically oriented than Farnell. Deals also exclusively with the long-wave limit. Formulation based on the surface Green function matching method.

A. A. Maradudin, R. F. Wallis, and L. Dobrzynski, "Surface phonons and polaritons," in: *Handbook of Surfaces and Interfaces*, Vol. 3, Garland STPM Press, New York (1980). Very extensive. Elastic waves and, mostly, discrete lattice phonons. Treats in detail formal aspects of surface lattice dynamics.

G. Benedek, "Dynamics of solid surfaces," *Physicalia Magazine*, Vol. 7, Suppl. 1 (1985). General and authoritative account of the state of the art up to 1985. *Physicalia Magazine* is published by the Belgian Physical Society, Boeretang 200, 2400 Mol, Belgium.

References

1. G. W. Farnell, in: *Physical Acoustics*, Vol. III (W. P. Mason, ed.), Academic Press, New York (1969).
2. J. R. Sandercock, *Solid State Commun.* **26**, 547 (1978).
3. V. R. Velasco and F. García-Moliner, *Surf. Sci.* **143**, 93 (1984),
4. F. García-Moliner, in: *Crystalline Semiconducting Materials and Devices*, ICTP International College, 1984 (P. N. Butcher, N. H. March, and M. P. Tosi, eds.), p. 483, Plenum Press, New York (1986).
5. V. R. Velasco and F. García-Moliner, *J. Phys. C* **13**, 2237 (1980).
6. W. Kress, F. W. de Wette, A. D. Kulkarni, and U. Schröder, *Phys. Rev. B* **35**, 5783 (1987).
7. A. A. Maradudin, R. F. Wallis, and L. Dobrzynski, in: *Handbook of Surfaces and Interfaces*, Vol. 3, Garland STPM Press, New York (1980).
8. V. R. Velasco and F. Yndurain, *Surf. Sci.* **85**, 107 (1979).
9. A. Fasolino, G. Santoro, and E. Tosatti, *Phys. Rev. Lett.* **44**, 1684 (1980).
10. G. Armand, *Solid State Commun.* **48**, 261 (1983).
11. T. S. Rahman, J. E. Black, and D. L. Mills, *Phys. Rev. B* **25**, 883 (1982).
12. V. Bortolani, A. Franchini, F. Nizzoli, and G. Santoro, *Phys. Rev. Lett.* **52**, 429 (1984).
13. J. Szefter and A. Khater, *J. Phys. C* **20**, 4725 (1987).
14. F. García-Moliner, G. Platero, and V. R. Velasco, *Surf. Sci.* **136**, 601 (1984).
15. F. García-Moliner and V. R. Velasco, *Prog. Surf. Sci.* **21**, 93 (1986).
16. L. V. Heimendal and M. F. Thorpe, *J. Phys. F* **5**, L87 (1975).
17. S. C. Yerkes and D. R. Miller, *J. Vac. Sci. Technol.* **17**, 126 (1980).
18. B. F. Mason and B. R. Williams, *Phys. Rev. Lett.* **46**, 1138 (1981).
19. M. Cates and D. R. Miller, Proc. Int. Conf. on Vibrations at Surfaces, Asilomar, California (Aug. 1982).
20. B. Feuerbacher and R. F. Willis, *Phys. Rev. Lett.* **47**, 526 (1981).
21. R. B. Doak, V. Harten, and J. P. Toennies, *Phys. Rev. Lett.* **51**, 578 (1983).
22. J. M. Szeftel, S. Lehwald, H. Ibach, T. S. Rahman, and D. L. Mills, *Phys. Rev. Lett.* **51**, 268 (1983).
23. G. Platero, V. R. Velasco, and F. García-Moliner, *Surf. Sci.*, **152/153**, 819 (1985).
24. R. E. Allen, G. P. Alldredge, and F. W. de Wette, *Phys. Rev. B* **4**, 1661 (1971).
25. S. Lehwald, J. M. Szeftel, H. Ibach, T. S. Rahman, and D. L. Mills, *Phys. Rev. Lett.* **50**, 518 (1983).
26. V. Bortolani, A. Franchini, F. Nizzoli, and G. Santoro, in: *Dynamics of Gas–Surface Interactions* (G. Benedek and U. Valbusa, eds.), Springer-Verlag, Berlin (1982).
27. V. Bortolani, A. Franchini, F. Nizzoli, G. Santoro, G. Benedek, V. Celli, and N. García, *Solid State Commun.* **48**, 1045 (1983).

28. G. Armand, *Solid State Commun.* **48**, 261 (1983).
29. N. Cabrera, V. Celli, and R. Manson, *Phys. Rev. Lett.* **22**, 346 (1969).
30. R. Manson and V. Celli, *Surf. Sci.* **24**, 495 (1971).
31. G. Brusdeylins, R. B. Doak, and J. P. Toennies, *Phys. Rev. B* **27**, 3662 (1983).
32. G. Brusdeylins, R. Rechsteiner, J. G. Skofronic, J. P. Toennies, G. Benedek, and L. Miglio, *Phys. Rev. Lett.* **54**, 466 (1985).
33. P. Brüesch, *Phonons: Theory and Experiments. I. Lattice Dynamics and Models of Interatomic Forces*, Springer-Verlag, Berlin (1982).
34. G. Benedek, *Surf. Sci.* **61**, 603 (1976).
35. G. Benedek, P. Brivio, L. Miglio, and V. R. Velasco, *Phys. Rev. B* **26**, 487 (1982).
36. A. A. Lucas, *J. Chem. Phys.* **48**, 3156 (1968).
37. G. Platero, V. R. Velasco, F. García-Moliner, G. Benedek, and L. Miglio, *Surf. Sci.* **143**, 243 (1984).
38. G. Benedek and U. Valbusa (eds.), *Dynamics of Gas–Surface Interactions*, Springer-Verlag, Berlin (1982).

Electronic Structure of Metal Surfaces

J. E. Inglesfield

5.1. Introduction

In this chapter we describe the electronic structure of clean metal surfaces, and the relationship between this and the physical properties. It is important, after all, to understand the physics of clean surfaces before we can understand their interaction with atoms and molecules. There is also a lot of inherently interesting physics in surface electronic structure, particularly its relationship with the bulk. We shall concern ourselves with the types of wave function which can occur at surfaces, localized surface states, as well as bulk wave functions reflected by the surface, the change in electronic charge distribution which contributes to the work function, and changes in bonding which lead to the surface energy and the displacement of atoms at the surface.

The surfaces considered here are ideal, that is, they can be related to crystal planes of the bulk crystal and have two-dimensional periodicity.[1] In reality, all surfaces have steps and other defects which are of great importance for adsorption, but modern preparation techniques make it possible to make very good surfaces, which stay clean by working in ultrahigh vacuum. In fact it is the development of modern vacuum equipment which has largely been responsible for the growth of surface science. The two-dimensional periodicity makes it possible to use diffraction techniques to study the surface atomic structure, in particular Low Energy Electron Diffraction (LEED). A proper understanding of LEED is based on understanding electrons at surfaces, and this will be developed later on in this volume. The occupied electronic states at surfaces, responsible for the bonding, can be studied by photoemission, in which light is shone on the surface, exciting electrons whose

J. E. Inglesfield ● SERC Daresbury Laboratory, Warrington, England. *Present address*: Department of Physics, Catholic University of Nijmegen, The Netherlands.

energy and angular distribution can be measured. The unoccupied electron energy levels can be studied by the reverse process to photoemission, usually called inverse photoemission, in which electrons are fired at the surface, drop down into unoccupied states, and emit light. In this chapter we shall relate this type of experiment to surface electronic structure. An excellent account of these and other experimental methods, with the underlying physics, is given in the book *Modern Techniques of Surface Science*, by Woodruff and Delchar.[2]

Experimental studies of surfaces have been matched (well, almost!) by the development of theoretical and computational techniques for calculating surface electronic structure and physical properties. This is in fact the author's own field, and he spends most of his life writing the big computer programs, thousands of lines long, which are needed in this work. The theoretical methods used are described toward the end of this chapter, and the structure of the computational problem outlined. Big computer programs require big computers to run them. However, there is still much interesting work to be done using simple models of surface electronic structure, such as the jellium model in which the atomic structure is smeared out into a uniform positive background. And there is still a great deal we do not understand about the interaction of electrons with each other at the surface—the many-body problem.

5.2. Surface Periodicity and the Two-Dimensional Bloch Property

An electron in a crystal moves in a potential field $V(\mathbf{r})$ originating from the electrostatic interaction with the atomic nuclei, and from the interaction with all the other electrons. These electrons have a static charge density, and the electron we are considering feels the electrostatic potential—the *Hartree* potential—of this charge density. But in addition the other electrons tend to get out of the way of our electron as it moves through the crystal, which lowers the energy of interaction. This correction to the interaction potential is called the *exchange-correlation* potential, $V_{xc}(\mathbf{r})$[3] and we shall see in later sections that its determination is important for accurate studies of surface electronic structure.

In a bulk crystal the total potential V has three-dimensional periodicity, but when the crystal is chopped in two to make a surface, the periodicity in the perpendicular direction is destroyed. However, the resulting semi-infinite crystal still has two-dimensional periodicity parallel to the surface—as long as there are no surface defects or adsorbates. This means that the potential at a point $(\mathbf{R}, z)$ ($\mathbf{R}$ is the vector parallel to the surface and z measures distance perpendicular to the surface) is the same as at $(\mathbf{R} + \mathbf{R}_I, z)$, where $\mathbf{R}_I$ is a vector in the two-dimensional surface lattice or mesh:

$$V(\mathbf{R}, z) = V(\mathbf{R} + \mathbf{R}_I, z) \tag{5.2.1}$$

This is illustrated in Figure 5.1, which shows potential contours on a plane through the Ag(001) surface. We see that the potential is indeed periodic parallel to the surface, but in the z-direction it goes from the constant potential in the vacuum to the bulk potential, and there is no overall periodicity.

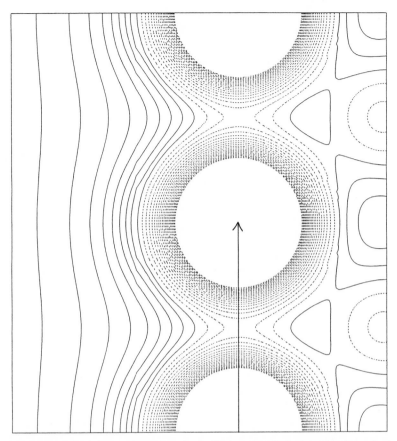

Figure 5.1. Potential felt by an electron at the Ag(001) surface (calculated by the author). The figure shows contours of constant potential on a plane passing through the surface layer of atoms. Solid contours represent a positive potential relative to the average potential between the atoms in the bulk, and dashed contours show a negative potential; the holes in the middle of the atoms are simply where the potential is too deep for the plotting program! The arrow shows a vector of the two-dimensional surface mesh. The figure clearly shows the periodicity parallel to the surface.

The potential outside the surface, in the vacuum (remember that we must keep a surface in ultrahigh vacuum to keep it clean), is actually longer range than the figure indicates. An electron outside a metal surface feels asymptotically an image potential of the form[4]

$$V(z) \sim -\frac{1}{4|z - z_0|} \qquad (5.2.2)$$

Here z_0 is close to the plane on which the solid was chopped in two to make the surface (Figure 5.2). The image potential results from the fact that there is perfect screening at a metal surface, and the electric field due to our electron cannot

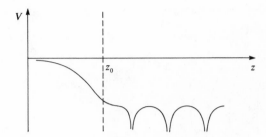

Figure 5.2. Potential as a function of z at the surface (schematic), along a line passing through the nucleus of a top-layer atom.

penetrate into the surface. The factor of 4 in equation (5.2.2) comes from the distance of $2z$ between the external electron and the image charge, $\times 2$ because the image charge is induced by the electron. The image potential can also be thought of as the limit of V_{xc} when the electron is outside the solid—after all, the image potential results from the electron–electron interaction. Figure 5.1 gives a misleading impression of the potential in vacuum because it has been calculated using an approximate form of V_{xc}, but the difference is not terribly important except when we are considering a special class of electron wave functions trapped by the image potential (Section 5.4.3).

As an electron goes into the solid through the surface, the potential goes smoothly from a flat potential a long way from the surface, through the image potential to the bulk potential (Figures 5.1 and 5.2). At a metal surface an electron feels the bulk potential beyond the first or second layer of atoms. There are relatively small changes in potential in the top layer or two, and this region is called the *selvedge*.[1] (If you look in the *Oxford English Dictionary* you will see that a selvedge is the edge of cloth, woven more tightly to stop it unraveling.) The changes in the static properties due to the surface, such as the charge density and the bonding, are almost entirely confined to the selvedge. The reason why the potential goes so rapidly to the bulk is, once again, perfect metallic screening, which means that any potential perturbation, such as a surface, is screened away within a Thomas–Fermi screening length or so. We note that even though the potential *does* go to the bulk value, overall the whole system has lost periodicity in the perpendicular direction.

The two-dimensional periodicity of the surface means that the electronic wave functions can be labeled by a two-dimensional Bloch wave vector.[5] An electronic wave function ψ satisfies the Schroedinger equation (in atomic units with $e = \hbar = m = 1$)

$$[-\tfrac{1}{2}\nabla^2 + V(\mathbf{R}, z)]\psi(\mathbf{R}, z) = E\psi(\mathbf{R}, z) \qquad (5.2.3)$$

Displacing $\mathbf{R}$ in equation (5.2.3) by the surface lattice vector $\mathbf{R}_l$ and using equation (5.2.1), we see that the *displaced* wave function satisfies just the same Schroedinger equation:

$$[-\tfrac{1}{2}\nabla^2 + V(\mathbf{R}, z)]\psi(\mathbf{R} + \mathbf{R}_I, z) = E\psi(\mathbf{R} + \mathbf{R}_I, z) \qquad (5.2.4)$$

If ψ is nondegenerate, the displaced wave function must be the same as the original wave function, apart from a phase factor of modulus unity, which can be written as $\exp{(i\mathbf{K} \cdot \mathbf{R}_I)}$:

$$\psi(\mathbf{R} + \mathbf{R}_I, z) = \exp{(i\mathbf{K} \cdot \mathbf{R}_I)}\psi(\mathbf{R}, z) \qquad (5.2.5)$$

If ψ is degenerate, it is also possible to choose the wave functions with this Bloch property. The two-dimensional Bloch wave vector $\mathbf{K}$ is thus a good quantum number which can be used to label the wave functions at a surface.

Just as in the case of a bulk crystal, where the wave functions are labeled by the three-dimensional Bloch wave vector, $\mathbf{K}$ is not defined to within a reciprocal lattice vector[5]—here a *surface* reciprocal lattice vector $\mathbf{G}$. We can add $\mathbf{G}$ to $\mathbf{K}$ without changing the phase factor in equation (5.2.5), because by definition $\mathbf{G} \cdot \mathbf{R}_I = 2\pi \times integer$. So

$$\exp{(i[\mathbf{K} + \mathbf{G}] \cdot \mathbf{R}_I)} = \exp{(i\mathbf{K} \cdot \mathbf{R}_I)} \qquad (5.2.6)$$

This lack of uniqueness in the Bloch wave vector means that we can choose $\mathbf{K}$ to lie in the surface Brillouin zone, and Figure 5.3 shows this for the (001) surface of a face-centered cubic solid like Ag.

The fact that we can add $\mathbf{G}$ to $\mathbf{K}$ without affecting the two-dimensional Bloch property of the wave functions is intimately connected with the diffraction properties of surfaces.[1,2] A low-energy beam of electrons fired at the surface in a particular direction has a free-electron wave function away from the surface, with a well-defined wave vector $\mathbf{K}$:

$$\psi(\mathbf{R}, z)_{incident} = \exp{(i\mathbf{K} \cdot \mathbf{R})} \exp{(ik_z z)} \qquad (5.2.7)$$

This is scattered by the surface into reflected waves traveling back from the surface, giving a total wave function

$$\psi(\mathbf{R}, z) = \exp{(i\mathbf{K} \cdot \mathbf{R})} \exp{(ik_z z)}$$
$$+ \sum_{\mathbf{G}} A_{\mathbf{G}} \exp{(i[\mathbf{K} + \mathbf{G}] \cdot \mathbf{R})} \exp{(-ik_G z)} \qquad (5.2.8)$$

The two-dimensional periodicity of the surface potential diffracts the electrons through reciprocal lattice vectors $\mathbf{G}$, without affecting the Bloch property of ψ, which still satisfies equation (5.2.5). As the reflected waves travel in directions determined by $\mathbf{K} + \mathbf{G}$, imaging the electrons on a fluorescent screen gives a map of the surface reciprocal mesh. The amplitudes $A_{\mathbf{G}}$ contain information about the location of the atoms themselves with respect to the corresponding real space surface mesh. Exactly the same analysis carries over to other surface diffraction techniques, such as helium atom diffraction. These techniques are described at length elsewhere in this book.

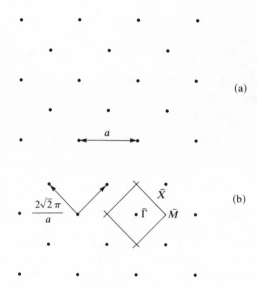

Figure 5.3. (a) Real space surface mesh for fcc (001) surface. (b) Corresponding reciprocal mesh, with arrows showing basis vectors; the first surface Brillouin zone is shown with some standard symmetry points.

5.3. Bulk States at the Surface

There are two types of electronic states with energy below the vacuum level, that is, confined to the semi-infinite solid by the surface potential barrier (Figure 5.2): bulk states which hit the surface and are reflected and which we consider in this section, and surface states which are localized at the surface and which we consider in Section 5.4.

Away from the surface we know that an electron experiences the same potential as in the bulk, so we can build up a solution of the Schroedinger equation in this region from bulk solutions.[6,7] The wave vector $\mathbf{K}$ is fixed parallel to the surface and we consider the bulk energy bands as a function of k_z, the wave vector component perpendicular to the surface. If, for example, we are working at $\mathbf{K} = 0\,(\bar{\Gamma})$ in the *surface* Brillouin zone for fcc (001), the bulk bands projecting onto this wave vector lie along $X\Gamma X$ in the *bulk* Brillouin zone, as shown in Figure 5.4. The state indicated by the minus sign in Figure 5.4 then represents a bulk wave function with vector component $\mathbf{K} = 0$, energy E, traveling toward the surface. In the same way that the LEED electron is scattered, this electronic state will be reflected back into the solid by the surface, into states with wave vector $\mathbf{K} + \mathbf{G}$ and energy E. All the intensity is reflected, because the energy lies below the vacuum zero, and the electron cannot escape, while energy is conserved in the elastic scattering by the surface potential.

The case in which the potential is one-dimensional, with no variation parallel to the surface, is examined first. The motion of electrons parallel to the surface is trivial in this case, giving a free-electron factor of $\exp(i\mathbf{K}\cdot\mathbf{R})$ to the total wave function which can be inserted at the end. With a one-dimensional potential there is only one reflected wave (as in Figure 5.4), so the total wave function away from the surface has the form

$$\psi = \phi^- + r\phi^+ \tag{5.3.1}$$

where ϕ^- is the wave traveling toward the surface (negative z-direction) and ϕ^+ is the reflected wave. A common approximation to the surface barrier shown in Figure 5.2 is to use a step potential (Figure 5.5), in which case equation (5.3.1) holds all the way up to the surface itself at $z = 0$. In the vacuum, the solution of the Schroedinger equation is then

$$\psi = t \exp(\hat{\gamma}z) \tag{5.3.2}$$

and the amplitudes r and t can be found by matching the amplitude and derivative of equations (5.3.1) and (5.3.2) at the surface. Coefficient r is a complex number with unit length to conserve flux, so equation (5.3.1) represents a standing wave.

In the real three-dimensional case, the band structure may be quite complicated, and there can be more than one reflected wave traveling away from the surface, with wave vector $\mathbf{K}$ and energy E. However, in the case shown in Figure 5.4 there is only one *traveling* wave reflected by the surface—the bands with surface wave vector $\mathbf{K} + \mathbf{G}$ are just the same and do not introduce any new states. However, in addition to the one *traveling* wave we must also consider *evanescent* waves,[8]

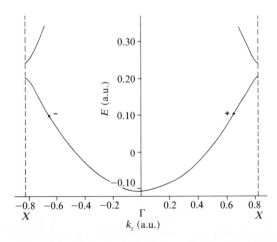

Figure 5.4. Bulk band structure of Al along $X\Gamma X$, corresponding to $\mathbf{K} = 0$ in the (001) surface Brillouin zone. The minus sign represents a wave traveling toward the surface, and the plus sign represents the traveling wave reflected by the surface.

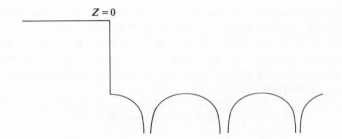

Figure 5.5. Step model of the surface potential barrier.

which in a nearly-free-electron system (like Figure 5.4) have the form

$$\phi^+_{\mathbf{K},\mathbf{G};E} \sim \exp\left(i[\mathbf{K}+\mathbf{G}]\cdot\mathbf{R}\right)\exp\left(-\gamma_G z\right) \qquad (5.3.3)$$

with

$$|\mathbf{K}+\mathbf{G}|^2 - \gamma_G^2 = 2E \qquad (5.3.4)$$

This wave function is not allowed in an infinite crystal, of course, but it is an allowed solution of the Schroedinger equation in the semi-infinite system because the surface stops it blowing up. As it has a reduced wave vector $\mathbf{K}$ and energy E, the surface can scatter the incident wave into this state, though because it is evanescent it does not carry flux away from the surface. There is clearly a one-to-one relationship between reflected traveling + evanescent waves like expression (5.3.3), and the surface reciprocal lattice vectors in the nearly-free-electron case. In other words, for each $\mathbf{G}$ there is either a wave traveling away from the surface of the form

$$\phi^+_{\mathbf{K},\mathbf{G};E} \sim \exp\left(i[\mathbf{K}+\mathbf{G}]\cdot\mathbf{R}\right)\exp\left(ik_G z\right) \qquad (5.3.5)$$

with

$$|\mathbf{K}+\mathbf{G}|^2 + k_G^2 = 2E \qquad (5.3.6)$$

or an evanescent wave like expression (5.3.3). This one-to-one property persists even when we turn on a strong crystal potential which distorts the bands from nearly-free-electron bands.[8]

With the full potential, the wave function away from the surface can now be written as

$$\psi = \phi^- + \sum_G r_G \phi^+_{\mathbf{K},\mathbf{G};E} \qquad (5.3.7)$$

namely, the incident traveling wave + reflected traveling and evanescent waves labeled by $\mathbf{G}$. In the vacuum, assuming a step potential, the solution has the form

$$\psi = \sum_G t_G \exp\left(i\mathbf{K}+\mathbf{G}]\cdot\mathbf{R}\right)\exp\left(\hat{\gamma}_G z\right) \qquad (5.3.8)$$

namely, waves decaying into the vacuum. As in the one-dimensional case, we must match amplitude and derivative across $z = 0$, and this can be done by expanding these functions in two-dimensional Fourier series made up of plane waves $\exp(i[\mathbf{K} + \mathbf{G}] \cdot \mathbf{R})$. If there are $\mathcal{N}$ $\mathbf{G}$'s in the Fourier series, this gives $2\mathcal{N}$ matching conditions from which we can find $\mathcal{N}$ quantities $r_{\mathbf{G}}$ and $\mathcal{N}$ quantities $t_{\mathbf{G}}$. We possess just the right number of conditions to find the unknown coefficients, and this means that the wave function can be determined uniquely. For every bulk state traveling toward the surface, we can thus find a solution for the semi-infinite solid with a surface, and at each value of $\mathbf{K}$ the bulk bands give continua of states at the surface, with energy gaps. This matching method for finding bulk states at the surface can be used to find the LEED wave function,[9] with the solution in vacuum given by equation (5.2.8) and the solution in the semi-infinite crystal given by equation (5.3.7), without ϕ^- traveling toward the surface.

5.3.1. Local Density of States

The most useful quantity to consider for the continua of states is the *local density of states*, defined as the charge density of states having a particular energy,[10]

$$\sigma(\mathbf{r}, E) = \sum_i \delta(E - E_i)|\psi_i(\mathbf{r})|^2 \tag{5.3.9}$$

This is a sum over states, in which the δ-function picks out those states with energy E. With surface symmetry this sum can be restricted to states with a particular wave vector $\mathbf{K}$, like equation (5.3.7), and Figure 5.6 shows the surface density of states with $\mathbf{K} = 0$ for Al(001)[11] (by surface density of states we just mean quantity

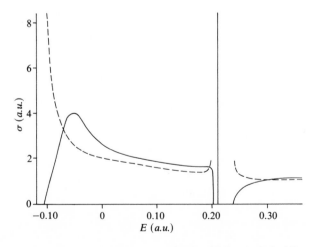

Figure 5.6. Solid line: surface density of states at Al(001), with $\mathbf{K} = 0$; dashed line: corresponding bulk density of states, with $\mathbf{K} = 0$ (after Benesh and Inglesfield[11]).

(5.3.9) integrated through the surface atoms). Let us first compare this with the density of states in the bulk, again states with $\mathbf{K} = 0$. The bulk density of states is just the number of states per unit energy range, and as the wave vector component parallel to the surface is fixed, we can find this immediately from the band structure (Figure 5.4):

$$\sigma_{\text{bulk}}(E) \propto \frac{1}{|dE/dk_z|} \tag{5.3.10}$$

This means that the bulk density of states at fixed $\mathbf{K}$ varies like $|E - E_0|^{-1/2}$ at a band edge, blowing up and giving the singularities shown in Figure 5.6. The continua of states at the surface have the same energy range as in the bulk, as is apparent in Figure 5.6. However, it is seen from the figure that the surface density of states is better behaved than the bulk at the band edges, varying like $|E - E_0|^{1/2}$. But in this figure an extra feature should be noted in the surface density of states which cannot arise from the bulk states as it lies in the energy gap—this is a discrete *surface state*, which we shall explore further in Section 5.4.

The surface density of states can be studied experimentally using angle-resolved photoemission,[2] in which the energy and angle distribution of photoemitted electrons is measured. Conservation of energy means that the energy of an emitted electron equals the energy of its initial state, when it was inside the solid, plus the energy of the absorbed photon $h\nu$. The photon carries negligible momentum, and the fact that $\mathbf{K}$ is a good quantum number means that $\mathbf{K}$ of the emitted electron equals $\mathbf{K}$ in its initial state, to within a surface reciprocal lattice vector, of course. So by measuring the energy and direction of the photoelectron we can find both E and $\mathbf{K}$ of the initial state. Now the photoelectron tends to originate from quite close to the surface, typically within 10 Å, because its mean free path between scattering events is rather short.[12] This means that the photoemission spectrum contains information about the local density of states, integrated over the top few layers of the sample. Figure 5.7 shows photoemission spectra from Al(001)[13] measured at normal emission; normal emission corresponds to $\mathbf{K} = 0$. We can see the clear relationship between these spectra and the surface density of states of Figure 5.6, with the band edges and the surface state all showing up very nicely.

5.3.2. Band Narrowing at the Surface

The reduction in weight in the surface density of states at band edges, compared with the bulk, is an example of so-called *band narrowing*.[10] This is a very general property, and we shall see how it affects the local density of states, summed over $\mathbf{K}$, in a system in which the electrons are tightly bound. A tight-binding picture of the electrons is the opposite extreme from the nearly-free-electron picture appropriate to s–p-bonded metals like Na, Mg, or Al, but it describes the d electrons in transition metals, with narrow bands arising from the limited overlap of atomic orbitals.[14]

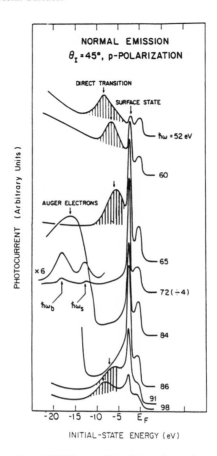

Figure 5.7. Photoemission from Al(001) normal to the surface, plotted as a function of initial state energy measured relative to the Fermi energy. Different spectra correspond to different values of $h\nu$. The bottom of the band is at -10.6 eV, and the bandgap lies between -2.83 and -1.15 eV (after Levinson et al.[13]).

In tight binding, the local density of states can be found directly in real space[14] without having to talk about reciprocal space or energy bands at all. To show this a bit of contour integration must be done, but the less mathematically inclined can omit this and proceed to the next-but-one paragraph. First, we express σ in equation (5.3.9) in terms of the Green function of the system, given by

$$G(\mathbf{r}, \mathbf{r}'; E) = \sum_i \frac{\psi_i(\mathbf{r})\psi_i^*(\mathbf{r}')}{E_i - E} \qquad (5.3.11)$$

so that

$$\sigma(\mathbf{r}, E) = \frac{1}{\pi} \text{Im } G(\mathbf{r}, \mathbf{r}; E + i\varepsilon) \qquad (5.3.12)$$

Now in tight binding, wave functions are expanded in terms of localized atomic orbitals α, β on atoms I, J:

$$\psi = a_{\alpha,I}\phi_{\alpha,I} + a_{\beta,J}\phi_{\beta,J} + a_{\gamma,K}\phi_{\gamma,K} + \cdots \tag{5.3.13}$$

with coefficients which form a vector satisfying the matrix form of the Schroedinger equation:

$$\sum_i H_{i,j}a_i = Ea_i \tag{5.3.14}$$

Here, $H_{i,j}$ is the matrix element of the Hamiltonian between orbitals $i(=\alpha, I$, say) and $j (=\beta, K$, say) which we write in Dirac notation as $\langle \alpha, I|H|\beta, J\rangle$. In this representation, we consider the matrix elements of the Green function *operator* $(H - E)^{-1}$:

$$G_{\alpha,I;\beta,J}(E) = \langle \alpha, I|(H - E)^{-1}|\beta, J\rangle \tag{5.3.15}$$

and the equivalent of equation (5.3.12) is

$$\sigma_{\alpha,I}(E) = \frac{1}{\pi} \text{Im} \langle \alpha, I|(H - E - i\varepsilon)^{-1}|\alpha, I\rangle \tag{5.3.16}$$

namely, the charge density in orbital α, I at energy E.

The shape of σ can be described in terms of its *moments*:

$$\mu_{\alpha,I}^{(n)} = \int_{-\infty}^{\infty} dE\, E^n \sigma_{\alpha,I}(E) \tag{5.3.17}$$

$$= \frac{1}{\pi} \text{Im} \int_{-\infty}^{\infty} dE\, E^n \langle \alpha, I|(H - E - i\varepsilon)^{-1}|\alpha, I\rangle \tag{5.3.18}$$

Contour integration enables this to be expressed in the form

$$\mu_{\alpha,I}^{(n)} = \frac{1}{2\pi i} \oint dE\, E^n \langle \alpha, I|(H - E - i\varepsilon)^{-1}|\alpha, I\rangle \tag{5.3.19}$$

where the contour encircles all the poles (at the energy eigenvalues) just below the real axis, in a clockwise direction. Then Cauchy's integral formula[15] gives

$$\mu_{\alpha,I}^{(n)} = \langle \alpha, I|H^n|\alpha, I\rangle \tag{5.3.20}$$

But this can be rewritten by inserting the sum over basis functions $\sum_{\beta,J}|\beta, J\rangle\langle\beta, J|$—the unit matrix—between the quantities H, giving

$$\mu_{\alpha,I}^{(n)} = \sum_{\beta,J} \sum_{\gamma,K} \cdots \sum_{\lambda,L} \langle \alpha, I|H|\beta, J\rangle\langle\beta, J|H|\gamma, K\rangle \cdots \langle \lambda, L|H|\alpha, I\rangle \tag{5.3.21}$$

So the nth moment is a sum over products of n matrix elements which take us from the orbital we are interested in, out into the crystal, and back again. The finer details of $\sigma_{\alpha,l}(E)$ correspond to the higher moments, which probe the environment more and more deeply.

This result can be used to find the mean-square width of the local density of states—the (width)^2, or moment of inertia of $\sigma_{\alpha,l}(E)$ is just the second moment. With one orbital per atom, and taking the atomic energy level as the zero of energy, we see from equation (5.3.21) that

$$\mu^{(2)} = nh^2 \tag{5.3.22}$$

where h is the matrix element coupling the orbital in which we are interested to orbitals on the n neighboring atoms. Now a surface atom has fewer neighbors than one in the bulk, so we immediately conclude that the local density of states is narrower—this is *surface band narrowing*. Figure 5.8 shows the local density of states in a simple cubic solid with a surface, with one s orbital on each atom.[16] We see that the density of states is indeed narrower at the surface than in the bulk, but even by the second layer σ is quite close to the bulk value. In this case, the surface atom has 5 neighbors compared with 6 in the substrate, so the surface width is $\sqrt{5/6} \times$ the bulk. Applying this to the d-bonded transition metals, the narrowing is particularly marked on the open (001) surfaces of bcc metals like W, where each atom has only 4 nearest neighbors compared with 8 in the bulk.[17] We see from Figure 5.9 that the main effect of the band narrowing is to produce a rather characteristic central peak in the surface density of states, coinciding with a minimum in the bulk density of states. The peak coincides with the Fermi energy in W, and this produces an instability responsible for the surface reconstruction.[18]

An important consequence of surface band narrowing is a shift in the potential on the surface atoms. Let us consider filling up the density of states shown in Figure 5.9 by moving across a transition metal row, varying the number of d-

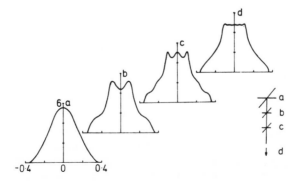

Figure 5.8. Local density of states on (a) a surface atom, (b, c) subsurface atoms, and (d) a bulk atom, in a simple cubic tight-binding s-band solid (after Haydock and Kelly[16]).

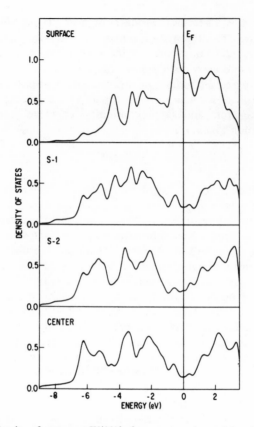

Figure 5.9. Local density of states on W(001), from a seven-layer slab calculation. Successive panels show the surface density of states, the density of states on the first $(S - 1)$ layer into the slab, the second $(S - 2)$ layer into the slab, and the density of states at the center of the slab. This is very close to the bulk density of states (after Posternak et al.[17]).

electrons from 0 to 10. The number of electrons in orbital α on atom I is given by

$$n_{\alpha,I} = \int^{E_F} dE \sigma_{\alpha,I}(E) \tag{5.3.23}$$

so changes in the density of states at the surface will lead to changes in the number of electrons on the surface atoms. The removal of weight from the band edges means that if the band is less than half-filled, with fewer than five electrons per atom, the surface atoms have a deficit of electrons compared with the bulk. But metallic atoms like to be almost exactly neutral, because of the short Thomas–Fermi screening length, so the potential on the surface atoms becomes more attractive, to attract extra charge and eliminate the charge deficit.[19] Conversely, if the band is more than half full, with more than five electrons per atom, the surface atoms would have a surplus of electrons compared with the bulk. In this case the potential becomes less attractive, to expel the charge surplus. These shifts in potential can be probed experimentally by measuring the energies of the localized core levels in photoemission.[20] It is indeed found that in the case of Ta, with less than five d electrons, the $4f$ level on the (111) surface is pulled down by 0.4 eV compared with the bulk.[21] And in the case of Ir, with more than five d electrons, the $4f$ level on the (001) surface is pushed up by 0.7 eV.[22] However, the whole story is a bit more complicated than we have implied because even with W, having a half-filled d band, there is an upward shift of 0.3 eV.[23]

5.4. Surface States

Figures 5.6 and 5.7 show that states can occur at the surface at energies for which there are no bulk states—within an energy gap of the bulk band structure. These are *surface states.* As the energy of a surface state lies in a bulk gap, its wave function decays exponentially into the solid, and as it lies below the vacuum zero of energy it also decays exponentially into the vacuum: the surface state wave function is localized at the surface. It has the surface Bloch property (5.2.5), with a well-defined wave vector **K** parallel to the surface, so it is delocalized over the surface. Surface states are important, first because they are a characteristic of surface electronic structure, second because they show up strongly in angle-resolved photoemission as discrete features (Figure 5.7), and third because they make a significant contribution to the surface density of states.

5.4.1. Shockley States

The first type of surface state to be considered here is typical of nearly-free-electron metals, occurring in the band gaps opened up by the crystal potential. This potential scatters the electrons relatively weakly in metals like Na, Mg, and Al, which is why they are rather free-electron-like, and it can be replaced by a weak *pseudopotential.*[24]

Let us consider one Fourier component of the pseudopotential in the bulk crystal:

$$V(z) = 2V \cos g_z z \qquad (5.4.1)$$

where $g_z = 2\pi/a$, a being the periodicity in the z-direction. This opens up an energy gap in the bulk band structure at $k_z = \pm\pi/a$, where k_z is the bulk wave vector component in the z-direction, and as before we set **K** = 0. Around $k_z = \pi/a$ the solutions of the bulk Schroedinger equation have the form[25]

$$\phi = a \exp(ik_z z) + b \exp(i[k_z - g_z]z) \qquad (5.4.2)$$

because the component of the potential given by equation (5.4.1) scatters the electrons though the bulk reciprocal lattice vector g_z. The coefficients can be found by substituting expression (5.4.2) into the bulk Schroedinger equation to yield the matrix equation

$$\begin{pmatrix} k_z^2/2 - E & V \\ V & (k_z - g_z)^2/2 - E \end{pmatrix} \begin{pmatrix} a \\ b \end{pmatrix} = 0 \qquad (5.4.3)$$

This has a solution when

$$E = \tfrac{1}{4}\{[k_z^2 + (k_z - g_z)^2] \pm ([k_z^2 - (k_z - g_z)^2]^2 + 16V^2)^{1/2}\} \qquad (5.4.4)$$

In the infinite bulk crystal k_z, the Bloch wave vector in the z-direction, must be real, and expression (5.4.4) gives the energy gap of $2|V|$ at $k_z = g_z/2$.

If we put E in the energy gap, equation (5.4.4) can still be solved, giving *complex k_z*. Now it has already been seen from equation (5.3.4) that solutions of the bulk Schroedinger equation with complex k_z are possible in the semi-infinite crystal, so let us study these decaying wave functions within the band gap in more detail. In the gap the wave vector can be written as

$$k_z = \kappa + i\gamma \qquad (5.4.5)$$

and substitution into equation (5.4.4) yields

$$E = \tfrac{1}{4}\{\kappa^2 + 2i\kappa\gamma + (\kappa - g_z)^2 + 2i(\kappa - g_z)\gamma - 2\gamma^2$$

$$\pm ([\kappa^2 + 2i\kappa\gamma - (\kappa - g_z)^2 - 2i(\kappa - g_z)\gamma]^2 + 16V^2)^{1/2}\} \qquad (5.4.6)$$

The imaginary part of this equation is satisfied by putting $\kappa = g_z/2$, and then the real part simplifies to

$$E = \frac{\kappa^2 - \gamma^2}{2} \pm (V^2 - \kappa^2\gamma^2)^{1/2} \qquad (5.4.7)$$

This can be written in the form

$$\gamma^2 = -2E - \kappa^2 + 2(V^2 + 2E\kappa^2)^{1/2} \qquad (5.4.8)$$

so γ goes to zero at the bottom and top of the gap and is maximum around the middle of the gap, where

$$\gamma \approx \frac{|V|}{\pi/a} \qquad (5.4.9)$$

In the gap, the wave function (5.4.2) becomes

$$\phi = \{a \exp(i\kappa z) + b \exp(-i\kappa z)\} \exp(-\gamma z) \qquad (5.4.10)$$

and from (5.4.3) the two equations for the coefficients are

$$\frac{a}{b} = \frac{-V}{(\kappa^2 - \gamma^2)/2 - E + i\kappa\gamma}$$

$$= \frac{(\kappa^2 - \gamma^2)/2 - E - i\kappa\gamma}{-V} \qquad (5.4.11)$$

By multiplying these equations it is seen that a/b is a complex number of modulus unity, and its variation as E moves through the gap is shown in Figure 5.10. Writing a/b as $e^{2i\chi}$, the wave function becomes

$$\phi \propto \cos(\kappa z + \chi) \exp(-\gamma z) \qquad (5.4.12)$$

This is a standing wave, decaying exponentially into the solid. If V is positive, the phase χ varies between $\pi/2$ at the bottom of the gap and 0 at the top, while if V is negative it varies between 0 at the bottom and $-\pi/2$ at the top (Figure 5.10).

In order to have a full solution of the Schroedinger equation, wave function (5.4.12) must be matched on to the solution in vacuum. As in Section 5.3, let us assume that the bulk potential given by equation (5.4.1) holds all the way up to the surface at $z = 0$, and that the vacuum potential is flat. The wave function in vacuum then looks like

$$\phi \propto \exp \hat{\gamma} z \qquad (5.4.13)$$

with a decay constant given by

$$\hat{\gamma}^2 = 2(V_{sb} - E) \qquad (5.4.14)$$

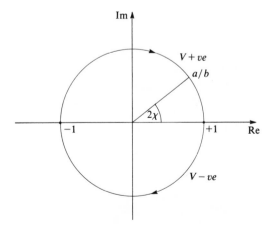

Figure 5.10. Variation of a/b in the complex plane as E moves up through the gap.

where V_{sb} is the step in the potential at the surface, the surface barrier. Wave functions (5.4.12) and (5.4.13) match in amplitude and derivative when their logarithmic derivatives at $z = 0$ are equal. The logarithmic derivative of function (5.4.12) is

$$\frac{1}{\phi}\frac{\partial \phi}{\partial z}\bigg|_{z=0} = -\gamma - \kappa \tan \chi \qquad (5.4.15)$$

and the logarithmic derivative of function (5.4.13) is $\hat{\gamma}$, a positive quantity. Clearly, if V is positive, the variation of χ through the gap makes the log derivative on the crystal side of the surface always negative and wave functions (5.4.12) and (5.4.13) can never match. *No solution of the full Schroedinger equation then exists in the gap.* On the other hand, if V is negative, derivative (5.4.15) varies between 0 at the bottom of the gap and $+\infty$ at the top, so that function (5.4.12) can be matched on to function (5.4.13) at some energy—this gives a valid solution of the Schroedinger equation at an energy in the gap, a *Shockley surface state.*[26]

A negative value of V with respect to an origin at the geometrical surface at $z = 0$, where the solid is chopped in two, means that the pseudopotential is positive with respect to an origin on the atoms—the topmost plane of atoms lies at $z = a/2$ in equation (5.4.1). We normally think of atoms as being very attractive to the electrons, so it might seem a bit unlikely that the pseudopotential can be positive, i.e., repulsive. However, we are talking here about a Fourier component of the pseudopotential, and we have absorbed the *average* pseudopotential in the material—which is indeed attractive—in the inner potential, with respect to which there is a step at the surface. In most s–p-bonded metals like Al, V is positive with respect to the atoms,[27] so we would indeed expect to see a Shockley surface state, as in Figures 5.6 and 5.7. We can see from Figure 5.6 that the surface state on Al (001) lies very close to the bottom of the gap, and indeed this is clear from the photoemission spectrum. From equation (5.4.8), this means that γ is rather small, and the surface state decays only slowly into the solid—calculation suggests that the charge density decays over about seven atomic layers by $1/e$. Figure 5.11 shows the charge density of this surface state, at $\mathbf{K} = 0$, and we see that charge is piled up on top of the surface atoms.[28] However, the figure also shows the state extending some way into the solid. It is interesting to see that the state has a node close to the atomic nucleus; in other words, the wave function is rather p-like. This is because positive V (with respect to the atoms) gives a bulk state which is p-like at the bottom of the energy gap, and one which is s-like at the top, and the surface state is very close to the bottom of the gap.

5.4.2. Tamm States

The characteristic surface state of tightly bound systems, like the d electrons in a transition metal, is the Tamm state.[29] This is pulled off the band by a shift in the potential at the surface, the sort of shift which we discussed in Section 5.3.2.

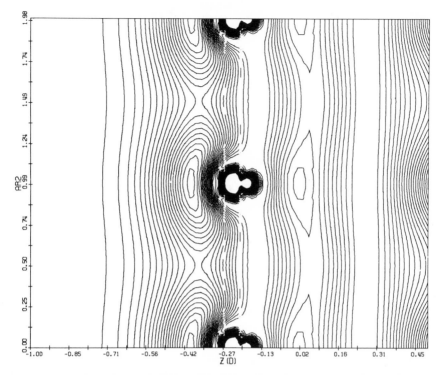

Figure 5.11. Charge density of Al(001) surface state at $\mathbf{K} = 0$, in the plane lying normal to the surface which intersects the top layer of atoms (after Inglesfield and Benesh[28]).

The conditions under which a Tamm state can occur are examined by considering a semi-infinite chain of atoms, each with a single orbital interacting with its nearest neighbors via the hopping integral h. The wave functions can then be written in the same form as equation (5.3.13):

$$\psi = a_0\phi_0 + \alpha_1\phi_1 + \alpha_2\phi_2 + \cdots \qquad (5.4.16)$$

where ϕ_0 is the orbital on the surface atom (0), ϕ_1 is the orbital on the subsurface atom (1), and so on into the solid. We note that this model can represent a real three-dimensional semi-infinite solid, if we think of ϕ_i as an orbital extended over the whole atomic plane i, with surface wave vector $\mathbf{K}$. The quantities a satisfy the matrix Schroedinger equation (5.3.14), which simplifies in this case to the tridiagonal form

$$\begin{pmatrix} v & h & 0 & 0 & \cdots \\ h & 0 & h & 0 & 0 \\ 0 & h & 0 & h & 0 \\ 0 & 0 & h & 0 & h \\ 0 & 0 & 0 & h & \end{pmatrix} \begin{pmatrix} a_0 \\ a_1 \\ a_2 \\ a_3 \\ \vdots \end{pmatrix} = E \begin{pmatrix} a_0 \\ a_1 \\ a_2 \\ a_3 \\ \vdots \end{pmatrix} \qquad (5.4.17)$$

The atomic energy level, corresponding to the diagonal elements of the matrix ($\langle i|H|i\rangle$ in the notation of Section 5.3.2), are set to zero in the bulk, but there is a shift of v on the surface atom.

A solution of this eigenvalue equation is now sought in which there is a simple factor relating orbital coefficients on adjacent atoms, that is,

$$a_1 = \alpha a_0$$

$$a_2 = \alpha a_1$$

and so on   (5.4.18)

On substituting (5.4.18) into (5.4.17), the first row of the matrix equation gives

$$\frac{E}{h} = \alpha + \frac{v}{h} \qquad (5.4.19)$$

while the second and subsequent rows give

$$\frac{E}{h} = \alpha + \frac{1}{\alpha} \qquad (5.4.20)$$

Here we have two equations for the two unknowns, E and α, which we can solve graphically. Figure 5.12 shows E/h versus α, and where the two curves corresponding to the two equations intersect, we have the solution. We are clearly looking for a solution in which $|\alpha| < 0$, so that the amplitude of the surface state wave function decays away into the solid. This occurs when:

- $v/h > 1$ (the case shown), giving a solution with positive α, with an energy $E > 2h$,
- or $v/h < -1$, giving a solution with negative α, with an energy $E < -2h$.

(It is assumed that h is positive here, but the argument is trivial to extend to negative h.) The bulk band extends between the energies:

$$-2h < E < 2h \qquad (5.4.21)$$

so a potential shift v *with magnitude greater than h* pulls a surface state either from the top or bottom of the bulk band.[30] This is a *Tamm surface state*.

A good example of a Tamm surface state is seen in photoemission from electron states at $\bar{M}$, the corner of the surface Brillouin zone (Figure 5.3), on the (001) surfaces of Cu and Ag.[31] The prominent feature marked S on the spectrum shown in Figure 5.13, from Ag (001), is a Tamm state pulled off the top of the d-electron bands. The actual band from which it is pulled shows up as state B_1 in the spectrum. The bulk states with the surface wave vector component at $\bar{M}$ lie along XW in the bulk Brillouin zone (inset of Figure 5.13), and we can see from

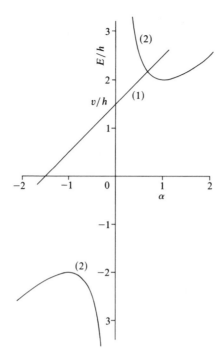

Figure 5.12. E/h as a function of α, given by (1) $E/h = \alpha + v/h$ and (2) $E/h = \alpha + 1/\alpha$.

the bulk band structure (Figure 5.14) that the topmost d band is very flat. This is made up of d_{xy} orbitals,[32] which lie parallel to the (001) plane and consequently interact only weakly in the perpendicular direction; in other words, the hopping integral h is very small. A small value of h means that a small shift in potential can pull off a Tamm state, and in the case of Ag(001) the surface atoms are slightly less attractive than the bulk. Hence the surface state is pulled off the top of the band.

The identification of B_1 with the bulk d_{xy} band, and S with the corresponding surface state, is consistent with the dependence of the photoemission on the polarization of the light.[31] Electrons are emitted from these states with s-polarized light (Figure 5.13), in which the electric field lies in the surface plane, perpendicular to the plane containing the directions in which the light is incident and the electrons are emitted. On the other hand, there is essentially no emission with p-polarized light (Figure 5.13), in which the electric field lies in the plane containing the incident light and emitted electrons. Now the matrix element for exciting an electron from initial state i to final state f is $\langle f | \mathbf{A} \cdot \mathbf{p} | i \rangle$, where $\mathbf{A}$ is the magnetic vector potential lying in the same direction as the electric field and $\mathbf{p}$ is the momentum operator;[33] the geometry for photoemission from the d_{xy} orbitals is shown in Figure 5.15. In this geometry, $\mathbf{A} \cdot \mathbf{p}$ is odd with respect to reflection in the xz-plane with s-polarized light, and even with p-polarized light. But the d_{xy} initial state is odd with respect to this reflection and the final state is even, so the matrix element

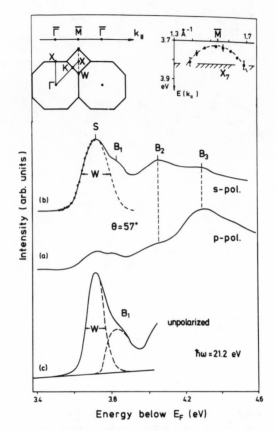

Figure 5.13. Photoemission spectra from $\bar{\text{M}}$ on Ag(001): (a) the spectrum with p-polarized light, (b) with s-polarized light, (c) with unpolarized light (after Goldman and Bartels[31]).

is finite with s-polarization and vanishes with p-polarization. This agrees with the experimental results.

5.4.3. Image Potential States

Electrons can be trapped by the image potential (5.2.2) if they have an energy coinciding with a bulk bandgap, so that they cannot propagate into the bulk solid. These are the image potential surface states, or Rydberg states,[34,35] which are normally unoccupied.

As the image-induced states have most of their charge density outside the solid, let us force their wave functions to go to zero at the geometrical surface at $z = 0$. The origin of the image potential z_0 in expression (5.2.2) will also be taken at the geometrical surface (it is actually just outside). In the vacuum, that is $z < 0$,

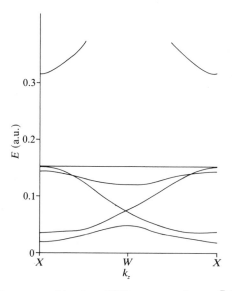

Figure 5.14. Bulk band structure of Ag along XWX, corresponding to $\bar{M}$ in the surface Brillouin zone.

the electron feels the potential $V(z) = -1/(4|z|)$, and its wave function has the form

$$\psi(\mathbf{r}) = \exp{(i\mathbf{K}\cdot\mathbf{R})}\phi(z) \tag{5.4.22}$$

where ϕ satisfies the Schroedinger equation

$$-\frac{1}{2}\frac{d^2\phi}{dz^2} - \frac{\phi}{4|z|} = \left(E - \frac{K^2}{2}\right)\phi \tag{5.4.23}$$

Compare this with the equation for $l = 0$ radial wave functions R in a hydrogenic atom with nuclear charge Z,

$$-\frac{1}{2r^2}\frac{d}{dr}\left(r^2\frac{dR}{dr}\right) - \frac{Z}{r}R = \varepsilon R \tag{5.4.24}$$

On substituting u/r for R this becomes

$$-\frac{1}{2}\frac{d^2u}{dr^2} - \frac{Z}{r}u = \varepsilon u \tag{5.4.25}$$

the same as our surface equation if we put $Z = \frac{1}{4}$ and $\varepsilon = E - K^2/2$. Now the eigenvalues of the hydrogenic equation are

$$\varepsilon = -\frac{Z^2}{2n^2} \tag{5.4.26}$$

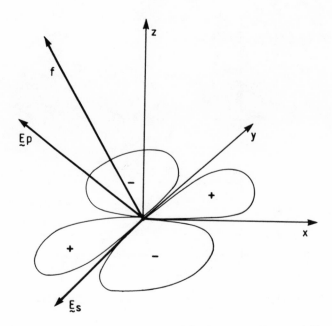

Figure 5.15. The initial state i is a d_{xy} orbital, and the final state f is essentially a plane wave traveling in the direction shown. E_s is the electric field with s-polarized light and E_p is the field with p-polarized light.

where n is an integer, corresponding to a solution R which stays finite as $r \rightarrow 0$; hence $u \rightarrow 0$. This is the behavior required of our surface state wave functions, so the eigenvalues of the Rydberg surface states are

$$E = -\frac{1}{32n^2} + \frac{K^2}{2} \tag{5.4.27}$$

These image-induced states have been studied particularly on noble metal surfaces. On Cu(001), for example, at $\bar{\Gamma}$ there is a bandgap in the bulk states at energies around the vacuum zero;[36] in other words, states with $K = 0$ and an energy close to zero cannot propagate into the solid. So we would expect to see the Rydberg series (5.4.27) of states localized outside the surface. As the vacuum zero lies 4.6 eV (the work function) above the Fermi energy, the image-induced surface states are unoccupied, but they can be measured in inverse photoemission— electrons fired in, light out. Figure 5.16 shows the inverse photoemission spectrum from Cu(001)[36] in which normal-incidence electrons drop down into the un-occupied states, so that the spectrum corresponds to states at $\bar{\Gamma}$. The Rydberg states are apparent just below the vacuum zero, and the resolution is such that the $n = 1$ state can be distinguished from the higher members of the series. Why study these apparently rather esoteric states—after all, there are normally no electrons in them? An important reason is that careful measurements of their energy, and its deviation

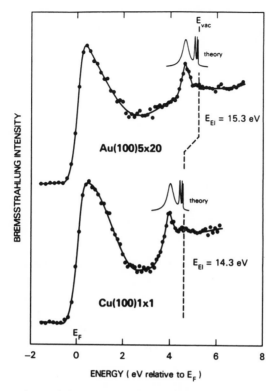

Figure 5.16. Inverse photoemission spectrum from Cu(001) and reconstructed Au(001), corresponding to unoccupied states with $\mathbf{K} = 0$. Energy is measured relative to E_F, and the vertical dashed line gives the vacuum zero (after Straub and Himpsel[36]).

from expression (5.4.27), give information about the exact form of the image potential—the value of the image plane z_0 and the way in which the asymptotic form (5.2.2) joins on to the selvedge potential.[37]

5.5. The Surface of Jellium

The surface states treated in the last section depend to a large extent on the atomic structure of the solid: the Shockley states depend on the scattering properties of the atoms, with a p state lying below an s state; the Tamm states originate very directly from the atomic orbitals; and even the image-induced states, lying outside the solid, depend on having a band-gap in the bulk band structure around the vacuum level. We shall now turn to a model of the surface in which the ions are smeared out into a uniform positive background, which is chopped off at $z = 0$ to create a surface. The electrons respond to the potential of the semi-infinite positive background, and to each other, but there are no ions to scatter them. This is the *jellium* model,[3,4] which is a reasonable first approximation for s–p-bonded metals like Na, Mg, or Al in which the electrons are scattered only weakly by the ions. It is the simplest model to give a reasonably accurate picture of the surface charge density, potential, and work function.

The electrons satisfy a Schroedinger equation in which the potential depends on their own charge density—in addition to the electrostatic potential of the positive background V_+, they feel the electrostatic potential due to the total electron density, the Hartree potential V_H, and the exchange-correlation potential V_{xc}.[3,38] The Hartree potential is given by

$$V_H(\mathbf{r}) = \int d^3r' \rho(\mathbf{r}')/|\mathbf{r} - \mathbf{r}'| \tag{5.5.1}$$

where ρ is the charge density of the occupied states:

$$\rho(\mathbf{r}) = \sum_{i<E_F} |\psi_i(\mathbf{r})|^2 \tag{5.5.2}$$

The exchange-correlation potential is actually a *functional* of the electron density, that is, it depends on the variation of the whole density as a function of $\mathbf{r}$. Unfortunately, the functional form is not known! The best we can do—and it is a very good approximation—is to put $V_{xc}(\mathbf{r})$ equal to the value of V_{xc} in a uniform free-electron gas of the same electron density as the actual density at point $\mathbf{r}$:[39]

$$V_{xc}(\mathbf{r}) \approx V_{xc}^{uniform}(\rho(\mathbf{r})) \tag{5.5.3}$$

The potential must be found self-consistently; this means that it gives wave functions which in turn give back the potential itself from equations (5.5.1)–(5.5.3). The potential is found by applying the cycle shown below, iterating until the input and output values of $V_H + V_{xc}$ are the same.

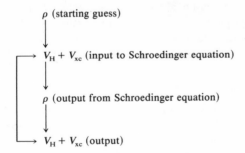

Figure 5.17 shows the form of potential in the jellium model,[4] given by

$$V(z) = V_+(z) + V_H(z) + V_{xc}(z) \tag{5.5.4}$$

The variation of the wave functions parallel to the surface is given by the free-electron factor $\exp(i\mathbf{K} \cdot \mathbf{R})$, because V depends only on z and is independent of

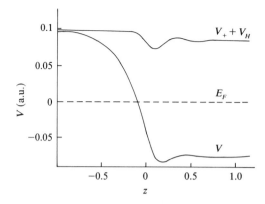

Figure 5.17. Self-consistent potential in the jellium model for K. Distance is measured in Fermi wavelengths from the positive background edge. Potential is measured in au from the Fermi energy. The two curves show the electrostatic part of the potential, $V_+ + V_H$, and the total potential V including V_{xc} (after Lang and Kohn[4]).

R. Deep in the solid, the potential becomes flat, so asymptotically the wave functions go to free-electron standing waves:

$$\psi(\mathbf{r}) \sim \exp(i\mathbf{K} \cdot \mathbf{R}) \sin(k_z z - \chi(k_z)) \qquad (5.5.5)$$

—standing waves as in equation (5.3.1), because the electrons bounce back from the surface. In the surface region, the Schroedinger equation must be integrated numerically, and the solution matched onto this asymptotic form to determine the phase shift χ. The charge density can then be constructed from the wave functions according to equation (5.5.2), and we enter the self-consistency cycle outlined above. In fact, self-consistency is very difficult to achieve at surfaces, because charge tends to flow across the surface in an unstable way—a small shift in charge gives a big change in potential, which in turn produces an even bigger change in charge. The standard way of coping with this instability in the self-consistency cycle is to mix only a small fraction of the output potential from the nth iteration into the input potential to give the input for the $(n + 1)$th iteration:

$$V_{input}^{(n+1)} = (1 - \alpha) V_{input}^{(n)} + \alpha V_{output}^{(n)} \qquad (5.5.6)$$

where α is typically between 1 and 10%.

The self-consistent surface charge density has been calculated in this way by Lang and Kohn[4] for electron densities corresponding to the s-p-bonded metals. This is in fact probably the most useful and influential surface calculation which has ever been performed. Figure 5.18 shows their charge densities for K and Al, and the main features are the overflow of charge at the surface, and small Friedel oscillations which extend into the bulk (decaying as $1/z^2$). Let us relate the form of the charge density with the corresponding potential (Figure 5.17). The charge

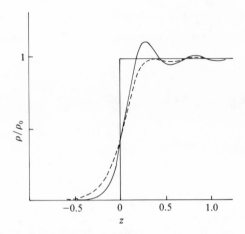

Figure 5.18. Self-consistent charge density in the jellium model for K (solid line) and Al (dashed line). Distance is measured in Fermi wavelengths from the positive background edge; the positive background itself is shown on this figure. Charge density is measured in terms of the bulk density ρ_0 (after Lang and Kohn[4]).

overflow sets up a surface dipole, with excess negative charge just outside the surface and excess positive charge just inside. This gives the electrostatic contribution to the surface potential barrier, shown in the results for $V_+ + V_H$ in Figure 5.17. The attractive exchange-correlation potential further deepens the inner potential—the inner potential is the potential well inside the solid compared with the vacuum. The part of the inner potential which comes from the mean potential inside the solid, additional to the surface dipole barrier, is sometimes called the internal contribution.

The work function can be found immediately from the self-consistent calculation.[3] This is the energy needed to remove an electron from the solid, and is given by the difference between the potential in the vacuum $V(z \to -\infty)$ and E_F. Figure 5.19 shows the calculated work function[40] as a function of bulk r_s, which is a measure of the electron density, the radius (in Bohrs) of a sphere containing one electron. This agrees quite well with the experimental values, and shows the trend for metals with a high electron density to have a larger work function. What is surprising is the fact that work functions vary so little in going from element to element—in most metals they lie in the range 2–5 eV, even when the bulk electron density itself varies by a factor of 20 in going from Al to Cs. The reason for this is that a high electron density gives a large E_F measured relative to the mean potential inside the solid, but the effect of this on the work function is offset by an increased surface dipole.

Work functions actually vary slightly from surface to surface,[10] and experimental examples of this are:[41-43]

fcc Al   (111) 4.24 eV   (100) 4.41 eV   (110)4.28 eV

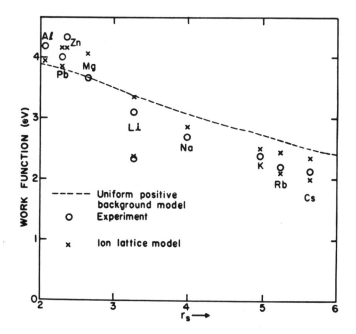

Figure 5.19. Dashed line shows the calculated work function in the jellium model as a function of r_s, compared with experimental values marked by circles.[40] Crosses show results with the ionic pseudopotential included (after Lang and Kohn[40]).

$$\text{fcc Cu} \quad (111) 4.94 \text{ eV} \quad (100) \ 4.59 \text{ eV} \quad (110) \ 4.48 \text{ eV}$$

$$\text{bcc W} \quad (110) \ 5.25 \text{ eV} \quad (100) \ 4.63 \text{ eV} \quad (111) \ 4.47 \text{ eV}$$

(We note that Cu and W, even though they are partly bonded by d electrons, still have work functions in the same range as our s–p-metals.) The work-function variation between surfaces of the same metal is due to changes in surface dipole, as the internal contribution stays the same. The jellium model does not contain any ionic lattice, so it cannot give this variation; however, the effect of the ionic pseudopotential can be included by perturbation theory, and this gives the observed surface-dependence.[40] There is a tendency for the work function to be biggest on surfaces with the greatest atomic density—(111) in the fcc structure and (110) in bcc—but we can see from the experimental results for Al that this is by no means completely true.

5.6. Calculating Surface Electronic Structure

The reason why surface electronic structure calculations are more difficult than bulk calculations is the lack of periodicity in the z-direction, as we have seen throughout this chapter. However, one way of avoiding this problem is to use a slab, typically 5–7 atomic layers thick, with the form of potential shown in Figure 5.20. The system is now finite in the z-direction, as far as an electron with energy below the vacuum zero is concerned, and standard basis set methods can then be used to solve the Schroedinger equation. If, for example, the ionic potentials are replaced by weak pseudopotentials, a plane wave expansion may be suitable.[44] As the system has two-dimensional periodicity, a state with two-dimensional Bloch wave vector $\mathbf{K}$ must be made up of plane wave components parallel to the surface of the usual form $\exp(i[\mathbf{K}+\mathbf{G}]\cdot\mathbf{R})$. In the z-direction we can assume that a bound-state wave function has dropped practically to zero at $z = \pm d$ (Figure 5.20), so it can be expanded in a Fourier series made up of cosines, $\cos(n+\frac{1}{2})\pi/d$, or sines, $\sin n\pi z/d$, for even and odd states, respectively. In other words, we expand even wave functions in the form

$$\psi_{\mathbf{K}}(\mathbf{r}) = \sum_{G,n} a_{G,n} \exp(i[\mathbf{K}+\mathbf{G}]\cdot\mathbf{R}) \cos\frac{(n+\frac{1}{2})\pi z}{d} \qquad (5.6.1)$$

and odd wave functions as

$$\psi_{\mathbf{K}}(\mathbf{r}) = \sum_{G,n} a_{G,n} \exp(i[\mathbf{K}+\mathbf{G}]\cdot\mathbf{R}) \sin\frac{n\pi z}{d} \qquad (5.6.2)$$

The coefficients are the eigenvectors of the corresponding matrix form of the Hamiltonian. This is almost exactly like a conventional band structure calculation.

If the full ionic potential is used, a plane wave expansion is not suitable, as numerous waves would be needed to describe the rapid oscillations of the wave functions inside the core region of the ions. Instead, a basis set "augmented" by atomic-like wave functions is often used. An example of this is the linearized augmented plane wave (LAPW) basis set;[45] each basis function looks like a plane wave as in expression (5.6.1) or (5.6.2) between the atoms, but inside the ion cores solutions of the radial atomic Schroedinger equation are used, chosen to match onto the plane waves. The LAPW basis functions can be used in a slab calculation just as easily as in a bulk calculation (perhaps it should be said, with no more difficulty).

A slab has two surfaces, separated by a few layers of bulk-like material. This means that the electronic structure is not exactly the same as for a single surface

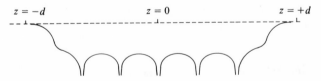

Figure 5.20. Schematic potential in a slab calculation as a function of z. If the potential is symmetric about $z = 0$ the wave functions can be classified as even (symmetric) or odd (antisymmetric).

of a semi-infinite substrate. First of all, the finite thickness of the slab means that the states at a particular **K** are discrete, below the vacuum zero of energy; this is because at fixed **K** we are essentially finding the bound states of a one-dimensional potential well, which occur at definite, discrete energies. This is unlike the energy spectrum of states at the surface of a semi-infinite solid, which consists of the continua of bulk states hitting the surface as well as the discrete surface states (Section 5.3). In other words a slab calculation cannot, in principle, distinguish between bulk and surface states, though in practice the surface states have their charge density fairly localized on the surface atoms. Another consequence of using a slab is that the surface state wave functions localized on each surface invariably mix, leading to a splitting of the surface state energy. These undesirable effects can be minimized by using as thick a slab as possible (the limitations come from computer time). In fact for quantities like the charge density, or total energy, which depend on integrations over all the occupied electronic states, slab calculations are quite satisfactory and have led to some impressive and accurate results.

As an example of a slab calculation, Figure 5.21 shows the discrete energy bands as a function of **K** for a nine-layer slab of Ni(001).[46] The circles show the

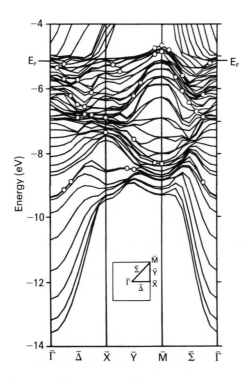

Figure 5.21. Two-dimensional energy bands as a function of **K** for a nine-layer Ni(001) slab. The open circles represent surface states (after Arlinghaus et al.[46]).

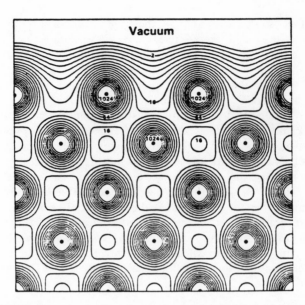

Figure 5.22. Ni(001) charge density from the nine-layer slab calculation. The plane shown passes through the centers of atoms in all the layers (after Arlinghaus *et al.*[46]).

states which are quite localized at the surface, and which are identified as surface states: at $\bar{M}$ the topmost state shown is the Tamm surface state, unoccupied in this case. The charge density is shown in Figure 5.22—the features to note here are the exponential decay of the charge density into the vacuum, and the fact that the charge density on the second layer of atoms is essentially bulk-like. The work function from this calculation is 5.1 eV, compared with the experimental value of 5.1–5.2 eV.

5.7. Plasmons and Many-Body Effects

So far we have treated the electrostatic interaction between the electrons through the Hartree potential entering the Schroedinger equation, corrected by the exchange-correlation potential to allow for the fact that the electrons get out of each other's way. There are, however, excited states of the system which are a consequence of the many-body electron–electron interactions, and the most important of these are the plasmons.[47]

We are all familiar with the most elementary derivations of the plasma frequency of a bulk electron gas, but let us consider it from the standpoint of the dielectric function.[5] In an electric field E, which includes the screening field as well as the external field, the equation of motion of classical electrons is

$$m\frac{d^2x}{dt^2} = eE \tag{5.7.1}$$

Therefore if the field has a time dependence $E_0 \exp i\omega t$, the amplitude of the displacement $x = x_0 \exp i\omega t$ is given by

$$x_0 = -eE_0/m\omega^2 \tag{5.7.2}$$

The polarization of the electron gas, the dipole moment per unit volume, is then given by

$$P = -\frac{ne^2}{m\omega^2} E \tag{5.7.3}$$

where n is the number of electrons per unit volume. This means that the electrical displacement D is given by

$$D = \left(1 - \frac{4\pi ne^2}{m\omega^2}\right) E \tag{5.7.4}$$

and the (frequency-dependent) dielectric constant is

$$\varepsilon(\omega) = 1 - \frac{4\pi ne^2}{m\omega^2} \tag{5.7.5}$$

If the frequency is such that $\varepsilon(\omega) = 0$, clearly we have finite E with zero D; but we know from Maxwell's equations that D is related to the *external* charge, so this condition corresponds to the electrons themselves producing a self-sustaining electric field and screening charge. This is a plasma oscillation, and $\varepsilon = 0$ occurs at the plasma frequency ω_p given by

$$\omega_p^2 = \frac{4\pi ne^2}{m} \tag{5.7.6}$$

The plasma oscillations have an energy which is quantized in units of $\hbar\omega_p$; the quantum is the plasmon.

At the surface, surface plasmons can occur—self-sustaining oscillations of the surface polarization charge.[48] Figure 5.23 shows the geometry, with a surface polarization charge density of σ per unit area. The electric field produced by this is $2\pi\sigma$ in each direction, or, counting a field *into* the metal as *positive*,

$$E = \begin{cases} -2\pi\sigma, & z < 0 \\ +2\pi\sigma, & z > 0 \end{cases} \tag{5.7.7}$$

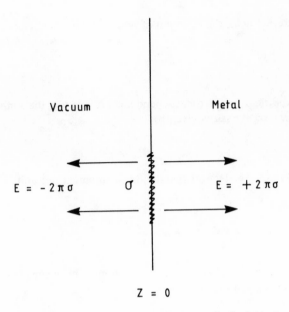

Figure 5.23. The surface polarization charge σ produces an electric field of $\pm 2\pi\sigma$, the $+$ field directed into the metal and the $-$ field directed into the vacuum.

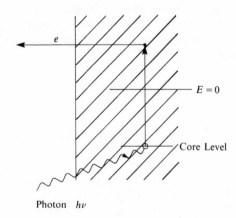

Figure 5.24. Core-level photoemission, in which a core electron absorbs a photon $h\nu$ and is excited from the solid. The outgoing photoelectron and the core hole can excite surface and bulk plasmons.

Hence the electrical displacement is given by

$$D = \begin{cases} -2\pi\sigma, & z < 0 \\ +2\pi\sigma\varepsilon, & z > 0 \end{cases} \qquad (5.7.8)$$

However, we know from Maxwell that the perpendicular component of D must be continuous across an interface; in other words, the two values of D around $z = 0$, from equation (5.7.8), have to be the same. This means that $\varepsilon = -1$; the frequency at which this occurs, when a self-sustaining surface charge can exist, is the surface plasmon frequency. It is seen from equation (5.7.5) that $\varepsilon = -1$ in a

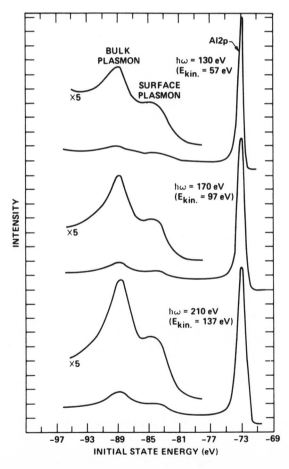

Figure 5.25. Photoemission intensity from the $2p$ core level in Al, with photons of energy 130, 170, and 210 eV, as a function of initial state energy. The peak at -73 eV is the no-loss line, and surface and bulk plasmon satellites are visible at -83 and -89 eV (after Flodström et al.[49]).

free-electron system at

$$\omega_s = \omega_p/\sqrt{2} \qquad (5.7.9)$$

For Al, to which the free-electron model of the plasma oscillations applies, the bulk plasmon energy $\hbar\omega_p$ is about 15 eV, and the surface plasmon has an energy of about 10 eV.[47]

Plasmons show up during energy-loss experiments in electron scattering, for example, as satellites in core-level photoemission.[49] A photon is absorbed, exciting a core electron into a state with energy above the vacuum zero such that it can leave the solid (Figure 5.24). As it passes through the gas of valence electrons, the outgoing electron can excite surface and bulk plasmons, losing energy as it does so. The ionized core can also excite plasmons, though this process tends to interfere destructively with the excitation by the outgoing photoelectron. Figure 5.25 shows the intensity of electrons photoemitted from the $2p$ core level in Al, as a function of energy. We can see a strong no-loss line corresponding to photoelectrons which have not excited any plasmons on the way out, together with satellites in which surface and bulk plasmons have been excited.

Surface plasmons are important (and interesting) because the dielectric response of the surface to an external charge can be described in terms of exciting the surface plasmons. Even when the electron gas does not contain any surface plasmon quanta, their zero-point energy contributes to the surface energy, and the shift in zero-point energy when two surfaces are brought up to one another gives rise to the van der Waals force.[50]

References

1. M. Prutton, *Surface Physics*, Oxford University Press (1983).
2. D. P. Woodruff and T. A. Delchar, *Modern Techniques of Surface Science*, Cambridge University Press (1986).
3. N. D. Lang, in: *Solid State Physics* (H. Ehrenreich, F. Seitz, and D. Turnbull, eds.), Vol. 28, p. 225, Academic Press, New York (1973).
4. N. D. Lang and W. Kohn, *Phys. Rev. B* **1**, 4555 (1970).
5. C. Kittel, *Introduction to Solid State Physics*, Wiley, New York (1976).
6. J. A. Appelbaum and D. R. Hamann, *Phys. Rev. B* **6**, 2166 (1972).
7. J. E. Inglesfield, *Prog. Surf. Sci.* **25** (1–4), 57 (1987).
8. V. Heine, *Proc. Phys. Soc.* **81**, 300 (1963).
9. J. B. Pendry, *Low Energy Electron Diffraction*, Academic Press, New York (1974).
10. J. E. Inglesfield, in: *Electronic Properties of Surfaces* (M. Prutton, ed.), Adam Hilger, Bristol (1984).
11. G. A. Benesh and J. E. Inglesfield, *J. Phys. C.* **19**, L539 (1986).
12. D. P. Woodruff, in: *The Chemical Physics of Solid Surfaces and Heterogeneous Catalysis* (D. A. King and D. P. Woodruff, eds.), Vol. 1, p. 81, Elsevier, Amsterdam (1981).
13. H. J. Levinson, F. Greuter, and E. W. Plummer, *Phys. Rev. B* **27**, 727 (1983).
14. V. Heine, in: *Solid State Physics* (H. Ehrenreich, F. Seitz, and D. Turnbull, eds.), Vol. 35, p. 1, Academic Press, New York (1980).
15. P. M. Morse and H. Feshbach, *Methods of Theoretical Physics*, McGraw-Hill, New York (1953).
16. R. Haydock and M. J. Kelly, *Surf. Sci.* **38**, 139 (1973).

17. M. Posternak, H. Krakauer, A. J. Freeman, and D. D. Koelling, *Phys. Rev. B* **21**, 5601 (1980).
18. C. L. Fu, A. J. Freeman, E. Wimmer, and M. Weinert, *Phys. Rev. Lett.* **54**, 2261 (1985).
19. C. Guillot, C. Thuault, Y. Jugnet, D. Chauveau, R. Hoogewijs, J. Lecante, T. M. Duc, G. Tréglia, M. C. Desjonquères, and D. Spanjaard, *J. Phys. C.* **15**, 4023 (1982).
20. W. F. Egelhoff, *Surf. Sci. Rep.* **6**, 253 (1987).
21. J. F. van der Veen, P. Heimann, F. J. Himpsel, and D. E. Eastman, *Solid State Commun.* **37**, 555 (1981).
22. J. F. van der Veen, F. J. Himpsel, and D. E. Eastman, *Phys. Rev. Lett.* **44**, 189 (1980).
23. T. M. Duc, C. Guillot, Y. Lassailly, J. Lecante, Y. Jugnet, and J. C. Vedrine, *Phys. Rev. Lett.* **43**, 789 (1979).
24. V. Heine, in: *Solid State Physics* (H. Ehrenreich, F. Seitz, and D. Turnbull, eds.), Vol. 24, p. 1, Academic Press, New York (1970).
25. R. O. Jones, *J. Phys. C.* **5**, 1615 (1972).
26. W. Shockley, *Phys. Rev.* **56**, 317 (1939).
27. M. L. Cohen and V. Heine, in: *Solid State Physics* (H. Ehrenreich, F. Seitz, and D. Turnbull, eds.), Vol. 24, p. 37, Academic Press, New York (1970).
28. J. E. Inglesfield and G. A. Benesh, *Surf. Sci.* **200**, 135 (1988).
29. F. Forstmann, in: *Photoemission and the Electronic Properties of Surfaces* (B. Feuerbacher, B. Fitton, and R. F. Willis, eds.), p. 193, (1978).
30. S. G. Davison and J. D. Levine, in: *Solid State Physics* (H. Ehrenreich, F. Seitz, and D. Turnbull, eds.), Vol. 25, p. 1, Academic Press, New York (1970).
31. A. Goldmann and E. Bartels, *Surf. Sci.* **122**, L629 (1982).
32. J. R. Smith, F. J. Arlinghaus, and J. G. Gay, *Phys. Rev. B* **22**, 4757 (1980).
33. R. H. Williams, G. P. Srivastava, and I. T. McGovern, in: *Electronic Properties of Surfaces* (M. Prutton, ed.), Adam Hilger, Bristol (1984).
34. P. M. Echenique and J. B. Pendry, *J. Phys. C* **11**, 2065 (1978).
35. N. V. Smith, *Phys. Rev. B* **32**, 3549 (1985).
36. D. Straub and F. J. Himpsel, *Phys. Rev. Lett.* **52**, 1922 (1984).
37. M. Weinert, S. L. Hulbert, and P. D. Johnson, *Phys. Rev. Lett.* **55**, 2055 (1985).
38. W. Kohn and L. J. Sham, *Phys. Rev.* **140**, A1133 (1965).
39. S. Lundqvist and N. H. March, *Theory of the Inhomogeneous Electron Gas*, Plenum Press, New York (1983).
40. N. D. Lang and W. Kohn, *Phys. Rev. B* **3**, 1215 (1971).
41. J. K. Grepstad, P. O. Gartland, and B. J. Slagsvold, *Surf. Sci.* **57**, 348 (1976).
42. P. O. Gartland, S. Berge, and B. J. Slagsvold, *Phys. Rev. Lett.* **28**, 738 (1972).
43. R. W. Strayer, W. Mackie, and L. W. Swanson, *Surf. Sci.* **34**, 225 (1973).
44. S. G. Louie, K. M. Ho, J. R. Chelikowsky, and M. L. Cohen, *Phys. Rev. B* **15**, 5627 (1977).
45. H. Krakauer, M. Posternak, and A. J. Freeman, *Phys. Rev. B* **19**, 1706 (1979).
46. F. J. Arlinghaus, J. G. Gay, and J. R. Smith, *Phys. Rev. B* **21**, 2055 (1980).
47. D. Pines, *Elementary Excitations in Solids*, Benjamin, New York (1964).
48. R. H. Ritchie, *Phys. Lett. A* **38**, 189 (1972).
49. S. A. Flodström, R. Z. Bachrach, R. S. Bauer, J. C. McMenamin, and S. B. M. Hagström, *J. Vac. Sci. Technol.* **14**, 303 (1977).
50. J. E. Inglesfield, in: *The Chemical Physics of Solid Surfaces and Heterogeneous Catalysis* (D. A. King and D. P. Woodruff, eds.), Vol. 1, p. 355, Elsevier, Amsterdam (1981).

Basic Structural and Electronic Properties of Semiconductor Surfaces

C. M. Bertoni

6.1. Introduction

In the last twenty years surface science has developed extensively. The preparation of crystal surfaces under ultrahigh vacuum conditions made possible the study of atomically clean surfaces with structural techniques sensitive to the first atomic layers. Moreover, the developments of surface spectroscopies allowed the electron band structure to be studied at the surface.

Semiconductor surfaces display a variety of phases with long-range order, showing relaxation of the atomic positions with respect to the bulk ideal sites or different kinds of reconstruction with new periodicity of the surface mesh. The surface electronic structure is affected strongly by these changes of the atomic geometry. Thus the structural and electronic properties were studied more in unison and enabled the properties of semiconductor surfaces to be understood at a microscopic level, both in the case of clean surfaces and in the first stage of chemisorption and ordered deposition of atoms, during the formation of an interface.

In this chapter the subject will be examined following a somewhat historical point of view; in particular we will treat together both experimental and theoretical aspects. Indeed, theory and experiment were developed together in the last two decades in order to describe systems with increasing level of complexity.

After a discussion of simple one-dimensional models to introduce concepts and terminology, we will review briefly the experimental methods and treat in

C. M. Bertoni • Dipartimento di Fisica, Seconda Università di Roma "Tor Vergata," 00173 Rome, Italy.

some detail the features of a band structure calculation for a surface, with special reference to the cleavage surface of III–V compounds. The fourth section deals with first-principles calculations, which evaluate the total energy as a function of the geometrical structure, allowing the prediction of surface structure. The subsequent section will be devoted to some specific cases.

6.2. Electronic Surface States: Simple Models

6.2.1. One-Dimensional Chain of Atoms

A simple model, like the one-dimensional (1D) linear chain, with a termination, can provide an elementary point of view on the formation of electronic surface states. The infinite chain in Figure 6.1a will be considered first. The simplest way to describe an electron moving in a 1D crystal is to examine a periodic cosine-type potential

$$V(z) = V_0 + V_g(e^{igz} + e^{-igz}) \tag{6.2.1}$$

where $g = 2\pi/a$, a being the lattice parameter of the linear chain. Near the boundary of the Brillouin zone (BZ), $k \simeq g/2$, the Bloch-type eigenfunctions can be approximated by the sum of two terms:

$$\psi(z) = c_1 e^{ikz} + c_2 e^{i(k-g)z} \tag{6.2.2}$$

The energy eigenvalues can be obtained by the secular system of equations

$$\begin{pmatrix} k^2 - E & V_g \\ V_g & (k-g)^2 - E \end{pmatrix} \begin{pmatrix} c_1 \\ c_2 \end{pmatrix} = 0 \tag{6.2.3}$$

where Rydbergs are energy units. Two bands $E = E_k^\pm$, separated by a gap, are obtained in the form

$$E_k^\pm = \tfrac{1}{2}\{k^2 + (k-g)^2 \pm \sqrt{[k^2 - (k-g)^2]^2 + 4|V_g|^2}\} \tag{6.2.4}$$

$$\psi_k^\pm(z) = e^{ikz} \pm \frac{E_k^\pm - k^2}{V_g} e^{i(k-g)z} \tag{6.2.5}$$

A gap is found at the border of the 1D BZ (see Figure 6.1c), where

$$E_{g/2}^\pm = \frac{g^2}{4} \pm |V_g| \tag{6.2.6}$$

$$\psi_{g/2}^\pm(z) = e^{igz/2} \pm \frac{|V_g|}{V_g} e^{-igz/2} \tag{6.2.7}$$

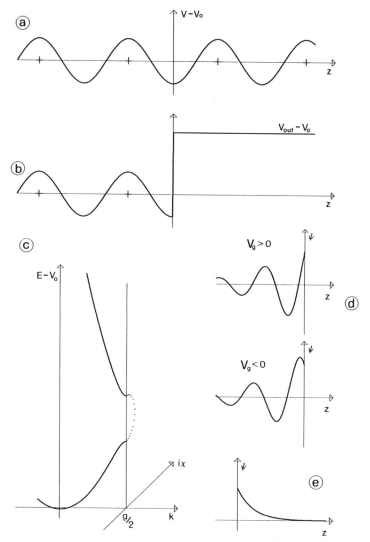

Figure 6.1. (a) Infinite chain potential (crosses indicate atomic positions); (b) truncated chain potential; (c) band structure at real and complex wave vectors; (d) evanescent waves; (e) outer solution.

We note that for $V_g < 0$ one has

$$\psi^+_{g/2}(z) = 2i \sin (gz/2) \qquad \text{and} \qquad \psi^-_{g/2}(z) = 2 \cos (gz/2)$$

and for $V_g > 0$ one obtains

$$\psi^+_{g/2}(z) = 2 \cos (gz/2) \qquad \text{and} \qquad \psi^-_{g/2}(z) = 2i \sin (gz/2)$$

The charge distribution of the lower-energy state has a maximum midway between the two atoms if $V_g < 0$, as the potential is attractive in the region of the bond. If $V_g > 0$ the lower-energy state is localized mainly at the atomic sites.

The approximate wave functions considered here are Bloch states which satisfy the conditions

$$\psi_k(z + a) = e^{ika} \psi_k(z) \tag{6.2.8}$$

$$|\psi_k(z + a)|^2 = |\psi_k(z)|^2 \tag{6.2.9}$$

They can be written as

$$\psi_k(z) = e^{ikz} u_k(z) \tag{6.2.10}$$

with $u_k(z + a) = u(z)$.

6.2.2. Evanescent Wave Functions, Truncated Chain, and Surface States

Before considering the case of a terminated lattice, we can extend the class of functions (6.2.5) to complex values of k, but with real values of the energy in equation (6.2.4). The condition (6.2.9) is not satisfied. If we write $k = g/2 - i\chi$ with $\chi > 0$, then

$$E^{\pm}_{g/2-i\chi} = \frac{g^2}{4} - \chi^2 \pm \sqrt{|V_g|^2 - g^2\chi^2} \tag{6.2.11}$$

The values of the energy are real if $\chi < |V_g|/g$, and they are found in the region of the forbidden gap (see Figure 6.1c). The corresponding wave functions are still eigenstates of the chain Hamiltonian and of the lattice translation operator. They satisfy equation (6.2.8), but not (6.2.9). Having complex k, they are of evanescent type. The phase factor in equation (6.2.10) becomes $e^{igz/2} e^{\chi z}$, vanishing when $z \to -\infty$ and diverging when $z \to \infty$. The wave functions with energy *in the gap* can be written as

$$\psi_{g/2-i\chi}(z) = e^{\chi z}(e^{igz/2} + e^{-2i\delta} e^{-igz/2})$$

$$= 2e^{\chi z} e^{-i\delta} \cos(gz/2 + \delta) \tag{6.2.12}$$

where δ is a phase, depending on the energy. The value of δ varies from 0 to $\pi/2$ when the energy goes from the lower to the upper edge of the gap, if $V_g < 0$. On the other hand, if $V_g > 0$, δ varies from $-\pi/2$ to 0 when passing from the lower to the upper edge of the gap. The two types of evanescent state are drawn in Figure 6.1d.

It is now assumed that the chain is terminated at the midpoint between two neighboring atoms at $z = 0$, so that the crystal is located in $z < 0$; this is the simplest way to describe the effect of the surface. The potential in the external region $z > 0$ is approximated by a constant term V_{out}, higher than the value of the energy at the upper edge of the gap. The wave functions inside the crystal must be matched at $z = 0$ with the external solution in the semiaxis $z > 0$, having the form e^{-qz}, with $q = \sqrt{V_{out} - E}$, at each energy E where the matching is possible. Continuity of the first derivative must also be imposed. In the region of allowed bulk bands the external solution is matched to a combination of $\psi_k(z)$ and $\psi_{-k}(z)$ at every E_k with real k. In the region of the gap, the matching condition, obtained by equating the logarithmic derivative of the evanescent solution and the external decaying function, gives

$$-q = \chi - \tan(\delta)\frac{g}{2} \tag{6.2.13}$$

where χ and q are positive and depend on energy. This equation has a solution only when δ is positive, i.e., when $V_g < 0$. It is known that V_g is negative when the lower-edge state $\psi_{g/2}^-$ has a maximum of the charge density distribution at the midpoint of the interatomic distance, which is the truncation point. The matching condition fixes the energy of a new state, obtained by matching an external exponentially decaying function and an internal evanescent wave with a decay length $1/\chi$. This state is a *surface state*, being *localized in the surface region*, and its energy eigenvalue lies *in the bulk energy gap*. In our simple model we find a single surface state if $V_g < 0$. This condition is a particular formulation of the Shockley theorem.[1]

The simple model employed here can roughly represent the case of a semiconductor terminated by a surface. If two electrons per atom are present, the gap separates the empty conduction band and the filled valence band. The abrupt termination of the crystal *can* create a localized state in the forbidden energy gap, if the existence condition involving the potential is satisfied. The surface, like other types of defect, can strongly modify the electronic structure of the semiconductor, giving rise to localized states.

6.2.3. Complex-k Energy Dispersion and Matching Conditions

In more general cases it is possible to follow the same procedure, namely, solving the band structure—also in a three-dimensional (3D) crystal—and obtaining evanescent solutions by removing the requirement of real $\mathbf{k}$ values, i.e., allowing complex values of k_z, the component of $\mathbf{k}$ normal to the surface.

These solutions must be found at every value of the energy and a linear combination of them must be matched, in addition to its derivative, with the external decaying function and its derivative. If the energy is in a forbidden gap, then the existence of a solution of the matching problem produces a *surface state*.

In the region of the energy values of a bulk state continuum, the *traveling* bulk states with components k_z and $-k_z$ are mixed, to be matched with the external solution. In this way a propagating wave is completely reflected by the surface. For the empty bands, with energy higher than the vacuum level ($E > V_{out}$), the traveling internal solutions can be combined and matched to an external traveling state. They correspond, in the 3D case, to the so-called LEED states, appearing in low-energy electron diffraction and in the final states of the photoemission process.

Moreover, forbidden gaps can exist also in the empty states above the vacuum level. In these region of *forbidden* energy for traveling waves, evanescent solutions match a combination of external propagating waves. These kinds of states can play an important role when they are used as final states for the excitation of an electron, in surface-sensitive spectroscopies.

6.2.4. Tight-Binding Model for the Linear Chain

Let us consider the linear chain in the tight-binding scheme. If we choose a chain with one atom per unit cell and describe the wave functions in terms of a linear combination of atomic-like orbitals, at each site l we have the set of functions $\{\phi_\alpha(z - la)\}$. For the infinite chain, we combine them to obtain Bloch functions:

$$\psi_k^{(\alpha)}(z) = \frac{1}{\sqrt{N}} \sum_l e^{ikla} \phi_\alpha(z - la)$$

$$\psi_k(z) = \sum_\alpha c_k(\alpha)\psi_k^{(\alpha)}(z) = \frac{1}{\sqrt{N}} \sum_\alpha \sum_l c_k(\alpha) e^{ikla} \phi_\alpha(z - la) \qquad (6.2.14)$$

From the relation $H\psi_k(z) = E\psi_k(z)$ one has the secular system

$$\sum_\alpha \{H_{\beta\alpha}(k) - E_k S_{\beta\alpha}(k)\} c_k(z) = 0 \qquad (6.2.15)$$

where

$$S_{\beta\alpha}(k) = \sum_l e^{ikla} \int \phi_\beta^*(z)\phi_\alpha(z - la) \, dz = \sum_l e^{ikla} S_{\beta\alpha}(0, l) \qquad (6.2.16)$$

and

$$H_{\beta\alpha}(k) = \sum_l e^{ikla} \int \phi_\beta^*(z) H\phi_\alpha(z - la) \, dz = \sum_l e^{ikla} E_{\beta\alpha}(0, l) \qquad (6.2.17)$$

If the crystal periodicity is lost owing to the presence of some defects or the termination of the chain, we can no longer use equation (6.2.14) with the phase factors containing the wave vector k, but we must replace $e^{ikla} c_k(\alpha)$ by $A_{l\alpha}$. In this case we set

$$\psi(z) = \sum_\alpha A_{l\alpha} \phi_\alpha(z - la) \qquad (6.2.14')$$

The secular equations are then given by

$$\sum_{l'} \sum_\alpha \{H_{\beta\alpha}(l; l') - E S_{\beta\alpha}(l; l')\} A_{l'\alpha} = 0 \qquad (6.2.15')$$

One of the simplest models is obtained with a set of two orbitals s and p_z. We can also put $S_{\beta\alpha}(l; l') = \delta_{\beta\alpha}\delta_{ll'}$, as in the *Slater–Koster* model.[2] The integrals $E_{\beta\alpha}(l; l')$ are included only for nearest neighbors. Equation (6.2.15) for the infinite chain becomes

$$\begin{cases} (H_{ss} - E_k)c_k(s) + H_{sp}c_k(p) = 0 \\ H_{ps}c_k(s) + (H_{pp} - E_k)c_k(p) = 0 \end{cases} \qquad (6.2.18)$$

with

$$H_{ss} = E_s - 2\gamma_{ss}\cos(ka), \qquad H_{pp} = E_p + 2\gamma_{pp}\cos(ka),$$
$$H_{sp} = -H_{ps} = 2i\gamma_{sp}\sin(ka)$$

where all quantities γ are positive.

The model can be further simplified by putting $\gamma = \gamma_{ss} = \gamma_{pp} = \gamma_{sp}$ and choosing the origin of the energy scale to have $(E_p + E_s)/2 = 0$. We write $\Delta = (E_p - E_s)/2$, where $\Delta > 0$. Hence

$$E_k^{\pm} = \pm\sqrt{\Delta^2 + 4\gamma^2 + 4\gamma\Delta\cos(ka)} \qquad (6.2.19)$$

This is the symmetrical two-band model of Figure 6.2. We note that at the center (the Γ point) and at the border $k = \pi/a$ (the X point) of the BZ we have $H_{sp} = 0$:

$$E_{\Gamma_1} = -(\Delta + 2\gamma) \qquad \text{and} \qquad E_{\Gamma_2} = +(\Delta + 2\gamma) \qquad (6.2.20)$$

At Γ, the lower and higher eigenvalues correspond to states composed of pure s and p orbitals, respectively, as follows from equation (6.2.18). At X one has

$$E_X = \pm|\Delta - 2\gamma| \qquad (6.2.21)$$

Also, the two states at the zone boundary are pure s or p. If $2\gamma < \Delta$, then the lower state is s-type (X_1) and the upper is p-type (X_2). If $2\gamma > \Delta$, then the two states have energies in the opposite order. The effect of large γ values is to create a *hybridization gap*. In the case $2\gamma > \Delta$, each band changes its orbital composition on going from Γ to X (see Figure 6.2). One says that the bands have *crossed*.

We now turn to the case of a terminated chain. The more general secular equation (6.2.15') is used and we investigate whether or not solutions exist at each energy. The last atom of the chain is taken at $l = 0$ and the first missing atom at $l = 1$. The function of type (6.2.14') is constructed by employing at each energy the solutions of equation (6.2.16), writing

$$A_{l\alpha} = \sum_j c(\alpha; k_j)\, e^{ik_j la} a_j \qquad (6.2.22)$$

where $k_j = k_j(E)$ are the solutions of the secular equations, as a function of E, obtained at any energy. They are real if E lies within one of the two bands (6.2.19), and in this case two values k and $-k$ are found.

If E is not in these regions, then one obtains complex solutions for k (*evanescent states*) which can give rise to localized states. One must also include appropriate

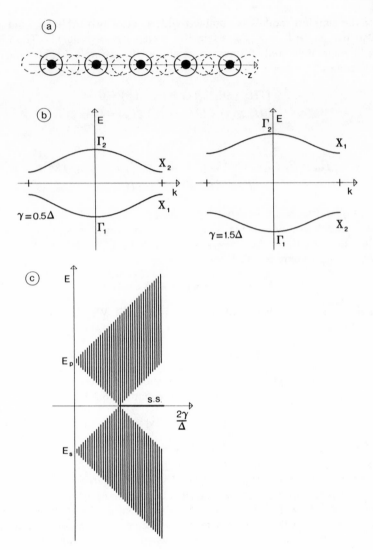

Figure 6.2. Tight-binding model for the linear chain. (a) s and p_z basis functions; (b) two-band models with uncrossed and crossed band; (c) bulk bandwidths and surface state (ss) as functions of $2\gamma/\Delta$.

boundary conditions to represent the missing terms in the Hamiltonian (6.2.15′) and ensure the appropriate behavior (no divergence) well inside the bulk.

In our case, real values of E in the gap are obtained with $k_{1,2} = \pi/a \mp i\chi$ and $\chi > 0$. We must accept only $k_1 = \pi/a - i\chi$, being the atomic positions of the semi-infinite row in $l \leq 0$ sites. Then there is only one term in the j-sum (6.2.22).

In our model, equations (6.2.14) and (6.2.22) yield that the only acceptable solution is $E = 0$, the center of the gap, with $c_{k_1}(s) = c_{k_1}(p)$, provided

$$\Delta + 2\gamma \cos(k_1 a) - 2i\gamma \sin(k_1 a) = 0 \tag{6.2.23}$$

However, because $2\cos(k_1 a) = -e^{\chi a} - e^{-\chi a}$ and $2i\sin(k_1 a) = -e^{\chi a} + e^{-\chi a}$, the condition reduces to

$$\Delta = 2\gamma e^{-\chi a} \tag{6.2.24}$$

which can be satisfied only for $2\gamma > \Delta$, i.e., in the case of a gap of hybridization type.

We find a state mainly localized at the surface and decaying inside the crystal with decay length $1/\chi$. Its orbital composition is given by a symmetric combination of $\psi_s(z)$ and $\psi_p(z)$, i.e., it is a *hybrid* orbital pointing in the positive direction of z (out of the surface).

It is noteworthy that it is possible to transform the basis of s and p orbitals in a basis of *hybrid sp* orbitals by their symmetric and antisymmetric combination. The two orbitals have directional lobes pointing respectively toward $z > 0$ and $z < 0$. The *surface state*, if it exists, comprises orbitals of the first type, with a major contribution coming from the surface atom at $l = 0$ (*dangling bond*).

We note that the surface state exists only if the lower bulk energy state at the X point is made up of p orbitals with a large value of the charge density at the middle of the bond, i.e., if $2\gamma > \Delta$. If we create the surface at this point and if the lower edge of the gap corresponds to a bonding state, then a surface state of the *dangling bond* type appears in the gap. We obtain a result consistent with that obtained in Section 6.2.2 using plane waves.

The energy of this state is zero, corresponding to $(E_p + E_s)/2$, the mean energy values calculated for the hybrid orbital. It is clear from equation (6.2.24) that this state decays rapidly inside the chain if 2γ is much larger than Δ.

The choice of a less symmetrical model with $\gamma_{ss} \neq \gamma_{pp} \neq \gamma_{sp}$ would give less simple equations, but essentially the same physical picture.

6.2.5. Classification of Surface States

By using two complementary approaches, we have seen that states localized at the terminating region of the linear chain with an exponential decay inside the bulk can arise under particular conditions. In both models the crystal is terminated abruptly and, roughly speaking, the condition of existence of surface states in the gap is linked to the possibility of accommodating electronic charge density in the region of the broken bond.

In both models the potential experienced by the electron is repeated periodically up to the termination and there is a sudden discontinuity at the surface. Reality should present a smoother profile. The potential at the last atomic sites is not necessarily the continuation of the perfect bulk potential, and in the external region it should smoothly reach the vacuum level. Additionally, the equilibrium position of the last atoms in the chain can change with a local relaxation of the interatomic distance.

These effects can be introduced in the tight-binding model by changing the values of the intra-atomic parameters and of the hopping integrals in the last atoms of the chain. Such changes can give rise to states localized at the surface and located in the energy near the edges of and inside the bandgap, also in the absence of Shockley states. These surface states are commonly classified as Tamm states.[3]

Similar states, which originate from the border of the band, are encountered in the study of ionic crystals. The 1D ionic crystal is a sequence of different atoms carrying different orbitals with $\Delta > 2\gamma$. The two bands are separated by a wide gap, not of hybridization type. Bands are *uncrossed*, but the last atom of the chain has a different environment with respect to the atoms of the same type inside the bulk. We can consider an *sp* chain like the one considered in Section 6.2.4, but with different sites for *s* and *p* orbitals. The value of Δ is reduced in the last two planes and a possible consequence is the presence of a state immediately above the valence band (1st band), if the termination is at an anion site, or immediately below the conduction (2nd band), if the terminating atom is the cation.[4] Both kinds of states can be found in a real surface where both atomic species are present. Their presence reduces the gap at the surface. Also, in the case of partially ionic semiconductors, the states arising in the hybridization gap can have different energy location as a function of the atom at the surface which mainly contributes to the state.

A review of one-dimensional models for the study of surface states is given by Davisson and Levine.[5]

6.3. Surfaces in Three Dimensions

6.3.1. Surface Periodicity: Ideal, Relaxed, and Reconstructed Surfaces

If a surface terminating a 3D crystal retains, up to the outermost plane of atoms, the geometrical arrangement of the corresponding bulk plane, the surface is named *ideal*. The surface plane has in two dimensions (2D) the periodicity of the bulk plane and the location of the atoms in the 2D unit cell is not changed with respect to the bulk planes.

If this is not the case, there are different kinds of deviation from ideality:

1. A *uniform displacement* of the atom of an outerplane inward or outward with respect to the bulk equilibrium position. This *relaxation* does not change the surface periodicity and the symmetry of the *ideal* configuration.
2. *In-plane* shift of the atomic positions and/or *different normal displacements* of the atoms inside the unit cell. The periodicity of the ideal surface is not lost, but symmetry can be changed.
3. Different shifts in the positions of atoms in neighboring cells or ordered sequences of defects and vacancies, with a sufficient long-range order to show a periodic arrangement, creating a new larger unit cell, i.e., a *reconstruction*.
4. Geometrical *disorder* with loss of the 2D periodicity.

The presence of *relaxation* and *reconstruction* is a common feature in most semiconductor surfaces, and we will see examples in the following paragraphs. Additionally, the *adsorption* (physisorption or chemisorption) of atoms of different species can give rise to new ordered structures, creating or modifying reconstructions of different types, or it can destroy the existing reconstruction of the clean surface, restoring the ideal geometry.

There are five types of 2D Bravais lattices: square, rectangular, centered rectangular, hexagonal, and oblique.[6] If the basis vectors of the direct mesh are

$$\mathbf{a}_1 = (a_{1x}, a_{1y}) \quad \text{and} \quad \mathbf{a}_2 = (a_{2x}, a_{2y}) \tag{6.3.1}$$

the unit cell in the reciprocal mesh is defined by the vectors

$$\mathbf{b}_1 = \frac{2\pi(a_{2y}, -a_{2x})}{a_{1x}a_{2y} - a_{2x}a_{1y}} \quad \text{and} \quad \mathbf{b}_2 = \frac{2\pi(-a_{1y}, a_{1x})}{a_{1x}a_{2y} - a_{2x}a_{1y}} \tag{6.3.2}$$

If a *reconstruction* is present, then the new basis vectors $\mathbf{a}_1^*$ and $\mathbf{a}_2^*$ describe the unit cell. In the most common notation the set of indices identifying the surface is followed by the expression $(m \times n)$, where $m = a_1^*/a_1$ and $n = a_2^*/a_2$; m and n are not necessarily integers or rational numbers, as the new unit cell can be rotated and can also correspond to a different mesh. In these cases additional information, like the value of the rotation angle or the letter p (primitive) or c (centered), is written explicitly.

The main experimental technique used in the study of surface structure is Low-Energy Electron Diffraction (LEED).[6] The surface periodicity and the presence of a reconstruction are observed directly via the diffraction pattern, which gives an image of the reciprocal mesh. The size and shape of the spots contain information about the extension of domains and the presence of surface steps or defects.

The atomic position inside the unit cell and the relaxation can be studied through the intensity profiles of the diffracted beams, i.e., by plotting the measured intensity of each diffraction spot as a function of the energy of the incoming electron. The information about the atomic positions is obtained by comparing the experimental LEED profiles with those obtained in a theoretical simulation of the electron diffraction in the crystal, where the atomic positions are the input data. Such computations are rather heavy, so only a limited number of atomic configurations can be compared with experiments, and some preliminary guesswork is necessary with respect to the nature of the reconstruction.

Other techniques employed to investigate the structural properties of semiconductors are angle-resolved light-ion backscattering,[7] diffraction of neutral atomic (He) beams,[8] and scanning tunnel microscopy.[9]

The first method uses beams of H^+ or He^{2+} ions with energies between one hundred and several hundred keV. The ions are scattered by the nuclei of the crystal following the dynamics of Rutherford scattering and also lose energy along straight trajectories through interaction with the electrons. By choosing the direction of the incoming beam and detecting the number of ions as a function of the energy and the outcoming directions of ions diffused backward, information on the composition and atomic displacements can be obtained in the surface layer and in the underlying atomic planes. With the choice of particular conditions of channeling and blocking geometry and with the help of computer simulations, it is possible to test the validity of a model of the surface structure.[7] This method also works for studying the geometry of interfaces buried some monolayers below the external surface.

The last two methods (atom diffraction and scanning tunneling microscopy) sample the corrugation of the last atomic plane and, in particular, of the external region fo the tails of electron density. As in the case of LEED, the interpretation of the data in order to obtain structural information is linked to a theoretical description which involves knowledge of the electronic structure (potential and charge density).

6.3.2. Observation of the Electronic Structure of Surfaces

It was pointed out in Section 6.2.5 that in the case of simple linear models, the electronic structure of the surface can depend strongly on the atomic geometry. The relaxation or reconstruction can very much change the electronic structure with respect to that of the ideally terminated crystal.

Among experimental techniques which allow one to inspect directly the band structure in the 2D BZ, we mention first *photoelectron spectroscopy*, especially with the use of synchrotron radiation. By choosing a photon energy so as to produce photoelectrons in the energy range where the escape depth has a minimum value (a few Å), it is possible to obtain a picture of the density of states at the last atomic planes, by considering the energy distribution curve (EDC). As bulk and surface states contribute to it, the surface features can be identified by removing or modifying them through the absorption of different atoms at the surface.

The *photoemission* with *angular resolution* allows one to separate the two contributions and derive information about the **k**-dispersion of the bands. One can distinguish the direct optical transition between filled and empty states in the bulk preceding the emission of the electron from the photoexcitations from surface filled states directly into the vacuum region. This can be done by recording, for a given direction of electron emission, the EDCs of the electrons at various photon energies. By plotting each EDC taken at each photon energy $\hbar\omega$ as a function of the initial energy E_i (the final energy minus $\hbar\omega$), one sees a discrete set of peaks, coming from bulk and surface transitions. The former connect states with the same 3D **k**-vector, thus the initial energy $E_i(\mathbf{k})$ varies with $\hbar\omega$ to fulfil the rule $E_f^{(n')}(\mathbf{k}) - E_i^{(n)}(\mathbf{k}) = \hbar\omega$ for a couple of bands in the bulk. The latter involves photoemission processes conserving only $\mathbf{k}_\parallel$, the 2D wave vector, between surface states and the continuum of states of the vacuum. For instance, let us consider photoelectrons emitted perpendicularly to the surface (normal photoemission). In the EDC one sees the peaks due to the bulk transitions, which move with $\hbar\omega$ due to the dispersion in k_z of initial and final state bands, and the peaks due to the surface states remaining at fixed E_i values. Away from the normal the analysis is more complex, but it is possible to map the bands for both the bulk[10] and the surface[11,12] with angular resolved photoemission. In the cited references, examples of band mapping are given for GaAs.

Empty surface bands can be studied by *inverse photoemission* (or Bremsstrahlung Isochromat Spectroscopy), detecting the energy of the photon emitted when an electron of an external beam of given E and $\mathbf{k}$ falls into an empty conduction surface band.

Additional means of studying surface electronic structure are *optical absorption* and *reflectivity* experiments, with the choice of appropriate conditions to enhance the contribution of transition involving surface states. This can be done in the infrared region, where it is possible to see transitions between surface bands, below the edge of the bulk bandgap.[13-15] In the ultraviolet range it is possible to study excitation from localized core levels and surface empty bands.[16]

A natural complement to optical spectroscopy is *electron energy loss spectroscopy* which, despite some complications in the interpretation of the data, turns out to be a surface-sensitive probe, particularly in the more recent version of high-resolution energy loss spectroscopy with low energy of the incoming beam.[17-19]

Scanning tunneling microscopy applied to semiconductors, by introducing a potential difference between the tip and the sample, allows one to have a spatial map of the wave function at different energies for both empty and filled states.[20,21]

The above list of experimental methods with which to study the surface electronic structure is far from complete. These and other methods will be recalled when discussing some specific cases in Section 6.5. We note meanwhile that the spectroscopies for the study of electronic structure are also an indirect test of the surface atomic structure, if the spectroscopic data are compared with the outcome of a computation of the surface band structure obtained with a geometrical model of the surface atomic structure. These theoretical calculations can be performed in semiempirical or *ab-initio* schemes. The second approach is more complex and can handle only the simplest cases. The first-principles methods, however, can attain a more ambitious goal: to obtain the total energies of the system at different geometrical configurations of the surface atoms and compare values in order to foresee the most stable structure.

In the following section we present general features of the surface band structure, starting from a comparison of bulk and surface band structure and continuing with special reference to empirical tight-binding methods. In Section 6.4 a first-principles calculation will be presented.

6.3.3. Remapping of the Bulk States: The Projected Bulk Band Structure

In order to identify where true surface states can exist, we must consider the values of $\mathbf{k}_\parallel$ where energy gaps in the bulk states distribution are present, $\mathbf{k}_\parallel = (k_x, k_y)$ being a vector of the two-dimensional Brillouin zone (2D BZ). The gaps are obtained by remapping the three-dimensional Brillouin zone (3D BZ) to a prismatic shape with the 2D BZ as its basis and k_z varying from $-\pi/d$ to π/d, where d is the bulk interplanar distance.

This construction can easily be carried out for the (001), (110), and (111) surfaces of the diamond and zinc-blende structures. In Figure 6.3, the 3D BZ for the face-centered cubic lattice is shown together with its prismatic rearrangement for the (001) surface. The 2D BZ is shown with high symmetry points and directions. The sequence of planes for this surface is of the type $AA'BB'AA'\ldots$, with atoms distributed at each plane on a square mesh. Two neighboring planes contain atoms of two different sublattices.

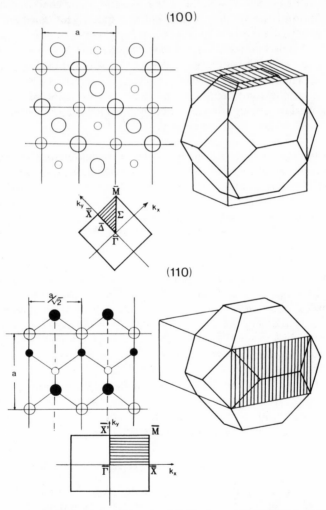

Figure 6.3. Atomic meshes and two-dimensional Brillouin zones (2D BZ) for ideal (100) and (110) surfaces of diamond and zinc-blende structures, and comparison of the 2D BZ with the 3D BZ of the fcc lattice. In both cases the 3D BZ is rebuilt into a prismatic shape according to the periodicity of the surface. The irreducible parts of the 2D BZ (hatched regions) and the labels of the high-symmetry points are shown.

The (110) surface, which is the cleavage surface of ionic semiconductors with zinc-blende structure, contains two atoms of different sublattices on the same plane in the ideal configuration. In Figure 6.3 both the direct mesh and the rectangular 2D BZ are shown; the usual form of the 3D BZ is also drawn. It should be rebuilt in a prismatic shape, to have the shaded area (2D BZ) as its basis.

At every $k_\parallel$ point of the 2D BZ one must plot the energy bands as a function of k_z and find the energy segments of existing bulk states. In symmetry directions, where eigenstates belong to different representations of the group of the $k_\parallel$ vector, one has to separate the energy intervals corresponding to states of different symmetries. In this way we obtain the *projected bulk band structure* (PBBS) associated with an ideal structure. It contains continua of bulk states separated by gaps, which do not extend necessarily all over the 2D BZ. As Figure 6.4 indicates that for the (110) surface of a III–V compound, in particular GaP(110), one obtains three kinds of gaps:

1. The gap between the valence and conduction bands, extending all over the BZ, having different width at different $k_\parallel$.
2. The gap between the lower part of the valence band (mainly the anion-derived s-band) and the upper part of the valence band; it extends from about -10 eV to about -6 eV. This gap closes at $\bar{\Gamma}$ in the diamond structure.
3. A lens or *relative gap* extending only on the external part of the 2D BZ some eV below E_v.

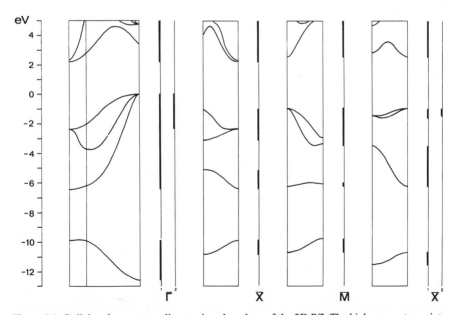

Figure 6.4. Bulk bands corresponding to given $k_\parallel$ values of the 2D BZ. The high-symmetry points of the GaP(110) surface are considered. The continua of states obtained from the bulk band dispersion along k_z are shown. At $\bar{\Gamma}$ and $\bar{X}'$ they are separated according to the different symmetries.

A further example across the high-symmetry directions of the 2D BZ is found in Figure 6.5, looking at the shaded area representing the continua of the bulk state for the case of GaSb(110).

All these gaps can accommodate bands of surface states, possibly dispersive as a function of $k_\parallel$. Usually, when a band of surface states crosses the border of a gap or a lens, it continues inside the bulk-state region as a *surface resonance*. A surface state of given symmetry can be degenerate with the bulk continuum of a different symmetry. This can occur only at high symmetry points or directions of the 2D BZ; when $k_\parallel$ becomes far from them, the surface state becomes again a surface resonance.

6.3.4. Finite-Slab Calculations. Surface Bands and Local Density of States

In order to calculate the electronic structure for a semi-infinite crystal, one must generalize the simple 1D model of Section 6.2 to a realistic 3D description. The $z = 0$ plane divides the region ($z < 0$) where the bulk potential is unperturbed, from a *selvage* region containing the last atomic planes and the vacuum ($z > 0$), where the potential can differ from the bulk and change continuously to reach the vacuum value.

For the infinite crystal, the one-electron problem is considered as completely solved: $E_n(\mathbf{k})$ and $\psi_{n\mathbf{k}}(\mathbf{r})$ are known across the whole 3D BZ, including the extension to complex values of k_z.

At every value of energy E, one has to match the external solutions ($z > 0$) with the internal ones at the matching plane. The 3D $\mathbf{k}$ vector is not a good label

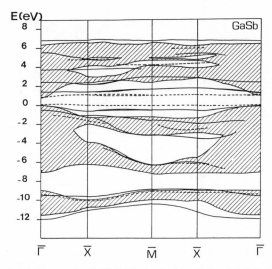

Figure 6.5. Surface band structure for GaSb(110) obtained in a finite-slab tight-binding calculation. The shaded areas correspond to the projected bulk band structure (PBBS). The dashed lines are the main surface bands in the *ideal* configuration of the surface. The full lines are the main surface bands in the *relaxed* geometry.

for the state, but $\mathbf{k}_\parallel = (k_x, k_y)$ is a good label and one must mix all the solutions with the same energy E corresponding to different k_z, including also the decaying solutions of the type $k_z = \kappa - i\chi$. Thus the bulk calculation must be extended to the complex values of k_z in order to include all bands of evanescent states, which branch out of maxima and minima of the real-$\mathbf{k}$ band structure. The internal solution can be a mixture of different (periodic and evanescent) states or, in the case of a gap, is composed of evanescent states only.

At a given $\mathbf{k}_\parallel$ one has

$$\psi_{\text{in}}(\mathbf{r}, \mathbf{k}_\parallel, E) = \sum_i A_i \psi(\mathbf{r}, \mathbf{k}_\parallel, k_z^i(E)), \qquad z < 0 \qquad (6.3.3)$$

and

$$\psi_{\text{out}}(\mathbf{r}, \mathbf{k}_\parallel, E) = \sum_n B_n \chi_n(\mathbf{r}, \mathbf{k}_\parallel, E), \qquad z > 0 \qquad (6.3.4)$$

A solution exists only if matching of the two functions and their derivatives with respect to z is possible at every point of the plane $z = 0$.

Among actual calculations of the electronic structure of surfaces, the method based on matching the wavefunctions and on calculation of the complex-$\mathbf{k}$ band structure is not widely used because of its computational difficulties, with the exception of approximate versions.[22] Another method, which allows one to work with a semi-infinite crystal, is based on matching the one-electron Green function.[23] This can avoid the need for evaluating the complex-$\mathbf{k}$ band structure. However, matching methods present additional difficulties if one wishes to undertake a *self-consistent* calculation, where the potential seen by an electron in the surface region is consistent with the electronic charge density distribution of the electrons in the occupied states.

In the surface calculation these methods have been overcome by the simpler use of *finite slabs*. Instead of describing a semi-infinite crystal one considers a finite number of atomic planes. The internal planes, preserving the geometry and the interplanar distance of the volume, represent the bulk while the outer planes are two equivalent surfaces where the atomic positions can be displaced in order to describe the surface relaxation or reconstruction. Along the planes the system is periodical; the unit cell size of the system has a total number of atoms given by the number of atoms in the 2D cell in the surface plane times the number of planes in the slab.

The slab must be sufficiently thick to avoid interaction between the two surfaces and to reproduce the bulk continua (represented here by a set of discrete bands extending across the 2D BZ) as well as the surface bands. The comparison between these bands and the continua of the PBBS (obtained through a bulk calculation in the rebuilt 3D BZ) is a tool to check the outcome of the calculation.

In Figure 6.5 the surface bands of states mainly localized at the surface and located inside the bandgaps are compared with the PBBS. The surface is the cleavage (110) surface of GaSb and two different calculations have been performed for a slab of 13 atomic planes, assuming an ideal termination of the surface (dashed lines) or a relaxed termination (full line). The details of the method will be discussed later.

Another source of information about the difference between surface and bulk electronic structures is obtained by looking at the *local density of states* (LDS). The electron density of states $n(E)$ can be resolved in real space by defining

$$n(E, \mathbf{r}) = \sum_{j}^{(occ)} \sum_{\mathbf{k}_\parallel} |\Psi_{\mathbf{k}_\parallel}^{(j)}(\mathbf{r})|^2 \delta(E - E_j) \tag{6.3.5}$$

Integration over $\mathbf{r}$ in the whole slab yields the total density of states $n(E)$. If the number of planes is increased up to infinity, $n(E)$ becomes the bulk density of states; if this number remains finite, $n(E)$ mixes bulk and surface features. By integrating on a plane normal to z in the surface unit cell area A_0 we obtain

$$n(E, z) = \sum_{j}^{(occ)} \sum_{\mathbf{k}_\parallel} \int_{A_0} d\mathbf{x}_\parallel |\Psi_{\mathbf{k}_\parallel}^{(j)}(\mathbf{x}_\parallel, z)|^2 \delta(E - E_j) \tag{6.3.6}$$

The contribution of different atomic planes can be identified by expressing, for instance, the *mth plane local density of states* as

$$n(E, m) = \int_{(z_m + z_{m-1})/2}^{(z_m + z_{m+1})/2} dz \, n(E, z) \tag{6.3.7}$$

and seeing how the LDS varies on passing from the outer plane to the inner planes. The central planes of the slab will have a function $n(E, m)$ closely resembling the bulk $n(E)$, while strong differences in shape will be found only at the first and second planes. Figure 6.6 shows LDS for the first, second, and sixth atomic planes for GaP(110). The latter closely resembles the bulk $n(E)$. It is now possible to point out the general features of LDS at the surface:

1. A reduction in the average width of the bands due to the reduced atomic coordination number at the surface plane.
2. Changes at the band edges. If a localized band exists all over the 2D BZ, it can originate step-like or logarithmic edges in the density of states because of the reduced dimensionality. Due to coupling with the substrate, the system is not strictly two-dimensional and this effect is smoothed in reality. However, sharp edges can be present.
3. New peaks and structures due to *surface states* and *resonances*; some of them can be localized at the second planes (*backbonds*).
4. A reduction of LDS in the planes below the surface with respect to the bulk density of states at the energies where resonances appear at the surface. This is due to orthogonality of the continuum states to the surface resonances.

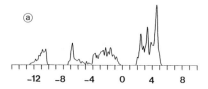

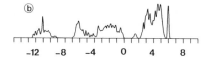

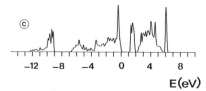

E(eV)

Figure 6.6. Local densities of states at the sixth (a), second (b), and first (c) atomic planes for GaP(110), obtained using the semiempirical tight-binding model. The surface is relaxed, according to the bond-rotation model, with maximum tilt angle $\theta = 34.8°$.

Study of the features of LDS at the first atomic planes has been a common tool for interpreting angular integrated photoemission spectroscopy. If the photon energy is chosen to minimize the escape depth of the electrons, the energy distribution curves of photoemitted electrons can be compared directly with the LDS of the first atomic planes.

6.3.5. Surface States in the Gap of Semiconductors, Band Bending, and Fermi-Level Pinning

The behavior of the LDS at the outer planes is particularly crucial in determining the surface properties of semiconductors. From a slab calculation, by filling the total density of states with the appropriate number of electrons per unit cell, we can fix the energy position of the last occupied state. The presence of surface bands in the gap between the bulk valence and conduction bands can give rise to two interesting cases:

1. If two nonoverlapping surface bands are present, their edges can be located inside the bulk gap producing a *surface gap* smaller than the bulk one.

2. A band of surface states not fully occupied can overlap the bulk gap entirely or partially, the Fermi-level position being fixed inside the gap.

The first possibility would occur for the cleavage surface (110) of partially ionic semiconductors if the surface plane had ideal atomic positions and no relaxation effects were present. In such a case of *ideal* termination of the surface, two bands (one completely occupied and the other completely empty) would appear at the two sides of the gap, partially overlapping it. The minimum gap between the two bands would be narrower than the bulk one.

As a consequence, the Fermi level can be located only inside this small segment, its position being determined by the defect levels, the doping, or any other charge transfer effect at the surface. At the edges of this surface gap, the large LDS due to the surface bands would strongly pin the position of the Fermi level.

It is meanwhile noteworthy that the two bands are related to the existence of two broken bonds per cell with wave functions of these two bands being composed mainly of the hybrid dangling-bond orbitals of the anion and cation, respectively. Putting two electrons on the anion dangling-bond band (the lower filled band), the other remains empty.

The other case arises if a single bond is cut in the surface unit cell; this would be the case of Si(111) in the ideal termination of the crystal with no reconstruction. This surface would comprise a hexagonal pattern of dangling bonds. A single Schockley-type dangling-bond band would be present in the gap and it would be half-filled, with the Fermi level in a region of high LDS, initiating a metallic electronic structure at the surface and a well-defined Fermi line in the 2D BZ. In this case the position of the Fermi level at the surface would be completely pinned in a given position of the bulk gap by the large density of states.

These two examples are rather academic. In the case of III–V compounds the relaxation of the surface, due to a shift of the anion outward and of the cation inward, pulls the surface states out of the gap in almost all compounds of the series. In the case of Si(111) a strong rearrangement of the atoms at the surface, giving rise to a (2×1) reconstruction, opens a gap between two surface bands, which is less than the bulk value (1.17 eV), optical measurement with surface sensitivity giving a value of 0.46 eV.[13]

In any case, even if the position of E_F is not completely pinned by the surface states, the existence of the surface bands can reduce the interval where E_F can vary inside the gap. As the position of E_F with respect to the band edges in the bulk depends on the bulk doping, it can differ from the position at the surface. In order to match the Fermi levels, an electric field must arise between the surface and the bulk, through a charge transfer between the surface and the impurity or defect level in the bulk. If the Fermi level is pinned, the charge transfer leaves the E_F position practically fixed, and if the bulk is not heavily doped, the field varies over a length which is much higher than the interplanar distance. We can then imagine that the band edges defining the bulk gap are rigidly shifted together as a function of the distance from the surface, giving rise to *band bending*. However, this happens over a wide distance deep inside the crystal, on a macroscopic scale (hundreds or thousands of Å); in the planes just below the surface, like those described in a slab calculation, the bands are practically flat, and this is so also at those planes which already possess the electronic structure of the bulk. The band bending region is much wider than that described in the slab.

6.3.6. The Empirical Tight-Binding Method

An application of the concepts introduced earlier can be illustrated by the empirical tight-binding calculations for finite slabs. This method was used extensively at the end of the seventies to study the electronic structure of semiconductor and transition metal surfaces, also in the case of relaxation and simple reconstructions. It is a natural extension to the 3D case of the simple model fo Section 6.2.4. The method is now reviewed with application to the surface of III–V compounds.

The eigenfunctions of the Schroedinger equations can be expressed as a combination of the orbitals $\phi_{\alpha\sigma}(\mathbf{r})$ centered at atomic sites, with α labeling the orbital (i.e., in our case four orbitals per atom of s, p_x, p_y, p_z type are sufficient) and σ the atom type ($\sigma = 1$ or 2 in binary compounds):

$$\Psi_{\mathbf{k}_\parallel}^{(n)}(\mathbf{r}) = \frac{1}{\sqrt{N}} \sum_{\sigma=1}^{2} \sum_{\alpha=1}^{4} \sum_{m=1}^{M} c_{\alpha\sigma m}^{(n)}(\mathbf{k}_\parallel) \sum_{\mathbf{l}} e^{i\mathbf{k}_\parallel \cdot (\mathbf{R}_\mathbf{l} + \mathbf{R}_{m\sigma})} \phi_{\alpha\sigma}(\mathbf{r} - \mathbf{R}_\mathbf{l} - \mathbf{R}_{m\sigma}) \quad (6.3.8)$$

where $\mathbf{R}_\mathbf{l}$ are the vectors of the 2D direct lattice and the 3D vectors $\mathbf{R}_{m\sigma}$ give the atomic position inside the unit cell of the slab at each plane m. As in the Slater–Koster method we do not need the explicit expression for the radial part of the orbital and we assume orthogonality also between orbitals on different sites. The coefficients $c_{\alpha\sigma m}^{(n)}(\mathbf{k}_\parallel)$ are obtained by solving the secular problem

$$\sum_{\sigma'=1}^{2} \sum_{\alpha'=1}^{4} \sum_{m'=1}^{M} \{H_{\sigma\alpha m,\sigma'\alpha'm'}(\mathbf{k}_\parallel) - E^{(n)}(\mathbf{k}_\parallel)\delta_{\sigma\sigma'}\delta_{\alpha\alpha'}\delta_{mm'}\}c_{\alpha'\sigma'm'}^{(n)}(\mathbf{k}_\parallel) = 0 \quad (6.3.9)$$

where the matrix element of the Hamiltonian is

$$H_{\sigma\alpha m,\sigma'\alpha'm'}(\mathbf{k}_\parallel) = \frac{1}{N} \sum_{\mathbf{l}} e^{i\mathbf{k}_\parallel \cdot (\mathbf{R}_\mathbf{l} + \mathbf{R}_{m\sigma} - \mathbf{R}_{m'\sigma'})} \int \phi_{\alpha\sigma}^*(\mathbf{r} - \mathbf{R}_\mathbf{l} - \mathbf{R}_{m\sigma}) H\phi_{\alpha'\sigma'}(\mathbf{r} - \mathbf{R}_{m'\sigma'})\, d\mathbf{r}$$

$$(6.3.10)$$

The integral can be expressed in the two-center approximation, as a function of the Koster–Slater parameters. They can include intra-atomic terms and hopping integrals between first and second neighbors, which are fitted to the bulk band structure of the perfect solids. A good fit of the band structure of the bulk (and consequently of the PBBS of the surface under consideration) can be obtained in a second-neighbor model; this model fitted to a bulk band calculation contains up to 17 parameters, in the case of binary semiconductor compounds, that can be *transferred* to the surface calculation, by introducing them into equation (6.3.10). If the geometry of the surface is modified and some change in the bond distances occurs, the interaction parameters can be varied according to a power law.[24]

The band structure can be obtained along the high symmetry direction of the 2D BZ, by finding the eigenvalues of equation (6.3.9) for a slab comprising a given number of planes. It is possible to identify surface states by looking at the energy location of the state at a given $\mathbf{k}_\parallel$ and comparing it to the bulk distribution of states in PBBS, and by considering the local density of states per plane

$$n(E, m) = \sum_{\sigma=1}^{2} \sum_{\alpha=1}^{4} \sum_n \sum_{\mathbf{k}_\parallel} |c_{\alpha\sigma m}^{(n)}|^2 \delta(E^{(n)}(\mathbf{k}_\parallel) - E) \qquad (6.3.11)$$

where the sum over $\mathbf{k}_\parallel$ is extended to a large mesh of points inside the irreducible part of the 2D BZ. The total orbital compositions and electronic charges at each atomic site can be obtained by similar expressions (avoiding sums over σ and α, integrating over E up to the last occupied state). Thus quantities $n(m, \alpha, \sigma)$ and $n(m, \sigma)$ can differ at the surface layers with respect to the bulk, giving charge transfers at the surface and change of orbital composition. This effect can alter the values of the potential at the surface and can be accounted for empirically by introducing shifts in the values of the intra-atomic parameters (a kind of effective atomic levels), similarly to changes of atomic ionization potentials with variation in configuration and charge. The calculation can be iterated to obtain consistency between these changes and the effective parameters in the Hamiltonian. This procedure has been adopted in calculations[25] and its effect is to reduce slightly the charge shifts at the surface obtained in the nonself-consistent scheme and to account for some additional state of Tamm type, due to the variation in the *atomic* potentials at the surface.

6.3.7. Application to the Cleavage Surfaces of III–V Compounds

The empirical tight-binding method has been widely applied to the study of cleavage surfaces of compound semiconductors of the III-V family. Today the (110) surface of these compounds is probably the best understood system, as far as the connection between atomic geometry and electronic structure is concerned, and the evolution of its knowledge fits well the needs of a tutorial presentation. The subject was not clear in the early seventies. At that time many experimental results on GaAs(110) seemed to give evidence of two bands of surface states partially overlapping the gap. Measurements of contact potential differences indicated the existence of pinning of the Fermi level at the surface in many samples, and the optical excitation of 3d to the empty states with synchrotron light seemed to confirm the picture of an empty surface band in the gap. These facts were consistent with the outcome of calculation based on the semiempirical tight-binding scheme, using a finite slab with ideal geometry of the surface.

A more accurate set of experiments showed that the states pinning the Fermi level were extrinsic and that no surface state is present in the gap for GaAs(110) if defects are removed or the surface is prepared in ultrahigh vacuum and kept clean from contamination. The problem of 3d-surface state excitation energy can be solved, if one accounts for the large exciton binding energy. As in the case of many other III–V compounds, GaAs(110) appeared free of surface states in the gap.

At the same period of time an accurate comparison between calculated and experimental LEED intensity profiles gave proof of the relaxation of GaAs(110). A *rotational model*,[26] which produced the displacement of As outward and Ga inward at the first plane without changing the nearest-neighbor distances, was proposed. A single parameter (the rotation angle θ between the plane of the Ga–As–Ga triangle at the first layer with respect to the ideal position, varying from 0° to 34.8°) can describe the geometry. In Figure 6.7 a perspective view is given of the relaxed surface with $\theta = 20°$. Some years later a more general model of relaxation,[17] which works for all III–V compounds, allowing also small shifts in the positions of the second-layer anion and cation (and different sign with respect to the first layer), was suggested. In this model a small variation of the bond length is obtained at the surface (about 2%). The parameters are indicated in the figure. An accurate review is given by Kahn.[27]

Calculations[25,28] were performed for the *bond-rotation* model in order to study the change in the electronic structure with the pseudopotential method and empirical tight-binding scheme. The effect of the relaxation is to induce a shift of the two surface bands, up to the almost complete removal of the states from the gap, with a change in the LDS of the first layer. The As-derived low band was moved below and the corresponding peak in the density of states overlaps the upper region of the valence band where it is still clearly visible. The upper bands, mainly formed of Ga dangling bonds, moves up, also changing in shape. This can be understood in terms of a partial dehybridization of the orbitals with respect to sp_3. Ga has three bonds in a quasi-planar coordination which favors sp_2 hybridization, thus the dangling bond becomes almost pure p_z.

This relaxation mechanism is common to the whole family of III–V compounds and has the same effects on surface electronic bands. Figure 6.5 gives a comparison between the main surface bands for GaSb(110) in the ideal geometry and for the relaxed configuration. The calculations were performed in the empirical tight-binding scheme. Figure 6.8 presents a kind of summary of the results obtained ten years ago[29] in the tight-binding scheme for the compounds containing Ga and In. On a scale of energies referred to the vacuum level, the edges of the valence and conduction bulk bands are shown together with the surface state regions (hatched regions). Cross hatches indicate maxima in the surface LDS. The only compound of the series which still has a peak of empty surface states in the gap is GaP. In spite of the approximations of the method this fact is confirmed by pseudopotential calculations and is consistent with the experimental findings,[30] which show a pinning of the Fermi level a few tenths of eV below the conduction band edge.

The theoretical description of surface states, also in terms of the simple tight-binding model, can then account for the features of the electronic structure which depend on the atomic displacements occurring in the surface relaxation, as the spectroscopic data are in overall agreement with the calculated LDS. The calculation of the electronic structure for different atomic configurations should provide also the evaluation of the total energy as a function of the relaxation parameters. If it were possible to reduce reasonably the number of configurations to be compared, we should be able to select the stable configuration as that of

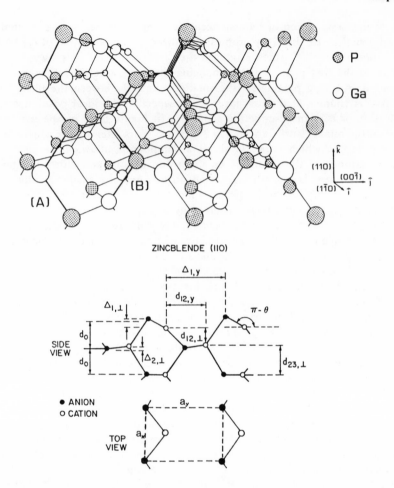

Figure 6.7. Views of the relaxed (110) surface of III–V compounds. The upper panel shows the geometrical arrangement of atoms in the $\theta = 20°$ bond rotation model (after Duke et al.[26]). The parameters describing a more general *two-layer* relaxation are shown in the second panel (after Kahn[27]).

minimum energy and then predict the surface relaxation. Although the empirical tight-binding scheme can be implemented[31] to provide information on total energy, this method is far from the reliability of a first-principles method. In order to achieve the most reliable results about the study of the minimum energy configuration and a first-principles determination of the atomic geometry of a relaxed or a reconstructed surface, it is necessary to calculate the electronic structure starting from an *ab-initio* description without parameters fitted to the bulk. In such a scheme it is possible, for instance, to obtain the total energy of the system as a function of the rotation angle θ in the case of III–V compounds, even if it involves

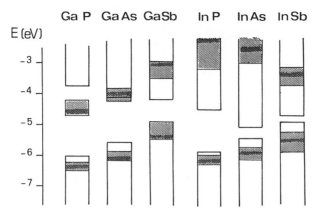

Figure 6.8. Energy level diagram of the surface states in the bandgap region for six III–V compounds, obtained in a tight-binding calculation. Surface states are indicated by hatching. Cross hatching shows the maxima in surface LDS.

very heavy computations.[32] This scheme is provided by the density functional theory in the local density approximation, where self-consistent slab calculations of band structure and total energy are possible.

6.4. First-Principles Description of the Surface Electronic Structure as a Function of the Atomic Configuration

6.4.1. Density Functional Theory and the Local Density Approximation

In the following we present the ingredients of a modern surface band structure calculation and consider the use of the self-consistent pseudopotential method in the framework of a density functional theory (DFT).[33] The electronic properties of the ground state of the many-electron system, composed of the valence electrons in the field of the ion cores frozen in a given configuration, can be described by this method and a set of *one-particle* equations can be derived[34] in order to obtain the particle density $n(\mathbf{r})$ and the total energy of the electron system.

The eigenvalues of these one-particle Kohn–Sham equations[34] are generally regarded (in analogy with the equations of the Hartree–Fock method) as the *quasi-particle* energies of the system. It has been pointed out in recent years that this approach is not completely adequate if the eigenvalues are seen as one-particle energies and we describe the excitation energies as differences between two of these eigenvalues corresponding to an empty and a filled state. This failure appears clearly in the case of a semiconductor, where the gap between filled and empty states turns out to be underestimated.[35,36]

However, the density functional theory is a powerful tool with which to describe ground state properties of the solid and, in particular, the electron charge density distribution, the total energy, the equilibrium structure and the lattice parameter, the formation energy of a defect in the bulk, as well as the energy changes induced by displacement of the atoms from their equilibrium positions. The eigenvalues of the one-electron equations provide a description of the band structure of a solid, even if some care must be taken in identifying them with the electron and hole *quasi-particle* energies.

By means of the slab method it is possible to treat the surface problem using the same scheme adopted in bulk calculations. In the *repeated* slab picture, an artificial periodicity is introduced in the direction normal to the surface by inter-calating a number of atomic planes (typically from 8 to 15) with a region with thickness a few interplanar distances (typically from 4 to 6) in order to describe the vacuum region. In this way the periodicity in the third dimension is restored and the methods used for calculation of bulk electronic structure, that work for compounds with many atoms per unit cell, are still useful. One has a very thin 3D BZ and a large number of basis functions must be employed to describe the eigenstates and the particle density, the only difficulties being of computational origin.

In the density functional theory the total energy is written as a functional of the electron density $n(\mathbf{r})$,

$$E[n] = \int d\mathbf{r} \, V(\mathbf{r})n(\mathbf{r}) + \frac{e^2}{2} \int d\mathbf{r} \int d\mathbf{r}' \frac{n(\mathbf{r})n(\mathbf{r}')}{|\mathbf{r} - \mathbf{r}'|} + T_0[n] + E_{xc}[n] \quad (6.4.1)$$

where $V(\mathbf{r})$ is the external potential experienced by the electrons due to the nuclei (in the case of an *all-electron* calculation) or to the ions (in a *pseudopotential* calculation). In the latter case $V(\mathbf{r})$ is a superposition of single ionic pseudopoten-tials which can be *nonlocal* operators. The quantity $T_0[n]$ is the kinetic energy of a system of noninteracting electrons with density $n(\mathbf{r})$ and $E_{xc}[n]$ is the exchange-correlation energy. This term is assumed to be a universal function of the density. The energy $E[n]$ achieves its minimum value for the ground state. To minimize it one can choose a form

$$n(\mathbf{r}) = \sum_i^{(occ)} |\psi_i(\mathbf{r})|^2 \quad (6.4.2)$$

The problem is then transformed into the solution of a system of N *one-particle* equations:

$$\left\{ -\frac{\hbar^2}{2m} \nabla^2 + V(\mathbf{r}) + e^2 \int d\mathbf{r}' \frac{n(\mathbf{r}')}{|\mathbf{r} - \mathbf{r}'|} + V_{xc}(\mathbf{r}) \right\} \psi_i(\mathbf{r}) = \varepsilon_i \psi_i(\mathbf{r}) \quad (6.4.3)$$

where the exchange-correlation potential $V_{xc}(\mathbf{r}) = \delta E_{xc}/\delta n$ is the functional deriva-tive of the exchange-correlation energy. As the last two terms in the curly brackets of equation (6.4.3) are functions of the density $n(\mathbf{r})$, the eigenvalues and eigenfunc-tions of equation (6.4.3) depend on the set of solutions via equation (6.4.2) and are determined through an *iterative* procedure, which evaluates at each step a total effective potential

$$V_{eff}(\mathbf{r}) = V(\mathbf{r}) + e^2 \int d\mathbf{r}' \frac{n(\mathbf{r}')}{|\mathbf{r} - \mathbf{r}'|} + V_{xc}(\mathbf{r}) \quad (6.4.4)$$

computed from the density that is constructed with the eigenfunctions. At each step of the procedure a new *output* potential is calculated and must be compared with the *input* potential in order to verify whether self-consistency is achieved.

The total energy of the system is given by

$$E = \sum_{i}^{(occ)} \varepsilon_1 - \frac{e^2}{2} \int d\mathbf{r} \int d\mathbf{r}' \frac{n(\mathbf{r})n(\mathbf{r}')}{|\mathbf{r} - \mathbf{r}'|} + E_{xc}[n(\mathbf{r})]$$

$$- \int d\mathbf{r} \, V_{xc}(\mathbf{r})n(\mathbf{r}) + E_{xc}[n(\mathbf{r})] + E_{\mathrm{I}} \tag{6.4.5}$$

where the sum is limited to the first N occupied states, taking into account spin degeneracy; E_{I} is the electrostatic energy of the ions (or of the nuclei in an all-electron calculation) and depends on the geometry of the surface, like the other terms.

The exchange-correlation potential is commonly assumed to be a local function of the density $n(\mathbf{r})$. This is the local density approximation (LDA), which is correct if the density varies slowly in space. The form of the local functional dependence of E_{xc} is chosen from the computed exchange-correlation energy density for a homogeneous electron system,

$$E_{xc}^{\mathrm{LDA}}[n] = \int d\mathbf{r} \, n(\mathbf{r})\varepsilon_{xc}[n(\mathbf{r})] \tag{6.4.6}$$

$$V_{xc}^{\mathrm{LDA}}(\mathbf{r}) = \varepsilon_{xc}[n(\mathbf{r})] + n(\mathbf{r}) \frac{d\varepsilon_{xc}}{dn} \tag{6.4.7}$$

The most often used form of $\varepsilon_{xc}[n(\mathbf{r})]$ has been given for different densities by Ceperley and Alder[37] through a quantum Monte Carlo calculation for an electron homogeneous system.

6.4.2. The Self-Consistent Potential in a Repeated-Slab Calculation

In a pseudopotential repeated slab calculation, with a plane wave basis set, the computation of the band structure is of the same difficulty as that of a bulk calculation with a large number of atoms in the unit cell. If L is the thickness of the repeated slab, the values of G_z are $2\pi n/L$. This G lattice is a dense mesh along z. A large number of G vectors must be included in the reciprocal space sphere used to construct the wave functions as a superposition of plane waves with wave vectors $\mathbf{k} + \mathbf{G}$. Typical cutoff values are $|\mathbf{k} + \mathbf{G}|^2 < 4$ or 5 Rydbergs for *soft-core* pseudopotentials with Fourier transforms which rapidly decay in G-space, or $|\mathbf{k} + \mathbf{G}|^2 < 10$ or 12 Rydbergs in the case of *hard-core* norm-conserving pseudopotentials.[38] In these two cases the wave functions must be obtained using approximately one thousand plane waves[28,39] or several thousand, respectively.

The sum (6.4.2) becomes, in the slab calculation,

$$n(\mathbf{r}) = \sum_{n,\mathbf{k}}^{(occ)} |\psi_{\mathbf{k}}^{(n)}(\mathbf{r})|^2 \tag{6.4.8}$$

where $\mathbf{k} = (\mathbf{k}_\parallel, k_z = 0)$ is limited to a set of special points inside the 2D BZ. This choice is made to optimize the evaluation of the density distribution $n(\mathbf{r})$ and of the total energy. In order to compute the band structure, one must use the final effective potential (6.4.4) obtained from the self-consistent procedure, and solve equation (6.4.3), getting $\varepsilon^{(n)}(\mathbf{k}_\parallel)$ and $\psi_{\mathbf{k}_\parallel}^{(n)}(\mathbf{r})$ along the high symmetry directions of the irreducible part of the 2D BZ. As an example, Figure 6.9 shows the bands, calculated by Manghi[40] for a slab of 9 atomic planes of GaP(110). The surface is relaxed. The bands can be obtained by joining the dots by a continuous line. The heavy lines are the bands of states localized at the surface or immediately below the first layer. Energies are referred to the valence band maximum E_v.

The consistency of the effective potential with the charge distribution in this case was obtained in 8 iterations. In slab calculations some care must be taken to avoid instabilities arising from the long-range behavior of the Hartree potential term. A technique of variable mixing of the input and output potentials at each iteration must be used in order to avoid such instabilities and to improve the speed in the last iterations.[41]

The total valence charge density for the relaxed GaP(110) surface is shown in Figure 6.10, together with the maps of $|\psi_{\mathbf{k}_\parallel}^{(n)}(\mathbf{r})|^2$ for selected surface states.

The density distributions are shown along $(1\bar{1}0)$ planes passing through the ionic sites along the chains shown in the figures. Two types of chain, ending with P or with Ga atoms, are present. The left edge of the panel corresponds to the

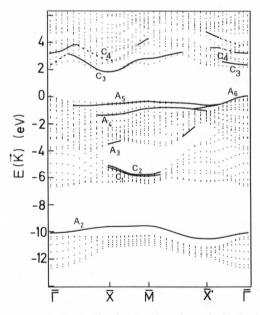

Figure 6.9. Band structure of the GaP(110) relaxed surface obtained with a self-consistent pseudopotential calculation using 9 atomic planes plus an empty region corresponding to 5 planes in the repeated-slab geometry. The labeling of the states is the same as in Figure 6.10.

center of the slab, while the right edge is at the center of the *vacuum* region. The height of the slab unit cell is twice the width of the panel.

It is seen that the total charge density distribution is not strongly affected by the presence of the surface, and only slightly by the presence of the relaxation. We must point out that the energy associated with the relaxation is just a few tenths of eV per surface atom and is much lower than the binding energy of the bulk, which depends on the pile up of the electronic charge along the bonds.

The surface states are labeled A (anion) or C (cation), depending on the atomic site where the wave function is mainly localized. The labeling is the same as in Figure 6.9. A_2 is a Tamm-type state arising at the lower edge of the ionic gap around 10 eV. It is localized at P sites of the first layer, and has mainly s character. The C_2 state appears in the lower part of the internal lens in the direction $\overline{MX'}$ of the 2D BZ. At $\overline{M}$ it is mainly localized on the first-layer Ga atoms, with an additional back-bond character toward the second plane. The C_1 state, not shown, is also localized in the second plane. Its dispersion can be seen in Figure 6.9 along $\overline{XM}$.

The anion dangling bond is denoted by A_5 and lies beyond E_v outside the bulk gap, but throughout almost the whole 2D BZ it is still in the relative gap, thus having the character of a true surface band. We must remember that the dispersion of this band, which depends of the relaxation of the surface, can be observed experimentally by means of photoemission spectroscopy with angular resolution.[11]

Finally, the last panel shows the particle distribution of the first empty surface band localized on the cation (C_3). In this and other pseudopotential calculations for GaP this state appears inside the gap only at $\overline{X}$. Among III–V (110) surfaces, this feature appears only in GaP(110). The possibility that an intrinsic surface state band can pin the Fermi level some tenths of eV below the conduction band (less than what is obtained in semiempirical tight binding) is thus confirmed in this compound.

Similar surface bands can be obtained in other III–V compounds. In the case of other Ga-containing compounds, the two surface bands are completely outside the bulk gap, but can be identified by surface-sensitive spectroscopies. In particular, for GaAs(110) the A_5 and C_3 states are outside the gap and their localization on the different sites at the surface is observed in the scanning images obtained at different voltages, with different signs, in tunneling microscopy.[21]

In this part of the discussion we have considered the one-electron band structure rather than the total energy obtained from the computed $n(\mathbf{r})$, which is the main goal of density functional calculations. It has already been pointed out that the gap is underestimated in a well-performed LDA calculation, although the total energy is described very well. If we are interested in the band structure, it is possible to adapt this scheme to reproduce values of the bulk gap closer to the experiments. One possibility is to replace equation (6.4.7) by the Slater form

$$V_{xc}(\mathbf{r}) = -\frac{\alpha 3 e^2}{2\pi}[3\pi^2 n(\mathbf{r})]^{1/3} \qquad (6.4.9)$$

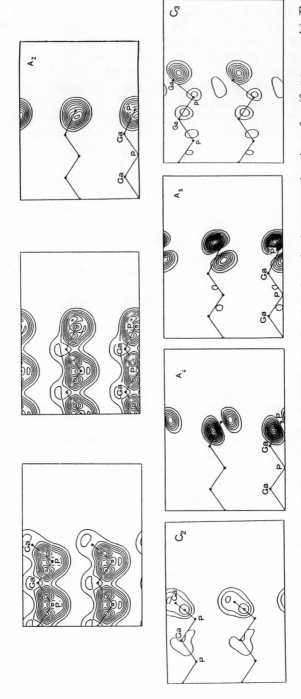

Figure 6.10. Valence charge density distribution for the GaP(110) relaxed surface shown in planes normal to the surface (first two panels). The atomic chains shown here, going from the bulk to the surface, are those indicated by (A) and (B) in Figure 6.7. The last five panels show the square modulus of the wave functions of the following states: A_2 at $\bar{M}$, C_2 at $\bar{M}$, A_4 at $\bar{X}$, A_5 at $\bar{X}$, C_3 at $\bar{X}$.

with α between 0.8 and 1. This choice, not justified rigorously in the density functional theory, reproduces the bulk excitation values more or less correctly, but gives an overestimate of the total energy. It is common in the literature to compute the band structure of filled and empty states with this method, in order to avoid more rigorous and complicated methods like the calculation of the *self-energy* corrections (see below).

If we are interested in total energy calculation, the problem of the gap evaluation is neglected and equation (6.4.7) is used. For example, a calculation for a slab of 9 atomic planes was performed by Pandey[32] for different relaxation angles θ in GaAs(110). Figure 6.11 gives the total energy per surface atom as a function of θ. It turns out that in the self-consistent calculation one obtains a curve having the minimum in approximately the same position as in a non-self-consistent case. In this latter case the energy value at the minimum is just double that in the former. The gain in energy of the relaxed surface is due to the lowering of the eigenvalues of the filled dangling bond band shifting downward in the relaxation. This shift lowers the sum of the energy values appearing in equation (6.4.5). The effect is partially compensated in the self-consistent calculation. The position of the minimum is at about $\theta = 24°$, which must be compared with the value $\theta = 27.2°$ given by LEED.[27] The ion crystallography[7] cannot handle the case of GaAs well, as the contribution of the two nuclei is not well resolved in the backscattering

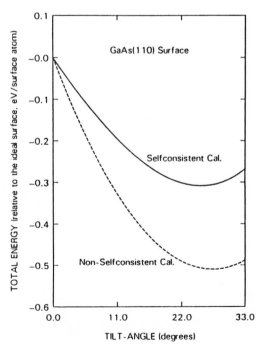

Figure 6.11. Total energy of the GaAs(110) surface as a function of relaxation tilt angle θ (after Pandey[32]).

spectra, because of the similar values of the two masses. In the most favorable case of GaSb(110) a best fit of the data is obtained[7] for $\theta = 29°$.

6.4.3. Limits of the Local Density Approximation and of the Density Functional Theory

To complete this short introduction to the calculation of the surface electronic structure, we mention briefly the possibility of improving the LDA calculations. A more appropriate description of E_{xc}, beyond LDA, would be a form of the type

$$E_{xc} = \int d\mathbf{r} \int d\mathbf{r}' \frac{n(\mathbf{r})n(\mathbf{r}')G(\mathbf{r},\mathbf{r}')}{|\mathbf{r}-\mathbf{r}'|} \tag{6.4.10}$$

where $G(\mathbf{r}, \mathbf{r}')$ is the two-particle correlation factor and $n(\mathbf{r}')G(\mathbf{r}, \mathbf{r}')$ is the exchange and correlation hole; this would introduce an intrinsic nonlocal dependence of $V_{xc}(\mathbf{r})$ on $n(\mathbf{r})$ is nonuniform electron systems.

Gunnarsson and Jones[42] proposed a model form for $G(\mathbf{r}, \mathbf{r}')$ which has the advantage of reproducing the correct behavior of the exchange and correlation potential in a system of electrons confined by a surface. Outside the surface, the effective potential must approach the vacuum level with the image form $1/4z$. For the case of a semiconductor having static dielectric constant ε, this behavior is

$$V_{xc}(z) \simeq -\frac{\varepsilon-1}{\varepsilon+1}\frac{e^2}{4z} \tag{6.4.11}$$

It can give rise to a new type of surface states, the *image* states, mainly localized in the region immediately outside the surface plane, if the wave function can match an internal decaying wave. The energy eigenvalues are located within 1 eV of the vacuum level, so that they are empty states and can be observed experimentally by inverse photoemission or tunneling spectroscopy.

The nonlocal approximation for V_{xc} is rather complicated to be used in actual calculations[43] and only recently has it been applied to surfaces.[44]

Another possible improvement of this scheme is to progress beyond the density functional theory. In describing the band structure we have assumed that equation (6.4.3) could give (through ε_i) the excitation energies of the system, i.e., the energy of an added particle (for empty states) and of a created hole (for filled states). This property is true if we write, instead of equation (6.4.3),

$$\left\{-\frac{\hbar^2}{2m}\nabla^2 + V(\mathbf{r}) + e^2 \int d\mathbf{r}' \frac{n(\mathbf{r}')}{|\mathbf{r}-\mathbf{r}'|}\right\}\psi_i(\mathbf{r}) + \int d\mathbf{r}' \Sigma(\mathbf{r},\mathbf{r}',E)\psi_i(\mathbf{r}') = \varepsilon_i\psi_i(\mathbf{r}) \tag{6.4.12}$$

where $\Sigma(\mathbf{r}, \mathbf{r}', E)$ is the *self-energy* and not the exchange-correlation potential. The eigenvalues ε_i are in principle complex: their real parts are here the quasi-particle energies while the imaginary parts are the inverse lifetimes. The evaluation of the real parts of ε_i can be done in the GW approximation;[35,36] starting from the wave functions obtained in LDA calculation, it requires rather heavy computational work. This method can be used in the bulk and gives good values for the gaps. An estimate of these effects can be derived by examining the gaps of Si and GaAs. Their LDA values, obtained by solving equation (6.4.3), are underestimated by about 50–60% in a fully converged calculation with a wide basis set; the use of a nonlocal functional for exchange and correlation reduces the discrepancy by only about 20%. The largest part of the remaining difference can be described[36] only with the aid of equation (6.4.12) in the GW approximation. In recent years both the nonlocal V_{xc} potential[44] and the GW approximation[45] have been applied also to surface problems. The future of the *ab-initio* surface approach seems to require both the density functional method to evaluate the total energy of the system in its ground state, and the self-energy calculation to obtain the quasi-particle band structure.

6.5. The Electronic States and the Structure of Semiconductor Surfaces

6.5.1. The Si(111) (2 × 1) Surface

The (111) surface of Si is created by cleaving the crystal with a cutting plane normal to the (111) direction. The surface can be maintained clean (without contamination) only in ultrahigh vacuum (about 10^{-10} Torr). The 2D mesh of the ideal surface would be hexagonal with lattice constant $a/\sqrt{2}$, where a is the bulk lattice parameters. In the ideal configuration each atom in the surface plane would be surrounded by six second neighbors, and (assuming that the cleave breaks one bond per atom) each surface atom is linked to three nearest-neighbor atoms in the second plane. Without relaxation or reconstruction the distance between the first and second plane would be $a/(4\sqrt{3})$ and between the second and third, $3a/(4\sqrt{3})$.

Actually, the simple unreconstructed surface is not observed and the LEED pattern shows a (2×1) reconstruction. Multiple reconstruction domains in the surface can be avoided by choosing appropriate conditions, e.g., by moving the cleaver along the $[\bar{1}, \bar{1}, 2]$ direction of the crystal or one of the other two equivalent directions. In this way it is possible to have a single domain reconstruction on the whole surface of the sample. In this case the LEED pattern shows extra spots as indicated in the lower panel of Figure 6.12 (smaller dots). This corresponds to a doubled periodicity along $\mathbf{a}_1 = a/2(0, 1, \bar{1})$ or, equivalently, to a new rectangular surface cell with new vectors $\mathbf{A}_1 = a/2(1, 1, \bar{2})$ and $\mathbf{A}_2 = a/2(\bar{1}, 1, 0)$. The new 2D BZ is obtained by folding the old one into a rectangular shape, as indicated in Figure 6.12.

A more comprehensive description of the properties of the (111) surfaces of diamond-structure semiconductors is given elsewhere;[46] we will summarize here the main points for Si. The 2 × 1 cleavage structure is metastable and, by heating at 500–700 K, one obtains a 7 × 7 pattern[47] with a unit cell 49 times greater than the ideal. This structure is more stable, thus the 2 × 1 structure cannot be restored by lowering the temperature. At 1150 K the 7 × 7 structure changes reversibly into a 1 × 1 structure.[48] The 1 × 1 pattern does not imply order and ideal periodicity

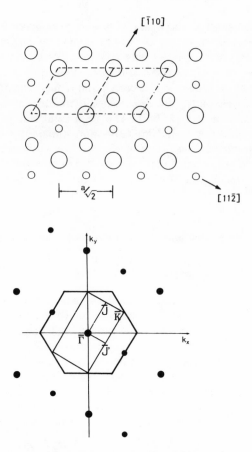

Figure 6.12. Top view of the ideal Si(111) surface and unit cell of the reconstructed (2×1) surface. The 2D BZ of ideal and reconstructed surfaces. The dots show schematically the single-domain LEED pattern with the extra spots (smaller dots) due to the reconstruction.

at the surface plane, but could be due to the order of the planes below the surface. The removal of the reconstruction and a 1×1 stable structure can be obtained in both 2×1 and 7×7 surfaces by deposition of different atomic species and adsorption also at submonolayer coverages.

The 7×7 structure will be treated in the next section. Let us consider first the 2×1 cleavage surface. It has been one of the most studied subjects in surface physics in the last twenty years. Different models have been proposed for the surface structure, as the interpretation of surface crystallographic methods (LEED, ion and atom scattering) requires comparison with complicated calculations based on structural models. At the same time, the investigation of electronic properties offered an indirect test of the structural models through comparison with surface band calculations.

Assuming the ideal termination fo the crystal, one expects a band of surface states mainly located inside the gap, due to dangling bonds. This band, approximately half-filled, would introduce a large LDS in the gap, strongly pinning the Fermi level. In the sixties, early measurements based on contact potential differences confirmed the presence of the pinning of the Fermi level. The effect was actually due to the existence of defect and impurity states arising from cleavage defects and contaminants.

The reconstruction of the surface, with displacements of the atoms from the bulk equilibrium positions, doubles the periodicity and folds the dangling-bond band into two bands in the new 2D BZ. Different models have been proposed for the reconstruction, based on different displacements of a couple of atoms at the first plane or on more complex changes involving the second plane too. Some of them are schematized in Figure 6.13. The displacements of the atoms open a gap between the two bands, creating a surface gap along the borders of the 2D BZ. If the two bands do not have a strong dispersion, they do not overlap and the lower band is completely filled while the higher is empty. The Fermi level is then pinned only at the edges of this surface absolute gap, if these edges lie inside the bulk bandgap.

The opening of the gap could be responsible for the gain in energy as in the case of III–V compound cleavage surfaces. This gap was observed by different methods in the late sixties, first by optical techniques based on the change of reflectivity observed in the infrared spectrum on passing from the freshly cleaved to the oxidized surface.[13] The oxidation suppresses surface states and inhibits optical transitions between them. To enhance surface sensitivity it is possible to use multiple internal reflections at the surface. A peak at about 0.45 eV was found for Si(111). In Figure 6.14 a new, recent version of that experiment is reported, where also the dependence of the differential reflectivity on polarization of the light is shown.[14]

Among the models proposed for the 2 × 1 reconstruction, for many years the *buckling model*[49] was considered a realistic explanation of the existence of the surface gap. It consists of alternate up and down shifts of atoms at the surface plane along the direction of the doubled periodicity. These displacements produce a charge transfer from one atom to the other in the new cell, creating a kind of ionic surface. Other models of pairing the atoms, through displacements *in the plane*, were also proposed,[50,51] with less or more drastic rearrangement of the atoms at the surfaces. Pandey pointed out that a gap with nonoverlapping bands cannot be obtained in a self-consistent LDA calculation, if the buckling model is assumed. In 1981 he proposed a new model[52] with a deep change of atomic positions and of the bonding geometry at the two outermost atomic planes at the surface (π-bonded chains). Two *zigzag* chains are obtained in this model at two different distances from the third atomic plane, which is left unchanged. They are directed along the $[\bar{1}, 1, 0]$ direction. This configuration can be attained by moving four atoms per cell in the two last planes along the $[\bar{1}, 1, 2]$ direction (the *cleave* direction to obtain a good single-domain 2 × 1 LEED pattern), followed by an appropriate upward or downward shift. A side view of this model is given in Figure 6.13b.

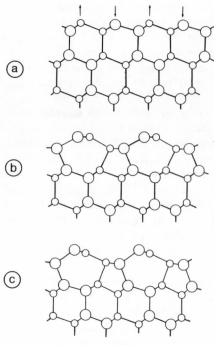

Figure 6.13. Side views of ideal and reconstructed Si(111) surfaces: (a) Ideal structure (arrows indicate atom displacements in the buckling model); (b) π-bonded chain model (the chains at the first two layers are directed normally to the plane of the figure); (c) chain model with tilted chains.

This model of reconstruction is consistent with many experimental results. If we consider the methods which directly involve the geometrical structure of the surface, we note that the angular resolved scattering of ions[53,7] is in favor of Pandey's model. Agreement between experiment and computer simulation is better if an additional tilt of the chains (as indicated in Figure 6.13c) is introduced. Scanning tunneling microscopy confirms the chain model,[21] while questions remain in the interpretation of the LEED data.

Surface total energy calculations[32] are in favor of the π-bonded chain model, if compared with previous models. However, another geometry leading to a more stable structure cannot be excluded.

The spectroscopic information on electronic structure compares rather well with the band structure of the π-bonded chain model. The buckling model can explain the spectroscopic data, although the use of different values for the buckling would be necessary to account for different features. In particular, the bands show the minimum gap along the border $\bar{J}-\bar{K}$ of the 2D BZ, as indicated in Figure 6.15, reporting the calculation by Northrup and Cohen.[54] The dispersion of the filled band is in agreement with angular resolved photoemission results. The value of the gap is lower than that seen in optical measurements. However, it should be noted that its value might be increased by the tilt of the chain, which can be present[53] in the reconstruction, and that it is underestimated in the calculation due to the limits of LDA.

The inverse photoemission technique, which samples empty states and maps the dispersion of the bands above the Fermi energy, also gives results in agreement with this model.

Stronger support of the π-bonded chain model comes from the results of differential reflectance with polarized light, shown in Figure 6.14. The symmetry, the k-space location, and the localization of the wave functions of the states involved in the optical transition give vanishing intensity when the A vector potential of the electromagnetic field is directed normally to the chains, while they produce the maximum intensity of the surface peak when A is along the chain direction $[\bar{1}, 1, 0]$.[14,15] This result is not compatible with the buckling geometry, which would give a ϕ plot of the intensity completely different from that observed in the experiments.

A similar effect is also shown in electron energy loss spectroscopy[18] with azimuthal resolution, where the contribution of the interband surface excitation

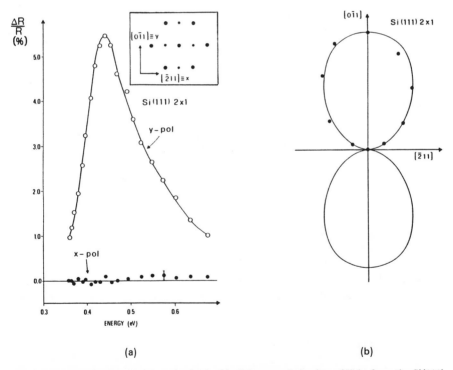

(a) (b)

Figure 6.14. Differential reflection obtained with different polarization of light from the Si(111) (2 × 1) surface: (a) Comparison between the two limiting cases: A—vector along the chains, A—vector normal to the chains; (b) ϕ plot of the intensity of the surface peak (after Chiaradia et al.[13]).

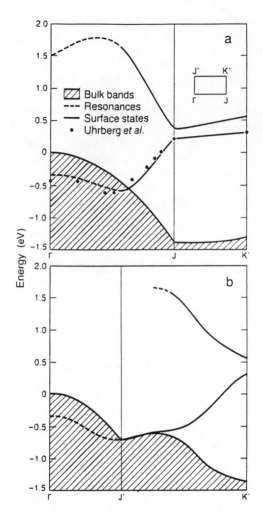

Figure 6.15. Dispersion of the surface bands of Si(111) (2 × 1) in the gap region, calculated for the π-bonded chain model (after Northrup and Cohen[54]).

is observed, for low-energy electrons, when the scattering plane is parallel to the chains, and vanishes almost completely when the plane is normal to the chain direction, as shown in Figure 6.16.

6.5.2. The Si(111) (7 × 7) Surface

The most stable structure of Si(111) is obtained by annealing at high temperature. It is the 7 × 7 structure. This surface has attracted the interest of physicists since its discovery, and many models were proposed before the direct observation of the 7 × 7 unit cell by scanning tunneling microscopy (STM) in 1983.[9] It was

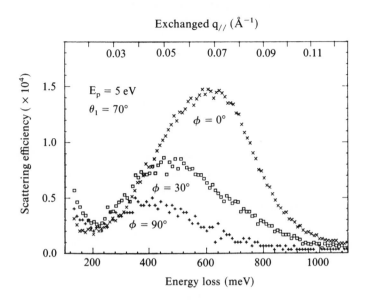

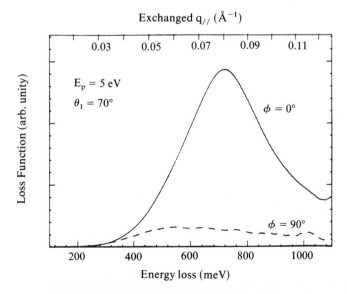

Figure 6.16. (a) Azimuthal dependence of the electron energy-loss scattering efficiency from Si(111) (2 × 1) and (b) loss function (obtained by dividing the scattering efficiency by the kinematic prefactor) in the two limiting cases ($\phi = 0°$ and $\phi = 90°$).

one of the first spectacular applications of this powerful method to surface science. The image contains nine minima and twelve maxima in each 7×7 cell.

The model presented in that paper[9] was a modification of the adatom model suggested by Harrison[55] in 1976. This one was a modification of another model, based on an ordered pattern of vacancies, proposed by Lander *et al.*[51] in 1963. The stability of the reconstruction is linked to the reduction of the number of dangling bonds and in this case 12 adatoms (added to the ideal substrate) saturate 36 dangling bonds, with a reduction of 24 in their total number in the 7×7 cell. At the four corners of the cell a deep minimum is found, surrounded by a pattern of local maxima with sixfold symmetry.

A model based on adatoms and a simplified total energy calculation in the tight-binding scheme were presented by Chadi[56] in 1984. The model contains adatoms, each bound to three atoms in the substrate. With additional shifts and exchanges in atomic positions, local configurations with small segments of π-chains are also obtained. This picture explains structures like 7×7 and 5×5. The latter is found in the case of Si deposited on Ge(111) and annealed.

Other models based on adatoms were proposed subsequently; in work by Tromp *et al.*[57] they are compared with the results of scanning tunneling spectroscopy. These authors obtained images of Si(111) (7×7) with different applied voltage ($+2$ and -2 V) between the tip and the sample in order to avoid masking effects arising from the localized states, mainly located near E_F, and to have a map of the charge distribution more directly comparable with the overall electron charge density distribution of the atoms at the surface. In this way one compares the experimental image with the superposition of electronic charge of the atoms calculated for the geometry of the different models. This comparison is in favor of the model by Takayanagi *et al.*,[58] proposed on the basis of a study using transmission electron microscopy and diffraction.

This model consists of 12 adatoms with their dangling bonds added to the external plane (plane 0) and 6 restatoms with dangling bonds in the first plane just below (plane 1). The restatom at the deep minimum at the corner of the unit cell has a dangling bond too, but is located at plane 3, which retains the ideal geometry. Nine dimers per cell are also present and the shorter diagonal of the unit cell divides the cell into two parts as a mirror symmetry line. This mirror symmetry can be see in the side view along the longer diagonal of the cell shown in Figure 6.17. Four adatoms are present on this line and two restatoms, together with the restatom at the corner in the third plane. As a consequence of the mirror symmetry in the left half of the unit cell a stacking fault is clearly visible between planes 1-2 and planes 3-4, the stacking sequence being of hexagonal–diamond structure instead of diamond structure, preserved in the right part of the cell.

Scanning tunneling spectroscopy can visualize the electronic states localized at the surface. In Figure 6.17 there are two schematic pictures[59] of the images showing the contribution to the tunneling current from the states with energy between E_F and -0.4 eV from E_F, and from other states with energy between -0.6 and -1.0 eV. They represent different kinds of surface states. The former are responsible for the metallic edge of the electronic structure of the 7×7 surface and are linked to the dangling bonds of adatoms. The latter give rise to other

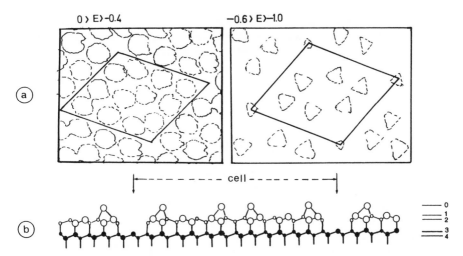

Figure 6.17. The Si(111) (7×7) surface: (a) Schematic drawings are presented of STM images of surface states localized at adatoms and restatoms (after Tromp *et al.*[59]); (b) side view of the reconstructed surface according the model of Takayanagi *et al.*[58]

surface bands at lower energy and correspond to the doubly occupied dangling bonds of restatoms. The image of these states shows clearly the mirror symmetry due to the stacking fault.

Calculation of the electronic structure and of the total energy of the reconstructed phase is not yet feasible in a first-principles approach, because of the too large extension of the unit cell. However, it is possible to handle smaller subunit models, with local environments similar to those of the large cell, and combine the results at the end. Further, it is possible to evaluate the energy per atom required to create the stacking fault from a bulk calculation for hexagonal–diamond structure. The fault does not need a large amount of energy and can be compensated by other energy gains. By using energy calculations for this simplified subunit model, Northrup[60] estimated the energy gain per surface atom of the 7×7 structure described above with respect to the ideal case. It ranges between 1.20 and 1.40 eV. This value must be compared with 1.27 eV per atom obtained for Si(111) (2×1). The possibility of an energy gain in the Takayanagi model is then justified.

6.5.3. Other Surfaces of Covalent Semiconductors

The Ge(111) cleavage surface shows, as in the case of Si(111), the 2×1 reconstruction as a metastable phase. This is a general feature of the diamond structure, including C(111). A mild annealing at 380 K is sufficient in Ge to reach the stable phase, corresponding to the $c(2 \times 8)$ LEED pattern, which reversibly changes at approximately 700 K into (1×1), with an intermediate multidomain phase between 560 and 700 K.

The surface of Si(100) has been widely studied. The stable structure is not the ideal geometry, described in Figure 6.3, but a reconstructed phase; some dependence of the reconstruction on the preparation and treatment of the surface can be found. The surface reconstruction is 2×1 and sometimes 4×2. Helium diffraction, LEED, and ion beam crystallography have been used to study the atomic geometry of Si(100) (2×1).[7,27] The model which compares most favorably with the structural data is the *dimer* model, obtained by shifting the positions of two atoms at the surface plane closer to each other, creating a bond and in this way saturating one of the two dangling bonds of each surface atom.

The main feature of the reconstruction is a lateral shift of the atoms with alternate signs and a gain in energy due to the new bond. In the cell two dangling bonds, one per atom, are still present and they would both be half-filled in the case of a symmetric dimer. An additional gain of energy of some tenths of eV can be obtained by tilting the dimers, producing a double filling of the *up-atom* dangling bond, while the *down-atom* surface state becomes empty. The electronic structure has been calculated by several authors[61,62] and the spectroscopic measurements are in agreement with the *asymmetric dimer* model.

Considerably lower is the number of studies on Si(110). Large reconstructions like (5×2), (16×2), and (32×2) can be found in this case, probably due to ordered arrangements of smaller subunits.

6.5.4. Polar Surfaces of Ionic Semiconductors

A large variety of surface reconstructions has been observed for the polar surfaces (100) and (111) of III–V compounds. These surfaces are important for technological reasons, as they are involved in the growth, obtained by molecular beam epitaxy (MBE), of microstructures and superlattices and in the fabrication of metal–semiconductor structures.

The complexity of these systems, the variability of the conditions of growth (temperature and molecular fluxes), and the possibility of different surface treatments do not allow one to understand the structural and electronic properties of these surfaces in terms of satisfactory microscopic models, even if they must be regarded as a challenging subject for present and future research.

The (100) surfaces of III–V compounds have many reconstructions. In the case of GaAs(100) prepared via MBE different structures are found, such as (4×6), (1×6), $c(8 \times 2)$, and $c(4 \times 4)$, associated in this order with an increasing excess of Ga over As in the composition of the last layer.

On the other hand, a general behavior is found for the cation terminated (111) surfaces and anion terminated $(\bar{1}\bar{1}\bar{1})$ surfaces, when they are prepared by ion bombardment with subsequent annealing. The 2×2 superstructure is obtained in the former case and the 3×3 in the latter, while many different structures can be obtained in the case of MBE growth in the directions (111) and $(\bar{1}\bar{1}\bar{1})$, as a function of the temperature of the substrate.

6.5.5. Overlayers and Interfaces

It is impossible to close this short review on semiconductor surfaces without mentioning some other important fields in this subject, i.e., atomic adsorption and chemisorption on semiconductor surfaces, chemical reactions and bonding at these surfaces, creation of ordered overlayers, formation of heterojunctions and of metal–semiconductor interfaces. All these topics correspond to an extensive literature and it would be difficult to select the papers to be included in a list of references of reasonable length. Only a small selection of some points, linked to the aspects of the surfaces already examined here, will be given, without discussion of the published results.

A first aspect is the removal of surface reconstruction due to atomic adsorption. This effect can be seen, for instance, in the adsorption of both H and Cl on Si surfaces. In these phenomena one problem is to understand whether the removal of the reconstruction corresponds to the destruction of long-range periodicity of the reconstructed phase or to the restoration of the ideal periodicity.

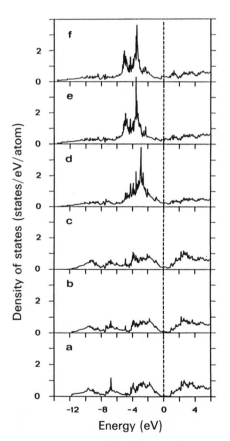
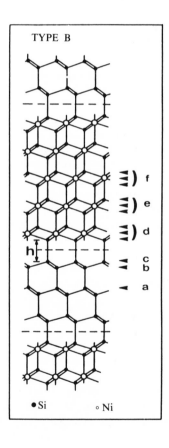

Figure 6.18.   Local density of states and interface atomic structure for an epitaxial type-B Si(111)–NiSi$_2$ interface (after Bisi and Ossicini[63]).

Another point is to study how the relaxation or the reconstruction, with the changes induced in the electronic structure, modifies the properties of the adsorption sites at the surface. In a cleavage surface of III–V compounds, the preferential site of adsorption seems to be the anion, which is relaxed outward.

The deposition of atoms of a different atomic species on the substrate can also stabilize a metastable phase or create a new one. As an example one can consider the $(\sqrt{3} \times \sqrt{3})$ reconstruction of Si(111) obtained by Ag deposition. Another important field of investigation is the study of metal–semiconductor interfaces. A variety of systems can be studied. A special case consists of the interfaces between alkali metals and semiconductors; if no intermixing takes place the model system consisting in a *jellium* slab over a semiconductor slab can account for many electronic properties; a more general and realistic theoretical model of an abrupt interface is obtained by considering repeated slabs with a number of planes of metal atoms plus a number of planes of the semiconductor.

This method can also work in order to describe reactive interfaces where a metallic compound grows on the semiconductor substrate, as in the case of the interface between silicon and silicides. This model is particularly suitable when dealing with the case of epitaxial growth, such as in the $Si–NiSi_2$ system.

As an example Figure 6.18 presents the results of a LDA calculation, based on the method of linear muffin tin orbitals (LMTO), for a repeated slab consisting of 6 planes of Si and 5 planes of $NiSi_2$. The interface can have two different geometrical arrangements of the atoms, usually called A and B type; in the first case there is ideal epitaxial orientation, in the second the structure of the silicide is rotated 180°. We are considering the latter structure, the most common at the first stage of the growth. The local density of states is shown at the different planes, passing from the internal plane of the silicide, having the density of states of bulk silicide, to the interface region, and to the internal planes of the silicon slab. At the interface the structures due to interface states can be identified. This kind of calculation can also describe the formation of the Schottky barrier, and the same method can be applied to semiconductor heterostructures to obtain the band offsets.

6.6. Conclusions

The investigation of the surface properties of semiconductors, presented in the preceding sections, shows how strongly the electronic surface structure is affected by the atomic geometrical arrangement. In the past two decades, the determination of the surface atomic structure and the study of the localized-state electron bands appeared clearly as related problems.

The discovery of new experimental methods for surface structure determination on the one hand, and the improvements of spectroscopic techniques on the other, offered mutual support to the advance of surface science.

The theory contributed to the increase of knowledge at two levels. At the first stage, qualitative information is obtained through semiempirical band structure calculations. More recently it was possible, through reliable first-principles methods, to describe more accurately all the features of the electronic structures and to compare the stability of the different surface phases, in the case of almost the simplest geometries.

The reliability of experimental and theoretical methods is particularly promising for future study in the field of semiconductor surfaces, with special reference to surfaces presenting a large variety of surface phases.

The knowledge and control of the atomic geometric arrangement and of the surface electron band structure are critically important in view of the design and fabrication of microstructures specifically tailored for a predefined behavior.

Acknowledgments

The author would like to thank Dr. Elisa Molinari for a critical reading of the manuscript. Financial support by Consiglio Nazionale delle Ricerche and Ministero della Pubblica Istruzione is acknowledged.

References

1. W. Shockley, *Phys. Rev.* **56**, 317–323 (1939).
2. J. C. Slater and G. F. Koster, *Phys. Rev.* **94**, 1498–1524 (1954); A. Nussbaum, *Solid State Physics*, **18**, 165–272 (1966).
3. I. Tamm, *Phys. Z. Sov. Un.* **1**, 317 (1939).
4. J. D. Levine and P. Mark, *Phys. Rev.* **144**, 751–763 (1966).
5. S. G. Davisson and J. D. Levine, *Solid State Physics* **25**, 1–149 (1970).
6. J. B. Pendry, *Low Energy Electron Diffraction*, Academic Press, New York (1974).
7. J. F. Van der Veen, *Surf. Sci. Rep.* **5**, 199–288 (1985).
8. M. J. Cardillo, *Phys. Rev. B* **23**, 4279–4282 (1981).
9. G. Binning, H. Rohrer, Ch. Gerber, and E. Weibel, *Phys. Rev. Lett.* **50**, 120–123 (1983).
10. T. C. Chiang, J. A. Knapp, M. Aono, and D. E. Eastman, *Phys. Rev. B* **21**, 3513–3522 (1980).
11. G. P. Williams, R. J. Smith, and G. J. Lapeyre, *J. Vac. Sci. Technol.* **15**, 1249 (1978).
12. G. J. Lapeyre and J. Anderson, *Phys. Rev. Lett.* **35**, 117–120 (1976).
13. G. Chiarotti, G. Del Signore, and S. Nannarone, *Phys. Rev. Lett.* **21**, 1170–1173 (1968).
14. P. Chiaradia, A. Cricenti, S. Selci, and G. Chiarotti, *Phys. Rev. Lett.* **52**, 1145–1148 (1984).
15. M. A. Olmstead and N. H. Amer, *Phys. Rev. Lett.* **52**, 1148–1151 (1984).
16. W. Gudat and D. E. Eastman, *J. Vac. Sci. Technol.* **13**, 831–837 (1976).
17. R. Math, H. Lüth, and H. Ritz, *Solid State Commun.* **46**, 343–346 (1983).
18. U. Del Pennino, M. G. Betti, C. Mariani, C. M. Bertoni, S. Nannarone, I. Abbati, L. Braicovich, and A. Rizzi, *Solid State Commun.* **60**, 337–340 (1986).
19. N. J. Di Nardo, W. A. Thompson, A. J. Schell Sorokin, and J. E. Demuth, *Phys. Rev. B* **34**, 3007–3010 (1986).
20. R. M. Feenstra, W. A. Thompson, and A. P. Fein, *Phys. Rev. Lett.* **56**, 608–611 (1986).
21. R. M. Feenstra, J. A. Stroscio, J. Tersoff, and A. P. Fein, *Phys. Rev. Lett.* **58**, 1192–1195 (1987).
22. J. A. Appelbaum, in: *Surface Physics of Materials*, Vol. 1 (J. M. Blakely, ed.), pp. 79–120, Academic Press, New York (1975).
23. F. García-Moliner, V. Heine, and J. Rubio, *J. Phys. C* **2**, 1797–1801 (1969).
24. W. A. Harrison, *Electronic Structure and the Properties of Solids*, Freeman, San Francisco (1980).
25. C. Calandra, F. Manghi, and C. M. Bertoni, *J. Phys. C* **10**, 1911–1927 (1977).
26. C. B. Duke, R. Lubinski, B. W. Lee, and P. Mark, *J. Vac. Sci. Technol.* **13**, 761–768 (1976).

27. A. Kahn, *Surf. Sci. Rep.* **3**, 193-300 (1983).
28. J. R. Chelikowsky and M. L. Cohen, *Phys. Rev.* B **13**, 826-834 (1976).
29. C. M. Bertoni, O. Bisi, C. Calandra, and F. Manghi, in: *Physics of Semiconductors 1978* (B. L. H. Wilson, ed.), Inst. Phys. Conf. Ser. **43**, 191-194 (1979).
30. A. Huijser, J. van Laar, and T. L. Roy, *Surf. Sci.* **62**, 472-481 (1977).
31. D. J. Chadi, *Phys. Rev. Lett.* **41**, 1062-1065 (1978).
32. K. C. Pandey, *Phys. Rev. Lett.* **49**, 223-226 (1982).
33. P. Hohenberg and W. Kohn, *Phys. Rev.* **136B**, 864-871 (1964).
34. W. Kohn and L. J. Sham, *Phys. Rev.* **140A**, 1133-1138 (1965).
35. M. S. Hybertsen and S. G. Louie, *Phys. Rev.* B **34**, 5390-5413 (1986).
36. R. W. Godby, M. Schlüter, and L. J. Sham, *Phys. Rev. Lett.* **56**, 2415-2418 (1986); *Phys. Rev.* B **37**, 10159-10175 (1988).
37. D. M. Ceperley and B. J. Alder, *Phys. Rev. Lett.* **45**, 566-569 (1980).
38. G. B. Bachelet, D. R. Hammann, and M. Schlüter, *Phys. Rev.* B **26**, 4199-4228 (1982).
39. F. Manghi, C. M. Bertoni, C. Calandra, and E. Molinari, *Phys. Rev.* B **24**, 6029-6042 (1981).
40. F. Manghi, unpublished (1988).
41. F. Manghi and E. Molinari, *J. Phys.* C **15**, 3627-3637 (1982).
42. O. Gunnarsson and R. O. Jones, *Phys. Scr.* **21**, 394-399 (1980).
43. F. Manghi, G. Riegler, C. M. Bertoni, and G. B. Bachelet, *Phys. Rev.* B **31**, 3680-3688 (1985).
44. F. Manghi, *Phys. Rev.* B **33**, 2554-2558 (1986).
45. M. Hybertsen and S. G. Louie, *Phys. Rev. Lett.* **58**, 1551-1554 (1987).
46. M. A. Olmstead, *Surf. Sci. Rep.* **6**, 129-252 (1987).
47. R. E. Schlier and H. E. Farnsworth, *J. Chem. Phys.* **30**, 917-925 (1959).
48. P. A. Bennett and M. B. Webb, *Surf. Sci.* **104**, 74-104 (1981).
49. D. Hanemann, *Phys. Rev.* **121**, 1093-1100 (1963); *Phys. Rev.* **170**, 705-718 (1968).
50. A. Selloni and E. Tosatti, *Solid State Commun.* **17**, 387-390 (1965).
51. J. J. Lander, G. W. Gobeli, and J. J. Morrison, *J. Appl. Phys.* **34**, 2298 (1963).
52. K. C. Pandey, *Phys. Rev. Lett.* **47**, 1913-1916 (1981).
53. R. M. Tromp, L. Smit, and F. J. Van der Veen, *Phys. Rev.* B **30**, 6235-6237 (1984).
54. J. E. Northrup and M. L. Cohen, *Phys. Rev. Lett.* **49**, 1349-1352 (1982).
55. W. A. Harrison, *Surf. Sci.* **55**, 1-19 (1976).
56. J. Chadi, *Phys. Rev.* B **30**, 4470-4480 (1984).
57. R. M. Tromp, R. J. Hamers, and J. E. Demuth, *Phys. Rev.* B **34**, 1388-1391 (1986).
58. K. Takayanagi, Y. Tanishiro, M. Takahashi, and S. Takahashi, *J. Vac. Sci. Technol.* A **3**, 1502-1506 (1985).
59. R. M. Tromp, E. J. van Loenen, R. J. Hamers, and J. E. Demuth, in: *Structure of Surfaces II (Proc. of ICSOS II)* (J. F. van der Veen and M. A. van Hove, eds.), Springer-Verlag, Berlin (1988).
60. J. E. Northrup, *Phys. Rev. Lett.* **57**, 154-157 (1986); J. E. Northrup, in: *18th International Conference on the Physics of Semiconductors, Stockholm 1986* (O. Engström, ed.), pp. 61-64, World Scientific, Singapore (1987).
61. J. A. Appelbaum and D. R. Hamann, *Surf. Sci.* **74**, 21-33 (1978).
62. M. T. Hin and M. L. Cohen, *Phys. Rev.* **24**, 2303-2306 (1981).
63. O. Bisi and S. Ossicini, *Surf. Sci.* **189/190**, 285-293 (1987).

Low-Energy Electron Diffraction

J. B. Pendry

7.1. Low-Energy Electron Diffraction Experiments

The essential elements are an ultrahigh vacuum (UHV) chamber to preserve surface cleanliness, an electron gun to produce a collimated beam of electrons in the energy range 0 to 500 eV, a crystal holder and manipulator, and some means of observing the diffracted electrons, typically a fluorescent screen. Further details may be found elsewhere.[1-3] The major difficulty is common to all surface experiments, namely, to keep the surface clean. The UHV chamber will normally contain an array of techniques for cleaning the surface (provision for heating the sample, ion bombardment) as well as some means of detecting impurities at the surface, usually by detection of Auger signals from adsorbed atoms. LEED is very sensitive to cleanliness of the surface and small amounts of contaminant can produce quite spurious results. Experiments conducted on clean, perfect, surfaces can produce a large amount of structural information of high precision. Obviously it is only possible to produce precise data for surfaces which are well defined in the first place.

The state of mechanical perfection of the surface can conveniently be monitored by the LEED experiment itself. Strangely enough LEED can be very tolerant of mechanical imperfections, perhaps because only perfect areas of surface contribute to the sharp diffraction features which are typically measured.

There are two sorts of observations to be made with LEED. The easiest sort is simply to take photographs of the screen. If the surface has been well prepared and is in a well ordered state, it will behave just like an optical diffraction grating. The incident beam is diffracted into a discrete set of beams which can be displayed on a fluorescent screen. The resulting pattern gives information about the unit cell of the surface structure. Sometimes, in simple cases this can be inferred from the bulk structure of the crystal, in other cases the surface may form a new structure due to interactions between adsorbed atoms or instabilities of the clean surface

J. B. Pendry • Imperial College, London SW7 2BZ, England.

itself. Such observations provide important information about elemental surface processes which has enabled us to greatly expand our understanding of the surface in the past few years.

The more difficult sort of experiment is to observe the intensities of the diffracted beams. While the spot profile on the screen tells us about the size and shape of the surface cell, the intensities are far richer in their information content: they tell us about how the contents of the cell are arranged. For example, if we wish to know the positions of atoms at surfaces then we must measure LEED intensities, not merely the positions of the spots.

In the early days of the subject diffracted intensities were measured by catching the electrons in a device called a Faraday cup. This was accurate, but very slow. Locating the beam presented difficulties that greatly taxed the patience of the experimenter and only very limited amounts of data have been acquired in this way. An easier but less accurate method is to measure the intensity of photoluminescence of the spots on the fluorescent screen. Here one can actually see what is being measured and guide the measuring device by eye. Nevertheless, even this method is time consuming when we wish to acquire the very large datasets needed for really precise work. It also lacks sensitivity, requiring large currents in the incident electron beam which can cause problems when chemically delicate species are present on the surface: the molecule CO can be desorbed and decomposed by an electron beam.

The most sophisticated solution to these problems developed so far is the "DATALEED II" apparatus developed in Erlangen, West Germany, by Müller and Heinz. The screen image is amplified by an image intensifier and digitized in real time by a TV camera working in conjunction with sophisticated electrons. Having acquired the data in electronic form it can be processed by computer so as to rapidly extract the information required.

With this apparatus very low beam currents can be used and data acquired in seconds rather than hours. It will be essential to the feasibility of the new methods for extracting structural information from diffraction data.

7.2. Typical Experimental Data

In principle the intensities could be measured as functions of other variables, such as the angle of incidence, but this is rarely done. Several points are worth noting. First, the $I(E)$ data are highly structured. Let us consider the curves from an information theory point of view: how many different numbers would one need to quote in order to describe the curves? In a typical $I(E)$ trace for the Cu(001) surface taken in the range 0 eV to 250 eV, it is possible to identify something like 11 peaks or shoulders in the data. For each of these one would need to specify the energy and intensity, 22 numbers in all. That is a great deal of information content. Furthermore, we can easily generate many more curves by looking at different beams, or by shifting the angle of incidence. This is the special power of LEED: it has an extraordinarily high information content. Even for the copper (001) surface an investigation of atomic positions will involve specifying the position of the second layer relative to the first (one atom per unit surface cell), and the third relative to the second. Sensitivity to the fourth layer is probably not great. Thus we need information on 6 independent coordinates, 3 for each displacement. Also we would expect to double check on each coordinate, hence the 11 peaks are only just enough to give a determination of the structure of this simple surface of clean copper!

7.3. Theoretical Problems

The experimental data contain information, but how can we extract that information? Somehow we have to relate the raw data, peak positions and intensities, to the arrangement of atoms at an atomic level.

The intensities are generated because electrons scatter from atoms in the surface. The electron wave function at the detector is given by a sum over all possible scattering events that the electron can make within the surface. Each scattering path will change the phase of the wave function by a different amount, depending on the path length and the wavelength of the electron. As the energy is varied, the wavelength changes and the interference conditions change: hence the succession of maxima and minima seen in a typical $I(E)$ spectrum. The problem for interpretation of the data is that the scattering paths contributing to the intensities are rather complex. In general an electron will scatter off more than one atom before it leaves the surface. This multiple scattering problem held up the development of LEED until we had developed an adequate theoretical base for calculating its effects. Today the theory is very well understood and we have access to far more sophisticated computers than was the case only 10 years ago. The theoretical calculations needed to interpret $I(E)$ data in terms of locations of copper atoms can now easily be done on a personal computer costing less than $1000.

What ingredients control this theory? the structure in the $I(E)$ curves tends to be rounded: none of the peaks is very sharp. In fact the most narrow peaks seen have a width given by

$$\delta E \approx 8 \text{ eV} \tag{7.3.1}$$

This uncertainty in the energy of a feature can be related by fundamental quantum mechanics to the lifetime of electrons in the surface. The uncertainty principle tells us that

$$\delta E >= \hbar/\tau \tag{7.3.2}$$

Taking the narrowest peaks we deduce that the lifetime of electrons is

$$\tau \approx \hbar/\delta E \approx 10^{-15} \text{ second} \tag{7.3.3}$$

during which time a 100 eV electron travels about 5 Å.

This short penetration depth explains the surface sensitivity of electrons with energies of the order of 100 eV. It is just enough to encompass the first few atomic layers of the solid in which all the interesting surface effects on atomic structure usually take place.

Now let us turn our attention to the intensities. During time τ the atoms have to scatter the electron strongly enough to give intensities of the order of 1%. If the rate of back scattering is given by a matrix element $|T_b|^2/\hbar$, then the scattered intensity will be given by

$$|T_b|^2 \tau/\hbar = |T_b|^2/\delta E = 0.01 \qquad (7.3.4)$$

hence

$$|T_b|^2 = 0.01\delta E \approx 0.1 \text{ eV} \qquad (7.3.5)$$

One further important parameter is missing from our picture. Scattering of electrons by atoms is by no means isotropic. The atoms are much better at scattering the electrons in the forward direction. Crudely speaking we can distinguish between forward and back scattering by defining a separate matrix element for forward scattering, $|T_f|^2$. It has already been noted that the electron scatters from several atoms before it leaves the surface. How do we know this? If the electron only scattered from one atom, i.e., scattering was very weak, then we can take over the theory used to describe X-ray diffraction. Here Bragg's law tells us that weak scattering waves diffract from planes of atoms. In the case of the 00 beam from a copper (001) surface, the relevant planes are the (001) planes. Bragg's law predicts peaks when the condition

$$\text{wavelength} = 2d \cos(\Theta) \qquad (7.3.6)$$

is satisfied, where d is the interplanar spacing and Θ is the angle of incidence of the waves measured relative to the normal to the planes. It is commonly observed in LEED data that Bragg's law fails to predict the existence of the majority of the peaks. The only way in which Bragg's law can break down is if the conditions for its validity are violated: the scattering cannot be weak.

We have demonstrated that the back-scattering matrix element is in fact small relative to the lifetime of the electron, therefore we can expect only one back-scattering event in general. Any multiple scattering must be due to forward scattering. Since this must occur in order that Bragg's law is violated, we deduce that

$$|T_f|^2 \approx \delta E \approx 8 \text{ eV} \qquad (7.3.7)$$

7.4. Kinematic Theory

So much for estimates of the nature of electron scattering at a surface. New let us try to do a better job, first by concentrating on how a single atom scatters an electron. Later the multiple scattering problem will be examined.

In order to simplify the writing of equations it is usual to adopt a simplified system of units. Atomic units will be used for which

$$\hbar = e^2 = m = 1 \tag{7.4.1}$$

In this system the units work out as follows:

$$\text{unit of energy} = 1 \text{ Hartree} = 27.2 \text{ eV} \tag{7.4.2}$$

$$\text{unit of length} = 1 \text{ Bohr radius} = 0.5292 \text{ Å} \tag{7.4.3}$$

and the Schroedinger equation becomes

$$-\tfrac{1}{2}\nabla^2\Phi + V\Phi = E\Phi \tag{7.4.4}$$

Conventionally we divide the potential into two parts,

$$V(\mathbf{r}) = V_0 + V_a(r) \tag{7.4.5}$$

where the constant potential

$$V_0 = V_{or} + iV_{0i} \tag{7.4.6}$$

comprises a real part which represents a shift of all electron energies in the crystal relative to the vacuum, and an imaginary part which represents the lifetime of the electron inside the solid. Typically,

$$V_{0r} \approx -10 \text{ eV}, \qquad V_{0i} \approx -4 \text{ eV} \tag{7.4.7}$$

In fact V_0 is only constant inside the surface, outside it gradually falls away to zero. However, it does so very smoothly and it is usual to neglect any reflection from this step in potential at the surface. Its effects are confined solely to changing the wave vector of the electron inside the surface.

That an imaginary potential reproduces the effect of finite lifetime can be seen as follows: neglect V_a in the Schroedinger equation. Then we can solve for Φ in a constant complex potential,

$$\Phi = \exp(i\mathbf{K} \cdot \mathbf{r}) \exp(-iEt) \tag{7.4.8}$$

so that,

$$|\mathbf{K}|^2 + V_{0r} + iV_{0i} = E \tag{7.4.9}$$

Evidently if we choose $\mathbf{K}$ to be a real vector, then E becomes complex, so that

$$|\Phi(t)|^2 = \exp[2\text{Im}(E)] = \exp(2V_{0i}) \tag{7.4.10}$$

and since V_{0i} is negative, this represents decay of Φ with time. Alternatively we may require E to be real (it may be fixed by voltages in the electron gun, for example) in which case we are free to choose the components of **K** to be complex. If we define x, y axes to lie in the surface plane, and the positive z axis to point into the surface, then we can only choose the z component of **K** to be complex: complex $\mathbf{K}_x$ would result in the wave function growing unphysically to ∞ at $\pm x = \infty$, and similarly for $\mathbf{K}_y$. On the other hand, we are free to choose $\mathbf{K}_z$ to have a positive imaginary part so that Φ decays away into the surface. Thus absorption can result in a wave function that decays in time or in space depending on the experimental circumstances. In a LEED experiment the energy is fixed to be real in the electron gun, therefore, from equation (4.4),

$$\mathbf{K}_z = +\sqrt{(2E - 2V_{0r} - 2iV_{0i} - \mathbf{K}_x^2 - \mathbf{K}_y^2)} \qquad (7.4.11)$$

The x and y components of **K** are fixed by the direction of the incident beam.

Next we consider the atomic part of the potential. It is usually a good approximation in a LEED calculation to take this to be spherically symmetric. Its effect is to scatter the wave function Φ into a different state so that far outside the range of V_a itself we must add corrections to our wave function in order to account for the scattering by an atom located at R_j,

$$\Phi' = \exp(i\mathbf{K} \cdot \mathbf{r}) + f(\Theta) \exp(i\mathbf{K} \cdot \mathbf{R}_j + i|\mathbf{K}||\mathbf{r} - \mathbf{R}_j|)/(|\mathbf{K}||\mathbf{r} - \mathbf{R}_j|) \quad (7.4.12)$$

where Θ is the angle between **K** and **r**. The scattered wave is an outgoing spherical wave centered on the atom, the amplitude of which depends on the direction in which it is scattered. Function $f(\Theta)$ will be large for forward scattering and small for backward scattering, as discussed above. It is usual to express the scattering factor in terms of entities, δ_l, called phase shifts:

$$f(\Theta) = 4\pi|\mathbf{K}|^{-1} \sum_{l=0}^{\infty} \sum_{m=-l}^{+l} \exp(i\delta_l) \sin(\delta_l)(-1)^m Y_{lm}(\hat{\mathbf{K}}) Y_{l-m}(\hat{\mathbf{r}}) \quad (7.4.13)$$

where Y_{lm} is a spherical harmonic and $\hat{\mathbf{r}}$ denotes the angular coordinates of **r**.

In LEED theory we usually think of a surface in terms of layers of atoms, each layer parallel to the surface, and the scattering from the surface as a whole is built up in stepwise fashion: first considering the individual atoms, next the layers of atoms, and finally the combined effect of the layers is calculated.

If we have a complete layer of atoms, all of the same type, spaced on a Bravais lattice defined by

$$\mathbf{R}_j = m_j \mathbf{a} + n_j \mathbf{b} \qquad (7.4.14)$$

then the wave field corrected for the effect of the layer scattering can be expressed in the form

$$\Phi'' = \exp(i\mathbf{K} \cdot \mathbf{r}) + \sum_{\mathbf{g}} M(\mathbf{k}_{\mathbf{g}}^{\pm}, \mathbf{K}) \exp[i\mathbf{K} \cdot \mathbf{R}_n + i\mathbf{K}_{\mathbf{g}}^{\pm} \cdot (\mathbf{r} - \mathbf{R}_n)] \quad (7.4.15)$$

where

$$M(\mathbf{K}_g^{\pm}, \mathbf{K}) = 2\pi i f(\Theta_g)/[A|\mathbf{K}||\mathbf{K}_{gz}|] \tag{7.4.16}$$

A is the area occupied by each atom in the layer, Θ_g is the angle through which the gth beam is scattered, and the superscript $\pm$ refers to whether the relevant beam has been scattered to the $+z$ or $-z$ side of the layer. We note that, now we are scattering from a periodic object, the scattered amplitude has the form of a discrete set of beams defined by the reciprocal lattice vectors, $\mathbf{g}$. These in turn are defined in terms of the unit cell:

$$\mathbf{g} = h\mathbf{A} + k\mathbf{B} \tag{7.4.17}$$

where

$$\mathbf{A} = \frac{2\pi}{a_x b_y - b_x a_y}(b_y, -b_x) \tag{7.4.18}$$

and

$$\mathbf{B} = \frac{2\pi}{a_x b_y - b_x a_y}(-a_y, a_x) \tag{7.4.19}$$

The wave vectors of the scattered waves are given by

$$[\mathbf{K}_g^{\pm}]_x = \mathbf{K}_x + \mathbf{g}_x \tag{7.4.20}$$

$$[\mathbf{K}_g^{\pm}]_y = \mathbf{K}_y + \mathbf{g}_y \tag{7.4.21}$$

$$[\mathbf{K}_g^{\pm}]_z = \pm\sqrt{[2E - 2V_0 - (\mathbf{K}_y + \mathbf{g}_y)^2 - (\mathbf{K}_x + \mathbf{g}_x)^2]} \tag{7.4.22}$$

Equation (7.4.16) is an approximation to the exact expression for the layer scattering. We have omitted to take account of the multiple scattering terms. However, they enter as corrections to M, and the general form of the scattered wave is retained even in the multiple scattering regime.

Equations (7.4.15) and (7.4.16) can be simply interpreted as follows: in the single scattering approximation, an ordered array of atoms results in the scattering directions being quantized, the amplitude of waves in each quantized direction being given by the atomic scattering factor for that direction.

Next we add together scattering from several layers to calculate the scattered wave field just outside the surface. The layers are assumed to be equally spaced with no multiple scattering between layers:

$$\mathbf{R}_n = n\mathbf{c} \tag{7.4.23}$$

and

$$\Phi''' = \exp\left(i\mathbf{K}\cdot\mathbf{r}\right) + \sum_{n=0}^{\infty}\sum_{\mathbf{g}} M(\mathbf{K}_{\mathbf{g}}^-, \mathbf{K})\exp\left[i\mathbf{K}\cdot\mathbf{R}_n + i\mathbf{K}_{\mathbf{g}}^-\cdot(\mathbf{r}-\mathbf{R}_n)\right]$$

$$= \exp\left(i\mathbf{K}\cdot\mathbf{r}\right) + \sum_{\mathbf{g}}\Phi_{\mathbf{g}}\exp\left(i\mathbf{K}_{\mathbf{g}}^-\cdot\mathbf{r}\right) \qquad (7.4.24)$$

where

$$\Phi_{\mathbf{g}} = \frac{M(\mathbf{K}_{\mathbf{g}}^-, \mathbf{K})}{1 - \exp\left[i(\mathbf{K}-\mathbf{K}_{\mathbf{g}}^-)\cdot\mathbf{c}\right]} \qquad (7.4.25)$$

Expressions (7.4.24) and (7.4.25) constitute the so-called kinematic formula for diffracted amplitudes. It contains no multiple scattering at all and therefore rarely represents an accurate account of a realistic situation. It does illustrate some interesting aspects of LEED which do not depend on multiple scattering.

This simple model reproduces peaks of the correct sort of shape and width. Including the proper atomic scattering factor would give roughly the correct intensities, provided we correct them for thermal vibrations, a topic which we shall not have time to treat here. It can also be seen from equation (7.4.25) that the peak positions are determined by the energy and the interplanar spacing. This gives the structural sensitivity that we seek. Increasing the spacing between planes moves the peaks to lower energies. The shape of the peaks is modified if each pair of planes has a different spacing: in this way it is possible to resolve all relevant interplanar spacings.

7.5. Multiple Scattering

This subject is a technical one and those interested in conducting detailed calculations should read one of the relevant texts on LEED theory. The underlying principles can be demonstrated by a one-dimensional (1D) model. Let us consider a 1D array of atoms. We assume that if a wave

$$\exp\left[+ik(z-nc)\right], \qquad z < nc \qquad (7.5.1)$$

is incident on a layer located at $z = nc$, then the transmitted and reflected waves are, respectively,

$$t\exp\left[+ik(z-nc)\right], \qquad z > nc \qquad (7.5.2)$$

and

$$r\exp\left[-ik(z-nc)\right], \qquad z < nc \qquad (7.5.3)$$

Correspondingly, if a wave

$$\exp\left[-ik(z - nc)\right], \qquad z > nc \tag{7.5.4}$$

is incident on a layer located at $z = nc$, then the transmitted and reflected waves are, respectively,

$$t \exp\left[-ik(z - nc)\right], \qquad z < nc \tag{7.5.5}$$

$$r \exp\left[-ik(z - nc)\right], \qquad z > nc \tag{7.5.6}$$

For simplicity we have assumed that for each of the two different incident waves, the relevant transmission and reflection coefficients are equal. In general this is not true.

We are now in a position to show how to calculate the band structure of the system. Bloch's theorem tells us that if the system consists of a set of identical objects, equally spaced, then the electron wave functions can be classified by a wave vector, K. In this instance, between layers $n - 1$ and n we have forward and backward waves of amplitudes $a_n^\pm$, and between layers n and $n + 1$ waves of amplitudes $a_{n+1}^\pm$. Bloch's theorem says that we can find solutions of the Schroedinger equation which satisfy

$$a_{n+1}^+ = \exp\left(iKc\right)a_n^+ \tag{7.5.7}$$

and

$$a_{n+1}^- = \exp\left(iKc\right)a_n^- \tag{7.5.8}$$

The wave vector K can be calculated from the scattering properties of the layers. Equations (7.5.1)–(7.5.6) yield the following relationships:

$$a_{n+1}^+ = \exp\left(+ikc\right)ta_n^+ + ra_{n+1}^- \tag{7.5.9}$$

$$a_n^- = \exp\left(+ikc\right)ta_{n+1}^- + ra_n^+ \tag{7.5.10}$$

By defining

$$t' = \exp\left(ikc\right)t \tag{7.5.11}$$

these equations can be rearranged and expressed in matrix form:

$$\begin{bmatrix} 1 & -r \\ 0 & t' \end{bmatrix}\begin{bmatrix} a_{n+1}^+ \\ a_{n+1}^- \end{bmatrix} = \begin{bmatrix} t' & 0 \\ -r & 1 \end{bmatrix}\begin{bmatrix} a_n^+ \\ a_n^- \end{bmatrix} \tag{7.5.12}$$

Matrix inversion leads to

$$\begin{bmatrix} a_{n+1}^+ \\ a_{n+1}^- \end{bmatrix} = \begin{bmatrix} 1 - r^2/t'^2 & r/t' \\ -r/t' & 1/t' \end{bmatrix}\begin{bmatrix} a_n^+ \\ a_n^- \end{bmatrix} \tag{7.5.13}$$

On substituting for the left-hand side with the aid of the Bloch condition from equations (7.5.6) and (7.5.7),

$$\exp{(iKc)}\begin{bmatrix} a_n^+ \\ a_n^- \end{bmatrix} = \begin{bmatrix} 1 - r^2/t'^2 & r/t' \\ -r/t' & 1/t' \end{bmatrix}\begin{bmatrix} a_n^+ \\ a_n^- \end{bmatrix} \tag{7.5.14}$$

we see that the band structure, that is to say $\exp{(iKc)}$, can be found from the eigenvalues of the *transfer matrix*,

$$T = \begin{bmatrix} 1 - r^2/t'^2 & r/t' \\ -r/t' & 1/t' \end{bmatrix} \tag{7.5.15}$$

The most general way of solving the problem of reflection from surfaces involves using the eigenvectors of the transfer matrix, i.e., the Bloch waves. The rules by which the calculation is made are as follows:

1. Solve for the eigenvectors of the transfer matrix; there will be two of them: $[b_1^+, b_1^-]$ and $[b_2^+, b_2^-]$.
2. Choose the eigenvector with net current flowing *into* the surface, $\mathbf{b}_1$ say:

$$k(b_1^{+2} - b_1^{-2}) > 0$$

3. The reflection coefficient is then given by

$$R = b_1^-/b_1^+$$

Rule (1) simply states that the wave function inside the surface must solve Schroedinger's equation. Rule (2) means that the wave function must not correspond to any sources of current buried inside the surface. This leaves a wave function corresponding to an externally incident wave plus a reflected wave.

This general method is rarely used in practice and simpler methods are used, the most popular of which is the "layer doubling method." It works like this: a single layer has reflection and transmission coefficients,

$$t(1) = t', \qquad r(1) = r \tag{7.5.16}$$

Taking two layers and calculating the scattering from the pair gives, for the combined transmission coefficient,

$$t(2) = t(1) \cdot t(1) + t(1) \cdot r^2(1) \cdot t(1) + t(1) \cdot r^4(1) \cdot t(1)$$

$$+ t(1) \cdot r^6(1) \cdot t(1) + \cdots \tag{7.5.17}$$

The terms in this series can be interpreted as follows: the first term describes transmission through both the layers; the second term corresponds to transmission through the first layer, reflection from the second, another reflection from the first, and finally transmission through the second layer at the second attempt. This is what we mean by multiple scattering. The series is a simple geometric series which can be summed to give

$$t(2) = t^2(1)/[1 - r^2(1)] \qquad (7.5.18)$$

Similarly, the combined reflection coefficient of the pair of layers is given by

$$r(2) = r(1) + t(1) \cdot r(1) \cdot t(1) + t(1) \cdot r^3(1) \cdot t(1) + t(1) \cdot r^3(1) \cdot t(1) + \cdots$$

$$= r(1) + t^2(1) \cdot r(1)/[1 - r^2(1)] \qquad (7.5.19)$$

We can repeat the process to calculate $t(4)$ and $r(4)$ from $t(2)$ and $r(2)$, doubling the number of layers each time. In this way we can easily find the reflection coefficient of an effectively infinitely thick pile of layers representing the surface. In practice 8 layers are usually enough, though 16 may sometimes be needed.

These methods have been described for a simple one-dimensional example, but they can be generalized to real surfaces using the same principles, the only difference being that t and r are matrices. Some further applications of LEED theory can be found elsewhere.[4]

References

1. J. B. Pendry, *Low Energy Electron Diffraction*, Academic Press, London (1974).
2. M. A. Van Hove and S. Y. Tong, *Surface Crystallography by LEED*, Springer-Verlag, Berlin (1979).
3. K. Heinz and K. Müller, in: *Structural Studies of Surfaces*, Springer-Verlag, Berlin (1982).
4. P. J. Rous, J. B. Pendry, D. K. Saldin, K. Heinz, K. Müller, and N. Bickel, *Phys. Rev. Lett.* **57**, 2951 (1986).

Theoretical Aspects of Adsorption

B. I. Lundqvist

8.1. Introduction

Molecule–surface interactions are important for a large number of technical reasons, such as corrosion and heterogeneous catalysis. They involve a range of physical and chemical phenomena of great intrinsic interest. This chapter will deal, for the most part, with theoretical issues. Descriptions, either quantum-mechanical or classical, of atoms and molecules at surfaces, require a physically sound input of dissipative and reactive forces to give meaningful results. Potential-energy surfaces, from which the reactive forces can be derived, are the chapter's main theme.

An attempt is made to demonstrate that a coherent, conceptual framework of the electronic aspects of chemisorption and surface reactions is developing. This knowledge originates from an increased understanding of the adsorbate-induced electron structure and potential-energy surfaces of adsorbates. In particular, such results have been derived from the self-consistent Kohn–Sham calculations of total-energy differences, with the density, and densities of states, as intermediate steps, and as helpful quantities in interpreting results

This fact—that potential-energy surfaces of single particles at metal surfaces can be understood, in terms of electron-structure features, in a systematic way—is a theme of the chapter. It is possible to identify: substrate factors such as electron density, d-band location, as well as, the shape and symmetry of substrate orbitals; and adsorbate factors such as electronegativity, subshell closure, core orthogonalization, as well as the shape and symmetry of adsorbate orbitals. The weakening and breaking of molecular bonds on metal surfaces may be explained in terms of the lowering (raising), broadening and filling (emptying) of molecular affinity (ionization) levels, all of which occur for molecules on the surface. The qualitative rules for the making and breaking of bonds at metal surfaces are particularly apt for finding trends with respect to substrate and adsorbate parameters, as well as

B. I. Lundqvist • Institute of Theoretical Physics, Chalmers University of Technology, S-412 96 Göteborg, Sweden.

for qualitative considerations of the modification of potential-energy surfaces. This will be illustrated through examples.

It may be said that fascinating physical phenomena occur at surfaces. Many are technically important in, e.g., heterogeneous catalysis, microelectronics, and corrosion. In addition, there exists a richness in fundamental phenomena and issues. The surface separates a solid and a gas, thereby linking the two disciplines of molecular and solid-state physics. The relevant particles and fields are numerous. The natures of the chemisorption bond[1,2] and the dissipation of molecular energy at a surface,[3] as well as the dynamics of molecule–surface collisions,[4,5] are all scientific issues of key interest.

Several steps are involved in the interaction of molecules with surfaces: adsorption, migration across the surface and into the substrate, dissociation, recombination, and desorption. Adsorption of molecules on surfaces occurs through physisorption or chemisorption. When molecules impinge upon a surface, molecular bonds may be broken (dissociative chemisorption), or not (associative adsorption). As a consequence of such processes, new molecules appear, which process forms the basis of heterogeneous catalysis.

This chapter is primarily concerned with adsorption on metal surfaces. When reacting with molecules, metal surfaces frequently demonstrate a high reactivity. The abilities of these surfaces not only to break chemical bonds but also to chemisorb and recombine fragments, as well as to transfer energy efficiently have all been mentioned as factors affecting the chemical reactivity of metals. However, wide variations exist and specificity is a must. Characteristic to metals is the quasi-continuous spectrum of electronic excitations—in particular, the absence of a gap at the Fermi level. High reactivity may be linked to this fundamental metallic property.[6] Electronically adiabatic potential-energy surfaces, nonadiabatic processes, local densities of states, molecular orbitals, and finally, the Woodward–Hoffman reaction rules are all among the relevant concepts.[7]

Recently, theoretical contributions have progressively augmented the description, not only of the mechanistic but also of the dynamic aspects of surface processes. One aim of this chapter is to explain that now there exists a conceptual framework which can be utilized by both experimentalists and theorists for qualitative considerations and to facilitate communication. There is a present trend toward dynamic, "state-to-state" experiments, corresponding with quantitative theoretical results, on differential cross-sections for elastic, inelastic, and reactive gas–surface scattering processes; but, in the general case, there is still a long way to go.

The interaction between a molecule and a surface is governed by both conservative and dissipative forces. The former can be derived from the potential-energy surface (PES), based on the Born–Oppenheimer approximation and represented both adiabatically and diabatically.[8] This chapter will focus upon various features of PESs and on the electron structures of molecules in the surface region, as well as their relationships. However, dissipative effects are also important. Potential-energy surfaces might inform us of possible "intermediate states" in a surface reaction, but the magnitude of dissipation, expressed as, e.g., trapping probabilities or vibrational lifetimes, determines whether these intermediates will form for a sufficient period of time that would permit study.

Like gas-phase collisions, surface scattering emerges from mutual interactions via valence electrons, and can be either elastic, inelastic, or reactive. Electronic (E), vibrational (V), and rotational (R) states can be engaged; the collisions involve energy transfers, $T \rightarrow V$, $T \rightarrow R$, where T denotes the translatory motion, etc. There are many differences from gas-phase collisions, however, They are basically due to the large number of atoms involved in the surface case. Thus the surface becomes a momentum and an energy sink. It allows extended electron states with energy bands, continuous excitation spectra, screening and correlation effects. There are also extended vibrational effects, the phonons, with continuous energy bands, allowing single and multiphonon excitations. In addition, the surface has a very particular geometry. This allows properties to vary along the surface, with periodicity giving diffraction. It fixes asymptotic forms, e.g., of the van der Waals and image interactions. Finally, the surface provides new paths for the particle, thus permitting it to migrate, on the surface or into the interior, i.e., to absorb or dissolve, and to desorb. Each of these elementary processes has a complex dependence on a great number of molecular and solid state parameters.

The continuous nature of the solid-state excitations causes other differences.[9] Translational, rotational, vibrational, and electronic states are not discrete as in the gas phase. Since the substrate adds channels for decay, characteristic lifetimes are significantly shorter. For instance, vibrational lifetimes can be up to 6 to 8 orders of magnitude shorter than in the gas phase. This, of course, significantly affects the dynamics.

This chapter is planned as follows: In Section 8.2 some questions posed by experiments are formulated; Section 8.3, which is the major section, is devoted to adsorbate-induced electron structure and potential-energy surfaces, noting, for example, systematics, relationships, and some potentialities for "chemical engineering." In Section 8.4 applications to experimental situations are described, including vibrational spectroscopy and heterogeneous catalysis. Finally, Section 8.5 concludes with an outlook.

8.2. Questions Posed by Experiments

Adsorption has both static and dynamic aspects. There is a massive experimental program to measure static properties. LEED,[10,11] UPS,[12,13] and EELS[14] are examples of experiments which inform us of adsorbate geometries, electronic structure, chemisorption energies, and vibrational frequencies. The experimental methods for studying the dynamics of molecule–metal interaction are fewer, for instance, time-of-flight measurements,[15] laser-[16] and electron-induced[17] fluorescence, as well as infrared emission spectroscopy.[18] This section will give a few examples of the questions posed by experiments.

8.2.1. Static Experiments

Some experimental questions are too well known to be dwelt upon here, e.g., how low-energy electron-diffraction (LEED)[10,11] and high-resolution electron-energy-loss spectroscopy (EELS)[14] experiments inform us about potential-energy surfaces. From potential-energy surfaces, static properties such as bond lengths and atomic positions can be deduced. In particular, the adsorption sites and adlayer structures can be compared with data from LEED. Vibrational frequencies can also give information about electronic and atomic structure. Finally, PES can yield information about energetics; e.g., by comparing potential-energy curves for atomic and molecular adsorption, as in the famous picture of Lennard-Jones,[19] we can judge whether the adsorption of a particular molecule is either associative or dissociative.

Many questions can be raised about adsorption with a successively sharper focus. When a molecule hits a surface, is it adsorbed at all? Is the adsorption dissociative or associative? Is the associative adsorption activated? Is there a precursor state, i.e., a molecularly adsorbed state in which the molecule becomes trapped upon hitting the surface and, then, is free to move along the surface in order to seek out the preferred site of adsorption or dissociation? Why are metals so reactive? Why is chemisorption so specific, with respect to adsorbate, substrate, and individual substrate faces? Where in the adsorption process does the chemisorption bond override the molecular bond? Which electron orbitals are of key importance for the adsorption? And so on.

8.2.2. Molecular Vibrations and Rotations at Surfaces

Increased experimental sensitivity and resolution pose increasingly more intricate questions. This will be briefly illustrated with the example of vibrations and rotations at surfaces. A natural starting point is our knowledge of vibrations and rotations of free molecules.[20] Due to the interaction with a surface these are changed, for example:

1. *Characteristic frequencies vanish,* as for H_2, D_2, and HD chemisorbed on the Ni(100) surface, which is significant for dissociative adsorption.[21]
2. *Characteristic frequencies are added,* as for H_2O adsorbed on Pt(111), where a hindered translation normal to the surface indicates that H_2O is molecularly adsorbed.[22]
3. *Characteristic frequencies are unchanged,* as for H_2 and D_2 on Cu(100) at 15 K, where energy losses close to rotational and rotational–vibrational transitions of the free molecules show that H_2 and D_2 physisorb on Cu(100) at low temperatures.[23]
4. *Characteristic frequencies are shifted,* as for CO on Ni(100), where red shift of the C—O stretch mode shows that CO is adsorbed primarily on top (terminal) sites on the clean surface but on the bridge and center sites on Ni(100) $p(2 \times 2)$O.[24,25]
5. *Loss peaks are broadened,* as for CO on Cu(100), where the stretch-mode width of 0.5 meV, corresponding to a vibrational lifetime of about 3×10^{-12} s,[26] informs about energy dissipation from the vibration into excitations, electrons, or phonons, in the substrate.[27,28]
6. *Loss intensities are changed,* as for Cu(100) $c(2 \times 2)$CO, where the C—O stretch-vibrational loss intensity[29-31] signals electronic rearrangements on CO upon adsorption.[32]

7. *Collective vibrational modes show dispersion,* as for Cu(100) $c(2 \times 2)$CO, where the dispersion of the collective mode is found to be dominated by dipole-dipole interactions among the adsorbed molecules.[33,34]

8. *Characteristic frequencies have satellite structures,* as for the ordered over-layer $c(2 \times 2)$0 on Cu(100), which, together with symmetry analysis, can be used to derive adsorption sites.[35-36]

This list indicates the great variety in adsorbate-induced spectral changes. It also shows that probing the vibrations and rotations of molecules at surfaces gives information over a broad range about structural, electronic, and chemical properties of such molecules.

8.2.3. Catalytic Reactions

The surface-science approach to catalysis is increasingly successful in establishing the microscopic reaction steps.[37] The important examples of ammonia synthesis on iron, and hydrogenation of CO on nickel, are used to illustrate this fact, and also to pose some of the questions for which attempted answers are described in the later part of the chapter.

8.2.3.1. Ammonia Synthesis

The reaction $3 H_2 + N_2 \rightarrow 2NH_3$ has an enormous activation energy in the gas phase but can be performed industrially with promoted iron catalysts in an N_2/H_2 mixture of approximately 100 atm and 400 °C. Metallic iron is primarily responsible for the catalytic activity. The nitrogen adsorption, $N_2 \rightarrow N_{2,ad} \rightarrow 2 N_{ad}$ is the rate-limiting step. The basic role of the catalyst is to produce atomic nitrogen N_{ad} and then add atomic hydrogen.[37] The molecular nitrogen $N_{2,ad}$ is weakly adsorbed and the adsorption into this precursor state is fairly rapid, with a sticking probability of the order of 0.01. The dissociation of $N_{2,ad}$ (Figure 8.1) is very slow, however, with an effective sticking coefficient for the process $N_2 \rightarrow 2N_{ad}$ of the order 10^{-7}, due to an activation energy E^* (Figure 8.1) of approximately 0.3, 0.2, and 0.03 eV for Fe (100), (100), and (111), respectively at low coverages, and an unfavorable activation entropy for the process.[37,38] At temperatures above approximately 300°C

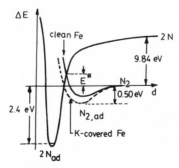

Figure 8.1. Lennard-Jones[19] potential-energy diagram for ammonia synthesis catalyzed by an iron surface (after Ertl[37]).

and under "normal" $N_2 : H_2$ partial-pressure ratios, this process is rate-limiting for the synthesis reaction.

The activity increases when potassium is added as an "electronic" promoter (Figure 8.1).[37] The promoter effect is strongest on the closely-packed (110) plane and weakest on the open (111), thus leveling off the pronounced face-dependence of the activity on the clean surfaces. It has been interpreted as a local effect, where at low K coverage each adsorbed K atom creates about one new state for $N_{2,ad}$ at a neighboring Fe site, and at higher K concentrations an increasing fraction of the surface is blocked by K atoms.[39]

8.2.3.2. Methanation

Surface-science experiments have also been shown to be relevant[40] for the methane synthesis over transition metals (Ni, Fe, Rh, and Ru), $3 H_2 + CO \rightarrow CH_4 + H_2O$. As with many reactions, the activity varies along the transition-metal series according to a "volcano"-shape curve.[41,42] This is commonly explained in terms of inhibitions caused by less suitable binding of reactants, too strong binding causing too slow desorption of products and too weak binding making the surface concentration of reagents too low.[43] A rather elaborate, but more realistic, picture will be indicated in Section 8.5.

A vital reaction intermediate is a surface carbide, i.e., an active carbon species adsorbed on the surface.[44,45] "Carbidic" carbon is distinguished from "graphitic" carbon,[40,46] which always accompanies the deactivation of the catalyst surface. The methanation rate is determined by a delicate balance between the formation and removal of surface carbide.[40] The carbide is formed from CO, probably by disproportion,[45] but possibly by dissociation[47] or hydrogenation.[48,49] The removal occurs by the reaction of the carbide with surface hydrogen atoms, formed by dissociative adsorption of H_2. Conversion to graphite occurs only when the carbide coverage and temperature have reached certain levels.

Electronegative adatoms (C, N, P, S, and Cl) on the Ni(100) surface cause a reduction of CO and H_2 adsorption rates and adsorption binding energies. This poisoning effect becomes stronger with increasing electronegativity. The initial sticking coefficients of CO and H_2 are affected almost as if there were a four-site blocking at low coverages.[50] Preadsorption of an electropositive atom, potassium, decreases the steady-state rate of methane formation but increases those for higher hydrocarbons, relative to clean Ni(100).[51] The surface carbide increases sharply in coverage upon the addition of 10% of a K monolayer. This is a result of a marked decrease in the activation energy for carbide formation from CO, affected by K[51]; cf the figure corresponding to Figure 1 in the work of Bonzel and Krebs.[47]

8.2.3.3. Potential-Energy Features to Be Explained

In addition to the list in Section 8.2.2, there are many theory questions raised by the heterogeneous-catalysis perspective, such as: Why are the surfaces so specific for certain reaction steps? What makes the catalysts selective, i.e., what facilitates the formation of one product molecule while inhibiting the production of other different, but thermodynamically allowed, molecules? How long-lived are internal excitations of adsorbates? What is the nature and efficiency of energy transfer among the reactants, the surface, and the products? Why is there a structural sensitivity in catalytic reactions on an atomic scale, i.e., an increased activity at, e.g., steps, kinks, and vacancies[52]? How do poisons and promoters act? How can the bonding and activity trends along the transition-metal series be explained (see also elsewhere[53,54])?

Complete answers to questions like these require solutions of equations of motions of atoms and molecules at surfaces. Such solutions have been obtained in a few cases for given assumptions about reactive and dissipative forces (see, e.g., Tully's contributions[55]).

8.3. Making Bonds at a Surface

The criterion for bond formation is that energy can be gained by leaving the constituting atoms in the bond configuration. The binding energy is one of the quantities that can be derived from the potential-energy function. The theme of this section is that *potential-energy surfaces can be understood in a systematic way in terms of electron-structure features.*[1,2,6,56,94]

The conceptual framework behind this ultimately emanates from the simple idea for two interfering atomic states, used in oversimplified accounts for the bond in, e.g., the H_2 and LiF molecules.[57] Interference results in shifts of the atomic energy levels into a lower, bonding molecular level and a higher, antibonding one, the strength of the bond being roughly a matter of relative occupation of bonding and antibonding states ("bond order"). A similar distinction can be made for energy bands of solids, with states in the bottom of the band being bonding and with the antibonding ones at the top.[58]

This section presents trends and results for the adsorbate-induced electron structure and the potential-energy surfaces of adsorbates, together with short accounts of methods to calculate the adsorbate–substrate interaction.

8.3.1. Separated Adsorbate and Substrate

In the limit of large separations, the electron structure of the adparticle is very little affected by the presence of the surface, and the interaction is weak. Yet three cases can be distinguished:

8.3.1.1. Neutral Adsorbate

A neutral particle, whose electron wave functions do not overlap those of a substrate, has a weak attraction to the surface. This van der Waals interaction emanates from the fluctuating dipoles of the adsorbate and substrate electrons. The corresponding potential energy has the limiting form

$$V_{vw}(d) = -c_3/(d - d_{vw})^3 \qquad (8.3.1)$$

where d is the distance to the surface, and c_3 and d_{vW} are constants that can be calculated from the polarizability of the adparticle and the dielectric behavior of the substrate.[59-61]

8.3.1.2. Charged Adsorbate

A charged particle is attracted to the surface by a stronger force. At large separations d, the classical image potential applies[62]:

$$V_{im}(d) = -q^2/4(d - d_{im}) \qquad (8.3.2)$$

where q is the charge of the particle and d_{im} the position of the image plane.[63] The affinity levels and ionization potentials of neutral particles vary with the distance d according to[64]

$$A(d) = A + e^2/4(d - d_{im}) \qquad \text{and} \qquad I(d) = I - e^2/4(d - d_{im}) \quad (8.3.3)$$

8.3.1.3. Charge Transfer

Already at rather large distances d, a strong electronegative, or electropositive, particle might cause an electron to exchange with the substrate. This is sometimes called harpooning.[65] In this way, an initially neutral particle can be strongly attracted to the surface. In the classical-charge limit, the potential energy is[66]

$$V(d) = \begin{cases} -A + \Phi - e^2/4(d - d_{im}) \\ I - \Phi - e^2/4(d - d_{im}) \end{cases} \qquad (8.3.4)$$

for electronegative and electropositive particles, respectively, where Φ is the work function of the substrate.

8.3.2. Pauli Repulsion

When the electron distributions of the adsorbate and substrate begin to overlap, the attractive forces between separated adsorbate and substrate are counteracted by a repulsive force. The Pauli principle states that only one electron can occupy each fully-specified quantum orbital. If the electrons of the adsorbate try to penetrate the substrate electron distribution, there is an energy penalty for that, in particular, an increase in kinetic energy. For a given adsorbate, this increase is stronger, the higher the density of the substrate. In other words, the higher the n_0 the further out does the repulsive wall meet the particle. This effect is called Pauli or kinetic-energy repulsion.

8.3.3. Physisorption

8.3.3.1. Inert Atoms and Molecules

Stimulated by the recent, significant progress in elastic atomic and molecular beam scattering experiments,[67] the theory of physisorption has made several important steps.

For an inert-gas atom on a simple model surface, the potential energy can be calculated by solving the fundamental equations self-consistently. In particular, this has been done in the so-called jellium model, which has the charges of the ions in the substrate lattice smeared out as a uniform, semi-infinite background charge.[68,69] Calculations in the Kohn–Sham scheme,[70] using a local density approximation for the exchange-correlation effects,[70-73] give realistic physisorption potentials for the rare-gas atoms close to the surface, but also indicate an increasing covalent character of the bond with atomic size.[74,75]

The repulsive potential, or Pauli repulsion, for, e.g., an He atom is proportional to the electron density $n_0(\mathbf{r})$ of the substrate at the classical turning point $\mathbf{r}$ of the He atom[76]; this is implied by the so-called effective-medium scheme[77,78] in its simplest version.[76,79]

In a refined approach to this problem one separates the interaction between atom and surface into two terms, V_{vW} according to equation (8.3.1) and improvements thereof, and V_R, of zeroth and first order, respectively, in the overlap between the He s-wave function and the metal wave function tails.[61] The Pauli-repulsion term V_R is due to the scattering of individual substrate Bloch electrons from the He atom; it is expressed as the sum of the shifts of the band energies due to the presence of the He atom.[61] For a general surface, these shifts can be calculated via perturbation theory in the nonlocal He pseudopotential.[80-83] In this way, the repulsive interaaction $V_R(\mathbf{r})$ can be written in terms of the local density of states $\rho(\varepsilon, \mathbf{r})$ of the substrate electrons at the He nuclear position $\mathbf{r}$ and a universal function of the energy, $g(\varepsilon)$,

$$V_R(\mathbf{r}) = \int_{-\infty}^{\varepsilon_F} d\varepsilon \, g(\varepsilon)\rho(\varepsilon, \mathbf{r}) \qquad (8.3.5)$$

In cases with little variation in $g(\varepsilon)$, like free-electron metals, this expression also gives a direct proportionality to the electron density $n_0(\mathbf{r})$.[82]

Helium-diffraction experiments give information about the corrugation of the surface, the dimensions of the attractive well, and the softness of the potential, i.e., its dependence on the energy of the incoming He atom.[67] Detailed comparisons between calculated and measured values of the quantities can now be made.[79,81-84]

Recent molecular-beam experiments with H_2 and D_2 scattering from a Cu(100) surface[85] show that there is still room for improvements in the theoretical description of physisorption. The well depth (30.9 meV) is found to be substantially larger than the commonly accepted value (around 22 meV).[85] The calculations described above are based on a division of the physisorption potential into one part due to van der Waals attraction, and one due to Pauli repulsion. No doubt, this approach has been successful in understanding the difference in well depth between He and H_2 on metal surfaces. More subtle details in the description of the physisorption interaction, such as the trends over the noble metals and different surface orientations, appear to require refined treatments of the van der Waals constants and the substrate–electron structure.[85]

8.3.3.2. Reactive Species with Frozen Electron Configurations

An inert atom is characterized by the prohibiting energy for changing its electron configuration. An electron configuration may stay frozen also for, e.g., a particle approaching the surface with a high speed with no time for it to change its configuration. It may remain on the *diabatic* potential-energy surface $(3p)^5$. For example, the $(3p)^5$ configuration of the Cl atom and the 1s configuration of the H atom correspond to excited states of the system, close to the surface. For a reactive species with a frozen electron configuration the results mentioned in Section 8.3.3.1 apply. For instance, the potential-energy surface of Cl $(3p)^5$ has a weak physisorption minimum and a repulsive wall that meets the atom rather distantly from the surface.[86]

8.3.4. Reactive Adsorbates

8.3.4.1. Chemisorption

When atoms are brought together, the electron energy levels are changed. Instead of having one center of force, an atomic valence electron gets two, or more, centers. This might allow the electron to be shared among the atoms. Such sharing has two major consequences.[87] One is interference (hopping, resonance, or mixing). Another is the implied likelihood to find more than one electron on the same nucleus, i.e., a penetration of the electrons. An increased Coulomb repulsion between the electrons is caused by this effect.

Reactive adsorbates are characterized by the fact that the new electronic configurations can be formed by sharing of electrons between adsorbate and substrate. This is the case of chemisorption.[88,89] An electronic configuration different from that of the free particle can allow the adsorbate electrons to benefit from the attractive potential of the nuclei,[88,90] which the adsorbate experiences as an attractive potential energy (Figure 8.2). New electron configurations can be present because the adsorbate one-electron energy levels broaden[91] and shift,[92,93] due to the chemisorption. An initially empty (filled) electron level can become partially or completely shifted to energies below (above) the Fermi level of the substrate and consequently change its occupancy, with the substrate acting as an electron reservoir.[6,94]

With the above perspective, the appearance of any attractive feature in a potential-energy surface, beyond the physisorption well, is very much a question about new configurations: whether, upon approaching the surface, an attractive well, associated with such a configuration, can be formed "before" the repulsive kinetic-energy wall of the same configuration becomes too strong. Figure 8.2[95]

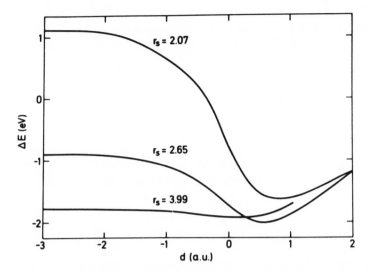

Figure 8.2. Potential-energy curves for atomic H chemisorbed on three different jellium substrates, modeling Al (r_s = 2.07 au), Mg (2.65), and Na (3.99). The distance to the jellium edge is d (after Hjelmberg[95]).

illustrates this point for H adsorbed on jellium surfaces. At all three indicated substrate-electron densities there is a "new" H-like configuration (cf. Section 8.3.4.3 below), giving attraction. The higher the density, the further out is the repulsive wall. The relative locations of the substrate-electron-density profile and the crossings between the adsorbate-induced levels and the Fermi level of the substrate are of key importance to understanding the general features of potential-energy surfaces, and in particular their specificity with respect to substrate metal, substrate face, and defect properties. The electron profile can be obtained from calculations on clean surfaces.[68,96,97] In addition, a systematics of factors that affect the adsorbate-induced electron structure are of great value. Some theoretical tools to derive such systematics will be described briefly in Section 8.3.4.2. The results show that, in the systematics (Section 8.3.4.3), one can distinguish between adsorbate and substrate factors.

a. Peak Widths due to Interference. The effects of interference are well known from many kinds of quantum systems. The simplest case is that of the two-state system, such as that of an electron in a diatomic molecule. As this model, and its immediate generalizations, has a bearing on the qualitative interpretation of several aspects of chemisorption,[6,94] the key results will be reviewed briefly.

It is assumed that there are two states possible for an electron, $|a\rangle$ on atom A and $|b\rangle$ on atom B, say, with the energy eigenvalues ε_a and ε_b, respectively. With some weak coupling V_{ab} between these states, the superposition $|\psi\rangle = c_a|a\rangle + c_b|a\rangle$ is a natural Ansatz for the combined system. The secular equation of the resulting linear equation system has the roots

$$\varepsilon_\pm = (\varepsilon_a + \varepsilon_b)/2 \pm [(\varepsilon_a - \varepsilon_b)^2/4 + |V_{ab}|^2]^{1/2} \qquad (8.3.6)$$

For a symmetrical case, i.e., when $\varepsilon_a = \varepsilon_b$, as the H_2 molecule, the interference breaks the original degeneracy into two separated energy levels, $\varepsilon_\pm = \varepsilon_a \pm |V_{ab}|$. Evaluation of the coefficients $c_{a\pm}$ and $c_{b\pm}$ shows the bonding and antibonding characters of the corresponding orbitals.

In the asymmetric case ($\varepsilon_a \neq \varepsilon_b$), equation (8.3.6) shows that the levels ε_a and ε_b are shifted with an amount that increases with $|V_{ab}^2/(\varepsilon_a - \varepsilon_b)|$. The corresponding expressions for c_a and c_b show that in this case the state derived from $|a\rangle$, say, has its major weight around atom A also with the interference.

For adatom energy levels the interference effects are of an asymmetric kind. A particular, and important, feature is the fact that the substrate has a macroscopic size. Consequently there is a macroscopic number of substrate states $|k\rangle$ to interfere with. The energy levels are the roots of the secular equation

$$\varepsilon - \varepsilon_a - \sum_k |V_{ak}|^2/(\varepsilon - \varepsilon_k) = 0 \qquad (8.3.7)$$

For the low ε_a values there is one level split off below the band of energies ε_k, and the other roots correspond to shifted band energies.

The character of the new state depends very much on the overlap and coupling between the orbital $|a\rangle$ and the $|k\rangle$ states. A rapidly oscillating Bloch state high up in the conduction band does not overlap much with a smooth atomic H or H^- orbital. Only Bloch states with tails stretching out sufficiently overlap with an adparticle outside the surface. In addition, the k-state has to have the proper local symmetry. Therefore, the effective band of k-states contributing to equation (8.3.7) differs from the bulk band. In particular, the split-off state should, according to this argument, follow the "local bottom of the band," which may differ from one type of adsorbate state to another.

With a final adsorbate level in an energy gap of the substrate, there is no possibility for an electron to propagate inside the substrate. Then the adsorbate electron state is localized to the adatom and a few of the nearest substrate atoms. Such a localized state has a sharp energy level. If, on the other hand, the adsorbate-induced level lies in an energy band, then there is a certain probability for propagation within the substrate. The electron then has a limited lifetime τ on the adsorbate, corresponding to an energy uncertainty $h/(4\pi\tau)$. The corresponding adsorbate-induced level has a nonzero width.[91] The Golden Rule of time-dependent perturbation theory gives the simple estimate

$$h/(2\pi\tau) = 2\pi \sum_k |V_{ak}|^2 \delta(\varepsilon_a - \varepsilon_k) \qquad (8.3.8)$$

b. *Shifts due to Coulomb Repulsion.* The effects previously mentioned are present in the one-electron approximation, i.e., when the Coulomb interaction between the electrons is ignored, or averaged out. The Coulomb repulsion between the electrons is an important effect.[98] The standard example is that of hydrogen chemisorbed on a transition-metal surface.[92] For the free H atom, the ionization potential I is 13.6 eV and the electron affinity A is 0.7 eV. The large difference $U = I - A$ for hydrogen is due to the Coulomb repulsion between the two electrons on H^-. For an adatom on a surface there is a new self-consistent setup at each position of the hydrogen atom which alters the atomic level to ε_{sc}. The size of the Coulomb repulsion depends upon the number of electrons on the adatom. Screening by the substrate conduction electrons should reduce the effective U from its free atom value.[99] Far outside the surface this can be expressed as an image-potential effect.

c. *Newns–Anderson Model.* For the conceptual development of chemisorption there is a model that has meant more than any other theoretical considerations. It originates from Anderson's description of magnetic impurities in metals.[98] In second-quantization language, the Newns–Anderson Hamiltonian[92] reads

$$H = \sum_{k,\sigma} \varepsilon_k n_{k\sigma} + \sum_{\sigma} \varepsilon_a n_{a\sigma} + U n_{a+} n_{a-} + \sum_{k,\sigma} (V_{ak} c_{k\sigma}^{+} c_{a\sigma} + \text{h.c.}) \qquad (8.3.9)$$

where c_i^{+}, c_i, and $n_i = c_i^{+} c_i$ are creation, destruction, and occupation-number operators, respectively, and h.c. denotes the Hermitian conjugate. The indices i stand for adatom, (a, σ) and substrate, (k, σ), orbital and spin indices.

In the Hartree–Fock approximation of equation (8.3.9), valid only in the limit of negligible correlation effects, $U = 0$, the adatom density of states can be expressed as

$$\rho_a(\varepsilon) = \pi^{-1} \Delta(\varepsilon)/\{[\varepsilon - \varepsilon_{sc} - \Lambda(\varepsilon)]^2 + \Delta^2(\varepsilon)\} \qquad (8.3.10)$$

The sums found in equations (8.3.7) and (8.3.8) are here generalized to a chemisorption function $q(\varepsilon + i\sigma) = \Lambda(\varepsilon) - i\Delta(\varepsilon)$, where

$$\Delta(\varepsilon) = \pi \sum |V_{ak}|^2 \delta(\varepsilon - \varepsilon_k) \qquad (8.3.11)$$

and the real part $\Lambda(\varepsilon)$ relates to $\Delta(\varepsilon)$ by a Hilbert transform. The metal properties appear through $\Delta(\varepsilon)$ and the screening effects on U. An analysis of equation (8.3.10) shows how the adatom electron structure is shifted, broadened, and further distorted due to the effects of V_{ak} and U.[88,89,92,93]

There are several classes of behavior: For weak metal–adatom coupling ($\Delta \ll B$, the substrate-electron bandwidth), sometimes called weak chemisorption, there are three typical cases: (1) a state that is localized near the adatom, with an energy below the band, (2) an atomic resonance state with energy in the band and a half-width Δ, and (3) a localized state with energy above the band. In the strong chemisorption regime ($\Delta \gg B$) there may be, in addition, a solution with a pair of levels embracing the band. This narrow-band solution exhibits a clear bonding-antibonding character. It is typically relevant for chemisorption on transition metals,[92,100] where the d-electron levels lie in a narrow band.

The problem has been treated completely only after self-consistency has been achieved. At a given value of ε_{sc}, the number of electrons $\langle n \rangle$ on the adatom is obtained by integrating equation (8.3.10) up to the Fermi level ε_F and summing over spins. The intersection of the corresponding curve for $\langle n \rangle$ as a function of ε_{sc} gives the self-consistent values of ε_{sc} and $\langle n \rangle$.[92]

While the Hartree–Fock solution gives one peak in the adatom density of states in the weak-chemisorption limit and two in the strong-chemisorption limit, correlation is expected to introduce additional structure.[101] For H chemisorption on Ni, in the equilibrium geometry, where the s–p band is very important, the qualitative changes in the spectrum due to these many-body effects seem to be small.[102] For a system with stronger effects of the d-electrons, however, the correlation might give observable effects.

8.3.4.2. Some Chemisorption Calculations

The Newns–Anderson and other model Hamiltonians require values of the ingoing parameters from other sources. There are methods which calculate the chemisorption properties, such as spectral changes upon adsorption and potential-energy surfaces, from more fundamental equations.[88] A frequently used basis is the Kohn–Sham scheme,[70] which is a simple and physically sound calculational scheme to manage the complex problem with a local-density approximation for exchange and correlation[70-72] or a nonlocal one.[103]

In this brief exposé, some methods that focus on the geometry with a single adsorbate and an extended substrate will be mentioned first. At the end, layer and cluster methods will be briefly accounted for.

a. Lang–Williams Method. Lang and Williams[69,104-106] have designed an efficient method of solving the Kohn–Sham equations, tailor-made for the adatom–jellium system. They utilize the fact that the metal electrons screen out the effects of the adatom on the charge density and potential (though not on the wave functions), except close to the adatom nucleus, and solve the Lippman–Schwinger equation equivalent to the one-electron equation of the Kohn–Sham scheme for the continuum states. Lang and Williams have calculated local densities of states, densities, dipole moments, potential-energy curves, equilibrium bond lengths, and chemisorption energies, in particular for H, Li, O, Si, and Cl on a high-density jellium substrate.[68,69,104-106]

b. Gunnarsson–Hjelmberg Method. The Gunnarsson–Hjelmberg method[90,107] to solve the equations of the Kohn–Sham scheme in the embedding configuration has been applied to the adatom–jellium geometry, too, but has a more general applicability. Inspired by Grimley's method[108] to treat the overcompleteness problem, it also utilizes the screening effects and describes the changes in the electron density locally around the adatom in terms of a *finite* set of N localized functions.

The adsorbate-induced Green function is calculated from an $N \times N$ matrix equation in the local representation

$$G_A = G_A^m (1 - V_A G_A^m)^{-1} \qquad (8.3.12)$$

where G^m is the clean-substrate Green function and V the potential energy due to the adsorbate. The approximations all improve, when the set of functions becomes more complete in the perturbed region, which usually occurs by increasing N. A simple but physically reasonable evaluation of G^m is obtained in the semi-infinite jellium model. Effects of the discrete ion lattice can be accounted for approximately in a perturbative way.[90,109] Originally designed for atomic substrates, the method has also been used for molecular hydrogen, defining the local basis set on a sphere with center in the midpoint of the molecular axis[110,111] and to dipoles.[112] It has also been extended to other atomic adsorbates[113] by using norm-conserving pseudopotentials.[114]

c. Effective-Medium Schemes.

c. Effective-Medium Schemes. The calculations described in Subsections a and b are self-consistent and concern the interaction of the atom or molecule with the substrate electrons described in the jellium model. They indicate clearly that the local electron density of the host material is a key parameter for the determination of several adsorbate-induced properties.[110,115]

The effective-medium scheme is an attempt to account for the effects of the local interactions in a simple quantitative way.[77,78,116,117] Its basic idea is to replace in the calculation the true, low-symmetric host by an "effective" host with higher symmetry and thus a simpler representation. In the simplest approximation, the chemisorption or, more generally, embedding energy ΔE, defined as the difference in energy between the combined atom–host system and the separated atom and host, would be

$$\Delta E_{\text{eff}}(r) \approx \Delta E_{\text{hom}}(\bar{n}_0(r)) \qquad (8.3.13)$$

where $\Delta E_{\text{hom}}(n)$ is the embedding energy of the atom in a homogeneous electron liquid of density n. $n_0(\mathbf{r})$ is the host electron density at the site $\mathbf{r}$ of the atom, and $\bar{n}_0(\mathbf{r})$ is a properly averaged value.[116] The host is characterized only by one quantity, $n_0(\mathbf{r})$. The properties of the atom are given by the immersion energy $\Delta E_{\text{hom}}(n)$ that can be calculated once and for all for each atom or molecule, immersed in a homogeneous electron gas.[118]

Equation (8.3.13) is the first version of the so-called effective-medium theory (EMT-1),[77,78] where a systematic expansion in deviations from the local density is employed. Keeping the first-order term in such an expansion gives potential-energy curves for, e.g., H and O on a jellium surface close to those of self-consistent calculations.[77]

In a second version of the effective-medium theory (EMT-2),[116,117] first, the true medium is approximated by an effective, approximate one within a small region **a** around the adsorbate. Then corrections to this are considered in a mixed first-order perturbation theory, with the atom regarded as a small perturbation outside region **a** and the deviations of the host from the effective medium assumed small inside region **a**. As a result one obtains an expression for the embedding energy ΔE with a simple structure,[116,117]

$$\Delta E = \Delta E_{\text{eff}} + \Delta E_{\text{es}} + \Delta E_{\text{cov}} \qquad (8.3.14)$$

where ΔE_{eff} is the embedding energy in the effective medium [equation (8.3.13)] and ΔE_{es} an electrostatic energy. The covalent term $\Delta E_{cov} = \delta\Delta(\sum_j \varepsilon_j)$ is a contribution not accounted for by the effective medium. It is the change in the atom-induced shifts of the one-electron energy parameters ε_j, when going from the effective host to the true host. The sum $\delta\Delta(\sum_j \varepsilon_j)$ is governed mainly by the possibility of finding a resonance between the atom- and host-derived one-electron energies. The covalent term requires a calculation of $\Delta\varepsilon_j$ in the real host for fixed potentials, i.e., not self-consistently. Often, for instance, for hydrogen, the first two terms of equation (8.3.14) give the dominant contribution to ΔE. These two facts make ΔE_{cov} accessible with rather simple methods. In the case of, e.g., chemisorbed or interstitial hydrogen and a transitional metal, the H 1s level interferes with the s-, p-, and d-valence electrons of the host. As a large part of the interference with the s–p electrons is accounted for by the effective medium, ΔE_{cov} expresses primarily the effects of the host d-electrons and can be calculated with rather simple methods.[119] In this way many useful results have been obtained, including explanations of trends along with the transition-metal series in chemisorption energies[116,117,120-124] and in rates for certain catalytic reactions,[122] and of electrostatic effects in the promotion of surface reactions.[125]

Starting from the variational property of the Hohenberg–Kohn[70] density functional for the total energy, using the local density approximation for the exchange-correlation energy,[70,73] a version of the EMT with interatomic forces (EMT-3) has been developed.[126] The electron density of the system under study is represented approximately by the sum of atomic densities, as calculated when the atom is embedded in a homogeneous electron gas of a density given by the first average of the densities from the neighboring atoms. This is a reasonable first approximation, which for instance includes the spherically symmetrical part of the screening effects. It also includes charge transfer, but again only in a spherically symmetric way. The variational property then insures that any errors in the Ansatz density relative to the true ground-state density will only show up to second order in the total energy of the system. The Ansatz allows us to write the total energy of the system as

$$\Delta E_{tot} = \sum_i E_{c,i}(n_i) + \Delta E_{1el} + \Delta E_{AS} \qquad (8.3.15)$$

where n_i is the density from the neighbors averaged in cell i:

$$n_i(S_i) = \sum_{j \neq i} \Delta n_j(S_i, r_{ij}) \qquad (8.3.16)$$

Δn_j being the average of the jth atom density over the Wigner–Seitz sphere i with radius S_i,

$$\Delta n_j(S_i, r_{ij}) = (4\pi S_i^3/3) \int_{S_i} \Delta n_j(\mathbf{r} - \mathbf{r}_j) \, d^3r \qquad (8.3.17)$$

The energy function $E_{c,i}(n)$ is related to the energy of ΔE_{hom} of atom i in a homogeneous electron gas. It can be calculated once and for all. For reactive atoms it shows a single minimum, indicating that atoms of this kind will tend to find surroundings providing a particular optimum density.[127] This term describes the dependence of energy of a closely packed metal on volume, and determines to a large extent the equilibrium lattice constant and bulk modulus.[126]

The term ΔE_{1el} is a one-electron energy difference, which is important in describing, e.g., the d-band formation in the transition metals corresponding to the ΔE_{cov} term in equation (8.3.14).

Finally, quantity ΔE_{AS}[126] describes an extra electrostatic energy that must be included for all systems, where the Wigner–Seitz cell cannot be well approximated by a sphere. This includes all situations, where the system is not in a perfect close-packed arrangement. For such systems one still chooses to work with atomic spheres and then corrects for the regions that are double-counted and those that are not included at all. The atomic spheres are chosen to be always neutral. In this way the long-range Madelung-type contributions to the total energy are avoided. The atomic-sphere correction is generally repulsive, when the neutral atomic spheres overlap. It is, for instance, this term which is responsible for the shear strength of a close-packed metal. Applications so far include surface energies and relaxation,[126,128] adsorption of simple gases on metal surfaces,[129] and reconstructions on surfaces.[130]

To allow the treatment of, e.g., H_2 dissociation on the N surface, the method has recently been extended to treat also covalent interactions between adsorbed atoms.[131]

d. Other Methods. To balance the emphasis of this chapter on certain methods, some examples with more pronounced molecular-physics and solid-state-physics approaches will be mentioned.

As an example of the far-reaching achievements of a molecular-physics approach, a cluster study of the interaction of a water molecule with an aluminum surface[132] can be mentioned. These calculations allow not only very explicit conclusions about the geometry of the water molecule, the identification of a dominant chemical component in the H_2O-surface interaction, and frequencies and intensities in good agreement with electron-energy-loss-spectroscopy (EELS) data, but also an identification of the nature of the H_2O-surface bond.[132] They also signal shortcomings of the cluster approach, in particular a substantial overestimate of the work-function change due to an H_2O overlayer.

As an example of a solid-state approach, a total-energy calculation for adsorbed hydrogen on a nickel surface $[p(1 \times 1) \, H/Ni(100)]$[133] can be mentioned. By forming a slab of up to seven layers of Ni, two outside layers of H, and four "layers" of vacuum, and repeating a supercell of this slab periodically in three dimensions, methods designed for the electron-structure problem of the ideal bulk crystal can be applied, in this case the linearized augmented-plane-wave (LAPW) method.[134] Extensive numerical calculations give the electron structure of the slab and the binding energy per hydrogen atom in the adsorbed layer. This represents an enormous achievement in computational accuracy. It is also interesting to note that the results[133] for the binding energy are in the same range as those from cluster[135,136] and embedding[116] methods.

The "superlattice" approach of, e.g., Umrigar and Wilkins[133] does not treat the ideally simplified adsorption problem. It accounts for a whole adlayer. To describe the making and breaking of chemical bonds at surfaces, the problem of a single adsorbate on a perfect surface is the ideal one, the problem focused on in the previous subsections. As a recent benchmark in solving this problem work by Feibelman[137] should be mentioned. Here a first-principles calculation for a single adatom on a crystal is performed without any other physical approximation than the Born–Oppenheimer approximation and the local-density approximation for exchange and correlation effects.[70-73] Based on the rapid screening of the adatom-induced one-electron potential, a localized-orbital basis is used to express the self-consistent Dyson equation as a finite matrix equation. With norm-conserving potentials[114] for the substrate and adsorbate ions, bond-charge densities and total energies have been calculated for an Si atom on an Al monolayer. These results are interesting, not only because they point forward toward future developments in this subfield, but also because the computed binding energies, force constants, and bond charges for two different bond geometries confirm more approximate calculations for the Si/Al adsorption system.[137] This matrix Green-function method is able to treat problems involving the displacement of substrate atoms. This has permitted studies of the force constants of a clean Al(001) surface, of the relaxation of the neighbors to an Al or a S adsorbed on Al(001), of the surface core level shift for Al(001) which includes the effects of final state screening.[138]

8.3.4.3. Systematics

From the calculations one can extract certain systematic features of chemisorption. Some of these relate to the adsorbate, others to the substrate.

a. Adsorbate Factors. The distinction between electropositive and electronegative adsorbates, present in the image-potential limit (Section 8.3.1.3), is present also in the chemisorption regime. Typical electronegative adsorbates are the Cl and O atoms. Figure 8.3a[106] illustrates how the p levels, or rather the corresponding Kohn–Sham energy parameters,[70] of these atoms are shifted to lower energies and broadened, when the atom is getting closer to the substrate. The more electronegative Cl atom has its 3p level cross the Fermi level of the substrate further out than the 2p level of the O atom.

As a matter of fact, the affinity of Cl is high enough to make the 3p level lie below the Fermi level of, e.g., the Na metal at all separations. Unlike the Na atom plus Cl atom system,[8] the Na *metal* plus Cl atom system thus shows no crossing of the covalent and ionic potential-energy curves. Rather, the Cl $3p^5$ configuration is an excited state, when Cl is in contact with an Na surface.[86] The de-excitation to the ground-state $3p^6$ configuration can be observed as chemiluminescence.[139] For oxygen, on the other hand, one can envision several potential-energy wells developing, a weak physisorption well for the p^4 configuration, a chemisorbed one

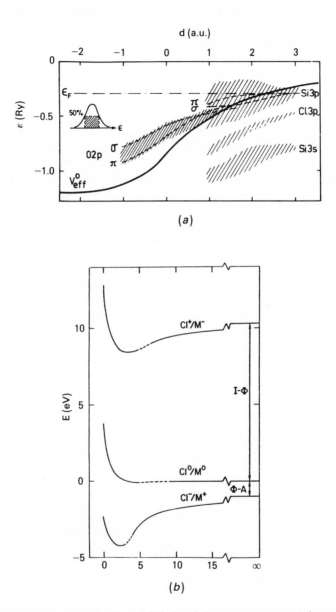

Figure 8.3. (a) Variation of the O-, Si-, and Cl-induced electron resonances vs. the metal-adatom separation d (measured from the jellium edge) for adsorption on a jellium substrate with $r_s = 2$. The shaded regions indicate the Kohn–Sham electron energy values within the full width at half maximum. The full curve is the effective electron potential of the bare jellium, and ε_F is its Fermi level.[69,94,104-106] (b) Potential-energy curves for atomic Cl chemisorbed on a jellium substrate modeling Na ($r_s = 4$) (after Lang et al.[86]).

for p^5, and a second chemisorption well for the p^6 configuration. The corresponding diabatic[8] potential-energy curves cross, and the adiabatic one goes smoothly[77] from one curve to another.

A case with curve crossing is shown by hydrogen adsorption (Figure 8.4[3,140]). At large separations d, the neutral H atom and metal M form the ground state. The charge transfer required to create $H^- + M^+$ gives a potential-energy curve [equation (8.3.4)] that crosses the H/M curve in the region of its weak physisorption well. In the limit with H inside the metal, the energy cost for having a vacancy in the H 1s level is high, and the stable configuration is H^-/M^+.[115,141]

For an electropositive adsorbate, the ionization potential is of key interest. Upon approach to the substrate the upward shift of energy, experienced already in the image limit [equation (8.3.4)], continues. At the equilibrium distance, an adsorbed alkali atom typically has a very broad valence-electron resonance, mainly occupied.[106,142]

The Group IIA elements (Be, Mg, Ca, Sr, ...) illustrate nicely the presence of chemisorption factors beyond the *electronegativity scale*.[113] They are next to the alkalis, with two valence electrons, and ought at first sight to resemble them. On the other hand, in their ground state they have a closed subshell $(ns)^2$ and ought to have some similarities with He. The self-consistent calculations show very weak chemisorption for Be and Mg (Figure 8.5). This can be explained in terms of the *subshell closure*. In analogy with the explanation for the weak binding energies of dimers of these elements (Be_2, Mg_2, ...) and the weak cohesive energies of the corresponding metals,[143] the $(ns)^2$ configuration would give essentially no bond, and the existence of any weak chemisorption bond is due to configurations with occupancy of 2p levels that are mixed in.[113,144] Plots of the adsorbate-induced density and density of states show the partial occupation of 2p-derived states (Figure 8.6).

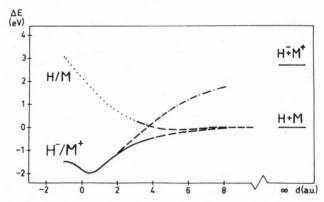

Figure 8.4. Potential-energy curves for H and H^- as functions of the distance d from the jellium edge, resulting from effective-medium [equation (8.3.13), dotted curve], image-potential [equation (8.3.4), dash–dotted], and self-consistent calculations (Subsection 3.4.2b, solid), and interpolation (dashed). The jellium substrate has $r_s = 3$, corresponding to the conduction-electron density in Ag metal (after Hellsing *et al.*[140]).

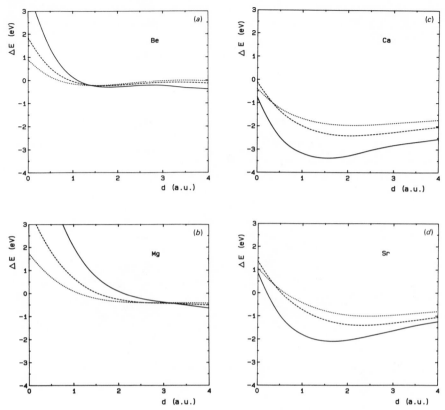

Figure 8.5. Potential-energy curves for (a) Be, (b) Mg, (c) Ca, and (d) Sr on jellium surfaces, modeling Al ($r_2 = 2$, full-drawn curve), Mg (2.65, dashed), and Li (3.28, dotted), d being the distance to the jellium edge (after Holmström[113]).

The calculated potential-energy curves for Mg on jellium (Figure 8.5) do not even show a chemisorption minimum. The additional weakness of the bond here is caused by another adsorbate factor: the *core orthogonalization*. Due to this effect the 3p electron in Mg is slightly more weakly bound than the 2p electron in Be and thus contributes less to the chemisorption bond.[113]

Figure 8.5 shows also that chemisorption of the heavier group IIA elements, Ca and Sr, is significantly stronger than for lighter ones. The density plots in Figure 8.7 illustrate the explanation, namely, the participation of adsorbate *d electrons* in the bond.[113] In the periodic system Ca and Sr are next to the left of the transition-metal series, thus with empty d states in the free atom. Obviously the jellium surface lowers the d-state energies sufficiently to allow some occupancy. The weaker chemisorption bond for Sr on jellium can be understood in terms of core orthogonalization for the 4d electrons.[113]

b. Substrate Factors. Figure 8.8 shows the variation of the H-induced resonance on three different substrates modeled by jellium.[95] For a substrate with a

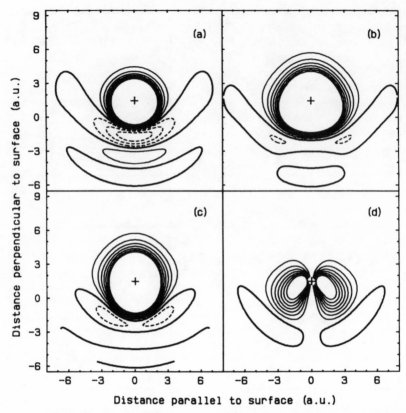

Figure 8.6. Level curves for the adsorbate-induced density $\Delta n(r)$ of (a) H, (b) Be, (c) Be, σ-symmetry, (d) Be, π-symmetry. The jellium, modeling an Mg substrate, has $r_s = 2.65$ au. The density step between the curves is 0.001 au. Dashed curves represent negative densities, full curves positive densities, and the fat curve is the zero-density curve (after Holmström[113]).

wide conduction band, like the high-density metal Al, the peak position varies by more than 10 eV per Å, while the variation for a low-density metal like Na is just 1–2 eV per Å. Obviously substrate factors affect the resonance position, the correlation with the effective electron potential of the clean surface being indicated, and the key parameter being the *conduction-electron density*.[95,145]

In spite of this large variation, the Fermi level does not cross the different resonance curves at very different distances from the surface. Therefore the attractive parts of the potential-energy curves do not differ very much for different values of r_s (Figure 8.2). The strong differences come from the Pauli repulsion. The repulsive wall meets the H atom further out on a high-density surface like Al.[95]

The behavior of the resonance peak in Figure 8.8 can be understood in terms of the interference picture of Subsection 3.4.1a. The interference is strongest with the band states having the proper s symmetry and spatial extent and form. This makes the resonances follow the "local bottom of the band," i.e., $V_{\text{eff}}^0(d)$. As such energies lie in the range of the bulk conduction band, the resulting levels broaden to resonances.[91] For adsorbate-induced levels with other symmetries (cf. Figure 8.3) other groups of substrate states are projected out, and other variations with d follow. Also in such cases, it is possible to infer the main features of the variation by just knowing the electron structure, i.e., *local symmetry of one-electron orbitals*, of the clean substrate.[142]

While the interference picture thus shows a high ability of metals to fill adsorbate affinity levels with electrons, the adatom Coulomb repulsion would imply a high energy cost for such increased occupancy (subsection 3.4.1b). However, the screening by the conduction electrons reduces the latter considerably. The high polarizability of the conduction electrons, and thus the strong screening, is ultimately due to the fundamental property of metals of having a quasi-continuous spectrum of electron excitations. The shifting and filling/emptying of adsorbate levels is the prerequisite for new configurations, which in turn form the basis for the reactivity. Thus the high reactivity of metals can be linked to the fundamental property of metals of having a gapless electron spectrum.[6]

The above results apply primarily to s and p electrons and thus directly to free-electron-like metals. For transition metals, in addition the *d electrons* have to be considered. The effective medium schemes here offer great possibilities for qualitative analysis. For instance, in the case of a hydrogen atom on a transition metal it gives a picture with the H-induced resonance state in the effective medium, which contains sp electrons, hybridizing with the d electrons.[116,117] Depending

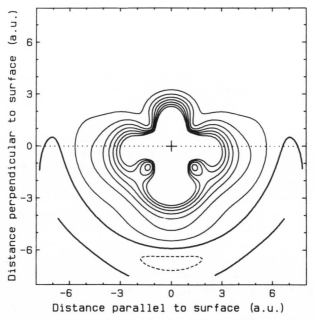

Figure 8.7. Level curves for the Ca-induced density $\Delta n(\mathbf{r})$ on a jellium surface modeling Mg ($r_s = 2.65$) with the Ca atom at a distance $d = 1.5$ bohr from the jellium edge (after Holmström[113]).

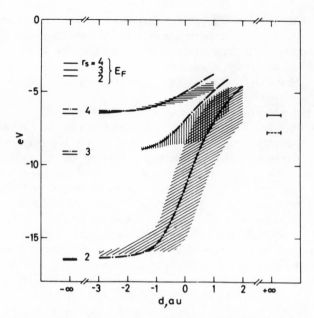

Figure 8.8. The hydrogen-induced resonance-peak position of the Kohn–Sham density of states (shaded areas), as a function of the distance d from the jellium edge, is shown for $r_s = 2$ (Al), 3 (Mg), and 4 (Na). the corresponding effective electron potential V_{eff}^0 is shown by dash–dotted curves (after Hjelmberg[95]).

on the nature of the transition metal and the degree of packing of substrate atoms in the surface region, the chemisorption type can be either weak or strong (cf. Subsection 3.4.1c).

The chemisorption bond for an adparticle on a transition metal surface can thus be considered to consist of a (major) contribution due to the interference of the adsorbate electrons with the sp electrons and an additional contribution from the interference with the d electrons. The former depends primarily on the local electron density of the host and does not vary very much along the transition-metal series. The latter depends on the relative occupation of levels that are bonding and antibonding with respect to the interaction between the effective-medium H 1s level and the d electrons. As the bonding levels lie in the lower part of the 1s d band, the filling of the d band, F, influences the magnitude of the binding energy.[117,120,146] The chemisorption energy is larger in the beginning of the transition-metal series than in the end, as shown in Figure 8.9.[120] This figure also gives an illustration of the usefulness of the extended effective-medium method as a quantitative tool. The same is shown by the good agreement between calculated and measured vibrational frequencies for an H-adsorbed atom oscillating on a number of transitional-metal surfaces,[120] and by the good account of the trend along the series Rh, Pd, and Ag of the wag-mode frequency of adsorbed OH.[147-149]

Figure 8.10 illustrates the dependence of chemisorption properties on local geometric factors.[116,117] These EMT results for H on Ni(100) can be compared

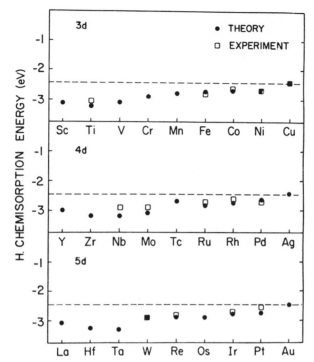

Figure 8.9. Chemisorption energies for atomic hydrogen on the most closely-packed surface of each of the transition metals. The dashed line shows the result of equation (8.3.13) (after Norlander et al.[120]).

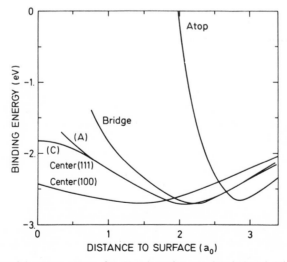

Figure 8.10. Potential-energy curves for H adsorption on low-indexed Ni surfaces (after Nørskov[116,117]).

with those of cluster[135,136] and layer methods.[133] Although the methods are addressing slightly different questions, it is still interesting to note that the resulting chemisorption energies agree within 0.5 eV and that the H–N distances agree within 10%.

8.4. Breaking Molecular Bonds at Surfaces

The adsorption of a molecule can be either associative or dissociative. Whether the breaking of, e.g., the H_2 molecular bond on a surface is thermodynamically favored or not can be seen by comparing the potential energies of the adsorbed H_2 and H.[19] Possible activation energies for dissociative adsorption can also be read off from the potential-energy curves.

The original Lennard-Jones picture,[19] with the potential-energy curve of a physisorbed molecule compared with that of chemisorbed fragments, has recently been extended and supplemented.[6,94,110,111,150,151] Often several electronic configurations are of interest and the discussion proceeds naturally in terms of the potential-energy surfaces for each configuration. The general trends and features described in the previous sections are, in this section, applied on molecular adsorbates and compared to results from a first-principles calculation.

8.4.1. Application of Simple Rules

When a molecule, say H_2, approaches a metal surface, it might experience several different potential-energy surfaces (Figure 8.11). If the quasi-continuum of electronic excitations of the metal were accounted for in the figure, there would be (quasi-infinitely) many more potential-energy surfaces of the whole system.[152] For simplicity, only the most stable energies of different configurations of the molecules will be considered here.

A molecule, like H_2, that approaches a metal surface may stay in its original

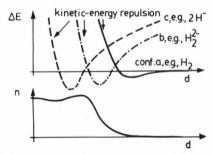

Figure 8.11. Schematic molecular potential-energy curves showing how different electronic configurations (a, b, and c, successively less inert) may give rise to different wells in the potential-energy surface. The lower figure illustrates the clean metal electron density (after Kasemo and Lundqvist[9]).

configuration $H_2 + M$. Then the major features of the potential-energy surface are a shallow physisorption well and a kinetic-energy-repulsion wall that meets the molecular far outside the surface.

As the adsorbate–substrate distance diminishes and the electron overlap increases, new electronic configurations might become energetically favorable. For instance, a configuration of the type $H_2^{-\delta}/M^{+\delta}$, where δ is not greater than 2, which has a high energy at large separations, might give an attractive well closer to the surface. This corresponds to a certain filling of the antibonding $2\sigma^*$ resonance when, due to interference with and screening by the substrate electrons, the latter is shifted down in energy to embrace the Fermi level of the substrate. However, with stronger electron overlap, there might be yet other configurations giving attractive potential-energy surfaces. Of particular interest is the potential-energy surface for the dissociated products, e.g., which could schematically be conceived as $(H^- + H^-)/M^{+2}$. With increased overlap between substrate and adsorbate electron states, chemisorption bonds build up and intramolecular bonds weaken or break.[110,111] In the hydrogen case, the latter is associated with the filling of the antibonding $2\sigma^*$ molecular-orbital resonance.[6,94,110,111,150,151,153] For other molecules, there are similar antibonding affinity levels, e.g., the $2\pi^*$ orbitals of N_2 and CO. For electropositive molecules, like Na_2, the emptying of the bonding ionization level is at focus.

8.4.2. Hydrogen Adsorption on an Mg Surface

Self-consistent calculations for chemisorbed H_2 illustrate these points clearly.[110,111,150] The dissociation energy of a free H_2 molecule is about 4.5 eV, while that of an adsorbed H_2 is one order of magnitude smaller, or even negative, depending on the substrate-metal electron density. This drastic reduction of the intermolecular forces comes about through the occupation of the antibonding $2\sigma^*$ molecular-orbital resonance. Not only the depth of the H–H potential well but also its curvature, and thus the molecular vibrational frequency, depend on the electron structure of the substrate (cf. subsection 3.4.3b) and on the adsorption site. This forms the basis of experimental probing of molecular vibrations at surfaces to inform about geometrical and electronic strucures (Section 2.2[14]).

Adiabatic potential-energy surfaces calculated for H_2 on the densely packed Mg(0001) surface are, as a model case, interesting as they show three different stages of hydrogen adsorption with almost the same energies (Figure 8.13): physisorbed H_2, associatively (molecularly) chemisorbed H_2, and dissociatively (atomically) chemisorbed $H + H$. Figure 8.12, which is closely related to Figure 8.11, is a schematic of the potential-energy contour plot in Figure 8.13. The wells P, M, and B are separated by activation-energy barriers that are about half an eV high, one for associative adsorption (A) and one for dissociation (D). From the self-consistent calculations the electronic structure can be obtained at any point of interest on the potential-energy surface. The conceptual picture described above is well supported, and the features of the potential-energy surface can be understood in these electron structure terms.[6,94,151]

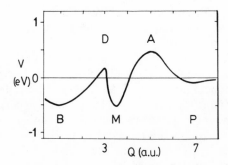

Figure 8.12. Schematic potential-energy curve for hydrogen adsorption on Mg(0001). The letters denote features in the contour plot of Figure 8.13. Q is the reaction coordinate, connecting the local potential-energy minima (after Kasemo and Lundqvist[9]).

This understanding is then a good aid to analyze how potential-energy surfaces on other substrate surfaces differ. For H_2 on a more open Mg surface, the outer activation barrier is thus expected to be lower or even absent, as the attractive potential from the substrate ions here can better reach the electrons close to the protons.[94] The same is expected at defects on the surface, like steps and vacancies.[151] On transition and noble metal surfaces, the d electrons in the spirit of Subsection 3.4.3b give a reduction of the outer potential barrier for H_2. On close-packed surfaces the barrier is reduced to about a quarter of an eV on Cu, and a complete absence on Ni,[6] in qualitative agreement with experimental findings. The difference is due to the different fillings of the d bands in these metals.[6,131] Further elaborations along these qualitative lines help us to understand differences in reactivity between different metals and surfaces.

Figures 8.13 and 8.14 also illustrate that the simple bonding–antibonding picture (Section 3.4.1) can be used to describe the dissociation of adsorbed molecules. The decrease in binding energy from point M to point D in Figures 8.12 and 8.13 is due to reduced overlap between the atomic orbitals with increasing separation between the protons, just as for a free molecule. Unlike the free molecule, however, upon further increase of the separation the chemisorbed molecule can get further stabilized, due to attraction of adsorbed H atoms to substrate ions. This is the case in the bridge position B in Figure 8.13, where the overlap is so small that there is no bonding–antibonding shift (Figure 8.14). Qualitative discussions of this kind provide a vehicle for discussing factors affecting the activation energy for dissociation.

8.4.3. Conceptual Picture of Dissociative Adsorption

The conceptual picture of molecular adsorption described above was presented already in 1978[110] and further refined thereafter.[6,110,111,150,151] Since then it has proven very useful for qualitative considerations, as will be described in, e.g., Section 8.5.

Recently another picture[154] has been advocated with emphasis.[155] It is natural to ask, whether this really is a different picture, describes the physics more adequately or, from a pragmatic point of view, has a better predicting power.

It addresses hydrogen dissociation at metal surfaces, in particular the marked difference between simple and noble metals on the one hand and transition metals on the other. Judging from observed dissociation probabilities for incident H_2 molecules, from observed excess translational energies of H_2 molecules desorbing from hot surfaces, and from theoretical work, the former have high entrance-channel energy barriers for dissociative adsorption. The latter display low entrance channel barriers (or no barriers whatsoever) and have relatively large binding energies. In particular, while H_2 molecules physisorb on Cu, they dissociate on Ni.

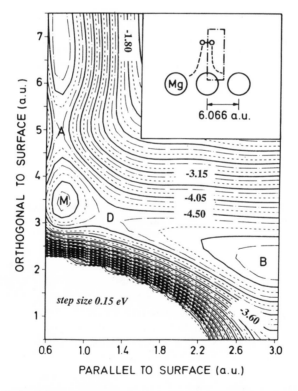

Figure 8.13. Equipotential-energy contours for H_2 on Mg(0001), with H_2 parallel to the surface, dissociating over the atop site into the bridge sites (see geometry in insert). Energies (eV) are relative to those of free atoms. The vertical distance is measured from the first atomic layer, and the distance parallel to the surface, which equals one half of the H–H separation, is measured from the atop site toward the bridge site (after Nørskov et al.[151]).

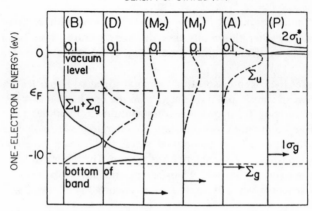

Figure 8.14. Development of the one-electron density of states along the reaction path in Figures 8.12 and 8.13. First the antibonding (Σ_u) and bonding (Σ_g) H_2 states are shifted down, as the distance d to the first Mg layer decreases for an almost constant H–H distance. Two spectra are shown for the molecular minimum (M) in Figure 8.13, for $d = 3.5$ (M_1) and 2.5 au (M_2). As the antibonding resonance gets filled, the H–H interaction energy E_{intra} goes to zero. As the molecule dissociates, M → D → B, the bonding–antibonding splitting decreases and vanishes (after Nørskov et al.[151]).

As for the repulsion of a He atom from the surface, so-called Pauli repulsion (cf. Section 8.3.2), the activation barrier is considered to be due to the closed-shell electronic structure of gas-phase H_2 together with the compactness of the molecule's $1\sigma_g$ orbital, which has its full complement of electrons.[154] As the H_2 molecule penetrates the outer tails of the wave functions of the metal electrons, these must deform so as to orthogonalize to the $1\sigma_g$ orbital. With increasing penetration, the electronic energy of the system increases, and this continues until the metal-molecule separation is small enough for a drastic change in the configuration to occur, e.g., dissociation. It is basically the Pauli principle that requires the metal wave functions to orthogonalize (Section 8.3.2).

The transition metals are characterized by unfilled d bands. The d wave functions are relatively well localized about the atoms of the substrate. As the H_2 molecules pass through the "physisorption region," they encounter first the s electrons of the metal, whose wave functions penetrate the surface region most effectively. This is the origin of the Pauli repulsion. The role of the d electrons, or rather the d holes, is to let the s electrons avoid the penetration into the region "occupied" by the H_2 $1\sigma_g$ orbital by transferring to the d band.[154] As a result the Pauli repulsion between a H_2 molecule and a transition metal is weaker than that of a simple or noble metal having a similar s electron density, and the corresponding activation barrier is therefore lowered and may suffice to remove the barrier entirely.[154]

This picture has been extracted from cluster calculations on the interaction of H_2 with Cu_2 and Ni_2.[156] Extrapolations to real systems of results from model calculations, whether jellium or clusters, have to be done with care and good judgment. It is therefore gratifying that both pictures claim a key role for the "d-hole" count[154] and "d-band filling factor" of the transition-metal electron structure,[6,120] respectively, in the lowering of the entrance-channel barrier. Obviously the same physical effect is accounted for, but it is articulated differently, as a perturbation to the Pauli repulsion[154] and an interference between molecular orbitals on weakly chemisorbed H_2 and d electrons,[6] respectively. Definite theoretical conclusions about molecularly adsorbed H_2 (a stable H_2 state has been observed on a stepped Ni surface presaturated with atomic hydrogen[158]) must await detailed, self-consistent calculations to be performed.

There is a major difference, however, and that concerns the account of the surface modified molecular orbitals (MOs). In both approaches the bonding 1σ MO of H_2 is accounted for. The antibonding $2\sigma^*$ MO, however, is considered to be of secondary importance in the cluster approach,[154] while the surface modification of the $2\sigma^*$ MO of the H_2 in particular, and of the HOMOs (Highest Occupied Molecular Orbital) and the LUMOs (Lowest Unoccupied Molecular Orbital)[157] of adsorbed molecules in general, is a big issue in the conceptual framework accounted for in this chapter. It has an intrinsic interest to explain the difference between Cu and Ni as H_2 chemisorbers in terms of, e.g., an increased interference between the LUMO $2\sigma^*$ and the d electrons. In addition, this interference picture has, indeed, predictive power and extensive practical·usefulness.

As a matter of fact, surface modification and repopulation of adsorbate MO resonances are concepts that have very broad applicability: lowering of stretch-mode frequency of, e.g., adsorbed CO by filling the antibonding $2\pi^*$ MO resonance,[159] and by the same mechanism many other effects, including formation of chemisorption bond,[159] weakening of molecular bond,[6,94,110,111,150,151,160] surface enhanced Raman scattering,[161] and energy dissipation by charge transfer, for vibrational[3,162,163] and transitional[3,164] adsorbate motions. The next section gives further applications of this concept.

8.5. Applications

In this section some applications of the rules from Sections 8.3 and 8.4 will be briefly described. The rules are particularly apt for finding trends with respect to substrate and adsorbate parameters and for qualitative considerations of modification of potential-energy surfaces.[56]

A trend with respect to an adsorbate parameter is shown in surface chemiluminescence from halogen gases on alkali metal surfaces.[165,166] The observed spectral variation for Cl_2, Br_2, and I_2 impinging on a sodium surface follows the trend given by electronegativity, in accordance with systematics described in Sections 8.3 and 8.4.[139] Recent measurements with potassium substrate give further support to the model.[167]

8.5.1. Modification of Potential-Energy Surfaces

The concepts and systematics of Sections 8.3 and 8.4 give guidance about how to modify potential-energy surfaces of adparticles.[56] The basic idea is to do this by manipulating electron energy levels of the adparticle, in particular the HOMOs, e.g., the antibonding $2\sigma^*$ orbital of H_2 and the $2\pi^*$ orbitals of N_2 and CO and LUMOs. The manipulation can be made in, for example, the following ways.[56]

8.5.1.1. Change of Substrate or Surface

Different substrates have different electron structures, as have different faces of one substrate. Thus they affect adsorbate electron orbitals differently and so give different potential-energy surfaces (Section 8.3.4). For instance, the $2\sigma^*$ molecular-orbital resonance of H_2 is affected differently by the d electrons on Cu and Ni surfaces, giving a further stabilization of the molecularly chemisorbed state, more stable the higher the transition-metal d-band energy.[6,56] Thus the external energy barrier for H_2 on Mg(0001) (Figures 8.12 and 8.13), which separates the physisorbed and chemisorbed molecular states, should be lowered on Cu(100) and absent on Ni(100).[6] This effect might explain, e.g., the inhibiting effect of Cu on the methanation reaction on Cu–Ni alloys,[45] suggesting the H_2 adsorption step to be rate-limiting in the Cu-rich regions.

The difference between close-packed and open surfaces has also been discussed. Basically, the denser the surface, i.e., the higher the electron density $n_0(\mathbf{r})$, the further out the repulsive kinetic energy wall meets the particle; the particle is thereby prevented from taking advantage of the increased attractive interaction due to the interference with the substrate at shorter distances. For instance, explicit calculations for H_2 on an open Mg surface show the potential barrier toward associative adsorption of H_2 to be absent there.[94,151] The tendency should be the same at defects, which might be an explanation for the observed[52] increased reactivity at steps on surfaces.[153]

8.5.1.2. Change of Site and Overlap

The energy shifts of molecular-orbital resonances may be different at different adsorption sites. Thus the $2\pi^*$ resonance of CO is lowered, when the CO molecule is moved from the top to the bridge position, and then to the center position on, e.g., the Ni(100) surface, as observed by, e.g., vibrational spectroscopy.[24] This may be due to the increased overlap between the $2\pi^*$ orbital and the d electrons of the substrate, associated with the increased coordination.[159] By manipulating the electron overlap one should then be able to affect the potential-energy surface. An example of this is given by the case of coadsorption of O and CO on Ni(100). On the clean surface, CO is primarily on top sites, while when O is preadsorbed in an open $p(2 \times 2)$ structure, the CO adsorbs in sites with higher coordination.[24] In an early verification of the electronic factor in heterogeneous catalysis,[24] the increased filling of the $2\pi^*$ molecular-orbital resonance in the high coordination sites has been suggested to explain the fact that the reaction rate for disproportionation of CO is higher on Ni(100) $p(2 \times 2)0$ than on clean Ni(100).

Behind this kind of analysis, where electronic information is extracted from vibrational spectra, is the insight that vibrations may depend on the occupation of certain electron energy levels. Such a correlation between the stretch frequency and the filling of the antibonding $2\pi^*$ resonance has been demonstrated in calculations for CO in carbonyl complexes[170] and for H_2 ($2\sigma^*$ resonance) chemisorbed on jellium.[111]

8.5.1.3. Application of Electric Field

In an electrolytic cell almost the whole potential drop occurs very close to the metal electrode. In an experiment on CO adsorbed on an Ag electrode, a lowering of the C—O stretch-mode frequency as a function of the applied potential has been observed by infrared measurements.[169] This implies an increased filling of the $2\pi^*$ molecular-orbital resonance, which in turn should imply increased reactivity.[170] The possibility of manipulating electrode reactions by varying the applied potential is indicated.

8.5.1.4. Change of Work Function

As discussed in Section 8.3.1.3, a reduced work function Φ stabilizes a state that results from a charge transfer from the substrate to the adsorbate. An illustration that this can happen at large separations is the reactive adsorption of the Cl_2 molecule on the Na surface.[139,165] Also at shorter distances, in the chemisorption region, a reduced work function increases the binding energy of a molecularly adsorbed state, associated with an increased filling of the affinity level. This is illustrated by calculated potential-energy curves[110,111,150] for H_2 on jellia with different electron densities and thus different work functions.[68] Measured from the position of the repulsive wall, the $2\sigma^*$ resonance peak crosses the Fermi level further out in the case with the lower substrate-electron density and thus lower work function. This allows an increased filling of the antibonding resonance. The work function of a surface can be changed by, e.g., adsorbing electropositive or electronegative atoms, which reduce and increase the work function, respectively.

8.5.1.5. Change of Electrostatic Potential

A change in the work function reflects a change in the macroscopic electrostatic potential. There can also be local electrostatic effects. Self-consistent calculations of the electrostatic potential around single atoms chemisorbed on jellium[125] show the potential to be screened out efficiently toward the interior of the metal and to be quite sizable (around 0.5 eV) outside the nearest-neighbor position (about 0.1 eV). The effects are particularly pronounced for strongly electronegative and electropositive atoms. Close to the adatom, coadsorbed species should be affected by such an electrostatic potential. This has been observed for CO on Cu(100), where coadsorbed Na reduces the frequency of the C—O stretch mode considerably for the CO molecules neighboring to a Na atom.[171] Implications for poisons and promoters in heterogeneous catalysis will be discussed in the next section.

8.5.2. Equilibrium Properties

Section 8.2 lists a number of important experimental results and questions posed by them. The examples are borrowed from, in particular, vibrational spectroscopy and heterogeneous catalysis. The theoretical methods reviewed in Section 8.3 are primarily aimed at conceptual and qualitative results. Nevertheless, there are several results on static properties derived from the potential-energy surfaces that are in good quantitative agreement with the experimental findings. In particular, the extended effective-medium scheme results[120] for chemisorption energies, bond lengths, and vibrational frequencies for hydrogen on transitional metals (Subsection 8.3.4.3b) should be noted. Other types of results are illustrated in Figure 8.15, indicating that chemisorbed O on Al(111) might stabilize in a subsurface position,[172,173] as well as in more recent results,[174] from which one may conclude that relaxation effects are important.

8.5.3. Molecular Vibrations and Rotations at Surfaces

Many of the spectral features listed in Section 8.2.2 have been explained qualitatively in Section 8.3. The absent stretch-mode frequency of H_2 on Ni(100),

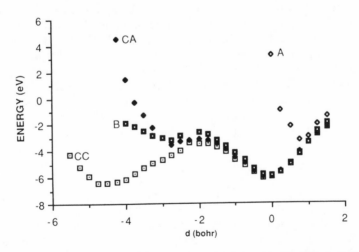

Figure 8.15. Potential-energy curves for an O atom chemisorbed on an Al(111) surface, calculated self-consistently on a jellium ($r_2 = 2$) surface, with lattice corrections added. The results indicate the possibility of a subsurface adsorption site (after Idiodi *et al.*[172-174])

signaling dissociation of H_2 on this surface,[21] is completely consistent with the calculated absence of any external potential-energy barrier on Ni (Section 8.5.1.1). The predicted presence of such a barrier on Cu, on the other hand, is a prerequisite for the unchanged rotational frequencies of H_2 observed on Cu(100),[23] as a pronounced physisorption well is provided by the barrier (cf. Figure 8.12). The shifted frequencies of CO[24,25] have been discussed in Section 8.4.3.

The extra frequencies observed for H on Ni(111) and possibly on Ni(100) have been interpreted as a manifestation of the quantum character of chemisorbed H on Ni.[175] The extended-effective-medium potential-energy surface is very anharmonic and has rather extended shallow regions connected with channels requiring little activation. The three-dimensional Schroedinger equation for the H atom in the calculated potential. The resulting energy bands for H on the Ni surface imply vibrational losses both in the range of the perpendicular vibrational frequency of a harmonic analysis and close to the "extra" frequency.[175]

8.5.4. Electronic Factor in Catalysis

Several questions raised by heterogeneous catalysis are the same as, or analogous to, those raised by adsorption. Here we will focus on the roles of different transition metals and of poisons and promoters.

8.5.4.1. Role of Transition Metals

Both the N_2 adsorption in the ammonia synthesis and the carbide formation in the methanation process give examples of chemisorption via a precursor, a molecularly adsorbed species, $A_2 \Leftrightarrow A_{2,ad} \rightarrow 2 A_{ad}$. If in a simple Lennard-Jones picture (cf. Figure 8.1) ΔE ($=|\Delta E|$) denotes the binding energy of the molecular species $A_{2,ad}$, and E_a is the energy barrier between the molecular and atomic potential-energy curves, measured from the lowest energy of $A_{2,ad}$, simple rate equations imply that the energy $E^* = E_a - \Delta E$ is one of the parameters that govern the sticking coefficient s_0 into A_{ad}.[37,125] Small and negative values of E^* favor a high rate of A_{ad} formation. This can be achieved by stabilizing, i.e., lowering, the molecular potential-energy curve E_{A2}.

Both atomic and molecular chemisorption come from overlap and interference between adsorbate and substrate electron states. The overlap affects atomic and molecular adsorption differently. For molecular chemisorption, the hybridization between the affinity level and the substrated electrons is an important marginal factor. In the common case, where the corresponding resonance is only partially filled, it gives an attraction ΔE^{hyb}, which is larger the closer A_2 is to the surface atoms. The kinetic energy ΔE^{rep} prevents A_2 from coming close. The potential energy $\Delta E_{A2} = \Delta E^{rep} + \Delta E^{hyb}$ and activation energy E^* are thus given by competition between ΔE^{rep} and ΔE^{hyb}.

For transition-metal substrates equation (8.3.14) forms a useful starting point for a discussion. In comparing similar surfaces of transition metals in the same series, the first term that primarily depends on the conduction-electron density does not vary significantly along the transitional-metal series. The covalent term, ΔE_{cov}, which expresses the hybridization with the d electrons, on the other hand, depends strongly on the location of the d band.[120,121] This is analogous to what is found for H adsorption in Section 8.3.3.[120,146] Starting to the right in the end of the transition metal series, one finds the molecular well depth ΔE to increase toward the left. This is a trend that does not need to apply systematically on a detailed level, since local geometrical factors may be important. Because of the variation of the position of the d band relative to the Fermi level along the transition-metal series, the binding energy of the molecular precursor A_2 increases toward the left of the series. This means a general decrease in the activation energy E^* and thus an increase in the adsorption rate and sticking into A_{ad}.[122] "Volcano plots" for the variation of the activity of, e.g., the methanation process along the transition-metal series should then be a result of a competition between the increase in dissociative adsorption rate as the rate-limiting step and the decrease in the desorption rate for the product, when going to the left in the transition-metal series (Figure 8.16).[122]

For some model reactions, in particular the ammonia synthesis,[176,177] the relevance of the surface science approach to catalysis has recently been demonstrated very explicitly. On the basis of the conceptual picture discussed in this chapter and on experimental data for single crystals under ultrahigh vacuum, the reaction rate of the ammonia synthesis at 149–300 bar and 375–500 °C has been calculated and found to agree closely with experimental findings.[176,177] This and similar works show that the so-called pressure gap between the ultrahigh vacuum of surface science and the high pressures of applied catalysis can be bridged, at least in some cases.

8.5.4.2. Poisons and Promoters

Section 8.2 describes how electronegative adatoms act as poisons and electropositive ones as promoters. The shifts in the electrostatic potential caused by such adatoms (Section 8.5.1.5) change adsorbate levels, e.g., the $2\pi^*$ molecular-orbital resonance of N_2 and CO and $2\sigma^*$ of H_2, and affect the molecular binding energy ΔE_{A2}, electropositive atoms stabilizing it.

This is most easily seen by applying the effective-medium equation (8.3.14) to the case, where the host is just changed by introducing the coadsorbed atom. In this approximation, the change in the binding energy of the molecule is

$$\delta E_B = \int_a d^3 r\, \delta\phi_0(\mathbf{r}) + \delta\Delta\left(\sum_j \varepsilon_j\right) \qquad (8.5.1)$$

where $\delta\phi_0(\mathbf{r})$ is the coadsorbate-induced change in the electrostatic potential.[125] The stabilization gives a decrease in the activation energy E^* and thus an increased rate for the reaction $A_2 \rightarrow 2\,A_{ad}$.

According to these results, electropositive adatoms, like Na and K, act as promoters for reactions, where adsorption is rate determining. Electronegative adatoms, on the other hand, shift the mentioned levels upward, increase E^*, and

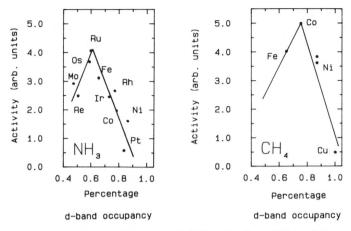

Figure 8.16. Semilogarithmic plots of the ammonia (left) and methane (right) activities as functions of occupancy of the substrate d-band (after Holloway *et al.*[(122)]).

thereby act as poisons in adsorption processes. The limited range of the electrostatic potential from charged adsorbates (Section 8.5.1.5) implies that these promoter and poison effects are local,[(125)] in agreement with the experimental findings mentioned in Section 8.2.

In addition to this kind of electrostatic interaction, poisons and promoters may act through a direct electronic interaction.[(178)]

8.6. Concluding Remarks

This chapter has dealt with the interaction of atoms and molecules with metal surfaces. While the emphasis has been on conservative forces, derived from potential-energy surfaces for chemi- and physisorption, it should be remembered that dissipative forces also are important, and that low-energetic phonon and electron-hole pair excitations of the surface are important dissipative channels.[(3)]

A central theme of the paper is the claim that a coherent picture of electronic aspects of adsorption and surface reactions, including heterogeneous catalysis, is developing. It originates from an increased understanding of the adsorbate-induced electron structure and potential-energy surfaces of adsorbates. For instance, the weakening or breaking of molecular bonds on metal surfaces get their explanation in terms of the lowering (raising), broadening, and filling (emptying) of molecular affinity (ionization) levels, occurring for molecules on the surface.[(1,6,56,94,110,150,151)] Potential-energy surfaces of atoms and molecules on metal surfaces can be sketched on the basis of existing systematics of electron energy levels.[(2,56)] By the effective-medium theory and similar schemes they can even be relatively simply calculated with a practically useful accuracy. Trends in chemisorption properties can be explained, for example, for substrates along the transition-metal series.

The qualitative picture can be used to analyze key features of potential-energy surfaces for surface reactions, in particular adsorption/desorption reactions. In this way the variation of reaction rates along the transition-metal series,[122] variations between different faces of the same substrate,[6] and the roles of defects, like steps, promoters, and poisons,[56,94,125,151] can be understood.

It should be stressed, however, that this type of approach has limitations.[6,56,94,151] For instance, in the calculations described in Section 8.2 the jellium model gives a very strong emphasis on the extended nature of the substrate electron states and the substrate lattice is introduced only by the lowest-order perturbation theory.[90,106,109] Further, the effective-medium theory is based on the double perturbation theory. However, comparisons with other models[133,135-138] give support to the methods and the qualitative picture.

References

1. B. I. Lundqvist, in: *Many Body Phenomena at Surfaces* (D. Langreth and H. Suhl, eds.), p. 93, Academic Press, New York (1984).
2. B. I. Lundqvist, *Chem. Scr.* **26**, 423 (1986).
3. B. I. Lundqvist, in: *Many Body Phenomena at Surfaces* (D. Langreth and H. Suhl, eds.), p. 453, Academic Press, New York (1984).
4. G. Benedec and U. Valbusa (eds.), *Dynamics of Gas–Surface Interaction*, Springer-Verlag, New York (1982).
5. R. B. Gerber, *Chem. Rev.* **87**, 29 (1987).
6. B. I. Lundqvist, O. Gunnarsson, H. Hjelmberg, and J. K. Nørskov, *Surf. Sci.* **89**, 196 (1979).
7. E. Drauglis and R. L. Jaffee, *Physical Basis for Heterogeneous Catalysis*, p. 476, Plenum Press, New York (1975), aired many of these aspects at an early stage.
8. T. F. O'Malley, in: *Advances in Atomic and Molecular Physics*, (D. R. Bates and I. Easterman, eds.), Vol. 7, p. 223, Academic Press, New York (1971).
9. B. Kasemo and B. I. Lundqvist, *Comments At. Mol. Phys.* **14**, 229 (1984).
10. J. B. Pendry, *Low Energy Electron Diffraction*, Academic Press, London (1974).
11. M. van Hove, *Surf. Sci.* **80**, 1 (1979).
12. B. Feuerbacher, B. Fitton, and R. F. Willis, *Photo-emission and the Electronic Properties of Surfaces*, Wiley, New York (1978).
13. E. W. Plummer and W. Eberhardt, in: *Advances in Chemical Physics* (I. Prigogine and S. Rice, eds.), Vol. 49, p. 533, Wiley, New York (1982).
14. H. Ibach and D. L. Mills, *Electron Energy Loss Spectroscopy and Surface Vibrations*, Academic Press, New York (1982).
15. See, e.g., G. Brusdeylins, R. B. Doak, and J. P. Toennies, *Phys. Rev. Lett.* **44**, 1417 (1980).
16. See e.g., F. Frenkel, J. Häger, W. Krieger, H. Walther, C. T. Campbell, G. Ertl, H. Kuipers, and J. Segner, *Phys. Rev. Lett.* **46**, 152 (1981).
17. See, e.g., R. P. Thorman, D. Andersson, and S. L. Bernasek, *Phys. Rev. Lett.* **44**, 143 (1980).
18. See, e.g., D. A. Mantell, S. B. Ryali, B. L. Halperin, G. L. Haller, and J. B. Fenn, *Chem. Phys. Lett.* **81**, 185 (1981).
19. J. E. Lennard-Jones, *Trans. Faraday Soc.* **28**, 333 (1932).
20. G. Herzberg, *Spectra of Diatomic Molecules*, 2nd ed., van Nostrand, New York (1959).
21. S. Andersson, *Chem. Phys. Lett.* **55**, 185 (1978).
22. B. A. Sexton, *Surf. Sci.* **88**, 299 (1979).
23. S. Andersson and J. Harris, *Phys. Rev. Lett.* **48**, 545 (1982).

24. S. Andersson, B. I. Lundqvist, and J. K. Nørskov, in: *Proc. 7th Int. Vacuum Congr. and 3rd Int. Conf. on Solid Surfaces*, p. 815, Vienna (1977).
25. S. Andersson, *Solid State Commun.* **21**, 751 (1979).
26. R. Ryberg, *Surf. Sci.* **114**, 627 (1982).
27. B. N. J. Persson and M. Persson, *Surf. Sci.* **97**, 609 (1980).
28. M. Persson and B. Hellsing, *Phys. Rev. Lett.* **49**, 662 (1982).
29. B. N. J. Persson, *Solid State Commun.* **24**, 573 (1977).
30. S. Andersson, *Solid State Commun.* **21**, 751 (1979).
31. B. N. J. Persson, *Surf. Sci.* **92**, 265 (1980).
32. B. N. J. Persson and R. Ryberg, *Solid State Commun.* **36**, 613 (1980).
33. S. Andersson and B. N. J. Persson, *Phys. Rev. Lett.* **45**, 1421, 3659 (1980).
34. S. Lewald, J. M. Szeftel, H. Ibach, T. S. Rahman, and D. L. Mills, *Phys. Rev. Lett.* **50**, 518 (1983).
35. S. Andersson and M. Persson, *Phys. Rev. B* **24**, 3659, (1981).
36. M. Persson and S. Andersson, *Surf. Sci.* **117**, 352 (1982).
37. G. Ertl, *Catal. Rev., Sci. Eng.* **21**, 201 (1980).
38. G. Ertl, S. B. Lee, and M. Weiss, *Surf. Sci.* **114**, 515 (1982).
39. G. Ertl, S. B. Lee, and M. Weiss, *Surf. Sci.* **114**, 527 (1982).
40. D. W. Goodman, R. D. Kelley, T. E. Madey, and J. T. Yates, Jr., *J. Catal.* **63**, 226 (1980).
41. M. A. Vannice, *J. Catal.* **37**, 449; **37**, 462 (1975).
42. M. A. Vannice, *J. Catal.* **50**, 288 (1976).
43. G. L. Bond, *Heterogeneous Catalysis. Principles and Applications*, Oxford Chemistry Series, Clarendon Press (1974).
44. P. R. Wentrak, B. J. Wood, and H. Wise, *J. Catal.* **43**, 363 (1976).
45. M. Araki and V. Ponec, *J. Catal.* **44**, 439 (1976).
46. R. D. Kelley and D. W. Goodman, *Surf. Sci.* **123**, L743 (1982).
47. H. P. Bonzel and H. J. Krebs, *Surf. Sci.* **117**, 639 (1982).
48. S. V. Ho and P. Harriott, *J. Catal.* **64**, 272 (1980).
49. S. V. Ho and P. Harriott, *J. Catal.* **71**, 445 (1981).
50. M. Kiskinova and D. W. Goodman, *Surf. Sci.* **108**, 64; **109**, L555 (1981).
51. C. T. Campbell and D. W. Goodman, *Surf. Sci.* **123**, 413 (1982).
52. I. Toyoshima and G. A. Somorjai, *Catal. Rev., Sci. Eng.* **19**, 105 (1979).
53. N. D. Spencer and G. A. Somorjai, *Rep. Prog. Phys.* **46**, 1 (1983).
54. G. A. Somorjai, in: *New Horizons of Quantum Chemistry* (P.-O. Löwdin and B. Pullman, eds.), p. 305, D. Reidel, New York (1983).
55. J. C. Tully, *Ann. Rev. Phys. Chem.* **31**, 319 (1980); in: *Many-Body Phenomena at Surfaces* (D. Langreth and H. Suhl, eds.), p. 372, Academic Press, New York (1984).
56. B. I. Lundqvist, *Vacuum*, **33**, 639 (1983).
57. J. C. Slater, *Quantum Theory of Molecules and Solids*, Vol. 7, McGraw-Hill, New York (1963).
58. See, e.g., V. Heine, in: *The Physics of Metals* (J. M. Ziman, ed.), Vol. 1, Cambridge University Press (1969).
59. E. M. Lifshitz, *Zh. Eksp. Teor. Fiz.* **29**, 94 (1955); *Sov. Phys. JETP* **2**, 73 (1956).
60. I. E. Dzyaloshinskii, E. M. Lifshitz, and L. P. Pitaevskii, *Adv. Phys.* **10**, 165 (1961).
61. E. Zaremba and W. Kohn, *Phys. Rev. B* **15**, 1769 (1972).
62. J. D. Jackson, *Classical Electrodynamics*, 2nd ed., Wiley, New York (1975).
63. N. D. Lang and W. Kohn, *Phys. Rev. B* **7**, 3541 (1973).
64. A. C. Hewson and D. M. Newns, *Jpn. J. Appl. Phys., Suppl. 2* **2**, 121 (1974).
65. W. S. Struve, J. R. Krenos, D. L. McFadden, and D. R. Herschbach, *J. Chem. Phys.* **62**, 404 (1975).
66. I. Langmuir, *J. Am. Chem. Soc.* **54**, 2798 (1932).
67. K. H. Rieder, *Europhys. News* **12**, No. 10, 6 (1981).
68. N. D. Lang, *Solid State Phys.* **28**, 225 (1973).

69. N. D. Lang, in: *Theory of the Inhomogeneous Electron Gas* (S. Lundqvist and N. H. March, eds.), p. 309, Plenum Press, New York (1983).

70. P. Hohenberg and W. Kohn, *Phys. Rev.* **136**, B864 (1964); W. Kohn and L. J. Sham, *Phys. Rev.* **140**, A1133 (1965).

71. L. Hedin and B. I. Lundqvist, *J. Phys. C* **4**, 2064 (1971).

72. U. von Barth and L. Hedin, *J. Phys. C* **5**, 1629 (1972).

73. O. Gunnarsson and B. I. Lundqvist, *Phys. Rev. B* **13**, 4274 (1976).

74. N. D. Lang, *Phys. Rev. Lett.* **46**, 842 (1981).

75. N. D. Lang and J. K. Nørskov, *Phys. Rev. B* **27**, 4612 (1983).

76. N. Esbjerg and J. K. Nørskov, *Phys. Rev. Lett.* **45**, 807 (1980).

77. J. K. Nørskov and N. D. Lang, *Phys. Rev. B* **21**, 2131 (1980).

78. M. J. Stott and E. Zaremba, *Phys. Rev. B* **22**, 1564 (1980).

79. M. Manninen, J. K. Nørskov, and C. Umrigar, *J. Phys. F* **12**, L7 (1982); M. Manninen, J. K. Nørskov, and C. Umrigar, *Surf. Sci.* **119**, L393 (1982).

80. J. Harris and A. Liebsch, *Phys. Rev. Lett.* **49**, 341 (1982).

81. J. Harris and A. Liebsch, *J. Phys. C* **15**, 2275, (1982).

82. P. Nordlander, *Surf. Sci.* **126**, 675 (1983).

83. P. Nordlander, C. Holmberg, and J. Harris, *Surf. Sci.* **152/153**, 702 (1985); **175**, L753 (1986).

84. J. Harris and A. Liebsch, *Phys. Scr. T* **4**, 14 (1983).

85. S. Andersson, L. Wilzén, and M. Persson, *Phys. Rev.* **38**, 2967 (1988).

86. N. D. Lang, J. K. Nørskov, and B. I. Lundqvist, *Phys. Scr.* **34**, 77 (1986).

87. K. Ruedenberg, *Rev. Mod. Phys.* **34**, 326 (1962).

88. See, e.g., B. I. Lundqvist, H. Hjelmberg, and O. Gunnarsson, in: *Photo-emission and the Electronic Properties of Surfaces* (B. Feuerbacher, B. Fitton, and R. F. Willis, eds.), Wiley, New York (1978).

89. J. P. Muscat and D. M. Newns, *Prog. Surf. Sci.* **9**, 1 (1978); R. Hoffmann, *Rev. Mod. Phys.* **60**, 601 (1988); P. J. Feibelman, *Annu. Rev. Phys. Chem.* **40**, 261 (1989).

90. O. Gunnarsson, H. Hjelmberg, and B. I. Lundqvist, *Phys. Rev. Lett.* **37**, 292 (1976).

91. R. W. Gurney, *Phys. Rev.* **47**, 479 (1935).

92. D. M. Newns, *Phys. Rev.* **178**, 1123 (1969).

93. J. W. Gadzuk, in: *Surface Physics of Materials* (J. M. Blakely, ed.), p. 339, Academic Press, New York (1975).

94. P. K. Johansson, B. I. Lundqvist, A. Houmøller, and J. K. Nørskov, in: *Recent Developments in Condensed Matter Physics* (J. T. Devreese, ed.), p. 605, Plenum Press, New York (1981).

95. H. Hjelmberg, *Phys. Scr.* **18**, 481 (1978).

96. D. R. Hamann, *Phys. Rev. Lett.* **46**, 1227 (1981).

97. O. Jepsen, J. Madsen, and O. K. Andersen, *Phys. Rev. B* **26**, 2790 (1983).

98. P. W. Anderson, *Phys. Rev.* **124**, 41 (1961).

99. C. Herring, in: *Magnetism* (G. T. Rado and H. Suhl, eds.), Vol. 4, Academic Press, New York (1966).

100. D. M. Newns, *Phys. Lett.* **33A**, 43 (1970).

101. W. Brenig and K. Schönhammer, *Z. Phys.* **267**, 201 (1974).

102. K. Schönhammer, *Int. J. Quantum Chem., Symp.* **11**, 517 (1977).

103. R. O. Jones and O. Gunnarsson, *Rev. Mod. Phys.* **61**, 689 (1989).

104. N. D. Lang and A. R. Williams, *Phys. Rev. Lett.* **34**, 531 (1975).

105. N. D. Lang and A. R. Williams, *Phys. Rev. Lett.* **37**, 212 (1976).

106. N. D. Lang and A. R. Williams, *Phys. Rev. B* **18**, 616 (1978).

107. O. Gunnarsson and H. Hjelmberg, *Phys. Scr.* **11**, 97 (1975).

108. T. D. Grimley, in: *Dynamic Aspects of Surface Physics* (F. O. Goodman, ed.), p. 197, Editrice Compositori, Bologna (1974).

109. H. Hjelmberg, *Surf. Sci.* **81**, 539 (1979).

110. B. I. Lundqvist, J. K. Nørskov, and H. Hjelmberg, *Surf. Sci.* **80**, 441 (1979).

111. H. Hjelmberg, B. I. Lundqvist, and J. K. Nørskov, *Phys. Scr.* **20**, 192 (1979).

112. S. Holmström and S. Holloway, *Surf. Sci.* **173**, L647 (1986).

113. S. Holmström, *Phys. Scr.* **36**, 529 (1987).

114. G. B. Bachelet, D. R. Hamann, and M. Schlüter, *Phys. Rev. B* **26**, 4199 (1982).

115. J. K. Nørskov, *Solid State Commun.* **24**, 691 (1977).

116. J. K. Nørskov, *Phys. Rev. B* **26**, 2875 (1982).

117. J. K. Nørskov, *Phys. Rev. Lett.* **48**, 1620 (1982).

118. M. J. Puska, R. M. Nieminen, and M. Manninen, *Phys. Rev. B* **24**, 3037 (1980).

119. A. R. Mackintosh and O. K. Andersen, in: *Electrons at the Fermi Surface* (M. Springford, ed.), Cambridge University Press (1980).

120. P. Nordlander, S. Holloway, and J. K. Nørskov, *Surf. Sci.* **136**, 59 (1984).

121. B. Chakraborty, S. Holloway, and J. K. Nørskov, *Surf. Sci.* **152/153**, 660 (1985).

122. S. Holloway, B. I. Lundqvist, and J. K. Nørskov, in: *Proceedings of the 8th International Congress on Catalysis*, Berlin, BRD (1984); J. K. Nørskov and P. Stoltze, *Surf. Sci.* **189/190**, 91 (1987).

123. P. Norlander, J. K. Nørskov, and F. Besenbacher, *J. Phys. F* **16**, 1161 (1986).

124. J. K. Nørskov, F. Besenbacher, J. Bøttiger, B. B. Nielsen, and A. A. Pisarev, *Phys. Rev. Lett.* **49**, 1420 (1982).

125. N. D. Lang, S. Holloway, and J. K. Nørskov, *Surf. Sci.* **150**, 26 (1984).

126. K. W. Jacobsen, J. K. Nørskov, and M. J. Puska, *Phys. Rev. B* **35**, 7423 (1987); K. W. Jacobsen, *Comments on Condensed Matter Physics* **14**, 129 (1988).

127. J. K. Nørskov, *Europhys. News*, **19**, 65 (1988).

128. K. W. Jacobsen and J. K. Nørskov, in: *The Structure of Surfaces*, 2 (van der Veen and van Hove, eds.), p. 118, Springer-Verlag, Berlin (1988).

129. K. W. Jacobsen and J. K. Nørskov, *Phys. Rev. Lett.* **59**, 2764 (1987).

130. K. W. Jacobsen and J. K. Nørskov, *Phys. Rev. Lett.* **60**, 2496 (1988).

131. J. K. Nørskov, *J. Chem. Phys.* **90**, 7461 (1989).

132. J. E. Müller and J. Harris, *Phys. Rev. Lett.* **53**, 2943 (1984).

133. C. Umrigar and J. W. Wilkins, *Phys. Rev. Lett.* **54**, 1551 (1985).

134. O. K. Andersson, in: *The Electronic Structure of Complex Systems* (W. Temmerman and P. Phariseau, eds.), NATO Advanced Study Institute, Plenum Press, New York (1982).

135. T. H. Upton and W. A. Goddard, III, *Phys. Rev. Lett.* **42**, 472 (1979).

136. T. H. Upton and W. A. Goddard, III, *CRC Crit. Rev. Solid State Mater. Sci.* **10**, 261 (1981).

137. P. J. Feibelman, *Phys. Rev. Lett.* **54**, 2627 (1985).

138. P. J. Feibelman, *Phys. Rev. Lett.* **63**, 2488 (1989).

139. J. K. Nørskov, D. M. Newns, and B. I. Lundqvist, *Surf. Sci.* **80**, 179, (1979).

140. B. Hellsing, M. Persson, and B. I. Lundqvist, *Surf. Sci.* **126**, 147 (1983).

141. C. O. Almbladh, U. von Barth, Z. D. Popovic, and M. J. Stott, *Phys. Rev. B* **14**, 2250 (1976).

142. O. Gunnarsson, H. Hjelmberg, and J. K. Nørskov, *Phys. Scr.* **22**, 165 (1980).

143. R. O. Jones, *J. Chem. Phys.* **71**, 1300 (1979).

144. S. Holmström, C. Holmberg, and B. I. Lundqvist, *J. Electron Spectrosc. Relat. Phenom.* **39**, 319 (1986).

145. H. Hjelmberg, O. Gunnarsson, and B. I. Lundqvist, *Surf. Sci.* **68**, 158 (1977).

146. C. M. Varma and A. J. Wilson, *Phys. Rev. B* **22**, 3795 (1980).

147. C. Holmberg, T. Fondén, A. Mällo, and B. I. Lundqvist, *Phys. Scr.* **35**, 181 (1987).

148. C. Nyberg and C. G. Tengståhl, *J. Chem. Phys.* **80**, 7 (1984).

149. G. B. Fisher and B. A. Sexton, *Phys. Rev. Lett.* **44**, 683 (1980).

150. P. K. Johansson, *Surf. Sci.* **104**, 5120 (1981).

151. J. K. Nørskov, A. Houmøller, P. K. Johansson, and B. I. Lundqvist, *Phys. Rev. Lett.* **46**, 257 (1981).

152. G. P. Brivio and T. B. Grimley, *Surf. Sci.* **89**, 226 (1979).

153. B. I. Lundqvist, in: *Vibrations at Surfaces* (R. Caudano, J. M. Gilles, and A. A. Lucas, eds.), p. 541, Plenum Press, New York (1982).
154. J. Harris, *Appl. Phys. A* **47**, 63 (1988).
155. A. Zangwill, *Physics on Surfaces*, Cambridge University Press, Cambridge (1988).
156. J. Harris and S. Andersson, *Phys. Rev. Lett.* **55**, 1583 (1985).
157. I. Fleming, *Frontier Orbitals and Organic Chemical Reactions*, Wiley, Chichester (1976).
158. A.-S. Mårtensson, C. Nyberg, and S. Andersson, *Phys. Rev. Lett.* **57**, 2045 (1986).
159. G. Blyholder, *J. Phys. Chem.* **68**, 2772 (1964).
160. A. B. Anderson and R. Hoffman, *J. Chem. Phys.* **61**, 4545 (1974).
161. B. N. J. Persson, *Chem. Phys. Lett.* **82**, 561 (1981).
162. B. N. J. Persson and M. Persson, *Solid State Commun.* **36**, 175 (1980).
163. M. Persson and B. Hellsing, *Phys. Rev. Lett.* **49**, 662 (1982).
164. J. K. Nørskov and B. I. Lundqvist, *Surf. Sci.* **89**, 251 (1979).
165. B. Kasemo and L. Walldén, *Surf. Sci.* **53**, 393 (1975).
166. B. Kasemo, E. Törnqvist, and L. Walldén, *Mater. Sci. Eng.* **42**, 23 (1980).
167. D. Andersson, B. Kasemo, and L. Walldén, *Surf. Sci.* **152/153**, 576 (1985).
168. E. J. Baerends and P. Ros, *Mol. Phys.* **30**, 1735 (1975).
169. B. Beden, A. Bewich, and C. Lamy, *J. Electroanal. Chem. Interfacial Electrochem.* **148**, 147 (1983).
170. S. Holloway and J. K. Nørskov, *J. Electroanal. Chem. Interfacial Electrochem.* **161**, 193 (1984).
171. L. Walldén, *Surf. Sci.* **134**, L513 (1983).
172. J. O. A. Idiodi, to appear.
173. B. I. Lundqvist, T. Fondén, J. Idiodi, P. Johnsson, A. Mällo, and S. Papadia, *Prog. Surf. Sci.* **25**, 191 (1987).
174. U. Yxklinten, M. Limbäck, and B. I. Lundqvist, in: *Many-Atom Interactions in Solids* (R. M. Nieminen, M. J. Puska, and M. Manninen, eds.), Springer Proceedings in Physics, Springer, Berlin (1990).
175. M. J. Puska, R. M. Nieminen, M. Manninen, Bulbul Chakraborty, S. Holloway, and J. K. Nørskov, *Phys. Rev. Lett.* **51**, 1081 (1983).
176. P. Stoltze and J. K. Nørskov, *Phys. Rev. Lett.* **55**, 2502 (1985); P. Stoltze, *Phys. Scr.* **36**, 824 (1987).
177. J. K. Nørskov and P. Stoltze, *Surf. Sci.* **189/190**, 91 (1987).
178. P. Feibelman and D. Hamann, *Phys. Rev. Lett.* **52**, 61 (1984).

9

Electronic Theory of Chemisorption

D. Spanjaard and M. C. Desjonquères

9.1. Introduction

In surface physics, one calls adsorption the accumulation at the solid–vapor interface of atoms or molecules coming from the gas phase. One usually classifies adsorption phenomena as two domains according to the energy E_B involved in the bonding:

1. When $|E_B| \lesssim 0.5$ eV, the adsorbate is said to be physisorbed on the surface. In physisorption, the adsorbate binds to the substrate via van der Waals forces which are due to dipole–dipole interactions. A typical example is the rare-gas adsorption.
2. When $|E_B| \gtrsim 0.5$ eV, the adsorbate is said to be chemisorbed. The bond between the adsorbate and the substrate is of chemical type, i.e., it involves sharing (covalent bond) or transfer (ionic bond) of electrons. This is the case for O, N, H, ... on transition metals. This domain is obviously the most important in view of its practical applications (catalysis, corrosion, ...) and we will limit ourselves to this phenomenon.

The adsorption theory can be tackled using three complementary approaches:

1. The macroscopic or thermodynamical approach which gives useful relationships between measurable quantities of the system, such as the relation between the amount of adsorbates and the pressure.
2. The microscopic approach which aims at calculating the interaction between the adsorbate and the substrate within the quantum theory.
3. The statistical methods which link the above approaches by relating macroscopic and microscopic physical quantities.

D. Spanjaard • Laboratoire de Physique des Solides, Université Paris-Sud, F91405 Orsay Cédex, France. **M. C. Desjonquères.** • IRF/DPhG/PAS, Centre d'Etudes Nucléaires de Saclay, F91191 Gif-sur-Yvette Cédex, France.

In this chapter, we will limit ourselves to the microscopic approach. Within this approach a complete knowledge of chemisorption phenomena requires the determination of

- The geometrical structure of the system (adsorption site, bond length, ...)
- The adsorbate binding and diffusion activation energies
- The charge transfer
- The electronic structure of the adsorbate and substrate
- The vibration frequencies

The physical quantities can be measured more or less directly by a variety of experimental techniques. Examples are[1]:

1. Field ion microscopy (FIM), low-energy electron diffraction (LEED), surface extended X-ray absorption fine structure (SEXAFS), surface core level spectroscopy (i.e., electron spectroscopy for chemical analysis—ESCA—at the surface) for the geometrical structure.
2. Microcalorimetry, thermal desorption, FIM for binding and diffusion energies.
3. Work function measurement, surface core level spectroscopy for charge transfers.
4. UV and X-ray photoemission (UPS, XPS), field emission microscopy (FEM), ion neutralization spectroscopy (INS),... for the electronic structure.
5. Electron energy loss spectroscopy (EELS), neutron diffraction, Raman spectroscopy,... for the vibration frequencies.

Although these experimental data are sometimes rather spread for a given system, they undoubtedly exhibit systematic trends which we will now summarize:

1. The bond length (i.e., the distance between the adsorbate and its nearest neighbors) increases with the coordination number Z of the adsorbate as shown in Table 9.1 for O, S, and Se on Ni(111) and (100).[2] Other examples can be found elsewhere.[3] The same trend is also obeyed by surface atoms which, due

Table 9.1. Experimental Variations of
the Bond Lengths of O, S, and Se on Ni
as a Function of the Coordination Number
(After Various Authors[2,3])

	Ni(111) $Z = 3$	Ni(100) $Z = 4$
d_{O-Ni}	1.88 Å	1.98 Å
d_{S-Ni}	2.02 Å	2.19 Å
d_{Se-Ni}	2.31 Å	2.35 Å

to their broken bonds, come closer to the subsurface plane, i.e., there is a contraction of the first interlayer spacing[4] which decreases or even vanishes in the presence of adsorbed atoms. Similarly one observes a dilatation of the interatomic spacing a for transition metals exhibiting a phase transition from a body-centered cubic structure (8 nearest neighbors) to a compact structure (12 nearest neighbors); for instance, $a = 2.48$ Å for bcc Fe and 2.52 Å for fcc Fe.

2. The most stable adsorption position is usually the site with the largest coordination number available on the surface.[3] It is noteworthy, however, that some chemisorbed atoms do not occupy the site with maximum coordination; it is known that, on the (100) face of W, H sits at the bridge rather than at a centered position.[5]

3. The variation of binding energy with the crystallographic orientation of the surface plane (anisotropy) is always smaller than expected from simple arguments such as the number of bonds even though there is a large straggling in experimental data.[6]

4. The variation of the binding energy has been studied either for a series of adsorbates on a given substrate or for a given adsorbate on a series of substrates. Figure 9.1 shows the results obtained for 5d adatoms on W(111), W(112), and Ir(111)[7] which follow a parabolic behavior similar to that of the cohesive energies. Figure 9.2 presents the variation in the binding energies of N, O, F and H[6,8] along a transition series which decreases continuously when the substrate d band fills.

5. The surface diffusion activation energies have only been measured for 5d adatoms on various faces of W.[9] They exhibit a maximum near the middle of the series similarly to their binding energy (Figure 9.3). More qualitatively it is known that transition adatoms deposited on the pole of a FIM tip are most often reflected by the edge of the tip.[10] One also observes that the diffusion of adatoms along a smooth ledge step is little perturbed by the presence of the step.[11]

6. Finally, the ability of transition metal surfaces to dissociate simple molecules (N_2, O_2, NO, ...) decreases along a transition series (see Table 9.2[12]).

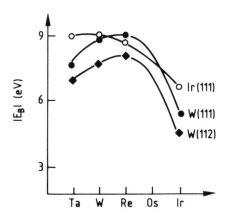

Figure 9.1. Experimental binding energies of 5d adatoms on W(111), W(112), and Ir(111) (after Menand and Gallot[7]).

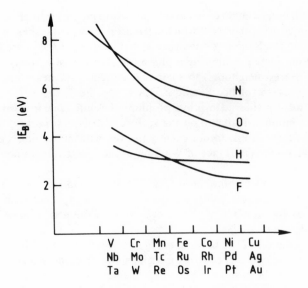

Figure 9.2. Variation in the binding energies of N, O, F, and H along the transition series (after Toyoshima and Somorjai[6] and Bolbach[8]).

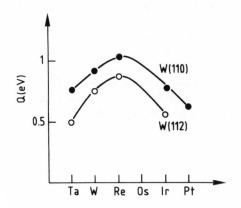

Figure 9.3. Experimental activation energies for surface diffusion of 5d adatoms on W(110) and W(112) (after Bassett[9]).

Table 9.2. Borderline between Dissociative (Left-Hand Side) and Molecular (Right-Hand-Side) Chemisorption at Room Temperature Thick line refers to N_2 and double line to NO (After Brodén et al.[12])

Sc	Ti	V	Cr	Mn	Fe	Co	Ni	Cu
Y	Zr	Nb	Mo	Tc	Ru	Rh	Pd	Ag
La	Hf	Ta	W	Re	Os	Ir	Pt	Au

One needs therefore a theory which can rationalize all these results and make good predictions for as yet uninvestigated systems. This is actually a very difficult task since, ideally, one should determine the geometrical and electronic structures self-consistently by minimizing the total energy of the system. Two main approaches have been followed based either on quantum chemistry methods or on solid state physics techniques. However, none of these lines of attack is fully appropriate, since the former most often deals with a finite number of atoms while the latter usually treats systems with three-dimensional periodicity. We will develop here mainly the methods derived from solid state physics and we will show how these methods can be modified to deal with nonperiodic systems. We will also give a brief outline of the most usual quantum chemistry methods. Finally, we will only consider the adsorption on metals of isolated atoms or molecules, i.e., we will assume that the adsorbate coverage is so small that the interaction between adsorbates can be neglected.

9.2. Outline of the Problem

Let us consider an atom X approaching the surface of a metal, the Fermi level and the work function of the latter being respectively E_F and Φ. Quantities I and A are termed, respectively, the ionization and affinity energies of the free atom (i.e., at an infinite distance from the surface) and are usually employed to define an effective Coulomb integral U which is the energy involved in the reaction

$$2X \to X^+ + X^- \qquad (9.2.1)$$

i.e.,

$$U = I - A \qquad (9.2.2)$$

(U is positive since $A < I$).

When the atom is far from the metal surface, the main interaction originates from classical electric image effects and leads to a shift of ionization and affinity levels (Figure 9.4) which can be easily calculated.[13]

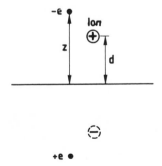

Figure 9.4. Influence of the image potential on the ionization and affinity levels of an adatom.

9.2.1. Ionization Level

The image force acting on the outgoing electron is (Figure 9.5)

$$F = -\frac{e^2}{4z^2} + \frac{e^2}{(d + z)^2} \tag{9.2.3}$$

the first and second terms are respectively due to its own image and to the ion image. The value of z is large, so the resulting force is repulsive and the ionization energy is decreased by an amount given by

$$W = \int_d^\infty F \, dz = \frac{e^2}{4d} = V_{\text{Im}} \tag{9.2.4}$$

and therefore the effective ionization energy is

$$I_{\text{eff}} = I - \frac{e^2}{4d} = I - V_{\text{Im}} \tag{9.2.5}$$

9.2.2. Affinity Level

Similarly, when approaching an additional electron toward the adsorbate, only its image contributes to the image force. This force is attractive and the modification of the affinity energy is

$$W = \int_\infty^d -\frac{e^2}{4z^2} \, dz = \frac{e^2}{4d} = V_{\text{Im}} \tag{9.2.6}$$

and thus

$$A_{\text{eff}} = A + V_{\text{Im}} \tag{9.2.7}$$

Consequently we can define an effective Coulomb integral U_{eff} by

$$U_{\text{eff}} = U - 2V_{\text{Im}} \tag{9.2.8}$$

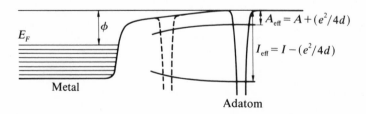

Figure 9.5. Electric image of an adatom.

In a very qualitative way and taking into account that physically we must have $A_{eff} < I_{eff}$, three cases can be distinguished according to the value of the metal work function:

1. $I - e^2/4d > \Phi > A + e^2/4d$, i.e., the Fermi level lies between the effective affinity and ionization levels. In these conditions the adatom tends to remain neutral (see Figure 9.4) since the transfer of an electron to or from the metal would cost some energy.
2. $(I + A)/2 < I - e^2/4d < \Phi$, i.e., the effective ionization level is above the Fermi level of the metal. The adatom tends to become a positively charged ion since one gains the energy $\Phi - I_{eff}$ when transferring an electron from the adatom to the metal.
3. $\Phi < A + e^2/4d < (I + A)/2$, i.e., the effective affinity level is below the Fermi energy of the metal. The adatom tends to become a negatively charged ion since one gains the energy $A_{eff} - \Phi$ when transferring an electron from the metal to the adatom.

When the adatom comes near the surface, the classical image theory starts to break down and, in addition to their shifts, the levels begin to broaden and therefore the charge transfer may be a fraction of an electron. It will be seen in the following that the exact determination of the charge transfer is very difficult, since it involves the treatment of electronic interactions. This is a many-body problem for which only very approximate solutions exist up to now. This range of distances is actually the most interesting, since it corresponds to the distances involved in chemisorption phenomena. The aim of this chapter is to show that, in spite of the difficulties arising from the charge transfer problem, one can set up simple models from which experimental trends can be understood.

9.3. The Anderson–Grimley–Newns Hamiltonian

This Hamiltonian was first introduced by Anderson[14] to treat dilute impurity problems in bulk metals. It was adapted by Grimley[15] and Newns[16] to study the chemisorption of adatoms on a metal surface. Since these early works, this type of Hamiltonian has been used by many others.

The Hamiltonian can be written in the form

$$H = H_{ads} + H_{subs} + H_{coupling} \tag{9.3.1}$$

where H_{ads}, H_{subs}, and $H_{coupling}$ are respectively the adsorbate, substrate, and coupling Hamiltonians. For the sake of simplicity, it is assumed that only one valence orbital of the adatom is involved in the bonding as in the hydrogen case. Then

$$H_{ads} = \sum_{\sigma} \left(\varepsilon_a^0 n_{a\sigma} + \tfrac{1}{2} U n_{a\sigma} n_{a-\sigma} \right) \tag{9.3.2a}$$

$$H_{\text{coupling}} = \sum_{k\sigma} [V_{ak} c_{a\sigma}^+ c_{k\sigma} + V_{ka} c_{k\sigma}^+ c_{a\sigma}] \qquad (9.3.2b)$$

$$H_{\text{coupling}} = \sum_{k\sigma} [V_{ak} c_{a\sigma}^+ c_{k\sigma} + V_{ka} c_{k\sigma}^+ c_{a\sigma}] \qquad (9.3.2c)$$

where ε_a^0 is the atomic level of the considered atomic orbital $|a\rangle$ and ε_k are the eigenvalues of H_{subs}; $c_{a\sigma}^+$ ($c_{a\sigma}$) and $c_{k\sigma}^+$ ($c_{k\sigma}$) are respectively creation (annihilation) operators in the spin-orbital $|a\sigma\rangle$ and metal state $|k\sigma\rangle$, the corresponding occupation number operators being

$$n_{a\sigma} = c_{a\sigma}^+ c_{a\sigma} \qquad (9.3.3a)$$

$$n_{k\sigma} = c_{k\sigma}^+ c_{k\sigma} \qquad (9.3.3b)$$

Quantity U is an effective Coulomb integral on the adatom ($U = I_{\text{eff}} - A_{\text{eff}}$) and is usually taken as a parameter. Finally

$$V_{ak} = \langle a|H|k\rangle = V_{ka}^* \qquad (9.3.4)$$

is the coupling matrix element. We note that this Hamiltonian does not take into account the core–core repulsion between the adatom and substrate atoms.

9.3.1. Hartree–Fock Treatment

In the Hartree-Fock approximation the two-body operator $n_{a\sigma} n_{a-\sigma}$ is replaced by an effective one-body operator:

$$n_{a\sigma} n_{a-\sigma} \Rightarrow \langle n_{a\sigma}\rangle n_{a-\sigma} + \langle n_{a-\sigma}\rangle n_{a\sigma} - \langle n_{a\sigma}\rangle \langle n_{a-\sigma}\rangle$$

In this approximation, the expression for H becomes

$$H = \sum_{\sigma} [H_{HF}^\sigma - \tfrac{1}{2}U\langle n_{a\sigma}\rangle\langle n_{a-\sigma}\rangle] \qquad (9.3.5)$$

with

$$H_{HF}^\sigma = \varepsilon_{a\sigma} n_{a\sigma} + \sum_k \varepsilon_k n_{k\sigma} + \sum_k (V_{ak} c_{a\sigma}^+ c_{k\sigma} + \text{h.c.}) \qquad (9.3.6a)$$

and

$$\varepsilon_{a\sigma} = \varepsilon_a^0 + U\langle n_{a-\sigma}\rangle \qquad (9.3.6b)$$

One sees at first glance that spins are now decoupled and therefore each spin can be treated separately. However, since $\varepsilon_{a\sigma}$ depends on $\langle n_{a-\sigma}\rangle$, the final solution should obey a self-consistency condition.

9.3.1.1. Local Density of States on the Adsorbate

Let us consider the states with spin σ and assume that $|a\sigma\rangle$, $|k\sigma\rangle$ is a complete orthonormal basis. The local density of states on the adsorbate can be obtained from the corresponding diagonal matrix element of the Green operator:

$$\rho_a^\sigma(E) = -\frac{1}{\pi} \lim_{\varepsilon \to 0} \text{Im}\, G_{aa}^{\sigma\sigma}(E + i\varepsilon), \qquad \varepsilon > 0 \qquad (9.3.7a)$$

with

$$G_{aa}^{\sigma\sigma}(E + i\varepsilon) = \langle a\sigma| \frac{1}{E + i\varepsilon - H_{HF}^\sigma} |a\sigma\rangle \qquad (9.3.7b)$$

$$(E + i\varepsilon - H_{HF}^\sigma)^{-1} = \begin{bmatrix} E + i\varepsilon - \varepsilon_{a\sigma} & -V_{ak_1} & -V_{ak_2} & \cdots \\ -V_{ak_1}^* & E + i\varepsilon - \varepsilon_{k_1} & 0 & \cdots \\ -V_{ak_2}^* & 0 & E + i\varepsilon - \varepsilon_{k_2} & \cdots \\ \vdots & \vdots & \vdots & \end{bmatrix}^{-1} \qquad (9.3.8)$$

Here $G_{aa}^{\sigma\sigma}(E + i\varepsilon)$ is the first element of this matrix, i.e.,

$$G_{aa}^{\sigma\sigma}(E + i\varepsilon) = \frac{\prod_k (E + i\varepsilon - \varepsilon_k)}{(E + i\varepsilon - \varepsilon_{a\sigma}) \prod_k (E + i\varepsilon - \varepsilon_k) - \sum_k |V_{ak}|^2 \prod_{k' \neq k} (E + i\varepsilon - \varepsilon_{k'})} \qquad (9.3.9)$$

or equivalently

$$G_{aa}^{\sigma\sigma}(E + i\varepsilon) = \frac{1}{E + i\varepsilon - \varepsilon_{a\sigma} - \sum_k |V_{ak}|^2/(E - \varepsilon_k + i\varepsilon)} = \frac{1}{E + i\varepsilon - \varepsilon_{a\sigma} - S(E + i\varepsilon)} \qquad (9.3.10)$$

which defines the S function.

Using the well-known identity

$$\lim_{\varepsilon \to 0} \sum_k \frac{|V_{ak}|^2}{E - \varepsilon_k + i\varepsilon} = \mathscr{P}\left(\sum_k \frac{|V_{ak}|^2}{E - \varepsilon_k}\right) - i\pi \sum_k |V_{ak}|^2 \delta(E - \varepsilon_k) \qquad (9.3.11)$$

in which $\mathscr{P}$ means the principal part, we can define two functions:

$$\Lambda(E) = \mathscr{P}\left(\sum_k \frac{|V_{ak}|^2}{E - \varepsilon_k}\right) \qquad (9.3.12a)$$

$$\Delta(E) = \pi \sum_k |V_{ak}|^2 \delta(E - \varepsilon_k) \qquad (9.3.12\text{b})$$

$$\lim_{\varepsilon \to 0} S(E + i\varepsilon) = \Lambda(E) - i\Delta(E) \qquad (9.3.12\text{c})$$

where $\Lambda(E)$ and $\Delta(E)$ are not independent but are related through a Hilbert transform:

$$\Lambda(E) = \frac{\mathscr{P}}{\pi} \int_{-\infty}^{+\infty} \frac{\Delta(E')}{E - E'}\, dE' \qquad (9.3.13)$$

Consequently

$$\lim_{\varepsilon \to 0} G_{aa}^{\sigma\sigma}(E + i\varepsilon) = \frac{1}{E - \varepsilon_{a\sigma} - \Lambda(E) + i\Delta(E)} \qquad (9.3.14)$$

is fully determined by the knowledge of $\Delta(E)$, which is usually called the *chemisorption function*. Equation (9.3.14) yields

$$\rho_a^\sigma(E) = \frac{1}{\pi} \frac{\Delta(E)}{[E - \varepsilon_{a\sigma} - \Lambda(E)]^2 + \Delta^2(E)} \qquad (9.3.15)$$

As already stated $\varepsilon_{a\sigma}$, and thus $\rho_a^\sigma(E)$, are functions of $\langle n_{a-\sigma}\rangle$ so therefore

$$\langle n_{a\sigma}\rangle = \int^{E_\mathrm{F}} \rho_a^\sigma(E)\, dE = N(\langle n_{a-\sigma}\rangle) \qquad (9.3.16)$$

Therefore the self-consistent function $\rho_a^\sigma(E)$ should be such that

$$\langle n_{a-\sigma}\rangle = N(\langle n_{a\sigma}\rangle) = N(N(\langle n_{a-\sigma}\rangle)) \qquad (9.3.17)$$

The features of $\rho_a^\sigma(E)$ will now be examined assuming that the self-consistent value of $\langle n_{a-\sigma}\rangle$ is known. Quantity $\rho_a^\sigma(E)$ has a continuous spectrum which comes from $\Delta(E)$ and thus coincides with the substrate energy band. In addition, one or two bound states may appear when the equation

$$E - \varepsilon_{a\sigma} - \Lambda(E) = 0 \qquad (9.3.18)$$

has roots $E_{l\sigma}$ outside the continuous spectrum $[\Delta(E) = 0]$. The weight of this bound state is given by

$$\langle n_{l\sigma}\rangle = [1 - \Lambda'(E_{l\sigma})]^{-1} \qquad (9.3.19)$$

where $\Lambda'(E) = d\Lambda(E)/dE$. This last quantity is always negative outside the band so that

$$0 < \langle n_{l\sigma}\rangle < 1$$

To proceed further we adopt a specific form of function $\Delta(E)$. Following Newns,[16] $|V_{ak}|^2$ is replaced by an average value $|V_{ak}|^2_{av}$:

$$\Delta(E) \simeq \pi |V_{ak}|^2_{av} \sum_k \delta(E - \varepsilon_k) \qquad (9.3.20a)$$

$$\simeq \pi |V_{ak}|^2_{av} \rho(E) \qquad (9.3.20b)$$

$\rho(E)$ being the (bulk) substrate density of states assumed to possess a semielliptic shape. It is known that this is actually the local density of states on the first atom of a semi-infinite linear chain in the tight-binding approximation with nearest-neighbor hopping integrals (see Appendix 9.1). Hence functions $\Lambda(E)$ and $\Delta(E)$ are given respectively by the real and imaginary parts of the corresponding Green function multiplied by $|V_{ak}|^2_{av}$. This Green function is derived in Appendix 9.1:

$$G(z) = \frac{z + \eta\sqrt{z^2 - W^2/4}}{W^2/8}, \qquad \eta = \pm 1 \qquad (9.3.21)$$

where W is the bandwidth, the value of η being chosen so that function $G(z)$ has the correct behavior at infinity and has a negative imaginary part. Thus:

if $E < -W/2$

$$\Lambda(E) = |V_{ak}|^2_{av} \frac{E + \sqrt{E^2 - W^2/4}}{W^2/8} \qquad (9.3.22a)$$

$$\Delta(E) = 0 \qquad (9.3.22b)$$

if $-W/2 < E < W/2$

$$\Lambda(E) = |V_{ak}|^2_{av} \frac{8E}{W^2} \qquad (9.3.23a)$$

$$\Delta(E) = |V_{ak}|^2_{av} \frac{8}{W^2} \sqrt{\frac{W^2}{4} - E^2} \qquad (9.3.23b)$$

if $E > W/2$

$$\Lambda(E) = |V_{ak}|^2_{av} \frac{E - \sqrt{E^2 - W^2/4}}{W^2/8} \qquad (9.3.24a)$$

$$\Delta(E) = 0 \qquad (9.3.24b)$$

The variations of $\Lambda(E)$ and $\Delta(E)$ with E are shown in Figure 9.6. The two limiting cases will now be discussed.

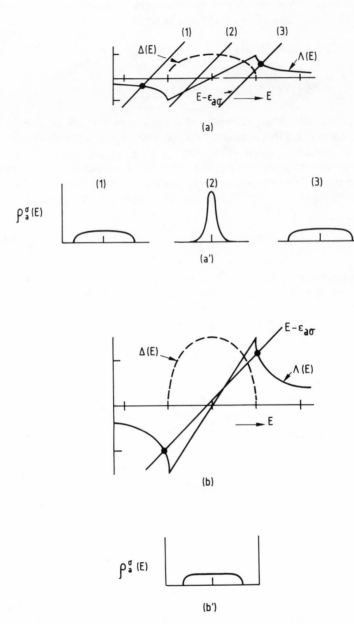

Figure 9.6. Solution of the Anderson–Grimley–Newns Hamiltonian in the weak (a, a') and strong (b, b') coupling limits: (a, b) present the intersections of $E - \varepsilon_{a\sigma}$ with $\Lambda(E)$ which define the localized states, (a', b') present the corresponding local densities of states on the adsorbate.

a. Weak-Coupling Limit. This is the case when $|V_{ak}|^2_{av}$ is small compared to W^2, in which case $\Delta(E)$ and $\Lambda(E)$ are small compared to W. The value of $\rho_a^\sigma(E)$ reaches a high value only when $E - \varepsilon_{a\sigma} - \Lambda(E)$ vanishes. This can be solved graphically (Figure 9.6a); one obtains a unique solution

$$E \simeq \varepsilon_{a\sigma} + \Lambda(\varepsilon_{a\sigma}) \tag{9.3.25}$$

If this energy falls within the energy band of the metal, Δ is not zero and can be approximated by $\Delta(\varepsilon_{a\sigma})$, in which case the adatom level can be thought of as a virtual bound state with a half-width at half-maximum $\Delta(\varepsilon_{a\sigma})$. If this value of energy falls outside the metal band, $\rho_a^\sigma(E)$ will be composed of a true bound state $E_{l\sigma}$ with a weight $\langle n_{l\sigma} \rangle$ given by equation (9.3.19) and a continuous contribution extending all over the metal band with a weight $1 - \langle n_{l\sigma} \rangle$ (Figure 9.6a′).

b. Strong Coupling Limit. In this case $|V_{ak}|^2_{av}$ is large relative to W^2, in which case $\Lambda(E)$ and $\Delta(E)$ are large compared to W. One sees from Figure 9.6b that equation (9.3.18) has generally three roots (at least when $\varepsilon_{a\sigma}$ is not too far from the center of the metal band): one in the metal band and the others on both sides of it. Quantity $\rho_a^\sigma(E)$ has thus two bound states and between them a weak continuous part extending over the metal band (Figure 9.6b′). These two bound states can be regarded as the bonding and antibonding states of a surface molecule formed by the adatom and its neighbors. This will be seen clearly below in the tight-binding formalism.

c. Magnetic versus Nonmagnetic Solutions. Two types of solution are possible for equation (9.3.17):

1. $\langle n_{a\sigma} \rangle = \langle n_{a-\sigma} \rangle$; this solution always exists and corresponds to a nonmagnetic adsorbate.
2. $\langle n_{a\sigma} \rangle \neq \langle n_{a-\sigma} \rangle$; in this case the adsorbate has a magnetic moment.

When the two types of solution exist, in principle, the correct solution has the lowest energy. For hydrogen, this solution turns out to be magnetic when the coupling is weak enough. However, the Hartree–Fock scheme for the adatom becomes questionable in this limit since U is larger that $\Delta(\varepsilon_{a\sigma})$. Then electronic correlations should be taken into account and it is well known that they decrease the tendency to magnetism, which is overestimated in the Hartree–Fock scheme.

d. Adsorption on a Substrate with a Narrow Band. When the substrate has a narrow band (transition metals) one can describe it in the tight-binding approximation. In this case, one obtains more physical insight into the chemisorption function which can be calculated exactly (i.e., without replacing V_{ak} by an average value). For the sake of simplicity, let us first consider a substrate with one s orbital per site. In the tight-binding approximation we can write

$$|k\rangle = \sum_i a_i(\varepsilon_k)|i\rangle \tag{9.3.26}$$

where $|k\rangle$ is a band state of the semi-infinite crystal and $|i\rangle$ is the atomic orbital centered at site i;

$$V_{ak} = \langle a|H|k\rangle = \sum_i a_i(\varepsilon_k)\beta'_{ai} \tag{9.3.27}$$

where $\beta'_{ai} = \langle a|H|i\rangle$ is the hopping integral between the adsorbate and substrate site i. Thus

$$\Delta(E) = \pi \sum_{i,j,k} \beta'_{ai}\beta'^*_{aj}a_i(\varepsilon_k)a_j^*(\varepsilon_k)\delta(E - \varepsilon_k) \tag{9.3.28}$$

The interaction can be limited to the nearest neighbors of the adsorbate, and when the adsorption geometry is such that all these neighbors are equivalent then $\beta'_{ai} = \beta'$ for any i nearest neighbor of a and $\beta'_{ai} = 0$ otherwise. Hence

$$\Delta(E) = \pi\beta'^2 \underset{\substack{i,j \\ \text{nearest neighbors of } a}}{\sum} \sum_k a_i(\varepsilon_k)a_j^*(\varepsilon_k)\delta(E - \varepsilon_k) \tag{9.3.29}$$

Let us now consider several simple geometries occurring, for instance, on a surface with a square lattice.

1. *On top position* (Figure 9.7a). The adatom has only one neighbor denoted by 1 and

$$\Delta(E) = \pi\beta'^2 \sum_k a_1^*(\varepsilon_k)a_1(\varepsilon_k)\delta(E - \varepsilon_k) \tag{9.3.30a}$$

$$= \pi\beta'^2 n_1(E) \tag{9.3.30b}$$

where $n_1(E)$ is the surface local density of states of the clean substrate. Hence function $\Delta(E)$ is proportional to this quantity.

2. *Bridge site* (Figure 9.7b). The adatom has two neighbors denoted by 1 and 2. Let us first introduce the matrix elements G^0_{ij} of the clean substrate Green function:

$$G^0_{ij} = \sum_k \frac{\langle i|k\rangle\langle k|j\rangle}{E + i\varepsilon - \varepsilon_k} = \sum_k \frac{a_i(\varepsilon_k)a_j^*(\varepsilon_k)}{E + i\varepsilon - \varepsilon_k} \tag{9.3.31}$$

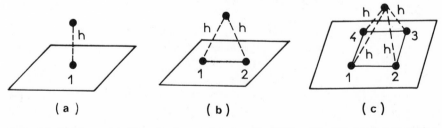

(a)   (b)   (c)

Figure 9.7. The three adsorption sites for an adatom on the (100) surface of a simple cubic lattice: (a) on top, (b) bridge, (c) centered.

Consequently, if real atomic wave functions are employed, the matrix elements of the tight-binding Hamilton are real and

$$\lim_{\varepsilon \to 0} \text{Im } G_{ij}^0 = -\pi \sum_k a_i(\varepsilon_k) a_j^*(\varepsilon_k) \delta(E - \varepsilon_k) \tag{9.3.32}$$

and

$$\Delta(E) = -\beta'^2 \text{ Im } (G_{11}^0 + G_{12}^0 + G_{21}^0 + G_{22}^0) \tag{9.3.33}$$

which can be simplified by introducing the group orbital[18]

$$|g_s^b\rangle = \frac{1}{\sqrt{2}} (|1\rangle + |2\rangle) \tag{9.3.34}$$

$$\Delta(E) = -2\beta'^2 \text{ Im } \langle g_s^b | G^0 | g_s^b \rangle \tag{9.3.35a}$$

$$= 2\pi\beta'^2 n_{g_s^b}(E) \tag{9.3.35b}$$

where $n_{g_s^b}(E)$ is the local density of states of the clean substrate associated with the group orbital $|g_s^b\rangle$.

3. *Centered site* (Figure 9.7c). Similarly one obtains

$$\Delta(E) = 4\pi\beta'^2 n_{g_s^c}(E) \tag{9.3.36}$$

with $|g_s^c\rangle = \frac{1}{2}(|1\rangle + |2\rangle + |3\rangle + |4\rangle)$.

Therefore, in the general case in which the adsorbate has Z_s equivalent nearest neighbors, $\rho_a^\sigma(E)$ is given by[17]

$$\rho_a^\sigma(E) = \frac{Z_s \beta'^2 n_{g_s}(E)}{[E - \varepsilon_{a\sigma} - Z_s \beta'^2 R_{g_s}(E)]^2 + \pi^2 Z_s^2 \beta'^4 n_{g_s}^2(E)} \tag{9.3.37}$$

$R_{g_s}(E)$ being the Hilbert transform of $\pi n_{g_s}(E)$, i.e.,

$$R_{g_s}(E) = \text{Re}\langle g_s | G^0 | g_s \rangle \tag{9.3.38}$$

The state $|g_s\rangle$ is the group orbital associated with the adsorption site, i.e., the symmetric linear combination (normalized) of substrate atomic orbitals centered on the nearest neighbors of the adsorbate.

In order to illustrate the preceding calculations we consider a nonmagnetic adatom at on top position on the (100) plane of a simple cubic lattice and assume, for simplicity, that $\varepsilon_{a\sigma} = \varepsilon_a$ coincides with the center of the substrate band.

The corresponding function $\rho_a(E)$ is given in Figure 9.8 as a function of β'/β, β being the substrate hopping integral. One sees that:

1. When $\beta' \lesssim \beta$, $\rho_a(E)$ has the shape of a virtual bound state centered at ε_a [since $R_{g_s}(E)$ is an odd function] with a half-mean-width of order $\pi\beta'^2 n_{g_s}(0)$.

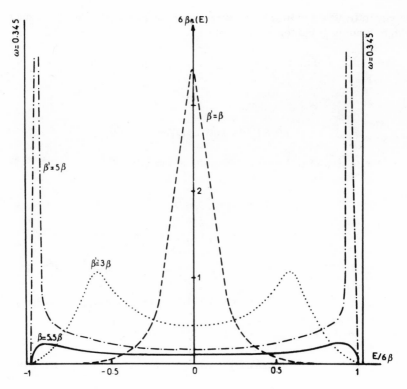

Figure 9.8. Local densities of states of an on top adatom on the (100) surface of a simple cubic lattice for various values of the adatom–substrate coupling.

2. When $\beta' \gtrsim \beta$, $\rho_a(E)$ is composed of two virtual bound states which become true bound states beyond a critical value of $\beta'/\beta \simeq 5.35$. These two bound states are actually the bonding and antibonding states of the molecule formed by the adatom and its substrate neighbor weakly perturbed by the "indented" solid.

The case of adsorption on bridge and centered sites is slightly more intricate, since the corresponding densities of states associated with the group orbital are asymmetric. The qualitative behavior is the same, except that bonding and anti-bonding bound states appear successively and for lower values of β'/β.

The concept of group orbital can be easily generalized to the case of substrates with degenerate atomic levels and when the atomic level of the adatom is itself degenerate. A group orbital must be associated with each adatom atomic orbital, this group orbital being defined as[15,18]

$$|g_{s\lambda}\rangle = A \sum_{i,\nu} \beta_{ia}^{\nu\lambda} |i\nu\rangle, \qquad \begin{array}{l} \lambda = 1, L_a \\ \nu = 1, L_s \end{array} \qquad (9.3.39)$$

where L_a and L_s are respectively the degeneracy of the atomic level of the adatom and the substrate, $\beta_{ia}^{\nu\lambda}$ are the hopping integrals

$$\beta_{ia}^{\nu\lambda} = \langle i\nu | H | a\lambda \rangle \tag{9.3.40}$$

between the adatom and its substrate neighbor i, and A is the normalization factor given by

$$A = \left(\sum_{i,\nu} |\beta_{ia}^{\nu\lambda}|^2 \right)^{-1/2} \tag{9.3.41}$$

However, the partial density of states on the adatom, $\rho_a^\lambda(E)$, corresponding to orbital λ does not usually take a simple form similar to equation (9.3.37). Nevertheless, its general shape is still governed not only by the values of the hopping integrals, but also by the adatom coordination number and the surface density of states $n_s(E)$ around the adatom atomic level. Consequently, for a given adsorbate-substrate system, the effective coupling may be weak or strong according to the surface crystallographic orientation. A typical example is shown in Figure 9.9,[19] where the coupling of a Mo adatom is weak on Mo(110) $[Z_s = 2, n_s(\varepsilon_a)$ small] and strong on Mo(100) $[Z_s = 4, n_s(\varepsilon_a)$ large].

We remark that within the tight-binding basis, H_{coupling} takes a simple form. Indeed, for an adsorbate with only one orbital (for simplicity)

$$V_{ak} = \sum_{j\lambda} \langle a|H|j\lambda\rangle\langle j\lambda|k\rangle \quad \text{and} \quad c_k = \sum_{i\nu} \langle k|i\nu\rangle c_{i\nu} \tag{9.3.42}$$

$$\sum_k V_{ak} c_a^+ c_k = \sum_{j\lambda} \langle a|H|j\lambda\rangle \underbrace{\sum_{k,i\nu} \langle j\lambda|k\rangle\langle k|i\nu\rangle}_{\displaystyle \sum_{i,\nu} \delta_{ij}\delta_{\lambda\nu}} c_a^+ c_{i\nu}$$

$$= c_a^+ \sum_{i\nu} \langle i\nu|H|a\rangle^* c_{i\nu}$$

$$= A^{-1} c_a^+ c_{g_s} \tag{9.3.43}$$

where c_{g_s} is the annihilation operator of the group orbital $|g_s\rangle$. Thus

$$H_{\text{coupling}} = A^{-1}(c_a^+ c_{g_s} + c_{g_s}^+ c_a) \tag{9.3.44}$$

9.3.1.2. Binding Energy

The binding energy E_B of the adsorbate is the difference between the total energy of the semi-infinite crystal with the adsorbed atom and that of the same system with the adatom infinitely far from the surface.

It will now be shown that this binding energy is fully determined by the knowledge of the chemisorption function $\Delta(E)$. If one neglects the core-core repulsion between the adsorbate and the substrate atoms and assumes that the adatom valence orbital is occupied in the free state by one electron, this energy can be expressed in the form

$$E_B = \int_{-\infty}^{E_F+\delta E_F} E \mathcal{N}'(E)\, dE - \int_{-\infty}^{E_F} E \mathcal{N}(E)\, dE - \varepsilon_a^0 - U\langle n_{a\sigma}\rangle\langle n_{a-\sigma}\rangle \tag{9.3.45}$$

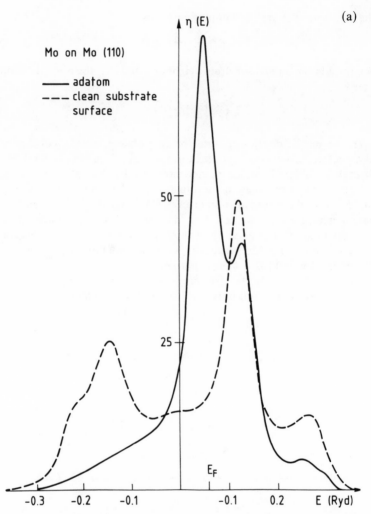

Figure 9.9. Local density of states of a Mo adatom on Mo compared with the surface density of states. The adatom is assumed to occupy a lattice site: (a) (110) surface, (b) (100) surface.

where $\mathcal{N}'(E)$ is the total density of states of the semi-infinite crystal with the adsorbed atom and $\mathcal{N}(E)$ the total density of states of the clean semi-infinite crystal. Thus $\mathcal{N}'(E)$ is normalized to $N+1$ atoms and $\mathcal{N}(E)$ to N atoms. The first three terms account for the variation of the total one-electron energy when the adatom is brought from infinity and the last term avoids the double counting of the electronic interactions in the Hartree scheme [see equation (9.3.5)]. In order to ensure the conservation of the total charge, it is necessary to allow a small (but unphysical) variation of the Fermi level. This point will be discussed later in more detail.

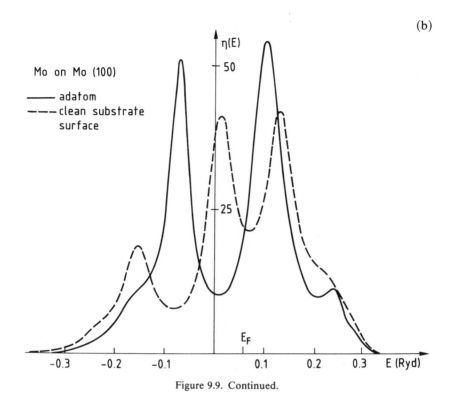

Figure 9.9. Continued.

Equation (9.3.45) can now be transformed into the form

$$E_B = \int_{-\infty}^{E_F} E \Delta \mathcal{N}(E)\, dE + E_F \mathcal{N}'(E_F) \delta E_F - \varepsilon_a^0 - U \langle n_{a\sigma} \rangle \langle n_{a-\sigma} \rangle \qquad (9.3.46)$$

with $\mathcal{N}'(E) - \mathcal{N}(E) = \Delta \mathcal{N}(E)$.

The charge conservation can be written as

$$\int_{-\infty}^{E_F + \delta E_F} \mathcal{N}'(E)\, dE - \int_{-\infty}^{E_F} \mathcal{N}(E)\, dE = 1 \qquad (9.3.47)$$

or

$$\int_{-\infty}^{E_F} \Delta \mathcal{N}(E)\, dE + \mathcal{N}'(E_F) \delta E_F = 1 \qquad (9.3.48)$$

The binding energy thus becomes

$$E_B = \int_{-\infty}^{E_F} (E - E_F) \Delta \mathcal{N}(E)\, dE + E_F - \varepsilon_a^0 - U \langle n_{a\sigma} \rangle \langle n_{a-\sigma} \rangle \qquad (9.3.49)$$

As shown in Appendix 9.2, the variation in the total density of states when the Hamiltonian H_0 is changed to $H_0 + V$ is given by

$$\Delta n(E) = \mp \frac{1}{\pi} \lim_{\varepsilon \to 0} \text{Im} \frac{d}{dE} \log \text{Det} (1 - VG^{0\pm}) \tag{9.3.50}$$

where $G^{0\pm} = (E - H_0 \pm i\varepsilon)^{-1}$ and $H_0 = \sum_\sigma \varepsilon_{a\sigma} n_{a\sigma} + \sum_{k,\sigma} \varepsilon_k n_{k\sigma}$. If V is the coupling between the adatom and the substrate, then

$$\Delta n(E) = \Delta \mathcal{N}(E) - \sum_\sigma \delta(E - \varepsilon_{a\sigma}) \tag{9.3.51}$$

Let us calculate Det $(1 - VG^{0\pm})$. This determinant factorizes into two determinants, one for each spin, having the following form:

$$\begin{bmatrix} 1 & -V_{ak_1} G^{0\pm}_{k_1 k_1} & -V_{ak_2} G^{0\pm}_{k_2 k_2} & \cdots \\ -V_{k_1 a} G^{0\pm}_{aa} & 1 & 0 & \cdots \\ -V_{k_2 a} G^{0\pm}_{aa} & 0 & 1 & \cdots \\ \vdots & \vdots & \vdots & \ddots \end{bmatrix} \tag{9.3.52}$$

Thus

$$\text{Det} (1 - VG^{0\pm}) = \prod_\sigma \left(1 - \sum_k |V_{ak}|^2 G^{0\pm}_{kk} G^{0\sigma\sigma\pm}_{aa} \right) \tag{9.3.53}$$

On replacing quantities $G^{0\pm}_{kk}$ and $G^{0\sigma\sigma\pm}_{aa}$ by their corresponding expressions one obtains

$$\text{Det} (1 - VG^{0\pm}) = \prod_\sigma \left(1 - \sum_k \frac{|V_{ak}|^2}{(E \pm i\varepsilon - \varepsilon_{a\sigma})(E \pm i\varepsilon - \varepsilon_k)} \right) \tag{9.3.54a}$$

$$= \prod_\sigma \left(1 - \frac{1}{E \pm i\varepsilon - \varepsilon_{a\sigma}} S(E \pm i\varepsilon) \right) \tag{9.3.54b}$$

and equation (9.3.50) yields

$$\Delta n(E) = \sum_\sigma \mp \frac{1}{\pi} \lim_{\varepsilon \to 0} \text{Im} \frac{d}{dE} \log [E - \varepsilon_{a\sigma} - S(E \pm i\varepsilon)] - \delta(E - \varepsilon_{a\sigma}) \tag{9.3.55}$$

thus

$$\Delta \mathcal{N}(E) = \sum_\sigma \mp \frac{1}{\pi} \lim_{\varepsilon \to 0} \text{Im} \frac{d}{dE} \log [E - \varepsilon_{a\sigma} - S(E \pm i\varepsilon)] \tag{9.3.56a}$$

$$= \sum_\sigma \mp \frac{1}{\pi} \lim_{\varepsilon \to 0} \text{Im} \frac{d}{dE} \log [G^{\sigma\sigma\pm}_{aa}(E)]^{-1} \tag{9.3.56b}$$

$$= \sum_\sigma \mp \frac{1}{\pi} \lim_{\varepsilon \to 0} \text{Im} \left[\left(1 - \frac{dS(E \pm i\varepsilon)}{dE} \right) G_{aa}^{\sigma\sigma\pm}(E) \right] \quad (9.3.56c)$$

If the function $\Delta N(E)$ is defined by

$$\Delta N(E) = \int_{-\infty}^{E_F} \Delta \mathcal{N}(E) \, dE \quad (9.3.57)$$

then integration by parts using equation (9.3.56b) with $G_{aa}^{\sigma\sigma+}(E)$ leads to

$$\int_{-\infty}^{E_F} (E - E_F) \Delta \mathcal{N}(E) \, dE$$

$$= - \int_{-\infty}^{E_F} \Delta N(E) \, dE \quad (9.3.58a)$$

$$= \sum_\sigma \left(E_{l\sigma} + \frac{1}{\pi} \int_m^{E_F} \text{Im} \log \left[E - \varepsilon_{a\sigma} - \Lambda(E) + i\Delta(E) \right] dE \right) \quad (9.3.58b)$$

in which $E_{l\sigma}$ is the occupied bound state with spin σ (when existing) and m is the bottom of the substrate band.

Finally

$$E_B = \sum_\sigma \left[E_{l\sigma} + \frac{1}{\pi} \int_m^{E_F} \arctan \frac{\Delta(E)}{E - \varepsilon_{a\sigma} - \Lambda(E)} \, dE \right]$$
$$+ E_F - \varepsilon_a^0 - U \langle n_{a\sigma} \rangle \langle n_{a-\sigma} \rangle \quad (9.3.59)$$

with $0 < \arctan x < \pi$ when there is an occupied bound state and $-\pi < \arctan x < 0$ otherwise.

With this formula Newns[16] has calculated the chemisorption energy of hydrogen on Ti, Cr, Ni, and Cu as a function of the coupling strength and the corresponding charge transfer. Although the sign of the charge transfer agrees with experiments (variation of the work function during adsorption or, more recently, variation of core level binding energy of the substrate surface atoms), its order of magnitude is too large. This formalism had also previously been used by Grimley[15] to study the indirect (i.e., via the substrate) interaction between two adatoms.

9.3.1.3. Discussion

Two weak points of this formalism have already been mentioned: one must assume a nonphysical variation of the Fermi level to ensure the total charge neutrality and magnetic solutions are too easily obtained. Actually, a proper treatment should obey the Friedel sum rule,[20] i.e.,

$$\int_{-\infty}^{E_F} \Delta \mathcal{N}(E) \, dE = 1 \quad (9.3.60)$$

which is generally not fulfilled by $\Delta\mathcal{N}(E)$ given by equation (9.3.56). Concerning the second point, the Hartree-Fock treatment implicitly assumes that U is small compared to the mean width of $\rho_a^\sigma(E)$; it allows any charge fluctuation as in any molecular orbital (MO) approach. The effect of Coulomb electronic interactions is only treated to the first order and thus an obvious improvement would be to proceed at least to second order. Such improved models have been proposed by several authors; they will be discussed briefly in the next section. On the other hand, for some transition metals, the effective Coulomb integral of the substrate may be important.

9.3.2. Beyond the Hartree–Fock Treatment

Since the first works based on the Hartree-Fock approximation described in the preceding sections, many attempts have been made to improve this treatment by including many-body effects. Many of them aimed at a better description of the one-particle spectrum of the adsorbate.[21-26] As far as the total energy is concerned, although the general formula derived by Kjöllerström et al.[27] for the dilute impurity problem can be used to compute the binding energy of a chemisorbed atom, few authors actually calculated this quantity.[28,29]

We have seen that in the Hartree-Fock treatment all the interesting physical properties of the chemisorbed system can be derived from the knowledge of the Green function $G_{aa}^{\sigma\sigma}$ of the adsorbate [see equations (9.3.56b) and (9.3.59)]. This fact remains true when many-body effects are taken into account. This chapter will not present the exact derivation, which can be found in Kjöllerström et al.,[27] but will rather generalize in an intuitive way equation (9.3.14) for the Green function and equation (9.3.49) for the binding energy.

In the expression for $G_{aa}^{\sigma\sigma}$, the introduction of the adsorbate Coulomb interaction gives rise to a self-energy $\Sigma_{a\sigma}$ which, in the Hartree-Fock scheme, is nothing but $U\langle n_{a-\sigma}\rangle$ [equation (9.3.6b)]. In the general case $\Sigma_{a\sigma}$ becomes a complex function of the energy and equation (9.3.14) now becomes

$$\lim_{\varepsilon\to 0} G_{aa}^{\sigma\sigma}(E+i\varepsilon) = \lim_{\varepsilon\to 0} \frac{1}{E - \varepsilon_a^0 - \Sigma_{a\sigma}(E+i\varepsilon) - \Lambda(E) + i\Delta(E)} \qquad (9.3.61)$$

from which the one-particle excitation spectrum of the adsorbate can be derived with the aid of equation (9.3.7a).

The expression for the binding energy can be generalized by rewriting equation (9.3.49). The first term can be transformed into a contour integral (see Appendix 9.3) in which the causal Green function on the adsorbate appears,

$$^c G_{aa}^{\sigma\sigma}(E) = \frac{1}{E - \varepsilon_a^0 - U\langle n_{a-\sigma}\rangle - S(E)} \qquad (9.3.62)$$

with

$$S(E) = \sum_k \frac{|V_{ak}|^2}{E - \varepsilon_k + i\eta \, \text{sgn}\,(\varepsilon_k - E_F)}$$

Then

$$\int^{E_F} (E - E_F) \, \Delta \mathcal{N}(E) \, dE = \frac{1}{2\pi i} \sum_\sigma \int_C (z - E_F)\left(1 - \frac{dS}{dz}\right) {}^c G_{aa}^{\sigma\sigma}(z) \, dz \quad (9.3.63)$$

where C is the contour consisting of the real axis and a semicircle at infinity in the upper half-plane.

The last term in equation (9.3.49) can also be written as a contour integral:

$$U\langle n_{a-\sigma}\rangle\langle n_{a\sigma}\rangle = \frac{1}{4\pi i} \sum_\sigma \int_C \Sigma_{a\sigma}^{HF} {}^c G_{aa}^{\sigma\sigma}(z) \, dz \quad (9.3.64)$$

where

$$\Sigma_{a\sigma}^{HF} = U\langle n_{a-\sigma}\rangle$$

Finally, the expression for the adsorption energy in the Hartree-Fock approximation becomes

$$E_B = \frac{1}{2\pi i} \sum_\sigma \int_C (z - E_F)\left(1 - \frac{dS}{dz}\right) {}^c G_{aa}^{\sigma\sigma}(z) \, dz$$

$$- \frac{1}{4\pi i} \sum_\sigma \int_C \Sigma_{a\sigma}^{HF} {}^c G_{aa}^{\sigma\sigma}(z) \, dz - (\varepsilon_a^0 - E_F) \quad (9.3.65)$$

One can show that this equation remains valid in the general case if $\Sigma_{a\sigma}^{HF}$ is replaced by $\Sigma_{a\sigma}(z)$ in equation (9.3.65). Thus the problem reduces to the determination of $\Sigma_{a\sigma}(z)$. Expressions for $\Sigma_{a\sigma}(z)$ exist only in different limits:

(1) If $U \to 0$, one can use a second-order perturbation theory in U/W,[23,30] W being the broadening of the adatom level. One finds

$$\Sigma_{a\sigma}(z) = \Sigma_{a\sigma}^{HF} + U^2 \int_{-\infty}^{E_F} dE_2 \int_{E_F}^{+\infty} dE_3 \int_{E_F}^{+\infty} dE_4 \frac{\rho_a^{-\sigma}(E_2)\rho_a^{-\sigma}(E_3)\rho_a^{-\sigma}(E_4)}{z + E_2 - E_3 - E_4}$$

$$+ U^2 \int_{E_F}^{+\infty} dE_2 \int_{-\infty}^{E_F} dE_3 \int_{-\infty}^{E_F} dE_4 \frac{\rho_a^{-\sigma}(E_2)\rho_a^{-\sigma}(E_3)\rho_a^{-\sigma}(E_4)}{z + E_2 - E_3 - E_4} \quad (9.3.66)$$

(2) When the coupling $V = 0$, one can write an exact expression for ${}^c G_{aa}^{\sigma\sigma}$:

$$ {}^c G_{aa}^{\sigma\sigma}(z) = \frac{\langle n_{a-\sigma}\rangle}{z - \varepsilon_a^0 - U} + \frac{1 - \langle n_{a-\sigma}\rangle}{z - \varepsilon_a^0} \quad (9.3.67)$$

which is rather intuitive, since the first term has a pole at the affinity level with weight $\langle n_{a-\sigma}\rangle$ and the other at the ionization level with weight $1 - \langle n_{a-\sigma}\rangle$. The above equation defines $\Sigma_{a\sigma}(z)$ since ${}^c G_{aa}^{\sigma\sigma}$ can be identified with the expression

$$ {}^c G_{aa}^{\sigma\sigma}(z) = \frac{1}{z - \varepsilon_a^0 - \Sigma_{a\sigma}(z)} \quad (9.3.68)$$

thus

$$\Sigma_{a\sigma}(z) = \Sigma_{a\sigma}^{\text{HF}} + \frac{\langle n_{a-\sigma}\rangle\langle n_{a\sigma}\rangle U^2}{z - \varepsilon_a^0 - U\langle n_{a\sigma}\rangle} \qquad (9.3.69)$$

In the general case one should find an interpolation formula between these two limits. Such a formula has been proposed by Martin-Rodero et al.[31] and Baldo et al.[24,25] Previously several treatments, based on the equations of motion of the Green function, were performed by Brenig and Schönhammer,[21] Schuck,[32] and Bell and Madhukar.[22] Recently, an extensive calculation of the chemisorption energy of H on Ni(100), Ni(111), and W(110) was conducted by Piccitto et al.[29] This work uses an approximate self-energy and takes into account the repulsion energy between the hydrogen adatom and the metal. The adatom-substrate coupling parameters are determined as a function of distance through the overlap integrals between the adatom and the metal orbitals as in the extended Hückel method. Finally, the variation of U with the adatom-surface distance is parametrized. These authors were then able to minimize the chemisorption energy as a function of the position of the adatom. Although the treatment of the electronic structure of the metal is greatly simplified (semielliptic density of states), the results seem to be in fair agreement with more sophisticated schemes.

In conclusion, although some improvements have been made to account for the effect of electronic correlations on the adsorbate, nothing is known about the effect of the electronic correlations of the substrate in this context, at least to our knowledge.

9.4. A Simple Chemisorption Model for Tight-Binding Systems

The Anderson-Grimley-Newns Hamiltonian has been mainly used to study the chemisorption of hydrogen on metals, however very few systematic studies of chemisorption on transition metals varying the nature of the adsorbate and substrate can be found in the literature. Moreover, save for the work of Piccitto et al.,[29] the position of the adsorbate is assumed to be known. Nevertheless, if our goal is to explain all the experimental trends discussed in the introduction, one needs to calculate the binding energy of the adsorbate as a function of its three coordinates from which it is possible to derive the most stable adsorption site and its energy, the bond length, the activation energy for surface diffusion (extrapolated at 0 K), and also vibration frequencies. This calculation should be conducted for a very large number of chemisorption systems and is obviously only possible practically if one uses simplified models pointing out the essential physical parameters. Such models have been derived in the framework of the tight-binding formalism for chemisorption on transition metals.[33-35] In this section we will describe these models and show that they explain, at least semiquantitatively, all the trends observed experimentally.

9.4.1. Models

9.4.1.1. General Characteristics

The binding energy is written in the form

$$E_B(x, y, z) = \Delta E_b + \Delta E_{rep} + \Delta E_{cor} \qquad (9.4.1)$$

where ΔE_b, ΔE_{rep}, and ΔE_{cor} are respectively the variations of the band, repulsive, and Coulomb correlation total energies when the adatom is brought from infinity to the point with coordinates (x, y, z) relative to the surface. These three contributions are now examined briefly.

a. The Band Contribution. The band contribution ΔE_b is calculated in a tight-binding formalism assuming that the adsorbate-substrate hopping integrals decrease exponentially with distance. The other parameters are the atomic levels of the atoms involved in the chemisorption bond. It is clear that the adsorbate-substrate interactions modify the potential, and then the atomic levels of these atoms. Ideally, these potentials and the charge on each atom should be calculated in a self-consistent manner. The simplest assumption is to limit the perturbation of the atomic levels to the adsorbate site. It is also assumed, for simplicity, that the adatom is not magnetic. Then two methods have been used to determine the effective atomic level ε_a^* of the adsorbate.

1. The Anderson-Grimley-Newns model in which, for an H adatom in the Hartree-Fock scheme,

$$\varepsilon_a^* = \varepsilon_a^0 + U\langle n_a \rangle \qquad (9.4.2)$$

where $\langle n_a \rangle$ is the occupation number per spin orbital. As seen previously, we are obliged in this method to introduce an unphysical variation of the Fermi level in order to ensure the total charge conservation.

2. One can also state that the Fermi level should remain unchanged in the chemisorption process, and then ε_a^* is determined from the Friedel sum rule which can be expressed as (the global neutrality condition)

$$\Delta N(E_F, \varepsilon_a^*) = N_a \qquad (9.4.3)$$

where N_a is the number of valence electrons of the adsorbate involved in the bond. From the knowledge of ε_a^*, the charge of any atom of the system can be derived. However, numerical calculations have shown that this often leads to a charge transfer on the adatom nearest neighbors that is of the same order of magnitude as on the adatom itself. In these conditions it seems unphysical to neglect the variation in the effective atomic levels of these neighbors. Moreover, in some particular cases unphysical charge is found on the adatom and should be strongly reduced if one extends the perturbation of the potential to the nearest neighbors. If this is done on p nearest neighbors, one needs $p + 1$ equations to derive the atomic levels of the adsorbate and its nearest neighbors. Since the Friedel sum rule only introduces one equation, new physical conditions should be sought.

The simplest condition obeying the Friedel sum rule is to assume that each atom involved in the chemisorption system remains neutral. This condition seems reasonable for many systems, at least at low coverage, since work function measurements and 4f surface core level spectroscopy (on 5d elements) are inconsistent with charge transfers larger than 0.1–0.2 electron. This implies that the adsorbate effective atomic level falls in the substrate band, since otherwise the charge transfer would be very large.

As in an Anderson–Grimley–Newns model two limits are usually considered:

- The weak coupling limit for which the adsorbate–substrate hopping integrals are much smaller than the substrate–substrate ones.
- The strong coupling limit in the opposite case.

In both cases some band energy is gained ($\Delta E_b < 0$). This contribution is attractive since it is roughly proportional to the adsorbate–substrate hopping integrals, which decrease with distance. We will see that it cannot be written as a sum of pair interactions.

b. *The Repulsive Contribution.* The repulsive contribution comes mainly from the compression of inner shells. It is expressed as a sum of phenomenological Born–Mayer pair potentials decreasing exponentially with distance.[36]

c. *The Electronic Correlation Contribution.* The contribution of the electronic correlation ΔE_{cor} is obtained from perturbation theory up to the second order in the band limit (intra-atomic Coulomb integral small compared to the bandwidth). It includes not only the contribution of the adsorbate as in the Anderson–Grimley–Newns Hamiltonian, but also the contribution of the substrate. We will see in the following that, to a good approximation, it can be written as a sum of local terms: it corresponds to an energy gain on the adsorbate and to an energy loss on its neighbors.

It is noteworthy that we thus take into account the interaction between the outer valence orbitals of the adsorbate and the d valence electrons of the metal neglecting the role played by the sp electrons of the metal. Indeed, these latter electrons give rise to both attractive and repulsive contributions to the binding energy. The repulsive contribution is phenomenologically taken into account in the Born–Mayer potential. Concerning the attractive contribution, we assume that the free sp electrons serve primarily to renormalize the hopping parameters. In any case, we believe that the sp contribution is rather small in transition metals and thus a treatment which includes only the d band of the metal is satisfactory.

9.4.1.2. Analytical Models

We will first use schematical densities of states and neglect the electronic correlation term, in order to be able to derive analytically simple expressions for the chemisorption energy.

a. Weak Coupling Limit. The band contribution can be split into two terms,

$$\Delta E_b = \Delta E_{ba} + \Delta E_{bs} \tag{9.4.4}$$

where ΔE_{ba} is the variation of energy due to the broadening of the adsorbate level and ΔE_{bs} is the variation of energy of the substrate due to the adsorption. In the weak coupling limit this last term can be neglected. Then, provided that the charge neutrality is ensured, i.e.,

$$N_a = L_a \int_{-\infty}^{E_F} n_a(E, \varepsilon_a^*) \, dE \tag{9.4.5}$$

quantity ΔE_b takes the form

$$\Delta E_b = L_a \int_{-\infty}^{E_F} E n_a(E, \varepsilon_a^*) \, dE - N_a \varepsilon_a - N_a(\varepsilon_a^* - \varepsilon_a) \tag{9.4.6a}$$

$$= L_a \int_{-\infty}^{E_F} E n_a(E, \varepsilon_a^*) \, dE - N_a \varepsilon_a^* \tag{9.4.6b}$$

where N_a, L_a, n_a, ε_a^*, and ε_a are respectively the number of electrons and the degeneracy in the valence shell of the adsorbate, its local density of states, and its atomic level in the adsorbed and free states. The first and second terms of equation (9.4.6a) represent the variation in the one-electron energy between the adsorbed state and the free state. The third term represents the variation in the average electronic Coulomb interaction which is counted twice in the first term (see Appendix 9.4). If we assume that the local density of states on the adsorbate is shifted rigidly when ε_a^* varies (i.e., when E_F varies), it is straightforward to show that ΔE_b remains constant and is thus independent of the substrate band filling. Indeed if ε_a^* varies by $\delta \varepsilon_a^*$, all one-electron energies in the first term vary by $\delta \varepsilon_a^*$, thus the total variation is $N_a \delta \varepsilon_a^*$, which is exactly cancelled by the variation of the second term.

In the weak coupling limit, the broadening of the adatom level is small and the local density of states on the adsorbate exhibits a single peak, which can be schematized by a rectangle of width W and centered at ε_a^*. The width W_a is chosen so that the centered second moment of the rectangle is equal to the centered second moment of the exact local density of states, i.e.,

$$\mu_{2a}^c = \int_{-\infty}^{+\infty} (E - \varepsilon_a^*)^2 n_a(E, \varepsilon_a^*) \, dE \tag{9.4.7a}$$

and, taking ε_a^* as the reference energy,

$$\mu_{2a}^c = \frac{1}{W_a} \int_{-W_a/2}^{W_a/2} E'^2 \, dE' = \frac{W_a^2}{12} \tag{9.4.7b}$$

so that

$$W_a = 2\sqrt{3}\sqrt{\mu_{2a}^c} \tag{9.4.8}$$

Quantity μ_{2a}^c can be easily obtained in the tight-binding approximation, since one can show that

$$\mu_{2a}^c = \frac{2}{L_a} \sum_{\substack{\lambda j\mu \\ j\neq i}} \langle i\lambda|H|j\mu\rangle\langle j\mu|H|i\lambda\rangle \tag{9.4.9}$$

If we assume that the adatom has Z equivalent nearest neighbors, we find that

$$\mu_{2a}^c = Z\beta'^2 \tag{9.4.10}$$

with

$\beta'^2 = sd\sigma^2$ for an adsorbate with an s orbital

$\beta'^2 = \frac{1}{3}(pd\sigma^2 + 2pd\pi^2)$ for an adsorbate with p orbitals

$\beta'^2 = \frac{1}{5}(dd\sigma^2 + 2dd\pi^2 + 2dd\delta^2)$ for an adsorbate with d orbitals

where $sd\sigma$, $pd\sigma$, $pd\pi$, $dd\sigma$, $dd\pi$, and $dd\delta$ are the usual Slater–Koster[37] hopping integrals between the adatom orbital and the d orbitals of the transition metal substrate. The variation of these hopping integrals with the interatomic distance R can be approximated by an exponential law,

$$\mu_{2a}^c = Z\beta_0'^2 e^{-2qR} \tag{9.4.11}$$

Equation (9.4.5) yields, using the rectangular density of states,

$$E_F = \frac{W_a}{L_a}\left(N_a - \frac{L_a}{2}\right) \tag{9.4.12}$$

Then from equation (9.4.6) we obtain

$$\Delta E_b = \frac{W_a}{2L_a} N_a(N_a - L_a) \tag{9.4.13}$$

$$= -\sqrt{Z}B(N_a)e^{-qR} \tag{9.4.14}$$

with

$$B(N_a) = \sqrt{3}\frac{|\beta_0'|}{L_a} N_a(N_a - L_a) \tag{9.4.15}$$

On the other hand, the variation ΔE_{rep} of the repulsive term is given by

$$\Delta E_{\text{rep}} = ZA\, e^{-pR} \tag{9.4.16}$$

where A and p are some given constants.

Finally, the binding energy of the adatom with bond length R is given by

$$E_B(R) = ZA\, e^{-pR} - \sqrt{Z}\, B(N_a)\, e^{-qR} \tag{9.4.17}$$

The value of E_B reaches a minimum if $p > q$ for an equilibrium bond length equal to

$$R_0 = \frac{1}{(2p-q)} \operatorname{Log} Z + \frac{1}{p-q} \operatorname{Log} \frac{pA}{qB} \tag{9.4.18}$$

thus R_0 increases with Z. This fact can be easily understood. Let us consider an adatom with coordinence Z, the bond length of which is R_0. If we increase Z keeping R_0 constant, the repulsive force overcomes the attractive force since this latter force varies less rapidly with Z and thus the bond length increases.

If equation (9.4.18) is substituted into equation (9.4.17), we obtain the binding energy at equilibrium:

$$E_B(R_0) = \left(\frac{q}{p} - 1\right) B \left(\frac{pA}{qB}\right)^{-q/p-q} Z^{(p-2q)/2(p-q)} \tag{9.4.19}$$

Therefore

$$E_B(R_0) \propto Z^{\alpha} \tag{9.4.20}$$

with $\alpha = (p - 2q)/2(p - q)$.

For real systems, $p > 2q$ and the binding energy increases with Z, therefore the most stable position corresponds to the site with the largest coordination number available on the surface. The latter number being larger on open surfaces than on closed-packed ones, we expect $|E_B(R_0)|$ to increase when the density of surface atoms decreases. In realistic cases $\alpha \lesssim \frac{1}{3}$ and consequently the anisotropy of E_B is very much reduced from the value predicted by a broken bond model ($\alpha = 1$) or in a tight-binding scheme with rigid bond lengths ($\alpha = \frac{1}{2}$).

In this model the binding energy is seen to be independent of the filling of the substrate band for given parameters A, p, β'_0, and q.

On highly symmetrical surfaces (with low index), the diffusion path can be easily guessed, the bottom of the well and the saddle point corresponding to special symmetry points. Surface diffusion activation energies Q at 0 K are thus given by the difference of binding energy between two sites which differ by their coordination number, the smaller one Z_s corresponding to the saddle point. Then

$$Q = E_B(Z_s) - E_B(Z) = [1 - (Z_s/Z)^{\alpha}]\, |E_B(Z)| \tag{9.4.21}$$

One sees that Q follows the same behavior as $|E_B|$.

 b. *Strong Coupling Limit.* In the weak coupling case, we have assumed a rigid local density of states on the adsorbate shifting with E_F in order to conserve the number of electrons on the adatom. In the strong coupling limit, this assumption is no longer valid since the adatom local density of states exhibits two peaks corresponding to the bonding and antibonding states of the "surface molecule." One can predict that their relative weight should be a function of the substrate band filling if the adatom remains neutral. This is pictured in Figure 9.10 in which one sees that the weight of the bonding state decreases in favor of the antibonding state when the number of d electrons of the substrate increases. As a consequence and contrary to the weak coupling case, the contribution of the adsorbate to ΔE_b is expected to depend on the position of the Fermi level.

 In the following we use a very simple model to show that this explains the decrease in the binding energy of N, O, F, and H along a transition series. We mimic the adatom local density of states by two δ functions of weights α and $1 - \alpha$ at energies $-X$ and X (the energy reference is arbitrary and can thus be chosen at the midpoint between the two δ functions):

$$n_a(E) = \alpha\, \delta(E + X) + (1 - \alpha)\, \delta(E - X) \qquad (9.4.22)$$

 The first and second moments of this distribution should be equal respectively to the effective atomic level of the adatom ε_a^* and to the second moment of the exact adatom local density of states (i.e., $\mu_2 = Z\beta'^2 + \varepsilon_a^{*2}$). These relations fix α and X:

$$\alpha = \frac{1}{2}\left(1 - \frac{\varepsilon_a^*}{\sqrt{Z\beta'^2 + \varepsilon_a^{*2}}}\right) \qquad \text{and} \qquad X = \sqrt{Z\beta'^2 + \varepsilon_a^{*2}} \qquad (9.4.23)$$

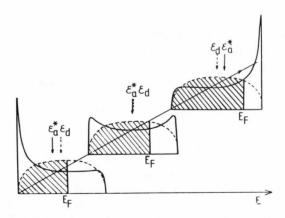

Figure 9.10. Deformation of the local density of states on the adatom when the substrate band filling (i.e., E_F) varies in the strong coupling limit (ε_a^* and ε_d are respectively the effective adatom atomic level and the center of the substrate d band).

The energy levels are then filled with the N_a electrons of the adsorbate and one finds that the one-electron contribution of the adatom to the binding energy is

$$|E_B| = N_a \sqrt{Z\beta'^2 \frac{1 - \alpha}{\alpha}} \qquad (9.4.24)$$

when all the adatom electrons are in the bonding state ($\alpha > N_a/L_a$), and

$$|E_B| = (L_a - N_a) \sqrt{Z\beta'^2 \frac{\alpha}{1 - \alpha}} \qquad (9.4.25)$$

when the antibonding state is partially occupied ($\alpha < N_a/L_a$). From these results, one can deduce easily the behavior of $|E_B|$ as a function of the substrate band filling, knowing that α is a decreasing function of this quantity. One finds that $|E_B|$ reaches a maximum when the bonding state is completely filled and the antibonding state is empty, which occurs for $\alpha = N_a/L_a$, i.e., at the beginning of the series when $N_a/L_a > \frac{1}{2}$, at the end when $N_a/L_a < \frac{1}{2}$, and around the middle of the series when $N_a/L_a = \frac{1}{2}$ (Figure 9.11). This explains the observed decrease in the binding energy of N, O, F, and H on substrates starting from the V, Nb, and Ta column to the end of the corresponding transition series (Figure 9.2).

9.4.1.3. Improved Models

We have just shown that the trends followed by the adsorbate binding energies can be explained using very simple models. It should be verified that these trends remain valid in a more realistic calculation and for more realistic systems for which the coupling can be intermediate between the two discussed limits. Actually, one should use a technique which does not make any assumption on the coupling strength in the calculation of the density of states. The continued fraction technique fulfills this requirement.

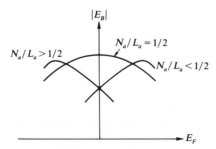

Figure 9.11. Schematic variation of the binding energy of an adsorbate with N_a/L_a electrons per orbital as a function of the substrate band filling (i.e., E_F).

In this technique, any diagonal Green function is written as a continued fraction[38,39]

$$g_{ii}(z) = \cfrac{1}{z - a_1^i - \cfrac{b_1^i}{z - a_2^i - \cfrac{b_2^i}{z - \cdots}}} \tag{9.4.26}$$

in which the first coefficients a_n^i, b_n^i are calculated exactly. When the spectrum presents no gap, these coefficients converge rapidly and consequently an excellent approximation is obtained by replacing the unknown coefficients by their asymptotic values a_∞, b_∞. This is equivalent to terminating the continued fraction by a complex function $F(z)$,

$$g_{ii}(z) = \cfrac{1}{z - a_1^i - \cfrac{b_1^i}{z - a_2^i - \cfrac{b_2^i}{\cfrac{\ddots}{z - a_N^i - b_N^i F(z)}}}} \tag{9.4.27}$$

where $F(z)$ obeys the quadratic equation

$$F(z) = \frac{1}{z - a_\infty - b_\infty F(z)} \tag{9.4.28}$$

and has an imaginary part when the discriminant

$$D = (z - a_\infty)^2 - 4b_\infty$$

is negative; a_∞ and b_∞ are then related to the band limits m, M by

$$a_\infty = (m + M)/2 \tag{9.4.29}$$

$$b_\infty = \frac{(M - m)^2}{16} = \frac{W^2}{16} \tag{9.4.30}$$

where W is the bandwidth.

We note that the continued fraction coefficients and the moment of the corresponding local density of states are related and that a continued fraction with N exact levels has $2N$ exact moments. In particular a_1^i is the center of gravity of the local density of states (i.e., the effective atomic level of atom i) and b_1^i is the centered second moment [see equation (9.4.9)]. Actually this technique, when applied to obtain the adatom local density of states, does not make any assumption as to the coupling strength (which is measured by $\sqrt{b_1}$). Consequently, it can describe both the weak ($b_1 \ll b_\infty$) and strong ($b_1 \gg b_\infty$) coupling cases but also all intermediate cases.

We have then developed an improved model in which the local densities of states are obtained from a continued fraction expansion of the Green function exact to the third moment (i.e., with exact a_1, b_1, and a_2). This has the advantage of giving the correct band limits and asymmetry. On the other hand, we now take into account the modification of the substrate electronic structure due to the adsorption. Finally, the calculation is conducted for a real surface crystalline structure and we are no longer restricted to equivalent bonds.

In practice, the perturbation due to the adsorbate extends only over a few sites. Consequently

$$\Delta E_{\text{rep}} = \sum_i A \exp(-pR_i) \tag{9.4.31}$$

where R_i is the distance between the adatom and the neighbor i;

$$\Delta E_{\text{b}} = \Delta E_{\text{ba}} + \sum_i \Delta E_{\text{b}}^i \tag{9.4.32}$$

where

$$\Delta E_{\text{ba}} = L_a \int_{-\infty}^{E_F} E n_a(E, \ldots, \varepsilon_j^*, \ldots)\, dE - N_a \varepsilon_a^* \tag{9.4.33}$$

$$\Delta E_{\text{b}}^i = 10 \int_{-\infty}^{E_F} E\, \delta n_i(E, \ldots, \varepsilon_j^*, \ldots)\, dE - N_s \delta \varepsilon_i^* \tag{9.4.34}$$

are respectively the adsorbate and substrate atom i contributions to the band part of the binding energy. The sum over i in equations (9.4.31) and (9.4.32) is limited to the perturbed atoms. All the effective atomic levels ε_j^* are determined by requiring that all atoms (adatom and substrate neighbors) remain neutral. Finally, δn_i and $\delta \varepsilon_i^*$ are respectively the variation in the local density of states and effective atomic level on atom i due to the adsorption.

The electronic correlation term is now examined. If the electronic correlations in the substrate are neglected, then we could start with equation (9.3.65) using expression (9.3.66) for the self-energy, but since we have assumed that the effective atomic levels can be fixed by a local charge neutrality condition and not by equation (9.4.2), such a treatment would be inconsistent. Moreover, since we want to take into account the Coulomb interaction in the metal, equation (9.3.65) can no longer be used. To describe the Coulomb correlations in the metal, let us add to the Anderson–Grimley–Newns Hamiltonian a Hubbard term:

$$H_{\text{Hub}} = \frac{U_s}{2} \sum_{\substack{i \\ \nu\sigma,\nu'\sigma'}} (1 - \delta_{\nu\nu'}\delta_{\sigma\sigma'}) n_{i\nu\sigma} n_{i\nu'\sigma'} \tag{9.4.35}$$

where $n_{i\nu\sigma}$ is the occupation number operator of orbital ν centered at site i with spin σ. Since we are interested in the chemisorption energy which, up to now, has been expressed as a sum of contributions from each perturbed atom [equations (9.4.33) and (9.4.34)], it is easier to compute directly the variation in local correlation energies instead of computing the self-energies. A local approximation and a second-order perturbation theory in the band limit[30,40] can be employed to show that the contribution of electronic correlations on atom i to the total energy is given by (see Appendix 9.5)

$$
E^i_{cor} = - \frac{L_i(L_i - 1)}{2}
$$

$$
\times U_i^2 \int_{E_F}^{+\infty} dE_1 \int_{E_F}^{+\infty} dE_2 \int_{-\infty}^{E_F} dE_3 \int_{-\infty}^{E_F} dE_4 \frac{n_i(E_1)n_i(E_2)n_i(E_3)n_i(E_4)}{E_1 + E_2 - E_3 - E_4}
$$

$$(9.4.36)$$

where L_i, U_i, and $n_i(E)$ are respectively the number of spin orbitals, the effective Coulomb integral (U on the adatom, U_s in the substrate), and the local density of states at site i. It is noteworthy that this correlation term accounts for the variation in Coulomb energy due to instantaneous fluctuations of the total number of electrons on each site. This clearly cancels in a free atom. An order of magnitude of this quantity is given by[41]

$$
E^i_{cor} \simeq - \frac{L_i(L_i - 1)}{2} \frac{U_i^2}{W_i} \left(\frac{N_i}{L_i}\right)^2 \left(1 - \frac{N_i}{L_i}\right)^2 \tag{9.4.37}
$$

where N_i and W_i are respectively the number of electrons and an effective band width at site i.

Similarly to the band contribution [equation (9.4.32)] one has, with obvious notation,

$$
\Delta E_{cor} = E^a_{cor} + \sum_i \Delta E^i_{cor} \tag{9.4.38}
$$

where E^a_{cor} is negative [equation (9.4.36)] and ΔE^i_{cor} is positive, since the presence of the adsorbate increases the effective bandwidth and therefore decreases E^i_{cor} [equation (9.4.37)]. As a consequence the sign of the correlation contribution may depend on the adsorbate or on the substrate.

The model can be used to study chemisorption on flat or stepped surfaces and also the adsorption of simple molecules. In the next section we will review the results obtained in all these cases.

9.4.2. Adsorption of Simple Elements on bcc Transition Metal Surfaces

The results obtained with the improved model for the adsorption of simple elements on bcc surfaces are now presented.

9.4.2.1. Low-Index Faces[33-35]

Let us consider a semi-infinite bcc crystal limited by a (110) or (100) surface. This surface is assumed to be the perfect termination of the bulk metal (i.e., there is neither perpendicular relaxation nor reconstruction). The number of d electrons of the substrate will be varied between 3 and 7 electrons per atom, since the bcc transition metals correspond to these band fillings. The adsorption sites are expected to be on high symmetry positions (see Figure 9.12). Calculations have been performed for various adsorbates: transition atoms,[33] N, O, F,[34] and H[35] which interact with the metal d orbitals through their d (transition atoms), p (N, O, F), and s (H) valence orbitals.

The semi-infinite crystal hopping integrals ($dd\lambda$, $\lambda = \sigma, \pi, \delta$) and their variation with distance are obtained from interpolation schemes.[42] Although they vary slightly between $N_s = 3$ and 7 d electrons per atom, we can with a good approximation neglect this variation. These hopping parameters are also used for adsorbate–substrate interactions in the case of transition adatoms. Adsorbate–substrate hopping integrals depend on two parameters, $pd\sigma$ and $pd\pi$, with $pd\pi \simeq -pd\sigma/2$ for N, O, F, and only one parameter, $sd\sigma$, for H adsorption. They can be determined either directly from their definition[43] [similar to equation (9.3.40)] or from an interpolation scheme on the band structure of the corresponding covalent compound when it exists. The tight-binding parameters being fixed, the parameters of the Born–Mayer potential are fitted to reproduce known experimental quantities. In the case of transition adatoms, they are determined from the values of cohesive energy and bulk modulus and they must satisfy the bulk equilibrium equation. In the case of O and H, they have been chosen to give reasonable values of the binding energy, bond length, and stretch frequency of O and H on W(110). It is assumed that these parameters do not vary rapidly from an element to its neighbor in the Periodic Table, thus we take the same values for N and F as for oxygen.

The last parameters of the model are the Coulomb integrals. In atoms, the Coulomb repulsion U is given by the difference between the ionization and affinity energies I and A [see equation (9.2.2)]. Usually A is much smaller than I, and U is therefore of the order of several eV. In transition metals the value of U is strongly reduced from the free atom value by screening. It has been shown that

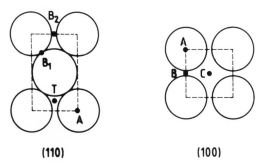

(110)   **(100)**

Figure 9.12. Labeling of adsorption sites on the (110) and (100) faces of a bcc metal.

$U_s \sim 1\text{-}3$ eV.[41,44,45] For a chemisorbed atom, the Coulomb integral is also lowered by screening which, far from the surface of the metal, is due to the image potential [see equation (9.2.8)]. Therefore a precise determination of the Coulomb integral is extremely difficult, because this parameter depends on the distance from the adsorbate to the surface. Since our goal in this work is to determine trends, we have introduced the simplifying assumption of taking U as a constant around the equilibrium position. For N, O, and F we have taken $U_a = 2U_s$; this relation is roughly true in the atomic state and therefore it has been assumed that the Coulomb interactions are screened to the same extent in the metal and the adsorbate. For H we have chosen $U = 2$ eV.[35] These values of the Coulomb integrals may seem somewhat small, but it is known that the second-order perturbation theory tends to exaggerate correlation effects.[40]

    a. Adsorption of Transition Adatoms. The bond lengths and binding energies of a transition adatom on a substrate of the same chemical species are shown in Figures 9.13 and 9.14 for the different adsorption sites shown in Figure 9.12. One sees that the bond length increases with the coordination number. It has a minimum near the middle of the series for a given site, similarly to the variation in the atomic volume along the transition series. In addition, taking into account the electronic correlation contribution without changing band and repulsion parameters increases the bond length, since in this case the main term in ΔE_{cor} is E_{cor}^a, which is negative and increases in absolute value with distance at least around the equilibrium position. On the other hand, binding energies increase with the coordination

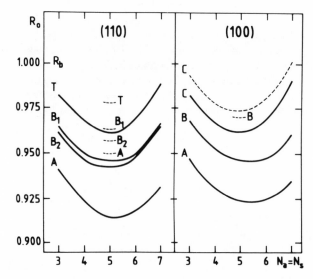

Figure 9.13. Adatom-substrate bond lengths R_0 for several sites on (110) and (100) bcc surfaces for an adatom on a substrate of the same chemical species as a function of the number of d electrons (full curves: $U = 0$, broken curves: $U = 1.2$ eV); R_b is the bulk interatomic distance.

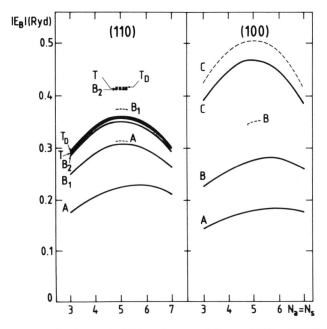

Figure 9.14. Adatom–substrate binding energy for several sites on (110) and (100) bcc surfaces for an adatom and a substrate of the same chemical species as a function of the number of d electrons (full curves: $U = 0$, broken curves: $U = 1.2$ eV). The curves labeled T_D refer to the most stable site between B_2 and T.

number and the most stable site corresponds to the most coordinated site available on the surface: the centered site on (100) and the almost perfect ternary site on (110). They reach a maximum for an almost half-filled band similarly to the cohesive energy. It is noteworthy that we have neglected the variation in exchange energy which should be introduced, since the free atom is magnetic and usually the adsorbed atom is not. It would produce a cusp in the middle of the series which corresponds to the special stability of the half-filled d atomic shells, due to a maximum of their exchange interaction. As in the case of cohesive energy[41] this cusp should be especially marked in the 3d metals, where these effects are important. The anisotropy between the (100) and (110) faces is small and is even reduced when correlation effects are taken into account (with $U = 1.2$ eV[44,45]). The correlation contribution is seen to stabilize the adatom, since $|E_{cor}^a|$ prevails.

We have also performed binding energy calculations of 5d adatoms on W surfaces, since experimental values of binding energies and surface diffusion activation energies on this substrate are available from FIM experiments.[7,9] Results are given in Figures 9.15 and 9.16. Experimental values of $|E_B|$ are rather dispersed and thus not conclusive about the anisotropy. However, the order of magnitude of $|E_B|$ is in agreement with our calculations. The easiest diffusion paths on (110) and (100) surfaces, and the corresponding diffusion activation energies Q (extrapolated at 0 K) are given in Figure 9.16 and compare favorably with

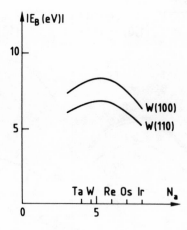

Figure 9.15. Calculated binding energies of 5d adatoms on W(110) and W(100).

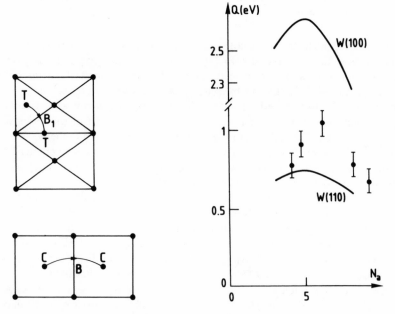

Figure 9.16. Easiest diffusion paths and their activation energies (extrapolated at 0 K) of 5d adatoms on W(100) and W(110) compared with experiments.[9]

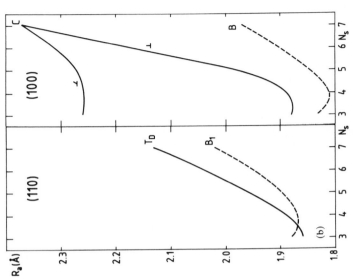

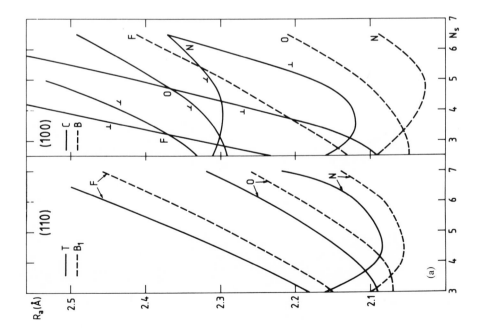

Figure 9.17. Adsorbate–substrate bond lengths for different sites on (110) and (100) bcc surfaces as a function of the number of d electrons of the substrate. For the centered site (C) on (100) two bond lengths are given, where ⊥ corresponds to the bond perpendicular to the surface and ⊥ to the bonds with the four metal atoms in the surface plane: (a) N, O, F adatoms; (b) H adatoms.

experimental data. In particular one sees that $|E_B|$ and Q have the same behavior when the adatom scans a transition series.

b. Adsorption of N, O, F, and H. First, the correlation energy term will be neglected. The variation in the bond lengths of N, O, F, and H on different sites of bcc (110) and (100) faces is given in Figure 9.17 as a function of the substrate band filling. They increase with the coordination number and with the number of d electrons of the substrate, at least when $N_s \geq 5$. The corresponding binding energies are given in Figure 9.18. The most stable site is the most coordinated site available on the surface save for H where one obtains, for example on the (100) face, an inversion of stability between the bridge and centered sites for $N_s \simeq 5$. This inversion of stability seems to occur between Ta(100) and W(100) according to surface core level shift data.[46] On the other hand, since the adatom–substrate coupling is rather strong and as expected from the simple model developed in subsection 9.4.1.2b, we find that the binding energy of N, O, F, and H decreases with the filling of the substrate d band at least for $N_s \geq 4$ and, for a given substrate, when going from N to O and F due to the filling of the antibonding state of the "surface molecule."

If we now take into account the contribution to the binding energy due to electronic correlations, we find that it changes sign along the transition series for

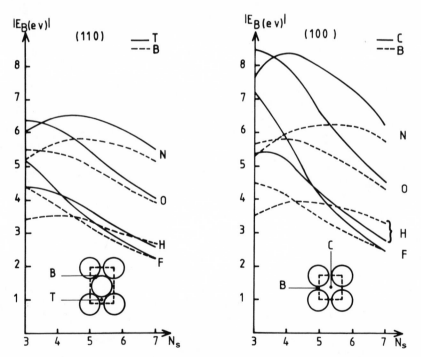

Figure 9.18. Binding energies of N, O, F, and H for different sites on (110) and (100) bcc surfaces as a function of the number of d electrons of the substrate.

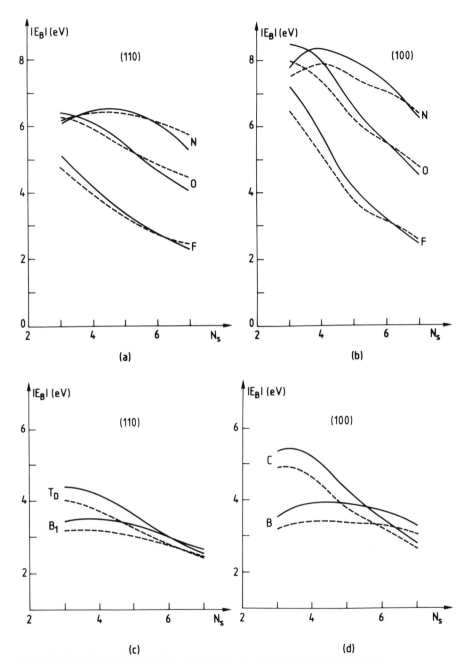

Figure 9.19. Influence of electronic correlations on the binding energies of N, O, F (a, b), and H (c, d) at their most stable position on (110) and (100) bcc surfaces (full curves, $U_a = U_s = 0$; broken curves, $U_a = 2U_s = 2.4$ eV for N, O, F and $U_a = 2$ eV, $U_s = 1.2$ eV for H) as a function of the number of d electrons of the substrate.

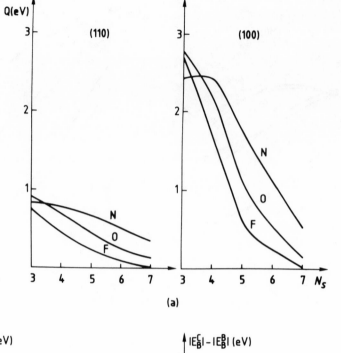

(a)

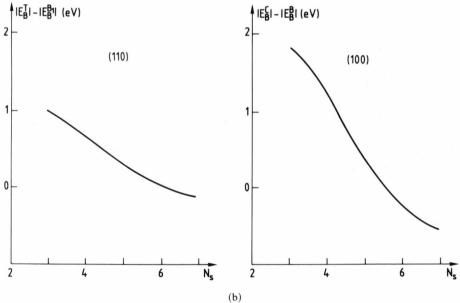

(b)

Figure 9.20. (a) Surface diffusion activation energies at 0 K of N, O, and F on (110) and (100) bcc surfaces as a function of the number of d electrons of the substrate. (b) Surface diffusion activation energies at 0 K of H on (110) and (100) bcc surfaces as a function of the number of d electrons of the substrate.

N, O, and F and decreases the value of $|E_B|$ for H (see Figure 9.19). This can be understood easily, since E_{cor}^a is proportional to the number of pairs of spin orbitals in the adatom valence shell which is reduced to 1 in the case of H, so that the variation in the correlation energy of nearest neighbors prevails.

Surface diffusion activation energies are shown in Figure 9.20a for N, O, and F. They decrease with the substrate band filling and are much larger on the (100) than on the (110) close-packed face. Similar results are obtained for H (Figure 9.20b). Therefore, the metals at the end of the d series are those for which chemical species, at least N, O, F, and H, diffuse easily. Finally, our results are in good agreement with the experimental data for O and H on W(110) (0.58 eV for O[47] and 0.21 eV for H[48]).

9.4.2.2. Stepped Surfaces[49-51]

Our model is simple enough to be applied to the adsorption on stepped surfaces. Since in this case the symmetry is rather low, we cannot limit ourselves to symmetrical sites. Therefore the adatom binding energy $E_B(x, y, z)$ is minimized with respect to the coordinate z (Oz perpendicular to the terraces) for given values of x and y. The minimization is repeated after a small translation of the adatom until a unit cell of the surface is scanned. From these results, a contour map of the binding energy is drawn from which we can deduce all the interesting physical quantities: the most stable adsorption sites and their energy; diffusion energies along any direction.

Practical calculations have been made on a stepped $[m(110) \times (0\bar{1}1)]$ bcc transition metal surface drawn schematically in Figure 9.21 for the adsorption of transition atoms, O and H. A typical binding energy contour map is shown in Figure 9.22 for a W adatom on W. The labeling of remarkable sites is given in Figure 9.23a (the U and L indices correspond, respectively, to upper and lower terrace sites). The most stable adsorption site is D_S for transition adatoms and oxygen and T'_U for H although, in this last case, the difference with D_S is small. Possible diffusion paths are shown in Figure 9.23a. In all cases, the diffusion parallel to the ledge is fairly insensitive to the presence of the step, even for an adatom diffusing in the ledge. The diffusion across the step, as is evident from the profiles of the potential energy of the adatom along this direction, is strongly perturbed in the case of transition adatoms (Figure 9.23b). This perturbation gets smaller and smaller when going to O and H. We point out that a transition adatom moving on the upper terrace is reflected by the outer edge of the step due to the occurrence of an extra barrier height, which does not exist for O and H. This effect has been seen in FIM experiments[10] as stated in the introduction. Moreover, the site T'_U close to the ledge is slightly more stable than other ternary sites. This type of behavior has been observed using FIM for a W adatom on the (211) pole of a W tip.[10]

In the case of transition adatoms, calculations have also been performed for the same terrace orientation but for different orientations of the ledge (Figure 9.24a, b). The corresponding potential energy profiles for diffusion across the ledge are drawn in Figure 9.24a', b'). Although this calculation confirms the trends

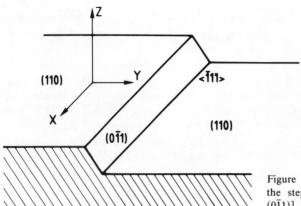

Figure 9.21. Schematic geometry of the stepped bcc surface $[m(110) \times (0\bar{1}1)]$.

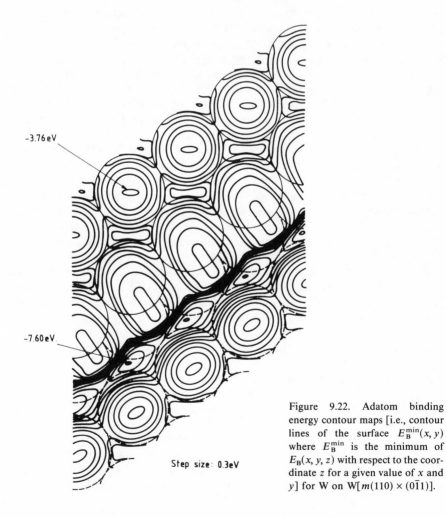

Figure 9.22. Adatom binding energy contour maps [i.e., contour lines of the surface $E_B^{min}(x, y)$ where E_B^{min} is the minimum of $E_B(x, y, z)$ with respect to the coordinate z for a given value of x and y] for W on W$[m(110) \times (0\bar{1}1)]$.

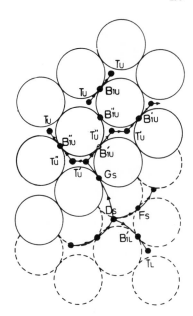

Figure 9.23a. Labeling of sites and diffusion channels of the $[m(110) \times (0\bar{1}1)]$ bcc stepped surface. The atoms of the upper (lower) terrace are drawn as full (broken) circles.

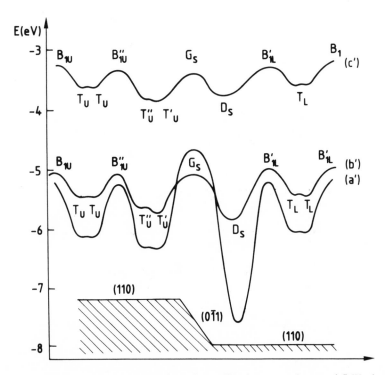

Figure 9.23b. Profile of the potential energy of an adatom diffusing across the step: (a') W adatom on W, (b') 0 on a metal with 5d electrons per atom, (c') H on a metal with 5d electron per atom.

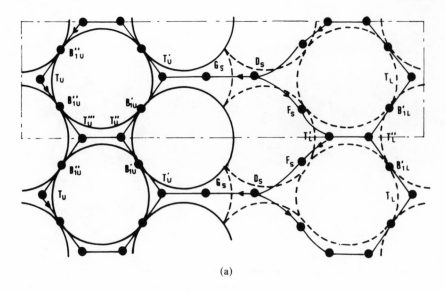

(a)

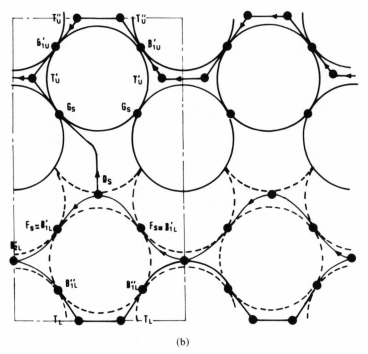

(b)

Figure 9.24a, b. Labeling of sites and diffusion channels of the $[m(110) \times (1\bar{1}0)]$ (a) and $[m(110) \times (001)]$ (b) bcc stepped surfaces. The atoms of the upper (lower) terraces are drawn as full (broken) circles.

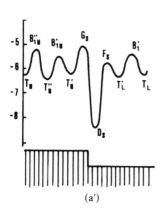
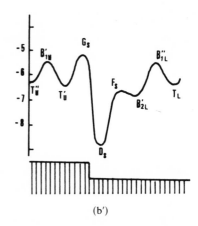

Figure 9.24a', b'. Profile of the potential energy of a W adatom diffusing across a $W[m(110) \times (1\bar{1}0)]$ (a') or $[m(110) \times (001)]$ (b') step.

discussed above, one sees that the potential-energy profile varies significantly with the roughness of the ledge.

9.5. Brief Summary of Other Methods

9.5.1. Effective Medium Theory[52-54]

Due to its underlying assumption (the starting point being the jellium model), this method is particularly suitable to the study of adsorption on simple and noble metals. However, it has also been used for transition substrates, the effect of d electrons (treated as a perturbation) being superimposed on the effective medium treatment.

The basic idea of the effective medium theory is to replace the metal by a simple effective medium. The simplest medium is obviously a jellium with a density $\bar{\rho}_0 = \rho_0(\mathbf{r}_a)$, $\rho_0(\mathbf{r}_a)$ being the substrate electron density at the point where the adatom is located. As a first approximation the adsorption energy can be taken as the difference in energy between the combined adatom–jellium system and a separated atom and jellium:

$$E_B^0(\mathbf{r}_a) = \Delta E^{\mathrm{hom}}(\bar{\rho}_0) \tag{9.5.1}$$

In this scheme, the host is characterized by $\bar{\rho}_0$ only and the quantity $\Delta E^{\mathrm{hom}}(\bar{\rho}_0)$ can then be calculated once and for all for each atom or molecule.[55] The first-order correction to this term, due to the inhomogeneity of the electron gas in the vicinity of the surface, can be written as

$$\Delta E_B^1(\mathbf{r}_a) = \int_a \Phi_0(\mathbf{r})\Delta\rho_a(\mathbf{r}) \, d^3\mathbf{r} \tag{9.5.2}$$

The integration is performed inside a region a, centered on the adatom, outside which the perturbation due to the adatom can be regarded as negligible. Function $\Phi_0(\mathbf{r})$ is the electrostatic potential of the host and $\Delta\rho_a(\mathbf{r})$ is the adatom-induced perturbation of the charge density in the homogeneous electron gas to which a suitable constant is added so that the volume a is rigorously neutral.

Finally, a third term should be added and can be expressed in the form

$$\Delta E_{\text{cov}} = \delta \int^{E_F} \Delta n(E) \, E \, dE \qquad (9.5.3)$$

This term is the difference in the sum of the one-electron energies of the adatom-induced states [density of states $\Delta n(E)$] between the effective medium and the real host.

In the case of transition metals, ΔE_{cov} contains the effect of the host d electrons which is treated to second order in perturbation. In our opinion this is the weak point of this theory since, for all transition metals except perhaps at the end of the series, d electrons provide the main contribution to the chemisorption energy.

For more details about this method the reader is referred to the chapter by B. Lundqvist in this volume.

9.5.2. Quantum Chemistry Methods[56,57]

Quantum chemistry methods most usually replace the semi-infinite substrate by a limited number of atoms, the cluster having the same symmetries as the adsorption site. The chemisorption system is studied either by *ab initio* (e.g., Xα, Hartree–Fock) or semiempirical methods (e.g., Hückel, extended Hückel). The obvious question that arises is whether the small aggregates mimic correctly the metal. This is questionable when the number of atoms is small, since a discrete energy spectrum may be a poor approximation of the true density of states of the metal. Moreover, it is not obvious that the small cluster should have the same electronic configuration as the bulk metal.

Here we restrict ourselves to a short discussion of the methods closest to those developed in the previous section, i.e., the Hückel and extended Hückel methods, which both rely on an expansion of the electron wave functions into linear combinations of atomic orbitals. In the Hückel method the overlap between neighbouring atomic orbitals is neglected, so that this method is altogether similar to the tight-binding approximation of physicists. In extended Hückel theory this overlap is not neglected and the energy levels are the solutions of the equation

$$\text{Det} \, |H_{ij} - ES_{ij}| = 0 \qquad (9.5.4)$$

(For simplicity we have assumed that there is only one orbital $|i\rangle$ per site.) The overlap integrals are calculated from their definition using atomic orbitals, for instance, of Slater type. The intra-atomic matrix elements H_{ii} of H is the energy of orbital $|i\rangle$ (referred to the vacuum level); they can vary with the charge of atom i and its neighbors (Madelung energy). Finally, one finds in the literature many expressions for the interatomic matrix elements H_{ij} as a function of S_{ij} and H_{ii}, the most popular one being the Wolfsberg–Helmholz formula [58]

$$H_{ij} = KS_{ij}(H_{ii} + H_{jj})/2 \qquad (9.5.5)$$

This method has been used extensively by R. Hoffmann.[56]

9.6. Adsorption of Homonuclear Diatomic Molecules on Metals

9.6.1. Outline of the Problem

We have seen that a considerable amount of work has been carried out in the case of single-atom chemisorption from which some general trends have been put forward, for instance, concerning the variation of binding energies with the nature of adsorbates and substrates.

On the other hand, present day knowledge about the chemisorption of molecules has not yet reached a similar level. In particular, it is of primary importance to know whether a molecule approaching a surface will dissociate or not.

Qualitatively, when a diatomic molecule X_2 approaches a metal surface, it may encounter three types of potential wells corresponding to a physisorbed state (far from the surface). Then, when its orbitals begin to mix with the metal ones, a molecular chemisorbed state may exist followed by a last well corresponding to dissociative chemisorption. If we disregard the physisorbed state, which is beyond the scope of this study, the potential energy diagram may be of the Lennard–Jones type[59] (see Figure 9.25) with the existence or not of an activation barrier ΔE^*

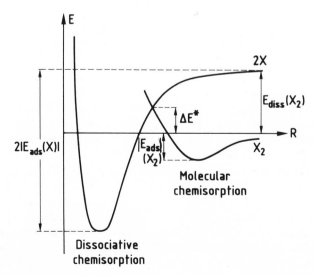

Figure 9.25. Lennard-Jones potential-energy diagram for the chemisorption of an X_2 molecule when the molecular and dissociative wells exist; ΔE^* is the activation barrier for dissociation.

for dissociation. From this diagram one sees that dissociative chemisorption is the most stable situation if there is an energy gain when going from the molecular chemisorbed state to the dissociated one, i.e., when

$$2|E_{\text{ads}}(X)| > E_{\text{diss}}(X_2) + |E_{\text{ads}}(X_2)| \qquad (9.6.1)$$

where $E_{\text{ads}}(X)$ is the atomic adsorption energy (<0) of atom X, while $E_{\text{diss}}(X_2)$ and $E_{\text{ads}}(X_2)$ are respectively the dissociation energy (>0) of the free molecule and its molecular chemisorption energy (<0). For X = N, O, F, and H, $|E_{\text{ads}}(X)|$ being a decreasing function when going across a transition series, this inequality may not be satisfied above some critical band filling. This explains qualitatively that the dissociative chemisorption of N_2 and NO, for instance, occurs for metals on the left of transition series, while on the right one observes a tendency to molecular adsorption[12] (Table 9.2).

However, the question of whether or not a molecule dissociates at the surface is a very complicated problem. The answer will depend on the dynamical motion of the molecule impinging on the surface and on the molecule–substrate energy transfers. Any attempt to treat this problem should start from the calculation of potential-energy surfaces. The potential energy is the total ground state energy of the chemisorption system minus that of the constituents calculated as a function of the coordinates of the involved atoms. The determination of these potential-energy surfaces can be investigated by different techniques derived either from quantum chemistry or from solid state physics theories. In the former approach, one treats the adsorbed atoms and a small number of neighboring solid atoms as a surface molecule, while the latter starts from a semi-infinite crystal on which the effect of the adatoms is superimposed.

Among quantum chemistry methods, one of the most popular for this type of problem is the LEPS (London–Eyring–Polanyi–Sato) approach, which is basically a valence bond treatment derived from the determination of the potential energy of a system of four one-electron atoms.[60] The LEPS potential has the advantage of being easy to evaluate numerically for any arrangement of atoms and can therefore be used as input in dynamical calculations.[61] However, in this approximation the electrons are assumed to be strongly localized, which is questionable for chemisorption on metal surfaces. On the other hand, the delocalization of the valence electrons in the metal has been taken into account for strongly delocalized electrons only (i.e., for simple and noble metals), the most appropriate method being the local density functional (LDF) within the effective medium theory.[62] When the electrons of the metal are more localized (valence d electrons of transition metals) the tight-binding theory has been most effective when explaining the general trends seen in atomic adsorption. This method can be applied to compute potential-energy surfaces for diatomic homonuclear molecules interacting with a transition metal surface.[63] It takes into account the delocalized character of the electrons, and is rather simple when used to calculate potential-energy surfaces for molecules in many geometrical configurations. This approach includes correctly the existence of a continuous spectrum of energy levels in the solid, but it certainly gives a rather poor description of the free molecule. For most realistic systems an adequate description of the molecule–surface interaction should be intermediate between the localized picture (LEPS) and the delocalized one.

9.6.2. Adiabatic Potential-Energy Surfaces for Diatomic Molecules on Transition Metal Surfaces[63]

Let us consider a frozen substrate. In this case, the adiabatic potential-energy surfaces for molecules in many geometrical configurations. This approach includes spanned by the three coordinates of both adsorbates, but if we reduce the number of degrees of freedom of the molecule by assigning a given geometry of approach, the dimension of these hypersurfaces is lowered.

In the following, we apply the models previously developed for the atomic adsorption to the case of the chemisorption of diatomic homonuclear molecules on simple bcc transition metal surfaces while neglecting the electronic correlation term. We consider an X_2 molecule impinging normally on the surface with its axis parallel to it. Moreover, it is assigned to stay in a given plane, and the projection of its center of gravity on the surface is a high symmetry point so that the two X atoms are equivalent. Within these restrictions the potential-energy contour maps are only functions of two coordinates (the distance to the surface and the molecule interatomic distance).

In a first approach, all local densities of states are assumed to have a rectangular shape with width fitted to the exact second moment. In addition, the perturbation of the substrate is taken into account and, similarly to the atomic case, charge neutrality of each involved atom is assumed. Practical calculations have been made for a molecule with a half-filled p valence shell interacting with a (100) surface of a bcc transition metal with a half-filled d band. We have taken $p_{XX}/q_{XX} = p_{XM}/q_{XM} = 3$ and $q_{XX}d_\infty = q_{XM}R_1 = 3$, where d_∞ is the bond length of the free molecule and R_1 the bond length of a single X adsorbed atom, subscripts XX and XM referring respectively to the X—X and X—metal bond.

The remaining parameters A_{XX}, A_{XM}, β_{XX}, and β_{XM}, i.e., respectively the prefactors of the Born–Mayer potentials and the hopping integrals between two X atoms and between an X atom and a metal atom M, are deduced from the values of R_1, d_∞, $E_{ads}(X)$, and $E_{diss}(X_2)$. The considered geometries are given in the insets of Figure 9.26 and the corresponding potential-energy contour maps are drawn in Figures 9.26a, b, d, e. We have also plotted (Figure 9.26c and f) the minimum value $E_B^{min}(X_2)$ with respect to the molecular interatomic distance as a function of the distance to the surface for four values of $E_{diss}(X_2)$, the other parameters remaining fixed. In Figures 9.26a and b the potential-energy contour maps exhibit first a single well corresponding to a dissociative chemisorption, then a second well appears, farther from the surface, corresponding to the molecular chemisorption which, for large values of $E_{diss}(X_2)$, becomes the most stable adsorption site. On Figure 9.26c one sees that the energy of the saddle point is larger than the energy of the free molecule for $E_{diss}(X_2) = 9$ and 12 eV. In this case the molecule needs an activation energy ΔE^* to dissociate. In Figure 9.26d and e the potential-energy contour lines present two wells at similar distances from the surface corresponding to atomic and molecular chemisorption (the molecular well being deeper when $E_{diss}(X_2)$ is large). This case occurs when the two atoms separating to reach atomic adsorption sites pass by a saddle point for atomic adsorption. The plots $E_B^{min}(X_2)$ as a function of the distance to the surface are shown in Figure 9.26f. This computation shows clearly the influence of $E_{diss}(X_2)/|E_{ads}(X)|$. In particular, over a given range of values of this ratio (1.3–2) the existence of both atomic and molecular adsorption wells is expected.

From the above results it is seen that, when a molecular well exists, the interatomic distance of the molecule varies only slightly from the equilibrium value of the free state. This suggests setting up a more simplified model in which we compare the adsorption energy of a rigid molecule X_2 with that of two isolated X adatoms. Furthermore, if we neglect the substrate contribution to the binding energy and assume that all X-substrate bonds are equivalent, then the relevant energies can be expressed analytically as a function of $E_{diss}/|E_{ads}|$ and geometrical parameters. From these expressions we can derive many of the trends put forward in the preceding calculations and thus clarify the respective roles played by energetic and geometrical factors. In this model, the binding energy of the molecule corresponding to an interatomic distance d is approximated by

$$E_B^\infty(X_2) = A_{XX} \exp\left[-3q_{XX}\left(\frac{d}{d_\infty} - 1\right)\right] - 2B\beta_{XX} \exp\left[-q_{XX}\left(\frac{d}{d_\infty} - 1\right)\right] \quad (9.6.2)$$

in which B is a function of N_X only, N_X being the number of valence electrons of an X atom. The dissociation energy, taking the equilibrium condition into account ($3A_{XX} = 2B\beta_{XX}$), is given by

$$E_{diss} = -[E_B^\infty(X_2)]_{d=d_\infty} = \tfrac{4}{3}B\beta_{XX} = 2A_{XX} \quad (9.6.3)$$

In the same way the binding energy of an X isolated adatom with coordination number Z and bond length R is

$$E_B(X) = ZA_{XM} \exp\left[-3q_{XM}\left(\frac{R}{R_1} - 1\right)\right]$$

$$- B\sqrt{Z}\beta_{XM} \exp\left[-q_{XM}\left(\frac{R}{R_1} - 1\right)\right] \quad (9.6.4)$$

where B is the same function of N_X as above and R_1 is the equilibrium bond

---

Figure 9.26. Potential-energy curves of a homonuclear diatomic molecule X_2 (with $d_\infty = 1.2$ Å) for several values of $E_{diss}(X_2)$ and for the geometries of approach toward a (100) bcc surface shown in the insets. The lattice parameter is $a_0 = 3.16$ Å, which corresponds to W. The valence shell of X and the substrate d band are both half-filled. a (d) and b (e) are respectively the equipotential energy curves for $E_{diss}(X_2) = 5$ eV and 9 eV, while c and f give $E_B^{min}(X_2)$ as a function of the distance z to the surface, i.e., the minimum of $E_B(X_2)$ relative to the intramolecular distance d for a given z. The corresponding values of d (in Å) are also given along these curves; ΔE^* is the activation barrier for dissociation. The reference energy is the energy of two free X atoms.

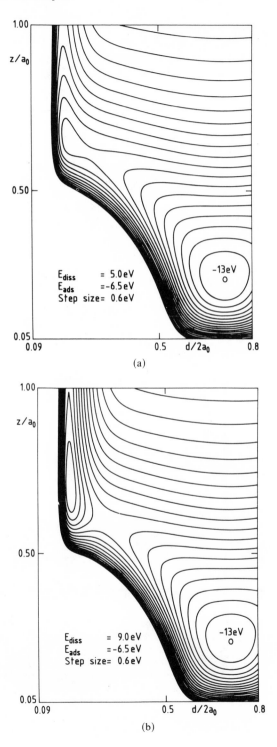

(a)

(b)

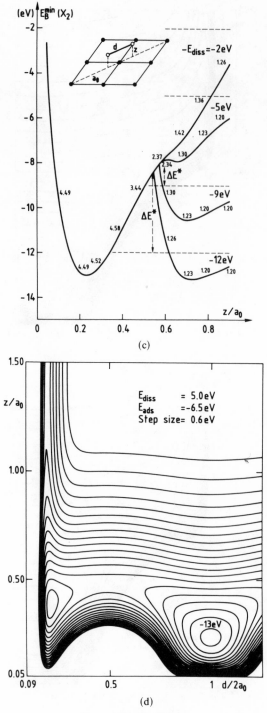

Figure 9.26. Continued.

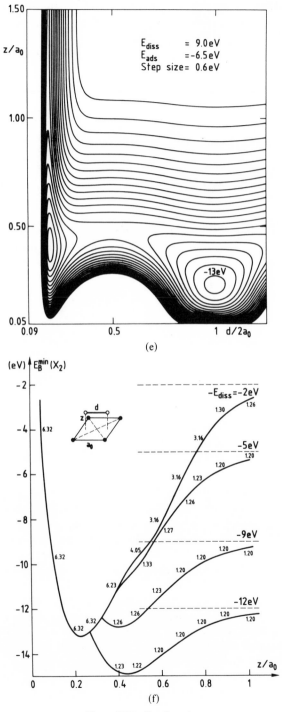

Figure 9.26. Continued.

length. On taking the equilibrium condition into account ($3ZA_{XM} = B\sqrt{Z}\beta_{XM}$), the adsorption energy is given by

$$E_{ads} = [E_B(X)]_{R=R_1} = -\tfrac{2}{3}B\sqrt{Z}\beta_{XM} = -2ZA_{XM} \qquad (9.6.5)$$

Finally, the energy of a flat rigid molecule ($d = d_\infty$) interacting with the substrate is

$$E_B(X_2) = 2Z'A_{XM}\exp\left[-3q_{XM}\left(\frac{R'}{R_1}-1\right)\right]$$

$$+ A_{XX} - 2B\sqrt{Z'\beta_{XM}^2}\exp\left[-2q_{XM}\left(\frac{R'}{R_1}-1\right)\right] + \beta_{XX}^2 \quad (9.6.6)$$

where Z' and R' are respectively the number of substrate neighbors of each atom X and the corresponding bond length. At the equilibrium position ($R' = R_1'$)

$$\exp\left[-2q\left(\frac{R_1'}{R_1}-1\right)\right] = \frac{1}{8}\frac{Z}{Z'}\left(\frac{E_{diss}}{E_{ads}}\right)^2(-1+S) \qquad (9.6.7)$$

with

$$S = \sqrt{1 + 64\frac{Z'}{Z}\left(\frac{E_{ads}}{E_{diss}}\right)^4}$$

from which we deduce

$$E_B^{R_1'}(X_2) = [E_B(X_2)]_{R'=R_1'}$$

$$= \frac{E_{diss}}{2}\left[1 + \frac{1}{8\sqrt{2}}\left(\frac{Z}{Z'}\right)^{1/2}\left(\frac{E_{diss}}{E_{ads}}\right)^2(-1+S)^{3/2} - \frac{3}{\sqrt{2}}(1+S)^{1/2}\right] \qquad (9.6.8)$$

In Figure 9.27, we have plotted $E_B^{R_1'}(X_2)/E_{ads}$ as a function of $|E_{diss}/E_{ads}|$; it is seen that $|E_B^{R_1'}(X_2)|$ increases with Z' for given values of Z, E_{diss}, and E_{ads}. On the other hand, the energy gain when adsorbing the molecule, $|E_B^{R_1'}(X_2) + E_{diss}|$, decreases continuously toward zero when E_{diss} increases. The molecular adsorption is favorable energetically when $E_B^{R_1'}(X_2) \le 2E_{ads}$, i.e., when $|E_{diss}/E_{ads}| \ge 1.7$ for $Z'/Z = \frac{1}{2}$ and $|E_{diss}/E_{ads}| \ge 1.9$ for $Z'/Z = \frac{1}{5}$, so the transition between atomic and molecular adsorptions is rather independent of the geometry.

Let us now consider the geometries of approach for which the distance to the surface of the molecular adsorption well is much larger than the atomic one. In these cases, we have seen that the saddle point between the two wells may be above the energy of the free molecule, giving rise to an activation energy for dissociation. This occurs for the particular geometry of Figure 9.26b when E_{diss} is larger than some critical value. In the framework of this simplified model we have to find the crossing point between the atomic and molecular potential curves and

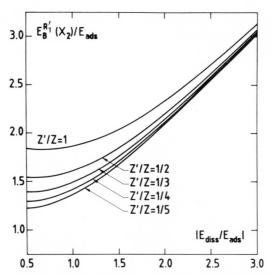

Figure 9.27. Variation in the equilibrium energy of the molecule interacting with the substrate as a function of its dissociation energy (in units of $|E_{ads}|$) and geometry.

examine whether the corresponding energy is larger or smaller than $-E_{diss}$. The critical value is thus given by

$$2[E_B(X)]_{z=z_c} = -E_{diss} \qquad (9.6.9a)$$

$$[E_B(X_2)]_{z=z_c} = -E_{diss} \qquad (9.6.9b)$$

In the physically interesting situation the crossing point is rather far from the z position of the bottom of the atomic well, i.e., in the attractive part, and the repulsive contribution can be neglected in $E_B(X)$. Under this condition equations (9.6.4), (9.6.5), and (9.6.9a) yield

$$\exp\left[-q_{XM}\left(\frac{R(z_c)}{R_1} - 1\right)\right] = -\frac{E_{diss}}{3E_{ads}} \qquad (9.6.10)$$

Similarly, equations (9.6.3), (9.6.6), and (9.6.9b) yield

$$\frac{Z'}{Z}Y_m^4 - 3\frac{E_{diss}}{E_{ads}}Y_m - 9 = 0 \qquad (9.6.11)$$

with $Y_m = \exp\left[-q_{XM}(R'(z_c)/R_1 - 1)\right]$.

By eliminating z_c between equations (9.6.10) and (9.6.11), we derive an implicit equation in $|E_{diss}/E_{ads}|$. Its solution gives the critical value of $|E_{diss}/E_{ads}|$ above which an activation barrier exists. This can be done quite easily on a computer in order to obtain the critical value η_c of $|E_{diss}/E_{ads}|$ as a function of d_∞ for various values of Z and Z'. The only physically interesting solution corresponding to dissociative activated adsorption ($\eta_c < 2$) occurs for the geometry of Figure 9.26b. One finds[63] that an increase in d_∞ (i.e., a decrease in the lateral distance between the atomic and molecular wells) leads to a slightly larger value of η_c while changing the geometry of approach (i.e., Z'/Z) is quite crucial.

As in the atomic adsorption case, we can perform more realistic calculations using local densities of states obtained from a continued fraction expansion of the local Green function on the atom i with three exact coefficients a_1^i, b_1^i, and a_2^i. We have carried out such computations for an O_2 molecule arriving on the (110) and (100) faces of a transition metal with a half-filled d band. The parameters describing the interaction between an oxygen atom and a substrate atom are the same as in Section 9.4. The potential-energy curve of the free O_2 molecule is assumed to be Morse-like ($p_{XX}/q_{XX} = 2$), and we assume $|pp\sigma|/|pp\pi| = 2$; the remaining parameters for the O–O interaction, namely $pp\sigma$, q_{XX}, and A_{XX}, are determined to fit $E_{diss}(O_2)$, d_∞, and the stretch vibration frequency.

Results are shown in Figures 9.28 and 9.29 with the corresponding geometry of approach. One sees in Figure 9.28 that on the (100) face the molecule dissociates.

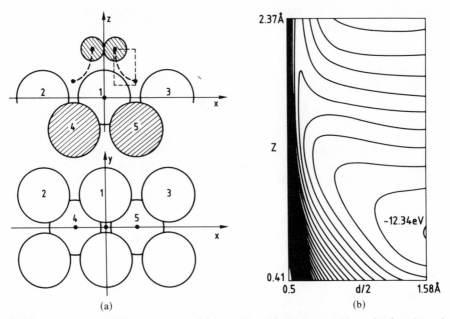

(a)   (b)

Figure 9.28. Equipotential energy curves (b) for an O_2 molecule approaching a (100) surface of a bcc transition metal with a half-filled d band (lattice parameter $a_0 = 3.16$ Å) according to the geometry of approach (a). The area limited by dotted lines shows the scanned region. The reference energy is the energy of two free O atoms.

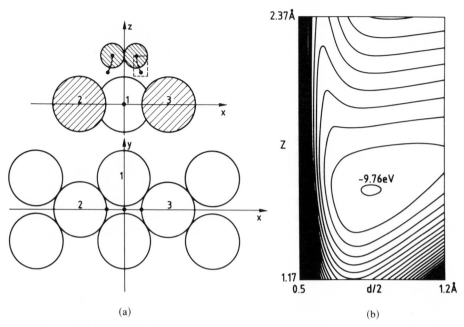

(a)   (b)

Figure 9.29. Equipotential energy curves (b) for an O_2 molecule approaching a (110) surface of a bcc transition metal with a half-filled d band (lattice parameter $a_0 = 3.16$ Å) according to the geometry of approach (a). The area limited by dotted lines shows the scanned region. The reference energy is the energy of two free atoms.

However, its final energy is not twice the atomic adsorption energy since there is a repulsive indirect (via the two substrate atoms labeled 1 in Figure 28a) interaction energy between the two atoms.

The case exhibited in Figure 9.29 is not so clear-cut, since the distance between the two atoms in the adsorbed state is such that there still exists a sizable direct interaction between them. Energy calculations are not sufficient to decide whether or not the molecule is dissociated. A more appropriate quantity would be the electron density between atom pairs which is connected to the bond order. This last quantity is quite familiar to quantum chemists. It can also be easily calculated using the Green function technique, since it is connected to the interatomic Green function. For example, between two atoms i and j with only one atomic orbital, it can be written in the form

$$P_{ij} = -\frac{1}{2\pi} \int_{-\infty}^{E_F} \lim_{\varepsilon \to 0} \text{Im} \left[G_{ij}(E + i\varepsilon) + G_{ji}(E + i\varepsilon) \right] dE \qquad (9.6.12)$$

Qualitatively speaking, one expects that making bonds between the molecule and the substrate usually weakens the intramolecular bond; this is indeed found using simple models.[63]

9.7. Conclusions

In summary we have shown that, despite its incompleteness, the theoretical description of chemisorption of simple elements on metals can provide some understanding of the main trends observed experimentally. In particular, for substrates at the end of the transition series one expects moderate binding energies, small activation energies for surface diffusion, and the possible existence of both molecular and dissociative chemisorption. These three factors are clearly favorable for surface chemical reactions and this should be one of the reasons why the best catalysts are found at the end of the transition series.

However, there is still a long way before we understand fully the dynamics of adsorption and surface reactions, since this involves the complete treatment of the energy transfers via phonons or electron–hole pairs from the adatom to the substrate, or *vice versa*. The comparison between molecular dynamics simulations and molecular beam experiments should provide useful information on this problem.

Appendixes

Appendix 9.1

Let us consider a semi-infinite linear chain with one orbital per site, $-\beta$ ($\beta > 0$) being the hopping integral between nearest neighbors.

We calculate the local density of states on the first atom using the Green function technique. In the basis of atomic orbitals

$$G = (z - H)^{-1} = \begin{vmatrix} z & \beta & & \\ \beta & z & \beta & \\ & \beta & z & \beta \\ & & \ddots & \ddots & \ddots \end{vmatrix}^{-1}$$

in which energies are referred to the atomic level

$$G_{11} = \frac{D_{n-1}}{D_n} = \frac{D_{n-1}}{zD_{n-1} - \beta^2 D_{n-2}} = \frac{1}{z - \beta^2(D_{n-2}/D_{n-1})}$$

D_{n-p} being the determinant obtained by suppressing the first p lines and columns.

Since the chain is semi-infinite, $D_{n-2}/D_{n-1} = G_{11}$. Thus

$$G_{11} = \frac{1}{z - \beta^2 G_{11}}$$

so

$$G_{11} = \frac{z + \eta\sqrt{z^2 - 4\beta^2}}{2\beta^2}, \qquad z = E + i\varepsilon \qquad \text{and} \qquad \eta = \pm 1$$

Quantity G_{11} has an imaginary part when $-2\beta < E < 2\beta$. Consequently the bandwidth is $W = 4\beta$ and

$$G_{11} = \frac{z + \eta\sqrt{z^2 - W^2/4}}{W^2/8}$$

The value of η outside the band should be such that G_{11} behaves like $1/z$ when E tends to infinity. Therefore

$$\eta = +1 \qquad \text{when } E < -2\beta$$

$$\eta = -1 \qquad \text{when } E > 2\beta$$

Inside the band, Im G_{11} should be negative to obtain a positive density of states

$$\eta\sqrt{z^2 - \frac{W^2}{4}} = -i\sqrt{\frac{W^2}{4} - E^2}$$

Appendix 9.2

Let us consider an unperturbed Hamiltonian H_0 and a perturbed Hamiltonian $H = H_0 + V$ and calculate the variation $\Delta n(E)$ in the density of states due to the perturbation. Hence

$$(E - H \pm i\varepsilon)\frac{1}{E - H_0 \pm i\varepsilon} = 1 - VG_0^{\pm}(E), \qquad \varepsilon > 0$$

$G_0^{\pm} = (E - H_0 \pm i\varepsilon)^{-1}$ is the unperturbed Green function. We thus obtain

$$\text{Det}\,[1 - VG_0^{\pm}(E)] = \prod_j \frac{E - E_j \pm i\varepsilon}{E - E_j^0 \pm i\varepsilon}$$

E_j and E_j^0 being respectively the eigenvalues of H and H_0,

$$\frac{d}{dE}\log\text{Det}\,[1 - VG_0^{\pm}(E)] = \sum_j \frac{1}{E - E_j \pm i\varepsilon} - \sum_j \frac{1}{E - E_j^0 \pm i\varepsilon}$$

The imaginary part of this expression is

$$\lim_{\varepsilon \to 0} \text{Im} \frac{d}{dE} \log \text{Det}\,[1 - VG_0^\pm(E)] = \mp \pi \left[\sum_j \delta(E - E_j) - \sum_j \delta(E - E_j^0) \right]$$

Hence

$$\Delta n(E) = \mp \frac{1}{\pi} \lim_{\varepsilon \to 0} \text{Im} \frac{d}{dE} \log \text{Det}\,[1 - VG_0^\pm(E)]$$

Appendix 9.3

We show that in the Hartree–Fock approximation

$$\int^{E_F} (E - E_F)\, \Delta \mathcal{N}(E)\, dE = \frac{1}{2\pi i} \sum_\sigma \int_C (z - E_F)\left(1 - \frac{dS}{dz} \right) {}^c G_{aa}^{\sigma\sigma}(z)\, dz$$

where C is the contour consisting of the real axis and a semicircle at infinity in the upper half-plane and ${}^c G_{aa}^{\sigma\sigma}(z)$ is the causal Green function. The integral on the semicircle vanishes and therefore the second member of the preceding equation can be written in the form

$$\lim_{\varepsilon \to 0} \frac{1}{2\pi i} \sum_\sigma \left[\int_{-\infty}^{E_F} (E - E_F)\left(1 - \frac{\partial S(E - i\varepsilon)}{\partial E} \right) G_{aa}^{\sigma\sigma-}(E - i\varepsilon)\, dE \right.$$

$$\left. + \int_{E_F}^{+\infty} (E - E_F)\left(1 - \frac{\partial S(E + i\varepsilon)}{\partial E} \right) G_{aa}^{\sigma\sigma+}(E + i\varepsilon)\, dE \right]$$

or

$$\lim_{\varepsilon \to 0} \frac{1}{2\pi i} \sum_\sigma \left[\int_{-\infty}^{E_F} (E - E_F)\left(1 - \frac{\partial S(E - i\varepsilon)}{\partial E} \right) G_{aa}^{\sigma\sigma-}(E - i\varepsilon)\, dE \right.$$

$$- \int_{-\infty}^{E_F} (E - E_F)\left(1 - \frac{\partial S(E + i\varepsilon)}{\partial E} \right) G_{aa}^{\sigma\sigma+}(E + i\varepsilon)\, dE$$

$$\left. + \int_{-\infty}^{+\infty} (E - E_F)\left(1 - \frac{\partial S(E + i\varepsilon)}{\partial E} \right) G_{aa}^{\sigma\sigma+}(E + i\varepsilon)\, dE \right]$$

The third integral can be transformed into a contour integral on C which vanishes, since the poles are below the real axis. Then, using equation (9.3.56c) of the main text, valid in the Hartree–Fock approximation, the equation to be proved is obtained immediately.

Appendix 9.4

In the Hartree approximation, the total energy is given by

$$E = \sum_{\varepsilon_k \leq E_F} \varepsilon_i - \tfrac{1}{2} \sum_{\substack{k \neq k' \\ \varepsilon_k \leq E_F \\ \varepsilon_{k'} \leq E_F}} \iint \frac{e^2}{|\mathbf{r} - \mathbf{r}'|} |\psi_k(\mathbf{r})|^2 |\psi_{k'}(\mathbf{r}')|^2 \, d^3\mathbf{r} \, d^3\mathbf{r}'$$

where ε_i are the one-electron energies, E_F is the Fermi level, and the second term represents the sum of Coulomb interactions between a pair of electrons in states ψ_k and $\psi_{k'}$, respectively, which are counted twice in the first term. This expression can also be expressed as

$$E \simeq \int^{E_F} En(E) \, dE - \tfrac{1}{2} \int \rho(\mathbf{r}) V(\mathbf{r}) \, d^3\mathbf{r}$$

where $n(E)$ is the total density of states,

$$\rho(\mathbf{r}) = \sum_{\varepsilon_k \leq E_F} |\psi_k(\mathbf{r})|^2$$

$$V(r) = \sum_{\varepsilon_{k'} \leq E_F} \int \frac{e^2 |\psi_{k'}(\mathbf{r}')|^2}{|\mathbf{r} - \mathbf{r}'|} \, d^3\mathbf{r}' = \int \frac{e^2 \rho(\mathbf{r}')}{|\mathbf{r} - \mathbf{r}'|} \, d^3\mathbf{r}'$$

Actually, the above two expressions for E only differ by the terms where $k = k'$, which are negligible when the number of electrons is large.

When the system is perturbed, the corresponding variation of energy is given by

$$\delta E = \delta \int^{E_F} En(E) \, dE - \tfrac{1}{2} \int \delta\rho(\mathbf{r}) V(\mathbf{r}) \, d^3\mathbf{r} - \tfrac{1}{2} \int \rho(\mathbf{r}) \, \delta V(\mathbf{r}) \, d^3\mathbf{r}$$

$$- \tfrac{1}{2} \int \delta\rho(\mathbf{r}) \, \delta V(\mathbf{r}) \, d^3\mathbf{r}$$

From the expression for $V(\mathbf{r})$, one sees at first glance that

$$\int \delta\rho(\mathbf{r}) V(\mathbf{r}) \, d^3\mathbf{r} = \int \rho(\mathbf{r}) \delta V(\mathbf{r}) \, d^3\mathbf{r}$$

Therefore

$$\delta E = \delta \int^{E_F} En(E) \, dE - \int \rho(\mathbf{r}) \, \delta V(\mathbf{r}) \, d^3\mathbf{r} - \tfrac{1}{2} \int \delta\rho(\mathbf{r}) \, \delta V(\mathbf{r}) \, d^3\mathbf{r}$$

In the tight-binding scheme one has

$$\psi_k = \sum_k a_i(\varepsilon_k)\varphi_i(\mathbf{r})$$

where $\varphi_i(\mathbf{r})$ is the atomic orbital centered at site i (we have considered s orbitals for the sake of simplicity). Thus

$$\rho(\mathbf{r}) = \sum_{\substack{i,j \\ \varepsilon_k \leq E_F}} a_i^*(\varepsilon_k)a_j(\varepsilon_k)\varphi_i^*(\mathbf{r})\varphi_j(\mathbf{r})$$

Then

$$\int \rho(\mathbf{r})\,\delta V(\mathbf{r})\,d^3\mathbf{r} = \sum_{\substack{i,j \\ \varepsilon_k \leq E_F}} a_i^*(\varepsilon_k)a_j(\varepsilon_k) \int \varphi_i^*(\mathbf{r})\,\delta V(\mathbf{r})\,\varphi_j(\mathbf{r})\,d^3\mathbf{r}$$

It is assumed that $\delta V(\mathbf{r})$ has only intra-atomic matrix elements $\delta\varepsilon_i$ so that

$$\int \varphi_i^*(\mathbf{r})\,\delta V(\mathbf{r})\varphi_j(\mathbf{r})\,d^3\mathbf{r} = \delta\varepsilon_i\,\delta_{ij}$$

$$\int \rho(\mathbf{r})\,\delta V(\mathbf{r})\,d^3\mathbf{r} = \sum_i \delta\varepsilon_i \sum_{\varepsilon_k \leq E_F} |a_i(\varepsilon_k)|^2$$

and since

$$\sum_{\varepsilon_k \leq E_F} |a_i(\varepsilon_k)|^2 = \int^{E_F} n_i(E)\,dE = N_i$$

where $n_i(E)$ is the local density of states at site i and N_i the corresponding number of electrons, we obtain:

$$\int \rho(\mathbf{r})\,\delta V(\mathbf{r})\,d^3\mathbf{r} = \sum_i N_i\,\delta\varepsilon_i$$

Similarly, if we set $\sum_{\varepsilon_k \leq E_F} a_i^*(\varepsilon_k)a_j(\varepsilon_k) = c_{ij}$ then

$$\frac{1}{2}\int \delta\rho(\mathbf{r})\,\delta V(\mathbf{r})\,d^3\mathbf{r} = \frac{1}{2}\sum_{ij} \delta c_{ij} \int \varphi_i^*(\mathbf{r})\,\delta V(\mathbf{r})\varphi_j(\mathbf{r})\,d^3\mathbf{r}$$

$$= \frac{1}{2}\sum_i \delta\varepsilon_i\,\delta c_{ii}$$

$$= \frac{1}{2}\sum_i \delta\varepsilon_i\,\delta N_i$$

Finally

$$\delta E = \delta \int E n(E) \, dE - \sum_i N_i \, \delta \varepsilon_i - \tfrac{1}{2} \sum_i \delta \varepsilon_i \, \delta N_i$$

If we use a local charge neutrality condition, the last term vanishes.

Appendix 9.5

We derive expression (9.4.36) of the main text giving the contribution of atom i to the correlation energy. Quantity H' denotes the electron Coulomb interactions. Within second-order perturbation theory, the total energy can be written in the form

$$E = E_0 + \langle 0|H'|0 \rangle + \sum_{|l\rangle \neq |0\rangle} \frac{|\langle 0|H'|l \rangle|^2}{E_0 - E_l}$$

where $|0\rangle$ is the fundamental state in the one-electron tight-binding approximation, i.e., a Slater determinant constructed with all occupied one-electron wave functions; E_0 is the corresponding energy and $|l\rangle$ are the excited states with energy E_l.

Quantity $\langle 0|H'|0 \rangle$ is the first-order contribution, i.e., the non-self-consistent Hartree–Fock correction. The only nonvanishing second-order terms are those involving excited states with one electron–hole pair $|l^{(1)}\rangle$ and two electron–hole pairs $|l^{(2)}\rangle$. One can show that the energy correction due to the $|l^{(1)}\rangle$ excited states are actually taken into account in a self-consistent Hartree–Fock approximation.

Therefore, for second order in H', the correlation contribution defined as the difference between the true energy and the self-consistent Hartree–Fock energy is given by

$$E_{\text{cor}} = \sum_{|l^{(2)}\rangle} \frac{|\langle 0|H'|l^{(2)}\rangle|^2}{E_0 - E_{l^{(2)}}}$$

We calculate $\langle 0|H'|l^{(2)}\rangle$, $l^{(2)}$ being the Slater determinant obtained from $|0\rangle$ by replacing two occupied states $(\psi_{m\sigma}, \psi_{n\sigma'})$ by two unoccupied states $(\psi_{p\sigma}, \psi_{q\sigma'})$,

$$\langle 0|H'|l^{(2)}\rangle = \left\langle \psi_{m\sigma}(\mathbf{r}) \psi_{n\sigma'}(\mathbf{r}') \left| \frac{e^2}{|\mathbf{r} - \mathbf{r}'|} \right| \psi_{p\sigma}(\mathbf{r}) \psi_{q\sigma'}(\mathbf{r}') \right\rangle$$

The spin of each electron is unchanged, since the interaction is spin-independent. In the tight-binding approximation

$$\psi_{m\sigma}(\mathbf{r}) = \sum_i a_{i\sigma}(\varepsilon_m) \varphi_\sigma(\mathbf{r} - \mathbf{R}_i) = \sum_i a_{i\sigma}(\varepsilon_m) \varphi_{i\sigma}(\mathbf{r})$$

hence

$$\langle 0|H'|I^{(2)}\rangle = \sum_{ijkl} a_{i\sigma}^*(\varepsilon_m) a_{j\sigma'}^*(\varepsilon_n) a_{k\sigma}(\varepsilon_p) a_{l\sigma'}(\varepsilon_q) \left\langle \varphi_{i\sigma}(\mathbf{r})\varphi_{j\sigma'}(\mathbf{r}') \left| \frac{e^2}{|\mathbf{r} - \mathbf{r}'|} \right| \varphi_{k\sigma}(\mathbf{r})\varphi_{l\sigma'}(\mathbf{r}) \right\rangle$$

Within the Hubbard model

$$H' = \tfrac{1}{2}\sum_{i\sigma} U_i n_{i\sigma} n_{i-\sigma}$$

where $n_{i\sigma}$, $n_{i-\sigma}$ are the occupation number operators, the only nonvanishing matrix elements in the expression for $\langle 0|H'|I^{(2)}\rangle$ are those with $i = j = k = l$ and $\sigma' = -\sigma$,

$$\langle 0|H'|I^{(2)}\rangle = \sum_i a_{i\sigma}^*(\varepsilon_m) a_{i-\sigma}^*(\varepsilon_n) a_{i\sigma}(\varepsilon_p) a_{i-\sigma}(\varepsilon_q) U_i$$

$$E_{\text{cor}} = \sum_{I^{(2)}} \frac{|\langle 0|H'|I^{(2)}\rangle|^2}{E_0 - E_{I^{(2)}}}$$

$$= \tfrac{1}{2}\sum_{\substack{i,j \\ m\sigma, n-\sigma \, \text{occ.} \\ p\sigma, q-\sigma \, \text{unocc.}}} U_i U_j \frac{a_{i\sigma}^*(\varepsilon_m) a_{j\sigma}(\varepsilon_m) a_{i-\sigma}^*(\varepsilon_n) a_{j-\sigma}(\varepsilon_n) a_{i\sigma}(\varepsilon_p) a_{j\sigma}^*(\varepsilon_p) a_{i-\sigma}(\varepsilon_q) a_{j-\sigma}^*(\varepsilon_q)}{\varepsilon_m + \varepsilon_n - \varepsilon_p - \varepsilon_q}$$

The factor $\tfrac{1}{2}$ in the latter expression must be introduced, since the interchange of p, σ with $q, -\sigma$ does not lead to physically different $|I^{(2)}\rangle$ states.

If real atomic wave functions are selected, the tight-binding Hamiltonian has real matrix elements and [see equation (9.3.32)]

$$n_{ij}^\sigma(E) = -\frac{1}{\pi} \lim_{\varepsilon \to 0} \text{Im } G_{ij}^\sigma = \sum_k a_{i\sigma}(\varepsilon_k) a_{j\sigma}^*(\varepsilon_k)\, \delta(E - \varepsilon_k) = n_{ji}^\sigma(E)$$

One can now verify easily that for a nonmagnetic system $[n_{ij}^\sigma(E) = n_{ij}^{-\sigma}(E)]$,

$$E_{\text{cor}} = \sum_{ij} U_i U_j \int_{-\infty}^{E_F} dE_1 \int_{-\infty}^{E_F} dE_2 \int_{E_F}^{+\infty} dE_3 \int_{E_F}^{+\infty} dE_4 \frac{n_{ij}(E_1) n_{ij}(E_2) n_{ij}(E_3) n_{ij}(E_4)}{E_1 + E_2 - E_3 - E_4}$$

It has been shown[30,40] that the contribution to E_{cor} of the nonlocal terms ($i \neq j$)

are small compared to the local ones and will be neglected. In this case

$$E_{\text{cor}} = \sum_i E_{\text{cor}}^i$$

$$= \sum_i U_i^2 \int_{-\infty}^{E_F} dE_1 \int_{-\infty}^{E_F} dE_2 \int_{E_F}^{+\infty} dE_3 \int_{E_F}^{+\infty} dE_4 \, \frac{n_i(E_1)n_i(E_2)n_i(E_3)n_i(E_4)}{E_1 + E_2 - E_3 - E_4}$$

where $n_i(E)$ is the local density of states on atom i.

If we now consider an atom i with L_i degenerate atomic spin-orbitals, then E_{cor}^i must be multiplied by the number of different pairs of atomic spin-orbitals, i.e., $L_i(L_i - 1)/2$. Equation (9.4.36) is obtained.

The self-consistent Hartree–Fock term is now examined. We have seen (Section 9.3.2) that in the case of the Anderson–Grimley–Newns model the adsorbate self-energy is $\Sigma_a^{HF} = U\langle n_a \rangle$. Obviously this is still valid in a self-consistent scheme. Similarly, it is easy to show that for a tight-binding Hamiltonian perturbed by the Hubbard term H', each atom i has a self-energy $\Sigma_i^{HF} = U_i\langle n_i \rangle$. Thus, in a self-consistent Hartree–Fock scheme, the atomic level of atom i is modified by this quantity, which is determined by solving a set of coupled self-consistency equations.

Hence this procedure is very similar to our renormalization of atomic levels based on the charge neutrality condition. Insofar as the self-consistent Hartree–Fock scheme leads to small charge transfers, the two approaches are equivalent.

References

1. See, for example, D. A. Woodruff and T. A. Delchar, *Modern Techniques of Surface Science*, Cambridge Solid State Science Series, Cambridge University Press, Cambridge (1986).
2. J. E. Demuth, D. W. Jepsen, and P. M. Marcus, *Phys. Rev. Lett.* **31**, 540 (1973); **32**, 1182 (1974); M. A. Van Hove and S. Y. Tong, *J. Vac. Sci. Technol.* **12**, 230 (1975); P. M. Marcus, J. E. Demuth, and D. W. Jepsen, *Surf. Sci.* **53**, 501 (1975); D. H. Rosenblatt, S. D. Kevan, J. G. Tobin, R. F. Davis, M. G. Mason, D. R. Denley, D. A. Shirley, Y. Huang, and S. Y. Tong, *Phys. Rev. B* **26**, 1812 (1982).
3. K. A. R. Mitchell, *Surf. Sci.* **149**, 93 (1985).
4. J. Sokolov, F. Jona, and P. M. Marcus, *Solid State Commun.* **49**, 307 (1984) and references cited therein.
5. M. R. Barnes and R. F. Willis, *Phys. Rev. Lett.* **41**, 1729 (1978).
6. I. Toyoshima and G. A. Somorjai, *Catal. Rev. Sci. Eng.* **19**, 105 (1979).
7. A. Menand and J. Gallot, *Rev. Phys. Appl.* **9**, 323 (1974).
8. G. Bolbach, Thesis, Paris (1982).
9. D. W. Bassett, *Surf. Sci.* **53**, 74 (1975); *J. Phys. C* **9**, 2491 (1976).
10. H. W. Fink and G. Ehrlich, *Surf. Sci.* **143**, 125 (1984).
11. J. Cousty, Thesis, Orsay (1980).
12. G. Brodén, T. N. Rhodin, C. Brucker, R. Benbow, and Z. Hurych, *Surf. Sci.* **59**, 593 (1976).
13. J. R. Schrieffer, Proceedings of the International School of Physics, Enrico Fermi, Course LVIII, *Dynamical Aspects of Surface Physics* (F. O. Goodman, ed.), Compositori, Bologna (1974).
14. P. W. Anderson, *Phys. Rev.* **124**, 41 (1961).

15. T. B. Grimley, *Proc. Phys. Soc.* **90**, 751; **92**, 776 (1967).

16. D. M. Newns, *Phys. Rev.*, **178**, 1123 (1969).

17. M. C. Desjonquères, Thesis, Grenoble (1976).

18. M. J. Kelly, *Surf. Sci.* **43**, 587 (1974).

19. M. C. Desjonquères and F. Cyrot-Lackmann, *Solid State Commun.* **26**, 271 (1978).

20. J. Friedel, *Philos. Mag., Suppl.* **3**, 446 (1954).

21. W. Brenig and K. Schönhammer, *Z. Phys.* **267**, 201 (1974).

22. B. Bell and A. Madhukar, *Phys. Rev. B* **14**, 4281 (1976).

23. K. Schönhammer, *Solid State Commun.* **32**, 51 (1977).

24. M. Baldo, F. Flores, A. Martin-Rodero, G. Piccitto, and R. Pucci, *Surf. Sci.* **128**, 237 (1983).

25. M. Baldo, R. Pucci, F. Flores, G. Piccitto, and A. Martin-Rodero, *Phys. Rev. B* **28**, 6640 (1983).

26. O. Gunnarsson and K. Schönhammer, *Phys. Rev. Lett.* **41**, 1608 (1978).

27. B. Kjöllerström, D. J. Scalapino, and J. R. Schrieffer, *Phys. Rev.* **148**, 665 (1966).

28. K. Schönhammer, V. Hartung, and W. Brenig, *Z. Phys. B* **22**, 143 (1975).

29. G. Piccitto, F. Siringo, M. Baldo, and R. Pucci, *Surf. Sci.* **167**, 437 (1986).

30. G. Tréglia, F. Ducastelle, and D. Spanjaard, *J. Phys. (Paris)* **41**, 281 (1980).

31. A. Martin-Rodero, F. Flores, M. Baldo, and R. Pucci, *Solid State Commun.* **44**, 911 (1982).

32. P. Schuck, *Phys. Rev. B* **13**, 5225 (1976).

33. M. C. Desjonquères and D. Spanjaard, *J. Phys. C* **15**, 4007 (1982).

34. M. C. Desjonquères and D. Spanjaard, *J. Phys. C* **16**, 3389 (1983).

35. C. Thuault-Cytermann, M. C. Desjonquères, and D. Spanjaard, *J. Phys. C* **16**, 5689 (1983).

36. F. Ducastelle, Thesis, Orsay (1972).

37. J. C. Slater and G. F. Koster, *Phys. Rev.* **94**, 1498 (1954).

38. R. Haydock, V. Heine, and M. J. Kelly, *J. Phys. C* **5**, 2845 (1972).

39. J. P. Gaspard and F. Cyrot-Lackmann, *J. Phys. C* **6**, 3077 (1973).

40. G. Tréglia, Thesis, Orsay (1983).

41. J. Friedel and C. M. Sayers, *J. Phys. (Paris)* **38**, 697 (1977).

42. H. Ehrenreich and L. Hodges, *Methods Comput. Phys.* **8**, 149 (1968).

43. Y. Boudeville, J. Rousseau-Violet, F. Cyrot-Lackmann, and S. N. Khanna, *J. Phys. (Paris)* **44**, 433 (1983).

44. G. Tréglia, M. C. Desjonquères, F. Ducastelle, and D. Spanjaard, *J. Phys. C* **14**, 4347 (1981).

45. G. Tréglia, F. Ducastelle, and D. Spanjaard, *J. Phys. (Paris)* **43**, 341 (1982).

46. C. Guillot, C. Thuault, Y. Jugnet, D. Chauveau, R. Hoogewijs, J. Lecante, Tran Minh Duc, G. Tréglia, M. C. Desjonquères, and D. Spanjaard, *J. Phys. C* **15**, 423 (1982); C. Guillot, P. Roubin, J. Lecante, M. C. Desjonquères, G. Tréglia, D. Spanjaard, and Y. Jugnet, *Phys. Rev. B* **30**, 5487 (1984).

47. M. Bowker and D. A. King, *Surf. Sci.* **94**, 564 (1980).

48. R. Di Foggio and R. Gomer, *Phys. Rev. Lett.* **44**, 1258 (1980).

49. J. P. Jardin, M. C. Desjonquères, and D. Spanjaard, *J. Phys. C* **18**, 1767 (1985).

50. J. P. Jardin, M. C. Desjonquères, and D. Spanjaard, *J. Phys. C* **18**, 5759 (1985).

51. J. P. Bourdin, J. P. Ganachaud, J. P. Jardin, D. Spanjaard, and M. C. Desjonquères, *J. Phys. F* **18**, 1801 (1988).

52. M. J. Stott and E. Zaremba, *Phys. Rev. B* **22**, 1564 (1980).

53. J. K. Nørskov and N. D. Lang, *Phys. Rev. B* **21**, 2136 (1980).

54. J. K. Nørskov, *Phys. Rev. B* **26**, 2875 (1982).

55. M. J. Puska, R. M. Nieminen, and M. Manninen, *Phys. Rev. B* **24**, 3037 (1981).

56. S. P. McGlynn, L. G. Vanquickenborne, M. Kinoshita, and D. G. Carroll, *Introduction to Applied Quantum Chemistry*, Holt, Rinehart, and Winston, New York (1972).

57. P. D. Offenhartz, *Atomic and Molecular Orbital Theory*, McGraw-Hill, New York (1970); J. A. Pople and D. L. Beveridge, *Approximate Molecular Orbital Theory*, McGraw-Hill, New York (1970).

58. M. Wolfsberg and L. Helmholz, *J. Chem. Phys.* **20**, 837 (1952).

59. J. E. Lennard-Jones, *Trans. Faraday Soc.* **28**, 333 (1932).

60. H. Eyring, J. Walter, and G. E. Kimball, in: *Quantum Chemistry*, Chap. XIII, Wiley, London (1967).

61. A. Gelb and M. J. Cardillo, *Surf. Sci.* **64**, 197 (1977); J. H. McCreery and G. Wolken Jr., *J. Chem. Phys.* **67**, 2551 (1977); G. F. Tantardini and M. Simonetta, *Chem. Phys. Lett.* **87**, 420 (1982); B. C. Khanra and S. K. Saha, *Chem. Phys. Lett.* **95**, 217 (1983).

62. J. K. Nørskov, H. Houmoller, P. K. Johansson, and B. I. Lundqvist, *Phys. Rev. Lett.* **46**, 257 (1981).

63. M. C. Desjonquères, J. P. Jardin, and D. Spanjaard, *Surf. Sci.*, **204**, 247 (1988).

10

Surface Vibration Spectroscopy

H. Ibach

10.1. Introduction

Surface science is an interdisciplinary field which encompasses the physics of solids, molecules, and atoms, their chemistry, as well as the material sciences. Applications of surface science branch into design and engineering of devices, micromechanics, sensorics, tribology, catalysis, and electrochemistry. Traditionally, surface physics has dealt with the solid vacuum interface, however, more recently we find increasing activity in solid–solid interfaces, solid–liquid interfaces, multilayered material, and also a revitalization of traditional fields like the physics of thin films, crystal growth, and epitaxy. The development of surface science has brought about an impressive number of ingenious methods for probing surfaces and interfaces. A common aspect of all these methods is that the tools need to be specific to surface properties, that is, they need to discriminate between surface and bulk properties and be less sensitive to the latter by a large margin. The use of nonpenetrating, strongly interacting particles with matter is therefore widespread. Traditionally, electrons have played a major role in the methodology of various tools. More recently, the use of atoms and ions have added to the arsenal of techniques.

This chapter will deal mainly with surface vibration spectroscopy and the instrumentation needed for this purpose. The physics of surface vibrations is the physics of chemical bonds at surfaces and of interatomic forces in general. The results of vibration spectroscopy thus relate to structural parameters as well, although vibration spectroscopy is not primarily a structural tool. Nevertheless, we shall see that surface vibration spectroscopy frequently allows for the analysis of the local symmetry of surface species, which is a valuable ingredient for any fine structure analysis. Furthermore, the study of interatomic forces near the surface

H. Ibach • Institut für Grenzflächenforschung und Vakuumphysik, Kernforschungsanlage Jülich, D-5170 Jülich, Federal Republic of Germany.

provides us with some valuable insight into the driving forces for structural changes at or near the surface.

The chapter is organized as follows. The next section contains a brief account of the various experimental techniques producing vibrational spectra of surfaces. We will then focus on three techniques: Fourier-transform infrared spectroscopy, scattering of helium beams, and electron energy loss spectroscopy. For the latter topic, we consider the electron optical principals for the design of electron energy loss spectrometers in greater detail in Section 3. Section 4 is concerned with the electron–solid interaction, and we shall learn about the different scattering mechanisms which one may exploit in order to produce information on surface vibrations. Section 5 is devoted to vibration spectroscopy of localized vibrations at surfaces. The use of selection rules for symmetry determination will be examined, as well as a few examples for the investigation of surface chemical processes by vibration spectroscopies. Section 6 deals with surface phonons and the determination of a surface force field using phonon spectroscopy. The chapter concludes with a more recent topic concerning epitaxial metal overlayers and phonons in such overlayers.

10.2. Experimental Techniques for Surface Vibration Spectroscopy

10.2.1. Overview

Various techniques have been developed to probe vibrational spectra near surfaces. Infrared transmission spectroscopy and neutron scattering have been used traditionally to probe vibrations of metal particles finely dispersed on insulating supports (Table 10.1).

In recent years the enhanced Raman effect has attracted much interest. It has become clear, however, that the 10^6 enhancement that one observes there requires particular surface conditions which are not typical for the general field of surface science. On specially prepared samples one can also apply tunnel spectroscopy. The scanning tunneling microscope in principle also offers the possibility for performing vibrational spectroscopy on a local base, although at present this method is still at its infancy and it is not clear whether it will become instrumental to surface science. By far the majority and the most general results have been produced by three techniques: inelastic helium scattering, infrared reflection absorption spectroscopy, and electron energy loss spectroscopy. Each of these methods has its own virtues and drawbacks, so a particular realm of applications will be discussed in connection with each technique. We begin with infrared reflection absorption spectroscopy.

10.2.2. Infrared Reflection Absorption Spectroscopy

The method of infrared reflection absorption spectroscopy was developed by Greenler[1] and Pritchard.[2] The technique exploits the effect that the metal surface becomes almost perfectly reflecting in the infrared for reflection near grazing incidence. If one places a molecule on the surface, the absorption due to the

Table 10.1. Comparison of Various Techniques for Surface Vibration Spectroscopy

	Sensitivity in monolayers	Surface area	Spectral regime	Resolution	
IR transmission	Many	m^2	$>1000 \text{ cm}^{-1}$	5 cm^{-1}	Finely dispersed material
Neutron scattering	Many	m^2	$<1000 \text{ cm}^{-1}$	Variable	
Enhanced Raman	Variable	10^{-2} cm^2	$>100 \text{ cm}^{-1}$	5 cm^{-1}	Special systems and surface conditions
Tunnel spectroscopy	10^{-2}	10^{-2} cm^2	$>200 \text{ cm}^{-1}$	$5\text{--}20 \text{ cm}^{-1}$	Specially prepared samples
He scattering	10^{-1}?	10^{-1} cm^2	$<300 \text{ cm}^{-1}$	2 cm^{-1}	Phonon-dispersion physisorbed molecules
IR reflection absorption	$1\text{--}10^{-3}$	1 cm^2	$>700 \text{ cm}^{-1}$ (today)	1 cm^{-1}	Ambient pressure, limited to the near infrared and dipole active modes
Electron energy loss	$10^{-1}\text{--}10^{-4}$	10^{-3} cm^2	$>40 \text{ cm}^{-1}$	$>15 \text{ cm}^{-1}$	Wide spectral range, phonon dispersion, largest versatility

vibrational eigenfrequency of that molecule will cause a dip in the reflectivity at that frequency. The dip can be as high as a few percent in the rather favorable case of a CO molecule on a metal surface. In general, however, the change in reflectivity will be much less than one percent. Modern instruments use Fourier-transform interferometers for that purpose. Figure 10.1 shows the layout of such a modern instrument as developed by Erley in this laboratory. The system consists of a Mattson Sirius 100 Fourier-transform interferometer which is evacuated to remove the background effects due to the absorption from gas-phase molecules. The principle of Fourier-transform spectroscopy is easily explained. Due to the different path length in the Michelson interferometer one has two different contributions to the field after reflection from the surface,

$$\mathscr{F}_1 = \sqrt{I_0 R(\omega)/2}\, e^{i\omega t} \quad \text{and} \quad \mathscr{F}_1 = \sqrt{I_0 R(\omega)/2}\, e^{i(\omega t + 2\omega x/c)} \quad (10.2.1)$$

where I_0 is the incident intensity, $R(\omega)$ the reflectivity of the sample, and x the position of the adjustable mirror in the Michelson interferometer. The detector measures the intensity as a function of the mirror position

$$I(x) = \int |\mathscr{F}_1 + \mathscr{F}_2|^2 \, d\omega$$

$$= I_0 \int R(\omega) \, d\omega + I_0 \int R(\omega) \cos(2\omega x/c) \, d\omega \quad (10.2.2)$$

which contains the Fourier transform of the reflectivity spectrum. Sample spectra are shown in Figures 10.2 and 10.3 demonstrating that the technique is fast, sensitive, and has an extremely good resolution.

While infrared spectroscopy leaves little to be desired in terms of resolution and sensitivity, it has some limitations. First of all it is sensitive only to those vibrations which have a perpendicular component of the dipole moment, since the electric field vector has only a perpendicular component at a metal surface. Second, the technique loses sensitivity very rapidly when one proceeds into the far infrared. Finally, one can only probe surface vibrations with nearly zero momentum transfer. The dispersion of vibrational modes is thus inaccessible. A technique that is ideally suited to the study of dispersion effects and still maintains a high resolution is inelastic helium scattering.

10.2.3. Inelastic Helium Scattering

Inelastic helium scattering became instrumental to surface vibration spectroscopy, in particular surface phonon spectroscopy, largely through the group of Toennies.[4] Monochromatic helium beams are produced by expanding helium from a high-pressure cell into vacuum and by "skimming off" all helium atoms which are not confined to the very narrow angular cone around the normal of the orifice of the expansion cell. By this procedure, which involves several steps of differential pumping, one discriminates against all helium atoms whose longitudinal

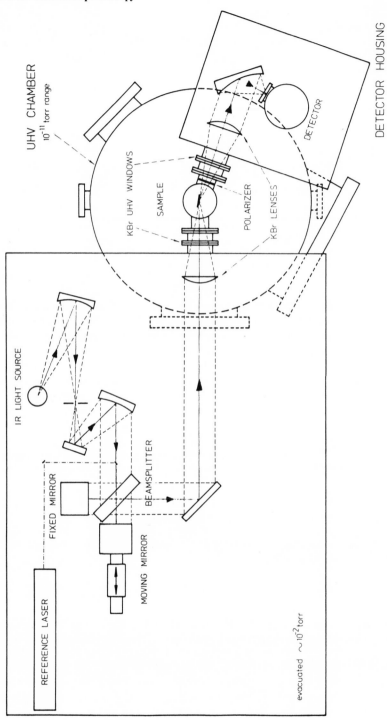

Figure 10.1. Fourier-transform infrared spectrometer for surface applications (after Erley[3]).

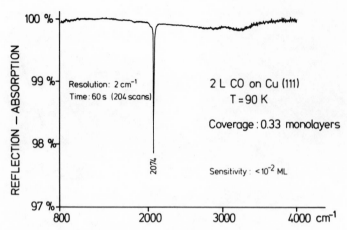

Figure 10.2. High-resolution spectrum of CO on Cu(111). Only 60 seconds of time are needed to produce a spectrum with a resolution of $2 \, \text{cm}^{-1}$ or $\frac{1}{4} \, \text{meV}$.

velocity differs from a certain mean value. The reason is that helium atoms not moving with a uniform speed will collide with each other in noncentral collisions which removes those atoms from the beam. Such monochromatic helium beams are then scattered from surfaces and energy losses are analyzed with time-of-flight spectroscopy. Resolution of a few tenths of 1 meV have been achieved with this method. Again the method is subject to some limitations which are best understood by considering a quasi-classical approach for the scattering mechanism. The scattering cross section for helium atoms in Born approximation is

$$\frac{d\sigma}{d\omega \, d\Omega} = \left(\frac{m}{2\pi\hbar^2}\right)^2 \frac{k'}{k} \left| \langle \varepsilon | e^{i\eta(r,t)} | \alpha \rangle^2 \delta(E' - E - \hbar\omega) \right. \tag{10.2.3}$$

Here α and ε denote the initial and final states of the surface and η is a classical phase of the helium atom given by the action integral

$$\eta = \frac{1}{h} \int L \, dt \tag{10.2.4}$$

where L is the Lagrangian to be evaluated along the classical atom trajectory. The remaining symbols in equation (10.2.3) have their usual meaning. To evaluate for phonon scattering one expands the Lagrangian in terms of the lattice displacements $\mathbf{u_n}$,

$$L = L_{\text{static}} + \sum_n \frac{\partial L}{\partial \mathbf{R_n}} \mathbf{u_n} \tag{10.2.5}$$

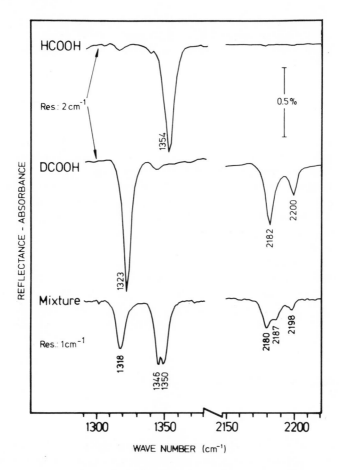

Figure 10.3. Spectrum of formic acidanhydride. The spectrum shows that reflectivity changes of much less than 0.1% can be detected (after Erley[3]).

where the derivatives of the Lagrangian with respect to the atom coordinates $\mathbf{R}_n$ represent the forces to which the helium atom is subjected by virtue of the motion of the atoms $\mathbf{n}$. If the classical point and time of reflection of a helium particle at the surface are denoted by $\mathbf{r}_\parallel$ and t, respectively, then the phonon contribution to the phase is

$$\eta(\mathbf{r}_\parallel, t) = -\frac{1}{h} \int \sum_\mathbf{n} \frac{\partial L}{\partial \mathbf{R}_n} (\mathbf{r}_\parallel, t) \cdot \mathbf{u}_n(t + t') \, dt' \qquad (10.2.6)$$

The frequency and momentum structure of the phase can be analyzed by working with the Fourier-transform representation. But even without employing such a detailed analysis we can see the following. Suppose the helium atom is approaching the surface rather slowly and the atoms are vibrating at rather high frequency, one is working in the limit where

$$\omega \gg v_{\text{He}}/\Lambda \tag{10.2.7}$$

where Λ is some characteristic interaction length of the helium atom with the surface. Then the integral in equation (10.2.6) vanishes and thus the phonon cross section also vanishes. The actual velocity of the helium beam is such that the cross section decays rather rapidly beyond 20–30 meV. High frequencies of vibration, which we have seen to be the domain of infrared spectroscopy, are thus inaccessible to helium scattering. One could, of course, enhance the sensitivity of helium scattering to higher frequencies by using higher beam energies, however, multiphonon processes are then prevailing to an extent that the technique becomes noninstrumental.

Helium atoms unlike light carry a large momentum. It is therefore possible to study energy losses as a function of a momentum transfer with helium atoms. Consequently the realm of helium scattering has been the study of surface phonons on clean and adsorbate covered surfaces, especially in the low-frequency regime. Examples of such studies are given in other chapters to which the reader is referred for further information.

10.3. Electron Energy Loss Spectroscopy

10.3.1. General Considerations

Electron energy loss spectroscopy of all vibration spectroscopies displays the greatest versatility with resolution nowadays down to 1.3 meV and an accessible frequency range that encompasses the far-infrared as well as the visible or even UV range. By introducing suitable scattering conditions, electron energy loss spectroscopy is sensitive to all types of vibrational modes. One may also use selection rules and thereby gain symmetry information. Finally, electrons also carry momentum and therefore the dispersion of vibrational modes is amenable to investigation.

In this section we examine some ideas about the proper design of electron spectrometers. Before proceeding further into the details it is important to realize what the design goals are for suitable instruments. Electron spectrometers should specular reflection. The angular resolution of the instrument should be adjusted to the desired momentum space that one wishes to explore, although usually it is more difficult to have a sufficiently large enough momentum space rather than having insufficient angular resolution. If one wishes to study localized vibrations as well as surface phonon dispersion, the instrument should feature a large range of impact energies at the sample ranging from 1–200 eV. Aside from these general features an obvious design goal is to have the maximum signal and the highest possible resolution. Unfortunately one has a certain universal power law linking the energy spread ΔE and the signal j_D at the detector, namely

$$j_D \sim \Delta E^n \tag{10.3.1}$$

where $n = 3\text{-}4$. This universal power law frequently sets a practical limit to the resolution with which one can work. Thus for weak scattering processes, such as interaction with surface phonons, one is usually confined to about 4 meV resolution while for localized vibrations 2 meV is feasible.

10.3.2. Cathode System

The task of the cathode system is to feed the monochromator with a beam subject to the following requirements.

1. One needs to have a homogeneous illumination of the entrance slit.
2. The cathode system should provide a horizontal focus at the slit. The focusing properties of the monochromator in that plane will then further transport the beam appropriately.
3. In the vertical plane one wishes to have a moderately converging beam so as to form a vertical image of the cathode somewhere between halfway in the monochromator and infinity. This different focusing is required, since the cylindrical condensers used as monochromators do not focus along the vertical plane.
4. The angular aperture in the horizontal plane should be confined to a few degrees. In order to fulfil these requirements a rather unusual shaping of the lens elements in the cathode system is necessary.

10.3.3. Cylindrical Deflectors

This section provides a brief account of the trajectories in the cylindrical deflector. Trajectories are best described by introducing cylindrical coordinates r, Θ. The Lagrangian then reads

$$L = m(\dot{r}^2 + r^2\dot{\Theta}^2)/2 + C \ln r \tag{10.3.2}$$

with m the electron mass and C a constant which determines the path energy E_0 of the device. The radial deviation from the center path is given by

$$r = r_0(1 + \rho) \tag{10.3.3}$$

and, after some algebra, the trajectory equation is obtained in the form

$$\rho'' + 2\rho - \frac{\Delta E}{E_0} = 2\rho'^2 - \rho^2 + \alpha_1^2 \tag{10.3.4}$$

Here quantity ρ' denotes the derivative of the reduced radial coordinate with respect to the deflection angle Θ and α_1 is the entrance angle measured with respect to the central path. This equation has a first-order solution

$$\rho = \frac{1}{\sqrt{2}} \alpha_1 \sin \sqrt{2}\,\Theta + \rho_1 \cos \sqrt{2}\,\Theta + \frac{1}{2}\frac{\Delta E}{E_0}(1 - \cos \sqrt{2}\,\Theta) \tag{10.3.5}$$

A first-order focus is achieved when the α_1 term vanishes, that is, at an angle

$$\Theta_f = \pi/\sqrt{2} = 127.28° \qquad (10.3.6)$$

At the position of the first-order focus one has an image equation

$$\rho_2 = -\rho_1 + \frac{\Delta E}{E_0} - \tfrac{4}{3}\alpha_1^2 \qquad (10.3.7)$$

where the term containing α_1^2 constitutes an angular aberration term originating from the right-hand side of equation (10.3.4). It may be noteworthy that the second-order aberration term appearing in equation (10.3.7) has no analog in the optics of electron or light lenses, because the second-order term there vanishes for symmetry reasons. Trajectories as well as the image equations are illustrated in Figure 10.4, which represents the result of a computer simulation. Equation (10.3.7)

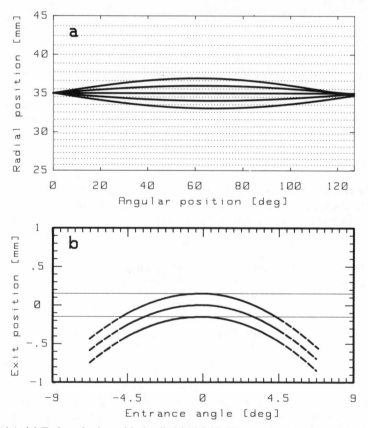

Figure 10.4. (a) Trajectories in an ideal cylindrical field. The angular coordinate Θ is displayed as a cartesian x-axis. (b) Position at the exit slit as a function of the entrance angle α_1 showing the second-order angular aberration [see equation (10.3.7)].

can be employed to easily calculate the maximum and minimum energy that the system will transport and the equation obtained for the base width of the transmitted beam reads

$$\Delta E_B / E_0 = 2s/r_0 + \tfrac{4}{3}\alpha_1^2 \tag{10.3.8}$$

with s the width of the entrance and exit slit. For the purpose of using cylindrical monochromators under conditions of space charge it is important to find out about the effect of space charge on the trajectories. This is a nontrivial issue that can be attacked only by relatively involved computer simulations. Within certain limits, however, the problem can also be approximately treated analytically. If one considers an ideal cylindrical field fed by a beam subject to the conditions

$$\Delta E_{in} / E_0 \ll \alpha_1, s/r_0 \tag{10.3.9}$$

then it becomes relatively straightforward to consider the space charge produced by such a feed beam and add an additional term to equation (10.3.4) that represents the space-charge effect. Such an analysis* yields that a prime effect of the space charge is to extend the focal length of the device which then becomes current dependent. While this is relatively straightforward to see even by visual inspection of a bundle of trajectories such as displayed in Figure 10.4, it seems that this idea has never been expressed before nor exploited. If one carries out the analytical calculation one obtains a relatively simple formula for the maximum input current as a function of the extension of the deflecting angle $\Delta\Theta_f$,

$$I_{input,max} = 4hkE_0^{3/2}\alpha_{1m}\Delta\Theta_f / r_0 \tag{10.3.10}$$

Here $k = 5.25 \times 10^{-6}$ A eV$^{-3/2}$ is a universal constant. By using equation (10.3.8) and noting that the maximum angle which still transmits electrons at the path energy is

$$\alpha_{1m} = \sqrt{\frac{3s}{4r_0}} \tag{10.3.11}$$

we can replace the path energy E_0 by the energy resolution and obtain for the monochromatic current

$$I_{mon} = I_{input}\Delta E_B / \Delta E_{in} \quad \text{and} \quad I_{mon} \approx \frac{2}{3}\frac{h}{s}k\frac{\Delta E_B^{5/2}}{\Delta E_{in}}\Delta\Theta_f \tag{10.3.12}$$

where ΔE_{in} is the energy width of the feed beam of the monochromator.

This equation seems to suggest that the deflection angle should be substantially extended for a monochromator. We must, however, keep in mind that our consideration has been confined strictly to two dimensions, and we did not examine the

---

* The author plans to give a detailed account of these considerations together with details on the experimental layout of optimized spectrometers in the near future (to appear in the Springer Series in Optical Sciences, 1990).

divergence of the beam in the vertical direction. It is in fact relatively straightforward to show that one loses transmission of the device owing to that beam divergence. Thus a practical extension is confined to 10°–20°. We note that cylindrical deflectors not featuring such an extension of the deflection angle can also produce monochromatic currents of significant amount. There, however, one uses a bundle of trajectories nonsymmetrical around the radial path. Such an approach leads to a current condition where the second-order angular aberration term and the space charge term in the imaging equations cancel. If one works with an instrument that is not specificly designed for space charge and tunes for optimum monochromatic current, then such operating conditions will be satisfied automatically.

10.3.4. Lens Systems and Performance of Spectrometers

The lens system between the monochromator and the sample, and the sample and the analyzer is of great importance for the performance of spectrometers. Since lens systems should properly account for the fact that the cylindrical deflector

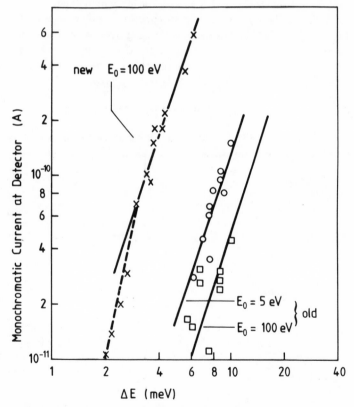

Figure 10.5. Comparison of the current at the detector for an older design and a new one with lenses calculated according to the procedures described in the text.

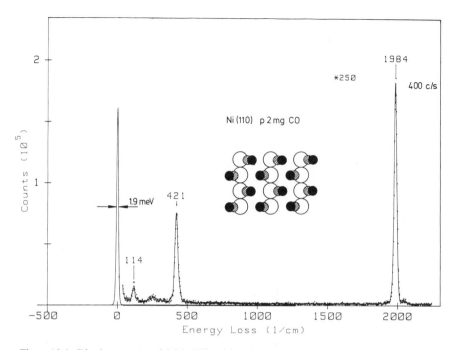

Figure 10.6. Dipole spectrum of CO in Ni(110) forming the p2mg structure. Since the CO molecule is tilted off the normal direction, a third mode, the hindered translation at 114 cm^{-1}, becomes dipole active. Despite the resolution of 15 cm^{-1} (or 1.85 meV) the count rate in the inelastic channels is quite substantial.

focuses only in one plane, radial symmetric lens systems* are excluded. Unfortunately, calculation of noncentrosymmetric lenses requires a fully three-dimensional solution of the Laplace equation as well as three-dimensional trajectory calculations. In order to perform such calculations it is necessary to use the remaining symmetry of the lenses (which is C_{2v}) and solve the Laplace equation on a relatively corse mesh, since otherwise one rapidly exceeds memory space. We have typically used a mesh of 0.5–1 mm length. For sufficient accuracy of the trajectory calculations one needs to employ dynamic interpolation schemes, which calculate the local field within the mesh effectively and rapidly. Optimized lenses using such calculation schemes show considerable improvement over previous designs (Figure 10.5). A spectrum that was obtained with such a spectrometer is shown in Figure 10.6. It refers to the p2mg structure of CO on Ni(110). The resolution in the spectrum is 1.9 meV.† Inspection of the spectrum shows that one is beginning to see the effect of the natural linewidths on the CO vibrations. Despite the high resolution one still has quite substantial count rates in the vibrational losses. The very low vibrational frequencies, at 114 cm^{-1}, arise from the hindered translation

* International patents are pending.
† In the meantime even a resolution of 1.3 meV has been achieved.

of the molecule. This mode becomes dipole active because of the canted position assumed by the carbon monoxide molecules on this surface.

10.4. Electron–Solid Interaction

10.4.1. Dipole Scattering—Dielectric Theory

In this section we investigate the interaction of electrons with the dipolar electric fields of elementary excitations at surfaces. The treatment is not confined to vibrational spectroscopy but applies equally to other elementary excitations such as electronic transitions and plasma excitations in the limit of small momentum transfer. In considering dipole scattering one basically has the choice of several approaches to the problem. One may either consider the Hamiltonian of the free electron and treat the dipole fields of the elementary excitations as a perturbation to this Hamiltonian. The advantage of this treatment is that it delivers the scattering kinematics, that is, energy and momentum conservation. The disadvantage, however, is that one has to focus on a particular elementary excitation. The second approach involves consideration of the Hamiltonian of the elementary excitation while the interaction with the electron is treated as a perturbation. As a result of the small momentum transfer involved in dipolar excitations, the perturbation can be treated as if the electron remains on a classical trajectory. This treatment easily provides for multiple losses. The disadvantage is that energy and momentum conservation have to be introduced *ad hoc*. A third alternative is also a classical derivation usually described as the dielectric theory. The advantage is that the formulation is independent of the special elementary excitation. The solid is simply described by a local or nonlocal dielectric function. The disadvantage of the latter treatment is that the quantum condition must be introduced *ad hoc*. All three derivations provide identical final equations in the realm of their applicability.

In the following we give the classical derivation. In classical electrodynamics the total energy loss is

$$W = \frac{1}{4\pi} \int_{-\infty}^{\infty} dt \int d\mathbf{r} \, \mathscr{F}(\mathbf{r}, t) \cdot \dot{\mathbf{D}}(\mathbf{r}, t) \tag{10.4.1}$$

which we equate with the loss probability through the equation

$$W = \int \hbar\omega P(\mathbf{q}_\parallel, \omega) \, d\mathbf{q}_\parallel \, d\hbar\omega \tag{10.4.2}$$

For surface losses one uses the two-dimensional Fourier expansion

$$\mathscr{F}_i(\mathbf{r}, t) = \int d\omega \, d\mathbf{q}_\parallel \, e^{-i\omega t} \, e^{i\mathbf{q}_\parallel \cdot \mathbf{r}_\parallel} \, e^{-q_\parallel |z|} \mathbf{P}_i(\mathbf{q}_\parallel, \omega)$$

$$= W = 2\pi^2 \int dz \, d\omega \, d\mathbf{q}_\parallel \, \omega \varepsilon_2(\omega) |\mathscr{F}_i(\omega, \mathbf{q}_\parallel, z)|^2 \, e^{-2q_\parallel |z|} \tag{10.4.3}$$

For a semi-infinite dielectric half-space the internal field $\mathscr{F}_i$ is screened by a factor satisfying

$$\mathscr{F}_i = \frac{2}{\varepsilon(\omega, \mathbf{q}_{\parallel}) + 1} \mathscr{F}_{ext} \tag{10.4.4}$$

where $\mathscr{F}_{ext}$ is the external field created by the electron.

It follows immediately that the dielectric losses are proportional to

$$\rho(\omega) \sim \frac{\varepsilon_2}{|\varepsilon + 1|^2} = \mathrm{Im} \frac{-1}{\varepsilon(\omega, \mathbf{q}_{\parallel}) + 1} \tag{10.4.5}$$

When $\varepsilon(\omega)$ is as for a free electron gas solid,

$$\varepsilon(\omega) = 1 - \frac{\omega_p^2}{\omega^2 + i\omega\tau} \qquad \text{with } \frac{1}{\tau} \ll \omega_p \tag{10.4.6}$$

ω_p being the plasmon frequency, then

$$\mathrm{Im} \frac{-1}{\varepsilon(\omega) + 1} = \frac{\pi\hbar\omega_{sp}}{4} \delta(\hbar\omega - \hbar\omega_{sp}) \tag{10.4.7}$$

where $\omega_{sp} = (1/\sqrt{2})\omega_p$ is the surface plasmon frequency. Thus we learn that the surface plasmons and all the so-called surface excitations are excited while the electron is still outside the material. Bulk plasmons on the contrary are excited while the electron is inside!

The case of a thin dielectric layer on top of a nonabsorbing substrate is now examined. The dielectric constants of the layer and substrate are denoted by ε_s and ε_b, respectively. If the layer is thin so that $|dq_{\parallel}\varepsilon_s| \ll \varepsilon_b$, the potential outside the material is as if there were no extra layer on top of the semi-infinite substrate (Figure 10.7):

$$\varphi_a = e\left(\frac{1}{r} + \frac{1 - \varepsilon_b}{1 + \varepsilon_b} \frac{1}{r'}\right) \tag{10.4.8}$$

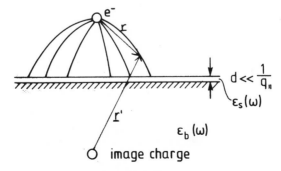

Figure 10.7. Dielectric losses in a thin surface layer on a nonabsorbing substrate.

We calculate the vertical electric displacement $D_{\nearrow}$ and the parallel electric field $\mathscr{F}_\parallel$ at the surface:

$$D_\perp\big|_{z=0} = -\frac{2\varepsilon_b}{1+\varepsilon_b} \nabla_z \frac{e}{r}\bigg|_{z=0} \quad \text{and} \quad \mathscr{F}_\parallel\big|_{a=0} = -\frac{2}{\varepsilon_b+1} \nabla_\parallel \frac{e}{r}\bigg|_{z=0} \quad (10.4.9)$$

We note that

$$\frac{\partial}{\partial z}\frac{1}{r}\bigg|_{z=0} = -\frac{\partial}{\partial z}\frac{1}{r'}\bigg|_{z=0} \quad \text{and} \quad \frac{\partial}{\partial x}\frac{1}{r}\bigg|_{z=0} = +\frac{\partial}{\partial x}\frac{1}{r'}\bigg|_{z=0} \quad (10.4.10)$$

The boundary conditions

$$\mathscr{F}_{i\parallel} = \mathscr{F}_{\text{ext}\parallel} \quad \text{and} \quad D_{i\perp} = D_{\text{ext}\perp} \quad (10.4.11)$$

are now employed to split the term under the integral sign in equation (10.4.1) using

$$\mathscr{F}_i \cdot \dot{\mathbf{D}}_i = \varepsilon_s \mathscr{F}_{\text{ext}\parallel} \dot{\mathscr{F}}_{\text{ext}\parallel} + \frac{1}{\varepsilon_s} \mathbf{D}_{\text{ext}\perp} \dot{\mathbf{D}}_{\text{ext}\perp} \quad (10.4.12)$$

where ε_s is the dielectric constant of the layer. The losses of the layer are then given by

$$W = d8\pi^2 \int \omega\, d\omega\, d\mathbf{q}_\parallel \left[\frac{\text{Im}\,\varepsilon_s}{(\varepsilon_s+1)^2} |\mathscr{F}_{\text{ext}\parallel}(\omega, \mathbf{q}_\parallel)|^2 - \frac{\varepsilon_b^2}{(\varepsilon_b+1)^2} \text{Im}\left(\frac{1}{\varepsilon_s}\right) |\mathscr{F}_{\text{ext}\perp}(\omega, q_\parallel)|^2 \right] \quad (10.4.13)$$

If one has a dilute layer of molecules on a metal surface, then only the second term survives and we may set

$$d \cdot \text{Im}\frac{1}{\varepsilon_s} = 4\pi n_s \, \text{Im}\, \alpha(\omega) \quad (10.4.14)$$

where $\alpha(\omega)$ is the polarizibility of the molecules and n_s is the surface density.

The Fourier components are now calculated by invoking

$$\frac{1}{r} = \frac{1}{2\pi} \int d\mathbf{q}_\parallel \frac{1}{q_\parallel} e^{-q_\parallel|z|} e^{-i\mathbf{q}_\parallel \cdot \mathbf{r}_\parallel} \quad (10.4.15)$$

in which case

$$-\nabla\frac{1}{r} = \frac{1}{2\pi} \int d\mathbf{q}_\parallel \, (ie_\parallel, 1)\, e^{-q_\parallel|z|} e^{-i\mathbf{q}_\parallel \cdot \mathbf{r}_\parallel} \quad (10.4.16)$$

When an electron is reflected at the surface, the vector $\mathbf{r}(t)$ must be replaced by (Figure 10.8)

$$\mathbf{r}(t) = \mathbf{r} + v_{\parallel}\mathbf{e}_{\parallel}t + v_{\perp}|t|\mathbf{e}_{\perp} \qquad (10.4.17)$$

where $\mathbf{r}$ denotes the position on the surface with respect to the point of reflection. Hence

$$\mathscr{F}_{\text{ext}}(\mathbf{r}, t) = -\nabla \frac{e}{|\mathbf{r} + v_{\parallel}\mathbf{e}_{\parallel}t + v_{\perp}|t|\mathbf{e}_{\perp}|}$$

$$= \frac{1}{2\pi} \int d\mathbf{q}_{\parallel} \, (i\mathbf{e}_{\parallel}, 1) \, e^{-i[\mathbf{q}_{\parallel} \cdot (\mathbf{r}_{\parallel} + \mathbf{v}_{\parallel}t)]} \, e^{-q_{\parallel}(z + v_{\perp}|t|)}, \qquad z > 0 \qquad (10.4.18)$$

Fourier expansion of the kernel yields

$$e^{-i\mathbf{q}_{\parallel} \cdot \mathbf{r}_{\parallel}t - q_{\parallel}v_{\perp}|t|} = \frac{1}{2\pi} \int d\omega \, e^{-i\omega t} \frac{2q_{\parallel}v_{\perp}}{(\omega - \mathbf{q}_{\parallel} \cdot \mathbf{v}_{\parallel})^2 + q_{\parallel}^2 v_{\perp}^2}$$

and

$$\mathscr{F}_{\text{ext}}(\omega, \mathbf{q}_{\parallel}) = (i\mathbf{e}_{\parallel}, 1) \frac{e}{(2\pi)^3} \frac{2q_{\parallel}v_{\perp}}{(\omega - \mathbf{q}_{\parallel} \cdot \mathbf{v}_{\parallel})^2 + q_{\parallel}^2 v_{\perp}^2} \qquad (10.4.19)$$

from which the loss probability is derived in the form

$$P(\omega, q_{\parallel}) = \frac{4e^2 n_s}{\pi^3} \frac{q_{\parallel}^2 v_{\perp}^2}{((\omega - \mathbf{q}_{\parallel} \cdot \mathbf{v}_{\parallel})^2 + q_{\parallel}^2 v_{\perp}^2)^2} \, \text{Im} \, \alpha_s(\omega) \qquad (10.4.20)$$

This loss probability possesses a sharp resonance when the "surfing condition" is matched,

$$\omega/q_{\parallel} = v_{\parallel} \qquad (10.4.21)$$

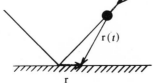

Figure 10.8. Illustration of the position vector of the electron.

If realistic numbers are inserted into this equation one finds that $q_\parallel$ is rather small. Dipole scattering is therefore confined to the vicinity of the $\bar{\Gamma}$ point in the surface Brillouin zone. As a result of momentum conservation, dipole losses are therefore also confined to a small angular cone around the specular reflected beam or around diffracted beams. We also see from the above derivation that only dipole moments oriented perpendicularly to the surface contribute to the losses when the surface is a metal. Thus, in that case, we have the same type of selection rule as for infrared reflection absorption spectroscopy.

10.4.2. Electron–Ion Core Scattering (Inelastic Electron Diffraction)

Even for a well-ordered surface considerable intensity is scattered in between the diffracted beam. This is inelastic diffuse scattering caused by the electron-phonon interaction. The total amount of this diffuse scattering is equivalent to the reduction of the intensities in diffracted beams, namely

$$I_{hk}(T) = I_{hk}^0 \exp \left\{ -\sum_{q\omega} \tfrac{1}{2} \langle [(\mathbf{k}_i - \mathbf{k}_f) \cdot \mathbf{u}(\mathbf{q}, \omega)]^2 \rangle_T \right\} \qquad (10.4.22)$$

The interaction between electron and solid is mediated via the crystal potential $V(\mathbf{r}_{el}, \{\mathbf{R}\})$, which depends *parametrically* on the coordinates $\mathbf{R}$ of all nuclei. In the following we describe briefly the theory of electron phonon scattering as developed by Tong et al.[5]

The initial and final states of the electrons and solid are described by

$$|i\rangle = |k_i\rangle |n_i(\mathbf{Q}_{\parallel_1}, j_1) n_i(\mathbf{Q}_{\parallel_2}, j_2) \cdots \rangle \qquad (10.4.23)$$

and

$$|f\rangle = |k_j\rangle |n_f(\mathbf{Q}_{\parallel_1}, j_1) n_f(\mathbf{Q}_{\parallel_2}, j_2) \cdots \rangle \qquad (10.4.24)$$

where $\mathbf{Q}_\parallel, j$ denotes the parallel momentum of the phonon and j the branch. With this notation it is assumed that the substrate is a slab of finite thickness.

The matrix element for the transition from $|i\rangle$ to $|f\rangle$ is

$$M = \langle f | T(E_i) | i \rangle \qquad (10.4.25)$$

where $T(E_i)$ is the transfer operator,

$$T(E) = V(\{\mathbf{R}\}) + V(\{\mathbf{R}\}) G(E) T(E) \qquad (10.4.26)$$

and $G(E)$ is the free electron propagator,

$$G(E) = \frac{1}{E - H_0 + i0} \quad \text{with } H_0 = \frac{\mathbf{p}^2}{2m} \qquad (10.4.27)$$

Function $T(E)$ depends also parametrically on the positions of the ion cores $\{\mathbf{R}\}$. The matrix element may therefore be expanded around the average atom positions, and thus also the scattering amplitude, which is proportional to the matrix element,

$$f(\mathbf{k}_i, \mathbf{k}_f, \{\mathbf{R}\}) \sim M \qquad (10.4.28)$$

$$f(\mathbf{k}_i, \mathbf{k}_f, \{\mathbf{R}\}) = f(\mathbf{k}_i, \mathbf{k}_f, \{\mathbf{R}^0\}) + \sum_{l_\parallel l_z \kappa j \alpha} \left(\frac{\partial f}{\partial R_{j\alpha}(l_\parallel l_z \kappa)} \right) u_\alpha(l_\parallel l_z \kappa) \quad (10.4.29)$$

The first term in the expansion describes the elastic scattering and the second term, the one-phonon scattering process. Quantity $l_\parallel$ in equation (10.4.29) labels the surface unit cell, l_z the layer, κ the atom in the unit cell, and α the cartesian coordinate. One has a kind of Bloch theorem for the derivatives:

$$\frac{\partial f}{\partial R_j(l_\parallel l_z \kappa)} = \frac{\partial f}{\partial R_j(0 l_z \kappa)} \exp\left[i R^0(l_\parallel l_z \kappa) \cdot \Delta \mathbf{k}_\parallel \right] \qquad (10.4.30)$$

As an illustration we conduct a kinematic approximation for a simple cubic lattice:

$$f = f_0 \sum_{l_\parallel l_z} \exp\left[i R(l_\parallel l_z) \cdot \Delta \mathbf{k} \right] \qquad (10.4.31)$$

hence

$$\frac{\partial f}{\partial R_j(0 l_z)} \bigg|_0 = f_0 i \Delta \mathbf{k}_i \exp\left[i \Delta \mathbf{k} \cdot \mathbf{R}^0(0 l_z) \right] \qquad (10.4.32)$$

Therefore a product $\mathbf{u} \cdot \Delta \mathbf{k}$ appears in the cross section. When the crystal has a mirror plane, the modes divide into even and odd modes. Odd modes are polarized perpendicular to the σ-plane when the plane runs through an atom. When Δk is aligned with σ, the cross section vanishes. This constitutes an important *selection rule* for electron–core scattering. This selection rule can be derived quite generally using time-reversal symmetry. It therefore applies also for multiple scattering.

The kinematic approach is not an adequate representation of the scattering problem. In fact it would be rather misleading to compare cross sections with the kinematic theory. Figure 10.9 presents a comparison of the calculated cross section in the kinematic approach (dotted line) and in the full dynamical theory. The eigenvectors of these phonons are also illustrated in Figure 10.9 for two particular phonons of the Ni(100) surface. It is seen that, owing to multiple scattering, one has rather rapid oscillations of the cross section as a function of the energy. These

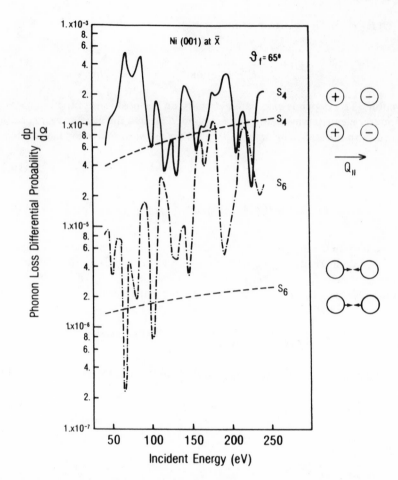

Figure 10.9. Cross section for inelastic electron scattering from the S_4 and S_6 phonon of Ni(100). The dotted line represents the result of a kinematic calculation (after Mu-Liang Xu *et al.*[6]).

rapid oscillations allow one to detect phonons that would be kinematically invisible. Frequently, the oscillations in the cross section also allow for a separation of phonons, which could not be resolved by the energy resolution of the spectrometer. An example is shown in Figure 10.10, which refers to the Ni(110) surface.

10.4.3. Resonance Scattering

It is well known from scattering of electrons with gas-phase molecules that the electron can form short-lived negative ion states with the molecules which manifest their presence by a resonance in the cross section as a function of the kinetic energy of the electron. It is a characteristic of this resonance scattering that the electron is captured for a brief period within the molecule so as to lose the information on the scattering parameters of the incident beam. The emission angle of the scattered electron is thus entirely defined by the orbital structure of the resonance. This effect can be used to determine the orientation of the weakly absorbed molecules and surfaces. An example has recently been given by Palmer et al.[8] Resonance scattering is also subject to particular selection rules in terms of the vibrational losses which are excited. These selection rules can be derived by considering the fact that the extra electron in a shape resonance gives rise to

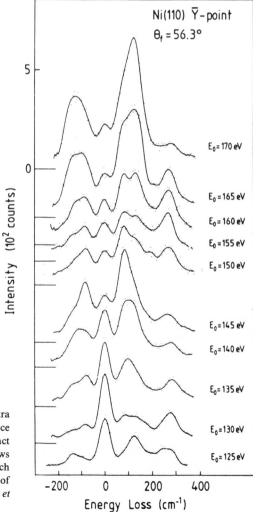

Figure 10.10. A set of phonon spectra taken at the same point in the surface Brillouin zone, but at different impact energies. A set of spectra like this allows for the separation of phonons, which could not be resolved by the resolution of the instrument alone (after Lehwald et al.[7]).

a short force pulse on the molecule. This force pulse has the symmetry of the charge density of the extra electron in the orbit. It can be shown that the force field, and thus the vibrations excited, are totally symmetric when the wave function of the electron belongs to a nondegenerate representation. Degenerate presentations are dealt with elsewhere.[9]

10.5. Localized Vibrations at Surfaces

10.5.1. Vibrations and the Chemical Bond

In the case of adsorbed molecules on surfaces it is useful to distinguish between localized vibrations and extended vibrations (surface phonons). For adsorbed molecules the frequency of internal molecular vibrations is typically much higher than the substrate vibrations. This means that with respect to the inner molecular vibrations the substrate is essentially rigid. To the extent that one may neglect direct coupling between the molecules (which will always be the case for a dilute overlayer), one may regard these molecular vibrations as being localized on the individual molecule. We have already noted that vibrations probe the interatomic

Table 10.2. Carbon–Carbon Bond Stretching Frequencies

Molecule	$\nu(C-C)$	$\nu(C=C)$	$\nu(C\equiv C)$
Ethane CH_3-CH_3	995		
Ethylene $CH_2=CH_2$		1623	
Acetylene $CH\equiv CH$			1974
Propane $CH_3-CH_2-CH_3$	$\left.\begin{matrix}869\\1054\end{matrix}\right\}962$		
Propylene $CH_2-CH=CH_2$	919	1647	
Allene $CH_2=C=CH_2$		$\left.\begin{matrix}1071\\1956\end{matrix}\right\}1513$	
Methylacetylene $CH_3-C\equiv CH$	931		2142
Butane $CH_3-CH_2-CH_2-CH_3$ (trans)	$\left.\begin{matrix}837\\1009\\1059\end{matrix}\right\}968$		
1,3-Butadiene $CH_3=CH=CH=CH_2$	1196	$\left.\begin{matrix}1596\\1630\end{matrix}\right\}1613$	
2-Butyne $CH_3-C\equiv C-CH_3$	$\left.\begin{matrix}725\\1152\end{matrix}\right\}1024$		
Butadiyne $CH\equiv C-C\equiv CH$	874		$\left.\begin{matrix}2020\\2184\end{matrix}\right\}2102$
Average frequency	~950	~1600	~2100

forces. In general, however, all atoms within the molecule participate in a particular vibration and thus make some contribution to a frequency. Nevertheless, one is often justified in associating a particular vibrational frequency with a particlar bond in the molecule, because the vibration is mainly localized at the two atoms which form the bond. This is immediately evident for the stretching vibrations of hydrogen within a hydrocarbon molecule which are practically entirely localized on the hydrogen atoms. By and large it also holds for the carbon–carbon stretching vibration. This situation is illustrated by Table 10.2, where the frequencies of carbon–carbon stretching vibrations of various molecules are tabulated. With the exception of the highly symmetric molecule allene, all the carbon–carbon frequencies fall into a certain realm characteristic for the carbon–carbon bond strength. For the symmetric molecule allene, one has a strong splitting between symmetric and antisymmetric combinations of the two carbon–carbon stretching frequencies which split the carbon modes to an extent that the vibrational frequencies fall into the realm of single-bonded carbon and triply-bonded carbon. Such highly symmetric situations, however, occur rarely on surfaces where, typically, the symmetry is additionally broken by the presence of the surface. Therefore one may safely regard the carbon–carbon stretching frequency of a hydrocarbon on a surface as a good indicator of the bond character. Along with the bond character goes the bond distance, as indicated in Figure 10.11. Observation of a bond stretching frequency therefore indicates also the parameters of the structure of a molecule on the surface. The results of surface vibration spectroscopy hence provide for important constraints on the number of different structures one may discuss for a particular molecule on a surface. While such qualitative arguments are of importance, one should however bear in mind that vibration spectroscopy is never a structural tool. Unfortunately, vibration spectroscopy and the analysis of vibrational spectra with some force constant models have sometimes been abused in order to determine structural parameters. Vibration spectroscopy should be used to obtain qualitative insight into a particular structure, while a quantitative determination of structural parameters should be left to genuine structural tools. Extremely powerful arguments about a structure also arise from symmetry considerations, which we discuss next.

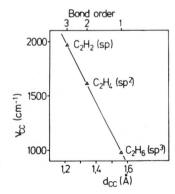

Figure 10.11. Correlation between the carbon–carbon bond stretching frequency and bond length or band order for hydrocarbons.

10.5.2. Selection Rules and Local Symmetry

We have already studied two selection rules in electron energy loss spectroscopy, one applicable to dipole scattering and the other to impact scattering. For dipole scattering we have seen that EELS is only sensitive to those modes which allow for a perpendicular dipole moment, at least on a metal surface. This is equivalent to saying that the vibrational mode should belong to the totally symmetric representation of the local point group. This argument applies to excitation from the ground state to the first excited state, which is the typical situation. For impact scattering we have seen that there is sensitivity only to even modes when the scattering plane is aligned with a mirror plane of the system, while no selection rule exists for other geometries. The full group theoretical analysis of a local vibration at surfaces has been discussed in detail before.[10] Here we only present two examples of recent studies. The first example refers to the dipole selection rule with the $p2mg$ structure of CO on Ni(110). The corresponding spectrum has already been shown in Figure 10.6, where it was noted that the spectrum displays three vibrational losses. Only two of them would be dipole active (the 420 cm^{-1} and 1985 cm^{-1} modes, which are the metal–carbon and CO stretching modes) if

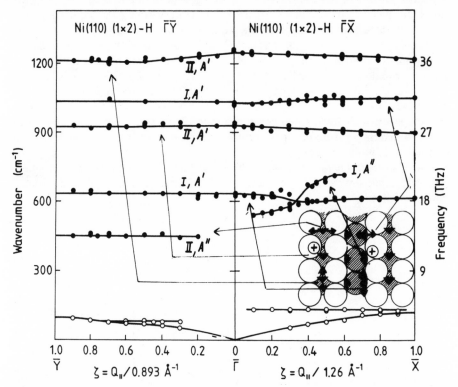

Figure 10.12. Dispersion of the hydrogen modes on the Ni(110) 1×2 reconstructed surface (after Ibach *et al.*[11]).

CO were sitting in an on-top position with a local symmetry of C_{2v} or in a bridging position with the same symmetry. We know, however, from the frequency of the CO stretching vibration that the site should be an on-top site.[10] The appearance of the mode at 115 cm^{-1}, which is the frustrated translation of the molecule at the surface, tells us that the symmetry is broken down to C_s.

The other example refers to hydrogen on Ni(110) at 1.5 monolayer coverage, where the surface has reconstructed to form a (1×2) reconstructed phase. The construction is assumed to be of the pairing-row type. Figure 10.12 shows the complete set of hydrogen vibrations along the two high-symmetry directions. From the absence of the A″ modes along the $\bar{\Gamma}\bar{Y}$ and $\bar{\Gamma}\bar{X}$ directions, respectively, it was determined that the two hydrogen species sit in threefold sites, but with mutual perpendicular orientation of the symmetry planes as indicated in the insert in the figure. A more detailed account of this analysis can be found elsewhere.[11] Here we merely note that the threefold site for the hydrogen within the paired row of nickel atoms has its mirror plane apparently along the $\bar{\Gamma}\bar{Y}$ direction ([110] direction). The symmetry of that site is such that it is only provided by a pairing-row reconstruction, not by a reconstruction of the missing-row type. This renders further strong evidence for the pairing-row as opposed to the missing-row reconstruction. These symmetry arguments are among the strongest that can be made about structure on surfaces, and so far they have not failed to provide a correct clue as to the nature of surface structures. A prerequisite for such analyses is, however, that one has performed a complete study of all the modes at different impact energies, to be absolutely certain that a particular mode has not been missed just because the cross section happened to be small at the experimental scattering parameters.

10.6. Surface Phonons: Force Constant Models

In Section 10.5.2 it was shown in the case of hydrogen that when localized vibrators are close to each other vibrational levels will interact; this is equivalent to saying that the mode acquires some dispersion with $q_\parallel$ when the surface is ordered and $q_\parallel$ is a good quantum number. A case of rather strong coupling is provided by the surface atoms themselves, which display a number of phonons localized near the surface (surface phonons), or so-called resonances when these surface phonons are not strictly localized to the surface, but couple to bulk phonons. The traditional instrument to relate such surface dispersion curves to interatomic forces is the theory of lattice dynamics.[12] In order to extract information on the force constant from the phonon dispersion curves, one must introduce some assumptions about the type and range of the force field. In general there is no uniqueness in this procedure, and different force fields may fit the data equally well. Physical intuition as well as guidance by total energy calculation are required. For some materials, such as copper and nickel, a relatively simple force field suffices to describe the short-range part of the phonon spectrum. An example is shown in Figure 10.13, where the neutron data for copper bulk phonon dispersion are fitted with a central force constant model between nearest neighbors, i.e., with

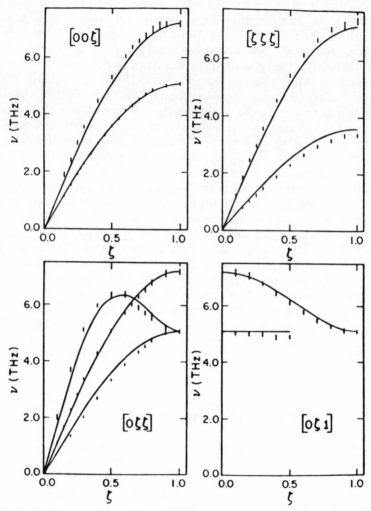

Figure 10.13. Fit of the experimental data of bulk phonon dispersion curves to a nearest-neighbor central force constant model (after Hall and Mills[13]).

a single parameter.[13] The overall agreement is quite good. However, the model does not give the correct account of the long-range part of the phonon spectrum, that is, of long-range correlations in the vibrations, since the elastic constants are not well reproduced. Surface phonons, however, are essentially of interest when they are localized to the surface and an accurate fit of the acoustic limit is less important. For materials like copper and nickel, it is therefore a useful strategy to start with this central force constant model and see what modifications are needed near the surface. The origin of such modifications lies in the rearrangement of the charge near the free surface as illustrated in Figure 10.14. Increased charge density

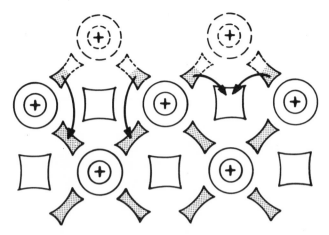

Figure 10.14. Illustration of the rearrangement of the electronic charge near a free surface.

between the first and second layer will give rise to the typically observed contraction of the interplanar spacing between first- and second-layer atoms. Increased charge density between the surface atoms gives rise to surface stress. In terms of the nearest-neighbor model of the force field near the surface, the surface stress is equivalent to introducing a first derivative of the potential between surface atoms. With such a surface stress one still has equilibrium for each atom, but nevertheless the surface as a whole is under stress. The only first-principles calculation of a surface stress was performed by Needs for aluminum.[14] The effect of surface stress is to raise the frequencies of vibrations polarized vertically to the orientation of the stress in the case of a tensile stress (like tuning a string in a violin). A complete analysis of the surface phonon spectrum on the (110) surfaces of nickel and copper was performed recently and the reader is referred to the relevant papers for further information.[7,15] Surface stress has also been held responsible for some of the adsorbate-induced reconstructions that have been observed.[16,17] Furthermore, surface stress appears to be important for an interesting reconstruction phenomenon that occurs with an epitaxial metal overlayer, to be examined in the next section.

10.7. Epitaxial Metal Overlayers

10.7.1. General Remarks

Epitaxial growth of metals on metals has received increased interest lately owing to the unusual properties of such epitaxially grown films. First of all, epitaxial growth offers the possibility of making perfect crystal surfaces with a low impurity level. In thin films quantum-size effects have been observed. Furthermore, one has the opportunity to produce metastable phases of material with unusual electronic, mechanical, and magnetic properties. Detailed studies also offer a unique possibility

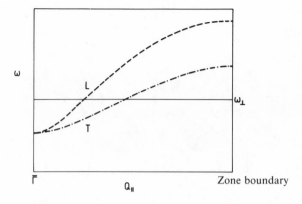

Figure 10.15. Schematic drawing of the dispersion curves of an adlayer with central force coupling in the limit of a rigid substrate.

for understanding the condition of epitaxial growth. We have recently studied the structure and phonon dynamics of Ag(111) overlayers on Ni(100) and Cu(100), as well as Fe(100) and Ni(100) overlayers on Cu(100) substrated.[18-20] In the following we discuss some principal features of the extra vibrations introduced by a monolayer metal film.

As a result of the extra three degrees of freedom for each adlayer atom, one has three additional branches. For a nearest-neighbor central force constant model, and in the limit that one has a flat adlayer, the perpendicularly polarized vibration has a flat dispersion branch (Figure 10.15), while the two parallel-polarized branches may display some dispersion. The extra dispersion branches introduced by the adlayer may lie below the maximum frequency of the Rayleigh wave, above it, or even above the entire bulk phonon spectrum. Figure 10.16 depicts the situation for the first case, which is characteristic of silver overlayers on copper and nickel. When the Rayleigh mode crosses the adlayer mode, one has a coupling between the two branches leading to coupling dispersion, which in some cases may extend over the entire Brillouin zone. This behavior was observed with silver overlayers.[18]

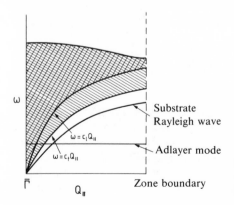

Figure 10.16. Perpendicularly polarized branch of an adlayer with frequency below the Rayleigh wave at the zone boundary.

The reverse situation is realized with the iron overlayer on Cu(100), to be discussed next.

10.7.2. Epitaxially Grown Iron Overlayers on Cu(100)

Iron overlayers on Cu(100) have already been studied in the number of publications. It was generally agreed that Fe grows in a layer-by-layer fashion up to five monolayers and that the structure of these monolayers corresponds to fcc Fe rather than bcc. It is noteworthy however, that the terminology "fcc" and "bcc" refers to three-dimensional space groups while an assembly of a few layers of "fcc"- or "bcc"-type layers belong to the same two-dimensional space group which is $p4m$. Consequently one cannot have a strict correlation between the lateral lattice parameter alongside the surface and the vertical distance of the layers. Nevertheless, the lateral lattice constant of iron overlayers on Cu(100) is very close to fcc iron while the vertical distance of the layers was found to be approximately in accordance with the fcc structure.[21] This fcc-type structure is evidently stable at room temperature only because the iron is forced into registry by the copper substrate, while otherwise at room temperature the bcc phase is stable. It was found recently[20] that the unusual stabilization of the fcc phase in an iron overlayer system gives rise to an interesting phase transition. For two monolayers of iron it was found that one has a (1×1) structure at room temperature. Lowering of the temperature leads to a gradual increase in half-order spots with systematic extinctions along the (01) and (10) axis for normal incidence of the electron beam. These systematic extinctions indicate a glide plane leaving only little choice for a structural model (Figure 10.17).

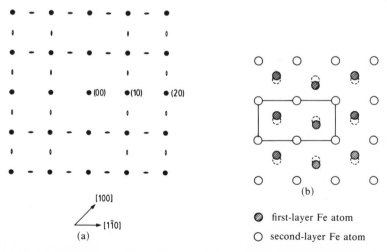

Figure 10.17. (a) Diffraction pattern for the monolayer system of Fe on Cu(100) with systematic extinctions along (01) and (10) axes indicating the existence of a glide plane. (b) Structural model for the (2×1) reconstructed phase. Two domains rotated by 90° of this structure produce the above diffraction pattern.

It is noteworthy that the (1×1) phase is also an ordered phase. This is indicated by the low elastic diffuse scattering observed in addition to the phonon scattering. An example of the phonon spectra of the Rayleigh wave for different monolayer coverages is shown in Figure 10.18. The figure also indicates that the system grows in a monolayer-by-monolayer fashion and that the monolayers are well ordered up to five monolayers. Beyond five monolayers one notes a sharp increase in the elastic diffuse intensity, indicative of the termination of layer-by-

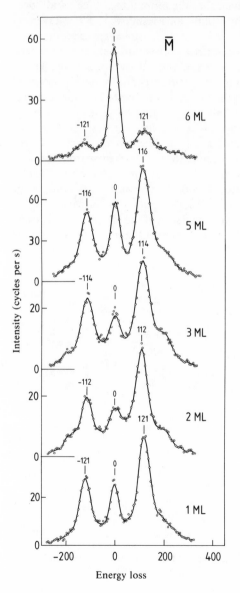

Figure 10.18. Phonon spectra for the 1–6 monolayer system of Fe on Cu(100). Note the sharp rise in the ratio of elastic diffuse scattering to phonon scattering beyond 5 monolayers.

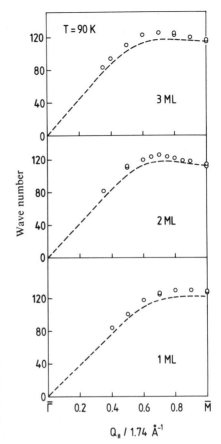

Figure 10.19. Phonon dispersion curves for the 1–3 monolayers of Fe on Cu(100). Data points refer to 90 K; the dotted line is the experimental result for 300 K. A lattice dynamical match is only obtained when a strong lateral surface stress is introduced (after Daum[18]).

layer growth. The phonon dispersion curves for the 1–3 monolayer system are shown in Figure 10.19.

The dispersion curve of the Rayleigh wave for the two monolayer system as well as for the rest of the 1–5 monolayer system can only be matched if one invokes a substantial repulsive stress between the iron atoms. The physical origin of this stress can be understood from the inherent instability of an fcc iron lattice at room temperature which would wish to expand laterally to become a bcc-type lattice. If one assumes a simple nearest-neighbor force field with additional stress between the surface atoms, one may calculate the frequency of the S_1 phonon and the $\bar{X}$ point in terms of the second derivative of the potential φ'' and the first derivative of the pair potential φ'. In the simple model the S_1 mode is strictly localized within the surface layer and a frequency of this particular mode is easily written in the form

$$M\omega_{S_1}^2 = \varphi'' + 4\varphi'/a \tag{10.7.1}$$

where a is the surface lattice constant.

One sees from this line that for a repulsive stress $\varphi' = 0.5\ \varphi''/a$ the phonon freezes. The displacement pattern of this phonon is precisely that proposed for the structure of the (2×1) reconstructed phase in Figure 10.16. Thus one has a good indication that the stress as perceived from the phonon dispersion curves is also the driving force for the phase transition. It would be interesting to establish more quantitative correlations between the theoretical consequences for such a model and a more detailed experimental study on the critical behavior of the phase transition itself. It would furthermore be of interest to establish correlations between structural properties of this important new system and the magnetic properties. Undoubtedly more work needs to be done. It is clear however, even at this point, that the system Fe on Cu(100) is a particular example of the possibility of growing materials at surfaces in a new state of matter with no equivalent in bulk solid-state physics.

References

1. R. G. Greenler, *J. Chem. Phys.* **44**, 310 (1966); **50**, 1963 (1969).
2. J. Pritchard, *J. Vac. Sci. Technol.* **9**, 895 (1972).
3. W. Erley, *J. Electron. Spectrosc. Relat. Phenom.* **44**, 65 (1987).
4. G. B. Brusdeylins, R. B. Doak, and J. P. Toennies, *Phys. Rev. B* **27**, 3662 (1983).
5. S. Y. Tong, C. H. Li, and D. L. Mills, *Phys. Rev. Lett.* **44**, 407 (1980); C. H. Li, S. Y. Tong, and D. L. Mills, *Phys. Rev. B* **21**, 3057 (1980).
6. Mu-Liang Xu, B. M. Hall, S. Y. Tong, M. Rocca, S. Lehwald, H. Ibach, and J. Black, *Phys. Rev. Lett.* **54**, 1151 (1985).
7. S. Lehwald, B. Voigtländer, and H. Ibach, *Surf. Sci.* **192**, 131 (1987).
8. R. E. Palmer, P. J. Rous, J. L. Wilkes, and R. F. Willis, *Phys. Rev. Lett.* **60**, 329 (1988).
9. H. Ibach, *J. Mol. Struct.* **79**, 129 (1982).
10. See also H. Ibach and D. L. Mills, *Electron Energy Loss Spectroscopy and Surface Vibrations,* Academic Press, New York (1982).
11. H. Ibach, S. Lehwald, and B. Voigtländer, *J. Electron Spectrosc. Relat. Phenom.* **44**, 263 (1987).
12. R. F. Wallis, *Prog. Surf. Sci.* **4**, Part 3 (1973).
13. M. B. Hall and D. L. Mills, Private communication.
14. R. J. Needs, *Phys. Rev. Lett.* **58**, 53 (1987).
15. P. Zeppenfeld, K. Kern, R. David, and G. Comsa *Phys. Rev.* **38**, 12 329 (1988).
16. J. E. Müller, M. Wuttig, and H. Ibach, *Phys. Rev. Lett.* **56**, 1583 (1986).
17. W. Daum, S. Lehwald, and H. Ibach, *Surf. Sci.* **178**, 528 (1986).
18. W. Daum, Thesis, RWTH Aachen, 1988.
19. W. Daum, *J. Electron Spectrosc. Relat. Phenom.* **44**, 271 (1987).
20. W. Daum, C. Stuhlmann, and H. Ibach, *Phys. Rev. Lett.* **60**, 2741 (1988).
21. A. Clarke, P. J. Rous, M. Arnott, G. Jennings, and R. F. Willis, *Surf. Sci.* **192**, L843 (1987).

Scanning Tunneling Microscopy

R. M. Feenstra

11.1. Introduction

Since its inception in 1982, the scanning tunneling microscope (STM) has proven to be a powerful tool in the sutdy of surfaces.[1-5] Ordered arrays of atoms and disordered atomic features have been observed on many metal and semiconductor surfaces. Clean surfaces as well as isolated adsorbates and thin overlayers have been studied. The STM has been used in a variety of environments including ultrahigh vacuum (UHV), air, and various liquids, and at temperatures ranging from liquid helium to above room temperature.

The power of the tunneling microscope lies in its ability to spatially and energetically resolve the electronic states on a surface. Spatially, the states can be observed with atomic resolution; 5 Å lateral resolution is routinely achieved and features on the 3 Å scale can be resolved under favorable circumstances. Energetically, states which lie within a few electron-volts on either side of the Fermi level can be observed with an energy resolution of a few kT. Details of the geometric arrangement of atoms on the surface are reflected in the spatial distribution of electronic states, and the STM thus provides a probe of the atomic structure of surfaces. The connection between the electronic states and the atomic structure depends on the type of system. For metals, the states generally follow the atoms in a uniform fashion, and the STM thus provides a direct "topographic" view of the atoms. For semiconductors, the electronic states often reflect "nontopographic" details of the surface. In that case, the interpretation of STM images in terms of their relationship to the geometric structure of the surface is a nontrivial task, and a major part of this chapter is devoted to a description of how such an interpretation can be accomplished.

In this chapter we mainly discuss STM data obtained from clean, ordered semiconductor surfaces in ultrahigh vacuum. This topic is just one of the many

R. M. Feenstra • IBM Research Division, T. J. Watson Research Center, Yorktown Heights, New York 10598, USA.

areas in which the STM has been applied. Aside from semiconductors, the surfaces of metals, semimetals (graphite), superconductors, insulators, organics, and other types of materials have been studied. One powerful aspect of the tunneling microscope is its ability to access a large number of variables, including current, voltage, tip-sample separation (z-position), and lateral (x, y) position on the sample. This ability leads to a large number of variations in the techniques used in acquiring the data, and these different techniques often emphasize different aspects in the interpretations of the data. No single acquisition method can be expected to solve all STM problems—the range of problems which can be addressed with the STM is just too broad to be encompassed by one single method of data acquisition and analysis. In this chapter we illustrate several of the basic methods which have been developed and used to date.

Following this introductory section, we present in Section 11.2 a brief description of the essential mechanical and instrumental aspects of the scanning tunneling microscope. In Section 11.3 we discuss the theoretical basis for tunneling, and we review work pertaining to the interpretation of STM images. The remainder of the paper is devoted to illustrating these principles of STM by the study of a particular system: the Si(111)2 × 1 surface. This surface is prepared by cleaving silicon in ultrahigh vacuum. We discuss both tunneling microscopy and spectroscopy of this surface. In the microscopy section, we present constant-current images of the surface, and discuss the interpretation of such images in terms of contours of constant state-density. Voltage-dependent imaging is used to obtain a complete picture of the structure of the surface. We then present spectroscopic measurements with the STM on the Si(111)2 × 1 surface. The constant-z method of spectroscopy, along with subsequent normalization of the acquired data, are used to build up a complete spectrum of the surface states. The relationship between this spectrum and the surface state-density is discussed.

11.2. Basic Principles

In Figure 11.1 we display a schematic view of the STM, taken from the original work of Binnig, Rohrer, Gerber, and Weibel.[1] In operation, the probe tip is brought to within 5–10 Å of the sample surface. Fine motion of the tip is accomplished with the piezodrives P_x, P_y, and P_z. A voltage V_T is applied between the tip and sample, producing the tunneling current J_T. The tunneling current is extremely sensitive to the separation between tip and sample, varying typically by one order of magnitude for each Å change in tip–sample separation. Thus, by scanning the tip across the sample, one can obtain an image of the sample topography. In practice, to avoid tip–sample contact, the scan is accomplished by continuously adjusting the tip height in order to maintain a constant tunneling current. The tip height as a function of lateral position, $z(x, y)$, thus constitutes a constant-current image of the surface. In the simplest case, such images reflect the topography of the sample, as illustrated in Figure 11.1 where the tip moves up over a surface step. More realistically, the tunnel current is also affected by local electronic effects on the surface. For example, a contamination spot may affect

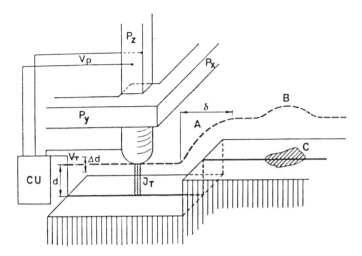

Figure 11.1. Schematic view of the scanning tunneling microscope. The piezodrives P_x and P_y scan the metal tip over a surface. The control unit (CU) applies the appropriate voltage V_p to the piezodrive P_z for a constant tunnel current J_T at tunnel voltage V_T. The broken line indicates the z displacement in a y scan at (A) a surface step and (B) a contamination spot, C, with lower work function (after Binnig et al.[1]).

the work function of the surface, and thus will change the tunnel current and subsequent tip height, as pictured in Figure 11.1. STM images thus contain a mixture of topographic and electronic information. In many cases, these types of information can be separated by acquiring images at a variety of tip–sample voltages.

Figure 11.2, also taken from the work of Binnig and Rohrer,[3] illustrates a number of other aspects of the STM. We see the probe tip mounted on a piezotripod, directed toward a sample which is mounted on an electrostatic walker or "louse." This walker, described elsewhere,[2] provides the coarse motion in the microscope, bringing the sample up to the tip within the scan range of the z-piezo (about $1\,\mu\text{m}$). The apex of the tip, in close proximity to the sample, is shown in two magnified views. At the highest magnification, dotted lines are used to indicate contours of constant electron density (state density) for the tip and sample. The tunnel current is determined by the overlap of these wave functions from the tip and sample, and the exponential dependence of the current comes from the exponential falloff of the wave functions into the vacuum region. Because of this exponential dependence of the tunnel current, only one or a few atoms right at the apex of the probe tip contribute significantly to the total current. The precise arrangement of atoms on the end of the tip affects the resolution of the microscope; this phenomenon is observed routinely by any operator of an STM as discrete changes in the resolution during the course of acquiring images. This dependence of the images on the sharpness and shape of the tip must always be kept in mind when analyzing STM images. Features which could result from asymmetric or multiple tips must be reproduced many times before they can be confidently

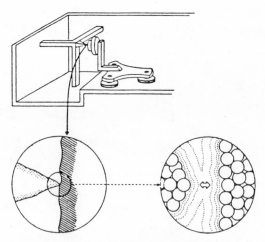

Figure 11.2. Schematic view of the physical principle and technical realization of the STM. In the upper part of the figure, the three piezodrives, the tip, and the sample holder with the sample on the louse are shown. The bottom part of the figure shows amplifications of the tip–sample region down to the atomic level. The dotted curves denote electron charge–density contours on a logarithmic scale (after Binnig and Rohrer[3]).

identified as a property of the sample itself. Nevertheless, many features of the images (e.g., the size and orientation of the unit cell) are relatively independent of tip structure, and thus form a most reliable database for structural determination with the STM.

We mention here a few additional technical features of the STM which are not illustrated in Figure 11.1 or 11.2. First, there is the issue of vibration isolation. Clearly the mechanical stability between probe tip and sample must be very high to avoid vibrational noise in the tunnel current. Typical stability levels are in the range 0.01–0.1 Å. In the original STM designs, this stability was accomplished using the simple, but effective, method of building the microscope as compact as possible and then suspending it on soft springs. These springs damp out vibrations with frequency greater than their resonance frequency (about 1 Hz). The resonant motion of the springs themselves can be suppressed by eddy-current damping using permanent magnets. This original design of the STM remains today one of the very best methods for designing and building the microscopes. Nevertheless, a number of other designs have been developed which simplify and improve on various aspects of the instrument. A review of these designs is presented elsewhere.[6]

Another important technical feature of the STM has to do with unintentional spatial drift between tip and sample. This drift typically amounts to a few Å per minute movement of the tip relative to the sample, in the x, y, or z direction. At this rate, the drift often amounts to a significant fraction of an image size in the time required to acquire the image. Sources of the drift include thermal expansion, and creep of the piezoelectric elements. The drift can be minimized by operating at low temperature, although this option is generally not available. The drift between tip and sample is a major factor in acquiring and analyzing data from the microscope. In the simplest case, the images themselves will be distorted. This distortion can be subsequently corrected either by measuring the drift using several successive images, or by using the known size of a unit cell on the surface. A more difficult problem arises when performing spectroscopic measurements at selected spatial locations. Typically it is difficult, or impossible, to go back precisely to a specified location and measure a spectrum after an image has been acquired. One way around this problem is to acquire the spectral information simultaneously with the image.[7-9]

11.3. Theory

Constant-current imaging is the original imaging method of the STM. This method has already been described in Section 11.2; essentially, the probe tip is scanned over the surface, with the z-position of the tip adjusted to maintain the tunnel current constant. The resultant image, $z(x, y)$ for constant tip-sample voltage V and current I, is often called a "topograph" of the surface. In this section we review the work which addresses the question of what exactly is measured in these constant-current images.

The tunnel current is determined by the overlap of wave functions between the tip and sample, as illustrated in Figure 11.3. This dependence is most clearly seen in the Bardeen tunneling formalism, where the tunneling current takes the form[10,11]

$$I = \frac{2\pi e}{\hbar} \sum_{\mu\nu} f(E_\mu)[1 - f(E_\nu - eV)]\delta(E_\mu - E_\nu)|M_{\mu\nu}|^2 \qquad (11.3.1)$$

with

$$M_{\mu\nu} = \frac{\hbar^2}{2m} \int d\mathbf{S} \cdot (\psi_\mu^* \boldsymbol{\nabla} \psi_\nu - \psi_\nu \boldsymbol{\nabla} \psi_\mu^*) \qquad (11.3.2)$$

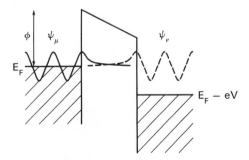

Figure 11.3. Energy diagram illustrating tunneling between two electrodes.

Here, $f(E_\mu)$ and $f(E_\nu - eV)$ are Fermi–Dirac occupation factors at the energies E_μ and $E_\nu - eV$ of the left- and right-hand electrodes, respectively. The delta function ensures energy conservation. Quantity $M_{\mu\nu}$ is called the "matrix element" for the process, and ψ_μ and ψ_ν are the wave functions of the left- and right-hand electrodes, respectively. The matrix element is evaluated over any surface lying entirely within the vacuum region separating the two electrodes.

Equations (11.3.1) and (11.3.2) can be used directly to illustrate some simple aspects of the tunneling problem. We consider a one-dimensional problem, where the electric field in the junction is neglected, the same work function ϕ is assumed for both electrodes, and all barrier rounding effects are neglected. The wave functions for the resulting rectangular-barrier problem are simple decaying exponentials. If the separation between electrodes is denoted as s, we have

$$\psi_\mu = \psi_\mu^0 \, e^{-\kappa z} \tag{11.3.3}$$

and

$$\psi_\nu = \psi_\nu^0 \, e^{-\kappa(s-z)} \tag{11.3.4}$$

where

$$\kappa = \sqrt{2m\phi / \hbar^2} \tag{11.3.5}$$

is the decay constant of the wave functions. For a typical work function of $\phi = 4.5 \text{ eV}$, the decay constant has a value $\kappa = 1.1 \text{ Å}^{-1}$. On substituting equations (11.3.3) and (11.3.4) into equation (11.3.2) we obtain

$$M_{\mu\nu} = \frac{\hbar^2}{2m} \int dS \, 2\kappa (\psi_\mu^0)^* \psi_\nu^0 \, e^{-\kappa s} \tag{11.3.6}$$

Hence equation (11.3.1) yields

$$I \propto \sum |\psi_\mu^0|^2 |\psi_\nu^0|^2 \, e^{-2\kappa s} \tag{11.3.7}$$

Here, using a shorthand notation, the summation sign denotes the sum over all relevant states which have appropriate energies and occupations to participate in the tunneling process. We note that in equations (11.3.6) and (11.3.7) the explicit dependence on z has dropped out, namely, the current can be evaluated along any surface separating the two electrodes.

Equation (11.3.7) illustrates several important features of the tunneling process. First, the factor $e^{-2\kappa s}$ gives the well-known exponential dependence of the current on separation. Using the above value of $\kappa = 1.1 \text{ Å}^{-1}$, this exponential dependence produces an order-of-magnitude change in the tunneling current for each Å change in the tip–sample separation. Second, within our simplified picture of a perfectly rectangular barrier, the quantities $|\psi_\mu^0|^2$ and $|\psi_\nu^0|^2$ give the state density (probability density) at the surface of the electrodes. In this picture then, we can consider the tunnel current to be a probe of the surface state-density, $\rho \equiv |\psi^0|^2$, and the exponential factor is regarded as the matrix element for the process.

Equation (11.3.7) was well known before the advent of the STM. In his treatise on tunneling, Duke arrives at essentially an identical formula displaying the proportionality between the tunneling current and the state density at the surface of the electrode(s).[12] However, it was then stated that "the [use of] plane wave basis functions, the neglect of irregularities in the actual (as opposed to the average) barrier potential, and the inapplicability of the WKB approximation to an abrupt junction conspire to strip this formal result of any simple physical interpretation."[12] Clearly, the situation changed after the STM was invented! The atomic-resolution images of surfaces,[1] and the oscillations of the tunneling current seen in the field emission range,[2] beautifully demonstrate the spatial localization and quantum coherence of the tunneling current in the STM. The observation of surface states in spectroscopic measurements with the STM demonstrates conclusively the close relationship between the tunnel current and the surface state-density.[7-9]

To understand the spatial resolution of the STM, it is necessary to go beyond the simple one-dimensional barrier problem discussed above and extend the discussion to three dimensions. In terms of the Bardeen formalism, this extension was accomplished by Tersoff and Hamann[11] and by Baratoff.[13] Tersoff and Hamann assume the probe tip to be a uniform sphere of radius R. In the limit $R \to 0$ the tip is replaced by a point probe, located at a position r_0. For the case of small voltage and temperatures (so that the relevant energy of the tunneling electrons is the Fermi energy E_F) they find that

$$I \propto \sum_\nu |\psi_\nu(\mathbf{r}_0)|^2 \delta(E_\nu - E_F) \qquad (11.3.8)$$

Thus the tunnel current is proportional to the local state-density of the sample evaluated at the point r_0. A constant-current STM image then simply corresponds to a surface of constant state-density. Tersoff and Hamann have shown that this same result can also be obtained for real probe tips with nonzero radius. In that case the point r_0 corresponds to the center of radius of curvature of the tip. This point is located a distance $R + d$ from the sample, where R is the tip radius and d is the distance from the tip apex to the sample. As the probe tip becomes blunter, R increases, and the STM effectively measures the state density at increasing distances away from the sample. As shown below, atomic features in the state density are attenuated as one moves away from the surface, and in this way the resolution of the STM decreases as the probe tips become blunter.

For metal surfaces, the local state-density at a surface roughly follows the corrugation of the surface atoms. For semiconductor surfaces the situation is more complicated since, at a given energy, the surface states tend to be localized on some atoms or bonds but not on others. In both cases, metals and semiconductors, we must also consider the changes in the state density as the distance from the surface increases. Generally, as the distance increases, contours of constant state-density tend to smooth out, with the higher-order Fourier components being more rapidly attenuated. This phenomenon is illustrated in Figure 11.4, where we show schematically the wave function ψ, state density $\psi^*\psi$, and the variation in state

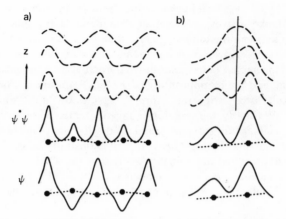

Figure 11.4. Schematic illustration of the wave function (ψ), the state density ($\psi^*\psi$), and the decay of state density with distance (z) from the surface (after Feenstra and Stroscio[14]).

density with increasing distance z from the surface.[14] Figure 11.4a illustrates the situation for a row of atoms on the surface, where the wave function is assumed to alternate between maximum and minimum values on subsequent atoms. The state density then has a maximum value on half the atoms and a minimum value on the other half. As the distance from the surface increases, the state density smooths out until eventually we are left with only a single sinusoidal corrugation component. Another situation is pictured in Figure 11.4b, in which we consider an isolated pair of atoms on the surface with the wave function concentrated more on one atom than the other. Now, as the separation from the surface increases, the state density will smooth out until it reaches a limiting form with a single maximum located somewhere between the two surface atoms. In both of these examples, the contours of constant state-density far from the surface cannot be interpreted in one-to-one correspondence with the surface atoms and thus, in general, one must exercise caution in the interpretation of STM images.

The smoothing out of the state density with increasing separation from the surface can be expressed in a simple analytic form, as derived by Stoll.[15] If we consider a Fourier component of the corrugation with period a and amplitude at the surface of h_s, then, for a distance d between surface and tip apex, the observed corrugation amplitude Δd is given by

$$\frac{\Delta d}{h_s} = \exp\left(\frac{-\pi^2(R+d)}{\kappa a^2}\right) \tag{11.3.9}$$

where R is the tip radius and κ is the electron decay constant defined above. This equation shows that, as the corrugation period a decreases, the attenuation of the corrugation amplitude increases rapidly. The effective resolution of the STM can be seen by plotting numerical values from equation (11.3.9), for fixed $R + d$, as shown in Figure 11.5. Smaller values for $\Delta d/h_s$ imply a reduced resolution of the microscope. We see that larger probe-tip radii, or greater tip–sample separations, both tend to reduce the resolution. This effect of the tip radius has been confirmed experimentally.[16] In practice, the ultimate resolution of the microscope is limited by the vibrational noise level in the instrument, below which a corrugation cannot be observed. For some value of the surface corrugation amplitude h_s, this noise limit corresponds to a particular value of $\Delta d / h_s$. Typically, corrugations which are attenuated by more than a factor of 0.1 will be difficult to observe. This cutoff value is shown by the dashed line in Figure 11.5, and the intersections of the various curves with this line specify the effective resolution of the microscope. For the very small $R + d$ value of 4 Å, we find a resolution of 3–4 Å, which is about the best that can be expected with the STM. A more realistic value for $R + d$ of 20 Å produces a resolution in the range 6–10 Å, which is easily attainable with the STM. These values for the tip radius refer to the *microscopic* radius at the apex of the probe tip; the *macroscopic* tip radius can be much larger than this value. Probe tips prepared by the usual methods of electrochemical etching apparently tend to have small "minitips" along their surface.[17]

Finally, we re-emphasize that values for the STM resolution given by equation (11.3.9) are only applicable to smoothly varying sinusoidal surface corrugation. For semiconductor (or semimetal) surfaces in which electronic effects can be large,

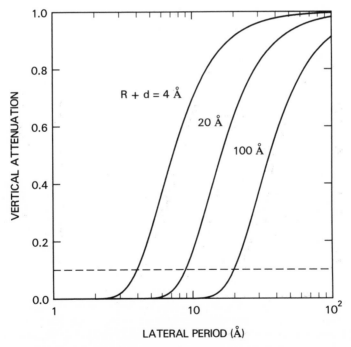

Figure 11.5. The ratio of observed corrugation amplitude to actual corrugation amplitude, as a function of the corrugation period. Results are shown for fixed values of $R + d$, where R is the probe-tip radius and d is the distance from tip apex to sample surface.

the resolution can be considerably enhanced over these values. A dramatic example of this effect has been presented by Tersoff for the surface of graphite,[18] in which it is argued that zeros on the surface wave function prevent the decay of the corrugation amplitude with increasing tip–sample separation. In that case, one may expect a resolution equal to the hexagonal spacing of graphite, which is 2.46 Å. Effects of this sort may also occur on semiconductor surfaces when only a few states are being sampled by the tunneling current.

11.4. Example: The Si(111)2 × 1 Surface

11.4.1. Microscopy

Figure 11.6 shows a typical STM image of the Si(111)2 × 1 surface.[14] Figure 11.6a shows the direct line scans of the data, and these are assembled into a top view in Figure 11.6b. In the top view a drift correction has been made to the data in order to achieve the known dimensions of the 2 × 1 unit cell, 6.65 × 3.84 Å. This top-view representation is then rotated in space to yield the perspective view shown in Figure 11.7. A small defect is seen near the center of that image. Such defects are relatively common on cleaved Si surfaces, although their exact nature is not known.

The image of Figure 11.7 can be compared with that shown in Figure 11.8, which shows a well-ordered region of the Si surface. Both images show a single topographic maximum per 2 × 1 unit cell. However, the shape of these topographic maxima appear quite different in the two images—the unit cells in Figure 11.7 are oblong, appearing to be longer in the $[1\bar{2}1]$ direction than the $[\bar{1}01]$ direction, while the unit cells in Figure 11.8 are more circular. The difference in the topographic shape arises from the magnitude of the corrugation in the two cases. In Figure 11.7 the $[\bar{1}01]$ corrugation amplitude is about twice as large as the $[1\bar{2}1]$ amplitude, while in Figure 11.8 the two corrugation amplitudes are almost equal. We attribute this difference between the images to the shape of the probe tip. As discussed in the previous section, a blunt tip is expected to yield a reduced corrugation amplitude, and this effect can vary for different lateral directions due to asymmetry in the tip structure. This dependence on tip geometry thus makes the corrugation amplitude itself an unsuitable quantity to use in a determination of the surface structure, and for that reason we turn to the dependence of the images on voltage.

Figures 11.9a and b show two STM images acquired at sample voltages of 1 V and −1 V, respectively.[14] The images are similar, each consisting of an array of topographic maxima with one maximum per unit cell. Cross hairs are superimposed on the images, located at identically the same surface locations in each image. Referring to the intersection of the cross hairs, we see in Figure 11.9a a topographic maximum, and in Figure 11.9b we find a saddle point in the topography. Thus, the topographic maxima have shifted by half a unit cell in the $[01\bar{1}]$ direction. This shift is seen more clearly in the cross section shown in Figure 11.10a. In the $[2\bar{1}\bar{1}]$ direction we observe a small shift in the corrugation, as shown

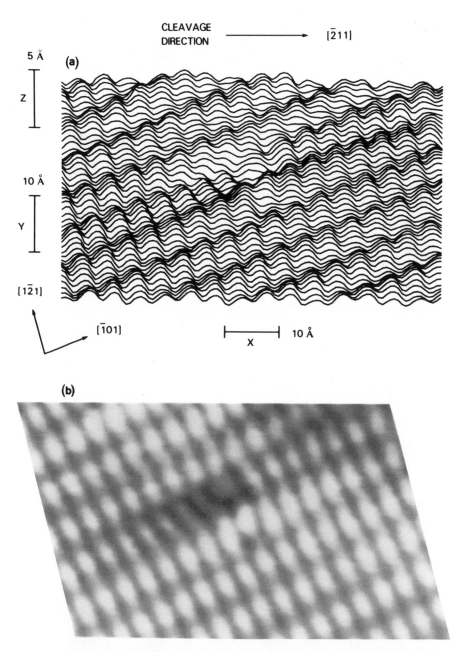

Figure 11.6. Constant-current STM image of the Si(111)2 × 1 surface. (a) Line scans of the surface. (b) Gray-scale image with drift correction (after Feenstra and Stroscio[14]).

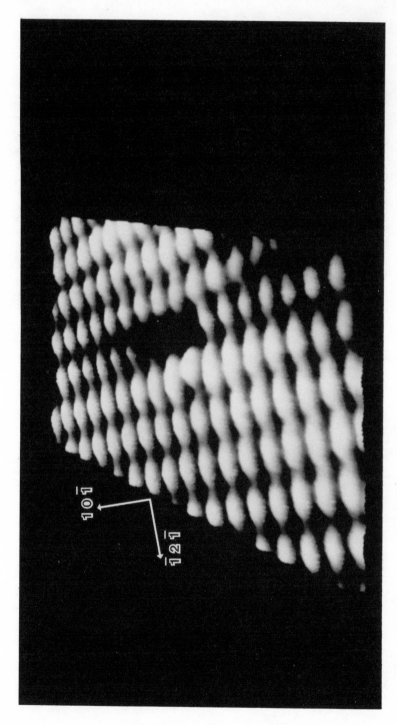

Figure 11.7. Three-dimensional view of the surface shown in Figure 11.6. The image extends over a lateral area of about 70×35 Å^2 with a vertical height variation of about 1 Å (after Feenstra and Stroscio[14]).

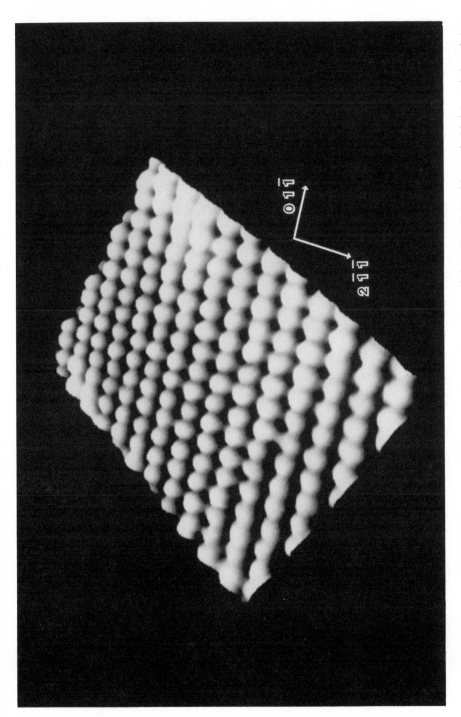

Figure 11.8. STM image of the Si(111)2 × 1 surface. The image extends over a lateral area of about 70 × 50 Å² with a vertical height variation of about 1 Å (after Feenstra and Stroscio[14]).

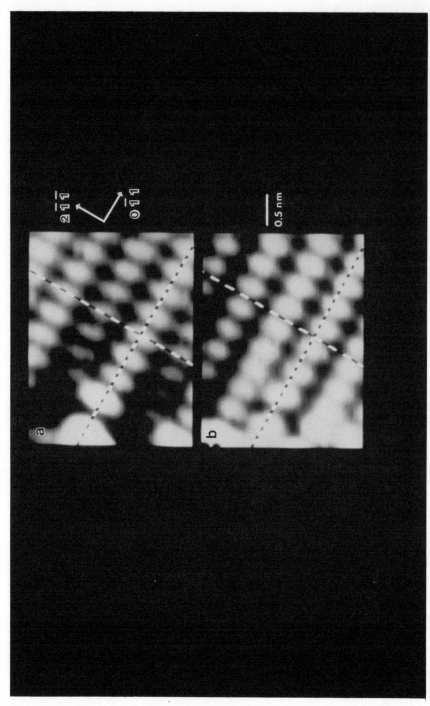

Figure 11.9. Two STM images of the Si(111)2 × 1 surface, acquired simultaneously at sample voltages of (a) +1 V and (b) −1V (after Feenstra nad Stroscio[14]).

in the cross sections of Figure 11.10b. The images of Figure 11.9 are analyzed by fitting the data to a surface consisting of the sum of two sinusoids, with adjustable periods, amplitudes, and phases. This procedure yields corrugation shifts between empty and filled states of about 2.0 ± 0.3 Å for the $[01\bar{1}]$ direction and 0.8 ± 0.2 Å for the $[2\bar{1}\bar{1}]$ direction.

To understand the observed voltage dependence of the images, it is necessary to consider the spatial location of the surface states. In principle one can perform a detailed numerical computation fo the surface states, although in practice such calculations are sufficiently difficult to perform so that they are rarely available for a particular surface of interest. Even for the case considered here of the

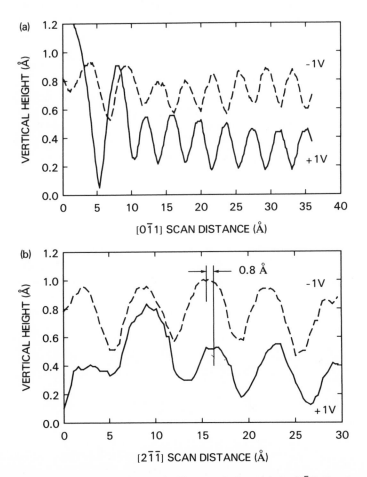

Figure 11.10. Cross sections of the images in Figure 11.9 along (a) the $[0\bar{1}1]$ direction and (b) the $[2\bar{1}\bar{1}]$ direction. Absolute height is arbitrary here. For the $[0\bar{1}1]$ direction the corrugation shifts by half a unit cell (1.92 Å) when the voltage changes from -1 to $+1$ V, and for the $[2\bar{1}\bar{1}]$ direction the corrugation shifts by about 0.8 Å (after Feenstra and Stroscio[14]).

Si(111)2 × 1 surface, which is one of the more extensively studied surfaces, detailed computations displaying results relevant to the STM experiments are not available for all the significant structural models. Thus, it is desirable to identify some simple, qualitative feature in the surface electronic structure which is independent of the details of a theoretical computation and can be used to distinguish between different structural models. For the Si(111)2 × 1 surface, and a number of other surfaces with two atoms per unit cell, the lateral separation between filled and empty states provides just such a quantity.

To understand the difference between empty and filled surface states, we introduce the concept of "buckling" on a semiconductor surface. We use this term here to refer to charge transfer between equivalent or inequivalent atoms on a semiconductor surface, as illustrated in Figure 11.11. We consider two atoms on the surface which may be two isolated atoms (e.g., forming a dimer) or two members of a chain of atoms on the surface. We suppose that in the nonrelaxed surface each of these atoms has a dangling bond, and each bond is occupied by a single electron, as shown in Figure 11.11a. A two-dimensional set of such bonds will form a half-filled band of states on the surface, as pictured in Figure 11.11b. On a semiconductor surface this band generally ends up lying somewhere within the bulk bandgap. We use the term "buckling" here to refer to the transfer of an electron from one of the dangling bonds to the other, creating one filled bond and one empty bond, as shown in Figure 11.11c. This transfer of charge results in a splitting of the band of states into a filled band and an empty band, separated by a bandgap, as pictured in Figure 11.11d. Accompanying the charge transfer there will be some relaxation in the position of the atoms on the surface. If the two atoms of the surface are initially inequivalent (such as a Ga and an As atom), then the energy of their bonds is necessarily different, and some buckling of the surface always occurs. If the two atoms are equivalent [such as two Si atoms in a symmetric dimer on the Si(100) surface], then buckling may or may not occur depending on the energetics of the situation. It is important to note that buckling does *not* require the transfer of an *entire* electron from one dangling bond to the other. In reality, all the states on the surface are composed of a linear combination of both dangling bonds. For the unbuckled surface this combination will include

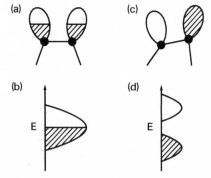

Figure 11.11. Schematic view of buckling on a semiconductor surface. Surface atoms with half-filled bonds (a) form a half-filled band of surface states (b). Charge transfer from one bond to the other (c) splits this band of states into a filled band and an empty band (d).

equal contributions from both bonds, while on a buckled surface one bond will dominate in the filled states and the other bond will dominate in the empty states.

We return now to the Si(111)2 × 1 surface and examine various models for the surface structure. Figure 11.12 shows the atomic positions and the wave functions for various structural models of the surface. The wave functions are schematic and are meant to highlight differences between empty and filled states on the surface, while neglecting details of the dispersion of the surface state bands. Each model has two dangling bonds per unit cell, leading to an empty and a filled band of states. At the center of the surface Brillouin zone, the empty states (antibonding) have nodes in the wave function, and the filled states (bonding) do not, as shown in Figure 11.12. The π-bonded chain model (Figure 11.12a) consists of zigzag chains of nearest-neighbor surface atoms. The two surface atoms in the unit cell are structurally inequivalent, so some buckling of the surface necessarily occurs. As discussed above, this buckling produces an asymmetry in the composition of the wave functions, with one dangling bond making a larger contribution in the filled states and the other dangling bond contributing more to the empty states. In the buckled model (Figure 11.12b) the surface atoms are second-nearest-neighbor. On an unrelaxed surface the two atoms in the unit cell are equivalent, although the basis for the model is to assume some buckling between the atoms

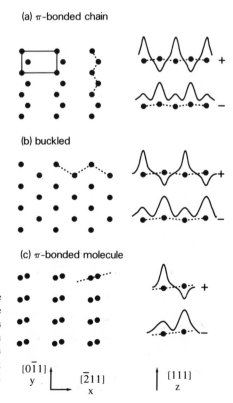

(a) π-bonded chain

(b) buckled

(c) π-bonded molecule

Figure 11.12. Schematic view of the surface atom locations (left side) and surface-state wave functions (right side) for various models of the Si(111)2 × 1 surface. A unit cell is shown by the solid rectangle in (a). The wave functions are shown for filled (−) and empty (+) states, along the spatial paths indicated by the dotted lines (after Feenstra and Stroscio[14]).

thereby forming the 2×1 unit cell. For the π-bonded molecular model (Figure 11.12c) the two atoms in the unit cell form a dimer. Again, the two atoms are inequivalent and some buckling necessarily occurs.

To go from the wave function pictured in Figure 11.12 to actual STM images, it is necessary to consider how the wave functions decay with distance from the surface. This behavior has already been discussed in the previous section and illustrated in Figure 11.4. Generally, as the separation between probe tip and sample increases, higher-order Fourier components in the state density are rapidly attenuated. For the dimensions of the Si(111)2×1 unit cell, usually only the lowest-order Fourier component is seen in the STM images. Generally, for semiconductor surfaces, peaks in the state density occur at the position of dangling bonds on the surface, which themselves occur near the location of surface atoms. In some cases, the state density is constrained by symmetry to have maxima or minima along particular crystallographic directions.

Given the surface wave functions, and the considerations for constructing the state density far from the surface, theoretical state-density contours can be determined qualitatively and compared with experiment. Such a comparison is made in Figure 11.13 for the Si(111)2×1 surface. The spatial locations of filled states are shown by the solid circles, and the spatial locations of empty states by open circles. The difference between the locations of filled and empty states is denoted by Δ_x and Δ_y in the figure. Experimentally, it was found above that $\Delta_x = 0.8 \pm 0.2$ Å and $\Delta_y = 2.0 \pm 0.3$ Å. In the $[01\bar{1}]$ direction (y-coordinate), the theoretical shift for the π-bonded chain and the buckled models is half a unit cell, 1.92 Å, in agreement with experiment. In the $[2\bar{1}\bar{1}]$ direction (x-coordinate), the theoretical shift is expected to be close to the difference between atomic coordinates, which for the π-bonded chain is 1.1 Å and for the buckled model is 3.3 Å. Thus, only the theoretical predictions for the π-bonded chain model are found to be in

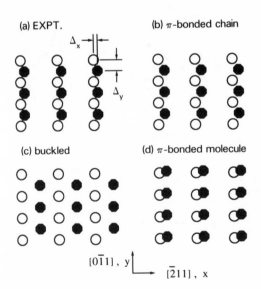

Figure 11.13. Schematic top view of the spatial location of filled states (solid circles) and empty states (open circles) for the observed STM experiment and various structural models. The difference in lateral position between filled and empty states is denoted by Δ_x and Δ_y (after Feenstra and Stroscio[14]).

agreement with experiment and, among the models considered here, the π-bonded chain model is found to be the best candidate for the structure of the Si(111)2 × 1 surface.

11.4.2. Spectroscopy

In addition to constructing images of a surface, the tunneling microscope can be used to directly probe the spectrum of surface states. On semiconductors, such investigations are generally limited in energy to a few eV on either side of the Fermi level. Practically all semiconductor surfaces possess an energy bandgap separating bands of filled and empty surface states. This surface gap may be larger or smaller than the bulk bandgap. Surface states which fall within the bulk bandgap are "strictly localized" at the surface, with wave functions which decay to zero inside the semiconductor. Surface states with energies equal to bulk conduction or valence band states are said to form "resonances" with those bulk states. Such resonant states have nonzero amplitude in the bulk, although they may have a substantial enhancement in their amplitude near the surface.

Spectroscopic measurement with the STM is performed by measuring current, or conductivity, versus tip-sample voltage. For semiconductors it is highly desirable to simultaneously measure both empty and filled states, which necessitates scanning through zero voltage. In that case the constant-current feedback loop cannot be used, since operation at zero volts would cause the probe tip to touch the sample surface. To overcome this problem, an "interrupted feedback" method was developed in which the probe tip is frozen at any particular (x, y, z) coordinate using a sample-and-hole circuit, and the current versus voltage $I(V)$ characteristic is measured.[19] To build up a large dynamic range in the current, the measurement can be repeated for several different z-values.

Figure 11.14 shows a series of $I(V)$ characteristics obtained from the Si(111)2 × 1 surface.[9] Each $I(V)$ curve is measured at a constant value of the tip-sample separation, and the separation is increased from curve a–m. A plot of the separation versus voltage, $s(V)$, for a constant current of 1 nA, is shown in the lower part of the figure. At high voltages, above 4 V, the $s(V)$ curve displays oscillations from an effect known as "barrier resonances," involving oscillations of the electron wave function in the vacuum region between tip and sample.[2] At lower voltages, between −4 and 4 V, surface state-density features show up as the various kinks and bumps occurring in the $I(V)$ curves

The state-density features seen in the $I(V)$ curves are obscured by the fact that the tunneling current depends exponentially on both separation and applied voltage. The dependence on separation arises, as discussed previously, from the exponentially decaying tails of the wave functions. The decay constant of the wave functions depends on the vacuum barrier height, and this in turn depends on the tip-sample voltage. Let us consider the case of two metals with equal work functions ϕ. The states at the Fermi level experience a barrier height of ϕ. We now apply a voltage V between tip and sample. States at the highest-lying Fermi level now see an *average* barrier of

$$\phi' = \phi - |eV|/2 \tag{11.4.1}$$

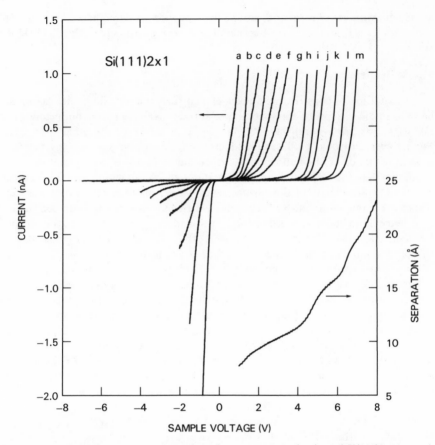

Figure 11.14. Tunneling current versus voltage for a tungsten probe tip and Si(111)2 × 1 sample at tip–sample separations of 7.8, 8.7, 9.3, 9.9, 10.3, 10.8, 11.3, 12.3, 14.1, 15.1, 16.0, 17.7, and 19.5 Å for the curves labeled a–m, respectively. These separations are obtained from a measurement of separation versus voltage, at 1 nA constant current, shown in the lower part of the figure (after Stroscio *et al.*[9]).

and in the WKB approximation their decay constant is given by

$$\kappa = \sqrt{2m\phi'/\hbar^2} \tag{11.4.2}$$

With the tunnel current proportional to $\exp(-2\kappa s)$, equations (11.4.1) and (11.4.2) show that, as the voltage approaches ϕ, the current will grow rapidly. This effect is clearly seen in the $I(V)$ curves of Figure 11.14, where a rapid turn-on at high voltages is seen in all the curves.

The exponential dependence of the tunneling current on separation and voltage can be practically removed by plotting the ratio of differential to total conductivity, $(dI/dV)(I/V)^{-1}$. This quantity provides a relatively direct measure of the surface density-of-states.[20,21] The results of such an analysis, applied to the data of Figure

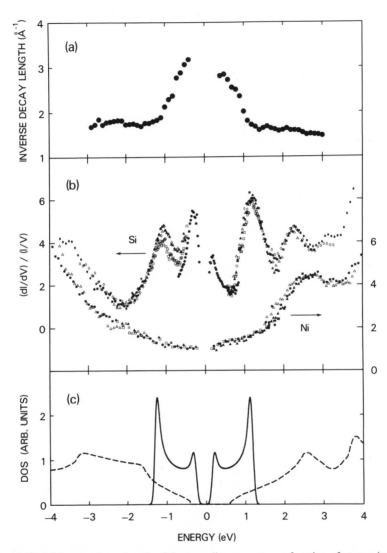

Figure 11.15. (a) Inverse decay length of the tunneling current as a function of energy (relative to the surface Fermi-level). (b) Ratio of differential to total conductivity for silicon and for nickel. The different symbols refer to different tip–sample separations. (c) Theoretical DOS for the bulk valence band and conduction band of silicon (dashed curve, after Chelikowsky and Cohen[22]), and the DOS from a one-dimensional tight-binding model fo the π-bonded chains (solid line) (after Stroscio et al.[9]).

11.14, are shown in Figure 11.15b. There, each type of data symbol refers to a different curve (i.e., a different tip–sample separation) from Figure 11.14. The measurements at different separations fall on top of each other in Figure 11.15, demonstrating that the quantity $(dI/dV)(I/V)^{-1}$ provides a convenient "invariant" quantity.

A number of peaks are evident in the spectrum of Figure 11.15b, at energies of −1.1, −0.3, 0.2, 1.2, and 2.3 eV, along with a small peak near 3.2 eV. To identify those peaks, Figure 11.15c shows a plot of theoretical results for the surface density-of-states (DOS). The dashed curve gives the bulk DOS for silicon,[22] and the solid curve gives the surface DOS for a one-dimensional tight-binding model of π-bonded chains.[23] By comparison of experiment and theory, some of the observed peaks can be immediately identified with known electronic features of the Si(111)2 × 1 surface. The peaks of −0.3 and 0.2 eV encompass the gap separating the filled and empty states. The peak at −1.1 eV corresponds to the bottom of the filled band, and the peak at 1.2 eV corresponds to the top of the occupied band. The peak at 2.3 eV may arise from the lowest-lying conduction band near the L point in the bulk band structure.[9]

It is important to note that, in principle, DOS features from the probe tip could contribute to spectroscopic observations. Theoretical studies by Lang for adsorbed atoms on two surfaces which are scanned over each other show that each atom contributes equally ot the spectrum (i.e., the tip and sample are indistinguishable in spectroscopy).[21] To evaluate the contribution of the probe tip to the above observations on the Si(111)2 × 1 surface, the same tips were used to perform spectroscopic measurements on evaporated nickel films. The resulting spectrum is also shown in Figure 11.15b. A single peak, near 2.7 eV, is seen in the spectrum, and this peak has been identified in other studies as being a Ni-related surface state. Thus, the tungsten probe tip apparently does not contribute to the observed spectra in Figures 11.14 and 11.15.

Another interesting quantity can be derived from the data of Figure 11.14. The relative separation of each $I(V)$ curve can be obtained by selecting the appropriate values from the observed $s(V)$ curve, thus constituting a measurement of current versus separation at various voltages. The current is found to decay exponentially with separation, and the measured decay constants are plotted in Figure 11.15a. These decay constants correspond to the quantity 2κ as used above. In the simplest approximation we expect a decay constant of $2\kappa = 2.2$ Å^{-1} for a work function of $\phi = 4.5$ eV. This value is close to what is observed in Figure 11.15a at energies with magnitudes greater than 1 eV. At energies closer to zero, the decay constant is observed to increase sharply. This increase is attributed to tunneling through states with large values of the parallel wave vector. Current arising from a state with parallel wave vector $k_\parallel$ will decay into the vacuum according to[11]

$$\kappa = \sqrt{2m\phi/\hbar^2 + k_\parallel^2} \qquad (11.4.3)$$

Using this formula, a value of $k_\parallel \simeq 1.1$ Å^{-1} is found for the states near zero energy in Figure 11.15a. This value is in agreement with the surface state band structure of the Si(111)2 × 1 surface, for which states near zero energy arise from the edge of the surface Brillouin zone, with maximum wave vector of 0.94 Å^{-1}.

11.5. Summary

In this chapter, we have discussed the interpretation of images obtained with the scanning tunneling microscope. For semiconductor surfaces, these images are dominated by the influence of dangling bonds on the surface. An understanding of the images thus translates into an understanding of the surface wave functions, expressed in terms of the state density $\rho \equiv |\psi|^2$. The state density may be peaked on certain atoms or bonds, but not on others, so that a one-to-one interpretation of the STM images in terms of surface atoms is not possible. However, by acquiring images at enough tip–sample bias voltages, one may succeed in imaging all of the surface dangling bonds and in this way build up a complete picture of the surface structure. In measurements of tunnel current versus voltage, the STM obtains an entire spectrum of the surface density-of-states. Effects of the tunneling matrix element can be partially, but not completely, removed from this spectrum. To lowest order the matrix element simply depends exponentially on tip–sample separation, with decay constant κ. Matrix-element effects can be expressed in terms of increasingly more accurate approximations for κ, including the dependence of κ on energy, tip–sample voltage, and parallel wave vector.

References

1. G. Binnig, H. Rohrer, Ch. Gerber, and E. Weibel, *Phys. Rev. Lett.* **49**, 57 (1982).
2. G. Binnig and H. Rohrer, *Helv. Phys. Acta* **55**, 726 (1982).
3. G. Binnig and H. Rohrer, *Surf. Sci.* **152/153**, 17 (1985).
4. G. Binnig and H. Rohrer, *IBM J. Res. Dev.* **30**, 355 (1986).
5. P. K. Hansma and J. Tersoff, *J. Appl. Phys.* **61**, R1 (1987).
6. Y. Kuk and P. J. Silverman, *Rev. Sci. Instrum.* **60**, 165 (1989).
7. R. S. Becker, J. A. Golovchenko, D. R. Hamann, and B. S. Swartzentruber, *Phys. Rev. Lett.* **55**, 2032 (1985).
8. R. J. Hamers, R. M. Tromp, and J. E. Demuth, *Phys. Rev. Lett.* **56**, 1972 (1986).
9. J. A. Stroscio, R. M. Feenstra, and A. P. Fein, *Phys. Rev. Lett.* **57**, 2579 (1986).
10. J. Bardeen, *Phys. Rev. Lett.* **6**, 57 (1961).
11. J. Tersoff and D. R. Hamann, *Phys. Rev. Lett.* **50**, 1998 (1983).
12. C. B. Duke, *Tunneling in Solids*, p. 253, Academic Press, New York (1969).
13. A. Baratoff, *Physica* **127B**, 143 (1984).
14. R. M. Feenstra and J. A. Stroscio, *Phys. Scr.* **T19**, 55 (1987).
15. E. Stoll, *Surf. Sci.* **143**, L411 (1984).
16. Y. Kuk and P. J. Silverman, *Appl. Phys. Lett.* **48**, 1597 (1986).
17. C. F. Quate, *Phys. Today* **39**, 26 (August, 1986).
18. J. Tersoff, *Phys. Rev. Lett.* **57**, 440 (1986).
19. R. M. Feenstra, W. A. Thompson, and A. P. Fein, *Phys. Rev. Lett.* **56**, 608 (1986).
20. R. M. Feenstra, J. A. Stroscio, and A. P. Fein, *Surf. Sci.* **181**, 295 (1987).
21. N. D. Lang, *Phys. Rev. B* **34**, 5947 (1986).
22. J. R. Chelikowsky and M. L. Cohen, *Phys. Rev. B* **10**, 5095 (1974).
23. R. Del Sole and A. Selloni, *Phys. Rev. B* **30**, 883 (1984).

Defect Structures at Surfaces

J. Lapujoulade

12.1. Introduction to Surface Defects

A crystal is characterized by a three-dimensional periodic atomic structure and a perfect crystal plane possesses the same property in two dimensions. Most of the theoretical studies of the properties of crystal surfaces deal with the properties of such crystal planes. A crystal plane is characterized by a unit cell defined by two unit vectors. The plane structure is invariant with respect to a translation by a multiple of these unit vectors.

But this is a very idealized situation and, indeed, real surfaces can never reach the structure of a perfect crystal plane. This can eventually be due to the method of preparation which generates defects but, more basically, it will be shown that the occurrence of defects is inherent to the thermodynamic equilibrium. So the study of surface defects is unavoidable for understanding the properties of surfaces which are readily observable in the laboratory. Moreover, knowledge of surface defect physics is of intrinsic interest since it is well known that many of the useful surface properties are rather a consequence of their defects than of their basic structure. This is especially true for chemical reactivity where the most perfect surfaces are known to be generally the less reactive.

A defect on a perfect crystal plane is a perturbation which breaks the translational invariance. A defect can be due either to atoms or groups of atoms which are misplaced, in which case we shall speak of a structural defect, or to atoms which are correctly located but have a foreign chemical nature, in which case we shall speak of a chemical defect. In the following we shall deal only with structural defects.

The easiest way to look at surface defects is to use the so-called TSK model (Terrace, Step, Kink; see Figure 12.1). In this model each atom of the crystal is depicted by a cube. For a perfect crystal plane all cubes have a face lying in this

J. Lapujoulade • Centre d'Etudes Nucléaires de Saclay, Service de Physique des Atomes et des Surfaces, 91191 Gif-sur-Yvette Cedex, France.

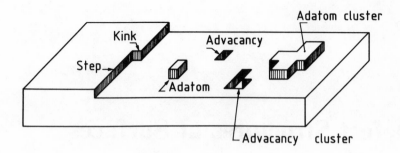

Figure 12.1. The TSK model.

common plane. Point defects like adatoms or advacancies can appear in this plane, but there exist also collective defects: groups of atoms or groups of advacancies defining a limited terrace which has the same structure as the basic plane but lies at a different level. This level is an integral multiple of elementary length c. The boundary of these defect clusters forms a step. A straight step itself may be considered as another kind of collective defect. A new kind of point defect then appears on a straight step: the kink.

Adatoms, advacancies, their clusters, steps, kinks are the only defects which can occur in the TSK model.

The scope of this chapter covers:

- Experimental characterization of surface defects.
- Physical properties of an individual defect: energy, entropy.
- Interactions between defects.
- Thermal equilibrium of defects.
- The roughening transition.
- The special case of vicinal surfaces.

12.2. Experimental Characterization of Surface Defects

The characterization of surface defects involves knowledge of their nature, concentration, and statistical distribution.

The identification of the defect nature is difficult and generally requires an image of the crystal structure on an atomic scale. This is the field of high-resolution microscopy techniques in the real space, such as scanning tunnel microscopy (STM), field ion microscopy (FIM), and high-resolution electron microscopy.

Once the nature of the defect is known (or guessed) the measurement of its concentration is easier, since both microscopic and macroscopic methods can be used. Indeed, any macroscopic property which is sensitive to defects can provide information about their concentration, e.g., work function, chemical reactivity, and so on. However, a calibration with a microscopic method is needed in order to obtain absolute determinations. Nevertheless, the most commonly used method is to monitor coherent diffraction peak intensities.

Diffraction methods (LEED, TEAS, glancing X-rays) operate in the reciprocal space and are thus sensitive to the Fourier transform of the correlation function of the defect distribution. Hence they give both the concentration and the statistical distribution of defects.

12.2.1. Surface Microscopy

Electron Microscopy. In some cases steps can be readily seen by ordinary electron microscopy using decoration techniques. This has been used for ionic crystals and also recently for molybdenum using gold atoms, which are preferentially adsorbed along steps.[1]

Field Ion Microscopy. In the field ion microscope individual atoms are directly seen, so that point defects like adatoms or advacancies can be readily observed. But there are many limitations to this technique:

1. It is restricted to refractory metals.
2. The presence of the strong field eventually modifies the distribution of defects.
3. A tip is needed where flat surfaces have necessarily a very small area (a hundred atoms).

In spite of these restrictions most of our knowledge about defect mobilities and their mutual interactions comes from this technique.

Scanning Tunnel Microscopy. This very powerful technique for imaging surfaces is, at least in principle, able to detect any kind of defect. Indeed steps were very soon observed. It must be emphasized that the defect is detected through the modification of the electron density in its vicinity, so it is easier to detect adatoms than advacancies on a close-packed surface. STM will be certainly in the near future one of the basic techniques employed to look at defects on surfaces. However, one must keep in mind some of its limitations:

1. In the present state of the technique, a scan over a surface needs some time (seconds or minutes). Thus STM imaging is restricted to frozen surfaces, i.e., low temperatures. Equilibrium steady states, when the mobility is high, fall out of the STM field.
2. The tip is sometimes able to displace an adatom, which is then not seen.

12.2.2. Diffraction Techniques

Diffraction techniques are very complementary to microscopy as they give information which is very tedious to obtain from a microscope image, namely the statistical distribution. Among the various diffraction techniques which can be used (LEED, X-Rays, etc.), thermal energy atom scattering (TEAS) is especially interesting for the study of surface defects due to its very high surface selectivity, and we shall concentrate here on this technique.

12.2.3. Thermal Energy Atom Scattering (TEAS)

We recall that helium atoms of thermal energy interact with the crystal through a repulsive potential due to the overlap of outer electron orbitals. The isopotentials are roughly unperturbed crystal electron isodensities in the range of 10^{-5} au. As a result helium is not very sensitive to the atomic structure of close-packed arrangements (in contrast to electrons or X rays). Defects on a surface introduce a very large perturbation of the electron density, thus TEAS will be very sensitive to them. The diffraction spectrum of helium on a close-packed surface is dominated by a strong specular peak and high-order diffraction peaks are negligible. Defects induce a weakening of the specular peak, possibly its broadening, and give rise to diffuse scattering. In order to go further into the analysis we must distinguish two classes of defects:

- Isolated point defects or isolated clusters of defects.
- Large domains delimited by steps.

a. Helium Scattering on a Plane Surface with Point Defects.[2-4] We suppose for simplicity that only one kind of defect is present on the surface: adatoms, advacancies, or identical small clusters of them. The probability that a defect is present on the surface is θ and the joint probability that sites i and j are affected simultaneously by a defect is P_{ij}. Hence $P_{ii} = \theta$ and, for sites very far apart, $P_{ij}^{\infty} = \theta^2$. We define a correlation coefficient C_{ij} by

$$C_{ij} = (P_{ij} - \theta^2)/\theta(1 - \theta) \tag{12.2.1}$$

such that $C_{ii} = 1$ and $C_{ij}^{\infty} = 0$.

Then one can show that the diffraction pattern comprises two parts:

1. *A coherent part* made up of δ-functions located at the nodes of the surface reciprocal lattice (Bragg peaks).

The specular intensity is (for a flat surface)

$$I_0 = 1 - 2\theta \, \text{Im} \, F(\mathbf{Q}) \tag{12.2.2}$$

where $\mathbf{Q}$ is the parallel momentum exchange; diffracted intensities appear due to the periodicity of the defect site lattice:

$$I_G = \theta^2 |F(\mathbf{Q})|^2 \tag{12.2.3}$$

2. *An incoherent part* due to the disorder in the site occupancy,

$$dP/d\Omega \approx |F(\mathbf{Q})|^2 \theta(1 - \theta) \sum C_{ij} \exp\left[-iQ(\mathbf{R}_i - \mathbf{R}_0)\right] \tag{12.2.4}$$

where the sum $\sum$ is extended over all lattice sites and $\mathbf{R}_i$ is a vector which defines the position of the lattice site i.

In these equations $F(\mathbf{Q})$ is the defect form factor, which only depends upon the variation of the He-surface potential in the vicinity of the defect. In the limit of low concentration ($\theta \ll 1$) $F(\mathbf{Q})$ is independent of defect distribution. The term Im $F(\mathbf{Q})$, which appears in the specular intensity, defines a cross section for incoherent diffusion by setting

$$\Sigma_d = A_{uc} \text{ Im } F(\mathbf{Q}) \tag{12.2.5}$$

where A_{uc} is the unit cell area; then

$$I_0 = 1 - 2\theta(\Sigma_d/A_{uc}) \tag{12.2.6}$$

The value of Σ_d is quite large (especially for adatoms, $\Sigma_d \approx 100$ Å), thus the specular peak measurement is a very sensitive method for measuring the defect concentration. The incoherent intensity contains information about the defect statistics through the Fourier transform of its correlation function C_{ij}. If the defects are completely randomly distributed over the lattice sites, then one obtains $C_{ij} = \delta_{ij}$ and the incoherent scattering is simply

$$dP/d\Omega = |F(\mathbf{Q})|^2 \theta(1 - \theta) \tag{12.2.7}$$

which reaches a maximum for $\theta = \frac{1}{2}$ as expected, but it must be emphasized that the above expression is only valid for low θ.

b. Helium Scattering on a Surface with Domains Limited by Monoatomic Steps. [5,6] Each domain has the same structure as the original surface, but they are shifted with respect to each other by an integral multiple of a constant vector $\mathbf{U}(\mathbf{R}, z)$, where $\mathbf{R}$ is the parallel displacement and z the normal displacement.

The phase shift between domains is

$$_f\phi_i = \mathbf{Q} \cdot \mathbf{R} + (k_i^z + k_f^z)z \tag{12.2.8}$$

Then, provided the domains are large enough in order that multiple scattering from one domain to another can be safely neglected, the scattered intensity can be written as

$$I_{scatt} = F(\mathbf{Q}) \sum_c \sum_{c'} \exp\left[i\mathbf{Q}(\mathbf{R}_c - \mathbf{R}_{c'})\right]\langle\exp\left[i_f\phi_i(d - d')\right]\rangle \tag{12.2.9}$$

where the summation is extended over all unit cells c, d is an integer which characterizes the unit cell shift, and $F(\mathbf{Q})$ is a form factor which only depends upon the scattering properties of the perfect surface, i.e., the corrugation of its unit cell. The domain distribution only appears in the second term, which can be called a structure factor.

If $(d - d')$ is not too large, $\langle\exp[i_f\phi_i(d - d')]\rangle$ can be developed in a cumulant expansion and only the first term is retained (Gaussian approximation):

$$\langle\exp\left[_f\phi_i(d - d')\right]\rangle \approx \exp\left[-\langle(d - d')^2\rangle(\pi^2/4)(1 - \cos_f\phi_i)\right] \tag{12.2.10}$$

Two important cases must be considered:

1. $_f\phi_i = 2n\pi$ "In-phase" Scattering. Then $\langle\exp[i_f\phi_i(d - d')]\rangle = 1$ and the peak shape is reduced to a δ-function as for the perfect surface case. Interferences between domains are constructive and the disorder is not seen.
2. $_f\phi_i = (2n + 1)\pi$ "Antiphase" Scattering. Then $\langle\exp[i_f\phi_i(d - d')]\rangle = \exp[-(\pi^2/2)\langle(d - d')^2\rangle]$. Interferences between domains are destructive and the effect of the disorder is maximum.

An example of this will be given later for the case of the roughening transition of stepped surfaces.

12.3. Physics of an Isolated Defect

The most important quantity associated with a defect is its free energy of creation:

$$F_d = E_d - TS_d \qquad (12.3.1)$$

One must be very careful in the calculation of F_d to specify correctly the initial and final states. For instance, if the initial state is a surface without defects and if an advacancy is created, the final state is necessarily an advacancy-adatom pair. But if the initial state is a surface having somewhere a step with a kink, one can speak of the creation of an advacancy alone by putting the removed atom on a kink site in the final state.

12.3.1. Energy Associated with the Creation of a Defect

The simplest way to estimate this energy is to count the number of broken bonds. Assuming a crystal with a coordination number Z ($Z = 8$ for a bcc and $Z = 12$ for a fcc) and only pairwise interactions between atoms, the energy per bond is simply

$$E_b = 2E_c/Z \qquad (12.3.2)$$

where E_c is the cohesive energy per atom.

Then the calculation of E_d is simply reduced to a count of the difference in bond number between the initial and final states. For instance, for the formation of an adatom-advacancy pair on a fcc (100) surface, 8 bonds are broken by removing the surface atom to form the vacancy while 4 new bonds are created by adsorbing it onto the surface. Thus the difference is $8 - 4 = 4$ bonds and the associated energy is

$$E_d = 4(2E_c/12) = 2E_c/3 \qquad (12.3.3)$$

Another example is the formation of a pair of kinks on a [110] step of a fcc (100) surface. In this operation only one bond is broken, so that we need $E_c/6$ for a pair of kinks, i.e., $E_c/12$ per kink.

This description is correct qualitatively but too naïve. Indeed interactions are not pairwise in usual crystals. In the case of metals the band structure must be taken into account. This can be achieved quite simply for transition metals using the tight-binding approximation. The bandwidth is proportional to the square root of the second moment of the d-state electron density, which is itself proportional to the number of its nearest neighbors. Thus for an atom near a defect with only Z_i nearest neighbors the energy is reduced with respect to a bulk atom with Z nearest neighbors by the ratio $\sqrt{Z_i/Z}$. An example of the results of such a calculation due to Allan and Wach[7] is indicated in Table 12.1, where the broken bond values are also reported for comparison. The calculation has been conducted for the creation energy of an advacancy with the removed atom placed in a kink position. The calculated values are always smaller than the broken bond values. In these estimates relaxations within the defect (see below) are not taken into account. These relaxations are expected to further decrease E_d, but this effect is generally assumed to be weak and can be neglected as a first approximation.

12.3.2. Entropy Formation of a Defect

The variation of entropy associated with the formation of a defect is due to the change in the vibration spectrum of the atoms involved in the defect. The calculation of the vibration spectrum of a solid with defects is a very difficult task owing to the lack of symmetry of the problem. However, Wynnblatt[8] has obtained a good estimate by using the quasi-harmonic Einstein approximation of the solid. Then every atom is allowed to vibrate independently in the potential well formed by the surrounding atoms. The results for the formation entropy of an advacancy on copper are shown in Table 12.2.

12.3.3. Relaxation Effects

Beside the main effect, which is a change in the surface energetics, defects also have local effects linked to relaxations. Crystal and surface structures result from the minimization of free energy. When a defect is present there is no reason

Table 12.1. Formation Energy of an Advacancy on a fcc Surface

	Number of broken bonds	E_d/E_C	
		Broken bond model	Allan and Wach[7]
Bulk	6	1.00	0.39
(111)	3	0.50	0.24
(100)	2	0.33	0.16
(110)	1	0.166	0.08

Table 12.2. Creation Entropy of a Point Defect on a
fcc Surface (After Wynnblatt[8])

Defect type	Face	S/k_B
Adatom	(111)	2.45
	(100)	1.46
	(110)	0.90
Advacancy	(111)	1.04
	(100)	2.82
	(110)	0.58

why the perfect periodicity of the surface lattice would be retained. On the contrary, it is expected that the electron density is rearranged introducing self-consistently a relaxation of neighboring atoms.

For a jellium the perturbation in the electron density due to a defect is well known to give rise to the so-called Friedel oscillations[9]:

$$\Delta\rho(r) \approx \cos(2k_p r + \tau)/r^3 \qquad (12.3.4)$$

An example of such a relaxation in the vicinity of a step is shown in Figure 12.2.[10]

12.4. Interaction between Defects

When defects on a surface are close together, they interact.[11] Different kinds of interaction may be involved.

12.4.1. Short-Range Interaction

This interaction has a chemical nature. It is either covalent or ionic. For adatoms or advacancies it is attractive, and its range is only a few angstroms. For adatoms or advacancies, it results in a tendency to form clusters. Steps under the action of these forces tend to make facets.

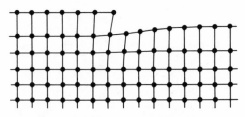

Figure 12.2. Relaxation within a step.

12.4.2. Long-Range Interaction

Direct interaction. If dipoles are associated with defects, then they can interact through dipole–dipole coupling. The interaction decreases like $1/r^2$.

Indirect interaction. This interaction originates from the overlap of the relaxations induced by every defect. The interaction between atomic relaxations brings about the so-called elastic interaction, which is repulsive and decreases monotonically as $1/r^4$. For the same reason two parallel steps of the same sign repel each other; this repulsion decreases with the distance d between the steps as $1/d^2$.

Before concluding this chapter devoted to the properties of defects on surfaces, it is important to note that very few experimental data are available. It is really a challenge for surface physicists to perform more experimental work on the physics of surface defects.

12.5. Statistics of Defects: The Roughening Transition

As soon as the individual properties of defects (free energy of creation) and their mutual interactions are known, we are in a position to look for their thermodynamic equilibrium. The surface without defects is the ground state of the system and, as temperature increases, fluctuations appear and more and more defects are present on the surface.

The equilibrium situation is governed by the free-energy excess F due to the defects,

$$F = \sum n_i F_i + F_{\text{int}} - TS \qquad (12.5.1)$$

where F_i is the free energy of formation of defect of type i, n_i the number of defects of type i, F_{int} the total interaction between defects, and S the entropy of configuration of defects.

There is a possibility that F becomes zero above some temperature T_R due to the increase of the entropy term; in this case defects will be created spontaneously on the surface. One gets an order–disorder phase transition.

A quantitative evaluation of the equilibrium structure needs knowledge of the partition function of the surface

$$Z = \sum \exp\left(-E_c/kT\right) \qquad (12.5.2)$$

where $\sum$ is the sum over all defect configurations and E_c is the energy of a configuration.

The probability for a particular configuration to appear is

$$P_c = \exp\left(-E_c/kT\right)/Z \qquad (12.5.3)$$

The partition function is related to the free energy by the relation

$$F = -kI \ln(Z) \qquad (12.5.4)$$

The expression for the partition function will be written for a special case of the TSK model, called the SOS model. In this model cubes are arranged into columns so that overhangs are forbidden (Figure 12.3). Only nearest-neighbor interactions are taken into account. Then the excess energy due to defects is simply obtained by counting the number of free vertical faces. A configuration is defined by an ensemble of integers $\{h_i\}$ defining the height of the column standing on each site i. If J is the energy associated with each bond, then the energy of a particular configuration is

$$E_{\{h_i\}} = \tfrac{1}{2} J \sum_i \sum_\delta f(h_i - h_{i+\delta}) \tag{12.5.5}$$

where δ indexes a site which is a neighbor of site i, and the partition function assumes the form

$$Z = \sum_{\{h\}} \exp\left[-(J/2kT) \sum_{i,\delta} f(h_i - h_{i+\delta}) \right] \tag{12.5.6}$$

If $f(h_i - h_{i+\delta}) = |h_i - h_{i+\delta}|$ one obtains the ASOS model, and if $f(h_i - h_{i+\delta}) = (h_i - h_{i+\delta})^2$ one obtains the DGSOS model. However, even with these simple models Z cannot be calculated analytically.

12.5.1. The BCF Argument

In 1951 Burton, Cabrera, and Frank[12] predicted the existence of a roughening transition on the basis of the following argument: if the height of the columns are restricted to two levels 0 and 1, the model reduces to a lattice gas which is equivalent to a 2D Ising model. This model is well known to undergo a phase transition at $kT_c = 1.14\,J$. For $T < T_c$ there is a ferromagnetic ordered phase, while for $T > T_c$ it is a nonmagnetic disordered phase. However, this argument is only approximate since additional levels are also present. Indeed, the transition falls into another class: the Kosterlitz–Thouless[13] transition that we shall now describe.

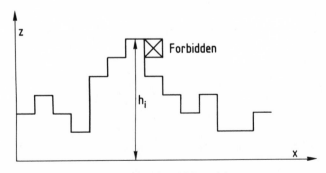

Figure 12.3. The SOS model.

12.5.2. A Renormalization Approach[14]

One considers a continuous solid. The surface energy is η per unit area. The surface is deformed by thermal fluctuations (Figure 12.4). The excess energy due to these fluctuations is

$$E = \iint dR(\eta\sqrt{(1 + \overline{\text{grad } z^2})} \tag{12.5.7}$$

The fluctuation $z(R)$ can be expanded as a Fourier series (wave vector k). The surface has size $L \times L$ and periodic boundary conditions are used. The average position is $\langle z \rangle = 0$. We define the operator $\sum_k$ as

$$\sum_k = (L^2/4\pi^2) \iint_{1/L}^{\Lambda_0} 2\pi k\, dk \tag{12.5.8}$$

where Λ_0 is a cutoff of the order of a^{-1} (a lattice parameter).
   If $|\text{grad } z| \ll 1$, then

$$E = \sum_k L^2\eta k^2\langle z_k\rangle^2/2 \quad \text{and} \quad z(\mathbf{R}) = \sum_k z_k \exp(ik\mathbf{R}) \tag{12.5.9}$$

At equilibrium, each mode contributes by $kT/2$ to the total energy so that

$$\langle z_k^2 \rangle = T/(\eta k^2 L^2) \tag{12.5.10}$$

Thus

$$\langle z^2 \rangle = \sum_k \langle z_k^2 \rangle = (T/2\pi^n) \int_{1/L}^{\Lambda_0} dk/k = (T/\pi\eta) \ln(\Lambda_0 L) \tag{12.5.11}$$

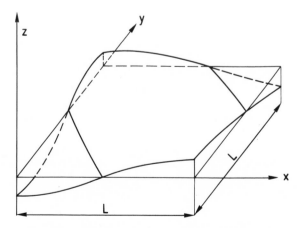

Figure 12.4. Fluctuations of a continuous solid surface.

and

$$\langle [z(\rho) - z(0)]^2 \rangle = \text{cte} + (T/\pi\eta) \ln(\rho) \qquad \text{for } \rho \text{ large} \qquad (12.5.12)$$

The conclusion is that fluctuation amplitudes diverge logarithmically on a free surface.

The atomic structure of the crystal can be taken into account by introducing a periodic potential which localizes the surface preferentially on horizontal parallel planes separated by a lattice parameter a (Sine-Gordon model). Then

$$E = \iint dR\eta\sqrt{1 - (\text{grad } z)^2} + U[1 - \cos(2\pi z/a)] \qquad (12.5.13)$$

One has to calculate the partition function:

$$Z = \iint dz_k \exp[-E(z_k)/k_B T] \qquad (12.5.14)$$

A renormalization procedure is applied to this equation. A new cutoff $\bar\Lambda$ is defined. Then a partial integration of equation (12.5.14) is conducted for short wavelengths ($\bar\Lambda < k < \Lambda_0$) and gives

$$\bar Z = \exp[-\bar E(k)/k_B T] \qquad (12.5.15)$$

where $\bar E$ is a function of the renormalized parameters $\bar\eta$ and $\bar V$. Hence

$$Z = \iint dz_k \bar Z \qquad (12.5.16)$$

Replacing $\bar\Lambda$ by $\bar\Lambda + d\Lambda$ leads to the renormalization equations

$$d\bar U/dl = \bar U[2 - (\pi T/\bar\eta a^2)], \qquad l = \ln(\bar\Lambda/\Lambda_0) \qquad \text{and} \qquad \bar U = \bar V/\bar\Lambda^2$$
$$(12.5.17)$$
$$d\bar\eta/dl = A(T)\bar U^2/Ta^2, \qquad A(T) \approx 0.2\text{-}0.4$$

This system defines the so-called Kosterlitz-Thouless transition. It is easy to integrate: the trajectories $\bar U/T$ as a function of $\bar\eta/T$ are displayed in Figure 12.5. Each different curve corresponds to a different temperature T. The line U_0, η_0 corresponds to the starting point of the renormalization $l = 0$. The renormalization curves are followed by increasing l from 0 up to infinity. There is clearly a fixed point at $T = T_R$ which separates two regimes:

For $T < T_R$ there is always a scale ξ above which $\bar U \to \infty$, in which case the surface is pinned at the potential $\bar U$.

When $T \to T_R$, then $\xi \approx \exp[C/(T - T_R)^{1/2}]$, and

$$\text{the singular part of } F \approx \exp[-C'/(T - T_R)^{1/2}] \qquad (12.5.18)$$

This is different from the Ising model, where $\xi \approx [(T - T_R)/T]^{-\nu}$.

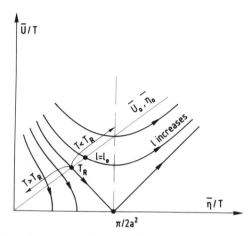

Figure 12.5. The renormalization trajectories for an isotropic surface.

For $T > T_R$, $\bar{U}$ vanishes at any large scale, hence the surface behaves for long-wavelength fluctuations as a free surface and ξ diverges for all $T > T_R$. We note again the difference with the Ising model, where ξ diverges only for $T = T_c$; here there is a line of critical points for all $T > T_R$. The previously derived equation for a free surface may still be used, provided η_0 is replaced by its renormalized value $\bar{\eta}$. For instance,

$$\langle [z(\rho) - z(0)]^2 \rangle = \text{cte} + (T/2\pi\bar{\eta}) \ln \rho \qquad \text{for } \rho \text{ large}$$

At $T = T_R$, $\bar{\eta}$ assumes the universal value $\bar{\eta}_R = \pi T_R / 2a^2$.

The singularity of ξ and F for $T = T_R$ is very weak: all their derivatives vanish at $T = T_R$. It is an infinite-order transition.

12.5.3. Application to Discrete Models: SOS, DGSOS, BCSOS[15]

In the previous renormalization calculation a development in powers of U has been used which is not valid for discrete models, since we then have $U_0 = \infty$. However, it can be shown directly that there exists also a roughening transition and that the previous results are still valid for $T \geq T_R$.

More precisely, the discrete Gaussian model (DGSOS)

$$E_{\{h\}} = (J/2) \sum (h_j - h_{j+\delta})^2 \tag{12.5.19}$$

has been shown by Chui and Weeks[16] to be equivalent to the Coulomb lattice gas which undergoes a transition from a dielectric to a conducting state.

The SOS models can be shown to be equivalent to planar XY models for spin systems which also undergo order–disorder transitions.[15,17] All these transitions fall into the Kosterlitz–Thouless class. The correspondence between these models is shown in Figure 12.6.

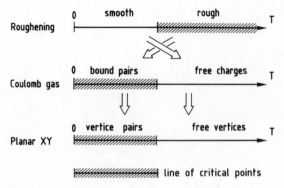

Figure 12.6. Correspondence between various equivalent models.

The relation between the energetic parameter J and T_R is not obtainable from renormalization calculations; Monte Carlo numerical simulations give[18]

ASOS model $kT_R = 1.24J$

DGSOS model $kT_R = 1.46J$

All these models describe a simple cubic crystal; a more realistic model has been proposed by van Beijeren[19] for fcc (100) faces, namely, the BCSOS model which is schematized in Figure 12.7. This model is formally equivalent to the 6-vertex model developed for ice crystals which is solvable analytically and presents also a Kosterlitz–Thouless transition.[20] In this case, for the BCSOS model

$$kT_R = (1/\ln 2)J \tag{12.5.20}$$

The BCSOS model can also be used for fcc (110) faces. In this case one must introduce two energies, J_y between nearest neighbors and J_x between next-nearest neighbors. The equivalent anisotropic 6-vertex model has also been solved giving the relation[21]

$$\exp\left(-J_x/k_B T_R\right) + \exp\left(-J_y/k_B T_R\right) = 1 \tag{12.5.21}$$

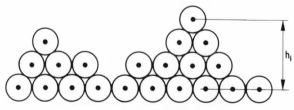

Figure 12.7. The BCSOS model.

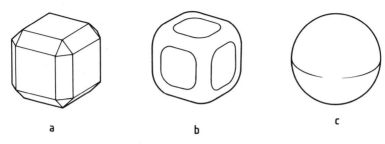

a   b   c

Figure 12.8. Crystal equilibrium shapes.

12.5.4. Consequences of the Roughening Transition for Crystal Shapes

The equilibrium crystal shape is due to the minimization of surface energy at constant volume. Far below T_R this is achieved when plane low-index facets delimited by sharp edges are developed (Figure 12.8a). As temperature is increased, steps appear and make the edges round (Figure 12.8b). Above T_R the free energy of creation of steps vanishes, at which point the surface-to-volume ratio is minimized when the crystal assumes a macroscopic quasi-spherical shape (in the absence of gravity) (Figure 12.8c). Thus the disappearance of facets is a signature of the roughening transition.

The first experimental evidence for a roughening transition was indeed the observation of this change of shape on a van der Waals crystal (C_2Cl_6, NH_4Cl)[22] but the most extensive studies were carried out at the liquid He–solid He interface.[23] Recently, roughening of metal faces [except the lowest-index faces: (111) and (100)] has been also observed (Pb,[24] Bi,[25] Au[25]).

For low-index faces of metals, such as fcc (111) and (100), the various SOS models previously discussed always predict a very high roughening temperature (larger than the melting point!), so they are not expected to roughen, in accordance with what is observed experimentally.

The cases of fcc (110) surfaces is less clear: there is some evidence that roughening occurs before melting in the case of Pb,[24] Bi,[25] and Ni[27]; for Ag[28] and Cu[29] X-ray experimentalists claim to have observed it around 650 K, but in the case of Cu their interpretation has been contested by a recent TEAS experiment.[30] The predictions of the anisotropic BCSOS model are inconclusive, owing to the uncertainties about the values of the energetic parameters to be introduced. This is certainly a very interesting field to be investigated both experimentally and theoretically.

We shall now develop in some detail the case of vicinal surfaces of fcc (100) metals, recently studied quite extensively by TEAS.

12.6. The Roughening Transition on Vicinal Surfaces

A vicinal surface is obtained by cutting a crystal along a plane making a small angle with a close-packed orientation. Figure 12.9 shows an example of the structure

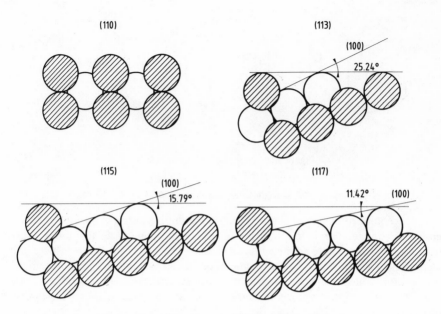

Figure 12.9. The structure of $(11n)$ vicinal surfaces.

of fcc (100) vicinal surfaces. They are composed of (100) terraces separated by parallel [110] monoatomic steps; their Miller index is $(11n)$ where n is an odd integer related to the terrace width l by

$$l = na/2$$

where a is the distance between nearest neighbors.

Before discussing the full problem of vicinal surfaces we examine first the case of an isolated step.

12.6.1. Statistical Equilibrium of an Isolated Step

As previously we initially deal with the case of a step on a continuous elastic solid (Figure 12.10). We can associate with the step a line tension β; the energy of the step is then simply

$$E = \beta \int_0^L dy \sqrt{1 + x'^2} \tag{12.6.1}$$

where L is the step length projected on the y-axis.

The same method as that developed above easily yields in two dimensions

$$\langle x^2 \rangle = LT/24\beta \tag{12.6.2}$$

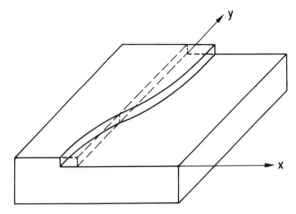

Figure 12.10. A free step on a continuous solid.

In one dimension the mean-square deviation varies linearly with the distance instead of logarithmically for two dimensions. Again the discreteness of the displacements may be taken into account by the introduction of a localization potential U. But in contrast with the two-dimensional case the renormalized value $\bar{U}$ is found to vanish at a large scale for any temperature. Thus an isolated step is always rough.

12.6.2. Statistics of a Vicinal Surface

LEED experiments have long shown that many vicinal surfaces are stable and can display very sharp diffraction patterns.[31] Thus some repulsive interaction must exist between steps. We have previously discussed some sources of repulsive interaction, such as elastic or dipole which decays as $1/d^2$. There is also a statistical repulsion arising from the impossibility for two steps to cross themselves; this leads also to a force decaying as $1/d^2$.

The roughening of such surfaces is quite different from the roughening of a close-packed surface. In the latter case one has to create steps on the surface, while in the former there is a first stage where the disorder is caused by the meandering of the steps without the creation of new ones. It is clear that it is easier to create kinks on a step than to create steps on a terrace. Terrace roughening can follow, but at a higher temperature. Pairs of kinks on the surface form detours (Figure 12.11) which, in order to minimize the interaction energy, have a tendency to form domains that retain the same structure as the original vicinal surface, but shifted. The lines of kink limiting these domains may be regarded as secondary steps.[32]

12.6.2.1. Renormalization Approach[33]

The step roughening can be treated as previously in a continuum picture. However, here the surface is anisotropic so two different excess surface energies must be considered, η_x and η_y, where η_y corresponds to the step rigidity as a result of the kink creation energy and η_x to the step-to-step interaction. Discrete displacements of the steps are favored by a localization potential U, so that the energy can be expressed in the form

$$E = \tfrac{1}{2}\sum L^2(\eta_x k_x^2 + \eta_x k_y^2)\langle u_k^2\rangle + U\{1 - \cos[\pi u_m(y)/a]\} \qquad (12.6.3)$$

where $u_m(y)$ is the displacement of the mth step at the ordinate y,

$$u_m(y) = \sum_k u_k \exp(ikR)$$

The same renormalization procedure leads to the same Kosterlitz–Thouless transition. Thus there is a transition temperature T_R above which $\bar{U}$ vanishes and the correlation function of step displacements is

$$\langle[u_m(y) - u_0(0)]^2\rangle = \text{cte} + (T/\pi\sqrt{\bar{\eta}_x\bar{\eta}_y})\ln(\rho) \qquad (12.6.4)$$

with $\rho^2 = y^2\sqrt{\bar{\eta}_y/\bar{\eta}_x} + m^2\sqrt{\bar{\eta}_x/\bar{\eta}_y}$.

The main difference between this and the previous result is contained in the distance ρ, which is now renormalized by the anisotropic surface energies.

12.6.2.2. Discrete Model[33]

It is possible to introduce a SOS description of the step roughening. It is assumed that the creation energy of a kink is W_0 and that the energy needed to displace a step by one atomic length between its two neighbors is w_n (Figure 12.12). This is completely analogous to a SOS model where the columns are lying flat on the surface. However, there is now an anisotropy in the interactions since

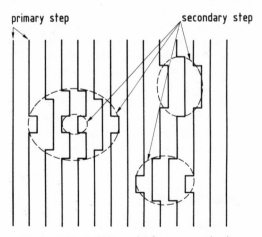

Figure 12.11. A TLK model for step roughening.

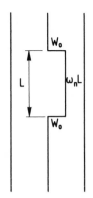

Figure 12.12. The VGL model.

w_0 is expected to be larger than w_n. Of course the previously defined parameters η_x and η_y are closely related to w_0 and w_n. An approximate relation has been given by Villain *et al.*[33] for $w_n \ll k_B T \ll W_0$:

$$\bar{\eta}_x = (k_B T/2) \exp(W_0/k_B T), \qquad \bar{\eta}_y = w_n \tag{12.6.5}$$

but the exact relation must be found in each particular case by a numerical Monte Carlo calculation.

12.6.3. Experimental Results

Vicinal surfaces Ni(113),[34] Ni(115),[35] Cu(113),[36] Cu(115),[37] and Cu(1, 1, 11)[38] have been studied by TEAS, and Cu(113) by X rays.[39] In the X-ray experiment the peak profiles were not analyzed, so we will concentrate our discussion on TEAS data. It is clear from Figure 12.11 that long-wavelength fluctuations of the steps create on the surface shifted domains, to which the analysis developed previously can be applied. Substitution of logarithmic correlation (12.6.4) in expression (12.2.9) yields the scattered intensity in the form

$$I = F(Q)f(\tau) \sum_{-\infty}^{+\infty} [a^2 q_y^2 \sqrt{\bar{\eta}_y/\bar{\eta}_x} + (Q_x l' + 2\pi)^2 \sqrt{\bar{\eta}_x/\bar{\eta}_y}]^{-1+\tau/2} \tag{12.6.6}$$

with

$$\tau = (T/\sqrt{\bar{\eta}_x \bar{\eta}_y})(\pi^2/4)(1 - \cos a q_x) \tag{12.6.7}$$

Here Q_x is the momentum exchange component parallel to the $(11n)$ surface and normal to the step, q_x and q_y are the momentum exchange in the (100) terrace plane, and l' is the step distance in the $(11n)$ plane.

Thus the diffraction pattern consists of peaks broadened according to power laws and centered on Bragg positions.

For $aq_x = 2n\pi$ (in-phase condition), $\tau = 0$ and the peak reduces to a δ-function.

For $aq_x = (2n + 1)\pi$ (antiphase condition), $\tau = (\pi^2/2)(T/\sqrt{\bar{\eta}_x \bar{\eta}_y})$, in which case the broadening is maximum. At $T = T_R$ one gets the universal value $\tau = 1$.

The peak shape is anisotropic; the anisotropy differs from the geometric anisotropy by a factor $\bar{\eta}_x/\bar{\eta}_y$.

Figure 12.13 shows the diffraction pattern for Cu(115) at two temperatures; broadening of the peaks in the antiphase condition is clearly observed. Figure 12.14 shows how the in-plane and out-of-plane profiles in the antiphase condition are fitted by the power law; $\bar{\eta}_x$ and $\bar{\eta}_y$ are the adjustable parameters. It is very important to note here that the profiles are found to be broader in-plane than out-of-plane, which is the reverse of the trend expected from the geometric anisotropy; this behavior emphasized the influence of the energetic anisotropy. The variations of τ and $\bar{\eta}_x/\bar{\eta}_y$ as a function of T are shown in Figure 12.15. The condition $\tau = 1$ in the antiphase yields: Cu(113), $T_R = 380$ K; Cu(115), $T_R = 720$ K.

Monte Carlo calculations[35,38] using a TSK model have been carried out for these surfaces. The parameters are W_0 and w_n; the correlation is computed and values for $\bar{\eta}_x$, $\bar{\eta}_y$ are deduced and can be compared with the experimental data. The best fit is obtained for:

$$\text{Cu(113),} \qquad W_0 = 800 \pm 100 \text{ K}, \qquad w_3 = 560 \pm 50 \text{ K}$$

$$\text{Cu(115),} \qquad W_0 = 900 \pm 100 \text{ K}, \qquad w_5 = 100 \pm 30 \text{ K}$$

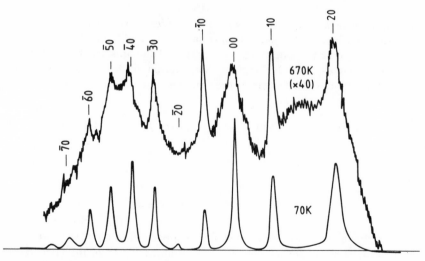

Figure 12.13. Diffraction pattern of He on Cu(115) for two temperatures, $T = 70$ K and $T = 670$ K.

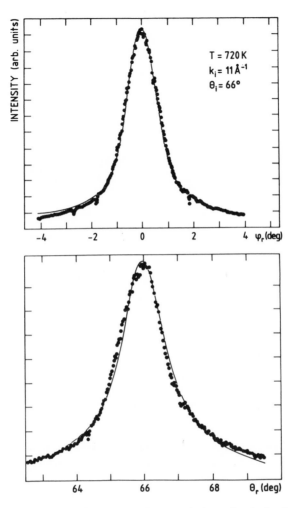

Figure 12.14. Fit of a power law to the experimental peak shapes for the in-plane (lower curve) and out-of-plane (upper curve) profiles.

The creation energy of a kink W_0 is, as expected, very similar for the two faces. However, its value is much lower than the estimated values (see above). Indeed, the latest theoretical calculations give $W_0 = 1500$ K. In fact, these calculations give the energy of creation at 0 K. The experimental values are rather free energies of creation at $T \geq T_R$. The vibrational entropy could eventually account for the difference.

The value of the interaction energy w_n is more delicate to discuss. The value of w_5 is in good agreement with a rough theoretical estimate:[40] $w_5 = 120$ K. As defined here (interaction of a step between its two neighbors) w_n is expected to vary as $1/l^4$. This dependence is not verified here, but it is not very surprising

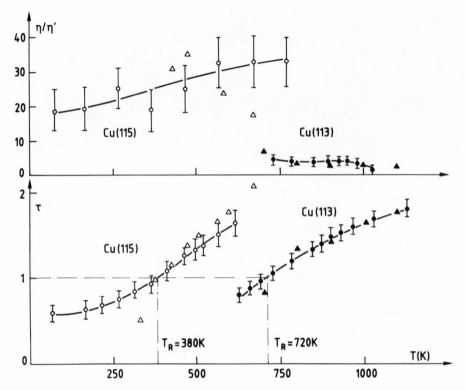

Figure 12.15. Plot of the exponent τ and the energetic anisotropy vs. T: ●○, experimental data; ▲△, calculated points.

since, in the case of Cu(113), the step-to-step distance is so short that direct interaction is certainly important. Unfortunately, in the case of Cu(1, 1, 11),[38] although the correct trend is observed qualitatively for the anisotropy, the peak profiles do not obey a power law. The origin of this discrepancy is still not clear.

12.7. Mobility of Defects

When the temperature is not zero, the various defects described here are able to diffuse on the surface. We shall not discuss the problem of diffusion on surfaces because it is developed elsewhere in this volume. However, we must point out that all that we have said about thermodynamical equilibrium of defects on a surface supposes that the mobility is high enough to insure that the equilibrium can be achieved in a reasonable time. This point has been carefully checked in the case of experiments conducted on copper vicinal surfaces.[41] Above room temperature perfect reversibility is observed, so that mobility is high on the scale of a few minutes. On the other hand, below 200 K mobility becomes negligible over the same time scale: the surface is frozen. This freezing is especially important when

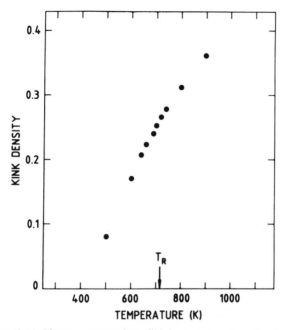

Figure 12.16. Linear concentration of kinks vs. temperature for Cu(113).

the roughening temperature T_R is lower than the freezing temperature [this is the case for Cu(1, 1, 11)]. Then, the surface can never be observed in a smooth state.

12.8. Conclusion

The main idea which emerges from this short review is that perfect plane crystal surfaces cannot occur in nature. Indeed, structural defects are always present and the thermodynamical equilibrium gives the lowest limit for their concentration. This limit is of course temperature-dependent. Figure 12.16 shows, for example, the kink linear concentration vs. temperature for the vicinal face Cu(113) as deduced from the previous measurements. The presence of these defects must always be kept in mind when surface properties like the chemical reactivity or the crystal growth behavior are studied. Not enough is known about the energetics and mobilities of surface defects, and much applied and basic research needs to be carried out.

Acknowledgments

The author is very grateful to his colleagues G. Armand, L. Barbier, F. Fabre, J. Perreau, and B. Salanon for many valuable discussions and critical reading of the manuscipt. The presentation of the roughening transition has been largely inspired by the lectures given at the College de France by Prof. P. Nozieres, to whom the author is very indebted. The treatment of the roughening of vicinal surfaces is mainly due to J. Villain and D. R. Grempel, to whom thanks are also due.

References

1. M. Mundshau, E. Bauer, and W. Swiech, *Phys. Rev. Lett.* **203**, 412 (1988).
2. G. Comsa and B. Poelsma, *Appl. Phys.* **A38**, 153 (1985).
3. G. E. Tommei, A. C. Levi, and R. Spadacini, *Surf. Sci.* **125**, 312 (1981).
4. G. Armand and B. Salanon, *Surf. Sci.* **217**, 341 (1989).
5. J. Lapujoulade, *Surf. Sci.* **108**, 317 (1981).
6. G. Armand and B. Salanon, *Surf. Sci.* **217**, 317 (1989).
7. G. Allan and J. Wach, in: Proc. 4th Int. Conf. on Solid Surfaces CANNES 1980 (D. A. Degras and M. Costa, eds.) p. 11, Le Vide, Les couches Minces Suppl. No. 201 (1980).
8. P. Wynnblatt, *Phys. Status Solidi* **36**, 797 (1969).
9. A. Haug, *Theoretical Solid State Physics*, Pergamon Press, Oxford (1972).
10. J. M. Blakely and R. L. Schwoebel, *Surf. Sci.* **26**, 321 (1971).
11. M. C. Desjonqueres, *J. Phys. (Paris), Colloq.* C3, Suppl. No. 4, **41**, C3-243 (1980).
12. W. K. Burton, N. Cabrera, and F. C. Frank, *Philos. Trans. R. Soc. London* **243**, 299 (1951).
13. J. M. Kosterlitz and D. J. Thouless, *J. Phys. C* **6**, 1181 (1973).
14. P. Nozieres and F. Galley, *J. Phys. (Paris)* **48**, 353 (1987).
15. J. D. Weeks, in: *Ordering in Strongly Fluctuating Matter* (T. Riste, ed.), p. 293, Plenum Press, New York (1980).
16. S. T. Chui and J. D. Weeks, *Phys. Rev. B* **14**, 4978 (1976).
17. J. M. Kosterlitz, *J. Phys. C* **7**, 1046 (1974).
18. W. J. Shugard, J. D. Weeks, and G. H. Gilmer, *Phys. Rev. Lett.* **41**, 1399 (1978).
19. H. van Beijeren, *Phys. Rev. Lett.* **38**, 993 (1977).
20. E. H. Lieb, *Phys. Rev. Lett.* **18**, 1046 (1967).
21. R. Youngblood, J. D. Axe, and B. M. McCoy, *Phys. Rev. B* **21**, 2512 (1980).
22. K. A. Jackson and D. E. Miller, *J. Cryst. Growth* **40**, 169 (1977).
23. S. Balibar and B. Castaing, *J. Phys. Lett. (Paris)* **41**, L329 (1980); J. E. Avon, L. S. Balfour, L. G. Kupfer, J. Landau, S. G. Libson, and L. S. Schumann, *Phys. Rev. Lett.* **45**, 814 (1980).
24. J. C. Heyraud and J. J. Metois, *Surf. Sci.* **128**, 334 (1983).
25. J. C. Heyraud and J. J. Metois, *Surf. Sci.* **177**, 213 (1986).
26. J. C. Heyraud, Thesis, Faculté des Sciences et Techniques de St-Jerome, Université d'Aix-Marseille, France (1987).
27. K. Yamashita, H. P. Bonzel, and H. Ibach, *Appl. Phys.* **25**, 231 (1981).
28. S. G. J. Mochrie, *Phys. Rev. Lett.* **59**, 304 (1987).
29. G. A. Held, J. L. Jordan-Sweet, P. M. Horn, A. Mak, and R. J. Birgeneau, *Phys. Rev. Lett.* **59**, 2075 (1987).
30. P. Zeppenfeld, K. Kern, R. David and G. Comsa, *Phys. Rev. Lett.* **52**, 63 (1988).
31. G. E. Rhead and J. Perdereau, in: *Colloque International sur la structure et les propriétés des surfaces des solides*, Editions du CNRS, Paris (1969).
32. H. J. Schultz, *J. Phys. (Paris)* **46**, 257 (1985).
33. J. Villain, D. R. Grempel, and J. Lapujoulade, *J. Phys. F* **15**, 804 (1985).
34. E. H. Conrad, L. R. Aten, D. L. Blanchard, and T. Engel, *Surf. Sci.* **187**, 1017 (1987).
35. E. H. Conrad, R. M. Aten, D. S. Kaufman, L. R. Allen, T. Engel, M. den Nijs, and E. K. Riedel, *J. Chem. Phys.* **84**, 1015 (1986) and Erratum **85**, 4756 (1986).
36. B. Salanon, F. Fabre, D. Gorse, J. Lapujoulade, and W. Selke, *J. Vac. Sci. Technol.* **A6**, 655 (1988); B. Salanon, F. Fabre, J. Lapujoulade, and W. Selke, *Phys. Rev. B* **38**, 7385 (1988).

37. F. Fabre, D. Gorse, J. Lapujoulade, and B. Salanon, *Europhys. Lett.* **3**, 737 (1987); F. Fabre, D. Gorse, B. Salanon, and J. Lapujoulade, *J. Phys. (Paris)* **48**, 2447 (1987).

38. F. Fabre, B. Salanon, and J. Lapujoulade, *Solid State Commun.* **64**, 1125 (1987).

39. K. S. Liang, E. B. Sirota, K. L. d'Amico, G. J. Hughes, and S. K. Sinha, *Phys. Rev. Lett.* **59**, 2447 (1987).

40. D. Gorse, Thesis, Université de Paris—Sud, Orsay, France (1986).

41. F. Fabre, D. Gorse, B. Salanon, and J. Lapujoulade, *Surf. Sci.* **175**, L693 (1986).

Molecular Scattering from Surfaces (Theory)

A. E. DePristo

13.1. Introduction and Scattering from Rigid Surfaces

In this chapter, the concepts and methods associated with the scattering of molecules from clean, perfectly periodic solid surfaces are introduced. Scattering from adsorbate covered and/or nonperiodic surfaces will not be discussed, since these areas are much more complex computationally and/or insufficiently advanced conceptually. Even with this restriction, such a wide variety of physical phenomena can occur in the scattering that it will not be possible to describe these within the confines of a single chapter. And, of course, in such a chapter no attempt will even be made to provide a comprehensive list of references. First the types of phenomena will be mentioned and then the restricted range will be considered.

A "cartoon" illustration of the various scattering processes is shown in Figure 13.1. In the simplest scattering event (Figure 13.1a), diffraction, a molecule reflects elastically from the solid, changing only the direction of its momentum. More complexity occurs when the molecule changes its internal rotational, vibrational, and/or electronic states (also possible in Figure 13.1a); this is denoted by inelastic scattering in this chapter. Even more complexity occurs in the dissociative chemisorption reaction (in Figure 13.1b), $AB(g) + M \rightarrow A(a) + B(a) + M$. [Throughout this chapter solid surfaces are referred to by the symbol "M," a gas phase species by "(g)," and an adsorbed species by "(a)."] It is also possible to have the scattering event disrupt the surface by ejecting surface atoms into the gas phase either directly by sputtering (in Figure 13.1c) or indirectly by forming volatile molecular species via reaction between the gas molecule and the surface atoms. And, in all of these processes, transfer of energy between the molecular and solid degrees of freedom may occur. The relevant solid's degrees of freedom for low-energy excitations involve motion of the nuclei (i.e., phonons), and for metals,

A. E. DePristo • Department of Chemistry, Iowa State University, Ames, Iowa 50011, USA.

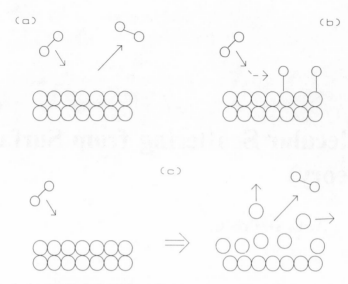

Figure 13.1. A schematic illustration of various collision processes between a gas atom or molecule and a solid surface. (a) elastic, diffractive, and inelastic scattering; (b) dissociative chemisorption; (c) sputtering.

excitation and de-excitation of electrons with energies close to the Fermi level [i.e., electron–hole pair or (e, h) processes]. High-energy excitations include collective excitation of the free electrons in a metal (i.e., plasmons), electronic excitation of core electrons of the surface atom, and the previously mentioned ejection of the surface atoms.

From the above list, only elastic, inelastic, and dissociative chemisorption events will be considered. For these cases, the treatment of low-energy excitation modes in the solid will also be described. Elastic and inelastic scattering have been subjected to intensive theoretical studies by many workers for over a decade, and for certain simple problems, much longer. A significant literature is available for such processes,[1-5] including a well-written (though slightly outdated) book, *Dynamics of Gas Surface Scattering*.[1] A number of reviews have appeared over the last decade.[2-5] Highly recommended is the recent comprehensive review of theory by Gerber,[4] who provides an introduction to a number of topics, along with a representative reference list.

High-energy scattering in which the surface atoms are ejected has been treated by a number of research groups. The interested reader is referred to the work of Garrison and co-workers for an introduction and relevant references.[6] To this author's knowledge, theoretical treatments of the dynamics of formation of volatile products, such as in the reaction of F with Si to form SiF_n, are not sufficiently advanced to warrant a description since the underlying potential-energy surface (PES) is essentially unknown, and must be extremely complex to describe the many arrangements occurring in such processes. There is little experience with multiarrangement PES in simpler gas-phase systems, not to mention in extended systems. Dissociation of gas-phase species via such processes as $AB(g) + M \rightarrow A(g) + B(g) + M$ or $AB(g) + M \rightarrow A(a) + B(g) + M$ will also not be described. This area is more complex than dissociative chemisorption, since one or two more arrangements must be included in a PES. These are higher-energy channels, since at most only one gas–surface bond is formed to compensate for the breaking of the molecular bond. At present, these have been treated using only the simplest PES, not allowing for multiple rearrangement channels. The reader is referred to the work of Gerber and co-workers as reviewed elsewhere[4] for more details.

Consideration has been restricted to the above-mentioned processes since these occur at kinetic energies less than 1 eV, which are important for thermally activated chemical processes. This energy range is also convenient for a single chapter since it naturally restricts the physical phenomena to a subset that can be described in a coherent manner.

We begin with the treatment of diffraction and ro-vibrationally inelastic scattering from rigid surfaces. Neglect of surface atom motion is rather severe unless the mass of the impinging molecule is much less than that of the solid's atoms. For solids in the third row, this limits the treatment to H_2 and perhaps D_2.

The solid is assumed to have perfect two-dimensional (2D) periodicity described by the surface reciprocal lattice vector G_{mn}. The fundamental quantum feature of the scattering is that the linear momentum of the scattered molecule parallel to the surface only changes by $\hbar G_{mn}$. This is simply the result of the scattering of a wave from a periodic array. If the initial and final momenta of the center of mass of the molecule are $\hbar k$ and $\hbar k$ and $\hbar k_f$ with the 2D components in the surface plane being K and K_f, respectively, and if the molecular mass is M and the internal energies are ε_i and ε_f, respectively, then conservation of energy and linear momentum yield

$$k_f^2 + 2M\varepsilon_f/\hbar^2 = k^2 + 2M\varepsilon_i/\hbar^2 \tag{13.1.1}$$

and

$$K_f = K + G_{mn} \tag{13.1.2}$$

where G_{mn} are 2D reciprocal lattice vectors. Such equations describe only the kinematics of the scattering, since they are independent of the interaction between the molecule and the solid.

The Hamiltonian is simply

$$H = T(r) + h(r_i) + V(r, r_i) \tag{13.1.3}$$

where $r = (x, y, z)$ specifies the position of the center-of-mass (CM) of the gas molecule and r_i denotes the internal molecular coordinates; $T(r)$ is the kinetic energy operator for the CM motion, $h(r_i)$ is the Hamiltonian for the internal motion of the free molecule, and $V(r, r_i)$ is the interaction potential between the molecule and the surface. Since the surface is rigid, V depends only upon r and r_i. Furthermore, the 2D periodicity of the potential implies the following expansion:

$$V(r, r_i) = \sum V_{mn}(z, r_i) \exp(iG_{mn} \cdot R) \tag{13.1.4}$$

where $\mathbf{R} = (x, y)$ specifies the 2D components of $\mathbf{r}$ in the surface plane. It is the periodicity in equation (13.1.4) that distinguishes molecule–surface scattering from other types of scattering, such as molecule–molecule scattering. This means that the standard quantum-mechanical scattering formalism can be used, but that the boundary conditions will be specific to the molecule–surface case.

The wave function is also expanded in products of 2D Bloch waves and internal wave functions,

$$\Phi(\mathbf{r}, \mathbf{r}_i) = \sum \exp\left[i(\mathbf{K} + \mathbf{G}_{mn}) \cdot \mathbf{R}\right]\phi_j(\mathbf{r}_i)g(z; m, n, j) \qquad (13.1.5)$$

The factor $\exp(i\mathbf{K} \cdot \mathbf{R})$ originates from the free molecule wave function,

$$\Phi_0(\mathbf{r}, \mathbf{r}_i) = \exp\left(i\mathbf{k} \cdot \mathbf{r}\right)\phi_j(\mathbf{r}_i) \qquad (13.1.6)$$

Equation (13.1.5) reduces to (13.1.6) for the $(m, n) = (0, 0)$ component when $g(z; 0, 0, j)$ satisfies a single-particle one-dimensional (1D) free wave equation [as it will in equation (13.1.7)]. Substitution of equation (13.1.5) into the time-independent Schroedinger equation, and use of orthogonality for both the 2D Bloch wave functions and functions ϕ_j, yields the set of close-coupled (CC) equations to be solved for the unknown z-dependent functions $g(z; m, n, j)$:

$$[d^2/dz^2 + [\mathbf{d}^2]_{mnj}]g(z; m, n, j)$$

$$- (2M/\hbar^2) \sum V(z; m, n, j, m', n', j')g(z; m', n', j') = 0 \qquad (13.1.7)$$

$$[\mathbf{d}^2]_{mnj} = (2M/\hbar^2)(E - \varepsilon_j) - (\mathbf{K} + \mathbf{G}_{mn})^2 \qquad (13.1.8)$$

$$V(z; m, n, j, m', n', j') = \langle\phi_j, \mathbf{G}_{mn}|V(\mathbf{r}, \mathbf{r}_i)|\phi_{j'}, \mathbf{G}_{m'n'}\rangle/A \qquad (13.1.9)$$

where A is the area of the unit cell. It is only necessary to specify the boundary conditions to finish the theoretical description. First, since the potential is assumed to become more and more repulsive as $z \to -\infty$, the wave functions $g(z \to -\infty; m, n, j)$ are forced to vanish. Second, since the potential is assumed to vanish as $z \to \infty$, the translational wave function must be a linear combination of incoming and outgoing plane waves. The incoming plane wave must have only a single component since equation (13.1.5) matches equation (13.1.6) with $\mathbf{k} \cdot \hat{z} < 0$. The net result is the boundary condition

$$g(z; m', n', j') \to (d_{mnj})^{-1/2} \exp\left(-id_{mnj}z\right)\delta_{m'n'j',mnj}$$

$$+ (d_{m'n'j'})^{-1/2} \exp\left(id_{m'n'j'}z\right)S(m', n', j' \leftarrow m, n, j) \qquad (13.1.10)$$

which defines the scattering or S-matrix. The factors $(d_{m'n'j'})^{-1/2}$ arise from enforcement of normalization of the square of the S-matrix. Because the equations reduce to those of an incoming plane wave with only the $(m, n) = (0, 0)$ channel, all the diffraction information is contained in the elements $S(m', n', j' \leftarrow 0, 0, j)$. Hence only one column of the S-matrix, for each $j' \leftarrow j$, is of interest for diffraction. In general, however, all columns must be solved for in this close-coupled set of differential equations, since that is the only way to ensure satisfying the proper boundary conditions.

It is currently possible to solve such sets of CC equations with high accuracy and numerical efficiency for perhaps a few hundred equations, with the computational time increasing like N_{eq}^3. If many calculations are required, as in adjustment of a PES for comparison to experimental data, fewer coupled equations can be handled. To get an idea of how many equations enter in a typical problem, consider $H_2/Ni(100)$ scattering and assume that no asymptotically closed channels, $[d^2]_{mnj} < 0$, must be retained. Then, the number of coupled equations is specified by $[d^2]_{mnj} > 0$. With $\mathbf{K} = 0$, $M = 2$ amu, $E = 0.1$ eV, $\mathbf{G}_{mn} = (m, n)2\pi/2.49$ Å, $\varepsilon_j = 0.00744j(j + 1)$ eV, we find $(\hbar\mathbf{G}_{10})^2/2M = 0.00665$ eV and thus that for $j = 0$, all $(m, n) = (0, 0), (1, 0), (1, 1), (2, 0), (2, 1), (2, 2), (3, 0), (3, 1), (3, 2)$ are allowed, and for $j = 2$, all $(m, n) = (0, 0), (1, 0), (1, 1), (2, 0), (2, 1), (2, 2)$ are allowed. Taking into account the symmetrical terms $(-1, -1), (0, 2)$, etc., we find the number of (m, n) channels is 1, 4, 4, 4, 8, 4, 4, 8, 8 for the $j = 0$ terms listed and 1, 4, 4, 4, 8, 4 for the $j = 2$ terms listed. Taking into account that each $j = 2$ rotational state is split into 5 m_j levels, we find that there are $45 + 125 = 170$ coupled equations. This number will decrease at lower E and higher incident angles. While symmetry can be used to lower the 170 dramatically in this case, it is clear that quantum CC approaches are inherently limited in the types of problems which can be treated.

A different type of quantum approach eliminates the use of a basis set entirely by solving the time-dependent Schroedinger equation (TDSE) directly.[4,7,8] One starts with a wave packet, $\Phi(t; \mathbf{r}, \mathbf{r}_i)$, which is localized at large z and with an average velocity directed toward the surface. The time-dependent Schroedinger equation is then solved over a short time interval $[t, t + \Delta t]$ via

$$\Phi(t + \Delta t; \mathbf{r}', \mathbf{r}_i') = \int \langle \mathbf{r}', \mathbf{r}_i' | \exp(-iH\Delta t/\hbar) | \mathbf{r}, \mathbf{r}_i \rangle \Phi(t; \mathbf{r}, \mathbf{r}_i) \, d\mathbf{r} \, d\mathbf{r}_i \qquad (13.1.11)$$

or via simpler finite-difference evaluation in time. To illustrate how the evaluation of the operator in the exponential is accomplished, we use a simpler notation suppressing the dependence upon r_i. Then a short time propagator is assumed (i.e., using $[T, V] = 0$), yielding

$$\exp(-iH\Delta t/\hbar) \approx \exp(-iT\Delta t/\hbar) \exp(-iV\Delta t/\hbar) \qquad (13.1.12)$$

Now T is a local operator in momentum space while V is a local operator in coordinate space. Putting a complete set of states $|\mathbf{k}\rangle\langle\mathbf{k}|$ in obvious places we find

$$\Phi(t + \Delta t; \mathbf{r}') = \int \langle \mathbf{r}' | \mathbf{k} \rangle \langle \mathbf{k} | \exp(-iT\Delta t/\hbar) | \mathbf{k} \rangle \langle \mathbf{k} | \mathbf{r} \rangle$$

$$\times \langle \mathbf{r} | \exp(-iV\Delta t/\hbar) | \mathbf{r} \rangle \Phi(t; \mathbf{r}) \, d\mathbf{k} \, d\mathbf{r} \qquad (13.1.13)$$

The transformation between coordinate and momentum space is accomplished via the fast Fourier transform (FFT) algorithm. This transforms the coordinate space evaluation of $\exp(-iV\Delta t/\hbar)\Phi(t)$ into momentum space; then the effect of $\exp(-iT\Delta t/\hbar)$ is determined in this space to yield the momentum-space wave function at $t + \Delta t$; and, finally, the momentum-space wave function is transformed back to coordinate space via another FFT. Proceeding to the next time step, this procedure eventually yields the wave function as $t \to \infty$ from which all scattering information can be extracted via projections onto plane waves.

A number of advantages can be attributed to such a time-dependent wave-packet approach. First, it allows computation of only one column of the S-matrix, exactly what is needed. Since a wave packet is built from a number of translational energy eigenstates, this one column can be determined for a number of energies in a single calculation. Second, the method can be applied to nonrigid surfaces by including a time-dependent classical motion for thee surface atoms. Third, the method is applicable to nonperiodic surfaces. The major limitation is computational: for each time step, the computational time is proportional to $N_g \ln N_g$ where N_g is the number of grid points used in the evaluation of the FFT. Assuming for simplicity an equivalent and low number of grid points of $2^4 = 16$ in each degree of freedom, then in M degrees of freedom we have $N_g = 16^M$ which limits current treatments to $M \leq 3$ on current supercomputers unless only a single calculation must be performed, in which case $M = 4$ may be feasible. However, even a rigid rotor–rigid surface collision entails $M = 5$ so that the FFT approach is currently only applicable to atom–surface scattering. However, in that case it should provide much more capabilities than the standard time-independent scattering theory presented previously. This argument is based upon the assumption that the number of time steps is not too large, which limits the approach to direct scattering.

Before leaving the simplest type of molecule–surface collision system, I will mention something about the interaction potential, $V(\mathbf{r}, \mathbf{r}_i)$. For simplicity, we ignore the structure of the surface for now. Only the term $V_{00}(z, \mathbf{r}_i)$ in equation (13.1.4) enters then. At distances far from the surface, the molecule–surface interaction is a van der Waals attraction, proportional to z^{-3}. At close distances, the overlap of the electronic distributions of the molecule and the surface leads to a repulsive interaction due to kinetic energy repulsion. For a nonreactive PES, the dependence of these interactions on the molecular bond length variation is quite weak, indicating that only the z and $\cos\theta = \hat{\mathbf{r}}_i \cdot \hat{z}$ dependence are important. The angular dependence is conveniently represented in a Legendre series:

$$V(z, \theta) = \sum V_l(z) P_l(\cos\theta) \tag{13.1.14}$$

where the terms with $l > 0$ give rise to rotational transitions. The more anisotropic the molecule solid PES, the larger the number of terms that must be retained in the expansion. The anisotropy in the long-range z^{-3} term can be determined from the anisotropy in the molecular polarizability, but the anisotropy at short range is much more difficult to determine.[9]

One method to determine the corrugated, $(m, n) > (0, 0)$, terms in the full PES involves simply assuming that higher terms are proportional to the $(0, 0)$ term. More sophisticated yet is the idea of using different proportionality constants for the long- and short-range parts separately, since it is unlikely that the long-range term varies significantly with position in the unit cell. The best determined potentials use many parameters in the PES form, and fit to accurate scattering data, but this has really only been done for atom scattering from rigid surfaces.

13.2. Low-Energy Excitations: Phonons and Electron-Hole Pairs

In this chapter, the problem of inclusion of the motion of the solid's atoms is examined. Formalisms for the quantum treatment of one-phonon events have been developed but these are limited in applicability to low-energy processes and light gas molecules, for which single quantum events are the only important process. These will be mentioned in some detail at the end of the chapter. The most general fully quantum treatment (using path integrals and influence functionals) has not been developed or implemented to a significant degree.

In the main part of this chapter we introduce the most common method for the treatment of surface atom motion in collision processes, at least within a classical mechanics description. The basic physical idea is simple: the colliding molecule interacts strongly with a few nearby solid's atoms, which then interact less strongly with their own neighbors, and so forth. Eventually, the energy exchange between the gas molecule and a few strongly interacting solid's atoms in the initial collision is dissipated throughout the entire infinite solid.

The difficulty is describing the exchange of energy between the localized and extended system. While a simple brute force procedure is always capable of arbitrary accuracy if enough atoms are followed using molecular dynamics, such an approach may not be feasible for complex events which require thousands of molecular dynamics calculations to ensure statistically reliable results. Besides, this might not be a wise use of computational resources.

To make progress, we note that while the motion of the strongly perturbed solid's atoms will sample anharmonic forces from the other solid's atoms, the less strongly perturbed can be described by simple-harmonic restoring forces. In fact, we *define* a less strongly perturbed solid atom as one that can be described by harmonic restoring forces. Furthermore, the PES between the impinging molecule and the solid will be rather short-ranged, and thus will not affect the less strongly perturbed solid's atoms directly.

With the above ideas in mind, we consider a harmonic solid, following closely the derivation due to Tully[10] based upon early work by Adelman and Doll[11] and even earlier developments by Zwanzig.[12] The equations of motion are

$$m_\alpha \, d^2 r_{\alpha i} / dt^2 = -\sum k(\alpha_i, \beta j)(r_{\beta j} - r_{\beta j}^0) \qquad (13.2.1)$$

where m_α and r_α are the mass and position of the αth solid atom, with $i = 1, 2, 3$ the three cartesian directions; $k(\alpha i, \beta j)$ is the force constant for the influence on the αth atom in the ith direction due to a distortion of the βth atom in the jth direction; $r^0_{\beta j}$ is the equilibrium position for the βth atom in the jth direction. Defining mass-weighted displacement coordinates,

$$(\mathbf{u})_{\alpha j} = m_\alpha^{1/2}(r_{\beta j} - r^0_{\beta j}) \tag{13.2.2}$$

and a frequency matrix,

$$(\mathbf{\Omega}^2)(\alpha i, \beta j) = k(\alpha i, \beta j)/(m_\alpha m_\beta)^{1/2} \tag{13.2.3}$$

Equation (13.2.1) can be rewritten in matrix-vector notation:

$$d^2\mathbf{u}/dt^2 = -\mathbf{\Omega}^2\mathbf{u} \tag{13.2.4}$$

Equation (13.2.4) is now partitioned into two sets, one for N_P primary atoms and the other for N_S secondary atoms. This is done via the definitions $\mathbf{y} = \mathbf{Pu}$, $\mathbf{z} = \mathbf{Qu} = (\mathbf{1} - \mathbf{P})\mathbf{u}$, $\mathbf{\Omega}^2_{PP} = \mathbf{P\Omega}^2\mathbf{P}$, etc. with the result

$$d^2\mathbf{y}/dt^2 = -\mathbf{\Omega}^2_{PP}\mathbf{y} - \mathbf{\Omega}^2_{PQ}\mathbf{z} \tag{13.2.5}$$

and

$$d^2\mathbf{z}/dt^2 = -\mathbf{\Omega}^2_{QP}\mathbf{y} - \mathbf{\Omega}^2_{QQ}\mathbf{z} \tag{13.2.6}$$

The idea is to eliminate the large set of $\mathbf{z}$, at least formally. This can be accomplished by solving equation (13.2.6) for the secondary atom motion and substituting this solution into equation (13.2.5). The result is the fundamental integrodifferential equation for the time evolution of the primary zone atoms within the generalized Langevin equation (GLE) approach:

$$d^2\mathbf{y}/dt^2 = -\mathbf{\Omega}^2_{PP}\mathbf{y} + \mathbf{M}(0)\mathbf{y} - \mathbf{M}(t)\mathbf{y}(0) - \int \mathbf{M}(t - t')\, d\mathbf{y}(t')/dt'\, dt' + \mathbf{R}(t) \tag{13.2.7}$$

The memory function and random force are defined by

$$\mathbf{M}(t) = \mathbf{\Omega}^2_{PQ} \cos(\mathbf{\Omega}_{QQ}t)\mathbf{\Omega}^{-2}_{QQ}\mathbf{\Omega}^2_{QP} \tag{13.2.8}$$

and

$$\mathbf{R}(t) = -\mathbf{\Omega}^2_{PQ} \cos(\mathbf{\Omega}_{QQ}t)\mathbf{z}(0) - \mathbf{\Omega}^2_{PQ} \sin(\mathbf{\Omega}_{QQ}t)\mathbf{\Omega}^{-1}_{QQ}\, d\mathbf{z}(0)/dt \tag{13.2.9}$$

The memory kernel in equation (13.2.7) involves the response of the full many-body system and thus retains "memory" of previous velocities. This is distinct from a standard Langevin-type equation, which replaces the memory function by a δ-function and thus a local friction. The random force, $\mathbf{R}(t)$, and the memory kernel obey the fluctuation-dissipation theorem:

$$\langle \mathbf{R}(t)\mathbf{R}(0)^{\mathrm{T}} \rangle = kT\mathbf{M}(t) \tag{13.2.10}$$

where the $\langle \ \rangle$ indicate an average over initial conditions of the secondary atoms. The derivation of equation (13.2.10) uses the equilibrium properties:

$$\langle \mathbf{z}(0)\mathbf{z}(0)^T \rangle = kT\mathbf{\Omega}_{QQ}^{-2} \tag{13.2.11a}$$

$$\langle d\mathbf{z}(0)/dt \, d\mathbf{z}(0)^T/dt \rangle = kT\mathbf{1} \tag{13.2.11b}$$

Before considering the relationship of the GLE and MD approaches, it is perhaps worthwhile to describe the point of the transition from deterministic to stochastic dynamics which appears to have occurred in reformulating equations (13.2.5) and (13.2.6) into (13.2.7)-(13.2.9). We note first of all that the GLE is not truly a stochastic dynamical technique. If the positions and velocities of the secondary and primary atoms are specified initially, then the sequence of forces generated by $\mathbf{R}(t)$ are completely determined. However, if there are a large number of the secondary atoms (i.e., $\mathbf{z}$ is a long vector), then a small change in the initial positions and velocities will generate a random force with a completely different sequence. It is this pseudorandomness due to the sensitivity of $\mathbf{R}(t)$ to the initial conditions of the large number of secondary atoms that is commonly termed stochastic when referring to the GLE technique. It is noteworthy that such a randomness depends upon having a large number of secondary atoms, so that reversing the solution in equations (13.2.5) and (13.2.6) to treat the $\mathbf{z}$ as the primary and the $\mathbf{y}$ as the secondary atoms would lead to formally identical equations to (13.2.7)-(13.2.9) with $(P, Q) \rightarrow (Q, P)$. These would not be stochastic, since the small number of secondary atoms would not give rise to a pseudorandom sequence of forces.

Equations (13.2.7)-(13.2.10) are equivalent to the original set of molecular dynamics equations in equation (13.2.4) or equations (13.2.5) and (13.2.6) in their ability to mimic an infinite solid. It appears that the GLE equations are limited in one sense: they cannot resolve energy transfer into the individual modes of the extended solid. However, this is not true since $\mathbf{z}(t)$ is given in terms of $\mathbf{y}(t)$ by the solution of equation (13.2.6). Nonetheless, the GLE equations are no easier to solve in their present form since evaluation of $\mathbf{M}(t)$ and $\mathbf{R}(t)$ via equations (13.2.8) and (13.2.9) is clearly impractical since the length of the Q-vectors is $3N_S$, which must be quite large to simulate the bulk solid correctly. In addition, evaluation of a convolution integral with a general time dependence in the kernel is also very time-consuming. Diagonalizing $\mathbf{\Omega}_{QQ}$ to $\mathbf{\tau}$, rewriting $\cos(\tau_i t)$ as separable exponentials, and accruing integrals of the form

$$\int \exp(\tau_i t') \, d\mathbf{y}(t')/dt' \, dt'$$

is not practical since $6N_S$ such integrals must be accumulated even before doing all the multiplications in equation (13.2.8).

What is then the advantage of the GLE approach? The answer is that it is rather easy to approximate the memory kernel and random force to provide a reasonably accurate description of both the short-time and long-time (actually long-wavelength) response of the primary zone atoms to an external perturbation (e.g., a collision). We now describe the common approximation to the memory function.[10,13] One expects a decaying and oscillating function on physical grounds. Using

$$\mathbf{M}(t) = \mathbf{M}_0^{1/2} \exp(-\boldsymbol{\tau}t)[\cos(\mathbf{w}_1 t) + \tfrac{1}{2}\boldsymbol{\tau}\mathbf{w}_1^{-1}\sin(\mathbf{w}_1 t)]\mathbf{M}_0^{1/2} \qquad (13.2.12)$$

which is the multidimensional generalization of the position autocorrelation function of a Brownian oscillator, provides such a function with a number of unknown parameters. Forcing equations (13.2.8) and (13.2.12) and their second time derivatives to agree at $t = 0$ (their first time derivative vanishes identically) yields a set of equations for the unknown coefficient matrices[13]:

$$\mathbf{M}_0 = \boldsymbol{\Omega}_{PQ}^2 \boldsymbol{\Omega}_{QQ}^{-2} \boldsymbol{\Omega}_{QP}^2 \qquad (13.2.13)$$

$$\mathbf{M}_0^{1/2}\mathbf{w}_0^2\mathbf{M}_0^{1/2} = \boldsymbol{\Omega}_{PQ}^2 \boldsymbol{\Omega}_{QP}^2 \qquad (13.2.14)$$

where

$$\mathbf{w}_0^2 = \mathbf{w}_1^2 + \tfrac{1}{4}\boldsymbol{\tau}^2 \qquad (13.2.15)$$

The exponential decay parameter is chosen to produce the correct long-wavelength limit for the density of states,

$$\boldsymbol{\tau} = (\pi w_b/6)\mathbf{1} \qquad (13.2.16)$$

where w_b is the bulk Debye frequency.

An alternate procedure due to Tully,[10] and which preceded the above method, is to choose the elements of the parameter matrices to mimic the phonon spectrum of the crystal. For a monatomic fcc crystal, this spectrum is related to the Fourier cosine transform of the velocity autocorrelation function. After considerable algebra, it can be shown that

$$g(w) = kTw^2\boldsymbol{\tau}\mathbf{A}\mathbf{U}\mathbf{B}\mathbf{U}^{\mathsf{T}}\mathbf{A} \qquad (13.2.17)$$

where $g(w)$ is the density of states at frequency w. This is a projection onto each atom. The new matrices in equation (13.2.17) are given by

$$\mathbf{A} = \{w^2\mathbf{1} - \boldsymbol{\Omega}_{PP}^2\}^{-1} \qquad (13.2.18)$$

and

$$\mathbf{B} = \{w^2\boldsymbol{\tau}^2 + (\mathbf{U}^{\mathsf{T}}\mathbf{A}\mathbf{U} + \mathbf{w}_0^2 - w^2\mathbf{1})\}^{-1} \qquad (13.2.19)$$

The two methods for choice of the parameters will agree if an accurate full force constant matrix, $k(\alpha i, \beta j)$, or frequency matrix, Ω, are available. If not, then it is a matter of convenience which method is preferable. Either one will capture the essential features of the GLE, namely, frictional energy loss from the primary atoms to the secondary atoms and thermal energy transfer from the secondary atoms to the primary atoms. And both will provide a reasonable description of the bulk and surface phonon density of states of the solid. This is illustrated in Figure 13.2, where a LJ(12, 6) model of an fcc crystal is used to generate Ω^2; in Tully's method, a lower surface normal frequency component was chosen but, in any case, the agreement between the methods is reasonable. Neither will provide the exact time-dependent response of the solid due to the limited number of parameters used to describe the memory function.

The approximate memory function in equation (13.2.12) allows for a replacement of the GLE equation (13.2.7) by

$$d^2 y / dt^2 = -\Omega^2_{PP} y + M_0^{1/2} w_0 s \tag{13.2.20}$$

and

$$d^2 s / dt^2 = w_0 M_0^{1/2} y - w_0^2 s - \tau \, ds/dt + f(t) \tag{13.2.21}$$

where $f(t)$ is a Gaussian white noise random force. The fictitious particles obeying the equations of motion of s are commonly referred to as "ghost" atoms, and thus this formulation is referred to as the GLE-"ghost" atom method.

To emphasize the physics again for this method, we note that the harmonic restoring force, $-\Omega^2_{PP} y$, provides the interaction among the primary zone atoms. The coupling between the primary and "ghost" atoms is incorporated by the terms $M_0^{1/2} w_0 s$ and $w_0 M_0^{1/2} y$. The frictional force, $-\tau \, ds/dt$, and the Gaussian random force are related via a simpler fluctuation dissipation than equation (13.2.10):

$$\langle f(t) f(0)^T \rangle = 2kT\tau\delta(t) \tag{13.2.22}$$

where $\delta(t)$ is a Dirac delta function. This balance ensures that in the long-time limit, the positions and velocities for the primary and "ghost" atoms become thermally distributed at a temperature T.

Finally, a few points are noted about the GLE formalism. First, it is easy to remove the assumption of harmonic interactions among the primary atoms simply by replacing $-\Omega^2_{PP} y$ in equation (13.2.7) or (13.2.20) by the exact forces. This does not cause any difficulties with any of the other equations. However, it is not possible to implement an (atomic-based) GLE in any *practical* and systematic way if the primary–secondary or secondary–secondary interactions are anharmonic. Second, a complication arises within the GLE formalism due to the localized nature of the primary zone atoms. In a molecule–surface collision, the initial localized interaction specifies a set of primary zone atoms as illustrated for a diatomic interacting with

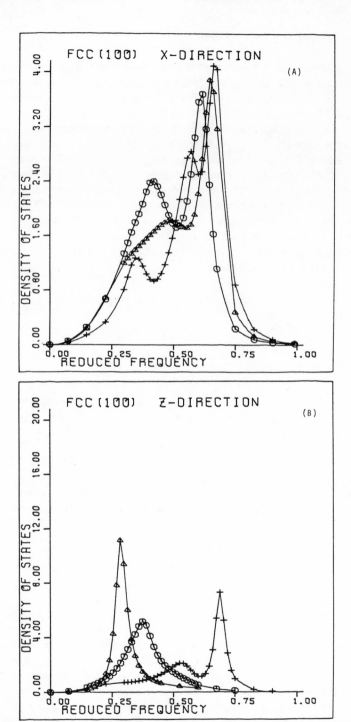

Figure 13.2A–D. Projected density of states for different faces, directions, and atoms using the parametrized memory function in equation (13.2.12). The curves are from DePristo.[13] The circles and plus symbols correspond to an atom in the first and second layers, respectively; the triangles are from Tully[10] for an atom in the first layer.

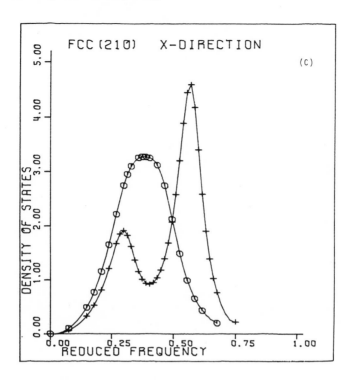

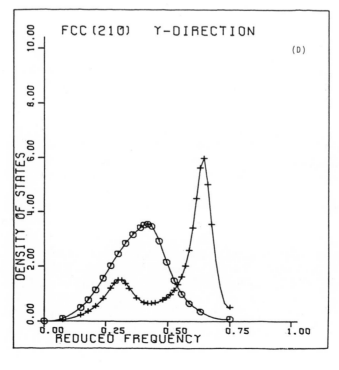

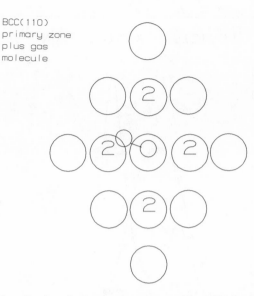

BCC(110)
primary zone
plus gas
molecule

Figure 13.3. Diatomic molecule and primary zone atoms used for GLE simulations of dissociative chemisorption on a bcc (110) surface. The atoms labeled by "2" are in the second layer.

a bcc (110) surface in Figure 13.3. (There is nothing fundamental about this surface or number of primary zone atoms; the same argument holds irrespective of the lattice and primary zone size.) When the molecule moves outside of these 13 atoms, a new set of primary zone atoms must be defined. This process is called switching of the primary zone. There is no *a priori* method to consistently define new primary zone atoms, since there is no information on the flow of energy from the original primary zone into the specific secondary zone atoms [i.e., the exact $M(t)$ is no longer available]. The currently used assumption is that the motion of the molecule across the surface is slow compared to thermalization of the surface atoms. Then the new primary zone atoms are reinitialized from a thermal distribution. Such a method will break down if the motion of the molecule is very fast and/or the distortion of the lattice is sufficiently great to inhibit thermalization on a fast enough time scale.

Before leaving this subject, we wish to mention that the GLE is not a particularly simple method to implement. One must choose the parameter matrices for the memory function, which will differ for each surface face and material, a procedure which is intensive of human time. Changing the number of primary zone atoms requires redetermination of these parameters, and major modifications of a computer code to implement the switching process. If a treatment of a particular solid and a surface face is to be the focus of investigation for an extended period of time, the investment in human time is definitely worthwhile. (If computational resources are limited to nonsupercomputers, it may be the only viable approach.) However, if a number of faces and materials are to be treated, still simpler methods combining LE and MD are desirable, although more expensive computationally.

Now we turn to a brief mention of the treatment of the other low-energy excitation modes, namely, electron–hole pair (denoted by e–h) creation and annihilation. It is our opinion that these will not be very important for translational to e–h pair processes since the coupling to phonons is so much stronger[14] as indicated in Figure 13.4. The exception may be atomic H and D.[15] The major role for e–h pair processes is likely in vibrational excitation and de-excitation for weakly interacting molecule–surface systems. It is emphasized that e–h pair processes are distinct from the adiabatic coupling of molecular and electronic degrees of freedom. For example, if the molecular bond length varies as the molecule approaches a surface, exchange of translational and vibrational energy is probable. This process should be relatively *independent* of the surface temperature. By contrast, experimental data[16] for the NO/Ag(111) system has shown that NO is vibrationally excited with a probability of the form

$$P = f(E_i \cos^2 \theta_i) \exp(-E_a/kT) \qquad (13.2.23)$$

which indicates that at $T = 0$ K there is no vibrational excitation. On the other hand, the probabilities are quite small even at $T = 760$ K, increasing from 0.01 to 0.06 as $E_i \cos^2 \theta_i$ increases from 0.01 eV to 1.2 eV as shown in Figure 13.5.

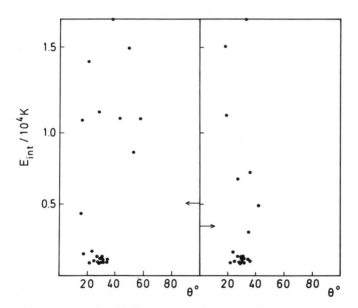

Figure 13.4. Energy transfer with the surface as a function of final scattering angle for CO/Pt(111). The initial kinetic energy, vibrational and rotational state are 100, 0, and 50 kJ mol^{-1}, respectively. The left- and right-hand panels show the results including only phonons and phonons plus electron–hole pair mechanisms, respectively (after Billing[14]).

The treatment of such processes is provided by Newns[17] along the lines developed by others.[18,19] The starting point is the Hamiltonian

$$H = \varepsilon_a(t)n_a + \sum \varepsilon_k n_k + V(t) \sum \{c_a^T c_k + \text{h.c.}\} + w_0 b^T b + \tau n_a (b^T + b) \quad (13.2.24)$$

Here $\varepsilon_a(t)$ is the energy level of the π^* level in NO and is dependent upon the position of the CM of NO from the surface (z), which provides the time dependence via motion of the NO CM, i.e., $z = z(t)$; $n_a = c_a^T c_a$ is the occupation number operator for this level; ε_k and n_k refer to the electronic eigenstates of the metal's electrons; $V(t)$ is actually $V(z(t)) = V^0 \exp(-\alpha z(t))$, where $z(t)$ is assumed to be known from motion on a PES. The frequencies of vibration in the NO and NO$^-$ are assumed to be the same, w_0. The operator "b" annihilates a vibrational quanta in either NO or NO$^-$. The electron-vibration coupling within the molecule is provided by the last term in equation (13.2.24). Thus, this equation contains two types of electron-vibration coupling, that between the e-h pairs of the solid and the π^* level of NO and that between vibrations of NO and NO$^-$ (i.e., $n_a = 1$).

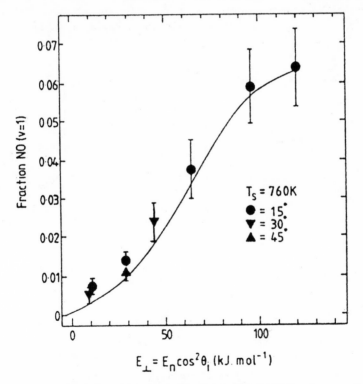

Figure 13.5. Vibrational excitation probability for the NO/Ag(111) system as a function of the initial normal kinetic energy. The curve is from Mowrey and Kouri[8] while the experimental data points are from Rettner et al.[16]

The solution of equation (13.2.24) for the operator "b" and then determination of $n_a(t)$ by first-order perturbation theory yields a formula similar to equation (13.2.23). A number of assumptions must be made to obtain quantitative agreement with experiment, including two critical ones. First, the energy of the π^* level as a function of the distance of the molecule from the surface is written as

$$\varepsilon_a(z) = \varepsilon_a(\infty) - \tfrac{1}{4}(z - z_0) \qquad (13.2.25)$$

Second, the lifetime broadening of the π^* level varies as

$$\Delta(z) = \Delta^0 \exp(-2\alpha z) \qquad (13.2.26)$$

The parameter z_0 controls the effect of the coupling with distance out into the vacuum, since it changes the energy gap between the excited and ground electronic level. Also, Δ^0 controls the rate of de-excitation from the π^* level. The particular values are $z_0 = -1.93$ Å and $\hbar/\Delta^0 = 6 \times 10^{-15}$ s. The sensitivity to these values is in accord with a different approach based upon use of an electronic friction, with the spatial extent of the electronic friction controlling the probability of energy dissipation to e–h pairs.[9] Finally, a recent article by Billing[14] is recommended for further reading since it provides a more collision-oriented approach to the treatment of e–h pair excitations for translational energy loss.

For completeness, we mention that inclusion of single phonon events can be accomplished within a time-independent theory simply by using a first-order Taylor series expansion of the interaction potential between the gas molecule and the solid surface. This is the generalization of equation (13.1.3) to include the solid's coordinates, $\{r_\beta\}$, which for the potential yields

$$V(\mathbf{r}, \mathbf{r}_i, \{r_\beta\}) = V(\mathbf{r}, \mathbf{r}_i, \{r_\beta^0\}) + \sum \nabla_{r_{\beta j}} V(\mathbf{r}, \mathbf{r}_i, \{r_\beta^0\})(r_{\beta j} - r_{\beta j}^0) \qquad (13.2.27)$$

The expansion of the wave function in equation (13.1.5) must then include eigenfunctions for all the relevant phonon modes, involving an integration over the continuous phonon degrees of freedom in the solid. This leads to a set of coupled integrodifferential equations in place of the coupled differential equations in equation (13.1.6). The effect of the phonons can then be included via a first-order perturbation treatment of the phonon coupling.

13.3. Scattering from Nonrigid Surfaces

We now turn to inclusion of surface atom motion into treatments of diffraction and inelastic scattering via the GLE technique. The collision specifies the primary zone atoms by the range of the interaction potential, adds an additional force on the primary zone atoms, and defines a time scale over which the response of the solid is required, namely, the collision time t_c (i.e., the duration of a "strong" interaction).

These GLE-"ghost" atom equations describe a set of coupled harmonic oscillators with damping and fluctuating forces. The rate of damping is determined by $|\boldsymbol{\tau}^{-1}|$ while the energy exchange between primary and "ghost" atom subsystems is controlled by $(\mathbf{w}_0 \mathbf{M}_0^{1/2})^{1/2}$. Typically, these two time scales are very similar since they both reflect the motions of the solid's atoms which are controlled by the forces between solid atoms. The other relevant time scale is t_c. When $t_c \ll |\boldsymbol{\tau}^{-1}|$, the solid response is limited to the motion of the primary zone atoms, with negligible effects due to the remainder of the solid, i.e., the "ghost" atoms. When $t_c \sim |\boldsymbol{\tau}^{-1}|$, the response of the solid is truly a *dynamical* many-body phenomena, since the energy exchange between the gas molecule and primary zone atoms is strongly influenced by the "ghost" atoms on the time scale of the collision duration. When $t_c \gg |\boldsymbol{\tau}^{-1}|$, the solid responds as a many-body *equilibrium* system, describable via either a classical Langevin treatment without memory or a temperature-dependent potential of mean force.[20] We note that $|\boldsymbol{\tau}^{-1}| = 1.5 \times 10^{-11}$ K-s/θ_D, where θ_D is the Debye temperature. A typical range is 100 K $< \theta_D <$ 400 K, yielding 3.8×10^{-14} s $< |\boldsymbol{\tau}^{-1}| <$ 15.0×10^{-14} s. A typical collision duration is 10^{-15}–10^{-11} s.

The most direct and accurate way to include energy transfer to the solid involves augmentation of the time-dependent scattering theory in equation (13.1.11) to include a time-dependent Hamiltonian[7,8]:

$$\Phi(t + \Delta t; \mathbf{r}', \mathbf{r}_i') = \int \langle \mathbf{r}', \mathbf{r}_i' | \exp\left(-iH(t)\,\Delta t/\hbar\right) | \mathbf{r}, \mathbf{r}_i \rangle$$

$$\times \, \Phi(t; \mathbf{r}, \mathbf{r}_i)\, d\mathbf{r}\, d\mathbf{r}_i \qquad (13.3.1)$$

where

$$H(t) = T(\mathbf{r}, \mathbf{r}_i) + V(\mathbf{r}, \mathbf{r}_i, \mathbf{y}(t), \{\mathbf{r}_\beta^0\}) \qquad (13.3.2)$$

$\{\mathbf{r}_\beta^0\}$ are the fixed positions of all the solid's atoms and must be included, since the PES will generally depend upon many more atoms than just those in the primary zone. In equation (13.3.2) variations in $H(t)$ over the time Δt have been neglected, since the lattice atoms are much heavier and move much more slowly than the impinging gas molecule. The time-dependence of $\mathbf{y}(t)$ is calculated via the GLE-"ghost" atom formalism as presented in Section 13.2, but with an additional force due to the interaction potential $V(\mathbf{r}, \mathbf{r}_i, \mathbf{y}(t), \{\mathbf{r}_\beta^0\})$. This can be done in a manner that would conserve energy in the absence of frictional and random forces, by using Ehrenfest's theorem:

$$d^2\mathbf{y}/dt^2 = \nabla_y V(\mathbf{y}, \{\mathbf{r}_\beta^0\}) - \boldsymbol{\Omega}_{PP}^2 \mathbf{y} + \mathbf{M}_0^{1/2} \mathbf{w}_0 \mathbf{s} \qquad (13.3.3)$$

$$d^2\mathbf{s}/dt^2 = \mathbf{w}_0 \mathbf{M}_0^{1/2} \mathbf{y} - \mathbf{w}_0^2 \mathbf{s} - \boldsymbol{\tau}\, d\mathbf{s}/dt + \mathbf{f}(t) \qquad (13.3.4)$$

where

$$V(\mathbf{y}, \{\mathbf{r}_\beta^0\}) = \langle \Phi(t; \mathbf{r}, \mathbf{r}_i) | V(\mathbf{r}, \mathbf{r}_i, \mathbf{y}(t), \{\mathbf{r}_\beta^0\}) | \Phi(t; \mathbf{r}, \mathbf{r}_i) \rangle \qquad (13.3.5)$$

This approach illustrates the power of the time-dependent solution of the Schroedinger equation, since the calculation of $y(t)$ and $s(t)$ involves a relatively small amount of extra time over that due to simple evolution of $\Phi(t; r, r_i)$ with a rigid surface. Even the integration in equation (13.3.5) is not particularly tedious since the wave function is already known on the grid of (r, r_i) values.

There is one computationally time-consuming feature however. For a nonzero T, the initial conditions of the primary and "ghost" atoms must be sampled. This implies that a considerable number of $\Phi(t; r, r_i)$ must be propagated and then averaged incoherently (or coherently) to determine the full wave function.

The above method is actually the time-dependent self-consistent field (TDSCF) approach.[4,21,22] In the TDSCF theory, a full wave function for two coordinates X, Y is approximated via

$$\phi(t; X, Y) = \Phi(t; X)\theta(t; Y) \qquad (13.3.6)$$

and the full TDSE

$$[T(X) + T(Y) + V(X, Y)]\phi(t; X, Y) = i\hbar\, \partial\phi(t; X, Y)/\partial t \qquad (13.3.7)$$

is replaced by the set of equations

$$[T(X) + \langle\theta(t; Y)|V(X, Y)|\theta(t; Y)\rangle]\Phi(t; X) = i\hbar\, \partial\Phi(t; X)/\partial t \qquad (13.3.8)$$

and

$$[T(Y) + \langle\Phi(t; X)|V(X, Y)|\Phi(t; X)\rangle]\theta(t; Y) = i\hbar\, \partial\theta(t; Y)/\partial t \qquad (13.3.9)$$

Equations (13.3.8) and (13.3.9) represent the best variational solution of equation (13.3.7) with the simple product wave function in equation (13.3.6). If the original equation (13.3.7) conserves energy, then the TDSCF equations (13.3.8) and (13.3.9) also conserve energy. The molecule–surface scattering description in equations (13.3.1)–(13.3.5) results from the identification of $X \to (r, r_i)$ and $Y \to \{r_\beta\}$. Then, the additional assumption that the wave function for Y is peaked around the classical value, $Y(t)$, is used to replace the TDSE in equation (13.3.9) by Hamilton's equations.

It is of course possible to simplify the molecule–surface scattering problem further by simplification of the solution of the TDSE in equation (13.3.1). This is equivalent to implementation of the multitude of approximations that have been developed in gas-phase scattering dynamics. For example, if the center-of-mass of the molecule is also assumed to follow a classical trajectory [i.e., $\Phi(t; r, r_i) \approx \theta(t; r_i)F(t; r)$ with $F(t; r)$ peaked around the classical trajectory $r(t)$], then the semiclassical stochastic trajectory (SST) approximation[23-25] results:

$$\theta(t; r_i) = \sum c_k(t)\phi_k(r_i) \exp(-i\varepsilon_k t/\hbar) \qquad (13.3.10)$$

$$dc_k(t)/dt = \sum \langle \phi_k(\mathbf{r}_i)| V(\mathbf{r}(t), \mathbf{r}_i, \mathbf{y}(t), \{\mathbf{r}_\beta^0\})|\phi_j(\mathbf{r}_i)\rangle \exp{(i[\varepsilon_k - \varepsilon_j]t/\hbar)}c_j(t)$$

$$\text{(13.3.11)}$$

$$M\, d^2\mathbf{r}/dt^2 = -\nabla_r V(\mathbf{r}, \mathbf{y}, \{\mathbf{r}_\beta^0\})$$

$$\text{(13.3.12)}$$

$$V(\mathbf{r}, \mathbf{y}(t), \{\mathbf{r}_\beta^0\}) = \langle \Phi(t; \mathbf{r}_i)| V(\mathbf{r}, \mathbf{r}_i, \mathbf{y}, \{\mathbf{r}_\beta^0\})|\Phi(t; \mathbf{r}_i)\rangle$$

$$\text{(13.3.13)}$$

The basis set expansion used in equation (13.3.10) is not necessary but is implemented to retain close correspondence with the original derivation. We could just as well have used the analog of equation (13.3.1) instead of equations (13.3.10) and (13.3.11) to determine the time evolution of $\theta(t; \mathbf{r}_i)$. The advantage in either case is the reduction of the number of quantal degrees of freedom which must be treated explicitly. For example, in a diatomic molecule–surface collision treated as in equations (13.3.1)–(13.3.3), there are six quantal degrees of freedom, which are too many for current computational facilities. However, using a SST technique there are only three quantal degrees of freedom, which is perfectly tractable computationally. Equivalently, in the basis set expansion (13.3.11), only a manageable set of basis functions is needed to span the rotation–vibration space.

The difficulty with the SST approach is replacement of the correlation between translational and internal energies of each internal state with a correlation between the *average* translational energy and all the states. The physical implication is that the prediction of the internal energy or state distribution for a particular angle of the product scattering is not possible. Instead, only the internal state distribution averaged over all final angles is provided. Although this is a much more severe limitation than that of separating molecular and phonon degrees of freedom in the TDSCF, the SST is much easier and faster to implement. Indeed, calculations of vibrational–rotational relaxation in CO_2/Pt collisions have been accomplished involving over 600 vibration–rotation states in the expansion.[25]

It is possible to retain some correlation between internal and translational degrees of freedom by replacing the exact translational wave function by a simpler function that is nearly but not totally classical. This is the idea of Gaussian wave-packet dynamics (GWD),[26,29] in which the full molecular wave function is expanded as

$$\Phi(t; \mathbf{r}, \mathbf{r}_i) = \sum c_{\alpha k}(t)[C_{\alpha k} G_{\alpha k}(t; \mathbf{r})]\phi_k(\mathbf{r}_i) \exp{(-i\varepsilon_k t/\hbar)}$$

$$\text{(13.3.14)}$$

where each translational Gaussian wave packet is of the form

$$G_{\alpha k}(t; \mathbf{r}) = \exp\{(i/\hbar)[(\mathbf{r} - \mathbf{r}_{\alpha k}(t)) \cdot \mathbf{A}_{\alpha k}(t) \cdot (\mathbf{r} - \mathbf{r}_{\alpha k}(t)) + \mathbf{p}_{\alpha k}(t) \cdot (\mathbf{r} - \mathbf{r}_{\alpha k}(t))]\}$$

$$\text{(13.3.15)}$$

The summation over α provides a linear combination of GWP to mimic an incident 2D plane wave, i.e., the $\mathbf{R}$ part of the initial wave function in equation (13.1.9). The coefficients $C_{\alpha k}$ are dependent on the index k only through the energy of the initial translational wave vector. The time-dependent coefficients $c_{\alpha k}(t)$ satisfy complicated differential equations.[29] The parameters in a Gaussian wave function obey (generic) equations of motion[26-29]:

$$d\mathbf{A}/dt = -(2/m)\mathbf{A}(t) \cdot \mathbf{A}(t) - \tfrac{1}{2}\mathbf{K} \qquad (13.3.16)$$

$$d\mathbf{r}/dt = (1/m)\mathbf{p}(t) \qquad (13.3.17)$$

$$d\mathbf{p}/dt = -\nabla_r\langle V\rangle \qquad (13.3.18)$$

where $\mathbf{K}$ is the matrix of expectation values of the second derivatives of V. The difficulty with this approach is the very large number of coupled differential equations which arise for each wave packet–internal state: 6 for the matrix $\mathbf{A}$, 3 for $\mathbf{r}$, 3 for $\mathbf{p}$, 1 complex for $c_{\alpha k}$. Since the $c_{\alpha k}$ couple all the wave packet–internal state combinations, the total number of real differential equations is $14N_g N_i$ where N_g and N_i are the total number of wave packets and internal states, respectively. A typical number is $N_g = 10\text{--}20$, which implies that about $200N_i$ coupled differential equations result from this approach. This poses a rather intractable problem since it is nearly always true that $N_i > 2$, with the typical case being $N_i > 10$.

To alleviate these problems, one may utilize a single translational function as in the TDSCF approach but not make the classical path approximation.[28] This replaces the multiple GWP expansion in equation (13.3.14) by the mean trajectory GWP method:

$$\Phi(t; \mathbf{r}, \mathbf{r}_i) = F(t; \mathbf{r}) \sum c_{\alpha k}(t)\phi_k(\mathbf{r}_i) \exp\left(-i\varepsilon_k t/\hbar\right) \qquad (13.3.19)$$

where the single translational wave function is given by

$$F(t; \mathbf{r}) = \sum C_\alpha G_\alpha(t; \mathbf{r}) \qquad (13.3.20)$$

Each GWP propagates independently, leading to equations nearly identical to those of the SST method but supplemented by equations for the time evolution of the matrix $\mathbf{A}$. The right-hand side of the coefficient differential equation (13.3.11) has an added term of the form $-(1/2m)p_\alpha(t) \cdot p_\alpha(t) - (i\hbar/m)\,\mathrm{Tr}\,\mathbf{A}(t)$ and utilizes an average over the appropriate Gaussian wave packet. The advantage of this approach is the coherence of the superposition, which enables the treatment of diffraction. However, it does not allow for any added correlation between the translational and internal degrees of freedom. In particular, it cannot be used to predict the simultaneous angular and internal energy distributions, although it can be used to predict the former in the absence of any internal energy changes.

The basic problem with all of the methods which utilize some TDSCF approximation is this destruction of the correlation between internal and translational degrees of freedom. The treatment of energy transfer with the lattice is quite tractable using the GLE-"ghost" atom formalism in equations (13.3.3)–(13.3.5) (which nevertheless destroys the correlation between individual phonon changes in the lattice and the energy changes in the molecule). Why then has there been so much work on the TDSCF and related approximations? It is the same reason that plagues all of dynamics: there are no general methods that can be used to treat more than three or at most four degrees of freedom via quantum mechanics. Hence, one is forced to develop simpler methods that allow *some* problems to be treated with *some* accuracy. The goal of treating these systems by consistent quantum-mechanical methods remains an elusive one.

In contrast to the considerable difficulties with either a full quantum mechanical or semiclassical treatment, the implementation of a full classical treatment of molecule–surface dynamics is not really difficult (but could still require substantial computational time depending upon the complexity of the PES and its derivatives). The Newtonian equations provide an exact (classical) correlation between internal and translational degrees of freedom. It is worthwhile to emphasize that the problem with a classical treatment is not implementation but *validity*. Classical simulations eliminate or at best treat poorly (and somewhat arbitrarily) the distinctly quantum-mechanical processes such as diffraction, and quantized internal energies. Unless these effects are negligible, classical dynamics is inadequate for describing the angular, translational, and internal state distributions of the scattered molecules. However, for heavier molecules, especially when state-to-state vibrational transitions are not desired, a classical trajectory-GLE-"ghost" atom approach is probably perfectly adequate. This has the advantage of focusing attention on the results of the calculation and not on the calculational method itself, precisely the way to understand real chemical and physical systems in this author's opinion.

Before considering some results, let us mention a possible solution to the implementation of quantum mechanics. Basically, we wish to treat many quantum dynamical degrees of freedom and shall denote these by the generic $\mathbf{x}$. Equation (13.3.1) can be rewritten in the more explicit form

$$\langle t'; \mathbf{x}'|\Phi \rangle = \int \langle \mathbf{x}'| \exp\left(-iH(t' - t)/\hbar\right)|\mathbf{x}\rangle\langle t, \mathbf{x}|\Phi \rangle \, d\mathbf{x} \tag{13.3.21}$$

which indicates that the necessary quantity is the propagator

$$U(t', \mathbf{x}'; t, \mathbf{x}) = \langle \mathbf{x}'| \exp\left(-iH(t' - t)/\hbar\right)|\mathbf{x}\rangle \tag{13.3.22}$$

For multidimensional problems, the only practical method to evaluate such a quantity is via path integrals:

$$\exp\left[-iH(t' - t)/\hbar\right] = \left[\exp\left(-iH\delta t/\hbar\right)\right]^N$$

$$\delta t = (t' - t)/N$$

$$U(t', \mathbf{x}'; t, \mathbf{x}) = \int \langle \mathbf{x}'| \exp\left(-iH\delta t/\hbar\right)|\mathbf{x}_1\rangle\langle \mathbf{x}_1| \exp\left(-iH\delta t/\hbar\right)|\mathbf{x}_2\rangle \cdots$$

$$\langle \mathbf{x}_{N-1}| \exp\left(-iH\delta t/\hbar\right)|\mathbf{x}\rangle \, d\mathbf{x}_1 \cdots d\mathbf{x}_{N-1} \tag{13.3.23}$$

By making each N large enough and each time interval $t' - t$ small enough, each effective short-time propagator $\langle \mathbf{x}_i| \exp\left(-iH\delta t/\hbar\right)|\mathbf{x}_{i+1}\rangle$ is approximated via use of some commutivity assumption $[T, V] = 0$. Then, the multidimensional integrals over $\{\mathbf{x}_i, i = 1, \ldots, N - 1\}$ are performed via Monte Carlo methods. Even quantum degrees of freedom for nuclear motion are unlikely to require more than 10–20 intermediate paths or expansion states. Thus a quantum system with N_Q degrees of freedom maps onto a classical-like Monte Carlo system with $\approx 10 N_Q$ degrees of freedom. This would allow for easy evaluation of even $N_Q = 100$. The bottleneck is the computation of an oscillatory function via Monte Carlo methods since $\exp(iF)$ does not provide a positive definite sampling function. A possible solution may be to add a small imaginary time component $t \rightarrow t - i\hbar\beta$, and to evaluate equation (13.3.23) either via newly developed stationary phase QMC methods[30,31] or via analytic continuation methods.[32]

Now, a few indications are given of the type of results which can be generated with the present methods. In Figure 13.6, results for the vibrational relaxation of CO_2 in collisions with Ag(111) are shown. These were generated using the SST technique[25] including all energetically accessible ro-vibrational states in the basis, over 600 in all! The surface was smooth to eliminate the m_j and diffraction levels, but was allowed to move in the z-direction to allow for energy exchange. The de-excitation probabilities are quite small, but very state specific.

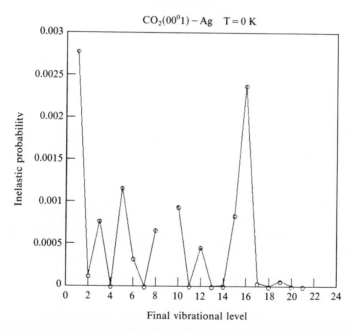

Figure 13.6. Semiclassical stochastic trajectory calculated results for the final vibrational distribution in direct scattering of $CO_2(00^01)$ from Ag(111) at a kinetic energy of 0.125 eV, normal incidence, and a surface temperature of 0 K. The curve is after DePristo and Greiger.[25] The final states are labeled using: $(1 = 00^00)$, $(2 = 01^10)$, $(3 = 02^00)$, $(4 = 02^20)$, $(5 = 10^00)$, $(6 = 03^10)$, $(7 = 03^30)$, $(8 = 11^10)$, $(9 = 00^01)$, $(10 = 04^00)$, $(11 = 04^20)$, $(12 = 02^00)$, $(13 = 04^40)$, $(14 = 12^20)$, $(15 = 20^00)$, $(16 = 01^10)$, $(17 = 05^10)$, $(18 = 05^30)$, $(19 = 13^10)$, $(20 = 13^30)$, $(21 = 21^10)$.

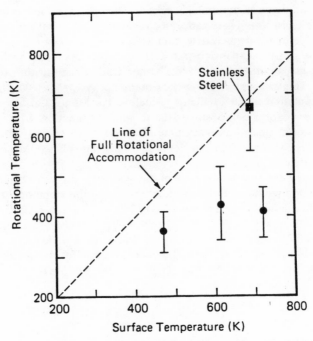

Figure 13.7. Rotational temperature of scattered NO molecules as a function of the Ag(111) surface temperature. The plot is from G. M. McClelland, G. D. Kubiak, H. G. Rennagel, and R. N. Zare, *Phys. Rev. Lett.* **46**, 831–834 (1981).

Next, Figures 13.7–13.9 present some results from experiments taken from a recent review by Rettner.[33] The original references, if available, are indicated on the figures. First, in Figure 13.7, the lack of rotational equilibration of NO scattering from Ag(111) is shown. Figure 13.8 illustrates the detailed information on the rotational state of the scattered NO molecules as a function of kinetic energy, while Figure 13.9 shows the rather peaked and moderately wide angular distributions for two different final rotational levels. We note that the difference between these curves is a measure of the correlation between translational and internal energy changes, precisely the information which is difficult to provide via quantum-mechanical dynamics. The detailed features of these distributions are duplicated by classical trajectory-GLE-"ghost" atom simulations.[34]

13.4. Dissociative Chemisorption: The Model

In this chapter we consider the description of the simplest type of chemical reaction involving a gas molecule and a solid surface, namely, the dissociative chemisorption of a diatomic molecule: $AB(g) + M \rightarrow A(a) + B(a) + M$. The treatment is limited to the case when the surface is clean. The dissociation probability of a molecule upon collision with a clean surface is referred to as the zero-coverage sticking coefficient, S_0 assuming that molecular trapping does not occur.

First it should be mentioned that it is now possible to measure S_0 by molecular beam scattering techniques[35-45] as a function of the molecule's incident kinetic energy, E_i, and direction as determined by the angle from the surface normal, θ_i. The initial azimuthal angle is fixed by orienting the crystal with respect to the incoming molecular beam. It is even feasible to excite a distribution of rotational and vibrational states of the incoming molecule.

For most of the systems studied so far, S_0 is found to be a function of the "normal" kinetic energy, $S_0(E_i, \theta_i, \phi_i) = S_0(E_i \cos^2 \theta_i, 0, \phi_i)$, but for a few systems it is found that S_0 is a function of only the total kinetic energy, $S_0(E_i, \theta_i, \phi_i) = S_0(E_i, 0, \phi_i)$. What is surprising about these findings is their simplicity, since in general one would not expect the combined effect of two variables to be represented so succinctly. Indeed, there are data which show neither type of scaling behavior.

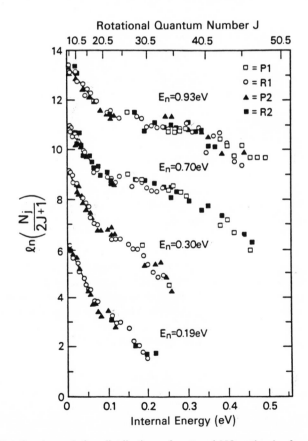

Figure 13.8. Rotational population distributions of scattered NO molecules for various normal kinetic energies. The plot is from A. W. Kleyn, A. C. Luntz, and D. J. Auerbach, *Phys. Rev. Lett.* **47**, 1169 (1981) and *Surf. Sci.* **117**, 33–41 (1982).

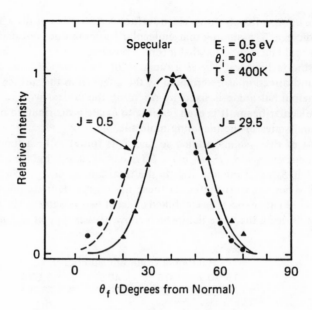

Figure 13.9. Final angular distributions for NO scattered from Ag(111) for an initial kinetic energy of 0.5 eV, angle of 30°, and surface temperature of 400 K. The plot is from Rettner.[33]

(Henceforth, we will suppress the explicit dependence upon ϕ_i for notational simplicity.) The explanation of such behavior is one goal of theoretical treatments of the dynamics. Another is to understand how factors such as the structure of the solid surface and the velocity and internal rotational–vibrational–electronic energy of the gas molecule influence the dissociation of the molecule on the surface. From a practical view, we note that with this knowledge it may be possible to indicate how such reactions can be made more efficient and selective via modification of either the surface or the gas molecule.

It is feasible to perform realistic simulations of *reaction* dynamics in molecule-solid surface systems under zero coverage conditions.[46-51] These utilize the stochastic trajectory, GLE-"ghost" atom, techniques developed in Section 13.2 to treat surface atom motion. The motion of the atoms in the gas molecule are also treated classically, by numerical integration of the Newtonian equations of motion. The particular system which results for a diatomic molecule–surface collision is as follows:

$$m_1 \, d^2\mathbf{x}_1/dt^2 = -\nabla_{x_1} V(\mathbf{x}, \mathbf{y}, \{\mathbf{r}_\beta^0\}) \tag{13.4.1a}$$

$$m_2 \, d^2\mathbf{x}_2/dt^2 = -\nabla_{x_2} V(\mathbf{x}, \mathbf{y}, \{\mathbf{r}_\beta^0\}) \tag{13.4.1b}$$

$$d^2\mathbf{y}/dt^2 = -\nabla_y V(\mathbf{x}, \mathbf{y}, \{\mathbf{r}_\beta^0\}) - \mathbf{\Omega}_{PP}^2 \mathbf{y} + \mathbf{M}_0^{1/2} \mathbf{w}_0 \mathbf{s} \tag{13.4.2}$$

$$d^2\mathbf{s}/dt^2 = \mathbf{w}_0 \mathbf{M}_0^{1/2} \mathbf{y} - \mathbf{w}_0^2 \mathbf{s} - \mathbf{\tau} \, d\mathbf{s}/dt + \mathbf{f}(t) \tag{13.4.3}$$

where $\mathbf{x} = (\mathbf{x}_1, \mathbf{x}_2)$ is the set of three-dimensional positions for the two atoms of the diatomic. The most crucial feature controlling the accuracy of such simulations is the adequacy of the PES, which must be able to describe the breaking of the molecular bond and the making of two new atom–surface bonds. Much will be said about such a PES later, but for now it is simply assumed that such a PES is available and we focus on the mechanics of such a calculation.

For a particular initial kinetic energy (E_i) and angles (θ_i, ϕ_i) of the center-of-mass (CM) of the impinging molecule, one equation constrains $\mathbf{x}$,

$$m_1 \, d\mathbf{x}_1/dt + m_2 \, d\mathbf{x}_2/dt = (m_1 + m_2) \, d\mathbf{X}/dt \equiv M \, d\mathbf{X}/dt \qquad (13.4.4)$$

where the CM velocity components are specified. The initial z-component of $\mathbf{X}$ is restricted to be large in order to specify a free molecule plus surface system initially. The in-plane components of the position $\mathbf{X}$ are constrained by the perfect 2D periodicity of the surface to lie within a given unit cell. And if the unit cell displays rotational symmetry, then these are constrained to sample only the irreducible part of the unit cell. In any case, the two in-plane components of $\mathbf{X}$ are only specified to within some part of the unit cell. If the initial molecular vibrational (ε_n) and rotational (ε_j) energies are also specified, then we have

$$\tfrac{1}{2}\mu|dr/dt|^2 + v(r) = \varepsilon_n \qquad (13.4.5)$$

and

$$\tfrac{1}{2}\mu r^{-2}|\mathbf{r} \times d\mathbf{r}/dt|^2 = \varepsilon_j \qquad (13.4.6)$$

where $\mathbf{r} = \mathbf{x}_1 - \mathbf{x}_2$ is the relative position vector, $v(r)$ is the interaction potential between the gas atoms in the absence of the surface [i.e., $V(\mathbf{x}, \mathbf{y}, \{\mathbf{r}_\beta^0\})$ as $|\mathbf{X}| \to \infty$), and μ is the reduced mass $m_1 m_2 / M$. These constraints imply that four internal variables must be averaged over: the vibrational phase, the two angles of the orientation $\hat{r}$, and the one angle of the orientation of the rotational angular momentum vector.

The net result of the above analysis is that six initial molecular variables are not specified exactly, but are only restricted to be sampled from some distribution at the beginning of each trajectory. In addition, for a nonzero surface temperature, the initial velocities and positions of the primary and ghost atoms must be sampled. This can be done efficiently by finding the normal modes of the $(\mathbf{y}, \mathbf{s})$ system and then sampling the velocities and positions from Gaussian random distributions. For a typical primary zone of 13 atoms with an associated 13 ghost atoms, there are 78 velocities and 78 positions to be sampled. The total number of variables to be sampled is thus 162. Physically meaningful results for S_0 are an average over all these variables of the form

$$S_0 = \int S_0(\mathbf{Q}) P(\mathbf{Q}) \, d\mathbf{Q} \qquad (13.4.7)$$

where Q symbolizes all 162 random variables and $P(Q)$ is the relevant distribution function. Such an average is accomplished by the Monte Carlo method with the number of Monte Carlo points (i.e., trajectories) reflecting the desired accuracy of the quantity to be computed. However, a few hundred trajectories are typically needed even for large values of $S_0 \approx 0.5$. If a less global quantity, such as the final angular-internal-kinetic energy distribution of the scattered molecules, is desired, then many more trajectories must be computed to provide statistically meaningful results.

The results of various dynamical calculations will be discussed later, but we now want to move on to a consideration of the PES in considerable detail. We shall focus on the basic physics of the interaction as described in a simple, but very concise and useful, model due to Norskov and co-workers.[52] At far distances from the surface, the molecule–surface interaction is a van der Waal's attraction, proportional to Z^{-3}. This is of little interest for a reactive PES, since the electronic structure of the molecule is unchanged. However, at closer distances, the affinity level of the molecule begins to feel an image-charge type attraction, causing a shifting of its energy as

$$E_a(Z) = E_a(\infty) - \tfrac{1}{4}Z \qquad (13.4.8)$$

As the distance Z gets smaller, the affinity level eventually gets pulled below the Fermi level of the solid, and the affinity level fills. If this level is antibonding, then the molecular bond may be weakened sufficiently to rupture, thereby effecting dissociative chemisorption. There is a process which competes with the lowering of the affinity level, namely, the extra kinetic-energy repulsion due to the Pauli exclusion principle operating between the solid's and molecule's electrons. If the latter increases more quickly than E_a decreases, a kinetic barrier to dissociation results. Indeed, if the breakdown in the image-charge approximation occurs before the barrier is reached, the barrier may increase continually thereby leading to stable molecular physi- or chemisorption.

The above concepts indicate that electron transfer is a major cause of bond rupture at surfaces. With this in mind, Gadzuk and Holloway[53] have described the PES as the lowest eigenvalue of two diabatic curves, corresponding to the arrangements $AB + S$ and $AB^- + S^+$. This quantifies the above ideas somewhat. However, the possible basic features of a PES in any description of dissociative chemisorption are the same: (1) a weak molecularly physisorbed species at far distances (4–8 bohr) from the surface; (2) a barrier to the continued decrease in Z, separating a molecularly chemisorbed species; (3) a barrier to stretching of the molecular bond, separating the final atomic chemisorbed products. A schematic diagram of such a PES is shown in Figure 13.10.

The main difficulty with such a simple argument is the neglect of the dynamics of the other degrees of freedom. The molecule can rotate and translate parallel to the surface, and the surface atoms can move, thereby distorting the PES. When the molecule rotates, the various barriers and even the gross topology of the PES changes. Similarly, the PES at different locations in the surface unit cell will be

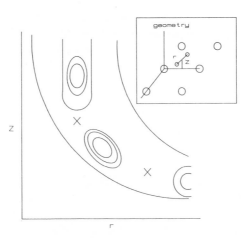

Figure 13.10. Schematic potential-energy surface for the dissociative chemisorption of a diatomic molecule. An arrangement of such high symmetry is chosen (as illustrated in the inset) such that only two distances can change. The height of the CM of the diatomic is Z while the bond length is r.

different. The dynamics in these other degrees of freedom may play a central role, and thus one must include them. This means that we must construct a full multidimensional PES to describe the reactive process.

It is worthwhile to emphasize that standard first-principles or *ab-initio* calculations of full PES for dissociative chemisorption are not feasible at present, given the accuracy ($1–2$ kcal mol^{-1} or $0.05–0.1$ eV) required by dynamical calculations. One generally must resort to a global representation with adjustable parameters at this time. While this leads to problems of uniqueness of the PES which will be addressed later, it does not lessen the importance of investigations of the dynamics since a general purpose is to provide insight into correlations between the topology of the PES and experimental data. We also mention that there are now becoming available high-quality calculations of at least a few points on the full PES[54] and these can be used to fix some of the parameters in a representation of the full PES.[55]

One successful method of attack on this problem, initially developed by McCreery and Wolken[56] in the mid-1970s and later modified and quantified by Lee, Kara, and DePristo,[46,50] is now examined. The representation of the PES which has been used is a modified four-body LEPS form. We consider a diatomic, AB, interacting with a surface, S. The basic idea is to utilize valence bond theory for the atom–surface interactions, V_{AS} and V_{BS}, and atom–atom interaction, V_{AB}, to construct the full molecule–surface interaction, $V_{AB,S}$. (We have been dealing with this PES throughout the chapter and are just using a more explicit notation here.) For each atom of the diatomic, we associate a single electron. In a three-body system, this allows only one bond. However, since the solid can bind both atoms simultaneously, two valence electrons are associated with the solid. The use of two electrons for the solid body is convenient mathematically, but this must not lead to a bond between these two electrons since this is nonsensical physically. Thus, the original four-body LEPS form must be modified to eliminate any unphysical electron–electron interactions based upon the rule that each electron can only interact with an electron on a different body. This leads to the modified four-body LEPS form. Alternatively, one may consider this as a parametrized form with a few parameters which have well-controlled effects on the global PES.

The explicit form is

$$V_{AB,S} = Q_{AS} + Q_{BS} + Q_{AB} - [J_{AB}(J_{AB} - J_{AS} - J_{BS}) + (J_{AS} + J_{BS})^2]^{1/2} \quad (13.4.9)$$

where Q and J are Coulomb and exchange integrals, respectively, for each constituent. The central feature of this form is the nonadditivity of the interaction potentials, which are $Q + J$ for each body–body interaction. It is the precise division into Q and J that will control the activation barriers in the reaction. An inkling of this can be seen by noting that if the terms with A–S and B–S vanish, then $V_{AB,S} = Q_{AB} - |J_{AB}| = Q_{AB} + J_{AB} = V_{AB}$. And, if the terms with A–B vanish, then $V_{AB,S} = (Q_{AS} + J_{AS}) + (Q_{BS} + J_{BS})$. This will become clearer later, but first bear with us through the details of each type of term in equation (13.4.9).

First, the A–B interaction will be examined since this is the simplest one given by

$$Q_{AB} + J_{AB} = V_{AB}$$

$$= D_{AB}\{\exp[-2\alpha_{AB}(r - R_{AB})] - 2\exp[-\alpha_{AB}(r - R_{AB})]\} \quad (13.4.10a)$$

$$Q_{AB} - J_{AB} = \tfrac{1}{2}[(1 - \Delta_{AB})/(1 + \Delta_{AB})]D_{AB}\{\exp[-2\alpha_{AB}(r - R_{AB})]$$

$$+ 2\exp[-\alpha_{AB}(r - R_{AB})]\} \quad (13.4.10b)$$

The A–B interaction potential is represented by the Morse potential in equation (13.4.10a). The parameters D_{AB}, α_{AB}, and R_{AB} are the bond energy, range parameter, and bond length, respectively, for the A–B molecule. These are specified by the A–B binding curve.[57] The A–B "antibonding" potential in equation (13.4.10b) is represented by the anti-Morse form. The Sato parameter, Δ_{AB}, is unspecified as yet. However, it clearly controls the division of the Morse–anti-Morse forms into Q_{AB} and J_{AB}.

The A–S interaction is considered next. This is more complicated and is given by

$$Q_{AS} + J_{AS} = V_{AS}$$

$$= D_{AH}\{\exp[-2\alpha_{AH}(r_{As} - R_{AH})] - 2\exp[-\alpha_{AH}(r_{As} - R_{AH})]\}$$

$$+ \sum D_{AS}\{\exp[-2\alpha_{AS}(R_{A\beta} - R_{AS})] - 2\exp[-\alpha_{AS}(R_{A\beta} - R_{AS})]\}$$

$$(13.4.11a)$$

$$Q_{AS} - J_{AS} = \tfrac{1}{2}[(1 - \Delta_{AS})/(1 + \Delta_{AS})]$$

$$\times D_{AH}\{\exp[-2\alpha_{AH}(r_{As} - R_{AH})] + 2\exp[-\alpha_{AH}(r_{As} - R_{AH})]\}$$

$$+ \tfrac{1}{2}[(1 - \Delta_{AS})/(1 + \Delta_{AS})]$$

$$\times \sum D_{AS}\{\exp[-2\alpha_{AS}(R_{A\beta} - R_{AS})] + 2\exp[-\alpha_{AS}(R_{A\beta} - R_{AS})]\}$$

$$(13.4.11b)$$

The summation extends over all of the atoms in the crystal including the moving primary atoms and the fixed atoms. The A–S bonding and "antibonding" potential are of similar form to those in equation (13.4.10). The distance between the atom A and the solid atom β is given by

$$R_{A\beta} = |\mathbf{R}_A - \mathbf{r}_\beta| \qquad (13.4.12)$$

The many more terms in equation (13.4.11) require further explanation.

First we consider the terms in D_{AH}, which represent the interaction between A and the valence electrons of the metal. This is modeled by the interaction of A with jellium of density provided by the metal's valence electrons at the position of A. The parameters are thus:

1. D_{AH} = strength of the interaction between atom A and jellium.
2. α_{AH} = range of the above interaction.
3. $R_{AH} = (3/4\pi n_0)^{1/3}$ where n_0 is the density at the minimum of the atom-jellium binding curve.

These are taken from the SCF-LD first-principles calculations of Puska et al.[58] on the embedding energy of an atom in jellium as a function of the density of the jellium. Their numbers are represented by a Morse-like form. (It is not strictly a Morse potential, since density is the variable instead of distance.) The variable, r_{As}, depends upon the density at the position of A:

$$r_{As} = (3/4\pi n)^{1/3} \qquad (13.4.13a)$$

$$n = \sum n(R_{A\beta}) \qquad (13.4.13b)$$

where $n(R_{A\beta})$ is the atomic density of the solid atom β at the center of the gas atom A and again the summation extends over all of the solid's atoms. It has been assumed that the density above the metal surface is well represented by the sum over the individual atomic densities. Since differentiation of the PES is needed, simple forms are used for the atomic densities for each shell, where only the outer or valence electrons need be considered. For example, in W, the $6s^2$ and $5d^4$ are used:

$$n(R_{A\beta}) = 2n_s(R_{A\beta}) + 4n_d(R_{A\beta}) \tag{13.4.14a}$$

where

$$n_s(r) = s_1[r - s_2]^{s_4} \exp(-s_3 r) \tag{13.4.14b}$$

$$n_d(r) = d_1 \exp(-s_2 r^{s_3}) \tag{13.4.14c}$$

The various parameters $\{s_k\}$, $\{d_k\}$ are determined by fitting to known atomic densities.

We now consider the terms in D_{AS}, which represent the interaction between A and the localized electrons and nuclear charge of the metal. This is modeled by a two-body interaction between A and each solid atom, β, which is represented by a Morse potential. The parameters are defined below:

1. D_{AS} = strength of the localized two-body interaction between atom A and the solid's atoms.
2. α_{AS} = range of the above interaction.
3. R_{AS} = position of the minimum of the two-body interaction.

These parameters must be determined from information on the atom–surface interaction potential which may come from either sufficient experimental or theoretical information, such as the binding energy, height, and frequency for different sites of adsorption or the full binding curves above each site.

The Sato parameter for the A–S interaction is Δ_{AS}. The B–S terms are exactly of the same forms but with different parameters, of course.

It is noteworthy that since $\mathbf{r}_\beta$ is related by mass-weighting to the primary zone displacement coordinates, the potential is dependent upon $\mathbf{y}$ and hence gives rise to forces, and energy exchange, between the gas molecule and the surface primary zone atoms.

13.5. Dissociative Chemisorption: Results

The results of dynamical simulations of dissociative chemisorption are now illustrated briefly. We focus in particular on the $N_2/W(110)$ system[41,49-51] and the $H_2/Ni(100)$ system.[38-40,46-48,54,55,59,60]

The $N_2/W(110)$ system will be considered first, spending considerable time on the construction of the PES along the lines detailed by Kara and DePristo.[50] Since the molecule is homonuclear, the parameters for the A=B=N atoms are identical. Explicit subscripts are used for the relevant bodies, but we note that the references employ a simpler notation which is also followed in the figures: (1) $\Delta_{NW} \equiv \Delta_{GS}$; (2) $\Delta_{NN} \equiv \Delta_{GG}$; (3) $D_{NW} \equiv D_2$; (4) $\alpha_{NW} \equiv \alpha_2$; (5) $R_{NW} \equiv R_2$.

Typical results of the embedding energy for an atom in jellium are shown in Figure 13.11 based upon SCF-LD calculations.[58] For the N atom, the Morse-like parameters modeling the embedding energy are found to be $(D_{AH}, \alpha_{AH}, R_{AH}) = (1.4\,eV, 0.78938\,bohr^{-1}, 3.7575\,bohr)$ with the last number corresponding to $n_0 = 0.0045\,bohr^{-3}$. From the atomic Hartree–Fock calculations for the W atom[61] it is found that the 6s and 5d electron densities per electron are represented accurately over the important distance range 3 bohr $< r <$ 10 bohr by the forms of equations (13.4.14b) and (13.4.14c) with the numbers given below (in atomic units):

$$n_s(r) = 0.1398(r - 1.6992)^{1.1043} \exp(-1.3967r) \qquad (13.5.1)$$

and

$$n_d(r) = 0.078513 \exp(-0.76027r^{1.5282}) \qquad (15.5.2)$$

This specifies the homogeneous part of the N–W potential.

The two-body part of the N–W potential is found by employing the theoretical potential[62] shown in Figure (13.12), which is clearly strongly dependent upon the binding site of the N atom. The lowest minimum is in very good agreement with the experimental value of 6.73 eV.[63] Fitting equation (13.4.11) to the data in Figure 13.12 yields the remaining N–W potential parameters, $(D_{NW}, \alpha_{NW}, R_{NW}) = (6.78\,eV, 1.089\,bohr^{-1}, 2.354\,bohr)$. We note the size of the two-body contribution to the N–W bond as compared to the homogeneous energy contribution, since the former has a minimum of -1.4 eV. Also noteworthy is the

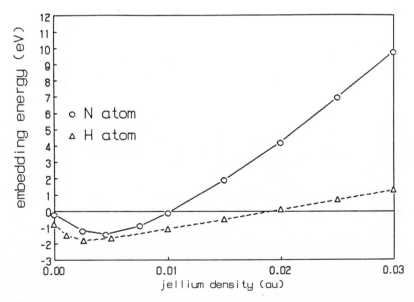

Figure 13.11. Embedding energy as a function of jellium density for two different atoms. The data are from Puska *et al.*[58]

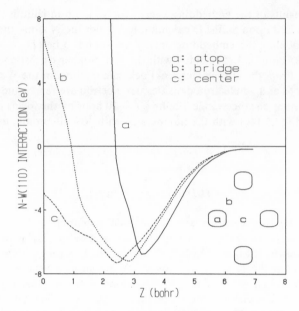

Figure 13.12. N–W(110) interaction, as calculated by the CEM-1 method,[62] as a function of the height of the N above the (rigid) W(110) surface for different sites. The plot is from Kara and DePristo.[50]

small value of R_{NW} and the moderate value of α_{NW}, which combine to make the two-body terms generally attractive (except close to the surface on an atop site) and relatively short-ranged. It should be emphasized that the parameter values can change by a few percent and still provide an excellent fit to the theoretical potential in Figure 13.12. This will be important in providing flexibility in adjustment of the NN–W potential.

In order to complete all individual interaction potentials, $(D_{NN}, \alpha_{NN}, R_{NN})$ must be specified. These are determined from the N_2 spectroscopic information[57] to be (9.9 eV, 1.42 bohr^{-1}, 2.06 bohr).

Next, the variety of topologies is illustrated for the full NN–W interaction potential that can be generated by varying mainly the Sato parameters, Δ_{NW} and Δ_{NN}, and slightly the two-body parameters, $(D_{NW}, \alpha_{NW}, R_{NW})$. Results will be shown for an N_2 molecule approaching a rigid W(110) surface with the N_2 molecular axis parallel to the surface, the center-of-mass above the bridge site, and the two atoms pointing to opposite center sites. For the best PES, described later, this is the most favorable site and configuration for dissociation, and it will be used for the illustration. For each set of parameters a PES is calculated, and a contour plot is made in the usual coordinates: bond length and height. For clarity, four critical features are extracted from each: the molecular well depth in the entrance channel; barrier energy; barrier location in both entrance channel (height) and exit channel (bond length). For the $N_2/W(110)$ system, this parameter variation never yielded

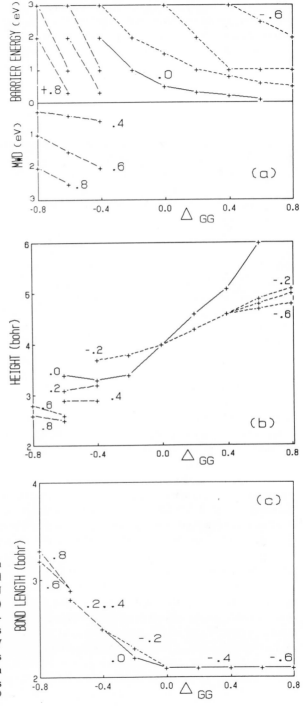

Figure 13.13. (a) Activation barrier and molecular well depth as a function of the N–N and N–W Sato parameters; (b) the height of the barrier location above the surface plane as a function of the N–N and N–W Sato parameters; (c) same as (b) but for the N–N bond length at the barrier location. The plots are from Kara and DePristo.[50]

two barriers since $(D_{NW}, \alpha_{NW}, R_{NW})$ were only allowed to vary slightly around the values specified above. (A different situation will exist for the $H_2/Ni(100)$ system discussed later.) An entrance channel barrier is merely a convenient name for an activation barrier in which the molecular bond length is "nearly unchanged" from its gas-phase value. By the same token, an exit channel barrier is one in which the molecular bond length is "significantly stretched" from its gas-phase value.

First, we consider the two Sato parameters which can vary as $-1 < \Delta \leq 1$ in principle. Figure 13.13 presents results for the range -0.8 to $+0.8$. All curves are shown for fixed Δ_{NW} as a function of Δ_{NN}. In the top panel, we note that as either of the Sato parameters approaches -1, the barrier energy increases and the molecular well depth decreases. This is easy to understand since as either $\Delta \rightarrow -1$ the "antibonding" interaction increases to ∞ and it is the separation between the "bonding" and "antibonding" interaction which controls the size of the dissociation barrier. From the top panel of Figure 13.13 two important points should be noted: (1) for any reasonable sized barrier, there are an infinite set of possible Sato parameters; (2) the molecular well depth can take on from small (≤ 0.1 eV) to large (≥ 1.0 eV) values. Thus, the parametrized PES is quite flexible with regard to the relevant energies.

Next, we consider the location of the barrier as shown in the middle and bottom panels of Figure 13.13. Some of the curves extend over a limited range because the barriers exist only over a small range. It is possible to place the barrier in either the entrance channel or the exit channel by varying the Sato parameters.

This investigation of the variation in $V_{NN,w}$ with parameters is concluded by demonstrating the effect of slight changes in the two-body parameters on the barrier height and molecular well depth in Figure 13.14. These changes have a very small effect on the location of the barrier, and therefore this is not shown. By adjusting $(D_{NW}, \alpha_{NW}, R_{NW})$ it is possible to effect substantial change in the size of the activation barrier without varying the topology of the PES or the molecular well depth. This allows for a "fine tuning" of the PES.

It should be emphasized that the original PES appears to lack sufficient flexibility, but it is clear from a detailed analysis that this is not true. Indeed, there is such great flexibility that one might even wonder how these parameters can be fixed in any meaningful way. This can be accomplished because each parameter has a clear effect on the PES, which implies that certain types of experimental or theoretical information can greatly restrict the possible ranges. This problem is now addressed in the context of the $N_2/W(110)$ system. For example, it is known[41] that the dissociation probability increases dramatically around $E_i = 0.5$ eV, and one may assume that this indicates some sort of barrier in the system; since there may be other effects, we simply use a reasonable range from 0–1 eV for such a barrier. From Figure 13.13 one can see that this restricts $\Delta_{NW} > -0.4$ but places no restrictions on Δ_{NN}. This information about the barrier energy alone does not reduce the range very much. It is also known[64] that the molecular well depth is approximately 0.27 eV, and we use a reasonable range of 0–0.4 eV. This restricts values to $0.2 < \Delta_{NW} < 0.6$ and $-0.8 \leq \Delta_{NN} \leq -0.2$. From both the barrier and well

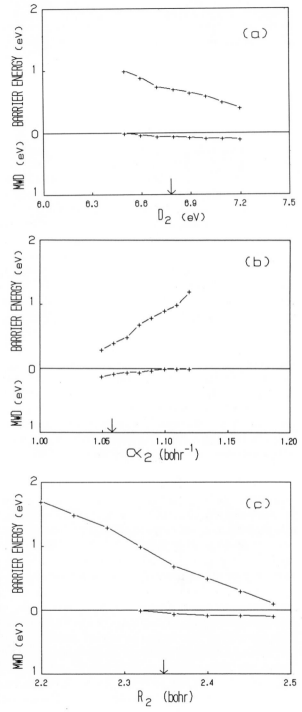

Figure 13.14. (a) Same as Figure 13.13b except that the strength of the N–W (110) two-body potential is varied; (b) same as (a) but as a function of the two-body range parameter; (c) same as (a) but as a function of the two-body interaction minimum. The plots are from Kara and DePristo.[50]

depth, the possible ranges are

$$0.2 \leq \Delta_{NW} \leq 0.6 \tag{13.5.3}$$

and

$$-0.4 \leq \Delta_{NN} \leq -0.2 \tag{13.5.4}$$

For these ranges, it is apparent from Figure 13.13 that the barrier lies in the exit channel (i.e., stretched N_2 bond length).

The exact values of the Sato and two-body parameters are determined by comparing molecular beam scattering data and GLE-"ghost" atom stochastic trajectory results. This iterative process requires human intervention to adjust the parameters, and is best accomplished by use of supercomputers to minimize the "turnaround time." Of course, it is also greatly facilitated by intelligence in the choice of the parameters (e.g., by systematically carrying out the same type of investigation in Figures 13.13 and 13.14 for the system under study). The $N_2/W(110)$ provided a difficult challenge because of the unusual experimental finding[41] that S_0 scales with the total kinetic energy, $S_0(E_i, \theta_i) = S_0(E_i, 0)$. It was almost impossible to find a PES which gave such behavior. This can be understood by inspecting Figure 13.15a, which shows a cut through the final PES, and contrasting it with the PES in Figure 13.15b, which is representative of essentially all PES that did *not* give total kinetic energy scaling.

We focus on the PES in the bottom panel first which, after full multidimensional GLE-"ghost" atom simulations, yields S_0 shown in Figure 13.16. The results labeled 45N are generated assuming "normal" kinetic energy scaling, $S_0(E_i, 45°) = S_0(E_i/2, 0)$, and are in excellent agreement with those from the dynamical simulation at 45°. The surprising feature about such a "normal" kinetic energy scaling is that the PES in Figure 13.15b exhibits an exit channel barrier. This illustrates the danger in simply focusing on the location of the barrier and not considering the topology of the PES at least in the general region of the barrier. In Figure 13.15b, the shape of the PES slightly on the entrance channel side of the barrier is such as to transform "normal" kinetic energy into vibrational energy. This makes an exit channel barrier act like an entrance channel one. We may conclude that observation of "normal" kinetic energy scaling is not particularly informative about the location of the PES barrier(s).

The distinguishing feature between the two PES in Figures 13.15a and b is the narrowness of the activation barrier region to motion perpendicular to the reaction path. The barrier region is extremely narrow in Figure 13.15a but much flatter in Figure 13.15b. In other words, the PES in the former enforces a much more stringent steric dependence on the molecule before reaction can occur. Although this PES also transforms "normal" kinetic energy into vibrational energy, such a process is too indiscriminate to enable just the correct bond length and height combination to surmount the low barrier. Instead, most trajectories lead to slightly wrong combinations of the height and bond length where the PES is much higher. Such a trajectory will not dissociate immediately but will transfer kinetic energy to the lattice phonons, molecular vibrations and rotations. If enough kinetic

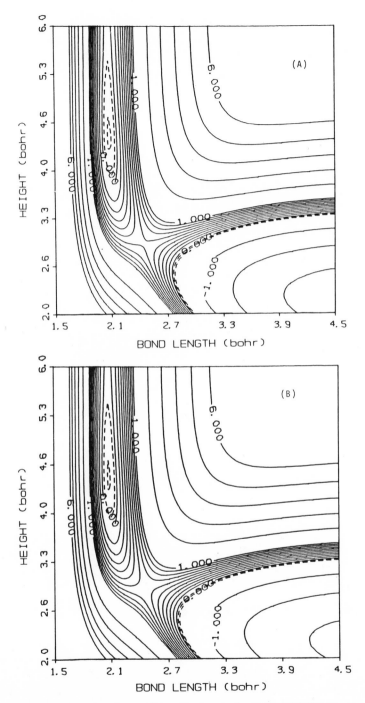

Figure 13.15. (a) Contour plot of a two-dimensional cut in the N_2/W (110) PES (for the configuration in Figure 13.13a) which yields total energy scaling for S_0; (b) same as (a) but for a typical PES which gave normal energy scaling. The plots are from Kara and DePristo.[50]

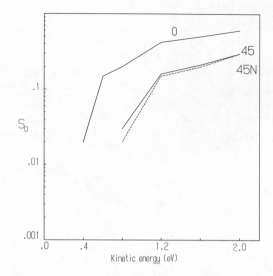

Figure 13.16. Theoretically calculated S_0 using the full N_2/W (110) PES illustrated in a 2D "cut" in Figure 13.15b. The curve for 45N corresponds to the assumption of normal energy scaling, $S_0(E_i, 45°) = S_0(\frac{1}{2}E_i, 0°)$. The plot is from Kara and DePristo.[50]

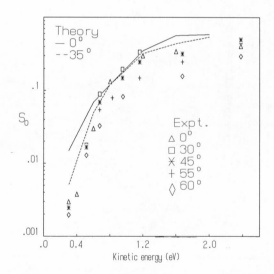

Figure 13.17. Experimental data from Pfnur et al.[41] for S_0 as a function of incident angle and kinetic energy. Theoretically calculated S_0 using the full N_2/W (110) PES illustrated in a 2D "cut" in Figure 13.15a. The plot is from Kara and DePristo.[50]

energy is transferred, the N_2 will not escape but will move out in height, move on the surface, and then recollide with the surface, attempting another barrier crossing. Eventually, some trajectories will sample just the correct height, bond length (and site), and will still have enough energy to surmount the barrier. If we wait long enough, all will sample the lowest barrier but not many will have enough energy to dissociate. In between, many of the N_2 molecules will transfer energy back to translations from rotations and vibrations, thereby desorbing. The question of the time scales for each of these processes can be addressed by the dynamical simulations, but has not been done so as yet. The results will depend upon the values of E_i, θ_i, and the surface temperature T_s.

It is clear, however, that the narrow barrier region makes the mechanism of the dissociation become rather indirect, thereby "scrambling" the translational velocity components, and leading to a total energy scaling as shown in Figure 13.17; the simulations show that the molecules often move 3–6 unit cells at $E_i = 1.2$ eV before dissociating, but are *not equilibrated* with the surface and thus are not what would be described as classical precursors. It is interesting to note another consequence of the PES. In recombinative desorption, the incipient molecule will be formed at the activation barrier and will then move into the entrance channel region which transforms vibrational energy to translational energy. This will yield a desorption pattern that peaks around the normal to the surface. Thus, there is no contradiction between desorption intensities peaked around the normal and total energy scaling of S_0, at least in the $N_2/W(110)$ system.

In the $N_2/W(110)$ system, S_0 remains less than unity even at kinetic energies well above those of the minimum barrier shown in Figure 13.15a. This occurs because many initial configurations of the N_2 with respect to $W(110)$ are repulsive and/or exhibit large activation barriers to dissociation and so are scattered back into the gas phase. These molecules are also interesting to analyze, and this is done in Figures 13.18 and 13.19 for the angular and energy distributions, respectively. The surprising feature of the angular distribution is its narrowness and peak at a slightly supraspecular angle. The reason is that the molecules which scatter back into the gas phase almost invariably only undergo a single impact with the surface, because of either an improper orientation or location during the initial collision. These scattered molecules do not sample the dissociative part of the PES but instead act just like the scattered molecules for inelastic scattering as described in section 13.3. Further evidence for this is seen in Figure 13.19 which shows the distribution of rotational, vibrational, kinetic and total energy for all the scattered molecules, each summed over all final angles. The surprising result, which corroborates the above idea, is the negligible change in the vibrational energy even though E_i is high enough to excite N_2 ($n = 4$). This demonstrates that the part of the PES which stretches the N_2 bond is not sampled by the scattered molecules. The scattering is not elastic, however. The energy loss to $W(110)$ is indicated by the average of the final TE being less than TE_i by about 0.3 eV. By contrast, the change in E_i is about 0.5 eV, which demonstrates that about 0.2 eV is transferred into rotations.

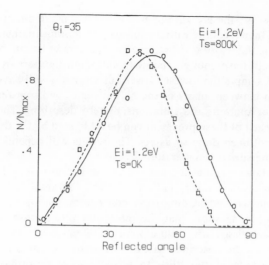

Figure 13.18. Number distribution of final angles for N_2 molecules scattered from the W(110) surface using the full PES illustrated in a 2D "cut" in Figure 13.15a. The plot is from Kara and DePristo.[50]

Another interesting feature of the dissociative chemisorption is the dependence upon initial vibrational and rotational excitation of the N_2. The results in Figure 13.20 demonstrate that translational and vibrational energy are equally efficient at increasing S_0. This may be expected for a system in which total energy scaling is found, especially in light of the explanation for such scaling in terms of the PES in Figure 13.15a. By contrast, a complicated dependence upon rotational state is exhibited in Figure 13.21. At each E_i there is a fast increase of S_0 at $j = 2$ followed by a nearly constant value. The increase is much larger than the variation of S_0 with E_i since $\varepsilon_{j=2} \approx 0.0015$ eV. The reason for the j-dependence can be seen in Figure 13.22 in terms of the variation in the dissociation probability with initial orientation of $\hat{j}$: larger dissociation occurs for $\hat{j}$ oriented parallel to the surface (i.e., rotating in the plane of the surface). This is due to both a geometric effect (i.e., rotations in the plane sample the dissociation barrier more fully) and a dynamical effect (i.e., small rotational energy perpendicular to the plane allows for significant translational to rotational energy transfer, increasing the surface residence time and thus S_0). It appears that the dependence of S_0 on j is a complicated and quite informative piece of data.

We now treat a different type of system, H_2 colliding with Ni and Cu surfaces. The light mass of H_2 precludes significant energy transfer to the lattice, and indicates that S_0 should proceed by a direct mechanism if possible. However, since H_2 is so light it is important to identify possible quantum-mechanical effects. For the dissociation of H_2 fixed parallel to the surface on a rigid linear chain of Ni atoms, spaced as along a row of Ni(100), the exact solution of the time-dependent Schroedinger equation is feasible as outlined in Section 13.1. This has been

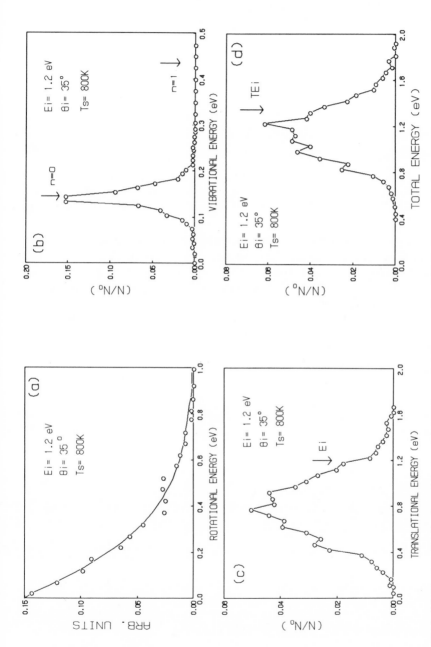

Figure 13.19. (a) Number distribution of final rotational energy for N_2 molecules scattered from the W(110) surface using the full PES illustrated in a 2D "cut" in Figure 13.15a; (b) same as (a) but for the vibrational energy; (c) same as (a) but for the translational energy; (d) same as (a) but for the total molecular energy.

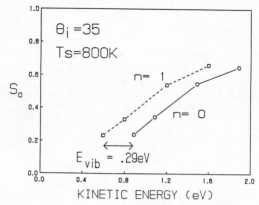

Figure 13.20. Comparison of the efficiency of vibrational and translational energy in the dissociative chemisorption of N_2 on W(110) using the full PES illustrated in a 2D "cut" in Figure 13.15a.

accomplished[59,60] using the FFT algorithm with the PES and dynamical results shown in Figure 13.23. The classical and quantum values of S_0 are in general agreement except at the lowest kinetic energies. Even for H_2, the classical results are not terrible, but are clearly not quantitative. This is due to both quantum-mechanical tunneling and reflection of the wave function at the second atop position as the H atoms separate on the surface. The latter effect will be negligible on a real 3D surface.

Mention should be made of a different approach to investigation of reactive scattering of H_2 on Ni and Cu, due to Halstead and Holloway.[65] They assume an activation barrier in the entrance channel that varies for different positions of

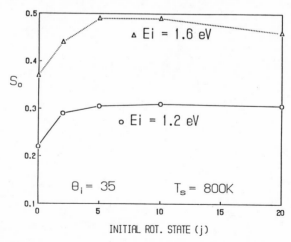

Figure 13.21. Initial rotational state dependence of the dissociative chemisorption probability of N_2 on W(110) using the full PES illustrated in a 2D "cut" in Figure 13.15a. The plot is from Kara and DePristo.[51]

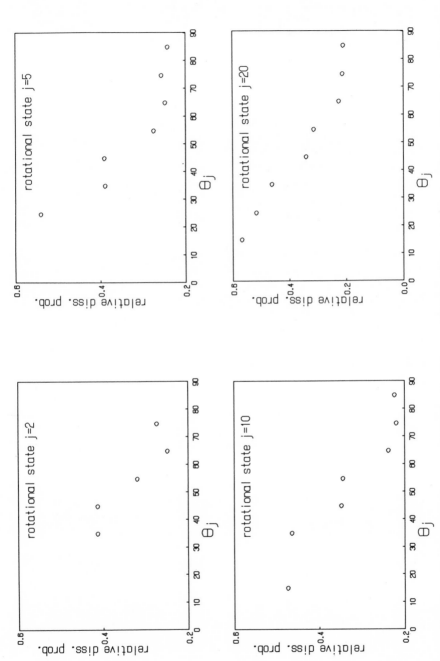

Figure 13.22. Initial rotational angular momentum orientation dependence of the dissociative chemisorption of N_2 on W(110) using the full PES illustrated in a 2D "cut" in Figure 13.15a. The plot is from Kara and DePristo.[51]

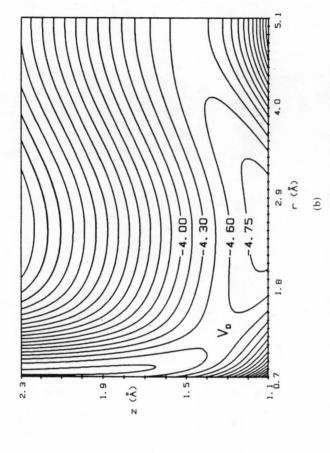

(b)

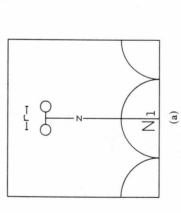

(a)

Figure 13.23. (a) Geometry and coordinate system for the H_2 dissociative chemisorption on Ni(100); (b) LEPS potential for (a); (c) variation of S_0 with kinetic energy for H_2 with the solid curve representing the quantum results and the circles and squares representing the quasi-classical trajectory values (with two different types of initial sampling of vibrational phase space); (d) same as (c) but for D_2; (e) same as (c) but for T_2; (f) same as (c) but for a fictitious H isotope of mass 7 amu. All plots are from Chiang and Jackson.[59]

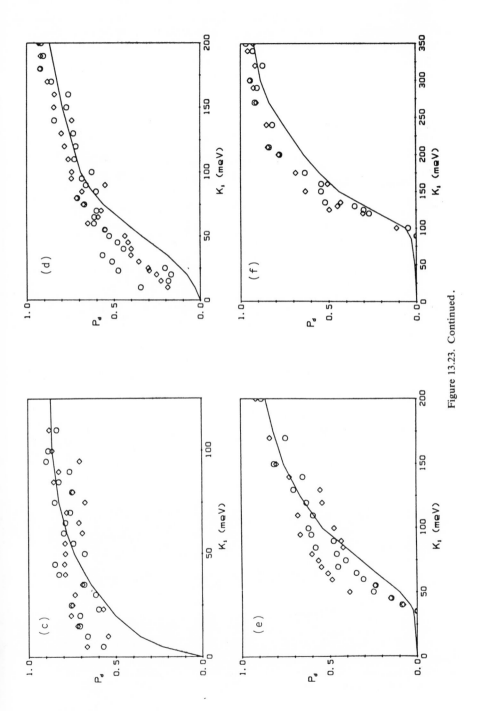

Figure 13.23. Continued.

the H_2 in the surface unit cell but which is independent of the orientation and bond length of the H_2. Under these approximations, the dynamical equations become identical to those of an atom scattering from a rigid, corrugated surface, and thus one can treat diffraction, tunneling, and dissociative chemisorption by accurate quantum techniques. In their paper, the equations are solved by the FFT procedure. The justification for the former approximation is the absence of rotational effects in the physisorbed species. The justification for the latter approximation is the absence of a stretched molecular bond at the position of the activation barrier(s). Further work will be needed to ascertain whether such assumptions are valid.

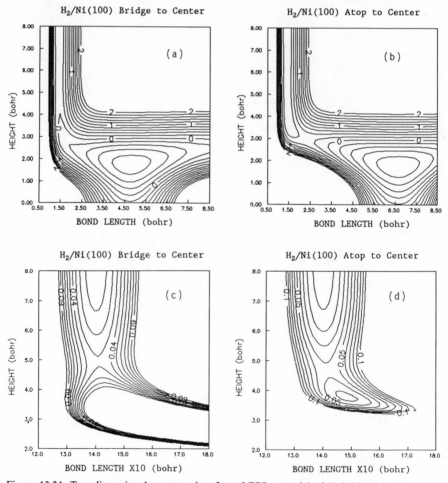

Figure 13.24. Two-dimensional contour plots for a LEPS potential of $H_2/Ni(100)$ based upon the *ab-initio* calculations of Siegbahn.[54] (a) and (c), bridge to center dissociation; (b) and (d), atop to center dissociation. The plots are from McCreery and Wolken.[56]

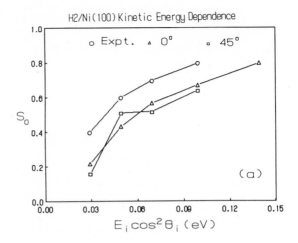

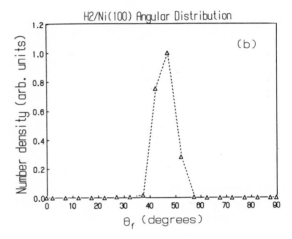

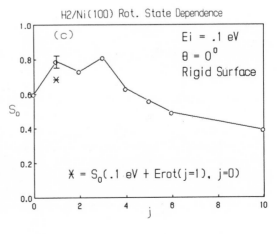

Figure 13.25. (a) Variation in S_0 with initial normal kinetic energy (circles represent experimental results from Hamza and Madix[40] while triangles and squares represent classical GLE calculations at 0° and 45°, respectively); (b) calculated number density of scattered H_2 vs. final polar angle; (c) calculated initial rotational state dependence of S_0. The PES is from Figure 13.24. The plots are from McCreery and Wolken.[56]

Finally, a new PES for the $H_2/Ni(100)$ system will be examined briefly.[55] This is based upon recent *ab-initio* calculations of Siegbahn[54] along with dynamically adjusted Sato parameters. Contour plots in Figure 13.24 demonstrate that there are small activation barriers of about 0.035 eV and 0.045 eV in the entrance channel of Figures 13.24c and d, respectively, but that at least for the atop → center dissociation in Figure 13.24b there is also a barrier in the exit channel. The orientation dependence of the entrance channel barrier has not yet been determined quantitatively, but the vibrational dependence is quite weak. In addition, the variation with position in the unit cell is also weak. However, the exit channel barrier obviously has a strong dependence upon bond length. Thus, the PES is more complex than assumed by Gadzuk,[19] but it is not apparent how the features other than the entrance channel barrier affect the dynamics.

In order to provide some understanding, we have performed classical simulations in which all degrees of freedom of H_2 can be included. The surface was allowed to move initially but this gave negligible effects, and all the results in Figure 13.25 are for a rigid surface. There must be some dynamical features of orientation and bond length since S_0 in Figure 13.25 does not become unity at $E_i > 0.1$ eV, which exceeds all the entrance channel barriers. On the other hand, S_0 is quite large (0.6–0.7) at such energies and so most of the molecules do dissociate after surmounting the entrance channel barrier. Furthermore, the angular distribution of scattered molecules is quite narrow, indicating small translational to rotational energy transfer. Perhaps the most interesting illustration of the importance of rotations is the dependence of S_0 on initial rotational state of the H_2. As for the $N_2/W(110)$ system, the dependence is quite complex but it is clear from the decrease of S_0 at high j that the dynamics after passage over the entrance channel barrier(s) must play a role since, by $j = 8$, S_0 decreases by 50% from the $j = 0$ result. These results show that the rotational dependence is quite substantial at *high kinetic energies*. However, it should be emphasized that at low kinetic energies the rotational dependence may be much less important than the neglect of all quantum effects inherent in the classical simulation. Hence, the model of Halstead and Holloway may have real utility at low kinetic energy.

Acknowledgments

The author's work on molecule/surface dynamics has been supported over the past six years mainly by the National Science Foundation, Division of Chemical Physics. Partial support by the Petroleum Research Foundation administered by the American Chemical Society has also been received. Generous amounts of supercomputer time were provided by NSF predominately at the Pittsburgh Supercomputer Center. Finally, much of the research on reactive scattering is the result of the cleverness, persistence, and physical insight of Drs. Abdelkader Kara and Chyuan-Yih Lee. By the same token, much of the work on semiclassical inelastic scattering is due to Drs. Ann M. Richard and Lynn C. Geiger.

References

1. F. O. Goodman and H. Y. Wachman, *Dynamics of Gas Surface Scattering*, Academic Press, New York (1976).

2. G. Wolken, Jr., in: *Dynamics of Molecular Collisions part A* (W. H. Miller, ed.), Plenum Press, New York (1976).

3. J. A. Barker and D. J. Auerbach, *Surf. Sci. Rep.* **4**, 1–99 (1984).

4. R. B. Gerber, *Chem. Rev.* **87**, 29–79 (1987).

5. R. B. Gerber and A. Nitzan (eds.), *Dynamics of Molecule-Surface Interactions*, special issue, Isr. J. Chem. **22**, No. 4, 283–413 (1982).

6. B. J. Garrison and N. Winograd, *Science* **216**, 805–812 (1982).

7. R. B. Gerber, R. Kosloff, and M. Berman, *Comput. Phys. Rep.* **5**, 59 (1986).

8. R. C. Mowrey and D. J. Kouri, *J. Chem. Phys.* **84**, 6466–6473 (1986).

9. A. E. DePristo, C. Y. Lee, and J. M. Hutson, *Surf. Sci.* **169**, 451–469 (1986).

10. J. C. Tully, *J. Chem. Phys.* **73**, 1975–1985 (1980).

11. S. A. Adelman and J. D. Doll, *Acc. Chem. Res.* **10**, 378–384 (1977); S. A. Adelman, *Adv. Chem. Phys.* **44**, 143–253 (1980) and references cited therein.

12. R. W. Zwanzig, *J. Chem. Phys.* **32**, 1173–1182 (1960).

13. A. E. DePristo, *Surf. Sci.* **141**, 40–60 (1984).

14. G. D. Billing, *Chem. Phys.* **116**, 269–282 (1987).

15. Z. Kirson, R. B. Gerber, A. Nitzan, and M. A. Ratner, *Surf. Sci.* **137**, 527–550 (1984).

16. C. T. Rettner, F. Fabre, J. Kimman, and D. J. Auerbach, *Phys. Rev. Lett.* **55**, 1904–1907 (1985).

17. D. M. Newns, *Surf. Sci.* **171**, 600–614 (1986).

18. B. N. J. Persson and M. Persson, *Solid State Commun.* **36**, 175–179 (1980).

19. J. W. Gadzuk, *J. Chem. Phys.* **79**, 6341–6348 (1983); **81**, 2828–2838 (1984).

20. J. C. Tully, private communication.

21. R. B. Gerber, V. Buch, and M. A. Ratner, *J. Chem. Phys.* **77**, 3022–3030 (1982).

22. N. Makri and W. H. Miller, *J. Chem. Phys.* **87**, 5781–5787 (1987).

23. A. M. Richard and A. E. DePristo, *Surf. Sci.* **134**, 338–366 (1983).

24. C. Y. Lee, R. F. Grote, and A. E. DePristo, *Surf. Sci.* **145**, 466–486 (1984).

25. A. E. DePristo and L. C. Geiger, *Surf. Sci.* **176**, 425–437 (1986).

26. G. Drolshagen and E. J. Heller, *J. Chem. Phys.* **79**, 2072–2087 (1983).

27. G. Drolshagen, *Comments At. Mol. Phys.* **17**, 47–63 (1985).

28. B. Jackson and H. Metiu, *J. Chem. Phys.* **84**, 3535–3544 (1986).

29. B. Jackson and H. Metiu, *J. Chem. Phys.* **85**, 4129–4139 (1986).

30. J. D. Doll, D. L. Freeman, and M. J. Gillan, *Chem. Phys. Lett.* **143**, 277–283 (1988).

31. N. Makri and W. H. Miller, *Chem. Phys. Lett.* **139**, 10–14 (1987).

32. A. E. DePristo, K. Haug, and H. Metiu, *Chem. Phys. Lett.* **155**, 376–380 (1989).

33. C. T. Rettner, *Vacuum* **38**, 295–300 (1988).

34. C. W. Muhlhausen, L. R. Williams, and J. C. Tully, *J. Chem. Phys.* **83** 2594–2606 (1985); a newer PES has now been developed which is much better in duplicating all the available data, J. C. Tully (private communication).

35. M. Balooch, M. J. Cardillo, D. R. Miller, and R. E. Stickney, *Surf. Sci.* **46**, 358–392 (1974).

36. M. J. Cardillo, M. Balooch, and R. E. Stickney, *Surf. Sci.* **50**, 263–278 (1975).

37. H. J. Robota, W. Vielhaber, M. C. Liu, J. Segner, and G. Ertl, *Surf. Sci.* **155**, 101–120 (1985).

38. H. P. Steinruck, K. D. Rendulic, and A. Winkler, *Surf. Sci.* **154**, 99–108 (1985).

39. H. P. Steinruck, M. Luger, A. Winkler, and K. D. Rendulic, *Phys. Rev. B* **32**, 5032–5037 (1985).

40. A. V. Hamza and R. J. Madix, *J. Phys. Chem.* **89**, 5381–5386 (1985).

41. H. E. Pfnur, C. T. Rettner, J. Lee, R. J. Madix, and D. J. Auerbach, *J. Chem. Phys.* **85**, 7452–7466 (1986).

42. C. T. Rettner and H. Stein, *J. Chem. Phys.* **87**, 770–771 (1987).

43. C. T. Rettner and H. Stein, *Phys. Rev. Lett.* **59**, 2768–2771 (1987).
44. M. B. Lee, Q. Y. Yang, S. L. Tang, and S. T. Ceyer, *J. Chem. Phys.* **85**, 1693–1694 (1986).
45. M. P. D'Evelyn, A. V. Hamza, G. E. Gdowski, and R. J. Madix, *Surf. Sci.* **167**, 451–473 (1986).
46. C. Y. Lee and A. E. DePristo, *J. Chem. Phys.* **85**, 4161–4171 (1986).
47. C. Y. Lee and A. E. DePristo, *J. Vac. Sci. Technol.* **A5**, 485–487 (1987).
48. C. Y. Lee and A. E. DePristo, *J. Chem. Phys.* **87**, 1401–1404 (1987).
49. A. Kara and A. E. DePristo, *J. Chem. Phys.* **88**, 2033–1035 (1988).
50. A. Kara and A. E. DePristo, *Surf. Sci.* **193**, 437–454 (1988).
51. A. Kara and A. E. DePristo, *J. Chem. Phys.* **88**, 5240–5242 (1988).
52. J. K. Norskov, A. Houmoller, P. Johansson, and B. I. Lundqvist, *Phys. Rev. Lett.* **46**, 257–260 (1981).
53. J. W. Gadzuk and S. Holloway, *J. Chem. Phys.* **84**, 3502–3508 (1986).
54. I. Paras, P. Liegbahn, and U. Wahlgren, *Theor. Chim. Acta* **74**, 167–184 (1988); P. Liegbahn, M. Blomberg, I. Panas, and U. Wahlgren, *Theor. Chim. Acta* **75**, 143–159 (1989).
55. A. Kara and A. E. DePristo, "On the concept and distribution of reactive sites in dissociative chemisorption," *J. Chem. Phys.* (to appear).
56. J. H. McCreery and G. Wolken, Jr., *J. Chem. Phys.* **67**, 2551–2559 (1977).
57. K. P. Huber and G. Herzberg, *Constants of Diatomic Molecules*, Van Nostrand, New York (1979).
58. M. J. Puska, R. M. Nieminen, and I. Manninen, *Phys. Rev. B* **24**, 3037–3047 (1981).
59. C. M. Chiang and B. Jackson, *J. Chem. Phys.* **87**, 5497–5503 (1987).
60. B. Jackson and H. Metiu, *J. Chem. Phys.* **86**, 1026–1035 (1987).
61. A. D. McLean and R. S. McLean, *At. Data Nucl. Data Tables* **26**, 197–381 (1981).
62. These were computed by the method detailed in J. D. Kress and A. E. DePristo, *J. Chem. Phys.* **87**, 4700–4715 (1987).
63. W. Ho, R. F. Willis, and E. W. Plummer, *Surf. Sci.* **95**, 171–184 (1980); P. W. Tamm and L. D. Schmidt, *Surf. Sci.* **26**, 286–296 (1971).
64. J. T. Yates, R. Klein, and T. E. Madey, *Surf. Sci.* **58**, 469–478 (1976).
65. D. Halstead and S. Holloway, *J. Chem. Phys.* **88**, 7197–7208 (1988).

14

The Structure of Molecules on Surfaces as Determined Using Electron-Stimulated Desorption

T. E. Madey, S. A. Joyce, and A. L. Johnson

14.1. Introduction

The structure of molecules on surfaces is an area of great importance in surface science, and a variety of surface-sensitive methods have been applied to structural problems. Many of these techniques [including Angle Resolved Ultraviolet Photoemission Spectroscopy (ARUPS), X-ray absorption near-edge structure (XANES), surface extended X-ray absorption fine structure (EXAFS), high-resolution electron energy loss spectroscopy (HREELS), and photoelectron diffraction (PD)] are discussed in other chapters in this volume.

The purpose of the present chapter is to discuss the use of electron-stimulated desorption (ESD) and photon-stimulated desorption (PSD) as probes of structure and bonding at surfaces. In contrast to many surface measurements, in which electron or photon bombardment-induced radiation damage is a nuisance to be avoided or minimized, we take advantage of beam damage to probe the structure of the surface layer.

In ESD and PSD, beams of energetic electrons or photons (generally ~10 to >1000 eV) incident on surfaces containing either adsorbed monolayers of atoms or molecules, or terminal bulk atoms, cause electronic excitations in the surface

T. E. Madey • Department of Physics and Astronomy, Rutgers, The State University, Piscataway, New Jersey 08855, USA.   S. A. Joyce and A. L. Johnson • National Institute of Standards and Technology (formerly National Bureau of Standards), Gaithersburg, Maryland, 20899, USA. NIST/NRC Postdoctoral Research Associates. The present address of Joyce is: Sandia National Laboratory, Albuquerque, New Mexico 87185, USA. The present address of Johnson is: Department of Chemistry, Cambridge University, Lensfield Road, Cambridge, England.

species. These excitations can result in desorption of ions, ground state neutral species, or metastable species from the surface.

ESD has been particularly useful in determining the structure of adsorbates due to the fact that ESD ions do not generally desorb isotropically. Instead, they desorb in discrete beams of emission, in directions determined by the orientation of the surface bonds that are ruptured by the excitations. As illustrated in Figure 14.1, ESD of CO bonded in a "standing up" configuration on a surface will result in desorption of O^+ in a direction perpendicular to the terrace plane. ESD of H^+ from "inclined" OH, or from NH_3 bonded via the N atom, will occur in directions away from the surface normal. Measurements of the electron-stimulated desorption ion angular distribution (ESDIAD) patterns can thus provide direct information about the structure of molecules oriented on surfaces.[1] The advantage of ESDIAD in determining the structure of molecules on surfaces is the fact that it is a direct method: the desorption angle is directly related to the angle of the ruptured bond. Angle-resolved PSD is more difficult experimentally, but provides similar information.

In recent years, a number of review articles and books have appeared in which the physics of ESD/PSD processes and the mechanisms and utility of ESDIAD are discussed. The DIET series (proceedings of Workshops on Desorption Induced by Electronic Transition) provides a useful and comprehensive summary of the entire field of ESD/PSD.[2-4] Other useful reviews of the mechanisms of stimulated desorption are by Knotek[5] and Menzel.[6] Detailed descriptions of the use of ESDIAD in determining the structure of surface molecules are given elsewhere.[1,7-9]

In the following pages we draw on work by Madey et al.[1,9] to summarize briefly the basic physics of ESD/PSD processes. Next, the experimental procedures used in ESDIAD measurements are described. Finally, we give a few examples of molecular structures identified using ESDIAD methods. While most of the measurements to date are of positive ions, we describe recent work on desorption of negative ions and indicate the new structural information available in these studies.

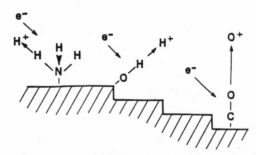

Figure 14.1. Schematic bonding configurations for adsorbed molecules, showing the relationship between surface bond angle and ion desorption angle in ESDIAD (electron-stimulated desorption ion angular distributions).

14.2. General Experimental Observations

As background for discussing the mechanisms of stimulated desorption, we summarize some basic experimental observations that characterize ESD and PSD measurements in the electron and photon energy range ~10 to 1000 eV. The following is summarized from Madey et al.[1] and based on other work.[1-6]

Desorption Products. The observed ESD and PSD desorption products include positive and negative ions as well as ground-state neutrals and vibrationally and electronically excited neutrals (metastables). Because of their relative ease of detection, positive ions have been the focus of most ESD and PSD studies; recent studies have characterized desorption of neutrals and negative ions. For adsorbed monolayers of small molecules on surfaces, the most common ions observed are atomic (such as $H^\pm$, $O^\pm$, $F^\pm$, and Cl^+), but substantial quantities of molecular ions (such as $OH^\pm$ and $CO^\pm$) are also seen. Generally, yields of positive ions are 10 to 100 times greater than yields of negative ions. Neutral desorption products include both atomic and molecular species; metastable atoms (Na^* and Li^*) and molecules (CO^* and NO^*) have been observed and characterized.[10-12] The desorbing molecules are often vibrationally hot. Desorbing ions appear to originate in the top one or two atomic layers of the solid surface.

Threshold Energies for Desorption. The threshold energy for desorption of neutral molecules[13] can be as low as 5 eV; this correlates with a one-electron valence excitation of the adsorbate. Valence and shallow-core excitations that lead to ion desorption have thresholds of 15 eV or greater. Deep-core excitations [for example, $C(1s)$ at 280 eV and $O(1s)$ at 530 eV] also correlate with ion desorption thresholds because new desorption mechanisms are available. Multiply charged ions are also seen at energies above deep-core hole ionization energies.

Cross Sections. The maximum cross sections for desorption of ions from surfaces ($\sim 10^{-20}$ to 10^{-23} cm^2) are generally smaller than those for the desorption of neutral species ($\sim 10^{-18}$ to 10^{-20} cm^2); both are smaller than typical cross sections for gas-phase dissociative ionization ($\sim 10^{-16}$ cm^2 for 100 eV electrons and diatomic molecules). For ESD, maximum ion yields are approximately 10^{-6} ions per incident electron; PSD ion yields are usually smaller, although a high PSD neutral yield of 10^{-2} atoms per photon has been reported for Li^* from LiF.[10] The cross sections for desorption of substrate ions are vanishingly small for metal surfaces but can be large for certain maximal valence oxides (such as TiO_2 and WO_3).[5]

Energies of Desorption Products. The most probable range of kinetic energies for ESD and PSD ions is 1 to 10 eV; energies as high as 15 eV have been reported. There are far fewer measurements of neutral products[13]; their most probable energies are significantly lower (≤ 1 eV). As indicated above, both vibrationally and electronically excited (metastable) species have been observed and characterized.

Sensitivity to Bonding Mode. ESD and PSD cross sections are sensitive to the mode of bonding of an atom or molecule to a surface. In general, the cross section for breaking an internal molecular bond in an adsorbed molecule is higher than that for breaking the bond to the substrate (for example, ESD of H^+ from OH bonded through the oxygen atom has a much higher cross section than ESD of H^+ from adsorbed atomic hydrogen).

Ion Angular Distributions: Relation to Structure. As indicated in the Introduction, the utility of ESD and PSD for determining the structure of surface species derives from the fact that ESD and PSD ions do not generally exhibit isotropic distributions. Instead, they desorb in directions determined by the orientation of the surface molecular bonds that are ruptured by electronic excitations. Figure 14.1 illustrates schematically the relation between the bond angle and the ion desorption angle in measurements of electron-stimulated desorption ion angular distributions (ESDIAD). Some factors which influence the final state ion trajectory (image force, reneutralization[14]) will be discussed below. Angle-resolved PSD contains information similar to ESDIAD[15] but because of experimental difficulties (primarily the lower excitation flux) it is less widely used. Angle-resolved desorption of neutrals and metastables has also been reported.[11]

ESD and PSD Similarities and Differences. ESD and PSD are thought to be initiated by essentially the same elementary electronic excitation of the surface.[5] The equivalence of ESD and PSD excitations has been demonstrated through similarities in desorption threshold energies, in ion energy distributions, in ion angular distributions, and in the nature of the surface species from which desorption occurs. There are differences, however, in the shapes of the ESD and PSD spectral yield curves (ion yield as a function of excitation energy) and in the magnitudes of the excitation cross sections for electron and photon excitation. A PSD ion yield curve generally has a sharp threshold that is followed by a maximum and a relatively abrupt decay of the signal. In contrast, ESD spectra have weak thresholds and rise smoothly above threshold. The origin of these spectral shapes and the relevant excitation physics have been discussed.[5]

In the following pages, we address the mechanisms of ESD and PSD, and then focus on the use of ESDIAD for determining the structure of surface molecules.

14.3. Mechanisms of ESD and PSD

The experimental observations described above indicate that stimulated desorption is initiated by electronic excitation of a surface molecular bond: the low threshold energies and the observation of massive ions with high kinetic energies are inconsistent with thermal effects, or with direct momentum transfer from bombarding electron to surface atom.

Various models have been proposed to account for stimulated desorption, including the Menzel–Gomer–Redhead model of desorption from covalent adsorbates[16] and the Knotek–Feibelman model of desorption from ionic adsorbates.[17] Although these (and other) models differ in detail, they have much in common. The essential features of stimulated desorption as described in all the models can be described approximately[2] as a sequence of three processes (Figure 14.2), as follows:

1. A fast initial electronic excitation ($\sim 10^{-16}$ s) (typically this is a valence or core excitation).

2. A fast electronic rearrangement ($\sim 10^{-15}$ s) to a repulsive electronic state having a lifetime of about 10^{-14} s (repulsive electronic energy is converted to nuclear motion).

3. A modification of the desorbing species (its energy, charge state, or trajectory) as it leaves the surface.

As illustrated in Figure 14.2, the surface bond is excited by electron or photon excitation through a valence or core hole ionization process (on a time scale of about 10^{-16} s).

It is widely believed[1-4,18] that ESD and PSD of ions from both covalently bonded and ionically bonded surface species proceed through multielectron excitations that produce two-hole (2h) or two-hole, one electron (2h1e) excited states. These excited states can be highly repulsive, with hole localization lifetimes on the order of 10^{-14} s, so that the repulsive electronic energy can be converted to nuclear motion. For example, an 8-eV O^+ ion will travel about 1 Å in 10^{-14} s, so that the surface bond is effectively broken. This process is an important route to ionic ESD and PSD products.

For valence excitations involving one-electron processes, excitation can be direct to a long-lived (10^{-15} to 10^{-14} s) antibonding repulsive state, from which desorption can occur. This can be a major route to desorption of ground-state and excited neutral species, as well as negative ions (dissociative attachment). Negative ions can also be formed via dipolar dissociation.[19]

The repulsive interaction in the excited electronic state (Figure 14.2) can be described as Coulombic in origin and is directed primarily along the direction of the bond that is ruptured by the excitation. Hence, the initial ion desorption angle in ESDIAD is determined by the ground-state surface bond angle. There are, however, final-state effects [process (3) above] that can influence ion desorption trajectories and yields; these include the surface image force and reneutralization

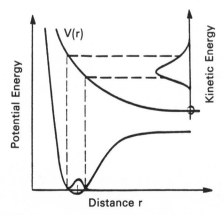

Figure 14.2. Schematic potential-energy diagram illustrating stimulated desorption of surface species. Electronic excitation from the attractive ground-state potential curve to the repulsive excited-state potential curve can lead to desorption of energetic species.

effects.[14] The image force invariably causes an increase in the polar desorption angle of an ion leaving a planar surface (that is, the trajectory is bent toward the surface). Reneutralization effects (electron hopping to the desorbing species by resonant tunneling or Auger neutralization) also influence the measured ion angular distributions and yields. Recent calculations indicate that, in addition, dynamical distortions of the substrate lattice after the initial excitation can influence the desorption processes.

14.4. Experimental Procedures

Our measurements of ESDIAD have been performed in a stainless steel ion-pumped ultrahigh vacuum chamber containing LEED/ESDIAD optics (Figure 14.3), a cylindrical mirror analyzer for Auger electron spectroscopy, a quadrupole mass spectrometer (QMS), an ion sputter gun, and an ancillary gas dosing system.[1,7,8] Samples are mounted on an XYZ rotary manipulator, and their temperatures are controlled from ~100 to ~1500 K. Samples are cleaned by Ar^+ ion bombardment and annealing. Surface purity is monitored by AES and surface order is checked using low-energy electron diffraction (LEED).

For ESDIAD measurements, a focused electron beam bombards a single crystal sample (Figure 14.3). The ESD positive ion beams pass through hemispherical grids and strike the front surface of a double microchannel plate assembly. The output signal from the assembly is accelerated to a phosphor screen, where it is displayed visually (the ESDIAD pattern). By changing potentials, the elastic LEED pattern from the sample can be generated and observed. The ESDIAD and LEED patterns can be photographed (as has been done in most ESDIAD studies), or they can be recorded with a high-sensitivity video acquisition system.

In our laboratory, a video data system was built around a high-sensitivity video camera, and an IP-512 video system in a LS1-11/73 microcomputer.* The video signal is digitized to 8 bits (256 levels) of gray scale. The images can be obtained directly from the phosphor using the video camera (in real time or off-line using video tape data storage), or from previously-obtained photographic negatives. The digitization of the data allows it to be displayed in various forms, including line contour and perspective plots, in black-and-white or false color.

Measurements of positive ion ESDIAD can be made in a dc mode, by appropriate biasing of the grids and microchannel plate entrance to reject secondary electrons.

The major technical challenge in negative ion ESDIAD is to separate the weak negative ion signal from the much larger (by ~10^7) secondary electron signal.[20] This is accomplished using time-of-flight techniques, by pulsing the electron beam, and "gating" a grid, as indicated in Figure 14.4. The gate pulse must be applied to an element of the analyzer to exclude the electron signal and

---

* Specification of a manufacturer is included to identify experimental conditions and does not imply an endorsement by NIST (formerly NBS) or Rutgers.

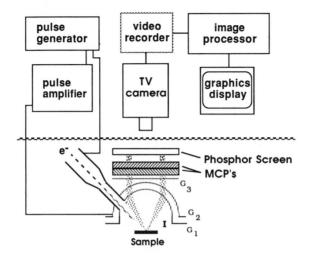

Figure 14.3. Imaging ESDIAD apparatus for measuring ion angular distributions. The high-transparency grids G_1, G_2, and G_3 are used to separate ions from electrons. The ion signal is amplified using the microchannel plates (MCP) and the secondary electrons from the MCPs are accelerated to the phosphor screen, where they produce light pulses. The resulting ion angular distribution pattern (the ESDIAD pattern) can either be photographed, or detected using a high-sensitivity video camera. The images are digitized and processed using the computer graphics system. The pulse generator and pulse amplifier are used for time-of-flight measurements.

still allow the transmission of the negative ions. The time-of-flight separation in our system is accomplished by pulsing the electron gun and the second grid (G_2) of the analyzer. The distance between the crystal and G_2 is ~2.6 cm. The electron gun is pulsed by applying a small voltage (~30 V) to the grid in the electron source triode assembly. The pulse width can be varied from 50 ns to several μs with a rise time of ~20 ns. Average currents range from 10^{-11} to 10^{-7} A, with a spot size of ~1 mm^2. G_2 is pulsed with a negative voltage from a high-voltage pulse generator. The advantage to pulsing G_2 as opposed to G_3 or the microchannel plates is that, as G_2 is hemispherical and centered about the crystal, all ion flight paths are the same length and there is therefore minimal temporal spreading of the signal. A four-channel digital delay/pulse generator controls the overall timing. A 10 KHz repetition rate is used. TTL pulses from the generator trigger both the high-voltage pulse generator and a low-voltage pulse generator whose output is used to pulse the electron gun. Using the present pulse amplifier configuration, G_2 is pulsed to more negative voltages, with a duty cycle of no greater than 2%. Since a negative voltage on G_2 is required to exclude negatively charge particles, the detector can only be gated off for short times and then returned to the "on" state for the majority of the duty cycle. It is important that G_2 be at ground during the "detector on" state to minimize the distortions of the ion energies. It is only possible with this arrangement to exclude lighter ions (shorter flight times) from heavier ions (longer flight times). As a result, TOF mass spectra for the negative ions are integrated

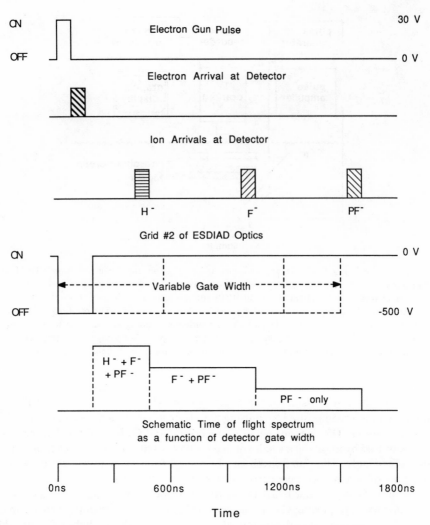

Figure 14.4. The schematic timing diagrams for mass-resolved negative-ion ESDIAD studies. The flight times for the ions are based on a bias potential at the crystal of 80 V.

from higher to lower masses as is shown schematically in Figure 14.4. This affects the sensitivity for the detection of small quantities of a lower mass ion when heavier ions are present. For typical signal levels, the lower mass ion must have an intensity of about 10% of all the higher mass ions to be detectable.

For positive ion detection, the negative high-voltage pulse is used to gate the detector on. By applying a positive offset voltage to G_2, positive ions are excluded from the detector. The pulse amplifier is then coupled to this offset in order to bring G_2 back to ground and thereby allow transmission of positive ions.

Typically in ESDIAD, a bias voltage is applied to the crystal in order to "compress" the ion trajectories so that the entire image can be viewed on the screen. For negative ion ESDIAD, the crystal bias also ensures complete separation of very-low-energy secondary electrons from the ions.

14.5. Examples of the Use of ESDIAD

14.5.1. ESDIAD of Positive Ions

There are a number of examples in which ESDIAD has been used to identify bonding structures of adsorbed molecules; many of these structures have been confirmed using other surface-sensitive methods.[1,7,8] Examples include "standing-up" CO and NO on metals, "inclined" OH and CO on surfaces,[21-23] NH_3 and H_2O adsorbed on many surfaces,[24] adsorbed fluorinated molecules,[25] and adsorbed hydrocarbons. Of particular interest are the many observations of impurity-induced ordering of H_2O, NH_3, and CO determined using ESDIAD.[1,7,26] Oxide surfaces of Ti and W have been characterized[27,28] and the bonding of small molecules to Si surfaces has been studied.[29,30]

ESDIAD is a useful, direct method for determining the structure of surface molecules. It is sensitive to the local bonding geometry; long-range order in the adsorbed layer is not necessary to obtain an ESDIAD pattern. Despite the apparent simplicity of the method, care is necessary in the measurements and the interpretation of the results. Some difficulties which can complicate measurements[31] include sensitivity to minority states (i.e., special sites or adsorbed fragments with unusually high ion yields), coadsorbed impurities which influence ion yields, sensitivity to steps and defects, beam damage, and final state effects (influence of image potential and reneutralization on ion yields and trajectories[14]). ESDIAD is most useful when used in conjunction with other surface-sensitive spectroscopic methods.

In the following paragraphs, we describe in more detail three applications of ESDIAD to surface molecular structure: chemisorption of NH_3 on metals, small molecules on silicon, and negative-ion ESDIAD of fluorinated molecules.

14.5.1.1. NH_3 on Metals

The ESDIAD method has been used to characterize the interaction of NH_3 with a number of metal surfaces[26,32,33] including Ni(111), Ru(0001), Ni(110), Ag(110), and Fe(100). For all of the planar surfaces studied, at fractional monolayer NH_3 coverages, a "halo" of H^+ ion emission is observed in ESDIAD. This pattern is consistent with bonding of the NH_3 via the N atom (see Figure 14.1) with the H atoms pointed away from the surface; this bonding configuration has been verified using other methods, including EELS and ARUPS. The continuous "halo" of H^+ has been interpreted as due to random azimuthal orientation of the NH bonds in adsorbed NH_3, or to free rotation of the NH_3 about the metal–nitrogen bond.

Recent studies of NH_3 on a stepped Fe(100) surface also provide evidence for a "tilting" of the NH_3 molecules due to bonding at step sites.[33] Figure 14.5 shows a series of ESDIAD patterns for NH_3 on Fe(s)(100), cut 3.2° off the [100] along the [110] direction; the average terrace width is ~13 atoms. Figure 14.5a is the H^+ "halo" pattern due to 0.2 monolayers of adsorbed NH_3 at 80 K. (The photos in the bottom row of Figure 14.5 are the ESDIAD patterns observed on the phosphor screen; the upper row contains computer-generated perspective intensity plots of the same patterns.) The "halo" is due to NH_3 adsorbed mainly on the planar terraces, with H atoms pointed away from the surface, as described above. Upon heating to 250 K, a substantial fraction of the NH_3 desorbs from the surface, and the ESDIAD pattern of Figure 14.5b results. The asymmetry is due to adsorption of NH_3 and NH_x fragments at step sites. At an even higher temperature, 300 K, a fraction of the asymmetric ESDIAD pattern disappears due to further desorption of molecular NH_3; in the schematic ESDIAD pattern in Figure 14.5c, the part which vanishes is designated as B. After this heat treatment we are left with two off-normal H^+ ESD spots and some fainter central emission (Figure 14.5d). This "two–spot" pattern is believed to be due to oriented NH_x fragments at step sites. Finally, heating to >350 K causes the H^+ ESD emission to disappear entirely.

Figure 14.6 shows structural models which Benndorf et al.[33] propose for NH_3 adsorption on (100) terraces and NH_3 and NH_x fragments at step sites. The proposed models are consistent with the observed ESDIAD patterns as well as thermal desorption data.

As indicated previously for the "halo" H^+ ESD pattern we suggest that NH_3 binds to Fe on (100) terraces via the N atom and that the molecular NH_3 axis is perpendicular to the surface so that the H atoms are pointed away from the surface (Figure 14.6a). This interpretation is consistent with recent cluster calculations of Bauschlicher for $NH_3/Ni(100)$[34] which demonstrated that the energy required for rotation along the threefold NH_3 axis (and also for small tilts of the NH_3 symmetry axis) is small. For $NH_3/Fe(100)$ we therefore assume that these energy barriers are comparably small.

We suggest that the two H^+ ESDIAD spots (designated A in Figure 14.5c) arise from NH_2 species (or, possibly, NH species) oriented with the NH bonds along [011] azimuths. The structural model proposed in Figure 14.6b shows two different NH_2 orientations, consistent with the two spot H^+ pattern observed in ESDIAD (Figure 14.5d). The existence of NH_2 (and NH) fragments has been reported for the atomically "rough" Fe(111) surface and these radicals have been identified by Grunze et al. using XPS (X-ray photoelectron spectroscopy).[35]

We suggest that the streaked spot B in the ESDIAD pattern (Figure 14.5b) is due to inclined molecular NH_3 adsorbed at step sites; the molecular NH_3 dipoles are believed to be tilted due to two effects: interactions with the electrostatic field associated with steps, and electronic effects due to the overlap of adsorbate levels with levels associated with the step edge. The electrostatic "dipole" field arises from the "smoothing" effect of the metal electron charge density at the step edge.

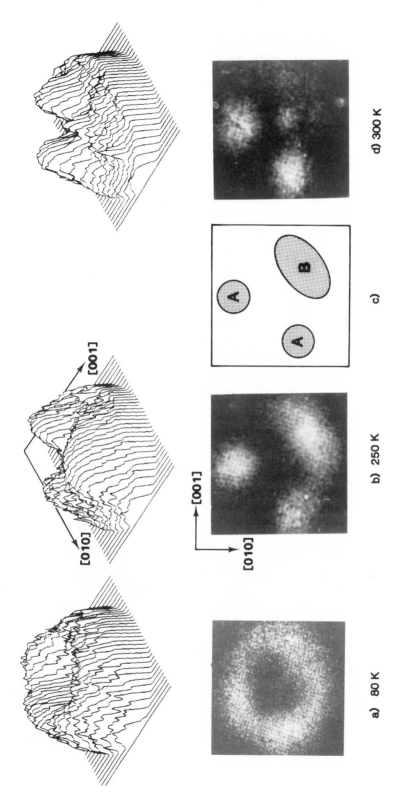

Figure 14.5. Effect of crystal heating on H$^+$ ESDIAD patterns for NH$_3$/Fe(s)(100). The lower row contains photos of the H$^+$ ESDIAD patterns as they appear on the phosphor screen, and the upper row contains computer-produced perspective intensity plots of the H$^+$ ESDIAD. (a) NH$_3$ "halo" for $\theta(NH_3) \approx 0.2$ at 80 K. (b) Asymmetric H$^+$ ESDIAD pattern after heating to 250 K. (c) Schematic ESDIAD pattern to demonstrate the experimental changes observed between 250 and 300 K. (d) Asymmetric H$^+$ ESDIAD pattern after heating to 300 K (from Benndorf et al.[33]).

a) 80 K

b) 250 K

c)

d) 300 K

[001]

[010]

[001]

[010]

A

A

B

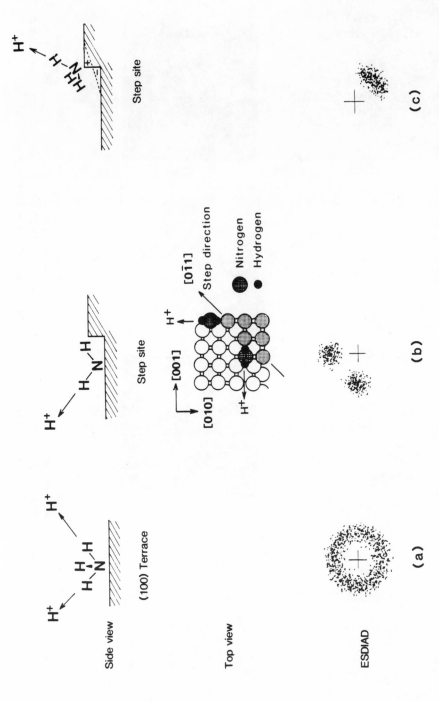

Figure 14.6. Structural models proposed for (a) NH_3 adsorption on Fe(100) terraces, (b) NH_x fragments on step sites, and (c) inclined NH_3 at step sites (from Benndorf et al.[33]).

Within this picture, the step ridge is positively charged and the "notch" is negatively charged. We suggest that this "dipole" field causes a reorientation of the NH_3 dipoles in a way that the positive end (with the H atoms) is tilted away from the step edge top and toward the step edge "notch."

In this model the three N—H bonds are now no longer equivalent for H^+ ESD emission. For H^+ ESD, N—H bond directions with an angle >60° with respect to the surface normal will be recaptured by the surface due to the attractive electrostatic interaction of the released H^+ ion with its image charge in the metal surface.[14] We conclude that a tilted NH_3 molecule would be associated with an "incomplete" H^+ ESDIAD halo pattern and the observed direction of H^+ ESD should be opposite to the direction of NH_3 tilting, as shown in Figure 14.5c. We note also that "inclined" species at step sites have been seen previously for both strongly chemisorbed and physisorbed systems, CO/W(s),[36] O/W(s),[37] and C_5H_{10}/Ag(s).[38]

14.5.1.2. H_2O, NH_3, and HF on Si(100)

ESDIAD has also been used to characterize adsorption on semiconductor surfaces, namely, adsorption of the first-row protic hydrides (H_2O, NH_3, and HF) onto planar and stepped Si(100) surfaces.[9,29,30] The LEED pattern from planar Si(100) indicates that the surface contains two domains of (2×1) reconstruction due to formation of Si—Si dimer bonds; the dimer bond axes are orthogonal in the two domains. In contrast, the surface of the stepped (100) sample cut 5° toward the (011) direction contains predominantly single domains of (2×1) reconstruction, with the dimer bonds parallel to the step edges. This distinction between planar and stepped Si(100) facilitates identification of adsorbate structures.

For adsorption temperatures between 125 and 300 K, we find evidence for dissociative adsorption for all three molecules ($H + OH$, $H + NH_x$, $H + F$) on Si(100). For monolayer coverages of each of the species, the LEED pattern shows no change from that of the clean (2×1) reconstructed surface. The ESDIAD patterns are dominated by emission in directions away from the surface normal over a wide range of coverages in each case.

The ESDIAD patterns of Figure 14.7 are due to desorption of F^+ following adsorption of HF onto stepped Si(100).[19] The two most intense beams of F^+ emission (Figure 14.7a) desorb along azimuths parallel to the step edges, i.e., parallel to the dimers (Figure 14.7b). The beams in the upstairs and downstairs directions in Figure 14.7a are due mostly to residual minority dimer domains. The ESDIAD patterns persist following heating to >600 K, showing only reversible beam broadening due to temperature-dependent vibrational effects. This behavior is expected for dissociative adsorption ($H + F$), in which the F^+ ion emission direction is controlled by the Si—F bond directions. The H^+ emission from Si—H is considerably less intense.

The correlation of polar ion emission angle with polar bond angle is complicated by image charge forces and ion reneutralization effects.[14] Calculations indicate that the maximum in the ion angular distribution is within a few degrees of the bond direction, when the emission direction is not too far from the surface normal (i.e., <30° to 40°), and the ion has a few eV of kinetic energy. We determine

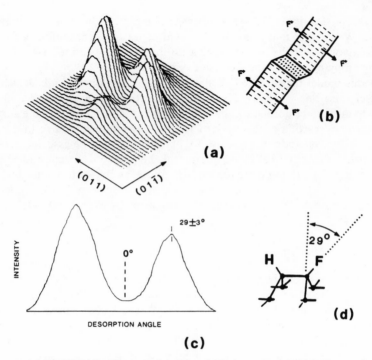

Figure 14.7. (a) F^+ ESDIAD of a saturation dose of HF on stepped Si(100) (perspective plot). (b) Model of a stepped Si(100) crystal showing that the F^+ emission is parallel to the surface dimers, along the azimuths of the Si—Si bonds. The Si—Si dimers are indicated by the symbol "–". The upstairs direction is toward the upper right of the drawing. (c), (d) The polar angle of the ion emission (and thus the Si—F bond) is $29 \pm 3°$, and the FWHM is $19°$ at 13 K.

a polar angle of $29 \pm 3°$ for the F^+ emission, with a full width at half maximum of $19°$ at 130 K (Figure 14.7c). This implies a polar angle of $\sim 29°$ for the Si—F bond (Figure 14.7d) which is along the azimuth of the dimer axis. Measurements of angle are made under field-free conditions, in which the crystal bias voltage normally used to "compress" the ESDIAD pattern is reduced to zero.

The inferred Si—F polar bond angle of $\sim 29°$ is greater than one would expect from tetrahedral bond angles and symmetric dimers, for which an angle of $19°$ to $21°$ is predicted. This difference can be rationalized either by an asymmetric dimer configuration, or by lengthening the dimer bond.

Adsorption of H_2O on Si(100) is also dissociative, with the formation of Si—H and Si—OH bonds.[30] The H^+ ESDIAD is dominated by emission from OH ligands. Unlike the F^+ described above the H^+ emission is along azimuths *perpendicular* to the dimer axes, indicating that the OH bonds are inclined nearly perpendicular to the Si—Si dimer bonds. We interpret this configuration as arising from a dative interaction between the oxygen lone-pair orbitals and the predominantly empty Si dangling bond on a neighboring dimer. Other possibilities were considered also.[30]

14.5.2. ESDIAD of Negative Ions

We have recently reported the first measurements of negative-ion ESDIAD[20,9] and are applying the new methodology to studies of small molecules on surfaces. There are several reasons for studying ESDIAD of negative ions. First, ESDIAD

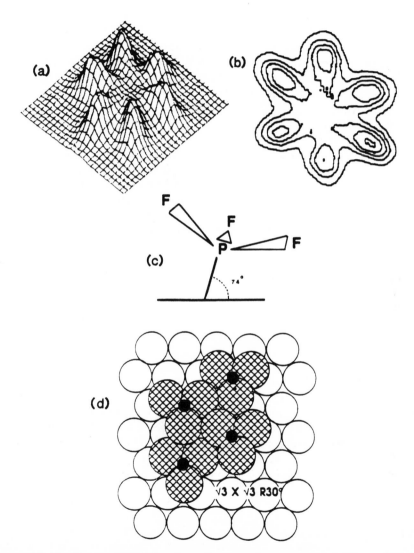

Figure 14.8. Negative-ion ESDIAD patterns for saturation coverages of PF_3 on Ru(001). (a) Perspective plot of F^- ESDIAD for saturation coverage of PF_3, annealed to 275 K, (b) line contour plot of same F^- ESDIAD, (c) model of "tilted" PF_3 consistent with the $\sqrt{3} \times \sqrt{3}$ R30° packing geometry shown in (d). The azimuthal orientations in (b) and (d) are coincident (from Johnson et al.[20]).

of positive ions has proven very useful in surface science and we anticipate new structural information about adsorbed molecules whose ion fragments include negative ions. Certain classes of adsorbed molecules (e.g., halogen-containing species) generate high yields of atomic negative ions (F^-, Cl^-) under electron bombardment. Other candidates include H^- and O^- from H_2O, CO, etc. Second, these studies are expected to provide new insights into the basic mechanisms of ion formation and the influence of surface parameters on ion trajectories, etc. Third, this is a new scientific and technical challenge. Such measurements have not been made before, and there may be unanticipated benefits.

The first experiments in negative-ion ESDIAD have involved the study of small halogen-containing molecules, including NF_3, $(CF_3)_2CO$, and PF_3 adsorbed on Ru(0001).[20] A ligand of particular interest is PF_3. Based on studies by Nitschke et al.[39] and Alvey et al.,[25] it is known that PF_3 is bonded to many metals [including Ru(0001) and Ni(111)] via the P atom, with the F atoms pointing away from the surface (Figure 14.8c).

Figure 14.8 contains two negative-ion ESDIAD patterns (Figure 14.8a is the perspective plot and Figure 14.8b the contour plot) corresponding to a saturation coverage of PF_3 on Ru(0001), dosed at 100 K and annealed to 275 K. Time-of-flight measurements demonstrate that the ions imaged in Figure 14.8 are F^-. The PF_3 coverage is 0.33 monolayers, as indicated by a sharp $(\sqrt{3} \times \sqrt{3})$ R30° LEED pattern. The six "petals" of the ESDIAD pattern confirm that PF_3 is azimuthally oriented on the hexagonal Ru substrate. The molecules, which rotate freely about the Ru—P bond at lower coverages, are constrained azimuthally at saturation coverage in a close-packed $\sqrt{3} \times \sqrt{3}$ R30° array; this is shown schematically in Figure 14.8d. Just as interlocking gears on a triangular lattice are unable to rotate, so the PF_3 molecules are "locked in" to the configuration shown. Although the molecules cannot rotate freely, they are able to "tilt" to relieve crowding; this results in the azimuthal elongation of each of the F^- ESDIAD beams (i.e., the beams are petal-like rather than circular).

The data of Figure 14.8 are complementary to positive-ion F^+ ESDIAD studies of PF_3 on Ru(0001), in that the F^+ patterns under the same experimental conditions are dominated by ion emission from dissociation products (PF_2, PF). A tentative conclusion based on our studies of adsorbed PF_3, NF_3, and $(CF_3)_2CO$ is that the undissociated adsorbed molecules produce large yields of F^-, while adsorbed molecular fragments generated by beam damage or thermal decomposition produce much larger yields of F^+ than F^-.[20] Thus, we are optimistic that negative-ion ESDIAD will provide new insights into molecular structures that are not accessible easily using positive-ion ESDIAD.

14.6. Summary and Conclusions

One of the main benefits of ESP/PSD as applied to studies of surfaces is the fact that direct information concerning the structures of surface molecules can be obtained from measurements of ESD ion angular distributions (ESDIAD). We can summarize some of the uses of ESDIAD in surface science as follows:[1]

1. ESDIAD provides *direct* information regarding surface molecular structures: ion desorption angles are related to surface bond angles. ESDIAD is not a diffraction technique; real-space images of bond directions are observed.
2. ESDIAD is sensitive to bond orientation, i.e., the local bonding geometry. Long-range order in the surface adlayer is not necessary to produce an ordered ESDIAD pattern.
3. ESDIAD is especially sensitive to the orientation of H atoms in surface molecular complexes. In contrast, low-energy electron diffraction (LEED) is relatively insensitive to H-atom positions in adsorbed molecules.
4. Negative-ion ESDIAD is complementary to positive-ion ESDIAD in the cases studied to date. Negative-ion ESDIAD is most sensitive to adsorbed molecular species, while positive-ion ESDIAD is sensitive to both adsorbed molecules and adsorbed molecular fragments.
5. Finally, ESDIAD is particularly helpful when used in conjunction with other surface-sensitive techniques. While bond directions are determined using ESDIAD, quantitative measurement of the bond length requires a technique like surface-extended X-ray absorption fine structure (SEXAFS). While ESDIAD is sensitive to the local order, low-energy electron diffraction (LEED) is sensitive to the long-range order in the surface layer. Vibrational spectroscopy, e.g., high-resolution electron energy loss spectroscopy (EELS), is extremely important for identifying the stoichiometry of surface molecular complexes whose structures are studied by ESDIAD (e.g., does H_2O dissociate into $H + OH$?).

Acknowledgment

This work was supported in part by the Office of Basic Energy Sciences of the U.S. Department of Energy.

References

1. T. E. Madey, *Science* **234**, 316 (1986); T. E. Madey, D. E. Ramaker, and R. L. Stockbauer, *Ann. Rev. Phys. Chem.* **35**, 215 (1984); J. J. Czyzewski, T. E. Madey, J. T. Yates, *Phys. Rev. Lett.* **32**, 777 (1974).
2. N. H. Tolk, M. M. Traum, J. C. Tully, and T. E. Madey (eds.), *Springer Ser. Chem. Phys.* **24**, 1 (1983).
3. W. Brenig and D. Menzel (eds.), *Springer Ser. Surf. Sci.* **4**, 1 (1985).
4. R. H. Stulen and M. L. Knotek (eds.), *Springer Ser. Surf. Sci.* **13**, 1 (1988).
5. M. L. Knotek, *Phys. Scr.* **T6**, 94 (1983); *Rep. Prog. Phys.* **47**, 1499 (1984); in Ref. 2, p. 139.
6. D. Menzel, *J. Vac. Sci. Technol.* **20**, 538 (1982); *Nucl. Instrum. Methods* **B13**, 507 (1986).
7. T. E. Madey, F. P. Netzer, J. E. Houston, D. M. Hanson, and R. L. Stockbauer, in Ref. 2, p. 120.
8. T. E. Madey, C. Benndorf, N. D. Shinn, Z. Miskovic, and J. Vukanic, in Ref. 3, p. 104.
9. T. E. Madey, A. L. Johnson, and S. A. Joyce, *Vacuum* **38**, 579 (1988).
10. N. H. Tolk, W. E. Collins, J. S. Kraus, R. J. Morris, T. R. Pian, M. M. Traum, N. G. Stoffel, and G. Margaritondo, *Springer Ser. Chem. Phys.* **24**, 156 (1983).

11. M. D. Alvey, M. J. Dresser, and J. T. Yates, Jr., *Phys. Rev. Lett.* **56**, 367 (1986).
12. A. R. Burns, *Phys. Rev. Lett.* **55**, 525 (1985).
13. P. Feulner, R. Treichler, and D. Menzel, *Phys. Rev. B* **24**, 7427 (1981); P. Feulner, *Springer Ser. Surf. Sci.* **4**, 142 (1985).
14. Z. Miskovic, J. Vukanic, and T. E. Madey, *Surf. Sci.* **169**, 405 (1986); *Surf. Sci.* **141**, 285 (1984).
15. T. E. Madey, R. L. Stockbauer, F. van der Veen, and D. E. Eastman, *Phys. Rev. Lett.* **45**, 187 (1980).
16. D. Menzel and R. Gomer, *J. Chem. Phys.* **41**, 3311 (1964); P. A. Redhead, *Can. J. Phys.* **42**, 886 (1964).
17. M. L. Knotek and P. J. Feibelman, *Phys. Rev. Lett.* **40**, 964 (1978); *Surf. Sci.* **90**, 78 (1979).
18. D. E. Ramaker, *Springer Ser. Chem. Phys.* **24**, 70 (1983); *Springer Ser. Surf. Sci.* **4**, 10 (1985); *J. Vac. Sci. Technol.* **A1**, 1137 (1983).
19. L. G. Christophorou, *Electron–Molecule Interactions and Their Applications*, Academic Press, New York (1984).
20. A. L. Johnson, S. A. Joyce, and T. E. Madey, *Phys. Rev. Lett.* **61**, 2578 (1988); S. A. Joyce, A. L. Johnson, and T. E. Madey, *J. Vac. Sci. Technol.* **A7**, 2221 (1989).
21. W. Riedel and D. Menzel, *Surf. Sci.* **163**, 39 (1985).
22. M. D. Alvey, M. J. Dresser, and J. T. Yates, *Surf. Sci.* **165**, 447 (1986).
23. C. Benndorf, C. Nobl, and T. E. Madey, *Surf. Sci.* **138**, 292 (1984).
24. P. A. Thiel and T. E. Madey, *Surf. Sci. Rep.* **7**, 211 (1987).
25. M. D. Alvey and J. T. Yates, Jr., *J. Am. Chem. Soc.* **110**, 1782 (1988).
26. F. P. Netzer and T. E. Madey, *Phys. Rev. Lett.* **47**, 928 (1981).
27. H. Niehus, *Surf. Sci.* **80**, 245 (1979).
28. R. Kurtz, R. L. Stockbauer, and T. E. Madey, *Nucl. Instrum. Methods* **B13**, 518 (1986).
29. A. L. Johnson, M. M. Walczak, and T. E. Madey, *Langmuir* **4**, 277 (1988).
30. C. U. S. Larsson, A. L. Johnson, A. Flodstrom, and T. E. Madey, *J. Vac. Sci. Technol.* **A5**, 842 (1987).
31. T. E. Madey, M. Polak, M. Walczak, and A. L. Johnson, *Springer Ser. Surf. Sci.* **13**, 120 (1988).
32. C. Benndorf and T. E. Madey, *Surf. Sci.* **135**, 164 (1983); *Surf. Sci.* **152/153**, 587 (1985).
33. C. Benndorf, T. E. Madey, and A. L. Johnson, *Surf. Sci.* **187**, 434 (1987).
34. C. N. Bauschlicher, Jr., *J. Chem. Phys.* **83**, 3129 (1985).
35. M. Grunze, F. Boszo, G. Ertl, and M. Weiss, *Appl. Surf. Sci.* **1**, 241 (1978); M. Grunze, *Surf. Sci.* **81**, 603 (1978).
36. T. E. Madey, J. E. Houston, and S. C. Dahlberg, Proc. ECOSS-3, Suppl. Le Vide, Les Couches, Minces **201**, 205 (1980).
37. T. E. Madey, *Surf. Sci.* **94**, 483 (1980).
38. M. D. Alvey, K. W. Kolasinski, J. T. Yates, Jr., and M. Head-Gordon, *J. Chem. Phys.* **85**, 6093 (1986).
39. F. Nitschke, G. Ertl, and J. Kuppers, *J. Chem. Phys.* **74**, 5911 (1981).

Photoemission from Adsorbates

A. M. Bradshaw

15.1. General Considerations

15.1.1. Introduction

Despite the plethora of techniques available to the surface scientist photoemission, or photoelectron spectroscopy, remains the most direct and effective probe of the bound electronic energy levels associated with adorbed species. In this experiment monochromatic radiation in the VUV or soft X-region causes photoexcited electrons (or photoelectrons) to be emitted into the vacuum continuum. Their kinetic energy E_f is determined with an electrostatic energy analyzer; their emission direction and spin state may also be important parameters. The energy balance is given by

$$E^N + h\nu = E^{N-1}(n_k = 0) + E_f \qquad (15.1.1)$$

where E_N is the total (N-particle) final state energy of the adsorbate–substrate system and $E^{N-1}(n_k = 0)$ is the energy of the ($N - 1$)-particle system with a hole in the kth level. The electron binding energy (also referred to as the ionization energy or ionization potential) is defined by

$$E_B = E^{N-1}(n_k = 0) - E^N = h\nu - E_f \qquad (15.1.2)$$

The different energies E_f correspond to different binding energies and thus to different ($N - 1$)-particle states. In a strictly one-electron picture, the binding energy may be identified with a single particle energy within the Hartree–Fock description (Koopmans' theorem—see Section 15.1.2). The photoelectron spectrum

A. M. Bradshaw • Fritz-Haber-Institut der Max-Planck-Gesellschaft, D-1000 Berlin 33, Federal Republic of Germany.

from an adsorbate-covered surface then consists of a series of lines corresponding to each discrete core and valence level of the system as well as a structured broad emission from the valence band of the substrate. When core levels are the main object of attention, the technique is often referred to as XPS (X-ray photoelectron spectroscopy) or ESCA (electron spectroscopy for chemical analysis), the latter being the original acronym proposed by K. Siegbahn. If the photon energy is low (say, $h\nu < 50$ eV), such that essentially only valence electrons are excited, then we normally refer to UPS (ultraviolet photoelectron spectroscopy) or simply photo-emission. At these low photon energies the photoionization cross section for valence (or outer-shell) electrons is also highest. The distinction between UPS and XPS has also been to some extent dictated by the choice of laboratory light source, i.e., the AlKα or MgKα X-ray emission lines in the soft X-ray region as opposed to the rare-gas resonance lines in the VUV. The more widespread use of synchrotron radiation now blurs the distinction between XPS and UPS and it is probably more sensible to refer to core-level photoemission and valence-level photoemission.

This chapter can only attempt to provide a very brief survey of what has become an extensive and still very topical area of surface science. Before proceeding with the various applications and examples, however, it is necessary to provide some more background information, particularly on the application of the technique to free molecules as well as on the specifically surface-related aspects of photoemission.

15.1.2. Photoelectron Spectroscopy of Free Molecules

Turner[1] in London and Viselov[2] in Leningrad performed the first valence-level experiments in the early 1960s. For a molecule in the gas phase equation (15.1.2) becomes

$$E_B = h\nu - E_f - \Delta E_{v,r} \qquad (15.1.3)$$

where $\Delta E_{v,r}$ represents the change in vibrational or rotational energy associated with the photoionization event.* The resolution of conventional spectrometers is not normally sufficient to be able to resolve rotational energy levels, but vibrational structure is strongly apparent as in the photoelectron spectrum of the N_2 molecule shown in Figure 15.1a. Each band corresponds to the removal of an electron out of the—in order of increasing ionization energy—$3\sigma_g$, $1\pi_u$, and $2\sigma_u$ orbitals, giving rise to the three ionic final states X $^2\Sigma_g^+$, A $^2\Pi_u$, and B $^2\Sigma_u^+$, respectively. With a photon energy of 21.2 eV (the HeI line) only three such states are accessible. It is immediately noticeable that the $^2\Pi_u$ band has a much more pronounced vibrational structure and that the spacing of the levels is smaller than for the $^2\Sigma_g^+$ and $^2\Sigma_u^+$

---

* As far as vibrational excitation is concerned most molecules will be in their vibrational ground state so that ΔE_v will be positive.

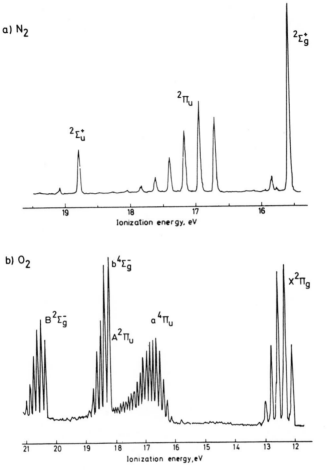

Figure 15.1. HeI photoelectron spectra of the (a) N_2 and (b) O_2 molecules (after Eland[7]).

bands. The $1\pi_u$ orbital is strongly bonding and the removal of an electron reduces the force constant in the ion, giving rise to a reduced vibrational frequency. The $2\sigma_u$ and $3\sigma_g$ orbitals are more weakly bonding and the reduction in frequency is not as great. For an antibonding orbital the frequency would increase relative to the neutral ground state. Furthermore, the $1\pi_u^{-1}$ ionization—as opposed to the $2\sigma_u^{-1}$ and $3\sigma_g^{-1}$ ionizations—is accompanied by an increase in bond length. Since according to the Franck–Condon principle we can regard the internuclear distance as remaining constant during the transition, higher vibrationally excited states of the ion can be reached as shown in Figure 15.2. Hence the strong vibrational fine structure on the $^2\pi_u$ band. The *adiabatic* ionization energy corresponds to the transition into the $v' = 0$ level, the *vertical* ionization energy to the transition of highest intensity in the vibrational envelope.

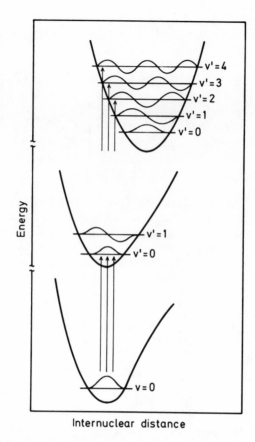

Figure 15.2. Potential energy curves illustrating the origin of vibrational excitation in molecular photoelectron spectroscopy (after Eland[7]).

The nitrogen spectrum of Figure 15.1a is also characterized by a one-to-one correspondence between the number of accessible orbitals and the number of bands in the spectrum. This is no longer true for open shell molecules such as oxygen, whose HeI photoelectron spectrum is shown in Figure 15.1b. Here, the electron configuration is $\ldots, 2\sigma_u^2, 3\sigma_g^2, 1\pi_u^4, 1\pi_g^2$. Ionization from each of the filled orbitals gives rise to quartet and doublet states due to parallel and antiparallel coupling with the two unpaired electrons in the $1\pi_g$ orbital. We note that the vibrational spacing on the $X\,^2\Pi_g$ band corresponds to a vibrational frequency higher than that of the neutral ground state because the $1\pi_g$ orbital is antibonding. Five bands are present in the spectrum at 21.2 eV photon energy but only three orbitals are ionized. The number of bands in the spectrum can also exceed the number of accessible orbitals for two other reasons. The first is a loss of degeneracy in the final ionic state due to spin–orbit coupling or to the Jahn–Teller effect, and the second is the presence of satellites due to relaxation and correlation. Spin–orbit splitting is normally only measurable in spectra of heavy atoms or of molecules containing heavy atoms. An example is provided by the photoelectron spectrum of adsorbed xenon found in Section 15.3 below. Spin–orbit splitting and satellite structures are also observed in core-level photoemission, a topic which we briefly summarize before going on to discuss relaxation effects in general.

The study of the photoionization of core levels in molecules is fascinating in itself, e.g., strong satellite structures (particularly for shallow levels), vibrational structure largely unresolved experimentally, and a very pronounced fragmentation behavior. Core-level photoelectron spectroscopy remains, however, primarily an analytical tool. This facility is based on the concept of the "chemical shift" introduced by Siegbahn and his Uppsala group,[3,4] who were the pioneers of core-level photoelectron spectroscopy. To a first approximation it neglects any complications resulting from relaxation effects and attributes shifts in the measured binding energy to changes in the orbital energy of the core level, which in turn is affected by the chemical environment of the atom. Siegbahn's work provides a highly illustrative example shown in Figure 15.3: the C1s region of the photoelectron spectrum of ethyl trifluoroacetate shows four peaks indicating four carbon atoms in distinctly different chemical environments.[4] These data were actually taken for the condensed species, not the free molecule. The linewidth is, however, determined

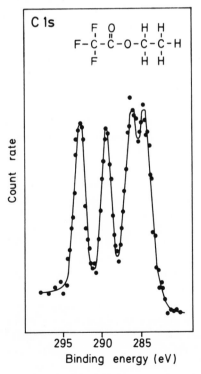

Figure 15.3. The C1s region of the photoelectron spectrum of ethyl trifluoroacetate (after Siegbahn *et al.*[4]).

here by the AlKα source. Natural linewidths in core-level photoelectron spectroscopy are determined largely by Auger decay rates for the lighter elements and are estimated to be about 0.1 eV. Vibrational structure may also be visible.[5]

The binding energy, or ionization energy, in photoelectron spectroscopy is an experimental quantity. When we assume in core-level photoelectron spectroscopy, for example, that changes in this quantity (the "chemical shift") reflect solely changes in the electronic structure of the neutral ground state, we are implicitly invoking Koopmans' theorem. This equates the measured binding energy with the orbital energy of the photoionized level:

$$E_B = -\varepsilon_k \qquad\qquad (15.1.4)$$

The approximation is, strictly speaking, only applicable to an orbital energy calculated within the Hartree–Fock self-consistent field model. Even then it remains an approximation: equation (15.1.2) shows that the binding energy is the difference between the total energy of the ionic final state and the total energy of the neutral ground state. Koopmans' theorem neglects relaxation and the change in correlation that occur upon photoionization. In the molecular orbital model each electron is considered as subject to an average electrostatic field owing to the Coulomb and exchange interactions. Upon creation of a hole these interactions among the remaining electrons will change, giving rise to a new average field. This will lead to an increased orbital energy of each bound electron and thus a more stable state of the final ion than would have been the case for "frozen orbitals." The use of Koopmans' theorem then gives an ionization energy that is too high by an amount termed the *relaxation energy*. The second point concerns the validity of the orbital model itself which cannot properly account for *correlation*, i.e., the dynamical interactions between electrons in atoms and molecules. The neglect of the change in correlation between ground state and final state works in the other direction: it tends to lower the ionization energy. In the valence region the two effects tend to cancel so that Koopmans' theorem is quite a reasonable approximation and, with one or two notable exceptions,[6] functions reasonably well. This is not the case in the core-level region, where the relaxation term dominates.

Owing to the dynamics of the relaxation process satellite structures may also be observed in the photoelectron spectrum. If the photoelectron is regarded as being removed instantaneously from the system (the sudden approximation), the other electrons may not have time to relax in response to the hole and there is a finite probability of the resulting ion being left in an excited state. This can give rise to a "shake-up" satellite corresponding to a two-electron process, i.e., the ionization of one electron and the simultaneous excitation of another. If the second electron is excited into the continuum, one refers to a "shake-off" process. Correlation in the initial state may also lead to the appearance of satellites corresponding to final states which have different symmetry and/or spin multiplicity from the initial state. To explain "relaxation" satellites, however, correlation does not necessarily have to be taken into account; it will nonetheless affect their intensity. In general, the shake-up excitation is often said to "screen" the photohole;

additional screening processes are possible when the atom or molecule is adsorbed on a surface.

Excellent introductions to photoelectron spectroscopy, particularly in its application to free atoms and molecules, are given by Eland[7] and by Siegbahn and Karlsson.[8]

15.1.3. The Surface Case

Without going into any details of the theoretical description of chemisorption at this point, it is first useful to consider what happens to the energy levels of an atom or molecule when it interacts with the surface of a metal. We take as an example the adsorption of an atom such as oxygen in a threefold hollow site on a single-crystal metal surface. The point group of this system is C_{3v}. The three 2p orbitals of the oxygen will give rise to adsorbate-induced states belonging to the A_1 and E irreducible representations. If we take the z-axis to be in the surface normal, the p_z-derived state belongs to A_1; the p_x,p_y-derived states distributed in the plane of the surface remain degenerate and belong to E. The lower symmetry of the surface thus only partially lifts the degeneracy of the oxygen 2p orbitals. Treating the semi-infinite substrate as a giant molecule, symmetry considerations tell us that only those substrate levels of A_1 and E symmetry can interact with the adsorbate levels and form the chemisorption bond. If we assume that only one substrate level of each representation participates in the bond, four molecular orbitals would be formed—two bonding and two antibonding—as in a molecule. The substrate, however, has a near-infinite number of electrons. The oxygen 2p orbitals thus interact with a semi-infinite continuum of levels and the simple picture of a molecular orbital-type scheme has to be extended. The former discrete levels of the adsorbate will shift toward lower energy owing to the attractive potential of the metal but, at the same time, are substantially broadened (Figure 15.4a). This interaction involves hopping of electrons between substrate and adsorbate orbital; the width of the level is now inversely proportional to the dwell time of the electrons on the adsorbate. In an equivalent description these broad levels are referred to as adsorbate-induced resonances: electrons incident from the bulk are scattered at the adsorbate. The states of the clean surface are interpreted as standing waves resulting from the incident electron wave and the wave reflected at the potential barrier with a node at the surface. An adsorbed particle modifies the reflection properties by shifting the phase of the reflected wave. On the low-energy side of the resonance the phase shift is such that charge is accumulated along the axis between adparticle and substrate, indicating that these states are bonding in character. On the high-energy side of the resonance the states are correspondingly antibonding in character. If the interaction with the substrate is very strong, however, the adsorbate-induced resonance will be pulled down below the valence band, becoming again discrete in nature (the "split off" state). A corresponding broad, antibonding resonance will appear at the top of the valence band (Figure 15.4b).

These considerations of the interaction of atoms and molecules with metal surfaces derive from the work of Lang[9] and others using the "jellium" model

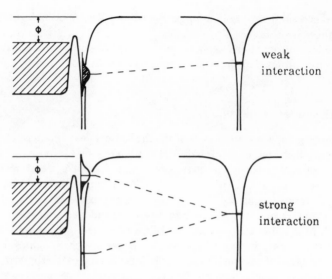

Figure 15.4. Schematic representation of the interaction of a discrete electronic energy level in an atom or molecule with the semi-infinite continuum of states at the surface of a free-electron-like metal.

and apply specifically to free-electron-like metals. In the case of transition metals, which are characterized by a narrow d-band in addition to the broad sp-band, strongly directional bonding may also take place. Such systems are good candidates for a description in terms of the so-called cluster approximation, where the metal substrate is treated as a small particle containing only a small number of atoms (typically 5–10). Molecular orbital theory can then be applied. At this point we should also mention the covalent interaction *between* adsorbed species which, in the case of ordered arrays, leads to two-dimensional (2D) band formation. This topic will be taken up in Section 15.3 below.

In summary, adsorbate-induced levels are likely to be broadened, shifted, and split due to the interaction with the surface. We expect this to be reflected in the photoelectron spectrum of the adsorbate. What other effects specific to the photoemission process itself should be expected?

First, the surface provides additional relaxation mechanisms, often referred to as extra-atomic screening. In the *image charge* screening process, the photoionization event outside the surface of the metal results in a movement of the valence electrons to screen the suddenly created hole. In a strictly classical model this interaction results in an induced surface charge and a corresponding interaction, or screening, energy of $e^2/4z$, where z is the separation between the hole and the surface of the electron gas. In a quantum description the "*image*" charge can be expressed as a superposition of various surface electronic excitations (plasma oscillations, electron–hole pairs). In core-level photoelectron spectroscopy of adsorbates surface plasmon satellites have in fact been observed.[10] In addition to the response of the bare metal surface, many-body effects associated with the metal–adsorbate interaction itself have to be considered. In the so-called "*charge transfer*" or "*unfilled orbital*" screening mechanism, an unoccupied adsorbate orbital is pulled down below the Fermi level due to the presence of the photohole and charge flows into it from the metal. This can lead to two peaks in the photoelectron spectrum, one corresponding to the well-screened situation where metal electrons have been transferred into the adsorbate orbital, and another at higher binding energy corresponding to the poorly screened situation with an empty adsorbate orbital in the final state. The strength of the well-screened feature depends on the strength of the metal–adsorbate interaction. Such effects are seen in both core- and valence-level photoelectron spectra of weak chemisorption systems[11,12] and are discussed again below, in both Section 15.2 and Section 15.4.

The second important effect of the surface on the photoelectron spectrum of a molecule is to remove the vibrational structure. The main reason for this is thought to be the finite lifetime of the photohole: it either hops into the solid or is filled via an interatomic Auger transition involving the metal valence-band electrons. If the vibrational period is longer than the hole lifetime, the vibrational fine structure will disappear, leaving a broadened line centered at the vertical ionization energy. The case of physisorbed or condensed molecules appears to lie somewhere between the chemisorption situation and the gas phase. The Franck–Condon envelope remains, but the individual vibrational lines are strongly broadened owing to the coupling to substrate phonon. The longer hole lifetime also promotes coupling to the even lower frequency external modes of the molecule itself.[13]

The third point concerns the general form of the photoelectron spectrum: the adsorbate-induced features are superimposed on the emission from the substrate. Figure 15.5 shows a schematic of the photoemission process for the metal adsorbate system. Since binding energies are conveniently referred to the Fermi level—in the case of metals it is always visible in the spectrum—equation (15.1.3) becomes

$$E_B^F = h\nu - E_f - \phi \tag{15.1.5}$$

where ϕ is the work function. The broadened adsorbate-induced features are visible superimposed on the substrate valence-band emission (or occur just underneath it in the case of split-off levels). At low kinetic energies there will be strong emission of secondary electrons which can cause high backgrounds for both valence- and core-level peaks depending on the photon energy. This may be a problem when working with the fixed photon energy of a laboratory source. In the case of valence levels it is often necessary to take difference spectra (spectrum of the clean surface subtracted from that of the adsorbate-covered surface) in order to clearly identify the features.

The last matter we address in this section concerns the surface sensitivity of photoelectron spectroscopy. When we remember that a fraction of a monolayer on a 1 cm^2 sample may contain less than 10^{14} atoms or molecules (compared to perhaps $\sim 10^{20}$ in a bulk sample), it is clear that the technique is extremely sensitive on an absolute scale. The reason lies in the low mean escape depth for the purely elastically scattered photoelectrons emitted from the substrate. A measure of this parameter is provided by the inelastic mean free path, λ, which has been measured for many solids. The dependence of λ on electron kinetic energy for the elements is shown in Figure 15.6[14] and indicates that there is a form of universal curve. The minimum of $\lambda = 5\text{-}10\ \text{Å}$ at 10-100 eV implies that, under these conditions, only the first few atomic layers of the solid are sampled. The long mean free paths at low kinetic energies are dictated by electron–phonon scattering. The minimum is given by a maximum in the cross sections for both plasmon and single-particle (valence) excitations. The curve shown in Figure 15.6 is the result of a least-squares fit to the empirical function[14]

$$\lambda = A/E^2 + BE^{1/2} \qquad (15.1.6)$$

where $E = E_f + \phi$, i.e., is measured relative to the Fermi level. It is clear from the figure that the use of a source of variable photon energy can be used to "tune"

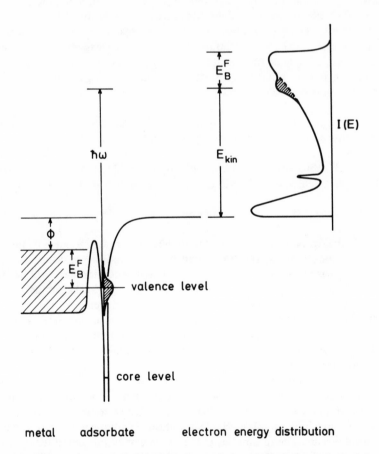

Figure 15.5. A schematic representation of the photoemission process in the single-particle picture.

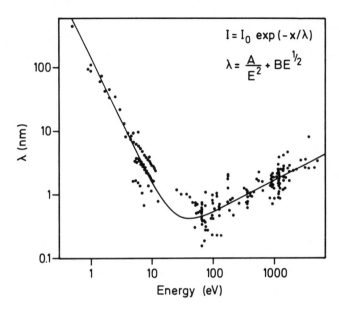

$$I = I_0 \exp(-x/\lambda)$$

$$\lambda = \frac{A}{E^2} + BE^{1/2}$$

Figure 15.6. The inelastic mean free path for electrons in solids (see text) (after Seah and Dench[14]).

the photoelectron kinetic energy to a position in or near the minimum and thus to optimize the surface sensitivity.

15.1.4. Experimental Aspects

The necessary experimental equipment for photoelectron spectroscopy is described in the two monographs cited above.[7,8] For more detailed information, the reader should consult the relevant texts on laboratory line sources,[3,15] electrostatic energy analyzers,[16] ultrahigh vacuum,[17] as well as single-crystal and adsorbate characterization techniques.[18,19] One important aspect should perhaps be singled out, namely, the increasing use of synchrotron radiation (SR). The facility of variable photon energy allows not only the tuning of the surface sensitivity as described above but also the selection of the optimal cross section. In addition, the intrinsically high degree of linear polarization allows the symmetry properties of the photoemission matrix element to be exploited in angle-resolved experiments.

The generation, properties, and uses of synchrotron radiation have been reviewed in many places. The reader is therefore referred only to a very recent book by Margaritondo[20] which serves as a useful introduction to the subject. Synchrotron radiation is produced when relativistic electrons (or positrons) are centripetally accelerated by the bending magnets of an accelerator or storage ring. It is emitted tangentially to the electron orbit as shown in Figure 15.7. In recent years a new generation of storage rings has been built whose sole function is to

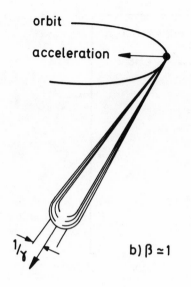

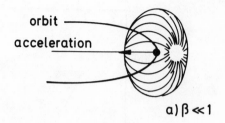

a) $\beta \ll 1$

b) $\beta \simeq 1$

$1/\gamma$

Figure 15.7. Schematic representations of the spatial distribution of electromagnetic radiation emitted by centripetally accelerated charged particles: (a) low energies; (b) relativistic energies (for electrons > 40 MeV).

serve as synchrotron radiation sources. Probably the most striking property of SR is the quasi-continuous spectrum extending from the infrared into the X-ray region. The characteristic spectral distribution curve for synchrotron radiation is shown in Figure 15.8. An SR source is characterized by its critical wavelength, λ_c, which corresponds approximately to the position of the maximum of this curve. Quantity λ_c is inversely proportional to the third power of the electron energy and directly proportional to the radius of curvature of the orbit. Thus, at the German XUV source BESSY in Berlin λ_c is 20 Å, while for the DORIS II storage ring ($E = 3.7$ GeV) in Hamburg (HASYLAB) $\lambda_c = 1.4$ Å. Experiments with hard X-rays are

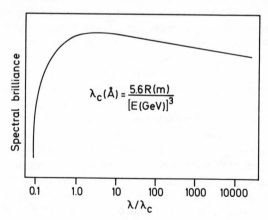

$$\lambda_c(\text{Å}) = \frac{5.6\,R(\text{m})}{[E(\text{GeV})]^3}$$

Figure 15.8. The characteristic spectral distribution curve for synchrotron radiation.

therefore also possible on high-energy rings such as DORIS. In addition, synchrotron radiation is characterized by its small source size and strongly directional emission. Both properties lead to a source of high spectral brilliance, a quantity which is defined as the number of photons per second in unit bandwidth, per unit solid angle and per unit area of source. Other important properties are the high degree of linear polarization (exactly 100% in the plane of the storage ring), the time structure (useful for certain time-resolved experiments), and the fact that its properties are entirely calculable via the Schwinger equations.[20]

Synchrotron radiation comprises "white" light and must therefore be used in conjunction with a suitable monochromator. The so-called normal-incidence region extends up to about 40 eV photon energy ($\sim$300 Å); as the name suggests, reflections at the optical components (gratings and mirrors) of the monochromator take place at normal, or near to normal, incidence. At higher photon energies the reflectivity at normal incidence falls drastically and it is necessary that reflections from the optical components take place at angles near to grazing. As a result, grating monochromators for the range 40–1000 eV are difficult to construct and the aberrations produced by the optical components are high. (Crystal monocromators can generally only be used above 800 eV.) These and other problems result in a resolution ($\lambda/\Delta\lambda$) much lower than that in the normal incidence region. Nonetheless, the advent of new designs of monochromator has recently given considerable impetus to the use of synchrotron radiation in this spectral region.

Before concluding this brief account it should be noted that undulators, produced by causing the electron beam to oscillate transversally in a periodic magnet structure inserted into the straight section of a storage ring, can further increase the intensity of synchrotron radiation. These devices produce pseudomonochromatic radiation due to constructive interference between the light emitted from the same point in successive excursions of the electron beam. The intensity can be as high as a factor 10 and the brilliance a factor 10^3 greater than that of the radiation emitted from the bending magnets on the same storage ring. Already storage rings are being planned or constructed where the main sources of radiation are provided by undulators. Three new facilities of this type—the ALS in Berkeley, Sincrotrone Trieste, and BESSY II—will provide considerable impetus in adsorbate studies, particularly for core-level photoemission and photoelectron diffraction.

15.2. Valence-Level Photoemission from Adsorbed Molecules

15.2.1. Selection Rules

Shortly after the appearance of Turner's book[1] on molecular photoelectron spectroscopy the first report of spectra from *adsorbed* species appeared in 1971: Eastman and Cashion[21] measured the valence-level spectrum of CO adsorbed on polycrystalline nickel using HeI radiation. This initially directed attention toward adsorbed *molecules* which, on account of the various bands observed, seemed somewhat more interesting than the spectra from atomic adsorbates. It was only later when dispersion effects were discovered in ordered overlayers that adsorbed atoms received closer scrutiny. We deal with this latter topic in Section 15.3. Only a few years after Eastman and Cashion's report the first experiments on adsorbed molecules using synchrotron radiation were performed at the Tantalus storage ring.[22-24] These investigations already used photoemission selection rules, although their full potential was perhaps not quite so well appreciated as it is today.

The total (elastic) photoelectron current at energy E_f from an isolated adsorbed molecule is given by

$$I \sim \sum_f \sum_i |\langle f|\tau|i\rangle|^2 \delta(E_f - E_i - h\nu) \qquad (15.2.1)$$

where $\tau = \mathbf{A} \cdot \mathbf{p} + \mathbf{p} \cdot \mathbf{A}$, with $\mathbf{A}$ the vector potential and $\mathbf{p}$ the momentum operator. The $|i\rangle$ are a complete set of single-particle states corresponding to the ground state configuration of the adsorbate–substrate system and the $|f\rangle$ represent the possible final states of the photoexcited electron. The particular $|i\rangle$ of interest are the adsorbate-induced levels corresponding to the molecular orbitals modified by the interaction with the substrate. For a given initial state $|i\rangle$ in the *angle-resolved* photoemission experiment; the final state $|f\rangle$ is uniquely characterized by its energy E_f and its momentum. The angle-resolved current is then given by

$$\frac{dj}{d\Omega} \sim |\langle f|\tau|i\rangle|^2 \delta(E_f - E_i - h\nu) \qquad (15.2.2)$$

The calculation of the matrix element is not trivial. In particular, the correct description of $|f\rangle$ must account for molecular effects (e.g., continuum orbitals giving rise to so-called shape resonances) as well as for the scattering of the photoelectron by the substrate atoms. Calculations for oriented molecules have been performed by Davenport[25] and, even though the interaction with the substrate is neglected, they show clearly the kind of information that can be obtained on orbital symmetry and orientation. As it turns out, there are certain conditions that can be placed on the angle-resolved experiment to extract this information without having to perform any calculations at all. They are essentially dipole selection rules based on the symmetry properties of the surface molecule. Strictly speaking they will only apply when the configuration of the molecule plus substrate still possesses at least one symmetry element, i.e., belongs to a point group. Invariably a lowering of the symmetry relative to the free, gas-phase species occurs.[26] Sometimes, however, the interaction with the substrate is weak and the full molecular point group may be assumed, at least for some orbitals. In such cases, the surface serves only to orient the molecule.

We consider an isolated molecule adsorbed on a metal surface with the point group C_{2v}. This could be, for example, CO in a so-called bridging site, or the surface formate species which is described in Section 15.2.4 below. The coordinates of the system are chosen so that the z-axis coincides with the twofold rotational axis in the surface normal. The Cartesian components of the matrix element are then given by

$$\left\langle f \left| A_x \frac{\partial}{\partial x} + \frac{\partial}{\partial x} A_x \right| i \right\rangle \tag{15.2.3}$$

with identical expressions for y and z. By placing symmetry constraints on $|f\rangle$, we obtain information on the $|i\rangle$.[27,28] For normal exit emission, for example, $|f\rangle$ must belong to the totally symmetric representation of the C_{2v} point group, namely a_1. This result follows from the requirement that the final-state wave function cannot have a node in the direction of emission; the final state must therefore be composed of $m = 0$ partial waves (s, p_z, d_{z^2}, etc.). In order that one of the three integrals represented by equation (15.2.3) be nonzero, the initial-state wave function must belong to the same representation of the point group as at least one of the components of the momentum operator. (This result derives from group theory and simply says that the integral must belong to or contain the totally symmetric representation in order that $\langle f|\tau|i\rangle \neq 0$, i.e., the integral must be an "even" function.) Since $\partial/\partial x$, $\partial/\partial y$, and $\partial/\partial z$ belong to the representations b_1, b_2, and a_1, respectively, it will not be possible to obtain normal emission from an a_2 state. Furthermore, for an arbitrary incident plane, s-polarized light ($A_x + A_y$) will produce emission from only b_1 and b_2 states whereas p-polarized light ($A_z + A_x + A_y$) will produce emission from b_1, b_2, and a_1 states. Were the incident plane to coincide with the xz-plane of the molecule, s-polarized light (A_y) would only excite b_2 states in normal emission. Similar considerations apply to emission in the mirror planes. When the detector is placed in the xz-plane, for example, the final-state wave function must be symmetric with respect to reflection in that plane (i.e., only partial waves without a node in xz will be detected). The a_1 and b_1 states can be excited by the A_z and A_x components, while a_2 and b_2 states can only be excited by the A_y component.

Before going on to discuss specific examples there are two general points to be made. Selection rules only give yes or no answers; they say nothing about the strength of a particular interaction which might, for example, induce a symmetry lowering. For a weak interaction, there may be neither a clear "yes" nor a clear "no." Furthermore, the answer might be inconclusive for another reason: a particular band might be intrinsically weak at an angle other than that at which it is expected to disappear as a result of a photoemission selection rule. This is simply due to the numerical value of the differential cross section as given by equation (15.2.2) under the particular conditions of the experiment. The second point concerns the overinterpretation of data. The preceding paragraph shows that by selecting final states of a particular symmetry and using the polarization vector, it is possible—knowing the orientation of the molecule—to assign the various photoemission bands. Conversely, if we are reasonably confident of the assignment, the selection rules can be used to determine the molecular orientation and/or the point group. Only in certain circumstances will enough information be available to do both.

15.2.2. CO Adsorption

We will describe here data from two adsorption systems: Cu{100}-CO, which is particularly interesting because it exhibits charge transfer screening and Cu{100}-CO/K, where the direct interaction with the coadsorbed alkali atom is apparent in the spectrum.[29] These and several other examples in this chapter are taken from the work of the author's own group. This has been done, however, purely for convenience and should not imply there is a shortage of other good examples in the literature. Numerous investigations of the adsorption of CO on copper surfaces have been performed with a variety of techniques. LEED investigations have shown that ordered overlayer structures are formed with unit meshes that are simply related to the unit mesh of the substrate. The first ordered structure on Cu{100} is designated $(\sqrt{2} \times \sqrt{2})R45°$ at a coverage of $\theta = 0.5$. A LEED structural analysis of this overlayer by Andersson and Pendry[30] has indicated that the CO molecules are bonded through the C atom in a vertical geometry directly over the Cu atoms (on-top site). The Cu-C and C-O distances were found to be 1.90 and 1.15 Å, respectively. These results have been confirmed by the photoelectron diffraction experiments of McConville et al.[31]

The photoemission spectrum of CO adsorbed on a *transition metal* surface is characterized by two peaks at 7-8 eV and 11-12 eV below the Fermi level, E_F. The first one is due to overlapping emission from the 5σ and 1π orbitals and the second to emission from the 4σ orbital. (We use the same symbols to refer to these orbitals as in the free molecule. It should be remembered, however, that they have been modified by the interaction with the metal surface.) The use of selection rules to arrive at this assignment is discussed below. The corresponding spectrum for CO on *copper* is more complicated, exhibiting more than two features below the Cu d band, as shown in curve a of Figure 15.9. The third peak at ~13.5 eV is a consequence of charge transfer screening: the main 4σ peak corresponds to the well-screened final state and the satellite to the poorly screened final state. The screening orbital is the $2\pi^*$. Another interesting feature of the spectrum of adsorbed CO is the observation of a shape resonance in the partial photoionization cross section for both σ-derived orbitals. For symmetry reasons this can only be observed for σ initial states and is strongly directed along the C—O bond axis.[23]

The use of the selection rules for assignment purposes (based on a known structure) may be understood by considering the three spectra in Figure 15.9. The spectra a and b were taken in normal emission ($\theta = 0°$) with p-polarized light. This configuration is often referred to as $k_\parallel \parallel E$, where $k_\parallel$ is the parallel component of the photoelectron momentum and E is the electric vector ($E = -\partial A/\partial t - \nabla\phi$). Under these conditions orbitals of both π and σ symmetry will be visible in the spectrum. By increasing the angle of incidence, α, from 25° to 80° the emission from σ orbitals is at its strongest, giving the binding energies of the 4σ and 5σ features fairly accurately. On the other hand, in the geometry of spectrum c ($k_\parallel \perp E$, often referred to as the forbidden geometry) only emission from states of π symmetry is allowed by the selection rules, thus enabling us to locate the 1π level. The binding energies (relative to E_F) in the system Cu{100} $(\sqrt{2} \times \sqrt{2})R45°$-CO are thus 8.4 eV ($1\pi$), 8.7 eV ($5\sigma$), and 11.5 eV ($4\sigma$).

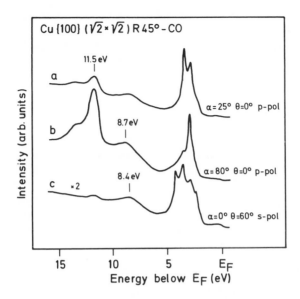

Figure 15.9. Photoemission energy distribution curves recorded for various geometries at $h\nu = 35$ eV for the $(\sqrt{2} \times \sqrt{2})R45°$ structure formed by CO on Cu{100} at 110 K. α is the angle of incidence of the incoming photons and θ the polar angle of emission (after Surman *et al.*[29]).

It is interesting to consider the position of these features relative to the corresponding ionization energies in the free CO molecule. The latter have values, relative to the Fermi level of the clean Cu surface, of 9.5 eV (5σ), 12.5 eV (1π), and 15.2 eV (4σ). Clearly a stabilization of the 5σ orbital relative to the 1π orbital takes place, while the 1π–4σ separation remains virtually unaltered. This reflects a strong interaction between 5σ and the substrate, although theory shows that the so-called π back-bonding interaction is in fact just as important (e.g., Bagus *et al.*[32]). The expected changes in the Cu d band corresponding to this interaction with the $2\pi^*$ orbital have not yet been studied in detail. At the same time an overall upward shift of ~3 eV has occurred due to the additional relaxation processes discussed in Section 15.1.3. Since we know that all the orbitals will also be pulled down due to the attractive potential of the metal, it is clear that this relaxation shift is actually larger than is at first obvious. This example immediately illustrates the problem of interpreting photoemission data from adsorbed molecules: while relative shifts of adsorbate-derived features are easy to explain qualitatively, it is difficult to compare the data with absolute values from calculated ground-state energy-level schemes. Comparison of relative values from adsorbate/metal cluster calculations should give, however, reasonable agreement with experiment, provided that the problems of charge transfer screening are recognized.

The coadsorption of CO with alkali metals produces substantial changes in the vibrational and electronic properties of the molecule. For example, a strong interaction between K and CO on a Cu{100} surface can lead to a lowering of the C—O stretch to 1530 cm^{-1} [33] indicative of substantial weakening of the C—O bond. Photoemission from the system Cu{100}-CO/K has been investigated by Heskett *et al.*,[34,35] Somerton *et al.*,[36] and Surman *et al.*[29] Representative spectra for high potassium predoses, corresponding to the relative coverage regime where a strong CO/K interaction is observed, are shown in Figure 15.10. Immediately apparent is the fact that the CO molecule retains its molecular integrity. Since we know from X-ray absorption measurements[37] that the C–O axis also remains perpendicular to the surface, the selection rules may be applied in the same way. The forbidden geometry of spectrum c shows that the 1π peak is split by the interaction into two features at 7.7 and 8.5 eV binding energy. Independent of the details of the model used in explanation, this is a manifestation of the symmetry lowering produced by a short-range direct interaction. The point group is reduced from C_{4v} to C_s, characterized by only one symmetry plane, and the degeneracy of the π states lifted. In the coadsorbed state there is also a strong suppression of the satellite structure, as shown in Figure 15.10. At the very highest K pre-coverages ($\theta_K \sim 0.33$) the "third peak" in the photoemission spectrum at $\sim$13.5 eV is no longer visible above the background. The bonding situation thus resembles that of CO on transition metals where stronger coupling of the $2\pi^*$ orbital to the metal states occurs and the charge-transfer screening mechanism becomes more

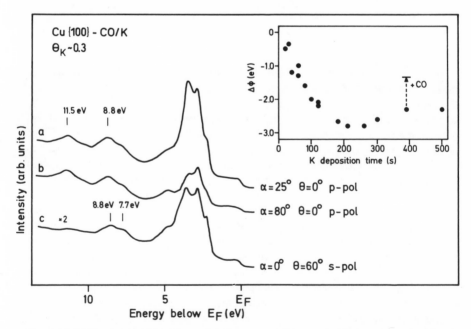

Figure 15.10. Photoemission spectra under the same conditions as in Figure 15.9 but for CO + K on Cu{100} at 100 K. $\theta_K \sim 0.3$ and CO dose 2×10^{-5} mbar · s. Inset: Change in work function as a function of deposition time as K is dosed to the clean Cu{100} surface (after Surman *et al.*[29]).

effective. This is compatible with the strongly reduced C—O stretch frequency and the increase in binding energy as measured with thermal desorption.[33]

15.2.3. Benzene Adsorption

Molecular adsorption of benzene with the ring plane oriented parallel to the surface has been established with photoemission for a number of metal surfaces (e.g., Nyberg and Richardson[38] and Netzer and Mach[39]). These results and corresponding NEXAFS data (e.g., Horsley et al.[40]) suggest that this is the usual bonding geometry, although further symmetry lowering[41] and even a tilt angle[42] have recently been proposed. Although the chemisorption bond is not particularly weak, only a minor perturbation of the molecular orbitals of the benzene takes place. Assignment is therefore not a major problem and the selection rules may be used for the determination of the point group, and thus the orientation of the molecule.

The adsorption of benzene on a Pd{100} surface results in a $(2 \times 2)R45°$ ordered overlayer.[43] An intuitive structural assignment would place the benzene molecules "flat" on the surface, i.e., with loss of the σ_h symmetry plane and thus the inversion center. In such an orientation the surface molecule group is either C_{6v}, if there is no strong influence of surface site on the molecular wave functions, or C_{2v}, if the surface site and/or the lateral interactions are important.[26] (We assume that distinct surface sites are involved because of the observation of a commensurate ordered overlayer.) The alternative structural model would be an edge-on "standing" benzene molecule as proposed for adsorption on graphite surfaces.[44] This also gives rise to a C_{2v} point group, which must be designated with a prime to distinguish it from the first case.

Seven valence levels of the free molecule (D_{6h}) are accessible with a photon energy of 21.2 eV. The HeI photoelectron spectrum of the free molecule is shown at the top of Figure 15.11; the assignment follows von Niessen et al.[45] It is now necessary to determine the irreducible representations of the benzene molecular orbitals in the point groups C_{6v}, C_{2v}, and C'_{2v} using a correlation table. This simply involves appropriate comparison of the character tables; the result is shown in Table 15.1. For normal emission the initial-state wave function must belong to the same irreducible representation as at least one of the Cartesian coordinates, i.e., A_1 and E_1 in C_{6v}; A_1, B_1, and B_2 in C_{2v} and C'_{2v}. We immediately see from Table 15.1 that normal emission from only four orbitals is allowed for the C_{6v} case, but from all orbitals for the C_{2v} and C'_{2v} cases. In fact only four orbitals are observed in normal emission with unpolarized or p-polarized light, indicating C_{6v} (see Figure 15.11; note that difference spectra are shown here). By moving the detector to an off-normal position the selection rule is relaxed, and emission from the former $2e_{2g}$, $1b_{1u}$, and $1b_{2u}$ orbitals is also observed. This indicates that the molecule takes up the "flat" orientation and that there is apparently no influence of surface site, i.e., the sixfold rotational axis is retained. The comparison of spectra taken with s- and p-polarized light, however, makes this orientational assignment quite conclusive. Of the four allowed orbitals in normal emission the former $1a_{2u}$ and $2a_{1g}$ orbitals (both a_1 and C_{6v}) should disappear with s-polarized radiation. This is

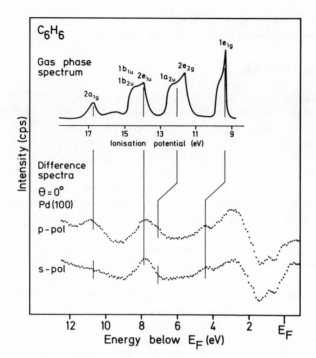

Figure 15.11. Top: Gas-phase photoemission spectrum of benzene (after Turner *et al.*[1]). Bottom: Normal emission difference spectra with p- and s-polarized light for benzene adsorbed on Pd{100}; $h\nu = 21.2$ eV (after Hofmann *et al.*[43]).

indeed observed in the lower difference spectrum of Figure 15.11. Furthermore, we observe that the ionization energies of the π orbitals, $1e_{1g}$ and $1a_{2u}$ in D_{6h}, are increased by about 1.1 eV relative to the σ orbitals [also relative to the $e_{2g}(\sigma)$ orbital which is forbidden in normal emission]. This was first noted by Nyberg and Richardson.[38] In view of the "flat" configuration of the molecule, we indeed expect chemical-bond formation to lead to a stronger interaction of the π orbitals with the surface and thus to their relative stabilization. That the chemical bond is

Table 15.1. Correlation Table for the Molecular Orbitals of Benzene

D_{6h}	C_{6v}	C_{2v}	C'_{2v}
$1e_{1g}$	e_1	$b_1 + b_2$	$a_2 + b_2$
$2e_{2g}$	e_2	$a_1 + a_2$	$a_1 + b_1$
$1a_{2u}$	a_1	a_1	b_2
$2e_{1u}$	e_1	$b_1 + b_2$	$a_1 + b_1$
$1b_{2u}$	b_2	b_2	a_1
$1b_{1u}$	b_1	b_1	b_1
$2a_{1g}$	a_1	a_1	a_1

indeed quite strong is borne out by the desorption temperature of ~500 K, where decomposition takes place. In the case of nondissociative adsorption and first-order desorption kinetics this temperature would correspond to an adsorption energy of ~120 kJ mol^{-1}. There is a certain contradiction involved here: a strong chemisorption bond to a distinct surface site should lower the symmetry further than C_{6v}. It is probably the case that the σ orbitals are only very weakly perturbed by the interaction with the surface. The behavior of the π orbitals with p and s light will be identical for C_{6v} and C_{2v} (see Table 15.1).

15.2.4. The Surface Formate Species

The decomposition of formic acid on metal and oxide surfaces is a model heterogeneous reaction. Many studies have shown that it proceeds via a well-defined reaction intermediate—the surface formate species—which can be isolated and investigated with spectroscopic techniques. Adsorption of formic acid at low temperatures gives rise to a molecular species on the surface which deprotonates on warming. On a Cu{110} surface this occurs at ~270 K, giving the formate species, which in turn decomposes at ~450 K with evolution of H_2 and CO.[46] Various studies, in particular with vibrational spectroscopy, have indicated that the two C—O bonds are equivalent and that the symmetry is C_{2v}.[47] In an X-ray absorption study Puschmann et al.[48] have recently shown that the molecular plane of the surface formate species on Cu{110} is aligned along the ⟨110⟩ azimuth (the directon of the close-packed rows) and oriented perpendicular to the surface (Figure 15.12a). The recent photoelectron diffraction study of Woodruff et al.[49] has indicated that the so-called aligned bridge site is occupied; the same local site pertains on the Cu{100} surface. This is described in more detail in Section 15.4.

Having established the orientation of the formate species on the Cu{110} surface, the photoemission data can be examined. Figure 15.12b shows the effect of deprotonation of adsorbed formic acid which occurs on warming the surface to above ~270 K. Spectrum (a) can be assigned by comparison with the photoelectron spectrum of the free molecule. The formate species also gives rise to four spectral bands and, since the number of expected orbitals is the same, it is tempting to assume a one-to-one correspondence, allowing of course for the change in symmetry from C_s to C_{2v}. The application of selection rules proves, however, that such an assignment is incorrect.[50,51]

Figure 15.13 shows three spectra at $h\nu = 25$ eV with the **E** vector parallel to the surface and aligned along the ⟨110⟩ azimuth, i.e., oriented in the molecular plane of the formate species. Spectrum b was obtained at normal photoelectron emission, for which the selection rules tell us that *only* levels belonging to b_1 in C_{2v} will be observed. This immediately assigns two features in the spectrum at 4.8 eV and 9.6 eV below E_F. By moving the detector off-normal into the ⟨100⟩ azimuth (spectrum a, **E** $\perp$ **k**$_\parallel$) emission from a_2 states should be observed as well. While peak 3 remains in the same place, peak 1 shifts slightly to lower binding energy, indicating that it also contains a level of a_2 symmetry. Similarly, by moving the detector off-normal into the ⟨110⟩ azimuth (spectrum c, **E** $\parallel$ **k**$_\parallel$) a_1 and b_1 states

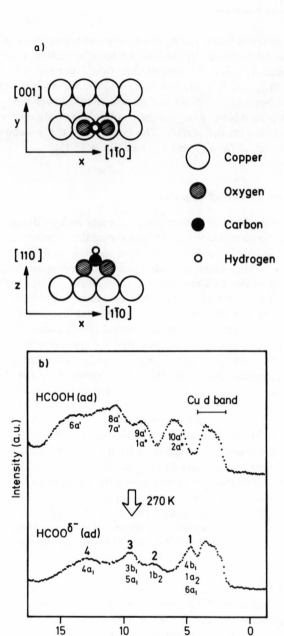

Figure 15.12. (a) Structural model for formate on Cu{110} deriving from X-ray absorption[48] and photoelectron diffraction data.[49] (b) Photoelectron spectrum of adsorbed formic acid on Cu{110} as well as the corresponding spectrum after formation of the formate species above ~270 K; $h\nu = 25$ eV (after Lindner et al.[50]).

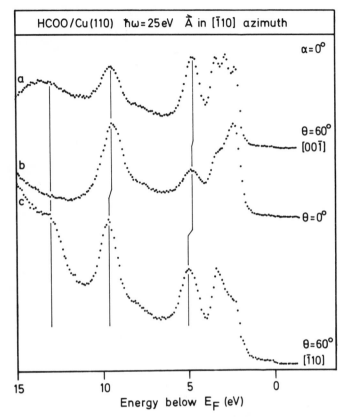

Figure 15.13. Photoelectron spectra from the system Cu{110}-HCOO for three different experimental geometries; $h\nu = 25$ eV (after Lindner et al.[50]).

are expected. Under these conditions peak 1 moves up in binding energy, as does peak 3. In addition, peak 4 is observed. Thus, three a_1 states are also present. Peak 2 is only visible with $E \perp k_{\parallel}$ for nonnormal incidence (not shown in Figure 15.13), indicating that it belongs to b_2. By performing further confirmatory experiments at other orientations of the E vector, in particular when it is aligned in the $\langle 110 \rangle$ azimuth, a complete assignment is possible. Peak 1 contains three bands due to $1a_2(\pi)$, $4b_1(\sigma)$, and $6a_1(\sigma)$ at 4.7, 4.8, and 5.1 eV below E_F, while peak 2 consists only of $1b_2(\pi)$ at 7.8 eV. Peak 3 contains $3b_1(\sigma)$ and $5a_1(\sigma)$ at 9.6 and 9.7 eV; peak 4 is due to $4a_1(\sigma)$ at 13.0 eV. These measured ionization energies have been compared with HF-SCF calculations for the formate ion[52] as well as with an INDO Cu{110}-HCOO cluster calculation.[53] The relative orbital energies from the latter, semiempirical treatment are in reasonable agreement with the measured binding energies although the absolute values, as expected, are very wrong.

The important result of these calculations by Rodriguez and Campbell[53] is the correct assignment of the photoelectron spectrum, in particular, the prediction that three levels are present in the first band and only one in the second. This comparison of orbital energies with measured binding energies is, of course, an even worse approximation than Koopmans' theorem: not only is the molecular orbital calculation non-self-consistent, but it is also assumed that the relaxation effects associated with the surface are the same for all orbitals. The calculation nonetheless delivers interesting additional information. An analysis of the percent formate character in the adsorbate-derived orbitals reveals that the $1a_2$, $4b_1$, and $6a_1$ orbitals are most strongly involved in the chemisorption bond. Relative to the formate ion, surface formate has both lower σ and π populations, but the σ population difference is the greater. The π donation occurs mainly via the $1a_2$ orbital; the strongest σ donor is the $4b_1$ orbital. Back-donation from the metal into the antibonding $\pi^*(2b_2)$ orbital is negligible because, unlike the situation in adsorbed CO, the latter is too high in energy.

15.3. Adsorbate-Induced Surface Band Structures

15.3.1. General Remarks

At sufficiently high adsorbate coverage, overlap of wave functions on adjacent species can occur either directly ("through space") or indirectly ("through substrate"). For two such species interacting through space, this overlap leads to the formation of two molecular orbital-type eigenstates, one symmetric with respect to the symmetry plane between the species and the other antisymmetric. The symmetric state is bonding and shifted to lower energy, the antisymmetric state antibonding and shifted to higher energy. Because the latter shift is greater, the total energy increases and the interaction is repulsive if both states are filled. In the case of a commensurate ordered overlayer on a single-crystal surface the 2D translational symmetry can be utilized for the description of this problem. The eigenfunctions are then 2D Bloch states defined by a reduced $k_\parallel$ vector in the surface Brillouin zone (SBZ). We show below how it is possible to determine the dispersion relationship $E(k_\parallel)$, or energy band structure, with photoemission. Figure 15.14 shows the surface Brillouin zones for a clean surface and for a $(\sqrt{2} \times \sqrt{2})R45°$ adsorbate overlayer on the {100} face of an fcc metal. The Bloch states of the overlayer can interact, or hybridize, with those of the substrate if they have the same $k_\parallel$ vector and are of similar energy. It is thus useful to be able to project the bulk band structure onto the surface band structure. This is performed by integrating over k_z inside the bulk Brillouin zone and shifting regions outside the first adsorbate 2D Brillouin zone by a reciprocal lattice vector into its interior. (Even for the clean surface it is usually found that the surface Brillouin zone is smaller than the projection of the bulk Brillouin zone.)

The dispersion of the adsorbate-induced bands, $E(k_\parallel)$, can be described for didactic purposes within a tight-binding, or LCAO, picture.[54] The simplest approach is to take the same square array as above and consider the p orbitals on each atom that have been suitably modified by the interaction with the substrate. Since the symmetry is C_{4v} at $\bar{\Gamma}$ and $\bar{M}$ these will be a_1 (p_z-derived) or e (p_x, p_y-

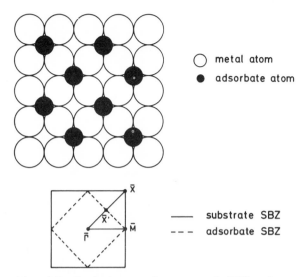

Figure 15.14. A $(\sqrt{2} \times \sqrt{2})R45°$ adsorbate overlayer on an fcc{100} surface together with the corresponding (substrate and adsorbate) surface Brillouin zones.

derived). Assuming that the Bloch states are described by

$$|n, \mathbf{k}_{\parallel}\rangle = \sum_{\mathbf{r}_j} \exp{(i\mathbf{k}_{\parallel} \cdot \mathbf{r}_j)}\phi_n(\mathbf{r} - \mathbf{r}_j) \tag{15.3.1}$$

where the $\mathbf{r}_j$ are the lattice points of the adlayer and $\phi_n(\mathbf{r} - \mathbf{r}_j)$ is the a_1 state or an appropriate linear combination of the two degenerate e states. The Bloch states at the three high-symmetry points of the surface Brillouin zone may be constructed as shown in Figure 15.15 (top). These are $\bar{\Gamma}$: $\mathbf{k}_{\parallel} = (0, 0)$, $\bar{M}$: $\mathbf{k}_{\parallel} = g/2(1, 1)$, $\bar{X}'$: $\mathbf{k}_{\parallel} = g/2(1, 0)$, where g is the length of the 2D reciprocal lattice vector of the adlayer. The sign of the wave functions at the nearest-neighbor sites around a given adspecies is given by the phase factor $\exp{(i\mathbf{k}_{\parallel} \cdot \mathbf{r}_j)}$, which in turn depends on the particular point in the SBZ. Consideration of these schematics already gives the qualitative band structure at the bottom of Figure 15.15. At $\bar{\Gamma}$ the a_1-derived Bloch state is completely bonding, while the e-derived states are mainly antibonding. At $\bar{M}$ the reverse is true. Thus, the a_1-derived band will disperse upward from $\bar{\Gamma}$ to $\bar{M}$ while the e-derived band will disperse downward. Because of the lower symmetry (C_s) the degeneracy of the e-derived band will also be lifted along $\overline{\Gamma M}$. This is also the case along $\overline{\Gamma X'}$, as well as at the $\bar{X}'$ point where the symmetry is C_{2v}. It will be seen in Section 15.3.3 that proper consideration of the interaction with the substrate gives rise to band broadening and to a mixing of bands due to symmetry lowering.

In photoemission from an ordered crystal, it is generally assumed that the parallel component of momentum is conserved as the electron crosses the surface. Since $\mathbf{k}_{\parallel}$ can be determined experimentally from

$$|\mathbf{k}_{\parallel}| = \frac{(2mE_f)^{1/2} \sin \theta}{\hbar} \tag{15.3.2}$$

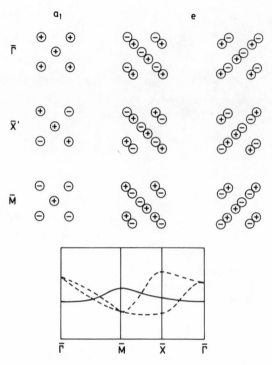

Figure 15.15. Top: Schematic representation of the 2D Bloch states formed from atomic wave functions belonging to A_1 (s or p_z) and E (p_x, p_y) at the three high-symmetry points of the surface Brillouin zone of Figure 15.14. Bottom: Resulting schematic band structure (after Bradshaw and Scheffler[54]).

it is clear that, unlike the case of the bulk band structure, the two-dimensional dispersion relationship can be determined directly. Such experiments have been performed for a variety of adsorption systems, even for adsorbed molecules. The band structures of rare-gas monolayers have proved particularly fruitful in testing some of the simple ideas above concerning dispersion in 2D arrays. Examples of these two types of system will be given below; in addition we discuss atomic chemisorption systems which are now also becoming accessible with "first-principles" electronic structure calculations.

Self-consistent slab calculations have in recent years produced very promising results for ordered adsorbate overlayers.[55-57] The slab is infinite in two dimensions but only a few atom layers thick. For ease of calculation it is also symmetrical about its central plane, i.e., there is an adsorbate layer on both sides. Using the density functional formalism in the local density approximation it has become possible not only to calculate total energy, work function, and $k_\parallel$-resolved densities of states, but also to predict the adsorption site by minimizing the total energy. Unfortunately, it appears difficult at present to calculate ordered overlayers other than (1×1), which restricts considerably the possibilities for comparison with experiment. An alternative approach is offered by the extended-layer KKR method,[58] which is not self-consistent but does allow the direct calculation of the photoemission spectrum. The comparison of slab calculations with photoemission data may not be quite straightforward because the matrix elements are not known. This is probably more important for 2D adsorbate band structures than for the isolated adsorbed molecules described in Section 15.2. In neither of these two theoretical approaches can one at present account for relaxation effects. Thus, the Koopmans' theorem discussion above applies equally well here, although valence-level relaxation probably occurs to a lesser extent in atomic adlayers than for adsorbed molecules.

Finally, we conclude these introductory remarks by noting that the symmetry discussion of Section 15.2.1 is also relevant for ordered overlayers. Thus in the application of the photoemission selection rules the initial states $|i\rangle$ are the 2D Bloch states which, at locations in the SBZ possessing symmetry, will belong to irreducible representations of the appropriate point group.[26]

15.3.2. Physisorption of Rare Gases

In physisorption systems the interaction of the adlayer with the substrate is so weak that a simple tight-binding model as described in the last section is found to describe reasonably well the 2D band structure.[59] Here, we concentrate on a specific adsorption system, Cu{110}-Xe.[60] The LEED data of Jaubert et al.[61] have indicated that a $(\sqrt{2} \times \sqrt{2})R45°$ ordered overlayer occurs at $\theta = 0.5$ followed by a commensurate-incommensurate transition occurring at slightly higher coverage. More recent LEED work[61] has indicated, however, that this picture is probably too simple and that the adlayer gives rise to a series of structures from just below $\theta = 0.5$ to saturation coverage which are all commensurate but with large unit cell dimensions. For present purposes it is sufficient to consider the adlayer at maximum coverage where the symmetry is very nearly hexagonal. A sequence of spectra from this system with increasing polar angle of emission is shown in Figure 15.16.

The occurrence of two main 5p emission peaks from adsorbed xenon—as in the free atom—is due to spin-orbit splitting. It is also another example of the breakdown of the rule "one band per orbital;" Koopmans' theorem is clearly not applicable either. The interesting feature of the *adsorbed* xenon spectrum is, however, the additional splitting of the $5p_{3/2}$ feature into two components with different magnetic quantum numbers, $m_j = \pm\frac{1}{2}$ and $m_j = \pm\frac{3}{2}$. Horn et al.[59] showed for the system Pd{100}-Xe that this effect is due to the same xenon-xenon interaction which leads to band formation: at the $\bar{\Gamma}$ point of the SBZ the symmetry is C_{6v} (or nearly C_{6v}!), for which there are only doubly degenerate representations of the corresponding double group. The $p_{3/2}$ states thus split. Tight-binding calculations and the dependence of normal emission intensity on angle of incidence[60] showed that the $m = \pm\frac{1}{2}$ component is at the higher binding energy. Following considerable controversy this original assignment has finally been confirmed in spin-resolved photoemission measurements using circularly polarized radiation.[64]

The effect of band formation in the adsorbed layer is clearly demonstrated in the set of spectra shown in Figure 15.16. As θ, and thus $\mathbf{k}_\parallel$, are increased, the binding energy of the adsorbate-induced features increases. The electron emission

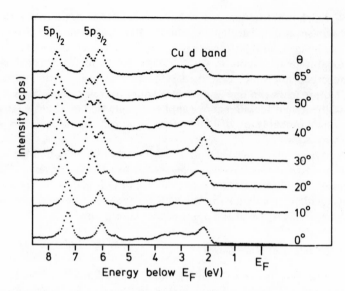

Figure 15.16. Photoelectron spectra for a pseudohexagonal xenon layer on Cu{110}; ⟨110⟩ azimuth ($\overline{\Gamma KM\Gamma'}$); $\alpha = 45°$ (unpolarized); $h\nu = 21.2$ eV (after Mariani *et al.*[60]).

direction is in the ⟨110⟩ azimuth of the substrate, corresponding to the $\overline{\Gamma KM}$ direction of the SBZ (see Figure 15.17, top). By plotting the energies of the three 5p-derived features as a function of $\mathbf{k}_{\parallel}$, the adsorbate-induced surface band structure is obtained (Figure 15.17, bottom). This has been done for both directions $\overline{\Gamma M\Gamma'}$ and $\overline{\Gamma KM\Gamma'}$. This experiment is only possible because of the twofold symmetry of the surface, which aligns the hexagonal array and enables the directions to be distinguished. On surfaces with three- or fourfold rotational symmetry, two domains are found and this information is lost. We note that in the direction $\overline{\Gamma KM\Gamma'}$ the 21.2 eV photon energy was not high enough to get back to the $\overline{\Gamma}$ point in the next zone [see equation (15.3.2)]. As expected, the dispersion curve along $\overline{\Gamma M\Gamma'}$ is symmetric about the $\overline{M}$ point. The experimental data turn out to be consistent in themselves: the $\overline{M}$ point can be reached by going along both $\overline{\Gamma M}$ and $\overline{\Gamma KM}$ and has the same binding energy in both cases. Another feature of the adsorbed xenon spectra is the presence of "satellites" on the main 5p-induced bands (these are, however, not readily apparent in Figure 15.16). More recent work shows that they are due to "density of states" (on non-**k**-conserving) emission from the flat parts of the bands at the critical points.[65] Thus in more than one respect these rare-gas adlayers may be regarded as model systems for photoemission studies.

15.3.3. Chalcogen Adsorption

As a first example of chemisorption it is instructive to take an adlayer with the same unit mesh as in Figures 15.13 and 15.14. It is known that sulfur chemisorbs on nickel{100} in both (2×2) and $(\sqrt{2} \times \sqrt{2})R45°$ phases at $\theta = 0.25$ and $\theta = 0.5$, respectively, and that in each case the fourfold hollow site is occupied. In the √2-structure, which is of immediate concern here, the S atom is situated 1.3 Å above the outermost Ni layer (see Stöhr *et al.*[66] and Section 15.4.4). The (√2 × √2)R45° overlayer exhibits much stronger dispersion of the S p-derived levels than the (2 × 2) overlayer, as might be expected at the higher coverage. Plummer *et al.*[67] have systematically investigated the position, width, and intensity of the S 3p bands as a function of both $\mathbf{k}_\parallel$ and $h\nu$. Their experimental band structure is shown in Figure 15.18a; there is certainly more than a superficial resemblance to Figure 15.15. The bands shown were obtained from a considerable number of experimental points; their symmetry (symmetric or antisymmetric with respect to reflection in $\overline{\Gamma M}$ or $\overline{\Gamma X'}$) was determined via the photoemission selection rules. The shaded area indicates the projection of the (symmetric) nickel sp band. The results of a KKR calculation are shown in Figure 15.18b: they reproduce the overall trends and the qualitative feature of the dispersion. The vertical hatching represents the broadening of the symmetric bands due to hybridization with the nickel sp band. The antisymmetric bands, p_y and $p_{x,y}$ (asym), do not mix with the symmetric nickel states. Outside the projection of the sp band all three sulfur bands form discrete "split-off" states. At $\overline{\Gamma}$ the p_z (a_1) level has a higher binding energy than the degenerate p_x, p_y (e) levels. Along $\overline{\Gamma M}$ the band splits as described in Section

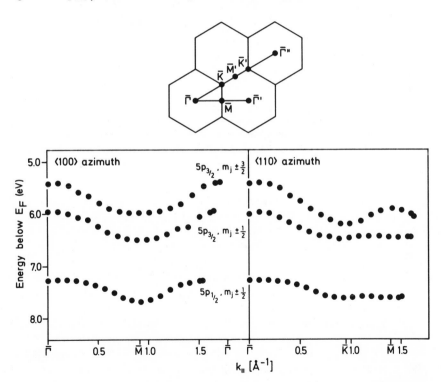

Figure 15.17. Top: Surface Brillouin zone for a hexagonal xenon layer on Cu{110}. Bottom: Experimental dispersion curve (after Mariani *et al.*[60]).

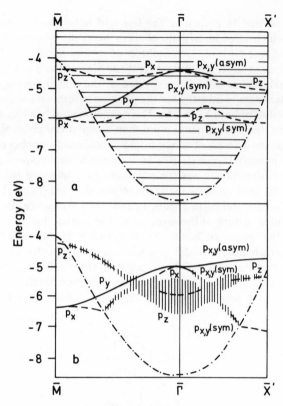

Figure 15.18. Comparison of (a) experimental and (b) theoretical energy band dispersion for the system Ni{100}($\sqrt{2} \times \sqrt{2}$)R45°–S. Solid lines: p states symmetric with respect to reflection in the symmetry plane. Dashed lines: antisymmetric p states. The hatched area in (a) denotes the projection of the symmetric Ni sp band. The hatched regions in (b) indicate the calculated one-electron width of the p levels (after Plummer *et al.*[67]).

15.3.1 but p_x and p_z are both symmetric and thus mix strongly. At $\bar{M}$ the p_x- and p_y-derived states are again degenerate because of C_{4v} symmetry. We note, however, the reversal of ordering of p_x, p_y and p_z relative to $\bar{\Gamma}$. Along $\bar{\Gamma}\bar{X}'$ the situation is somewhat different because the splitting of the symmetric and antisymmetric p_x, p_y bands is larger and, as explained in Section 15.3.1, remains at $\bar{X}'$. The symmetric band hybridizes with p_z in the same way as along $\bar{\Gamma}\bar{M}$. In the experiment, the $p_{x,y}$ band along $\bar{\Gamma}\bar{X}'$ is not observed which Plummer *et al.* ascribe to final-state multiple-scattering resonances.

As a second example of an atomic chemisorption system we choose oxygen on the Cu{110} surface. Various structural tools, particularly SEXAFS,[68] have shown that the oxygen atoms in the (2 × 1) overlayer on Cu{110} occupy the long bridge sites and that the accompanying reconstruction of the surface is of the missing row type. The corresponding angle-resolved photoemission experiments have been performed by DiDio *et al.*[69] and Courths *et al.*[70] Three largely oxygen

2p-derived bands are found below the metal d states as in the case of S 3p in the system Ni{110}($\sqrt{2} \times \sqrt{2}$)R45°-S above. In addition, there are two adsorbate-induced bands, largely Cu-derived, between the metal d bands and the Fermi level. As in the system Ag{110}-O,[71] the lower bands are bonding in character with respect to the metal–oxygen bond and the higher bands antibonding. A band structure derived from both sets of data is shown in Figure 15.19. The unoccupied adsorbate-induced band measured by Jacob et al.[72] in inverse photoemission is also included. The surface Brillouin zone is shown as an inset. It should be noted

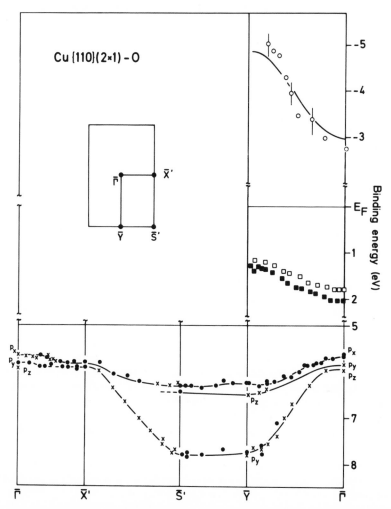

Figure 15.19. Measured O 2p-derived energy band dispersion obtained from both photoemission and inverse photoemission for the system Cu{110}(2 × 1)-O, after DiDio et al.[69] (filled circles and crosses), Courths et al.[70] (squares), and Jacob et al.[72] (open circles). The latter are inverse photoemission data.

that in reciprocal space the length of the substrate SBZ is halved in the $\langle 110 \rangle$ direction ($\overline{\Gamma X}$) by the formation of the adlayer. Intuitively we would expect little dispersion in this direction because the oxygen atoms are separated by 5.12 Å, and this is in fact borne out in practice. In the $\langle 100 \rangle$ direction they are separated by only 3.62 Å and the dispersion is correspondingly strong. It is thus the O p_y band with the p lobes oriented in this direction that disperses most strongly. Because of the C_{2v} symmetry the p_x- and p_y-derived bands are not degenerate at the $\overline{\Gamma}$ point, but split by about 0.3 eV. A comparison with Figure 15.18 reveals that, unlike the S p_z-derived band, the O p_z-derived band here disperses downward away from $\overline{\Gamma}$. This is immediately indicative of substrate-mediated effects: we saw in Section 15.3.1 that for the simple tight-binding picture of overlap of adjacent p_z orbitals the energy is expected to be lowest at the zone center. Further progress in understanding this band structure can only be made after the appropriate calculations have been performed. In particular, it would be useful to know why the accompanying surface reconstruction does not affect more dramatically the Cu d bands and the Cu sp surface state. DiDio et al.[69] argued at the time that these observations in fact favored the buckled row model in which the shift of the alternate Cu atoms rows is relatively small.

15.3.4. Band Formation in CO Layers

That dispersion effects are also observed in molecular adlayers was first noted by Horn et al.[73,74] for the systems Ni{100} and Pd{100}. Freund and Neumann have recently investigated the dependence of the extent of dispersion (bandwidth) on the CO–CO separation using data from a number of such systems.[75] For purposes of illustration, we take some recent data from the system Ni{110}(2 × 1)–CO.[76] In this structure, formed at $\theta = 1.0$, the C–O axes are tilted alternately to each side of the $\langle 110 \rangle$ Ni atom rows (Figure 15.20). The resulting zigzag chains give rise to a glide line and the space group is pmg. Such nonsymmorphic structures can be recognized in LEED by systematic absences of the expected, adsorbate-induced fractional-order features. The presence of a glide line means that there is necessarily more than one CO per unit cell. The repercussions of this for the band structure are shown schematically in Figures 15.20 and 15.21. For simplicity, only p_z (or s) orbitals are shown. Along the direction $\overline{\Gamma Y}$ and $\overline{\Gamma X'}$ the bands are doubled due to the formation of symmetric and antisymmetric combinations of the atomic wave function in the unit cell. At $\overline{\Gamma}$ all the orbitals in the symmetric combination have the same phase and the band is bonding in character relative to a noninteracting array of orbitals. The antisymmetric combination is accordingly antibonding. At $\overline{X'}$ on the zone boundary the bands become degenerate, which can be shown by using equation (15.3.1) and going to $\overline{X}'$ from both $\overline{\Gamma}_+$ and $\overline{\Gamma}_-$. This does not occur at $\overline{Y}$ since theory shows that a degeneracy is only possible perpendicular to a glide line.[77]

The photoemission spectra of Kuhlenbeck et al.[76] are quite complicated because even at $\overline{\Gamma}$ the degeneracy of the 1π orbital is lifted. Since the 4σ, 5σ, $1\pi_x$, and $1\pi_y$ levels are then further split due to the effect discussed above, we should expect eight bands along the two main symmetry directions $\overline{\Gamma X'}$ and $\overline{\Gamma Y}$, as well

Ni {110} pmg (2 × 1) - CO

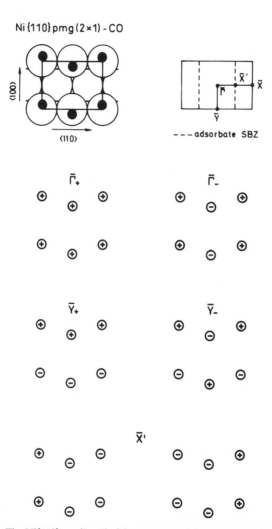

Figure 15.20. Top: The Ni{110}pmg(2 × 1)–CO structure and the corresponding surface Brillouin zone. Bottom: Schematic representation of the 2D Bloch states found from atomic wave functions (s or p_z) at the $\bar{\Gamma}$, $\bar{Y}$, and $\bar{X}'$ points of the SBZ (after Kuhlenbeck et al.[76]).

as considerable mixing where 5σ- and 1π-derived features overlap. The data are shown in Figure 15.21. We first note that measurable dispersion is indeed observed in such molecular adlayers and, second, that such band structures may also be calculated reasonably within the framework of the tight-binding model.[78] The figure shows results of calculations for a free, unsupported CO overlayer with a tilt angle of 17° between the C–O axis and the surface normal. In order to partially account for the interaction with the substrate the authors have calculated self-consistently a linear NiCO cluster, "cut off" the Ni, and renormalized the CO

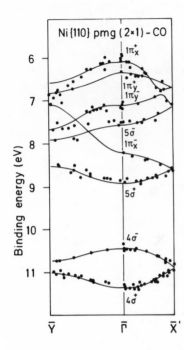

Figure 15.21. Comparison of experimental and calculated band structures for the system Ni{110}pmg(2 × 1)–CO along $\overline{\Gamma X}'$ and $\overline{\Gamma Y}$. See text (after Kuhlenbeck et al.[76]).

wave functions. This has the advantage of lowering the energy of the 5σ orbital relative to 1π (see Section 15.3.2), thus giving rise to the observed 5σ–1π mixing. We note that there is good agreement with experiment. In particular, the dispersion of the 4σ bands is described very accurately. The bandwidth of 0.8 eV (at $\overline{\Gamma}$) is in fact the largest reported so far, the reason being the very low intermolecular separation: the Ni–Ni distance is only 2.49 Å along ⟨110⟩. The less good agreement along $\overline{\Gamma Y}$ for the 5σ- and 1π-derived features is probably due to the fact that through-substrate lateral interactions cannot be properly accounted for. The band structure was in fact calculated for different tilt angles. For 17° (±2°) there was a best fit to the width of the $1\pi_x$-derived bands, agreeing well with the ESDIAD measurements of Riedl and Menzel.[79]

A further interesting feature of such nonsymmorphic structures is the operation of changed photoemission selection rules. As we have seen, these require that the integrand or part thereof in matrix elements of the type $\langle f|\tau|i\rangle$ belong to the totally symmetric representation of the appropriate point group before emission into a particular final state, $|f\rangle$, can be observed (Section 15.3.1). As Hermanson observed in his treatment of the photoemission selection rules,[27] it is sufficient to treat the final state as a plane wave in order to find the allowed transitions. This is a useful starting point for considering the selection rules for photoemission from nonsymmorphic structures. While for a symmorphic structure the symmetry of the plane wave is uniquely determined by the direction of its wave vector, in the case of a nonsymmorphic space group the symmetry of a plane wave is a function of both the *magnitude* and direction of its wave vector and is no longer unique. Thus,

Pescia *et al.*[80] have shown for normal photoemission from the basal plane of graphite that the final state could be classified as Δ_1 or Δ_2 depending on the kinetic energy of the photoelectron (i.e., on the length of its wave vector). The symmetry switching was shown to occur at the Brillouin zone boundaries. Prince *et al.*[81] have examined this phenomenon for the case of the Ni{100}p4g(2 × 2)–C structure; a detailed consideration of the topic lies, however, outside the scope of this chapter. The reader is referred to a recent review article by Prince.[82]

15.4. Core-Level Photoemission from Adsorbates

15.4.1. Introduction

Just as in the last two sections on valence-level photoemission, there is hardly space in this chapter to do full justice to the topic of core-level photoelectron spectroscopy of surfaces, and in particular of adsorbates. It is an area of increasing technological interest because of the applications of the technique in surface analysis. Core-level photoemission has lagged behind Auger spectroscopy in this respect, largely because of the poor spatial resolution, but does have several distinct advantages. These lie in the well-defined structure of the spectrum (one sharp peak per level, allowing of course for spin–orbit splitting) as well as in the chemical shift information, to which we will turn briefly again below. In addition, spatial resolution has recently been improved both by better focusing of the incident X-ray beam (this is actually much easier on a synchrotron radiation source) and by imaging the photoemitted electrons. This latter possibility—already offered by one instrument maker with a resolution of about 5μm—appears to be the most powerful approach for the future. "Small-spot" focusing means that the sample has to be scanned, while the imaging technique enjoys the multiplexing advantage.

In this section we will concentrate on the case of core-level spectroscopy in "*fingerprinting*," i.e., in the characterization of adsorbate species under well-defined conditions on metal surfaces, and in studying *surface reactions* as well as in *photoelectron diffraction*. In the first two of these areas—and in the third probably to a limited extent as well—the chemical shift plays an important role. In a simple picture of the chemical shift (referred to as the ground-state potential model) the energy required to remove a core electron is affected by the electrostatic potential at the core which in turn is affected by the charge density in the valence shell. Thus a strongly electronegative species such as a chlorine atom might form a chemical bond such that valence charge accumulates in the vicinity of its core. The potential energy of the core electrons would then be raised and the binding energy, or ionization energy, would decrease relative to the free atom. The discussion on free molecules in Section 15.1.2 has shown us, however, that this picture is actually highly simplified although it may work very well in certain situations (for example, in explaining the C 1s chemical shifts in the molecule ethyl trifluoroacetate shown in Figure 15.3). The ground-state potential model is based on Koopmans' theorem and neglects relaxation and the change in correlation. In the core-level region the two effects do not cancel as they tend to in the valence region.

Furthermore, on surfaces there are additional relaxation effects ("extra-atomic screening") as was briefly discussed in Section 15.1.3. Thus, instead of equation (15.1.4) we have

$$E_B = -\varepsilon_k + E_R + \delta E_C \tag{15.4.1}$$

where E_R is the relaxation energy and δE_C the change in correlation energy. The chemical shift is then

$$\Delta E_B = -\Delta \varepsilon_k + \Delta E_R + \Delta(\delta E_C) \tag{15.4.2}$$

The first term is usually described as the "initial state" contribution; the latter two constitute the final state contribution. The simplest way to proceed further is to set $\Delta(\delta E_c)$ to zero (probably a reasonable approximation) and to try and separate $\Delta \varepsilon_k$ and ΔE_R. This is only possible by calculation. For the chemical shift occurring upon adsorption Lang and Williams[83] have attempted this in a rigorous way for various atoms (O, Na, Si, and Cl) on a high-density jellium ($r_s = 2$) substrate. They have specifically calculated the change in binding energy of the 2s level between the free and adsorbed atom, ΔE_B^{ads}. In each case, the relaxation shift (typically about 5 eV) is larger than the corresponding initial state shift, as shown in Table 15.2. We note that, in keeping with the ground-state potential model, the initial state shift is such as to increase the binding energy of the 2s electron in the case of the electropositive atom sodium, but to decrease it in the case of oxygen, silicon, and chlorine.

This work notwithstanding, it is in practice not possible to distinguish between initial state and final state contributions to the chemical shift in surface-related experiments, although there are certainly cases where the effect of one or other may be ascertainable. A case in point is xenon photoemission, where the physisorbed atom acts as a probe of the local electrostatic potential at the surface.[84] From the observed binding energy shifts on the 5p levels (actually valence levels, but very core-like) the effect of crystal face, steps, and alloying in the initial state are discernible. Another interesting approach is the use of the equivalent core approximation to interpret semiquantitatively core-level binding energy shifts and to relate them to thermochemical data. This topic lies outside the scope of the

Table 15.2. Contributions to the Reduction in Core-Level Binding Energy upon Chemisorption of O, Na, Si, and Cl Atoms on a Jellium ($r_s = 2$) Substrate (eV) (From Lang and Williams[83])

	Initial state $\Delta \varepsilon_{2s}^{ads}$	Final state ΔE_R^{ads}	Sum ΔE_B^{ads}
O	3.1	7.8	10.9
Na	-1.0	4.4	3.4
Si	1.8	5.3	7.1
Cl	1.4	4.7	6.1

present chapter in which the examples require only a phenomenological treatment of the chemical shift. The reader is referred to an article of Egelhoff for a comprehensive review of core-level shifts at surfaces in general, as well as of the equivalent core approximation and the so-called reference level problem in particular.[85]

Before concluding this brief introduction it is worthwhile taking up again the subject of interatomic relaxation discussed in Section 15.1.3 and referring once more to the calculation of Lang and Williams.[83] The adsorption-induced relaxation energy, ΔE_R^{ads}, comes about because the positive charge of the "hole" (i.e., the reduced screening of the nucleus by the adsorbate electrons) polarizes the rest of the system, which induces a screening charge at the metal surface. This corresponds to the classical image effect. Another possible effect—as we have already seen—is that the hole will pull the density of states at the adsorbate to lower energies, followed by transfer of a substrate electron into an unfilled adsorbate orbital. The adsorbate is thus partially neutralized and the hole is screened. There is an important difference between these two mechanisms. While the metallic or dielectric screening process is determined by the substrate, the unfilled-orbital mechanism is largely determined by the electronic structure at the surface, in particular by the strength of the metal–adsorbate interaction. For the adsorption of a Na atom on a jellium substrate ($r_s = 2 \equiv$ Al), the latter effect is shown in Figure 15.22. To analyze whether the relaxation is more metallic (of the image type) or "charge transfer" in character, Lang and Williams calculated the spatial distribution of the induced charge due to a hole on the adsorbate. They find that, for adsorbed atoms which have partially filled valence resonances at the Fermi level, the model of Figure 15.22 describes the dominant relaxation mechanism. We thus see how the adsorbate density of states at the Fermi level, which depends sensitively on the nature of the chemical bond, might influence core-level spectra. However, if the valence levels of the adsorbate are filled and lie well below the Fermi level, as for Cl, an image type of screening appears to be dominant.

15.4.2. Fingerprinting: CO, NO, and N_2 Adsorption

Under experimental conditions which lead to more than one surface species in an adsorption experiment, the core-level spectrum can often be used for identifying different adsorption states and following their change in concentration as a function of temperature and/or coverage. To illustrate this possibility we take two examples from the work of the Menzel group who, together with Madey and Yates,[86] pioneered such studies on well-defined metal surfaces. Figure 15.23 shows the O 1s region of the core-level photoelectron spectrum when CO is adsorbed on W{110}.[87] Referring to the left-hand side (a) the middle spectrum corresponds to saturation coverage at room temperature. The sample is subsequently heated to 600 K (bottom) and then re-exposed to 10^{-6} torr CO (top). Since the various adsorption states of CO on W are well characterized, it is known that this sequence gives rise to the following CO states: $v + \beta$, pure β, and $\alpha + \beta$, respectively. By appropriately taking difference spectra, the O 1s bands of each species can be obtained (Figure 15.23b). The O 1s binding energy of β-CO (530.4 ± 0.2 eV) is

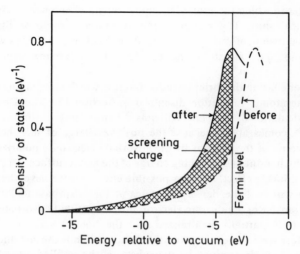

Figure 15.22. Density of states induced by adsorption of a Na atom on a jellium surface corresponding to Al ($r_j = 2$). The metal–adatom distance is that which minimizes the total energy of the ground state (after Lang and Williams[83]).

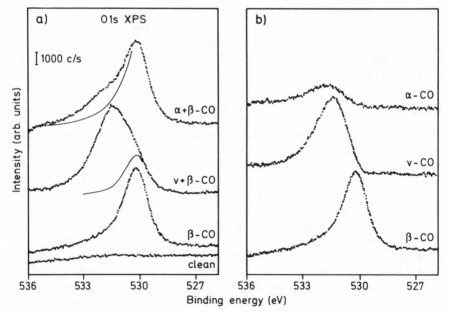

Figure 15.23. (a) O 1s photoelectron spectra from various CO layers on W{110} (see text). (b) Isolation of contribution from individual CO states (after Umbach et al.[87]).

identical with that of adsorbed oxygen on the same surface and supports the view that CO is dissociated in this state. Furthermore, the binding energies of v–CO and α–CO are indicative of molecular adsorption and show that they are probably very similar species. Similar conclusions may be drawn from the corresponding C 1s spectra (not shown).

Another example is provided by the adsorption of NO on Ru{001} at 83 K.[88] Figure 15.24 shows the O 1s and N 1s spectra from the surface after sequential exposure to NO as well as after heating a saturated layer of NO to various temperatures. There are two distinct peaks in the O 1s region following adsorption at 83 K but only one broad peak in the N 1s region; the latter shifts slightly with increasing coverage. These and other measurements indicate that the peaks at 530.3 and 531.8 eV are due to two molecular "virgin" states, which Umbach et al. have designated v_1 and v_2, respectively. Dissociated NO leads to peaks around 530 eV for O 1s and 397.0 for N 1s.[89] Heating to 318 K produces only minor changes in the spectra; at about 430 K, the peak at 531.8 disappears, the peak at 530.3 eV

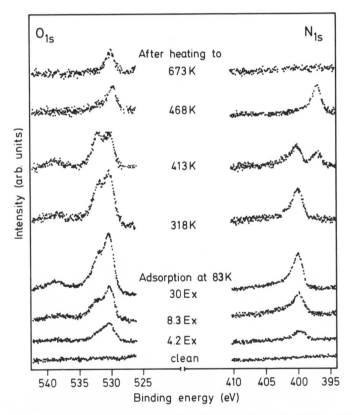

Figure 15.24. O 1s and N 1s core-level spectra for NO adsorption on Ru{001} at 83 K and for subsequent stepwise heating of the saturated layer. 1 Ex corresponds to an exposure of 2.7×10^{-7} torr · s NO (after Umbach et al.[88]).

shifts to 529.6, and the peak in the N 1s region at 399.9 eV is replaced by a peak at 369.8 eV. The latter two features thus indicate that dissociation into the β-NO state takes place. Similar measurements at room temperature show that NO is initially adsorbed in the β form followed by the formation of v_2, which is accompanied at higher exposures by v_1. On the basis of vibrational spectra for this system,[90] the v_1 state is assigned to a bridge-bonding species and the v_2 state to a linear species.

Just as in the gas phase, satellite lines can also be found in the core-level photoelectron spectra of adsorbed species. In fact, in the example quoted above—W{110}-CO—the satellite structure on the O 1s line from β-CO and v-CO are quite different, thus providing another fingerprinting possibility.[87] In Section 15.1.3 we have already briefly discussed this subject and in Section 15.2.2 we have seen that such effects also occur in the valence-level photoelectron spectrum of the Cu{100}-CO system. The calculations of Lang and Williams cited above were performed for simplicity in the so-called adiabatic approximation, where it is assumed that the $(N-1)$-particle system is always left fully relaxed, i.e., in its state of lowest energy with a hole in the kth orbital. The transition from the N-particle ground state to the ground state of the $(N-1)$-particle system cannot, however, proceed instantaneously: the time scale for the build up of a screening charge of the image type is given by the inverse of the surface plasma frequency, ω_s^{-1}. The time scale for filling an adsorbate orbital via the charge transfer mechanism is determined by the strength of the interaction between adsorbate and surface (i.e., by the width of the adsorbate-induced resonance or, where appropriate, by the bonding-antibonding splitting). The $(N-1)$-particle system will therefore, with some probability, be left in an excited (not fully relaxed) state. The observation of satellites thus depends on the dynamics of the screening processes. This is an important aspect of the characterization of adlayers; satellites may be of sufficient intensity to cause confusion in the assignment of a multipeak photoemission spectrum.

Gunnarsson and Schönhammer[91] have performed parameterized calculations to study dynamical effects in an adsorption system where both screening mechanisms pertain. The finite lifetime of the hole has been treated by adding an imaginary part to the energy. Figure 15.25 shows their core hole spectral function for a reasonable set of parameters. The sequence (a) to (c) corresponds to increasing adsorbate-substrate interactions ($V = 0.3$ for physisorption to $V = 0.75$ for strong chemisorption). The two curves in each case represent no coupling to surface plasmons ($\lambda = 0$) and a reasonable interaction of the hole with the surface plasmons ($\lambda = 0.4$). Figure 15.25a shows that for very weak adsorption, the unfilled orbital screening mechanism, although possible, is not important. The screening is mainly metallic in character because the hopping time for an electron from the substrate to the unfilled orbital at the adparticle is too long. Due to this metallic screening, the main peak is shifted to higher energy and a small surface plasmon satellite appears at $\hbar\omega_s$ below the main peak. This has been observed experimentally for the system Al{111}-O.[10] For a weak chemisorption system (Figure 15.25b), we find two peaks for $\lambda = 0$. The high-energy peak corresponds to the state where the unfilled orbital at the adparticle is filled, the hole charge is screened, and the

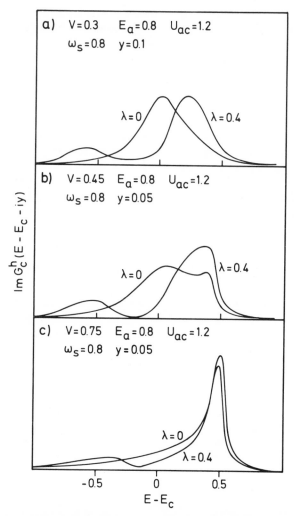

Figure 15.25. The core hole spectral functions for different values of the coupling strength of the hole to surface plasmons, λ, and of the adsorbate-substrate coupling strength, V. E_a is the energy of the unoccupied adsorbate orbital, $-U_{ac}$ the interaction between the hole in the core level of energy E_c and an electron at E_a, ω_s the surface plasmon energy, and y the imaginary part added to the energy to take into account the finite hole lifetime. Rydberg units (1 Ry = 13.6 eV) (after Gunnarsson and Schönhammer[91]).

$(N - 1)$-particle state is fully relaxed. The low-energy peak corresponds to the unrelaxed system, i.e., to an excited state (shake-up) of the $(N - 1)$-particle system. If metallic screening is also included, the unrelaxed peak is shifted to higher energy. The relaxed peak is, however, only little affected because the filling of the adparticle orbital essentially neutralizes the adparticle and the coupling to surface plasmons becomes very weak. A better resolution of the two peaks is therefore not achieved. For strong chemisorption, the metallic screening mechanism becomes less important (Figure 15.25c) because now the response time in which the lowered (initially unoccupied) adparticle level is filled is very short. The shape and the position of the peaks in the hole spectral function depend on the details of the hybridization of the unfilled adsorbate orbital with the substrate.

Umbach has analyzed the N 1s lineshapes for N_2 weakly adsorbed on three different metal single-crystal surfaces.[92] He comes to the conclusion that the spectra are similar, but not identical, and are composed of three to four peaks or shoulders, although only one chemical state of the N atom can be resolved. (For N_2 bonded perpendicular to the surface, two inequivalent N atoms are expected: the observed splitting of the well-screened peak by 1.0–1.4 eV is much larger, however, than the expected binding energy difference.) Umbach uses the Gunnarsson–Schönhammer (GS) model to understand most of the details of the observed structures.

To compare experimental data with the GS model it is, however, more convenient to look at the same species adsorbed with varying strength on different surfaces. This comparison has been made by Krause et al.[93] for the adsorption of CO on Ni{100}, Cu{100}, and Ag{110}, corresponding to strong chemisorption, weak chemisorption, and physisorption, respectively. Figure 15.26 shows the spectra in both the C 1s and O 1s regions. In the case of CO on Ni the well-screened peak is accompanied by some broad structure extending about 6 eV to higher binding energy. For CO on Cu there are three strong features: the change from strong to weak chemisorption indeed reflects the changes in going from (c) to (b) in the GS model (Figure 15.25). For CO on Ag the expected simple spectrum consisting of one poorly screened peak as in Figure 15.25a is not reproduced. More structure is apparent; it is clear that even for physisorption the spectrum will depend on the details of the adsorbate–substrate interaction. We note too that for CO on Cu and Ag the well-screened peak is relatively more intense in the O 1s spectra. This is probably due to the fact that the $2\pi^*$ screening electron is more strongly localized at the O end of the molecule.

15.4.3. Surface and Interface Reactions

Much has been published in recent years on *substrate* core-level shifts and the reader is referred again to the review article by Egelhoff.[85] The topic has assumed particular importance because photoelectron spectroscopy has become a very useful probe for investigating interface reactions that are of technological importance. These include silicon oxidation and nitridation, the metal–semiconductor interaction (Schottky barrier formation), and molecular beam epitaxy. Synchrotron radiation has played an important role in many of these studies.[20] Since the tenor of this chapter has been adsorbates on *metal* surfaces, however, we take as a single example a recent study of oxygen chemisorption and oxidation of an aluminum surface.

Figure 15.27 shows the photoemission spectra of McConville et al.[94] at $h\nu = 100$ eV in the Al 2p region during the exposure of an aluminum{111} surface to molecular oxygen. On the clean surface the spin–orbit split doublet is clearly

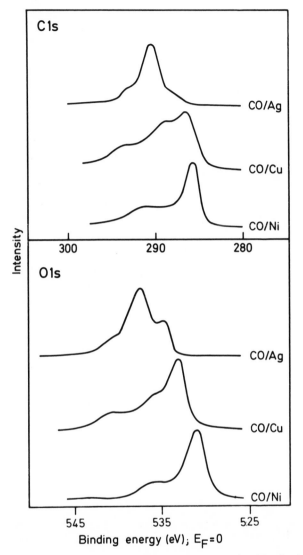

Figure 15.26. Comparison of C 1s and O 1s photoelectron spectra for CO adsorbed on Ag{110}, Cu{100}, and Ni{100} (after Krause *et al.*[93]).

resolved; the Al $2p_{3/2}$ binding energy is 72.5 eV. While the spectra show that at the highest oxygen coverages the Al 2p intensity becomes concentrated in an oxide state characterized by a chemical shift of about 2.6 eV, it is clear that one or more chemisorption states can also be distinguished. In fact, the authors found that the spectra in Figure 15.27 can be fitted reasonably satisfactorily by five spin–orbit split doublets, not all of which are resolved due to broadening. Apart from the metallic substrate emission and the oxide state, three distinct chemical shifts due

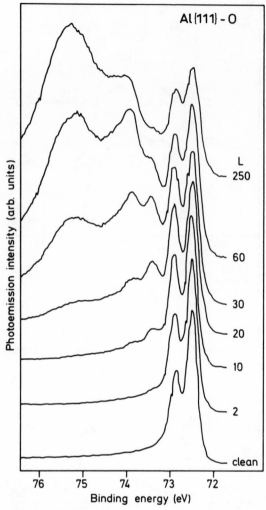

Figure 15.27. Photoelectron spectra taken at 100 eV photon energy in the vicinity of the Al 2p emission from Al{111} following different exposures to molecular oxygen (after McConville *et al.*[94]). 1 L = 1 Langmuir = 1 × 10⁻⁶ torr s.

to chemisorbed species are ascertainable as shown in Figure 15.28b for the 100 L spectrum. Similar analyses of all the spectra (with the same relative binding energies for all five components) show that the first chemisorption state grows in alone between 1 and 10 L but that the second and third both appear at about 10 L. Shortly after the appearance of the "oxide" peak the features due to chemisorption begin to decrease in intensity. The observation of three different chemisorption features can be explained with a single model. It is known that oxygen forms a 1 × 1 overlayer on Al{111} and that the atoms are adsorbed in threefold hollow

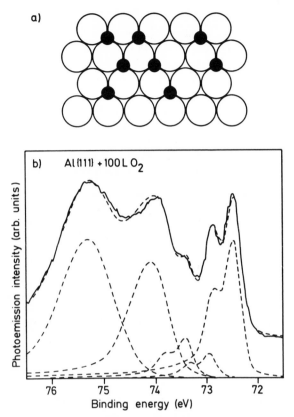

Figure 15.28. (a) Model for the incomplete (1 × 1) overlayer of oxygen on Al{111}. (b) Al 2p photoelectron spectrum from an Al{111} surface exposed to 100 L oxygen from Figure 15.27 and the fit to this (dashed curve) by a sum of the individual contributions shown underneath. These are (from right to left) the metallic state, the first, second, and third chemisorption states, and the oxide state (after McConville *et al.*[94]).

sites.[95] It is reasonable to suppose that the formation of this layer proceeds by random filling of sites and that there will be surface Al atoms bonded to either one, two, or three O atoms. This is shown schematically in Figure 15.28a. If this model is correct, the data also indicate that the oxide phase begins to form before the (1 × 1) overlayer is complete. In fact, the full development of the peak due to the highly coordinated aluminum only occurs in the presence of substantial oxide.

15.4.4. Photoelectron Diffraction

The last section of this chapter is concerned with a potentially very powerful *structural* technique in adsorbate studies: photoelectron diffraction (PhD). The experiment is none other than angle-resolved photoemission measured for a particular adsorbate core level. It can be performed in one of two ways, either by measuring the photoelectron current as a function of emission angle or by varying the photon energy (and thus the photoelectron kinetic energy) at constant emission angle. This latter, energy-scanned form of PhD is also known as constant initial-state spectroscopy (CIS). The interference effects which lead to the use of the term "diffraction" are shown schematically in Figure 15.29a. Photoelectron waves emitted from an adsorbate can reach the detector (in this case positioned so as to measure normal emission) either directly or after scattering at the substrate atoms. The resulting interference at the detector is determined by the path length differences, which in turn depend not only on the separation but also on the direction of emitter and neighboring scatterers. PhD is thus particularly sensitive to the adsorption site of the emitter. The current in each emission direction is given by equation (15.2.2). In the angle-scanned experiment $|f\rangle$ is changed by varying Ω and in the energy-scanned experiment by varying E_f. To obtain the desired structural information the resulting curves of photoelectron emission intensity plotted against angle (or energy) must be compared with calculated curves for various positions of the emitter atom on the surface.

A continuously tunable source of soft X-radiation is required for energy-scan PhD, which in practice can only be provided by suitably monochromatized synchrotron radiation. This version of the experiment is also related to surface EXAFS. In the latter, the total photoionization cross section is determined and corresponds to collecting the total photoelectron current in 4π solid angle (which in practice is not feasible). In energy-scan PhD a differential cross section is measured: as the photon energy is scanned the measured photoelectron current in a particular direction varies by up to $\pm 50\%$ as the emission is redistributed between the various $|f\rangle$. In SEXAFS these effects are entirely averaged out; what remains is the interference between the outgoing photoelectron wave and the waves back-scattered from the surrounding atoms toward the emitter.

Scattering factor amplitudes play an important role in both forms of photoelectron diffraction. In the energy-scan experiment, where the data are normally taken for photoelectron energies up to a few hundred eV, the most favorable geometry is for scattering angles close to 180°. Thus a scatterer which is directly *behind* the emitter, relative to the direction of the detector, is likely to produce pronounced structure in the PhD curves. In the angle-scan experiment, which is normally performed with higher-energy photons from laboratory AlKα and MgKα sources, a high scattering factor amplitude is also obtained for 0° scattering angle.[96] This enables bond directions to be measured in molecular adsorbates. The simplest example is perhaps CO: by measuring the angular profile of the C 1s emission the strong forward scattering along the C—O bond direction readily gives the orientation of the molecular axis relative to the surface. In this section we show two examples of energy-scan PhD; this is probably the form of the experiment with most potential for the future (providing that a synchrotron source is available!). For examples of angle-scan PhD the reader is referred to the work of Fadley.[96]

The Ni{100}($\sqrt{2} \times \sqrt{2}$)R45°–S system is one of the better understood chemisorption systems and therefore ideally suited as a model system for structural analysis (see also Section 15.3.3). All investigations have shown that the sulfur atoms are adsorbed in fourfold hollow sites, while the vertical positions and thus

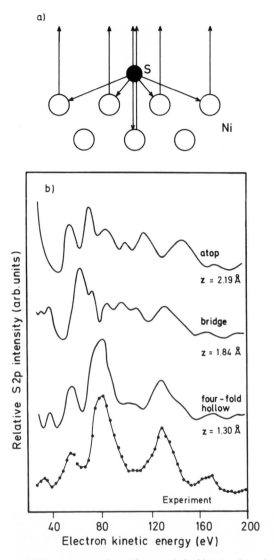

Figure 15.29. (a) Schematic representation of normal incidence photoelectron diffraction. (b) Normal emission PhD curve for S 2p photoelectrons from the adsorption system Ni{100} $(\sqrt{2} \times \sqrt{2})\text{R}45°$-S compared with calculated curves for the on-top, bridge, and fourfold hollow sites (after Rosenblatt et al.[99]).

the nearest-neighbor distances differ only slightly from technique to technique. LEED studies[97] showed that sulfur adsorbs at a distance $z = 1.30 \pm 0.1$ Å above the surface, corresponding to a nearest-neighbor S–Ni distance of $r = 2.19 \pm 0.06$ Å while SEXAFS[66] and LEIS[98] gave distances of $r = 2.23 \pm 0.02$ Å and $r = 2.25 \pm 0.3$ Å, respectively. Within experimental error, there is thus excellent agreement. Rosenblatt et al.[99] have performed a scanned-energy PhD study of the

Ni{100}($\sqrt{2} \times \sqrt{2}$)R45°-S system. The intensity of the S 2p level (170 eV binding energy with respect to the vacuum level) was measured as a function of the photon energy in 3 eV steps up to electron kinetic energies of 200 eV. At each photon energy, angle-resolved photoemission spectra normal to the surface were taken in the region of the core-level peak. After background subtraction the peak area was determined and corrected for photon flux and analyzer transmission. The measurements were made after cooling the sample (prepared at 300 K) to 120 K to increase the peak-to-valley ratio of the S 2p intensity, which is plotted in Figure 15.29b versus the electron kinetic energy (lower curve). Also shown in the figure are theoretical curves for the fourfold hollow sites ($z = 1.30$ Å), the twofold bridge site ($z = 1.80$ Å), and the atop site ($z = 2.19$ Å). For the fourfold hollow site different z spacings were tested at intervals of 0.10 Å. The extensive multiple-scattering calculations were based on a LEED formalism. The best agreement between theory and experiment is found for the fourfold hollow site with a z value of 1.30 ± 0.04 Å. A later study by Barton et al.[100] using the S 1s level at 2472 eV binding energy comes to the same result.

To end this chapter we return to the surface formate species discussed in Section 15.2.4. The structure of this molecular fragment on Cu{110} and Cu{100} has been determined using photoelectron diffraction by Woodruff et al.[49] (Figure 15.12a). The important point about this study is its demonstration of the use of energy-scan PhD for investigating systems containing no long-range order, i.e., systems where the LEED method cannot be applied. Furthermore, it shows that multiple-scattering calculations are not always necessary for structure determination.[101] Spectra were collected from both the C 1s and O 1s levels along the surface normal by stepping the photon energy in typically 3 eV steps from 80 eV to 380 eV above the relevant core-level ionization threshold and recording the photoelectron energy distribution curves over a 30 eV "window" around the photoelectron peak. After linear background subtraction the peak areas were determined and normalized to the background on the high kinetic energy side of the peak. The O 1s core-level intensities obtained in this way are plotted in Figure 15.30 as a function of the photoelectron kinetic energy. The data for the Cu{110} and Cu{100} surfaces show a remarkable similarity, implying that not only are the O–Cu nearest-neighbor distances similar but also the relative positions of the nearest-neighbor Cu atoms. The data thus provide a strong indication that the adsorption site of the formate on the two surfaces is identical.

Model calculations were performed using a curved wave and double scattering on clusters of typically 500 substrate atoms. Scattering by the C and O atoms in the formate as well as vibrational effects via Debye–Waller factors[101] were also included. For the O 1s data the best agreement between theory and experiment is found for the aligned bridge site (Figure 15.12) with the oxygen atoms close to atop positions and with a nearest-neighbor O–Cu distance of 1.98 ± 0.04 Å. The model calculations for this adsorption site and for the aligned atop site established with SEXAFS[102,103] are also shown in Figure 15.30. Although detailed agreement with the C 1s photoelectron diffraction data (not shown) is not present in any of the calculations, a comparison between theory and experiment reveals that the main features are best fitted by a shorter O–Cu distance (1.90–1.94 Å). This

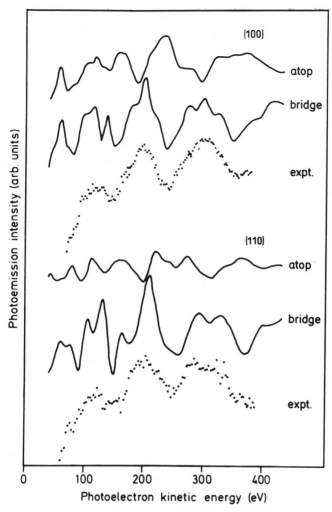

Figure 15.30. Comparison of experimental and theoretical O 1s PhD spectra for formate on Cu{100} for the two locally equivalent sites on these surfaces. O–Cu distances of 1.99 Å for {100} and 1.94 Å for {110} were used for the calculated curves (after Woodruff et al.[49]).

discrepancy need not be real and may suggest that the O—C—O bond angle is about 134° compared to 125° in inorganic formates.[104] Whereas the O–Cu nearest-neighbor bond lengths of 1.98 Å agree exactly with the SEXAFS results of 1.98 Å for Cu{110}[102] and 1.99 Å for Cu{100},[103] the adsorption sites determined are different. Since the structural conclusions drawn from the PhD data appear very reasonable and the local geometry is clearly the same on both surfaces, it must be concluded that the SEXAFS studies (see Puschmann et al.[102] and Crapper et al.[103] as well as earlier studies cited therein) most probably suffered from the single shell analysis and the limited data range.

In summary, photoelectron diffraction—particularly in the scanned-energy mode—is an extremely promising surface structural tool and represents an exciting application of adsorbate photoemission which has hitherto received very little attention.

Acknowledgments

The author would like to thank his colleagues and co-workers at the Fritz-Haber-Institut as well as Prof. D. P. Woodruff and his group at the University of Warwick. Without them, many of the adsorbate photoemission experiments used as examples in this article would not have been performed. The work described from the Fritz-Haber-Institut has been supported partly by the Deutsche Forschungsgemeinschaft through the Sonderforschungsbereich 6, by the Bundesministerium für Forschung und Technologie, and by the Fonds der chemischen Industrie.

References

1. Turner's experiments are summarized in: D. W. Turner, A. D. Baker, C. Baker, and C. R. Brundle, *Molecular Photoelectron Spectroscopy*, Wiley, London (1970).
2. F. I. Viselov, B. L. Kurbatov, and A. N. Terenin, *Dokl. Akad. Nauk SSSR* **138**, 1329 (1961).
3. K. Siegbahn, C. Nordling, A. Fahlman, R. Dordberg, K. Hamrin, J. Hedman, G. Johansson, T. Bergark, S.-E. Karlsson, I. Lendgren, and B. Lindberg, *ESCA-Atomic, Molecular and Solid State Structure Studied by Means of Electron Spectroscopy*, Almqvist and Wiksells, Uppsala (1967).
4. K. Siegbahn, C. Nordling, G. Johansson, J. Hedman, P.-F. Hedén, K. Hamrin, U. Gelius, T. Bergmark, L.-O. Werne, R. Manne, and Y. Baur, *ESCA Applied to Free Molecules*, North-Holland, Amsterdam (1969).
5. U. Gelius, S. Svensson, H. Siegbahn, E. Basilier, Å. Faxälv, and K. Siegbahn, *Chem. Phys. Lett.* **28**, 1(1974).
6. L. S. Cederbaum and W. Domcke, *Adv. Chem. Phys.* **36**, 205 (1977).
7. J. H. D. Eland, *Photoelectron Spectroscopy (2nd ed.)*, Butterworths, London (1984).
8. H. Siegbahn and L. Karlsson, *Photoelectron Spectroscopy* (Handbuch der Physik Vol. XXXI), Springer-Verlag, Berlin (1982).
9. N. D. Lang, in: *Theory of the Inhomogeneous Electron Gas* (S. Lundqvist and N. H. March, eds.), Plenum Press, New York (1983).
10. A. M. Bradshaw, W. Domcke, and L. S. Cederbaum, *Phys. Rev. B* **16**, 1480 (1977).
11. J. C. Fuggle, E. Umbach, D. Menzel, K. Wandelt, and C. R. Brundle, *Solid State Commun.* **27**, 65 (1978).
12. C. L. Allyn, T. Gustafsson, and E. W. Plummer, *Solid State Commun.* **24**, 53 (1977).
13. M. Šunjić and A. Lucas, *Chem. Phys. Lett.* **42**, 462 (1976); J. W. Gadzuk, *Phys. Rev. B* **14**, 5458 (1976).
14. M. P. Seah and W. Dench, *J. Surf. Interface Anal.* **1**, 1 (1979).
15. J. A. R. Samson, *Techniques of Vacuum UV Spectroscopy*, Wiley, New York (1967).
16. H. Ibach and D. L. Mills, *Electron Energy Loss Spectroscopy and Surface Vibrations*, Academic Press, New York (1982).
17. P. A. Redhead, J. P. Hobson, and E. V. Kornelsen, *The Physical Basis of Ultrahigh Vacuum*, Chapman and Hall, London (1968).

18. G. Ertl and J. Küppers, *Low Energy Electrons and Surface Chemistry* (*2nd ed.*), Verlag Chemie, Weilheim (1985).
19. D. P. Woodruff and T. A. Delchar, *Modern Techniques of Surface Science*, Cambridge University Press, Cambridge (1986).
20. G. Margaritondo, *Introduction to Synchrotron Radiation*, Oxford University Press, New York (1988).
21. D. E. Eastman and J. K. Cashion, *Phys. Rev. Lett.* **27**, 1520 (1971).
22. R. J. Smith, J. Anderson, and G. J. Lapeyre, *Phys. Rev. Lett.* **37**, 1081 (1976).
23. C. L. Allyn, T. Gustafsson, and E. W. Plummer, *Chem. Phys. Lett.* **47**, 127 (1977).
24. D. R. Penn, *Phys. Rev. Lett.* **28**, 1041 (1972).
25. J. W. Davenport, *Phys. Rev. Lett.* **36**, 945 (1976).
26. A. M. Bradshaw, *Z. Phys. Chem.* NF **112**, 33 (1978); N. V. Richardson and A. M. Bradshaw, in: *Electron Spectroscopy: Theory, Technique and Applications* (A. Baker and C. R. Brundle, eds.), Vol. 4, Academic Press, London (1980).
27. J. Hermanson, *Solid State Commun.* **22**, 9 (1977).
28. K. Jacobi, M. Scheffler, K. Kambe, and F. Forstmann, *Solid State Commun.* **22**, 17 (1977).
29. M. Surman, K. C. Prince, L. Sorba, and A. M. Bradshaw, *Surf. Sci.* **206**, L864 (1988).
30. S. Andersson and J. B. Pendry, *Phys. Rev. Lett.* **43**, 363 (1979).
31. C. F. McConville, D. P. Woodruff, K. C. Prince, G. Paolucci, V. Cháb, M. Surman, and A. M. Bradshaw, *Surf. Sci.* **166**, 231 (1986).
32. P. S. Bagus, C. J. Nelin, and C. W. Bauschlicher, Jr., *J. Vac. Sci. Technol.* **A2**, 905 (1984).
33. L. H. Dubois, B. R. Zogarski, and H. S. Luftman, *J. Chem. Phys.* **87**, 1267 (1987).
34. D. Heskett, I. Stratky, E. W. Plummer, and R. A. de Paola, *Phys. Rev. B* **32**, 622 (1985).
35. D. D. Heskett and E. W. Plummer, *Phys. Rev. B* **33**, 2322 (1986).
36. C. Somerton, C. F. McConville, D. P. Woodruff, D. E. Grider, and N. V. Richardson, *Surf. Sci.* **138**, 31 (1984).
37. G. Paolucci, M. Surman, K. C. Prince, L. Sorba, A. M. Bradshaw, C. F. McConville, and D. P. Woodruff, *Phys. Rev. B* **34**, 1340 (1986).
38. G. L. Nyberg and N. V. Richardson, *Surf. Sci. B* **5**, 335 (1979).
39. F. P. Netzer and U. Mach, *J. Chem. Phys.* **79**, 1017 (1983).
40. J. A. Horsley, J. Stöhr, A. P. Hitchcock, D. C. Newbury, A. L. Johnson, and F. Sette, *J. Chem. Phys.* **83**, 6099 (1985).
41. F. P. Netzer, H. H. Craen, H. Kuhlenbeck, and M. Neumann, *Chem. Phys. Lett.* **133**, 49 (1987).
42. F. P. Netzer, G. Raugelov, G. Rosina, H. B. Saalfeld, M. Neumann, and D. R. Lloyd, *Phys. Rev. B* **37**, 10399 (1988).
43. P. Hofmann, K. Horn, and A. M. Bradshaw, *Surf. Sci.* **105**, L260 (1981).
44. B. Boddenberg and J. Moreno, *J. Phys.* (*Paris*), *Colloq.* **38**, C4-52 (1977).
45. W. von Niessen, L. S. Cederbaum, and W. P. Kraemer, *J. Chem. Phys.* **65**, 1378 (1976).
46. D. H. S. Ying and R. Madix, *J. Catal.* **61**, 48 (1980).
47. B. E. Hayden, K. C. Prince, D. P. Woodruff, and A. M. Bradshaw, *Surf. Sci.* **133**, 589 (1983).
48. A. Puschmann, J. Haase, M. D. Crapper, C. E. Riley, and D. P. Woodruff, *Phys. Rev. Lett.* **54**, 2250 (1985).
49. D. P. Woodruff, C. F. McConville, A. L. D. Kilcoyne, Th. Lindner, J. Somers, M. Surman, G. Paolucci, and A. M. Bradshaw, *Surf. Sci.* **201**, 228 (1988).
50. Th. Lindner, J. Somers, A. M. Bradshaw, and G. P. Williams, *Surf. Sci.* **185**, 75 (1987).
51. P. Hofmann and D. Menzel, *Surf. Sci.* **191**, 353 (1987).
52. S. D. Peyerimhoff, *J. Chem. Phys.* **47**, 349 (1967).
53. J. A. Rodriguez and C. T. Campbell, *Surf. Sci.* **183**, 449 (1987).
54. A. M. Bradshaw and M. Scheffler, *J. Vac. Sci. Technol.* **16**, 447 (1979).
55. P. J. Feibelman and F. J. Himpsel, *Phys. Rev. B* **21**, 1394 (1980).
56. R. Richter and J. W. Wilkins, *Surf. Sci.* **128**, L190 (1983).
57. C. L. Fu, A. J. Freeman, W. Wimmer, and M. Weinert, *Phys. Rev. Lett.* **54**, 2261 (1985).

58. A. Liebsch, *Phys. Rev. B* **17**, 1653 (1978).

59. K. Horn, M. Scheffler, and A. M. Bradshaw, *Phys. Rev. Lett.* **41**, 822 (1978).

60. C. Mariani, K. Horn, and A. M. Bradshaw, *Phys. Rev. B* **25**, 7798 (1982).

61. M. Jaubert, A. Glachant, M. Bienfait, and G. Boato, *Phys. Rev. Lett.* **46**, 1979 (1981).

62. W. Berndt, *Surf. Sci.* **219**, 161 (1989).

63. M. Scheffler, K. Horn, A. M. Bradshaw, and K. Kambe, *Surf. Sci.* **80**, 69 (1979).

64. G. Schönhense, *Appl. Phys.* **A41**, 39 (1986).

65. T. Mandel, M. Domke, and G. Kaindl, *Surf. Sci.* **197**, 81 (1988).

66. J. Stöhr, R. Jaeger, and S. Brennan, *Surf. Sci.* **117**, 503 (1982).

67. E. W. Plummer, B. Tonner, N. Holzwarth, and A. Liebsch, *Phys. Rev. B* **21**, 4306 (1980).

68. M. Bader, J. Haase, A. Puschmann, and C. Ocal, *Phys. Rev. Lett.* **59**, 2435 (1987).

69. R. A. DiDio, D. M. Zehner, and E. W. Plummer, *J. Vac. Sci. Technol.* **A2**, 852 (1984).

70. R. Courths, B. Cord, H. Wern, H. Saalfeld, and J. Hüfner, *Solid State Commun.* **63**, 619 (1987).

71. K. C. Prince, G. Paolucci, and A. M. Bradshaw, *Surf. Sci.* **175**, 101 (1986).

72. W. Jacob, V. Dose, and A. Goldmann, *Appl. Phys.* **A41**, 145 (1986).

73. K. Horn, A. M. Bradshaw, and K. Jacobi, *Surf. Sci.* **72**, 719 (1978).

74. K. Horn, A. M. Bradshaw, K. Hermann, and I. P. Batra, *Solid State Commun.* **31**, 257 (1979).

75. H.-J. Freund and M. Neumann, *Appl. Phys.* **A47**, 3 (1988).

76. H. Kuhlenbeck, M. Neumann, and H.-J. Freund, *Surf. Sci.* **173**, 194 (1986).

77. F. Hund, *Z. Phys.* **99**, 119 (1936).

78. I. P. Batra, K. Hermann, A. M. Bradshaw, and K. Horn, *Phys. Rev. B* **34**, 2199 (1986).

79. W. Riedl and D. Menzel, *Surf. Sci.* **163**, 39 (1985).

80. D. Pescia, A. R. Law, M. T. Johnson, and H. P. Hughes, *Solid State Commun.* **56**, 809 (1985).

81. K. C. Prince, M. Surman, Th. Lindner, and A. M. Bradshaw, *Solid State Commun.* **59**, 71 (1986).

82. K. C. Prince, *J. Electron Spectrosc. Relat. Phenom.* **42**, 217 (1987).

83. N. D. Lang and A. R. Williams, *Phys. Rev. B* **16**, 2408 (1977).

84. K. Wandelt and J. E. Hulse, *J. Chem. Phys.* **80**, 1340 (1984).

85. W. F. Egelhoff, *Surf. Sci. Rep.* **6**, 253 (1987).

86. T. E. Madey, J. T. Yates, and N. E. Erickson, *Surf. Sci.* **43**, 257 (1984).

87. E. Umbach, J. C. Fuggle, and D. Menzel, *Surf. Sci.* **10**, 15 (1977).

88. E. Umbach, S. Kulkarni, P. Feulner, and D. Menzel, *Surf. Sci.* **88**, 65 (1979).

89. R. I. Masel, E. Umbach, J. C. Fuggle, and D. Menzel, *Surf. Sci.* **79**, 26 (1979).

90. G. E. Thomas and W. H. Weinberg, *Phys. Rev. Lett.* **41**, 1181 (1978).

91. O. Gunnarsson and K. Schönhammer, *Solid State Commun.* **23**, 691 (1977).

92. E. Umbach, *Solid State Commun.* **51**, 365 (1984).

93. S. Krause, C. Mariani, K. C. Prince, and K. Horn, *Surf. Sci.* **138**, 305 (1984).

94. C. F. McConville, D. L. Seymour, D. P. Woodruff, and S. Bao, *Surf. Sci.* **188**, 1 (1988).

95. See, e.g., I. P. Batra, and L. Kleinmann, *J. Electron Spectrosc. Relat. Phenom.* **33**, 175 (1984).

96. C. S. Fadley, *Phys. Scr.* **T17**, 39 (1987).

97. Y. Gauthier, D. Aberdam, and R. Baudoing, *Surf. Sci.* **78**, 339 (1978) and references therein.

98. Th. Fauster, H. Dürr, and D. Hartwig, *Surf. Sci.* **178**, 657 (1986).

99. D. H. Rosenblatt, J. G. Tobin, M. G. Mason, R. F. Davis, S. D. Kevan, D. A. Shirley, C. H. Li, and S. Y. Tong, *Phys. Rev. B* **23**, 3828 (1981).

100. J. J. Barton, C. C. Baker, Z. Hussain, S. W. Robey, L. E. Klebanoff, and D. A. Shirley, *Phys. Rev. Lett.* **51**, 272 (1983).

101. D. P. Woodruff, *Surf. Sci.* **166**, 377 (1986).

102. A. Puschmann, J. Haase, M. D. Crapper, C. E. Riley, and D. P. Woodruff, *Phys. Rev. Lett.* **54**, 2250 (1985).

103. M. D. Crapper, C. E. Riley, and D. P. Woodruff, *Surf. Sci.* **184**, 121 (1987).

104. G. A. Barclay and C. H. L. Kennard, *J. Chem. Soc.*, 3289 (1961).

16

Rate Equations, Rate Constants, and Surface Diffusion

G. Wahnström

16.1. Introduction

One of the themes in the present volume is the way the basic interaction potential for adparticles physisorbed and chemisorbed on solid surfaces relates to the dynamics of adsorbed particles. In this chapter we will try to make this connection in some simple cases.

We will mainly treat the situation of a single atom adsorbed on the surface, but the extension to finite coverages will be discussed. Most of the theoretical models presented here can equally well be applied to the adsorption–desorption problem and may also serve as a starting point in examining the chemical reaction catalyst by a solid surface. The restriction to classical mechanics for the motion of adatoms is often natural, but in connection with diffusion of hydrogen we will comment on possible quantum effects. It will be assumed that the substrate forms a perfect periodic lattice by neglecting all defects, steps, polycrystalline effects, etc. In order to obtain a detailed microscopic understanding of the dynamics, simple and well-defined models must be used.

The chapter is organized in the following way. In the first part we discuss two different phenomenological ways to treat surface diffusion. In Section 16.2 jump diffusion models are presented, while Section 16.3 deals with models where the continuous motion of the adatom or adatoms is considered. The second part of the chapter is devoted to the microscopic basis for these two different kinds of models. In Section 16.4 we derive and discuss the microscopic expressions for the rate constants introduced in connection with the jump diffusion models, and in Section 16.5 the microscopic basis for the models based on continuous motion is

G. Wahnström • Institute of Theoretical Physics, Chalmers University of Technology, S-412 96 Göteborg, Sweden.

presented. In both Sections 16.4 and 16.5 we will discuss actual numerical calculations, where the aim is to determine dynamical parameters from the basic interaction
potentials.

16.2. Jump Diffusion Models

Due to the periodic arrangement of the substrate particles, adsorbed atoms
experience a periodic potential along the surface. In many cases the activation
energy for diffusion, i.e., the difference between the maximum and minimum of
this potential energy along the surface, is large compared with the thermal energy.
This implies that the adatoms are mainly localized at different surface lattice sites,
while the change of sites, and consequently the diffusion, becomes thermally
activated. The slow diffusive motion can be expressed in terms of the rate of change
of the probabilities to be located at different sites. This is the basis for using jump
diffusion models.[1]

16.2.1. Noninteracting Adatoms

Let us consider first the simplest case: a single adsorbed atom. The following
description is also valid for an adsorbed layer of adatoms if the adatoms can be
treated as *completely* noninteracting.

The notation $P^s(\mathbf{s}_i, t)$ is introduced for the probability to find the single adatom
at site i ($i = 1, \ldots, N$) at time t. The different sites are defined by the vectors $\mathbf{s}_i$
while N is the number of sites. The discussion proceeds in terms of probabilities
and a statistical thermal average is assumed. By conservation of the number of
particles we have

$$\sum_{i=1}^{N} P^s(\mathbf{s}_i, t) = 1 \tag{16.2.1}$$

The probability to be located at a specific site i will decrease in time due to jumps
from that site, and increase in time due to jumps to the same site. Function $P^s(\mathbf{s}_i, t)$
obeys the rate equation

$$\frac{\partial}{\partial t} P^s(\mathbf{s}_i, t) = \sum_{j(\neq i)} [k_{j \to i} P^s(\mathbf{s}_j, t) - k_{i \to j} P^s(\mathbf{s}_i, t)] \tag{16.2.2}$$

where $k_{i \to j}$ is the rate at which the adatom is moving from site i to site j. The sites
i and j need not be adjacent in configuration space. We will call $k_{i \to j}$ the rate
constants. Equation (16.2.2) is also valid in the quantum regime, provided the
fluctuations in the surrounding is large enough to entirely eliminate coherent
motion over two or more lattice spacings. This is generally assumed to be the case
for diffusion of hydrogen and other heavier particles at not too low temperatures.[2]

At equilibrium $P^s(\mathbf{s}_i, t)$ does not vary in time. Hence $P^s(\mathbf{s}_i, t) \equiv P^{s;eq}(\mathbf{s}_i)$, and
we have the detailed balance condition

$$\sum_{j(\neq i)} k_{j \to i} P^{s;eq}(\mathbf{s}_j) = \sum_{(j \neq i)} k_{i \to j} P^{s;eq}(\mathbf{s}_i) \tag{16.2.3}$$

If all N sites are identical, then

$$P^{s;eq}(\mathbf{s}_i) = 1/N \tag{16.2.4}$$

and

$$k_{i \to j} = k_{j \to i} \tag{16.2.5}$$

This assumption will be used henceforth. We introduce a matrix, the transfer matrix, defined by

$$T(\mathbf{s}_i, \mathbf{s}_j) = \begin{cases} -k_{j \to i} & \text{if } i \neq j \\ \sum\limits_{m(\neq i)} k_{i \to m} & \text{if } i = j \end{cases} \tag{16.2.6}$$

so the rate equation can be written as

$$\frac{\partial}{\partial t} P^s(\mathbf{s}_i, t) = -\sum_j T(\mathbf{s}_i, \mathbf{s}_j) P^s(\mathbf{s}_j, t) \tag{16.2.7}$$

If the sites $\mathbf{s}_i$ form a Bravais lattice, the transfer matrix can easily be diagonalized. By introducing the intermediate scattering function

$$P^s(\mathbf{q}, t) \equiv \sum_i \exp(-i\mathbf{q} \cdot \mathbf{s}_i) P^s(\mathbf{s}_i, t) \tag{16.2.8}$$

the solution can be expressed in the form

$$P^s(\mathbf{q}, t) = \exp[-t/\tau(\mathbf{q})] P^s(\mathbf{q}, t = 0) \tag{16.2.9}$$

where the different relaxation times are defined by

$$\tau^{-1}(\mathbf{q}) \equiv T(\mathbf{q}) \equiv \sum_i T(\mathbf{s}_i, \mathbf{s}_j) \exp[-i\mathbf{q} \cdot (\mathbf{s}_i - \mathbf{s}_j)]$$

$$= 2 \sum_{j(\neq i)} k_{i \to j} \sin^2[\mathbf{q} \cdot (\mathbf{s}_j - \mathbf{s}_i)/2] \tag{16.2.10}$$

Quantity $P^s(\mathbf{q}, t = 0)$ is the initial condition for the probability distribution, and we have made use of the fact that the transfer matrix only depends on the relative distance, not on $\mathbf{s}_i$ and $\mathbf{s}_j$ separately. By transforming back to real space the solution becomes

$$P^s(\mathbf{s}_i, t) = \frac{1}{N} \sum_{\mathbf{q}} \exp(i\mathbf{q} \cdot \mathbf{s}_i) \exp[-t/\tau(\mathbf{q})] P^s(\mathbf{q}, t = 0) \tag{16.2.11}$$

The sum over $\mathbf{q}$ is over all wave vectors in the first Brillouin zone of the reciprocal lattice.

16.2.1.1. The Self-Diffusion Constant

The motion is conveniently characterized in terms of the self-diffusion constant, which is related to the mean-square displacement of the adatom. It is assumed that initially the adatom is located at site k with $\mathbf{s}_k = \mathbf{0}$. The mean displacement of the adatom is zero, namely

$$\langle \mathbf{R}(t) \rangle \equiv \sum_i \mathbf{s}_i P^s(\mathbf{s}_i, t) = \mathbf{0} \tag{16.2.12}$$

but the mean-square displacement

$$\langle \mathbf{R}^2(t) \rangle \equiv \sum_i \mathbf{s}_i^2 P^s(\mathbf{s}_i, t) \tag{16.2.13}$$

is nonzero. We can determine this quantity from equation (16.2.11), but it is simpler to directly use the rate equation. If we multiply both sides of equation (16.2.2) by $\mathbf{s}_i^2$, sum over i, and interchange the summation indices for the first term on the right-hand side, we obtain

$$\frac{\partial}{\partial t} \langle \mathbf{R}^2(t) \rangle = \sum_i \sum_{j(\neq i)} k_{i \to j}(\mathbf{s}_j^2 - \mathbf{s}_i^2) P^s(\mathbf{s}_i, t) \tag{16.2.14}$$

The relationship $\mathbf{s}_{i \to j} \equiv \mathbf{s}_j - \mathbf{s}_i$ is introduced:

$$\frac{\partial}{\partial t} \langle \mathbf{R}^2(t) \rangle = \sum_i \sum_{j(\neq i)} k_{i \to j}(\mathbf{s}_{i \to j}^2 + 2\mathbf{s}_i \cdot \mathbf{s}_{i \to j}) P^s(\mathbf{s}_i, t) \tag{16.2.15}$$

Both $k_{i \to j}$ and $\mathbf{s}_{i \to j}$ depend only on the relative distance between site i and site j, and if we use equations (16.2.1) and (16.2.12) we derive

$$\frac{\partial}{\partial t} \langle \mathbf{R}^2(t) \rangle = \sum_{j(\neq i)} k_{i \to j} \mathbf{s}_{i \to j}^2 \tag{16.2.16}$$

or

$$\langle \mathbf{R}^2(t) \rangle = \sum_{j(\neq i)} k_{i \to j} \mathbf{s}_{i \to j}^2 t. \tag{16.2.17}$$

The self-diffusion constant $\mathbf{D}^s$, which in the general case is a tensor, is defined by

$$\mathbf{D}^s = \tfrac{1}{2} \sum_{j(\neq i)} k_{i \to j} \mathbf{s}_{i \to j} \mathbf{s}_{i \to j} \tag{16.2.18}$$

For a lattice with cubic symmetry the diffusion is independent of the direction. Hence

$$\langle \mathbf{R}^2(t) \rangle = 2dD^s t \tag{16.2.19}$$

with

$$D^s = \frac{1}{2d} \sum_{j(\neq i)} k_{i \to j} s^2_{i \to j} \tag{16.2.20}$$

Here D^s is the self-diffusion constant, d the dimensionality associated with the diffusion process ($d = 1$ for linear motion, $d = 2$ for motion in a plane, etc.), and $s_{i \to j}$ is the distance between site i and site j. Equation (16.2.19) yields the important connection between the mean-square displacement for the adatom and the self-diffusion constant, and it can be viewed as a definition of D^s. This will be made more precise in Section 16.3. According to equation (16.2.20) the self-diffusion constant D^s is simply related to the rate constants $k_{i \to j}$, and by calculating $k_{i \to j}$ the self-diffusion constant can be determined.

16.2.1.2. Macroscopic Approximation

It is evident from equation (16.2.11) that the solution consists of a sum of modes that relax with different relaxation times $\tau(\mathbf{q})$. At sufficiently long times only the modes that correspond to the longest relaxation times will survive. Inspection of equation (16.2.10) implies that in this limit only small wave vectors contribute, in which case we can write

$$T(\mathbf{q}) = \tfrac{1}{2} \sum_{j(\neq i)} k_{i \to j} [\mathbf{q} \cdot (\mathbf{s}_j - \mathbf{s}_i)]^2 \tag{16.2.21}$$

In this limit the transfer matrix is related to the self-diffusion constant and, for a system with cubic symmetry,

$$T(\mathbf{q}) = D^s q^2 \tag{16.2.22}$$

If only small wave vectors are important, i.e., long wavelength, the lattice structure becomes unimportant and the difference equation (16.2.2) can be simplified to a differential equation. We call this the macroscopic approximation to indicate that the microscopic lattice structure is neglected. Expansion of $P^s(\mathbf{s}_j, t)$ around $\mathbf{s}_i$ yields

$$P^s(\mathbf{s}_j, t) = P^s(\mathbf{s}_i, t) + (\mathbf{s}_j - \mathbf{s}_i) \cdot \nabla P^s(\mathbf{s}_i, t) + \tfrac{1}{2}[(\mathbf{s}_j - \mathbf{s}_i) \cdot \nabla]^2 P^s(\mathbf{s}_i, t) + \cdots \tag{16.2.23}$$

Insertion of this expansion into the rate equation (16.2.2) leads to the standard diffusion equation,

$$\frac{\partial}{\partial t} P^s(\mathbf{r}, t) = D^s \nabla^2 P^s(\mathbf{r}, t) \tag{16.2.24}$$

We have assumed cubic symmetry and that $\mathbf{r}$ is a continuous variable with

$$P^s(\mathbf{s}_i, t) \equiv \int_{\Omega_i} d\mathbf{r}\, P^s(\mathbf{r}, t) \tag{16.2.25}$$

where the volume Ω_i defines the lattice site i.

The solution to equation (16.2.24) with initial condition $P^s(\mathbf{r}, t = 0) = \delta(\mathbf{r})$ is

$$P^s(\mathbf{r}, t) = (4\pi D^s t)^{-d/2} \exp\left(-r^2/4D^s t\right) \tag{16.2.26}$$

which is easily obtained by Fourier transforming equation (16.2.24). The probability distribution for the adatom is a Gaussian. Initially it is a delta function, and when time proceeds the distribution broadens, and the rate of this broadening is determined by the self-diffusion constant.

16.2.1.3. One-Dimensional Example

Direct experimental observation of details of the motion of adatoms is obviously desirable. In field ion microscopy (FIM) the diffusive motion of single metal atom is observed directly.[3,4] The self-diffusion constant can be extracted from the mean-square displacement of the adatom and also the actual probability distribution $P^s(\mathbf{s}_i, t)$ can be obtained. The picture that has emerged is that the adatom motion over low-index surfaces is quite well described in terms of uncorrelated jumps between nearest-neighbor sites.[5,6]

On some crystal surfaces the adatom moves along channels and the diffusion becomes one-directional in nature. One such case which has been experimentally studied is W adatoms on the W(211) surface.[7] If we assume that a single jump occurs only to two nearest-neighbour sites with rate α and that successive jumps are independent of each other, the probability distribution can be expressed in terms of the modified Bessel function I_j of order j according to

$$P^s(\mathbf{s}_j, t) = \exp\left(-2\alpha t\right) I_j(2\alpha t) \tag{16.2.27}$$

This result is obtained from equation (16.2.11) with initial condition $P^s(\mathbf{s}_j, t = 0) = \delta_{j,0}$. The self-diffusion constant is given by equation (16.2.20), i.e.,

$$D^s = \alpha a^2 \tag{16.2.28}$$

where a is the lattice spacing. If we now allow for jumps to both nearest sites at a rate α and to second-nearest sites at a rate β, the probability distribution is then given by

$$P^s(\mathbf{s}_j, t) = \exp\left[-2(\alpha + \beta)t\right] \sum_{k=\infty}^{\infty} I_{j-2k}(2\alpha t) I_k(2\beta t) \tag{16.2.29}$$

and the self-diffusion constant by

$$D^s = (\alpha + 4\beta)a^2 \tag{16.2.30}$$

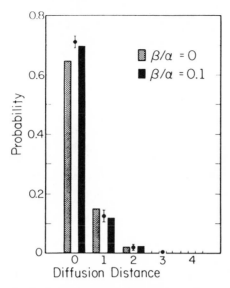

Figure 16.1. Probability distribution for a single tungsten atom diffusing on W(211). Points give results of experiment. Light bars indicate fit of experimental data to equation (16.2.27) and dark bars to equation (16.2.29) with $\beta/\alpha = 0.1$ [after G. Ehrlich and K. Stolt, in: *Growth and Properties of Metal Clusters* (J. Bourdon, ed.), Elsevier, Amsterdam (1980)].

We note that the contribution to the self-diffusion constant scales as the square of the jump distance, and it implies that even a small jump rate β makes a comparatively large contribution to the diffusion constant. Experimental observations of the probability distribution for W on the W(211) surface have been compared with predictions from the above two models.[7] Figure 16.1 shows the results at a time when the tungsten adatom has on average made half a jump. Points give results of the experiment, light bars indicate fit of experimental data to equation (16.2.27) and dark bars to equation (16.2.29). The best result is obtained on the assumption that $\beta/\alpha = 0.1$, i.e., there is a small tendency to correlated jumps.

16.2.2. Interacting Adatoms

So far we have restricted ourselves to a single adsorbed atom or, equally well, the case with completely noninteracting adatoms. The situation is much more complicated if we have an adsorbed layer of interacting adatoms. To describe this situation we introduce the notation $P(\mathbf{s}_i, t)$ for the probability to find one, or maybe several, adatoms at site i at time t. Conservation of number of particles implies that

$$\sum_{i=1}^{N} P(\mathbf{s}_i, t) = N_p \tag{16.2.31}$$

where N_p is the number of adsorbed atoms. The concentration of adatoms is defined by

$$c = N_p / N \tag{16.2.32}$$

16.2.2.1. Macroscopic Description

The phenomenological description of diffusion is embodied in Fick's law, which is an empirical statement that relates the diffusive flow of matter to concentration gradients. On a macroscopic level we neglect the lattice structure (cf. Section 16.2.1.2) and write for the probability distribution, or density of adatoms, $P(\mathbf{r}, t)$, where $\mathbf{r}$ is a continuous variable. Fick's law states that the flux of diffusing adatoms is proportional to the concentration gradient,

$$\mathbf{J}(\mathbf{r}, t) = -D\nabla P(\mathbf{r}, t) \tag{16.2.33}$$

where the proportionality factor D is the diffusion constant. If we add the continuity equation

$$\frac{\partial}{\partial t} P(\mathbf{r}, t) = -\nabla \mathbf{J}(\mathbf{r}, t) \tag{16.2.34}$$

to Fick's law, the standard diffusion equation is obtained:

$$\frac{\partial}{\partial t} P(\mathbf{r}, t) = D\nabla^2 P(\mathbf{r}, t) \tag{16.2.35}$$

Here D is sometimes called the chemical or collective diffusion constant and it is important to realize that D is *not* identical to the self-diffusion constant D^s. In Section 16.2.1 the motion of a single adatom was monitored. Also, when considering an adsorbed layer of interacting adatoms it can be useful to follow the motion of a specific, tagged, particle. The self-diffusion constant for the system of interacting adatoms is then related to the mean-square displacement of the tagged particle. We use a superscript s on $P^s(\mathbf{r}, t)$ and on D^s to indicate that these quantities are related to the self-motion of the adatom, in contrast to D and $P(\mathbf{r}, t)$ which are a measure of the collective behavior. Only when the adatoms can be treated as completely noninteracting are D and D^s identical.[8]

16.2.2.2. Hard Core Interaction

If we restrict ourselves to the case where double occupancy is prohibited but no other interaction between adatoms is present, some progress can be made. We call this type of interaction hard core interaction.

The rate equation takes the form

$$\frac{\partial}{\partial t} P(\mathbf{s}_i, t) = \sum_{j(\neq l)} \{k_{j \to i} P(\mathbf{s}_j, t)[1 - P(\mathbf{s}_i, t)] - k_{i \to j} P(\mathbf{s}_i, t)[1 - P(\mathbf{s}_j, t)]\} \tag{16.2.36}$$

which directly simplifies to

$$\frac{\partial}{\partial t} P(\mathbf{s}_i, t) = \sum_{j(\neq i)} [k_{j \to i} P(\mathbf{s}_j, t) - k_{i \to j} P(\mathbf{s}_i, t)] \qquad (16.2.37)$$

This equation is identical to equation (16.2.2), so all the results in Section 16.2.1 can be used. In the limit of small wave vectors a connection can be made with the diffusion equation and we have the relation

$$D = \frac{1}{2d} \sum_{j(\neq i)} k_{i \to j} s_{i \to j}^2 \qquad (16.2.38)$$

The rate constants are here related to the chemical diffusion constant D and not to the self-diffusion constant D^s. Surprisingly, the chemical diffusion constant is independent of the concentration c if hard core interaction is assumed.

An expression for the self-diffusion constant is obtained by solving for $P^s(\mathbf{s}_i, t)$. This is more complicated and no simple rate equation can be obtained.[9] Generally we have

$$\frac{\partial}{\partial t} P^s(\mathbf{s}_i, t) = \sum_{j(\neq i)} \{k_{j \to i} P^s(\mathbf{s}_j, t)[1 - P(\mathbf{s}_i, t)] - k_{i \to j} P^s(\mathbf{s}_i, t)[1 - P(\mathbf{s}_j, t)]\}$$

$$(16.2.39)$$

We can get a closed equation for $P^s(\mathbf{s}_i, t)$ by using a mean-field approximation, i.e., simply replacing $P(\mathbf{s}_i, t)$ by c, the average concentration. Within this approximation the self-diffusion constant is given by

$$D^s = (1 - c) \frac{1}{2d} \sum_{j(\neq i)} k_{i \to j} s_{i \to j}^2 = (1 - c)D \qquad (16.2.40)$$

The factor $(1 - c)$ is called the blocking factor. Quantity D^s decreases with increasing concentration, which is evident from the relation between D^s and the mean-square displacement of the tagged adatom. We note that D and D^s become identical at low concentrations when interaction between adatoms can be neglected.

16.2.2.3. General Interaction

Calculation of both static and dynamic properties becomes involved in the presence of nontrivial interaction between adatoms. This goes far beyond the scope of the present chapter. The jump diffusion models can also be extended to several species and chemical reactions between species can be included.

16.2.3. Limitations

The jump diffusion models automatically exclude a description of the oscillatory motion of the adatoms. In neutron scattering experiments both the vibrational motion and the diffusive motion are revealed, and for a proper interpretation of those experiments the vibrational motion must be included in the jump diffusion models. If the diffusive jumps are assumed to be uncorrelated to the vibrational motion, this can easily be done.[10]

When the thermal energy becomes comparable to the activation energy for diffusion along the surface, the adatoms are no longer mainly localized at different lattice sites. The time between two consecutive jumps becomes comparable to the duration of a jump, and the subdivision into individual jumps loses its meaning. A more precise description of the motion is then needed. In the next section models based on the continuous motion of the adatoms will be presented.

16.3. Continuous Diffusion Models

The surrounding substrate influences the motion of the adatom in a complicated way. In the previous section this influence was taken care of phenomenologically in two separate steps. First, we made use of the periodic arrangement of the surrounding substrate atoms and assumed that the adatom most of the time was located around stable lattice sites. Second, we incorporated into the rate constants the effect that the adatom can gain and lose energy to the substrate and thereby change site. In this section we will carry out a similar decomposition of the influence from the surrounding. The difference is that here we will treat the continuous motion of the adatom. This treatment is more general, and the corresponding rate equation is more complicated and cannot be solved analytically in the general case. We restrict ourselves to a single adsorbed atom, but the formalism can be generalized to incorporate interactions among adatoms. No attempts will be made to include quantum effects in the motion of the adatom.

If we know the force on the adatom, we can solve Newton's equation and obtain the motion of the adatom. In principle we could use the microscopic equation of motion for the total system and obtain the time dependence of this force. Here we approach the problem quite differently and more phenomenologically. We will try to describe the influence from the surrounding in some simple terms.

16.3.1. Brownian Motion Theory

First we restrict ourselves to a homogeneous medium. We disregard for a moment the static part of the force acting on the adatom and arising from the electrons and nuclei of the substrate in their average positions. The purpose is to present a phenomenological way to introduce the effect that the adatom can gain and lose energy to the substrate. Or, in other words, we want to introduce dissipation and fluctuation of the energy of the adatom.

16.3.1.1. Langevin's Equation

The force on the adatom is split into a smooth friction term, which is assumed to be proportional to the velocity of the adatom, and a rapidly fluctuating force, the stochastic force, arising from collisions with the individual surrounding particles. Newton's equation is expressed in the form

$$m\frac{d}{dt}\mathbf{V}(t) = -m\eta\mathbf{V}(t) + \mathbf{F}^{st}(t) \tag{16.3.1}$$

where $\mathbf{V}(t)$ is the velocity of the adatom, m its mass, η the friction coefficient, and $\mathbf{F}^{st}(t)$ the stochastic force. Equation (16.3.1) is called *Langevin's equation*. The stochastic force is assumed to have zero average value,

$$\langle \mathbf{F}^{st}(t) \rangle = \mathbf{0}, \tag{16.3.2}$$

and is assumed to be totally uncorrelated in time, i.e., $\langle \mathbf{F}^{st}(t)\mathbf{F}^{st}(t') \rangle = \mathbf{0}$ for $t \neq t'$. However, the average value of $\langle \mathbf{F}^{st}(t)\mathbf{F}^{st}(t) \rangle$ cannot be zero and we postulate that

$$\langle \mathbf{F}^{st}(t)\mathbf{F}^{st}(t') \rangle = \sigma\mathbf{1}\delta(t - t') \tag{16.3.3}$$

where $\mathbf{1}$ is the unit tensor and σ is a constant, which for the moment is unspecified. The motion of the adatom is treated in a probabilistic manner. Quantity $\mathbf{F}^{st}(t)$ is a stochastic variable and we will only be able to give statistical predictions for the position $\mathbf{R}(t)$ of the adatom and for its velocity $\mathbf{V}(t)$. The angle brackets in equations (16.3.2) and (16.3.3) denote a canonical ensemble average while $\langle \mathbf{R}(t) \rangle$ and $\langle \mathbf{V}(t) \rangle$ will be the average values over this ensemble, characterized by the temperature.

16.3.1.2. The Fluctuation-Dissipation Theorem

Direct solution of equation (16.3.1) leads to

$$\mathbf{V}(t) = \mathbf{V}(0)\exp(-\eta t) + \frac{1}{m}\int_0^t dt'\exp[-\eta(t - t')]\mathbf{F}^{st}(t') \tag{16.3.4}$$

where the *fixed* initial velocity of the adatom is denoted by $\mathbf{V}(0)$. It follows directly from equation (16.3.2) that the average velocity is

$$\langle \mathbf{V}(t) \rangle = \mathbf{V}(0)\exp(-\eta t) \tag{16.3.5}$$

and for times $t \gg 1/\eta$ there is no information left on the initial velocity, while the average velocity is zero. This does not imply that $\langle V^2(t) \rangle$ is zero. In fact we know from the equipartition theorem of classical statistical mechanics[11] that for a system in thermal equilibrium at temperature T,

$$\langle V^2 \rangle = d\frac{k_B T}{m} \tag{16.3.6}$$

where d is the dimension of the system and k_B is Boltzmann's constant. The formal solution of Langevin's equation enables us to obtain

$$V^2(t) = V^2(0)\exp(-2\eta t) + \frac{2}{m}\exp(-\eta t)\int_0^t dt' \exp[-\eta(t-t')]\mathbf{F}^{st}(t')\cdot\mathbf{V}(0)$$

$$+ \frac{1}{m^2}\int_0^t dt' \int_0^t t'' \exp[-\eta(t-t')]\exp[-\eta(t-t'')]\mathbf{F}^{st}(t')\cdot\mathbf{F}^{st}(t'')$$

$$(16.3.7)$$

By averaging over the thermal ensemble and using the two properties of the stochastic force, postulated in equations (16.3.2) and (16.3.3), we get

$$\langle V^2(t)\rangle = V^2(0)\exp(-2\eta t) + d\frac{\sigma}{2\eta m^2}[1 - \exp(-2\eta t)] \qquad (16.3.8)$$

We note again that the initial information fades away exponentially in time. At sufficiently long times the adatom is assumed to reach thermal equilibrium and its mean-squared velocity approaches the value given in equation (16.3.6). In order to fulfill this we must have

$$\sigma = 2m\eta k_B T \qquad (16.3.9)$$

The parameter σ refers to the magnitude of the fluctuations in the medium, while η refers to the magnitude of the collective friction force. They are evidently related to each other and that is by no means obvious. The above relation is a special case of a general one, called the *fluctuation-dissipation theorem*. Any system which shows dissipative effects, as friction above, will also show fluctuations and there exists a relation between the two. The constant σ in equation (16.3.3) is now specified in terms of the friction coefficient.

16.3.1.3. The Self-Diffusion Constant

The self-diffusion constant D^s was introduced in equation (16.2.18) by connecting it to the mean-square displacement of the adatom. In the jump diffusion model the mean-square displacement was linear in time for all times. This is not true in a more microscopic treatment. At very short times the adatom is moving as a free particle and

$$\langle[\mathbf{R}(t) - \mathbf{R}(0)]^2\rangle = \langle V^2(0)\rangle t^2 = d\frac{k_B T}{m}t^2 \qquad (t \to 0) \qquad (16.3.10)$$

Here we have conducted a thermal average over all possible initial velocities. Equation (16.3.10) is an exact result, and a model that is supposed to describe correctly the short-time behavior must fulfill this relation. If the motion becomes diffusive, the mean-square displacement will approach at long times a linear dependence on time. We *define* the self-diffusion constant by

$$D^s \equiv \lim_{t \to \infty} \frac{1}{2dt} \langle [\mathbf{R}(t) - \mathbf{R}(0)]^2 \rangle. \qquad (16.3.11)$$

The relation

$$\mathbf{R}(t) - \mathbf{R}(0) = \int_0^t dt' \, \mathbf{V}(t') \qquad (16.3.12)$$

can be used to relate the mean-square displacement to the velocity correlation function for the adatom. Hence

$$\langle [\mathbf{R}(t) - \mathbf{R}(0)]^2 \rangle = \int_0^t dt' \int_0^t dt'' \, \langle \mathbf{V}(t') \cdot \mathbf{V}(t'') \rangle$$

$$= 2 \int_0^t d\tau \, (t - \tau) \langle \mathbf{V}(\tau) \cdot \mathbf{V}(0) \rangle \qquad (16.3.13)$$

where we have made use of the property $\langle \mathbf{V}(t') \cdot \mathbf{V}(t'') \rangle = \langle \mathbf{V}(t' - t'') \cdot \mathbf{V}(0) \rangle$. We can now relate D^s to the velocity correlation function and a second and equivalent definition of D^s is obtained,

$$D^s \equiv \frac{1}{d} \int_0^\infty d\tau \, \langle \mathbf{V}(\tau) \cdot \mathbf{V}(0) \rangle \qquad (16.3.14)$$

Equations (16.3.11) and (16.3.14) can also be used as definitions of the self-diffusion constant for an adsorbed layer of interacting adatoms. The variables $\mathbf{R}(t)$ in equation (16.3.11) and $\mathbf{V}(t)$ in equation (16.3.14) are then the position and velocity for a single tagged adatom, respectively. The chemical diffusion constant D can be defined in terms of another velocity correlation function.[8] The correlation function then contains the velocities of all adatoms and not only the velocity of a tagged adatom as in equation (16.3.14).

Langevin's equation can be used to derive a connection between the self-diffusion constant and the friction coefficient. Equation (16.3.4) yields directly

$$\langle \mathbf{V}(t) \cdot \mathbf{V}(0) \rangle = d \frac{k_B T}{m} \exp{(-\eta t)} \qquad (16.3.15)$$

this is inserted into equation (16.3.13), we obtain

$$\langle [\mathbf{R}(t) - \mathbf{R}(0)]^2 \rangle = d \frac{2k_B T}{m\eta} \left\{ t - \frac{1}{\eta} [1 - \exp{(-\eta t)}] \right\} \qquad (16.3.16)$$

which is consistent with equation (16.3.10) at short times and in the limit $t \to \infty$ the identification

$$D^s = k_B T / m\eta \qquad (16.3.17)$$

can be made.

16.3.2. Adatom Dynamics

The substrate breaks the symmetry of the system and an adatom experiences a static potential arising from the electrons and nuclei of the substrate in their average positions. This part of the interaction with the substrate is crucial for the character of the adatom motion. It gives rise to the periodic force along the surface and to the attractive and repulsive forces perpendicular to the surface of the substrate. In order to properly describe surface dynamical problems as diffusion, adsorption, and desorption one must, besides the above conservative part of the forces, also include the dissipative part. It is the latter that is responsible for dissipation and fluctuation of the energy of the adatom and these effects must be included.

A starting point for treating adatom dynamics is the following generalized Langevin's equation:

$$m \frac{d}{dt} \mathbf{V}(t) = \mathbf{F}^{ad}(\mathbf{R}) - m\eta \mathbf{V}(t) + \mathbf{F}^{st}(t) \qquad (16.3.18)$$

with

$$\langle \mathbf{F}^{st}(t)\mathbf{F}^{st}(t') \rangle = 2m\eta k_B T \mathbf{1} \delta(t - t'). \qquad (16.3.19)$$

The static force $\mathbf{F}^{ad}(\mathbf{R}) = -\nabla V^{ad}(\mathbf{R})$ depends on the position $\mathbf{R}(t)$ of the adatom and we call it the adiabatic force. This is the conservative part of the interaction with the substrate. The dissipative part is represented by the friction coefficient and by the stochastic force, which are connected by the fluctuation-dissipation theorem in equation (16.3.19).

Equation (16.3.18) is a stochastic equation and one often converts it into an equation for the probability distribution $f(\mathbf{r}, \mathbf{p}, t)$ to find the adatom at time t at point $\mathbf{r}$ with momentum $\mathbf{p}$. This is the Fokker–Planck equation with a static force and it reads[12]

$$\left[\frac{\partial}{\partial t} + \frac{\mathbf{p}}{m} \cdot \nabla_{\mathbf{r}} + \mathbf{F}^{ad}(\mathbf{r}) \cdot \nabla_{\mathbf{p}} \right] f(\mathbf{r}, \mathbf{p}, t) = \eta \nabla_{\mathbf{p}} \cdot [\mathbf{p} + m k_B T \nabla_{\mathbf{p}}] f(\mathbf{r}, \mathbf{p}, t). \qquad (16.3.20)$$

The latter equation replaces the rate equation (16.2.2) and it was first given by Klein.[13] In the absence of the static force it can be solved analytically.[12] The incorporation of the static force complicates the situation considerably. Most discussions based on equation (16.3.20) and aiming at analytic results have been restricted to one-dimensional (1D) models and to situations where the friction coefficient is either small or large compared to some typical frequency. Here, we will discuss two such models: escape from a 1D potential well and diffusion in a 1D periodic potential.

16.3.2.1. Escape from a Potential Well

Kramers based a discussion of the *steady-state escape rate* from a 1D potential well on equation (16.3.20) in a seminal paper.[14] He was unable to solve the equation for an arbitrary value of η. He could, however, analyze the solution for small and large values of η.

For large values of η, so large that the distance a thermally drifting particle traverses in time η^{-1} is small on the scale of variation of $V^{ad}(x)$, equation (16.3.20) reduces to a diffusion equation for the density $\rho(x, t) = \int dp\, f(x, p, t)$ alone, the so-called Smoluchowski equation. Kramers solved it for the steady-state escape rate and under the additional assumption $\Delta E \gg k_B T$; he obtained the rate

$$k = \frac{\omega_A \omega^*}{2\pi\eta} \exp\left(-\Delta E / k_B T\right) \qquad (\eta \gg \omega^*) \tag{16.3.21}$$

The frequencies ω_A and ω^* are those associated with the second derivative of the static potential at the bottom and top of the barrier, respectively, and ΔE is the barrier height.

In the opposite limit, η small, the energy of the particle, not its position, will be slowly varying. Equation (16.3.20) can then be reduced, by using a canonical transformation, to a diffusion equation for the probability $P(E, t)$ to find the particle with a certain energy. Kramers solved this equation too and, employing the same assumption about the barrier height as above, i.e., $\Delta E \gg k_B T$, he found the rate

$$k = \eta \frac{\Delta E}{k_B T} \exp\left(-\Delta E / k_B T\right) \qquad (\eta \ll \omega_A k_B T / \Delta E) \tag{16.3.22}$$

The Kramers's treatment is an attempt to go beyond the transition state theory (TST) for the escape rate which, for the above situation, leads to the expression

$$k = \frac{\omega_A}{2\pi} \exp\left(-\Delta E / k_B T\right) \tag{16.3.23}$$

The TST will be discussed in Section 16.4. Here we only stress that there is no explicit dependence on the dissipative part of the forces in the TST. Kramers showed that by including dissipation and fluctuation through the use of the Brownian motion theory, large deviations from the TST result are obtained for small and large values of η. He also argued that TST applies for intermediate values of η, large enough to ensure replenishment of the equilibrium distribution but not so large as to inhibit the motion of escaping particles. It is the former effect that is responsible for the depletion of the rate in the low-friction limit and the latter effect gives rise to the reduction of the rate in the high-friction limit. From this it follows that TST always overestimates the true steady-state escape rate.

16.3.2.2. Diffusion in a Periodic Potential

We now consider the motion in a 1D periodic potential, $V(x)$. The diffusion constant is not determined solely by the escape rate, but also the jump distance is needed. This distance depends on how efficiently the energy of the diffusing atom is dissipated away.

For small values of η the jump distance can become large. If the coupling to the substrate is weak, i.e., η is small, adatoms with energies above the potential barrier, activated adatoms, will move rather freely and adatoms with energies less than the barrier energy will vibrate locally for a long time. To a first approximation in the low-friction limit only adatoms with energies above the potential barrier contribute to the diffusion constant. Their contribution is given approximately by equation (16.3.17). A more careful treatment leads to the result[15]

$$D^s = \frac{k_B T}{m\eta} \sqrt{\frac{2}{m\pi k_B T}} \int_{\Delta E}^{\infty} dE\, \bar{v}(E)^{-1} \exp\left(-E/k_B T\right)$$

$$\times \left\{ \frac{1}{a} \int_0^a dx\, \exp\left[-V(x)/k_B T\right] \right\}^{-1} \tag{16.3.24}$$

where $\bar{v}$ is the average velocity given by

$$\bar{v}(E) = \frac{1}{a} \int_0^a dx\, |v(x, E)|. \tag{16.3.25}$$

This is essentially equation (16.3.17) times the probability to find the adatom above the potential barrier. We note that a system in thermal equilibrium is considered and we always have adatoms with energies above and below the potential barrier. In the low-friction limit both activation and deactivation of adatoms becomes equally inefficient.

In the opposite limit, η large, the energy dissipation is efficient and the jump distance is equal to the lattice spacing a. The diffusion constant is then given by the Kramers' result in equation (16.3.21) times a^2. Again, the analysis can be made more precise and by solving the Smoluckowski equation one obtains[16-18]

$$D^s = \frac{k_B T}{m\eta} \left\{ \frac{1}{a} \int_0^a dx\, \exp\left[-V(x)/k_B T\right] \frac{1}{a} \int_0^a dx\, \exp\left[V(x)/k_B T\right] \right\}^{-1} \tag{16.3.26}$$

In the limit $\Delta E \gg k_B T$, equation (16.3.26) reduces to the Kramers result times a^2. The diffusion constant is seen to be proportional to η^{-1} in both limits, in contrast to the steady-state escape rate that depends linearly on η in the small-friction limit.

16.3.3. Limitations with the Brownian Motion Model

This section is concluded by pointing out some limitations with the Brownian motion model and making some comments on the applicability of the above analytic results.

16.3.3.1. Three-Dimensional Motion

The result for the escape rate in Section 16.3.2.1 and for the diffusion constant in Section 16.3.2.2 are all based on 1D treatments of the adatom motion. The results for the escape rate will not be qualitatively different if the full three-dimensional (3D) motion of the adatom is included.[19] However, it has been shown that in order to describe correlated jumps one must include the 3D character of the adatom motion.[20] This is particularly important for a light adatom which can move rather erratically on the surface. This cannot be simplified to a 1D motion without loss of qualitative features. The result in equation (16.3.24) is therefore of limited value in the case of surface diffusion. There is no problem, expect numerical, to proceed beyond the 1D treatment within the Brownian motion model. One has to solve the full 3D Fokker–Planck equation (16.3.20).

16.3.3.2. Position-Dependent Friction

We have only considered a constant η, but in many situations it would be more appropriate to use a friction coefficient which depends on the position of the adatom. Such a situation obviously occurs in thermal desorption, where $\eta(z)$ approaches zero with increasing distance z from the solid surface. But also in surface diffusion the friction coefficient can have a strong dependence on the position of the adatom as well as on the direction of the adatom motion.[21] At the bridge position on a fcc (001) surface the coupling to the vibrations of the substrate atoms is weak in the direction of the minimum energy path but strong in the other two directions. These effects can be incorporated into the Brownian motion model if one allows for a position-dependent friction, $\eta \to \eta(\mathbf{r})$, and if the tensor nature of the friction coefficient is included, $\eta(\mathbf{r}) \to \boldsymbol{\eta}(\mathbf{r})$.

16.3.3.3. Memory Effects

A more serious limitation with the above approach is connected to the basic assumption underlying the Brownian motion model. In writing the equation of motion in the simple Langevin's form, one has assumed that the fluctuations in the surrounding are fast with respect to the adatom motion. This is the case if the adatom is much heavier than the substrate particles, but it is certainly not true in the general case. For the latter cases Kubo has suggested that one should modify Langevin's equation to the form

$$m\frac{d}{dt}\mathbf{V}(t) = \mathbf{F}^{\mathrm{ad}}(\mathbf{R}) - m\int_0^t dt'\,\Gamma(t - t')\mathbf{V}(t') + \mathbf{F}^{\mathrm{st}}(t) \qquad (16.3.27)$$

where the simple friction term is replaced by a term that depends on the whole past history of the adatom. The function $\Gamma(t)$ is usually called the *memory function*. In Section 16.5 we will return to equation (16.3.27), the *generalized Langevin's equation*, which can be shown to be exact as it is merely a formal separation of the total force on the adatom into different components.

16.3.3.4. Quantum Effects

The Brownian motion model is based on the assumption that the motion of the adatom can be treated classically. In recent years many attempts have been made to generalize the model to the quantum regime. The quantum version of the Kramers problem has been solved[22,23] and results concerning the influence of a dissipative environment on the dynamics of a quantum system have been reviewed.[24]

16.4. Thermal Rate Constants

In Section 16.2 the jump diffusion model was examined. The important parameter in that model is the thermal rate constant $k_{i \to j}$. In this section we will relate $k_{i \to j}$ to the basic interaction potential between the constituents.

16.4.1. The Born–Oppenheimer Approximation

The dynamics of an atom adsorbed on a surface is fundamentally a complex quantum-mechanical problem. An important simplification is obtained by using the Born–Oppenheimer approximation[25] to separate the motion of the nuclei from the electronic degrees of freedom. The quantum-mechanical interactions of the electrons and nuclei are reduced to an interatomic potential-energy hypersurface,

$$V(\mathbf{R}_0, \mathbf{R}_1, \ldots, \mathbf{R}_N). \tag{16.4.1}$$

Here $\mathbf{R}_0$ denotes the position of the adatom (previously denoted by $\mathbf{R}$) and $\mathbf{R}_i$, $i = 1, \ldots, N$, denotes the positions of the N substrate atoms. The assumption of classical mechanics is usually appropriate and the motion of the atoms are obtained by solving Newton's equation,

$$m_i \frac{d^2}{dt^2} \mathbf{R}_i = -\nabla_{\mathbf{R}_i} V(\mathbf{R}_0, \mathbf{R}_1, \ldots, \mathbf{R}_N), \qquad i = 0, \ldots, N \tag{16.4.2}$$

with appropriate initial conditions. At low temperatures and for small masses m_i quantum-mechanical effects such as zero-point motion, tunneling, and inteference phenomena may be important. Newton's equation is then replaced by the Schroedinger equation.

There are cases when the Born–Oppenheimer approximation must be improved. In several surface dynamical problems, for instance, in ion neutralization and ionization at surfaces and, generally, in chemical reactions, more than one potential-energy hypersurface is involved. For such cases "surface-hopping" techniques have been developed.[26] Metal surfaces introduce new phenomena, compared with gas-phase reactions, due to the presence of the conduction electrons. The moving adatom may excite electron–hole pairs which can effectively dissipate excess energy. Electron–hole pairs can be provided with infinitesimal energy and the single Born–Oppenheimer potential hypersurface should be replaced by a continuous band of very nearly parallel hypersurfaces corresponding to different combinations of electron–hole pair excitations. This effect will be discussed in the next section in terms of a friction coefficient.

16.4.2. Transition State Theory

Diffusion is in essence a *dynamical problem*, but in an important class of theories the explicit dependence on the dynamics is bypassed and the diffusion constant obtained from an *equilibrium consideration*. These are based on the *transition state theory* (*TST*) for the escape rate,[27,28] which originates from the absolute rate theory of Eyring.[29,30] In applications to solids the theory was developed through the work of Zener and Wert[31] and Vineyard.[32]

If we assume that the diffusive motion consists of uncorrelated jumps between adjacent lattice sites, the diffusion constant is obtained from the relation

$$D^s = (1/2d)ka^2 \qquad (16.4.3)$$

where k is the escape rate from a lattice site. A key observation is that the change of sites is a thermally activated process. In the realm of classical mechanics the adatom must pass over the energy barrier that separates regions of lower energies from each other, the surface lattice sites. Thus, a barrier region is a transition state, a state that must be visited during the passage from one site to another and it acts as a bottleneck for the reaction.

16.4.2.1. One-Dimensional Example

To introduce the method of transition state theory (TST), we consider a 1D example, as shown in Figure 16.2. We wish to calculate the rate $k_{A \to B}$ at which the particle is moving from site A to site B. In moving from A to B the particle must pass the point S, the saddle point or the transition state. During a short time interval, say Δt, the particle will escape if it has positive velocity, $v > 0$, and if it is within the distance $v\Delta t$ of S, i.e., $x_0 - v\Delta t < x < x_0$. The time interval Δt is chosen small enough so that the potential energy can be assumed to be constant over the distance $v\Delta t$. We now assume that the system is in thermal equilibrium at the temperature T. The probability to find the particle within the distance Δx

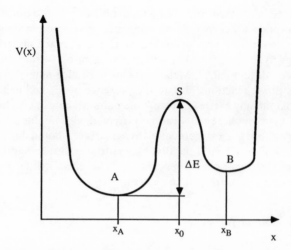

Figure 16.2. A one-dimensional potential $V(x)$ for the reaction coordinate x. An energy barrier ΔE separates the two sites A $(x < x_0)$ and B $(x > x_0)$ from each other.

of S is given by

$$P(\Delta x) = \Delta x \frac{\exp\left[-\beta V(x_0)\right]}{\displaystyle\int_{-\infty}^{x_0} dx \, \exp\left[-\beta V(x)\right]} \tag{16.4.4}$$

and the probability for the particle to have the velocity v is

$$P(v) = \frac{\exp\left(-\beta m v^2/2\right)}{\displaystyle\int_{-\infty}^{\infty} dv \, \exp\left(-\beta m v^2/2\right)} = \sqrt{\frac{m\beta}{2\pi}} \exp\left(-\beta \frac{m v^2}{2}\right) \tag{16.4.5}$$

where $\beta = 1/k_B T$. We note that in the expression for $P(\Delta x)$ we have normalized by integrating from $-\infty$ to x_0, i.e., equation (16.4.4) gives the probability for the particle to be located at the transition state, provided it is situated at site A $(x < x_0)$. The escape rate is now obtained by taking into account all positive velocities, v, i.e.,

$$\Delta t \, k_{A \to B}^{TST} = \int_0^\infty dv \, P(v) P(\Delta x = v \, \Delta t) \tag{16.4.6}$$

or

$$k_{A \to B}^{TST} = \frac{\exp\left[-\beta V(x_0)\right]}{\sqrt{2\pi m\beta} \displaystyle\int_{-\infty}^{x_0} dx \, \exp\left[-\beta V(x)\right]} \tag{16.4.7}$$

Another way to write this is

$$k_{A \to B}^{TST} = \langle v\delta(x - x_0)\theta(v)\rangle_A \tag{16.4.8}$$

where $\delta(x)$ is the Dirac delta function, $\theta(x)$ the Heaviside step function given by

$$\theta(x) = \begin{cases} 1 & \text{if } x > 0 \\ 0 & \text{if } x < 0 \end{cases} \tag{16.4.9}$$

and where subscript A in the bracket notation, $\langle \ldots \rangle_A$, indicates an ensemble average restricted to the configuration space of the site A only.

If the potential barrier is large compared to $k_B T$, then the integral in the denominator in equation (16.4.7) is dominated by values around $x = x_A$. If we expand the potential in the form

$$V(x) = V(x_A) + \tfrac{1}{2} m \omega_A^2 x^2 + \cdots \tag{16.4.10}$$

retain the quadratic term, and extend the integration limit to infinity, the well-known result

$$k_{A \to B}^{TST} = \frac{\omega_A}{2\pi} \exp\left(-\beta \Delta E\right) \tag{16.4.11}$$

is obtained [cf. equation (16.3.23)]. The prefactor $\omega_A/2\pi$, the so-called attempt frequency, comes from proper normalization and has nothing to do with kinetics. The usual physical intepretation, that $\nu_A = \omega_A/2\pi$ should be considered as the frequency with which the particle tries to escape, is therefore somewhat misleading in this context.

16.4.2.2. The Multidimensional Case

The above 1D expression for the TST escape rate can be extended to the multidimensional case. Figure 16.3 gives a 2D picture of the motion in the $(3N + 3)$-dimensional configuration space. It is useful to introduce new coordinates such

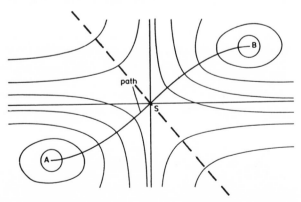

Figure 16.3. Two-dimensional picture of the $(3N + 3)$-dimensional configuration space. A and B are the minima and S is the saddle point. The dividing surface is defined by $s = s_0$ and it is shown as a dashed line in the figure.

that one coordinate, the reaction coordinate s, runs along the diffusion path between A and B. We denote the corresponding velocity by v_s. The value $s = s_0$ defines a $(3N + 2)$-dimensional surface, which we call the dividing surface or the TST surface. The TST value for the escape rate is now obtained by taking the probability to be located in the transition state region times the mean velocity in the reactive direction,

$$k_{A \to B}^{TST} = \langle v_s \delta(s - s_0) \theta(v_s) \rangle_A \tag{16.4.12}$$

where the average is over all degrees of freedom of the system. The escape rate can also be written as

$$k_{A \to B}^{TST} = \frac{k_B T}{h} \frac{Q_{cl}^*}{Q_{cl}^A} \tag{16.4.13}$$

where Q_{cl}^A is the classical mechanical partition function for the system when the particle is located in the region A,

$$Q_{cl}^A = \int \prod_{i=1}^{3N+3} \left(\frac{dq_i dp_i}{h} \right) \exp(-\beta H) \tag{16.4.14}$$

and Q_{cl}^* is the corresponding function when the particle is located in the transition state region,

$$Q_{cl}^* = \int \prod_{i=1}^{3N+2} \left(\frac{dq_i dp_i}{h} \right) \exp(-\beta H^*) \tag{16.4.15}$$

Quantity H is the total Hamiltonian for the system and H^* is defined in the transition state region, i.e., $s = s_0$ and v_s does not appear. Equations (16.4.12) and (16.4.13) are two equivalent expressions for the escape rate within the transition state theory. It is noteworthy that the integrals in equations (16.4.14) and (16.4.15) are equilibrium averages and can be evaluated exactly by means of numerical Monte Carlo methods.[33] The starting point is then an assumed form for a Born–Oppenheimer potential-energy hypersurface. For computational reasons one often uses a model with pairwise additive interaction potentials. This gives a connection between the escape rate within the transition state theory and the microscopic force laws.

In the same say as in the 1D example we can expand the potential energy and obtain an expression valid in the harmonic approximation. We write for the potential energy around the stable point A

$$V(\ldots) = V_A + \tfrac{1}{2} \sum_{i=1}^{3N+3} m_i \omega_i^2 q_i^2 + \cdots \tag{16.4.16}$$

and around the saddle point

$$V^*(\ldots) = V^* + \tfrac{1}{2} \sum_{i=1}^{3N+2} m_i (\omega_i^*)^2 q_i^2 + \cdots \tag{16.4.17}$$

where ω_i and ω_i^* are normal-mode frequencies and q_i normal-mode coordinates (not necessarily the same in the two expressions). The phase-space integrals in equations (16.4.14) and (16.4.15) can now be performed and, using the notation $\Delta E = V^* - V_A$, we arrive at the frequency-product formula of Vineyard,[32]

$$k_{A \to B}^{TST} = \frac{1}{2\pi} \frac{\displaystyle\prod_{i=1}^{3N+3} \omega_i}{\displaystyle\prod_{i=1}^{3N+2} \omega_i^*} \exp(-\beta \Delta E) \tag{16.4.18}$$

16.4.3. Quantum Transition State Theory

Classical concepts have been used in an essential way to obtain the transition state theory result in equation (16.4.13). Especially in cases when hydrogen atoms are involved one would like to incorporate quantum effects into the expression for the rate constant.

First, we make the obvious modification that the classical mechanical partition functions are replaced by their quantum counterparts,

$$Q_{cl}^A \to Q_{qm}^A = Tr_A \{\exp(-\beta H)\} \tag{16.4.19}$$

and correspondingly for Q_{cl}^*. This takes into account the discreteness of the excitation energies and it is important if a typical vibrational energy is large compared to the thermal energy. For hydrogen this can be important even at room temperature.

Second, we note that the factor $k_B T/h$ in equation (16.4.13) originates from the evaluation of the rate at which the particle crosses the dividing surface. The calculation is based on the assumption that velocity and position coordinates are independent, which is not true quantum mechanically. We must correct for tunneling through the potential barrier at energies below the classical threshold, as well as the possibility of reflection at energies above the classical threshold. To do this we look at the classical calculation,[27]

$$\frac{k_B T}{h} = \frac{1}{h} \int_0^\infty dp \, \frac{p}{m} \exp\left(-\beta \frac{p^2}{2m}\right) = \frac{1}{h} \int_0^\infty dE \, \exp(-\beta E)$$

$$= \frac{1}{h} \int_{-\infty}^\infty dE \, T_{cl}(E) \exp(-\beta E) \tag{16.4.20}$$

where $T_{cl}(E)$ is the classical probability of crossing the potential barrier at an energy E measured with respect to the barrier maximum, $T_{cl}(E) = 1$ if $E > 0$ and zero otherwise. This suggests that we replace $T_{cl}(E)$ by the quantum probability of crossing a suitably chosen 1D potential barrier at energy E. We define the quantum correction $\kappa(T)$, which takes into account tunneling and nonclassical reflection:

$$\kappa(T) = \beta \int_{-\infty}^{\infty} dE \, T_{qm}(E) \exp(-\beta E) \qquad (16.4.21)$$

where $T_{qm}(E)$ is the quantum-mechanical transmission coefficient and where the zero of energy E is located at the classical threshold. We can now write the rate constant within the quantum transition state theory (QTST) as

$$k_{A \to B}^{QTST} = \kappa(T) \frac{k_B T}{h} \frac{Q_{qm}^{\#}}{Q_{qm}^{A}} \qquad (16.4.22)$$

There are several weaknesses in this simplified QTST. One of the main approximations in equation (16.4.23) is the assumption of separability between the reaction coordinate and all the other degrees of freedom. Several attempts have been made to go beyond the assumption of separability.[27] The quantum transition state theory is, however, not as well founded as its classical counterpart. For applications of QTST to hydrogen diffusion on solid surfaces, see the work by Lauderdale and Truhlar,[34,35] Valone, Voter, and Doll,[36,37] and Jaquet and Miller.[38]

If we again use the harmonic approximation, the partition functions Q_{qm}^{A} and $Q_{qm}^{\#}$ can be evaluated analytically. We can also determine $\kappa(T)$ at the same level of approximation. For not too low temperatures the rate will be dominated by the shape of the potential barrier in the vicinity of the classical threshold. We can then use a parabolic approximation

$$V(s) = -\tfrac{1}{2} m_s (\omega^{*})^2 s^2 \qquad (16.4.23)$$

for which the transmission probability is known.[39] The rate constant can then be written in the symmetric form

$$k_{A \to B}^{QTST} = \frac{\displaystyle\prod_{i=1}^{3N+3} \frac{\sinh(\beta \hbar \omega_i / 2)}{(\beta \hbar \omega_i / 2)}}{\dfrac{\sin(\beta \hbar \omega^{*}/2)}{(\beta \hbar \omega^{*}/2)} \displaystyle\prod_{i=1}^{3N+2} \frac{\sinh(\beta \hbar \omega_i^{*}/2)}{(\beta \hbar \omega_i^{*}/2)}} k_{A \to B}^{TST} \qquad (16.4.24)$$

16.4.4. Dynamic Correction Factor

The TST is approximate and does not give the correct value for the escape rate. The basic assumption in TST is that each crossing of the dividing surface corresponds to a reactive site change event. This is not true. TST ignores dynamical effects such as recrossing effects. Another assumption that must be made in order to describe diffusion using TST is that a particular jump mechanism must be imposed. If one assumes single uncorrelated jumps between adjacent lattice sites, the diffusion constant is related to the escape rate through equation (16.4.3). The *dynamic correction formalism*[40,41] is a method by which both these assumptions can be relaxed simultaneously and formally exact expressions for the different rate constants are obtained.[42,43] They are expressed in terms of different *reactive flux correlation functions.* A major advantage with this approach is that the correct quantum version can be derived.[44,45] We consider first a two-site system and thereafter the multisite case, necessary in describing correlated multiple jumps on a surface.

Before we discuss the dynamic correction factor we comment on the use of the equilibrium configuration in the transition state region. The adatom is in constant contact with a nearly infinite "heat bath" and it is therefore perfectly appropriate to use the canonical ensemble to describe this system. This is the *basic assumption* we make and the corresponding rate constant is then the *thermal* rate constant. The use of the equilibrium distribution in the transition state region is a rigorous result that directly follows from the use of the canonical ensemble.[46] It does not depend on how easy or difficult the bottleneck (i.e., the transition state region) is to enter, or on how quickly the typical trajectory passes through. It is misleading to regard the relaxation of the surrounding substrate atoms when the adatom is approaching the transitions state solely caused by the approaching adatom. It can equally well be that a fluctuation of the substrate atoms, a relaxation, makes the adatom approach the transition state. The jump event is more properly treated as a fluctuation in a many-body system at thermal equilibrium; the presence of the adatom in the transition state region neither causes, nor results from, but is rather instantaneously correlated with a relaxation in the mean positions of the surrounding substrate atoms. Similar arguments imply that the velocity distribution of adatoms found in the transitions state region is Maxwellian. Although a jumping adatom will usually need more than average kinetic energy to approach the saddle point, all this excess kinetic energy has on average, been converted into potential energy at the saddle point, only to be recovered as kinetic energy during the descent.

16.4.4.1. Two-Site System—Heuristic Derivation

As shown in Figure 16.4, we consider a system with two stable configurations, A and B, separated by an energy barrier at $s = s_0$. The reaction coordinate $s(t)$ can, for instance, be the distance $z(t)$ between an adsorbed atom and the surface in the case of thermal desorption. The "site" B then extends to infinity. In isomerization $s(t)$ can be an angle $\alpha(t)$ characterizing two different stable configurations of the molecule. Five typical trajectories which all cross the dividing surface at $s = s_0$ one or several times are shown. Only two of these trajectories, (a) and (b), are true reactive events for the reaction A → B and only these should contribute when calculating the rate constant $k_{A\to B}$. If we examine a canonical ensemble of systems at a given instant of time, we find that each of the eleven crossings shown in Figure 16.4 are present. If we now use the TST approximation, defined in equation (16.4.12), we will find the following: trajectory (a) will contribute. In trajectory (b) both crossing 2 and 4 will contribute, but not crossing 3 due to the negative sign of the velocity. Trajectory (b) is a single reactive event, but in TST it will contribute twice. Trajectories (c), (d), and (e) are all nonreactive

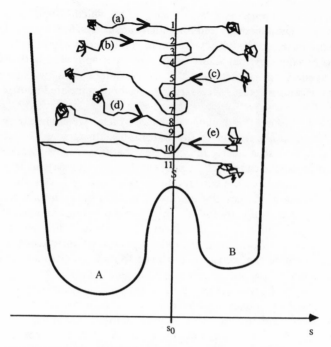

Figure 16.4. Five typical trajectories which all cross the dividing surface at $s = s_0$ one or several times. The arrows indicate the direction of the time. Only trajectory (a) and (b) correspond to true reactive events for the reaction $A \to B$.

but in TST crossings 6, 8, and 11 will contribute when calculating the rate constant. These "correlated dynamical events" cause the TST approximation to be an *upper bound* on the true rate constant, since each reactive event consists of at least one TST surface crossing.

To get the true rate constant, one must ensure that a crossing corresponds to a reactive event and that no multiple counting of a single reactive event is performed. This is done by generalizing the TST expression in equation (16.4.12) to[27,47]

$$k_{A \to B} = \langle \theta[s_0 - s(-\Delta t)] v_s \delta(s - s_0) \theta[s(\Delta t) - s_0] \rangle_A \qquad (16.4.25)$$

For each crossing of the TST surface we must run the trajectory backward in time to ensure that it started at site A and forward in time to ensure that it ended at site B. This is taken care of by the two Heaviside step-functions. The time Δt is somewhat arbitrary, but in a system with well-defined reactive events the time Δt will also be well-defined.[48] More precisely, we define τ_{corr} as the time scale on which the correlated dynamical events occur and define t_{rxn} as the average time between reactive events, i.e., the inverse of the true rate constant. Systems with well-defined reactive events will then be characterized by the time-scale separation $\tau_{corr} \ll \tau_{rxn}$ and the time Δt should be chosen longer than τ_{corr} but much shorter than τ_{rxn}. We now apply equation (16.4.25) to the trajectories in Figure 16.4.

Trajectory (a) still contributes to the rate. In trajectory (b) all three crossings contribute. However, crossing 3 gives a negative contribution ($v_s < 0$) and, in a perfect sampling of trajectories, crossings like 3 and 4 (or equally well 3 and 2) will exactly cancel each other. The net contribution from trajectory (b) will then be a single reactive event. Trajectories (c), (d), and (e) make no contribution to the rate due to one or both of the step-functions in equation (16.4.25).

We can further simplify the expression for the rate and write it as[48]

$$k_{A \to B} = \langle v_s \delta(s - s_0) \theta[s(\Delta t) - s_0] \rangle_A \qquad (16.4.26)$$

This is seen by first using the identity $\theta(s_0 - s(-\Delta t)) = 1 - \theta(s(-\Delta t) - s_0)$. Hence it follows that the difference between equations (16.4.25) and (16.4.26) is the quantity $\langle \theta(s(-\Delta t) - s_0) v_s \delta(s - s_0) \theta(s(\Delta t) - s_0) \rangle_A$, which is the average flux across the surface $s = s_0$ given that $s(-\Delta t) > s_0$ and $s(\Delta t) > s_0$. According to Liouville's theorem, that averaged flux is zero.

Next we derive equation (16.4.26) in a more formal way and put it on a more rigid basis. Before doing this we note that the TST result in equation (16.4.12) is obtained by taking the limit $\Delta t \to 0$ in equation (16.4.26). The TST result is obtained by assuming that all trajectories passing the dividing surface in the reactive direction correspond to a true reactive event.

16.4.4.2. Two-Site System—Formal Derivation

We consider again the situation in Figure 16.4. The rate at which the system is changing from A to B is required. To make the derivation more general it will be conducted fully quantum mechanically. We introduce the operator

$$A = \theta(s_0 - s) = \int_{-\infty}^{s_0} ds \, |s\rangle\langle s| \qquad (16.4.27)$$

which is unity if the system is in configuration A and zero if it is in configuration B. We also introduce the notation

$$B = 1 - A = \theta(s - s_0) = \int_{s_0}^{\infty} ds \, |s\rangle\langle s| \qquad (16.4.28)$$

Variables s, A, and B are now operators and we must keep track of the order. The Hamiltonian is assumed to possess the standard form

$$H = p_s^2/2m_s + \sum_i (p_i^2/2m_i) + V(s, q_1, q_2, \ldots) \qquad (16.4.29)$$

and the time evolution of A is given by

$$A(t) = \exp(iHt/\hbar) A \exp(-iHt/\hbar) \qquad (16.4.30)$$

The expectation value of A, given by

$$\langle A \rangle = Z^{-1} \operatorname{Tr} \{ A \exp(-\beta H) \} \tag{16.4.31}$$

with

$$Z = \operatorname{Tr} \{ \exp(-\beta H) \} \tag{16.4.32}$$

is the probability that the system is in configuration A.

The system is now perturbed slightly from equilibrium by applying an external time-dependent field that couples to the dynamical variable A. We turn on the perturbation slowly and, at $t = 0$, it is suddenly turned off,

$$\delta a^{\text{ext}}(t) = \begin{cases} \delta a \exp(\varepsilon t) & \text{if } t < 0 \\ 0 & \text{if } t > 0 \end{cases} \tag{16.4.33}$$

with δa and ε small and positive. The unperturbed Hamiltonian H is modified to

$$H'(t) = H - A \delta a^{\text{ext}}(t) \tag{16.4.34}$$

and at $t = 0$ the expectation value of A, namely $\langle A \rangle_{\text{ne}}$, will be slightly larger than the equilibrium value. The subscript "ne" denotes a nonequilibrium average. After $t = 0$, when the perturbation is turned off, the system will spontaneously relax back to the equilibrium configuration. By calculating the linear response to the weak external field $\delta a^{\text{ext}}(t)$ this relaxation is given by[49,50]

$$\delta \langle A(t) \rangle_{\text{ne}} = \beta C(t) \delta a \tag{16.4.35}$$

where

$$\delta \langle A(t) \rangle_{\text{ne}} = \langle A(t) \rangle_{\text{ne}} - \langle A \rangle \tag{16.4.36}$$

is the fluctuation from the true equilibrium value and $C(t)$ is the relaxation function

$$C(t) = \frac{1}{\beta} \int_0^\beta d\lambda \, \langle \delta A(-i\hbar\lambda) \delta A(t) \rangle \tag{16.4.37}$$

By taking the time derivative of equation (16.4.35), the following expression is obtained for the relaxation of the perturbed system:

$$\frac{d}{dt} \delta \langle A(t) \rangle_{\text{ne}} = \frac{\dot{C}(t)}{C(t)} \delta \langle A(t) \rangle_{\text{ne}} \tag{16.4.38}$$

We now *assume* that a simple linear rate equation describes the relaxation, i.e.,

$$\frac{d}{dt}\langle A(t)\rangle_{ne} = k_{B\to A}\langle B(t)\rangle_{ne} - k_{A\to B}\langle A(t)\rangle_{ne} \tag{16.4.39}$$

or

$$\frac{d}{dt}\delta\langle A(t)\rangle_{ne} = -k_{eff}\delta\langle A(t)\rangle_{ne} \tag{16.4.40}$$

with

$$k_{eff} = k_{A\to B}/\langle B\rangle \tag{16.4.41}$$

We expect the rate equation to be valid for times $t > \tau_{corr}$, where τ_{corr} is a microscopic time-scale that characterizes the rapid transient relaxation. We can then make the *identification*

$$k_{eff} = -\dot{C}(\Delta t)/C(\Delta t) \tag{16.4.42}$$

where the time Δt satisfies the condition

$$\tau_{corr} < \Delta t \tag{16.4.43}$$

By numerically calculating the time-dependent function $k(t) \equiv \dot{C}(t)/C(t)$ one can establish whether the proposed phenomenological rate equation (16.4.39) is valid. In that case $k(t)$ should reach a constant value for times $t > \tau_{corr}$, the plateau value behavior, and the value of the rate constant can be extracted. If the plateau value behavior is not observed, one must conceive of a different phenomenology. The correlation function $C(t)$ is changing on the time scale $\tau_{rxn} \sim 1/k_{eff}$ and as long as we are considering times $\Delta t \ll \tau_{rxn}$ we can replace $C(\Delta t)$ by $C(0)$ in the denominator in equation (16.4.42). In the classical limit we have the identity

$$C(0) = \langle A\rangle\langle B\rangle \tag{16.4.44}$$

In the quantum case this is an excellent approximation if the system is "well localized" in site A or in site B.[20,51] The rate constant can then be expressed in the form

$$k_{A\to B} = \frac{1}{\langle A\rangle}C_{fs}(\Delta t), \qquad \tau_{corr} < \Delta t \ll \tau_{rxn}, \tag{16.4.45}$$

where

$$C_{fs}(t) = Z^{-1} \frac{1}{\beta} \int_0^\beta d\lambda \operatorname{Tr} \{\exp(\lambda H) F_A \exp(-\lambda H) B(t) \exp(-\beta H)\}$$

$$(16.4.46)$$

Here F_A denotes the flux operator

$$F_A = -\dot{A} = \frac{i}{\hbar}[A, H] = \frac{1}{2m_s}[p_s \delta(s - s_0) + \delta(s - s_0)p_s] \qquad (16.4.47)$$

Equation (16.4.45) is the quantum version of the classical expression in equation (16.4.26). One can show that the initial value of the correlation function is zero in the quantum case, namely $C_{fs}(t = 0) = 0$,[52] in contrast to the classical limit where $C_{fs}(t \to 0^+)$ is finite and equal to the TST value.[48] This makes the problem of defining a quantum version of the classical TST apparent.

By using the fact that $C_{fs}(t = 0) = 0$ we can also write the rate constant as

$$k_{A \to B} = \frac{1}{\langle A \rangle} \int_0^{\Delta t} dt \, C_{ff}(t), \qquad \tau_{corr} < \Delta t \ll \tau_{rxn} \qquad (16.4.48)$$

where

$$C_{ff}(t) = Z^{-1} \frac{1}{\beta} \int_0^\beta d\lambda \operatorname{Tr}[\exp(\lambda H) F_A \exp(-\lambda H) F_A(t) \exp(-\beta H)]$$

$$(16.4.49)$$

The rate constant describes how the system evolves over macroscopic times. In equation (16.4.48) it is related to a time integral over a flux–flux correlation function, which decays on a microscopic time-scale. This kind of relation between a transport coefficient, the rate constant, and an equilibrium correlation function[53] is often called a Green–Kubo formula. In connection with reaction rates it was first given by Yamamoto.[44]

16.4.4.3. Multisite System

In the case of surface diffusion the adatom can be located in more than two sites, allowing for the possibility that an energized adatom will make a correlated multiple jump and thermalize in a nonadjacent lattice site. This kind of system can be treated in a fashion similar to the two-site system, using the formalism developed by Voter and Doll.[54] They simply generalize the expression for the rate constant in equation (16.4.26) to

$$k_{i \to j} = \langle v_i \delta_i \theta_j(\Delta t) \rangle_i, \qquad \tau_{corr} < \Delta t \ll \tau_{rxn} \qquad (16.4.50)$$

where v_i is the velocity normal to the dividing surface (defined as positive when the system is exiting from site i) and δ_i defines the location of the dividing surface. The function $\theta_j(t)$ is 1 if the adatom is located at site j at time t and zero otherwise. Equation (16.4.50) connects the phenomenological rate constant $k_{i\rightarrow j}$, introduced in equation (16.2.2), to the microscopic force law.

Voter and Doll have applied the formalism to a model system that is supposed to mimic the diffusion of a Rh-atom on a Rh(001) surface.[54] The Born–Oppenheimer potential-energy surface is represented as a sum of pair-potentials of the Lennard-Jones form. They combine a Monte Carlo approach for static properties with molecular dynamics to obtain the time dependence needed in equation (16.4.50). For computational reasons it is then crucial that the trajectories have to be followed only for relatively short times. In their case Δt is of the order 1 ps. For that particular system they conclude that at $T = 1000\,K$ and below the dynamical corrections are negligible, indicating that the TST value for the escape rate together with the assumption of single uncorrelated jumps is a very good approximation. They also claim that it is probably true that dynamic correction factors are quite minor in general for surface diffusion of adatoms, provided the TST surface is chosen properly.[43] That conclusion is supported by FIM experiments.[6]

The method with dynamic correction factors can be extended to include interactions among adsorbed adatoms.[55]

16.5. Friction Coefficient and Memory Function

The phenomenological treatment in Section 16.3 is based on the Langevin's equation, where the force on the adatom is divided into a friction force and a stochastic force. This is an approximation. In Section 16.3 we pointed out that in order to generalize Langevin's equation one must incorporate memory effects into the formalism.

16.5.1. Formal Exact Equation of Motion

It is more convenient to base a discussion on the generalization of the Fokker–Planck equation than on the generalized Langevin's equation (16.3.27). We wish to derive an equation of motion for the probability distribution $f(\mathbf{r}, \mathbf{p}, t)$ to find the adatom at time t at position $\mathbf{r}$ with momentum $\mathbf{p}$. To do this the microscopic phase-space density is introduced in the form

$$\rho(1t) = \delta[\mathbf{r}_1 - \mathbf{R}_0(t)]\delta[\mathbf{p}_1 - \mathbf{P}_0(t)] \qquad (16.5.1)$$

where 1 is a shorthand notation for the phase-space point $(\mathbf{r}_1, \mathbf{p}_1)$. The position and momentum of the adatom is, as previously, denoted by $\mathbf{R}_0(t)$ and $\mathbf{P}_0(t)$, respectively. In thermal equilibrium we have the probability distribution

$$\langle \rho(1) \rangle = \langle n(\mathbf{r}_1) \rangle (\beta/2\pi m)^{3/2} \exp(-\beta\, p_1^2/2m) \qquad (16.5.2)$$

where $\langle n(\mathbf{r}) \rangle$ is the mean density of the adatom. To discuss dynamic phenomena in an equilibrium system we form the phase-space correlation function

$$C^s(11't) = \langle \rho(1') \rangle^{-1} \langle \rho(1t)\rho(1') \rangle \tag{16.5.3}$$

where, as earlier superscript s indicates that we are treating the self-motion; $C^s(11't)$ is a mathematically well-defined quantity which satisfies the initial condition

$$C^s(11't = 0) = \delta(11') = \delta(\mathbf{r}_1 - \mathbf{r}'_1)\delta(\mathbf{p}_1 - \mathbf{p}'_1) \tag{16.5.4}$$

It is equal to the conditional probability, i.e., the probability of finding the adatom at time t at position $\mathbf{r}$ with momentum $\mathbf{p}$ if, at time $t = 0$, it was located at $\mathbf{r}'$ and moved with momentum $\mathbf{p}'$. On multiplying the phase-space correlation function by different powers of momenta and integrating, one can extract the more familiar density and current correlation functions. The latter can be used in determining the velocity correlation function and, from equation (16.3.14), one can then obtain the self-diffusion constant.

The derivation of formally exact transport equations is nowadays well established. One method is based on the projection operator technique[49,50,56] introduced by Zwanzig[57] and extended by Mori.[58] If we apply that formalism to the phase-space correlation function, then we can write[21]

$$\left[\frac{\partial}{\partial t} + \frac{\mathbf{p}_1}{m} \cdot \nabla_\mathbf{r} + \mathbf{F}^{\mathrm{ad}}(\mathbf{r}_1) \cdot \nabla_\mathbf{p} \right] C^s(11't)$$

$$= \int_0^t d\bar{t} \int d\bar{1} \nabla_\mathbf{p} \cdot \mathbf{L}(1\bar{1}t - \bar{t}) \cdot \left[\frac{\beta}{m}\mathbf{p} + \nabla_\mathbf{p} \right] C^s(\bar{1}1'\bar{t}). \tag{16.5.5}$$

This is a correct formal generalization of the Fokker–Planck equation. The influence from the surroundings on the adatom is split into two parts, one static and one dynamic. The static part gives rise to the conservative force $\mathbf{F}^{\mathrm{ad}}(\mathbf{r}) = -\nabla V^{\mathrm{ad}}(\mathbf{r})$, which is temperature-dependent. It is equal to the force acting on a fixed adatom at position $\mathbf{r}$ when the surrounding is in thermal equilibrium around that atom. We can define it by

$$\mathbf{F}^{\mathrm{ad}}(\mathbf{r}) = \langle \mathbf{F}(\mathbf{r}) \rangle_\mathbf{r} \tag{16.5.6}$$

where $\mathbf{F}(\mathbf{r})$ is the force on the adatom and $\langle \cdots \rangle_\mathbf{r}$ denotes an average over all the degrees of freedom of the substrate in the presence of a fixed adatom at $\mathbf{r}$. The corresponding potential $V^{\mathrm{ad}}(\mathbf{r})$ is the free energy and the mean density can be written as

$$\langle n(\mathbf{r}) \rangle = Z^{-1} \exp\left[-\beta V^{\mathrm{ad}}(\mathbf{r})\right] \tag{16.5.7}$$

where Z is the configurational part of the partition function,

$$Z = \int d\mathbf{r} \exp\left[-\beta V^{\mathrm{ad}}(\mathbf{r})\right] \tag{16.5.8}$$

On average the force on the adatom is given by $\mathbf{F}^{ad}(\mathbf{r})$, and in transition state theory (TST) only this part of the interaction is included while all fluctuations of the force around the static value are neglected. The dynamic part, represented by the *memory function* $\mathbf{L}(11't)$, is a direct generalization of the friction term and is responsible for fluctuation and dissipation of the energy of the adatom. Obviously, one cannot evaluate the memory function exactly for any real many-body system, but its formal expression can be the base for relevant approximations. It contains the fluctuations $\delta\mathbf{F} = \mathbf{F} - \mathbf{F}^{ad}$ of the force on the adatom from its adiabatic value and therefore depends on the simultaneous motion of this atom and the surrounding particles.

16.5.2. Fokker–Planck Approximation

In cases where a clear time-scale separation exists between the motion of the adatom and the motion in the surrounding, one can justify approximations of the memory function. If the fluctuations in the substrate have decayed before the adatom has moved appreciably, the motion of the adatom can be replaced by its initial value and the memory function reduces to

$$\mathbf{L}(11't) = \delta(11')\langle\delta\mathbf{F}(0)\delta\mathbf{F}(t)\rangle_{\mathbf{r}_1} \qquad (16.5.9)$$

Here $\delta\mathbf{F}$ is the fluctuating part of the force on the adatom, $\delta\mathbf{F} = \mathbf{F} - \mathbf{F}^{ad}$, and the subscript $\mathbf{r}_1$ in the bracket notation indicates that the force–force correlation function should be evaluated in the presence of a *fixed* adatom at position $\mathbf{r}_1$. The delta-function represents the effect that the adatom has no time to move or to change momentum during the time the force–force correlation function decays to zero. If we are only interested in times longer than the typical relaxation time of $\langle\delta\mathbf{F}(0)\delta\mathbf{F}(t)\rangle_{\mathbf{r}_1}$, then the equation of motion (16.5.5) reduces to the Markoffian equation[49,59]

$$\left[\frac{\partial}{\partial t} + \frac{\mathbf{p}_1}{m}\cdot\nabla_{\mathbf{r}} + \mathbf{F}^{ad}(\mathbf{r}_1)\cdot\nabla_{\mathbf{p}}\right]C^s(11't) = \nabla_{\mathbf{p}}\cdot\boldsymbol{\eta}(\mathbf{r}_1)\cdot\left(\mathbf{p} + \frac{m}{\beta}\nabla_{\mathbf{p}}\right)C^s(11't) \quad (16.5.10)$$

with

$$\boldsymbol{\eta}(\mathbf{r}_1) = \frac{\beta}{m}\int_0^{\infty} dt\,\langle\delta\mathbf{F}(0)\delta\mathbf{F}(t)\rangle_{\mathbf{r}_1} \qquad (16.5.11)$$

This is the Fokker–Planck equation but with a position-dependent friction tensor, in contrast to what was assumed in equation (16.3.20). Equation (16.5.11) gives the microscopic expression for the friction coefficient and is basically a property of the surrounding. The only dependence on the adatom is that the force–force correlation function should be evaluated in the presence of a fixed adatom.

16.5.2.1. Electron–Hole Pair Excitations

The obvious candidate to treat within the Fokker–Planck approximation is the coupling to the electronic degrees of freedom.[60,61] Since the electron mass is much smaller than the mass of any adatom, the fluctuations in the surrounding are fast with respect to the adatom motion, and the assumption underlying the Fokker–Planck equation should be valid. If the adatom is assumed to couple to the density of the surrounding electrons, then the force on the adatom can be written as

$$\delta F = - \int d\mathbf{r} \nabla_\mathbf{R} V(\mathbf{R}, \mathbf{r}) \delta n_e(\mathbf{r}) \tag{16.5.12}$$

where $\delta n_e(\mathbf{r}) = n_e(\mathbf{r}) - \langle n_e(\mathbf{r}) \rangle_\mathbf{R}$ is the fluctuating part of the density operator for the electrons, $\mathbf{R}$ is the position of the adatom, and $V(\mathbf{R}, \mathbf{r})$ describes the interaction between the electrons and the adatom. The friction coefficient can now be expressed in terms of the density–density correlation function for the electronic motion,

$$\boldsymbol{\eta}(\mathbf{R}) = \frac{\beta}{m} \int d\mathbf{r} \int d\mathbf{r}' \, \nabla_\mathbf{R} V(\mathbf{R}, \mathbf{r}) \int_0^\infty dt \, \langle \delta n_e(\mathbf{r}, 0) \delta n_e(\mathbf{r}', t) \rangle_\mathbf{R} \nabla_\mathbf{R} V(\mathbf{R}, \mathbf{r}')$$
$$\tag{16.5.13}$$

An estimate of the order of magnitude of this expression leads to[60]

$$\eta \sim \frac{m_e}{m} \frac{\varepsilon_F}{\hbar} \tag{16.5.14}$$

where m_e is the electron mass, m the mass of the adatom, and ε_F the Fermi energy.

The friction coefficient can be evaluated by calculating the density–density correlation function for the electronic motion in the presence of a fixed adatom. For a chemisorbed adatom this is very difficult to do from first principles. The density functional theory provides a one-electron scheme for calculating ground state properties and takes into account exchange and correlation effects.[62,63] The friction coefficient is, however, a dynamic property, but due to the time integral from zero to infinity only the low-lying excitations ($\omega \to 0$) are needed. It is plausible that the density functional scheme can be used.[64] Within a one-electron scheme equation (16.5.13) reduces to[65]

$$\eta^{\beta\beta'}(\mathbf{R}) = \frac{2\pi\hbar}{m} \sum_{\substack{\mathbf{kk'} \\ (\varepsilon_\mathbf{k} = \varepsilon_{\mathbf{k}'} = \varepsilon_F)}} \int d\mathbf{r} \, \psi_\mathbf{k}(\mathbf{r}) \psi_{\mathbf{k}'}(\mathbf{r})^* \nabla^\beta V(\mathbf{R}, \mathbf{r})$$
$$\times \int d\mathbf{r}' \, \psi_\mathbf{k}(\mathbf{r}')^* \psi_{\mathbf{k}'}(\mathbf{r}') \nabla^{\beta'} V(\mathbf{R}, \mathbf{r}') \tag{16.5.15}$$

where $\psi_{\mathbf{k}}(\mathbf{r})$ is a one-electron wave function with wave vector $\mathbf{k}$ and energy $\varepsilon_{\mathbf{k}}$, while β denotes a cartesian coordinate. The factor 2 comes from spin degeneracy. Actual calculations based on the density functional scheme with the local density approximation for the exchange-correlation potential have been performed.[66] Typically one finds $\hbar\eta$ of order 1 meV for hydrogen chemisorbed on a jellium surface.

The excitations of electron–hole pairs in metals provide an energy dissipation mechanism in addition to coupling to phonons.[67,68] In the language of Born–Oppenheimer potential-energy hypersurfaces the electron–hole pair excitations promote transitions among nearby hypersurfaces.[26] The strength of this non-adiabatic coupling can be obtained from the above *ab initio* calculations. They can also be the starting point for quantitative investigations of the relative importance of electron–hole pair excitations versus phonons in adatom dynamics.

16.5.2.2. Phonons

If coupling to phonons is examined instead, we can simply replace the electronic density in equation (16.5.13) by the density of the substrate atoms. It is then convenient to expand in lattice displacements $\mathbf{u}_l$ and the following formula for the friction coefficient can be derived[21]:

$$\eta^{\beta\beta'}(\mathbf{R}) = \frac{\beta}{m} \sum_{ll'} \sum_{\alpha\alpha'} v_{\mathrm{eff}}^{\alpha\beta}(\mathbf{R} - \mathbf{R}_l) \int_0^\infty dt \, \langle u_l^\alpha(t) u_{l'}^{\alpha'} \rangle v_{\mathrm{eff}}^{\alpha'\beta'}(\mathbf{R} - \mathbf{R}_{l'}) \quad (16.5.16)$$

In deriving the latter equation it is assumed that the adatom interacts with the substrate atoms through a pairwise additive potential $v(\mathbf{r})$. The subscript eff indicates that a Debye–Waller factor has been combined with the potential to an effective coupling term while α and β denote cartesian coordinates. In the presence of a fixed adatom the equilibrium positions of the substrate atoms are displaced and $\mathbf{u}_l$ are the displacements from these perturbed equilibrium positions, which are denoted by $\mathbf{R}_l = \mathbf{R}_l(\mathbf{R})$. The harmonic approximation for the lattice motion is assumed and only the one-phonon term has been retained. An estimate of the order of magnitude of this expression, valid when the adatom is located around its equilibrium position, leads to[69]

$$\eta \sim \frac{m}{m_s}\left(\frac{\omega_{\mathrm{A}}}{\omega_{\mathrm{D}}}\right)^4 \omega_{\mathrm{D}} \quad (16.5.17)$$

where m_s is the mass of the substrate atoms, ω_{A} is the vibrational frequency for the adatom, and ω_{D} is the Debye frequency of the substrate phonons.

Using the Fokker–Planck equation in this case is certainly highly questionable. Only when the adatom is much heavier than the substrate atoms is the friction description supposed to provide an accurate description of coupling to the phonons. However, actual calculations of the motion of an adsorbed atom show that the time-scale separation between the motion of the adatom and the substrate motion need not be large. If the Debye frequency is at least twice as large as the characteristic vibrational frequency for the adatom, the Fokker-Planck equation gives quite accurate results, provided one uses a proper position-dependent friction coefficient.[69]

16.5.3. Mode Coupling Approximation

For light adsorbates one must proceed beyond the Fokker-Planck approximation and treat the memory function in a more accurate way. The time extension of this function depends on how the excitations in the substrate propagate and decay, as well as on the motion of the adatom and the coupling in between. These effects are incorporated in an approximate way by the so-called mode coupling approximation, which takes into account explicitly the appropriate time scale of the fluctuating force through the inclusion of the dynamics in the surrounding. In the simplest approach the memory function is expressed as a product of $C^s(11't)$ and a part containing the substrate motion. The resulting equation of motion becomes nonlinear in $C^s(11't)$ and has to be solved self-consistently. The lack of translational symmetry makes the numerical work complicated and only studies where the adatom is restricted to move in one dimension have been performed.[21,70]

References

1. C. P. Flynn, *Point Defects and Diffusion*, Clarendon Press, Oxford (1972).
2. Y. Fukai and H. Sugimoto, Diffusion of hydrogen in metals, *Adv. Phys.* **34**, 263-326 (1985).
3. E. W. Müller and T. T. Tsong, *Field Ion Microscopy*, Elsevier, New York (1968).
4. G. Ehrlich, Wandering surface atoms and the field ion microscope, *Phys. Today* **6**, 44-53 (1981).
5. G. Ehrlich and K. Stolt, Surface diffusion, *Ann. Rev. Phys. Chem.* **31**, 603-637 (1980).
6. V. T. Bien (ed.), *Surface Mobilities on Solid Materials*, Plenum Press, New York (1983).
7. G. Ehrlich, Quantitative examination of individual atomic events on solids, *J. Vac. Sci. Technol.* **17**, 9-14 (1980).
8. G. Mazenko, J. R. Banavar, and R. Gomer, Diffusion coefficients and the time autocorrelation function of density fluctuations, *Surf. Sci.* **107**, 459-468 (1981).
9. K. W. Kehr, R. Kutner, and K. Binder, Diffusion in concentrated lattice gases. Self-diffusion of noninteracting particles in three-dimensional lattices, *Phys. Rev.* **B 23**, 4931-4945 (1981).
10. K. Sköld, in: *Hydrogen in Metals I, Basic Properties* (G. Alefeld and J. Völkl, eds.), pp. 267-287, Springer-Verlag, Berlin (1978).
11. F. Reif, *Fundamentals of Statistical and Thermal Physics*, McGraw-Hill, Tokyo (1965).
12. S. Chandrasekhar, Stochastic problems in physics and astronomy, *Rev. Mod. Phys.* **15**, 1-89 (1943).
13. O. Klein, Zur statistischen theorie der suspensionen und lösungen, *Ark. Mat. Astron. Fys.* **16** (5), 1-51 (1922).
14. H. A. Kramers, Brownian motion in a field of force and the diffusion model of chemical reactions, *Physica* **7**, 284-304 (1940).
15. H. Risken and H. D. Vollmer, Low friction nonlinear mobility for the diffusive motion in periodic potentials, *Phys. Lett.* **69A**, 387-389 (1979).
16. R. L. Stratonovich, *Topics in the Theory of Random Noise*, Gordon & Breach, New York (1967).
17. V. Ambegaokar and B. I. Halperin, Voltage due to thermal noise in the dc josephson effect, *Phys. Rev. Lett.* **22**, 1364-1366 (1969).
18. D. L. Weaver, Effective diffusion coefficient of a brownian particle in a periodic potential, *Physica* **98A**, 359-362 (1979).

19. J. R. Banavar, M. H. Cohen, and R. Gomer, A calculation of surface diffusion coefficients of adsorbates on the (110) plane of tungsten, *Surf. Sci.* **107**, 113–126 (1981).
20. G. Wahnström, Surface self-diffusion of hydrogen on a model potential: Quantum aspects and correlated jumps, *J. Chem. Phys.* **89**, 6996–7009 (1988).
21. G. Wahnström, Diffusion of an adsorbed particle: Theory and numerical results, *Surf. Sci.* **159**, 311–332 (1985).
22. P. G. Wolynes, Quantum theory of activated events in condensed phases, *Phys. Rev. Lett.* **47**, 968–971 (1981).
23. P. Hanggi, Escape from metastable state, *J. Stat. Phys.* **42**, 105–148 (1986).
24. A. J. Leggett, S. Chakravarty, A. T. Dorsey, Matthew P. A. Fisher, Anupam Garg, and W. Zwerger, Dynamics of dissipative two-state system, *Rev. Mod. Phys.* **59**, 1–85 (1987).
25. A. Messiah, *Quantum Mechanics, Vol.* 2, North-Holland, Amsterdam (1961).
26. J. C. Tully, Theories of the dynamics of inelastic and reactive processes at surfaces, *Ann. Rev. Phys. Chem.* **31**, 319–343 (1980).
27. P. Pechukas, in: *Dynamics of Molecular Collisions, Part B* (W. H. Miller, ed.), pp. 269–322, Plenum Press, New York (1976).
28. D. G. Truhlar, W. L. Hase, and J. T. Hynes, Current status of transition-state theory, *J. Phys. Chem.* **87**. 2664–2682 (1983).
29. H. Eyring, The activated complex in chemical reactions, *J. Chem. Phys.* **3**, 107–115 (1935).
30. S. Glasstone, K. J. Laidler, and H. Eyring, *The Theory of Rate Processes*, McGraw-Hill, New York (1941).
31. C. A. Wert and C. Zener, Interstitial atomic diffusion coefficients, *Phys. Rev.* **76**, 1169–1175 (1949).
32. G. H. Vineyard, Frequency factors and isotope effects in solid state rate processes, *J. Phys. Chem. Solids* **3**, 121–127 (1957).
33. A. F. Voter, A Monte Carlo method for determining free-energy differences and transition state theory rate constants, *J. Chem. Phys.* **82**, 1890–1899 (1985).
34. J. G. Lauderdale and D. G. Truhlar, Diffusion of hydrogen, deuterium, and tritium on the (100) plane of copper: Reaction-path formulation, variational transitions state theory, and tunneling calculations, *Surf. Sci.* **164**, 558–588 (1985).
35. J. G. Lauderdale and D. G. Truhlar, Embedded-cluster model for the effects of phonons on the hydrogen surface diffusion on copper, *J. Chem. Phys.* **84**, 1843–1849 (1986).
36. S. M. Valone, A. F. Voter, and J. D. Doll, The isotope and temperature dependence on self-diffusion for hydrogen, deuterium, and tritium on Cu(100) in the 100–1000 K range, *Surf. Sci.* **155**, 687–699 (1985).
37. S. M. Valone, A. F. Voter, and J. D. Doll, The influence of substrate motion on the self-diffusion of hydrogen and its isotopes on the copper (100) surface, *J. Chem. Phys.* **85**, 7480–7486 (1986).
38. R. Jaquet and W. H. Miller, Quantum mechanical rate constants via parth integrals: Diffusion of hydrogen atoms on a W(100) surface, *J. Phys. Chem.* **89**, 2139–2144 (1985).
39. D. L. Hill and J. A. Wheeler, Nuclear constitution and the interpretation of fisson phenomena, *Phys. Rev.* **89**, 1102–1121 (1953).
40. J. C. Keck, Variational theory of reaction rates, *Adv. Chem. Phys.* **13**, 85–121 (1967).
41. C. H. Bennett, in: *Diffusion in Solids: Recent Developments* (A. S. Nowick and J. J. Burton, eds.), pp. 73–113, Academic Press, New York (1975).
42. D. Chandler, Roles of classical dynamics and quantum dynamics on activated processes occurring in liquids, *J. Stat. Phys.* **42**, 49–67 (1986).
43. J. D. Doll and A. F. Voter, Recent developments in the theory of surface diffusion, *Ann. Rev. Phys. Chem.* **38**, 413–431 (1987).
44. T. Yamamoto, Quantum statistical mechanical theory of the rate of exchange chemical reactions in the gas phase, *J. Chem. Phys.* **33**, 281–289 (1960).
45. W. H. Miller, S. D. Schwartz, and J. W. Tromp, Quantum mechanical rate constants for bimolecular reactions, *J. Chem. Phys.* **79**, 4889–4898 (1983).
46. C. H. Bennett, in: *Algorithms for Chemical Computations* (R. E. Christoffersen, ed.), pp. 63–97, American Chemical Society, Washington D.C. (1977).

47. W. H. Miller, Importance of nonseparability in quantum mechanical transition-state theory, *Acc. Chem. Res.* **9**, 306-312 (1976).

48. D. Chandler, Statistical mechanics of isomerization dynamics in liquids and the transition state approximation, *J. Chem. Phys.* **68**, 2959-2970 (1978).

49. D. Forster, *Hydrodynamic Fluctuations, Broken Symmetry, and Correlation Functions*, W. A. Benjamin, Reading, Massachusetts (1975).

50. S. W. Lovesey, *Condensed Matter Physics: Dynamic Correlations*, The Benjamin/Cummings Publ., Menlo Park, California (1986).

51. G. Wahnström and H. Metiu, Numerical study of the correlation function expressions for the thermal rate coefficients in quantum systems, *J. Phys. Chem.* **92**, 3240-3252 (1988).

52. J. Costley and P. Pechukas, Short-time behavior of quantum correlation functions in rate theory, *Chem. Phys. Lett.* **83**, 139-144 (1981).

53. R. W. Zwanzig, Time-correlation functions and transport coefficients in statistical mechanics, *Annu. Rev. Phys. Chem.* **16**, 67-102 (1965).

54. A. F. Voter and J. D. Doll, Dynamical corrections to transition state theory for multistate systems: Surface self-diffusion in the rare-event regime, *J. Chem. Phys.* **82**, 80-92 (1985).

55. A. F. Voter, Classically exact overlayer dynamics: Diffusion of rhodium clusters on Rh(100), *Phys. Rev. B* **10**, 6819-6829 (1986).

56. B. J. Berne, in: *Statistical Mechanics, Part B: Time-Dependent Processes* (B. J. Berne, ed.), pp. 233-257, Plenum Press, New York (1977).

57. R. Zwanzig, in: *Lectures in Theoretical Physics, Vol. 3* (W. E. Brittin, B. W. Downs, and J. Downs, eds.), Interscience, New York (1961).

58. H. Mori, Transport, collective motion and Brownian motion, *Prog. Theor. Phys.* **33**, 423-455 (1965).

59. W. L. Schaich, Brownian motion model of surface chemical reactions. Derivation in the large mass limit, *J. Chem. Phys.* **60**, 1087-1093 (1974).

60. E. G. d'Agliano, P. Kumar, W. Schaich, and H. Suhl, Brownian motion model of the interactions between chemical species and metallic electrons: Bootstrap derivation and parameter evaluation, *Phys. Rev. B* **11**, 2122-2143 (1975).

61. A. Blandin, A. Nourtier, and D. W. Hone, Localized time-dependent perturbations in metals: Formalism and simple examples, *J. Phys. (Paris)* **37**, 369-378 (1976).

62. P. Hohenberg and W. Kohn, Inhomogeneous electron gas, *Phys. Rev.* **136**, B864-B871 (1964).

63. W. Kohn and L. J. Sham, Self-consistent equations including exchange and correlation effects, *Phys. Rev.* **140**, A1133-A1138 (1965).

64. B. Hellsing and M. Persson, Electronic damping of atomic and molecular vibrations at metal surfaces, *Phys. Scr.* **29**, 360-371 (1984).

65. A. L. Fetter and J. D. Walecka, *Quantum Theory of Many-Particle Systems*, McGraw-Hill, New York (1971).

66. M. Persson and B. Hellsing, Electronic damping of adsorbate vibrations on metal surfaces, *Phys. Rev. Lett.* **49**, 662-665 (1982).

67. V. P. Zhdanov, Diffusion of hydrogen in bulk or on surface of metal, *Surf. Sci.* **161**, L614-L620 (1985).

68. G. Wahnström, Role of phonons and electron-hole pairs in hydrogen diffusion on a corrugated metal surface, *Chem. Phys. Lett.* **163**, 401-406 (1989).

69. G. Wahnström, Diffusion of an adsorbed particle: Dependence on the mass, *Surf. Sci.* **164**, 449-463 (1985).

70. W. G. Kleppmann and R. Zeyher, Diffusion in a deformable lattice: Theory and numerical results, *Phys. Rev. B* **22**, 6044-6064 (1980).

Theory of Adsorption–Desorption Kinetics and Dynamics

S. Holloway

17.1. Introduction

The dynamics of gas–surface processes is one of the fastest developing areas within surface science. After the initial thrust to understand electronic, vibrational, and structural properties of clean and adsorbate covered surfaces, it is obvious that the next logical step is to understand the details of kinetic and dynamic phenomena. Following from the expertise developed in gas-phase studies (both experimental and theoretical), it is now possible to address a diverse range of fundamental problems which have considerable technological interest. The application of molecular beam scattering and laser diagnostic and pumping techniques to study the interaction of atoms and small molecules with well-characterized metal surfaces promises to revolutionize our understanding of the detailed dynamics involved in gas–surface interactions. For example, it is now possible to measure directly the vibrational and translational energy dependence of dissociative chemisorption at metal surfaces[1] and probe the internal energy distributions of scattered[2] or desorbed molecules.[3]

Figure 17.1 shows a breakdown of a stylized experimental configuration. A monoenergetic beam of molecules (here referred to as species A_2) is produced from a supersonic source. The expansion process used to generate the beam results in molecules having very low internal energies (≤ 10 K), a FWHM of speed distributions typically $\geq 10\%$ and variable translational energies in the range 0.05 to 10 eV. For studying state-to-state reaction cross sections, laser radiation may be employed to select the initial vibrational and rotational states and additionally to probe the scattered molecules. As in conventional studies, surface temperature,

S. Holloway • Surface Science Research Centre, University of Liverpool, Liverpool L69 3BX, England.

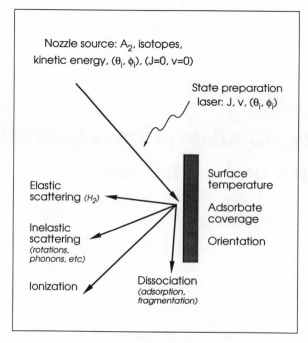

Figure 17.1. An idealized state-to-state gas–surface scattering experiment. Even if the initial beam were composed of delta-function probability distributions in each of the coordinates, the number of possible final states is still too large to contemplate an exact coupled channel solution similar to that encountered in gas-phase scattering.

crystallographic orientation, and adsorbate coverage may be controlled to provide a well-defined "target." As indicated in Figure 17.1, the number of possible "open channels" for the scattered particles is large. Broadly speaking the A_2 molecule can either be preserved intact or dissociate, with the products being trapped on the surface or scattered into the gas phase. Possible variations on this theme include inelastic scattering (vibrational, rotational excitation of the molecule; phonon, electron–hole pair excitations in the surface), ionization, and so on. For more details the reader is referred to the review articles.[4]

17.2. The Born–Oppenheimer Approximation and Potential-Energy Surfaces

Figure 17.2 presents a typical diagram which may be found in many elementary textbooks which deal with chemical rate processes. Before approaching the microscopic world of surface reaction dynamics, three questions concerning this diagram need answering: (1) what is the black bump and what are its origins? (2) what is a reaction coordinate?, and (3) why is the rate given by the Arrhenius expression?

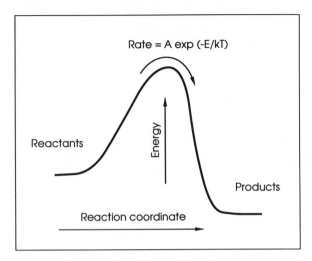

Figure 17.2. An infamous diagram showing the rate of reaction in a system having an activation barrier.

The black bump is a rudimentary potential-energy surface (PES) which contains an activation barrier separating reactants (e.g., A_2 in the gas phase) from products (e.g., A_2 or 2A adsorbed on a surface). Figure 17.3 shows a model PES for H_2 interacting with a Ni surface, which was first suggested by Lennard-Jones in 1932, and has since become the standard vehicle for discussing the dissociative

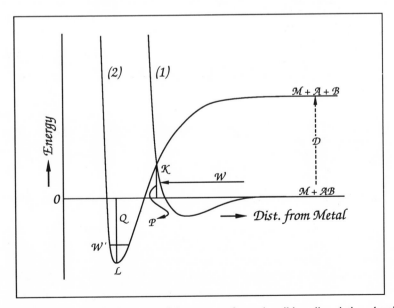

Figure 17.3. The Lennard-Jones potential-energy surfaces describing dissociative chemisorption,[5] originally captioned "The interaction of a molecule and a metal."

adsorption of molecules at metal surfaces.[5] In this PES, the activation barrier of magnitude P is shown at point K, D is the dissociation energy of molecular species AB, Q is the sum of chemisorption energies for species A and B, and L is the equilibrium chemisorption location. To obtain a "real" PES describing a reactive encounter of a molecule with a metal surface is a formidable task, but the solution of the Schroedinger equation is facilitated by the large disparity in mass of the electrons and nuclei.† As a molecule approaches a surface, we assume the coordinates of the electrons are given by $\mathbf{r} \equiv (\mathbf{r}_1, \mathbf{r}_2, \mathbf{r}_3, \ldots, \mathbf{r}_n)$ and the nuclei by $\mathbf{R} = (\mathbf{R}_1, \mathbf{R}_2, \mathbf{R}_3, \ldots, \mathbf{R}_N)$. The total Hamiltonian may be expressed as

$$\mathcal{H} = \mathcal{T}_{\mathbf{R}} + \mathcal{H}_0 \tag{17.2.1}$$

where

$$\mathcal{T}_{\mathbf{R}} = \sum_{M=1}^{N-1} -\left(\frac{\hbar^2}{2\mu_M}\right)\nabla_M^2 \tag{17.2.2}$$

is the nuclear kinetic energy operator and

$$\mathcal{H}_0 = \sum_{i=1}^{n} -\left(\frac{\hbar^2}{2m}\right)\nabla_i^2 + \sum_{i=1}^{n-1}\sum_{j>i}^{n} \frac{1}{|\mathbf{r}_i - \mathbf{r}_j|} - \sum_{i=1}^{n}\sum_{M=1}^{N} \frac{z_M}{|\mathbf{r}_i - \mathbf{R}_M|}$$

$$+ \sum_{M=1}^{N-1}\sum_{M'>M}^{N} \frac{z_M z_{M'}}{|\mathbf{R}_M - \mathbf{R}_{M'}|} \tag{17.2.3}$$

is the electronic Hamiltonian for *fixed* nuclear positions. By selecting a suitable basis set for the electronic wave functions, $\phi_k(\mathbf{r}; \mathbf{R})$, the total system wave function $\Psi(\mathbf{r}, \mathbf{R})$ may be expressed as the linear combination

$$\Psi(\mathbf{r}, \mathbf{R}) = \sum_k \phi_k(\mathbf{r}; \mathbf{R})\chi_k(\mathbf{R}) \tag{17.2.4}$$

The wave function $\chi_k(\mathbf{R})$ describes the motion of the nuclei on the PES associated with the electronic state k. On substituting equations (17.2.1)–(17.2.4) into the Schroedinger equation

$$[\mathcal{H} - \mathcal{E}]\Psi(\mathbf{r}, \mathbf{R}) = 0 \tag{17.2.5}$$

one obtains the infinite set of coupled equations

$$[\mathcal{T}_{\mathbf{R}} + U_{kk} - \mathcal{E}]\chi_k = -\sum_{k' \neq k} [\mathcal{T}'_{kk'} + U_{kk'}]\chi_{k'} \tag{17.2.6}$$

where

$$U_{kk'} = \langle \phi_k | \mathcal{H}_0 | \phi_{k'} \rangle \tag{17.2.7}$$

† The discussion here closely follows that in Tully.[6]

and

$$\mathscr{T}'_{kk'} = \sum_{M=1}^{N-1} -\left(\frac{\hbar^2}{2\mu_M}\right)\langle\phi_k|\nabla_M|\phi_{k'}\rangle \cdot \nabla_M \qquad (17.2.8)$$

It is the diagonal elements of $U_{kk'}$ *which are the potential-energy surfaces governing the nuclear motion in a particular electronic state,* k. The off-diagonal terms $\mathscr{T}'_{kk'}$ and $U_{kk'}$ give rise to transitions between the various electronic states and are usually referred to as nonadiabatic couplings. They depend critically upon the velocity of the nuclei as can be seen in equation (17.2.8).

Having arrived at the point where a set of PES have formally been defined, the problem now to be addressed is what choice of electronic basis is most applicable.

17.2.1. Adiabatic Representation

This is the most frequently used choice of representation in gas–surface problems and uses as basis states the eigenfunctions of $\mathscr{H}_0$

$$[\mathscr{H}_0 - V_k]\phi_k = 0 \qquad (17.2.9)$$

Then from equation (17.2.7) it is seen that the eigenvalues V_k themselves are the required PES. The off-diagonal elements $U_{kk'}$ are identically zero leaving only terms $\mathscr{T}'_{kk'}$ to couple motion between different electronic states. The states $\phi_k(\mathbf{r}; \mathbf{R})$ then satisfy the variational principle and may be calculated (sic) using, for example, density functional methods.[7]

17.2.2. Diabatic Representation

Electronic nonadiabaticity occurs mainly at regions in "$\mathbf{R}$-space" when potential-energy surfaces for different states k approach one another. From the point of view of bonding interactions, these regions correspond to a change in electronic configuration of the interacting system. To illustrate this, consider the interaction of an electronegative molecule with an electropositive surface. Figure 17.4 shows schematically a one-electron diagram and the accompanying PES for this problem. Essentially the bonding may be described by two simple diabatic states, covalent and ionic.[8,9] Far from the surface the affinity level on the adsorbate is above the Fermi level and the interaction may be thought of as being covalent in nature, $A_2/$Metal (here we ignore dispersion forces). Close to the surface, the affinity level is sufficiently lowered in energy (from image interactions) that it becomes occupied and the bonding then becomes ionic, $A_2^-/$Metal$^+$. In the region where the affinity and Fermi levels are degenerate, the two diabatic states cross and it is here that nonadiabatic effects are greatest. The adiabatic PES, which are suitable linear combinations of these diabatic states, in general will not cross[10] but will exhibit an "avoided crossing" as indicated in Figure 17.4. For molecular adsorbates there will be a continuous series of avoided crossings as, for example, the molecular

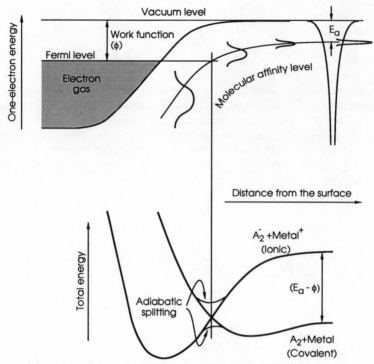

Figure 17.4. The one-electron picture and the corresponding total-energy curves for a diatomic molecule (A_2) approaching a metal surface. The molecule is taken to be of fixed orientaion and the abscissa measures the distance from a notional surface-plane. The A_2 diabatic[(8,9)] curve is for the electronic state where the affinity level is always empty. For large gas–surface separations this is a weakly attractive van der Waals interaction while, closer in, a strongly repulsive force is felt. The A_2^- curve is for the interaction of the molecular ion where the affinity level is always occupied. At large distances, an attractive image force is felt while past the curve-crossing point, a repulsion again obtains. The adiabatic potential curves have been indicated and constitute a "rounding off" of the diabatic curves near the crossing point.

bond is extended or the molecule rotates. The locus of these points gives rise to "crossing seams" in the PES, which will be discussed a greater length in the following sections.

Problematically, there exists no unique definition of the diabatic representation. One possibility of obtaining diabatic potentials is to simply extrapolate adiabatic potentials in the neighborhood of avoided crossings![(11)] Another possibility is to construct the basis functions in such a way as to minimize the $\mathcal{T}'_{kk'}$ terms, which has the consequence that the off-diagonal terms $U_{kk'}$ promote transitions between the various states.[(12)] The diabatic representation is of most use when fast-moving adsorbates have curve crossing points far from the surface (such as surface chemiluminescence),[(13)] since in this case the avoided crossing region remains well localized and the relative velocity is large [see equation (17.2.8)]. For the two-state system in Figure 17.4, an approximate criterion due to Massey[(9)] says that adiabatic behavior will predominate if $h\mathbf{V} \cdot \mathbf{d}_{kk'}/|V_k - V_{k'}| \ll 1$. Here $\mathbf{V}$ is the nuclear velocity, $\mathbf{d}_{kk'}$ $(=\langle \phi_k|\nabla_M|\phi_{k'}\rangle)$ is the nonadiabatic coupling, and V_k is the adiabatic PES defined in equation (17.2.9).

17.3. Desorption Rates and the Transition State Method

Having established the origin of the black line in Figure 17.2, it remains to determine an expression for the reaction rate. To make contact with the gas–surface problem, it is assumed that we are studying the desorption of a molecule, a topic to which we will return later. Then "reactants" will signify the chemisorbed species and "products" the bare surface plus desorbed molecule. The development follows that of transition state theory,[14] which takes as its starting point an equilibrium gas–surface system in which the probability of finding momenta $\{\mathbf{V}\}$ and positions $\{\mathbf{R}\}$ for the particles is

$$P(\mathbf{V}, \mathbf{R}) \sim \exp\left\{\frac{-[T_R(\mathbf{V}) + V(\mathbf{R})]}{kT_s}\right\} \tag{17.3.1}$$

If one now draws a line in phase space S at the top of the activation barrier, or more generally at the crossing seam (see Figure 17.5), separating "reactants" from "products," then at equilibrium the flux of particles crossing S in the forward and backward directions must be equal. The dividing line may be parameterized by an arc length s, with $\mathbf{R}_s$ a point on the curve having a local normal $\mathbf{n}_s$ pointing toward the products. The classical thermal rate constant is then proportional to the flux of trajectories passing through S from reactants to products,

$$k = Q^{-1} \int d\mathbf{V}\, ds\, (\mathbf{V} \cdot \mathbf{n}_s)\chi_r(\mathbf{V}, \mathbf{R}_s) \exp\left[\frac{-H(\mathbf{V}, \mathbf{R}_s)}{kT_s}\right] \tag{17.3.2}$$

where $H[= T_R(\mathbf{V}) + V(\mathbf{R})]$ is the Hamiltonian of the nuclear subsystem and Q is the partition function of the reactants; $\chi_r(\mathbf{V}, \mathbf{R}_s)$ is the characteristic function of reactive phase points and is unity if $(\mathbf{V}, \mathbf{R})$ lies on a reactive trajectory and zero otherwise. In transition state theory, the total rate constant k_{ts} is obtained by replacing $\chi_r(\mathbf{V}, \mathbf{R}_s)$ with $\chi_+(\mathbf{V}, \mathbf{R}_s, \mathbf{n}_s)$, which is defined by

$$\chi_+(\mathbf{V}, \mathbf{R}, \mathbf{n}) = \begin{cases} 1 & \text{if } \mathbf{V} \cdot \mathbf{n} > 0 \\ 0 & \text{otherwise} \end{cases} \tag{17.3.3}$$

Clearly k_{ts} will be larger than k because it includes (1) nonreacive trajectories which begin as reactants, cross S but then recross and end again as reactants (RPR), (2) trajectories that start and end as products, but cross S (PRP), and (3) true reactive trajectories with multiple crossings (either RP, ..., RP or PR, ..., PR). These trajectories are illustrated in Figure 17.6. Grimmelmann et al.[15] addressed this issue by writing the true rate constant as the product

$$k = k_{ts}F \tag{17.3.4}$$

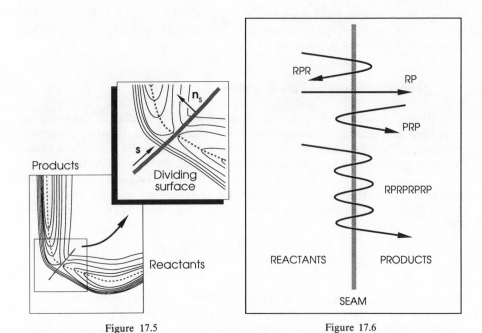

Figure 17.5   Figure 17.6

Figure 17.5. Schematic contour map for a reactive event showing equipotential (adiabatic) contours, a reaction path, and a dividing surface (seam). The reaction path is the line of steepest descent from the saddle point, which is orthogonal to the seam and may be obtained from the relationship $d\mathbf{R}/ds = \nabla V(\mathbf{R})/|\nabla V(\mathbf{R})|$ (after Pechukas[14]).

Figure 17.6. A set of representative trajectories near the seam in the PES for an equilibrium system.

where the factor F (<1) contains all of the recrossing dynamics in the problem. By choosing a dividing surface in the asymptotic region of the products, they invoked microscopic reversibility and equated F with the sticking probability. It has been shown[14] that by constraining S, one can obtain an upper bound to k_{ts}, a procedure often referred to as variational transition state theory.[16] If the transition state is located at a bottleneck which truly does not allow recrossings, the variational procedure yields an exact result and $k_{ts} = k$.

To make further progress, an explicit form for $V(\mathbf{R})$ is required. If the surface is modeled by a system of near-neighbor potentials, and the desorbing species interacts with a single site upon the surface via $V(\mathbf{R}_0 - \mathbf{R}_1)$, then it is possible to integrate equation (17.3.2) over the phonon states of the surface and derive a simple expression for the rate constant,[17]

$$k_{ts}^{-1} = \sqrt{\frac{M}{2\pi kT_s}} \frac{1}{r_{w-s}^2} \int d(\mathbf{R}_0 - \mathbf{R}_1)\left\{ \exp\left[\frac{-V(\mathbf{R}_0 - \mathbf{R}_1)}{kT_s} \right]^{-1} \right\} \quad (17.3.5)$$

Here M is the mass of the molecule and r_{w-s} is a surface Wigner–Seitz radius.

The desorption energy is approximately equal to the well depth and, for a "reasonable" parameter choice, this expression yields pre-exponents in the picosecond range.

17.4. The Interaction of H_2 with Cu Surfaces: A Brief Review

Rather than simply list scores of examples of results on a variety of gas–surface systems, we present a case study of H_2/Cu, which represents a prototypical system studied extensively both theoretically and experimentally.

17.4.1. Experimental

The notion of a kinetic activation barrier to H_2 adsorption on a Cu surface was first suggested to explain the absence of any chemisorption on polycrystalline films.[18] These results were at variance with volumetric measurements on reduced copper powders[19] where a $\Delta H_{ads} \sim 0.31$ eV heat of adsorption was found. It was not until a detailed kinetic analysis of the rate of parahydrogen conversion over Cu films was interpreted by Eley and Rossington,[20] using a similar value of ΔH_{ads}, that results from film and powders were reconciled. Alexander and Prichard[21] measured work function changes ($\Delta \phi$) for H_2 adsorption on Cu films in the surface temperature 242–337 K. With the assumption that adsorption was dissociative with coverage linear in $\Delta \phi$, a kinetic analysis based on Langmuir adsorption yielded an activation energy to adsorption of $E_{act} \sim 0.3$–0.4 eV.

Following the work of Van Willigen on transition metals,[22] Bradley and Stickney[23] measured the spatial distributions of H_2 molecules desorbing from polycrystalline Cu films using a permeation technique. They concluded that Cu was unlike the other surfaces studied (Fe, Pt, Nb, and stainless steel) since, even for a clean surface, the polar angle distribution did not tend toward a diffuse $\cos \theta$ distribution but, instead, was well represented by $\cos^n \theta$ with $n = 4$ (see Figure 17.7). This work was extended by Balooch and Stickney[24] who measured angular distributions from (100), (110), and (111) Cu single crystals. Their measurements yielded the expected anisotropy with $n = 5$, 2.5, and 6, respectively, but, rather surprisingly, no dependence on azimuthal angle was found for any of the surfaces. More recent measurements by Comsa and David[25] find a $\cos^8 \theta$ angular distribution for $D_2/Cu(100)$. In this work, the velocity distributions of H_2 and D_2 molecules desorbing from Cu(100) and Cu(111) surfaces were also obtained. In all cases, the measured time-of-flight spectra were extremely narrow with a mean energy of $\sim 8 kT_s$. This, again, was at variance with results for transition metal surfaces where H_2 velocity distributions obtained using the permeation method could be separated into "fast" and "slow" components.[26] For Cu surfaces the slow component was conspicuous only by its absence.

At this point it is useful to review the model of Van Willingen[22] for the associative *desorption* of H_2 from metal surfaces with an activation barrier to *adsorption*. His analysis for the angular distribution $N(\theta)$ of desorbed molecules

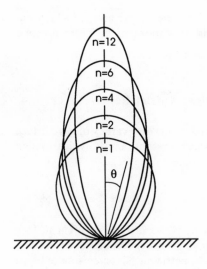

Figure 17.7. Angular distributions corresponding to the molecular desorption of H_2. While $n = 1$ corresponds to the Knudsen law $N(\theta) \propto \cos \theta$, the general function $\cos^n \theta$ has no theoretical basis but is a convenient way of representing experimental data.

was based upon the following assumptions: (1) the adsorbed and gas-phase molecules are in equilibrium with the surface, (2) the activation barrier is single valued over the entire surface with flat, parallel equipotential contours, (3) the rotational and vibrational molecular degrees of freedom are not altered by the PES, thus their respective distributions are thermal, at the surface temperature, and (4) incident molecules with a kinetic energy normal to the equipotentials in excess of the barrier energy have a unity sticking coefficient. Points (1) and (4) enable detailed balance arguments to be used in formulating $N(\theta)$, while (2) reduces the problem, effectively, to one dimension. Applying equilibrium statistical mechanics, Van Willigen obtained the following result:

$$N(\theta) = N(0)\left(\frac{E_{act} + kT_s \cos^2 \theta}{(E_{act} + kT_s) \cos \theta}\right) \exp\left(\frac{-E_{act}}{kT_s} \tan^2 \theta\right) \qquad (17.4.1)$$

Using the same model, Comsa and David[27] showed that the angular dependence of the mean energy of the desorbing molecules is

$$\langle E(\theta)\rangle = 2kT_s\left[1 + \frac{1}{2}\left(\frac{E_{act}^2}{(E_{act} + kT_s \cos^2 \theta)kT_s \cos^2 \theta}\right)\right] \qquad (17.4.2)$$

In the limit $E_{act} \ll kT_s$, equations (17.2.1) and (17.2.2) reduce to the familiar Knudsen laws, $N(\theta) \sim \cos \theta$ and $\langle E(\theta)\rangle = 2kT_s$. Although Van Willigen did not treat intramolecular degrees of freedom, his hypothesis of thermal equilibrium of *molecules* prior to desorption implies that the distributions of vibrations and rotations are those of a Boltzmann gas at the surface temperature,

$$N_J = N_0(2J + 1) \exp\left(-E_J/kT_s\right) \qquad (17.4.3)$$

and

$$N_n = N_0 \exp\left(-E_n/kT_s\right) \qquad (17.4.4)$$

where E_J and E_n are the rotational and vibrational energy levels. Hypothesis (3) ensures that the ro-vibrational distributions are unaffected by the presence or absence of a barrier. Using their measured angular distributions and equation (17.4.1), Balooch and Stickney[24] derived the following values for (one-dimensional) barriers: (100), 0.13 eV; (110), 0.08 eV; and (111), 0.26 eV. Comsa and David[15] approached the problem via equation (17.4.2) and obtained a value of $E_{act} \sim 0.56$ eV for $D_2/Cu(100)$ from their measured ($\theta = 0$) velocity distribution. When converted [via equation (17.4.1)] into an angular distribution, this barrier should result in $N(\theta) \sim \cos^{13}\theta$, which was significantly sharper than their measured $\cos^8\theta$. However, when compared to deviations between the measured and predicted θ-dependence of $\langle E(\theta) \rangle$, this difference was of minor consequence. Comsa and David[25] observed an *angular independent* $\langle E(\theta) \rangle$ which bears no resemblance to equation (17.4.2) and therefore casts doubt on Van Willigen's model, which predicts a steep increase of mean translational energy with θ.

Additional data pertaining to the H_2/Cu system were obtained by Balooch *et al.*[28] in a detailed set of scattering experiments. They measured the rate of HD production when H_2 was scattered from $Cu(100)$, (110), and (310) surfaces, precovered with dissociated D_2. By assuming that H_2 dissociation is rate limiting,[29] the overall HD reaction probability was taken as a measure of the H_2 sticking coefficient S_0. By varying the primary H_2 beam energy (E_i) and polar angle (θ_i) it was possible to probe the activation barrier from the gas phase. The value of S_0 varied in a sigmoidal fashion according to the normal projection of the primary beam energy ($E_\perp = E_i \cos^2\theta_i$), with $S_0 \sim 0.02$–03 for $E_\perp < 0.13$ eV increasing to $S_0 \sim 0.10$–15 for $E_\perp > 0.30$ eV, results depending upon the particular crystallographic face studied (Figure 17.8). Again these results are in strong disagreement with the Van Willigen model, which would predict a step-function jump in S_0 at E_{act}. Angular resolved thermal desorption measurements for HD performed in this study confirmed their earlier permeation results with $N(\theta) \sim \cos^{2.5}\theta$ for $Cu(110)$ and $\sim\cos^{5.5}\theta$ for both the $Cu(100)$ and (310) faces. These results were used by Cardillo, Balooch, and Stickney[30] to test the principle of detailed balance for surface reactions. By performing "quasi-equilibrium averages" over the measured $S_0(\theta_i, E_i)$, they were able to predict their measured angular distribution of desorbed molecules, $N(\theta)$. However, using this 0.13–22 eV barrier in equation (17.4.2), Comsa and David[26] pointed out that a derived $\langle E(\theta) \rangle \sim 1800$ K, at a

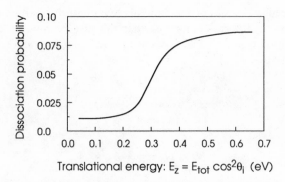

Figure 17.8. The experimental dissociation probability for $H_2/Cu(100)$ as a function of the initial kinetic energy normal to the surface (after Balooch *et al.*[28]).

surface temperature of 1000 K, is in strong disagreement with their measured 3950 K. To explain this discrepancy between their permeation data and the scattering results of Balooch *et al.*,[28] Comsa and David[26] invoked a PES having two atomic traps in the selvedge region, a chemisorption well, and a high-lying subsurface state, separated by a barrier $E_{abs} \sim 0.59$ eV. They postulated that diffusing atoms from the bulk do not equilibrate in the chemisorption well but migrate across the surface, find a partner, either equilibrated or not, and desorb as a molecule. Since the nascent molecules will have an energy so far in excess of E_{act}, they will not perceive its presence and appear at the detector with the observed $\langle E(\theta) \rangle \sim 3950$ K.

Although Balooch *et al.* noted that the majority of H_2 molecules were in fact *scattered* from the Cu surface, Lapujoulade and co-workers[31] were the first to present detailed information regarding this fraction for a Cu(110) surface. In this work diffraction data for H_2 beam energies $76 \le E_\perp \le 218$ meV were presented. This is a particularly interesting range, since at the higher end it should probe regions of the surface where dissociative chemisorption occurs. The data were analyzed using a "Morse corrugated potential" representation of the PES and the findings may be summarized: (1) at low energies a relatively large value of the surface corrugation (a sixfold increase by comparison to He scattering from the same surface) was obtained, and (2) although impossible to measure, significant out-of-plane scattering was inferred from the high-energy data where no satisfactory model for the surface corrugation function could be found.

Kubiak, Sitz, and Zare[3] have measured the ro-vibrational state distribution of desorbing H_2 and D_2 from Cu(110) and (111) using the permeation technique in conjunction with REMPI detection of the molecules. They found that the rotational distributions were not well described by equation (17.4.3) and that mean rotational energies were only 80–90% of kT_s. The vibrational distributions were also highly non-Boltzmann with N_1/N_0 ratios ~50–100 times greater than those expected from equation (17.4.4). While these results implicitly condemn the Van Willigen model, the authors went beyond this and cast doubts upon the detailed balance arguments of Cardillo *et al.*[30] by reanalyzing their scattering data in the light of the observed vibrational enhancement in desorption.[3] Within a detailed balance framework, these data imply that $S_0(n = 1) \sim 50$ times $S_0(n = 0)$, a result which was not inherent in any of the results of Balooch et al.[28] On this basis, Kubiak et al. concluded that the translational and vibrational dynamics in the scattering and permeation experiments are dissimilar, with results from the latter technique being consistently in excess of those obtained using the former.

In conclusion, much data have been accumulated on this prototypical "barrier" system. It remains to question how much of the data can be understood with conventional models for the dynamics. The "normal energy scaling" results obtained by Balooch et al. for S_0 coupled with the angular desorption distributions would appear to indicate that a one-dimensional model similar to Figure 17.3 may suffice. Unfortunately, the diffraction data and the measured internal energy distributions are in conflict with this Van Willigen model and suggest that the surface presents a range of activation barriers to the incident molecule and, further, that the transition state geometry does not correspond to that of the gas-phase species.

17.4.2. Theoretical

Adopting the terminology of Lennard-Jones (Figure 17.3), the interaction of H_2 with a Cu surface can be described by the mixing of the two diabatic states, H_2/Cu and $2H/Cu$.[5]

The physisorption state is described by the diabatic state H_2/Cu and is itself composed of two parts.[32] For large separations there is no overlap of the molecule electrons with those of the surface and a van der Waals attraction dominates. Specifically, taking R_1 and R_2 to be the position of the H nuclei and defining R ($= X, Y, Z$) to be the location of the bond centre with molecular orientation (θ, ϕ), then the van der Waals potential is

$$V_{vw}(Z, \theta) = -\frac{C_{vw}[1 + 0.05 P_2(\cos \theta)]}{(Z - Z_{vw})^3} f[k_c(Z - Z_{vw})] \qquad (17.4.5)$$

where C_{vw} and Z_{vw} are the van der Waals constant and origin, $P_2(\cos \theta)$ is a Legendre polynomial, and $f(x)$ is a function which accounts for the saturation of the interaction.[32] The bracketed term in the numerator describes the (weak) variation of $V_{vw}(Z, \theta)$ with molecular orientation. Closer to the surface the electronic overlap increases and a strong repulsive interaction is experienced which arises from the cost in orthogonalizing the surface "band" states to the $H_2(1\sigma_g)$ molecular orbital. This interaction, which is approximately proportional to the charge density of the surface electrons at the molecular bond center $\rho(R)$,[33] may be modeled by

$$V_{rep}(R, \theta) = V_0 e^{-\alpha Z}[1 + \alpha h_{rep}(R)][1 + 0.18 P_2(\cos \theta)] \qquad (17.4.6)$$

where

$$h_{\text{rep}}(\mathbf{R}) = \tfrac{1}{2}e^{-\beta(Z-Z_0)}\left[h_X\left(1 + \cos\frac{2\pi X}{a_X}\right) + h_\gamma\left(1 + \cos\frac{2\pi Y}{a_Y}\right)\right] \quad (17.4.7)$$

The terms h_X and h_Y describe the lateral corrugation of the potential, which is considered to decay exponentially with increasing distance from the surface; a_X and a_Y are the unit cell lengths in the X and Y directions. Values of the various parameters occurring in equations (17.4.5)–(17.4.7) have been discussed elsewhere.[34] The total molecular potential is then given by the sum, $V_{H_2}(\mathbf{R}, \theta) = V_{\text{vw}}(Z, \theta) + V_{\text{rep}}(\mathbf{R}, \theta)$, which for Cu results in a physisorption well of depth ~ 20 meV.

The second diabatic state is 2H/Cu and corresponds to the chemisorption interaction. There exist many electronic structure calculations for the interaction of an isolated H atom with a surface. The effective medium approximation[35,36] affords an intuitively appealing description of the binding of atomic adsorbates to metal surfaces. This scheme, based upon the density functional scheme for solving equation (17.2.9), in its most primitive form equates the binding energy of an adsorbate at some position $V_H(\mathbf{R}_1)$, with the embedding energy of that species in a *homogeneous* electron gas of density equal to $\rho(\mathbf{R}_1)$. This describes the binding of the adsorbate with the s–p electrons in the surface and additional terms are required to describe the (covalent) interaction with the d electrons.[37] The electronic structure of the adsorbate consists of a doubly occupied H(1s) state whose energetic position is approximately given by the *local* value of the effective potential, upon which is *superimposed* a screening charge $\Delta\rho_{\text{ads}}$. The net effect is to render the adsorbed species as an almost neutral particle with a total charge density not terribly dissimilar from the isolated atom![38] The application of this method to treat hydrogen interacting with a wide range of metals has been successful in reproducing experimental chemisorption energies E_{chem}, vibrational frequencies $\hbar\omega$, diffusion barriers E_{diff}, and absorption energies.[39] Typically for a Cu surface $E_{\text{chem}} \sim -2.45$ eV, $\hbar\omega \sim 80$ meV, and $E_{\text{diff}} \sim 150$ meV, with the most favorable adsorption location usually corresponding to the highest coordinated site on the particular crystal face in question. With use of the effective medium approximation it is possible to estimate the additional interaction which arises when two H atoms approach one another on the surface $V_{2H}(\mathbf{R}_1, \mathbf{R}_2)$.[40] Since the energies are so large for the chemisorbed species, $V_{2H}(\mathbf{R}_1, \mathbf{R}_2)$ will show a strong dependence upon the bond orientation (θ, ϕ), a feature absent from $V_{H_2}(\mathbf{R}, \theta)$.

To complete the story of the H_2/Cu PES, the interaction between the two diabatic states is required, which involves following the detailed electronic configuration switch from H_2/Cu to 2H/Cu. This has been discussed at length by Harris and co-workers in a series of papers where the adiabatic PES for the reaction $H_2 + Cu_2 \to 2$ HCu has been obtained using Kohn–Sham local density calculations.[41] On the basis of the above discussion, they key question concerning the mixing of the diabatic states is whether the curves cross with positive or negative energy, since this determines the sign of E_{act}. This is determined by the competition between an energy lowering, arising from the filling of the $H_2(1\sigma_u)$ antibonding

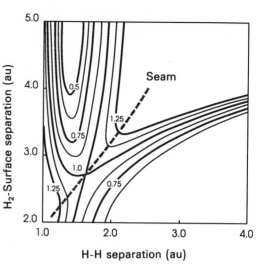

Figure 17.9. The potential-energy surface for the interaction of H_2 approaching the bridge site of Cu(100). This surface is adapted from a cluster calculation for H_2/Cu_2.[32] There is a clear bottleneck occurring which inhibits both the desorption and absorption of H_2 molecules. The calculations in Sections 17.5 and 17.6 are based upon this surface.

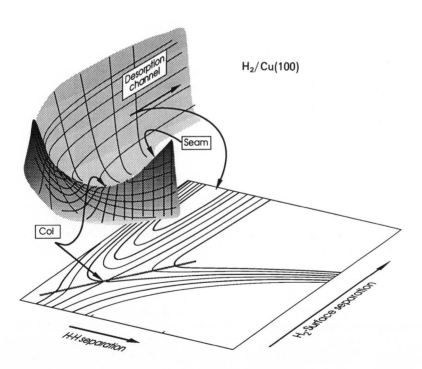

Figure 17.10. Three-dimensional plot and two-dimensional projection of model potential in Figure 17.9. The figure displays the "col" or saddle point in the PES and the "seam" that extends to higher energies on either side.

resonance,[42] and the kinetic energy cost in orthogonalizing the surface electronic states to the $H_2(1\sigma_g)$ "core." The PES shown in Figure 17.9 is based upon the calculation by Harris for a H_2-Cu_2 cluster having a planar C_{2v} geometry. The energy diagram separates naturally into product and reactant regions, separated by a seam on either side of which the energy falls off rapidly. In simple terms a barrier to H_2 dissociation arises because of the relative inertness of the Cu surface arising due to the complete occupation of the d states. Going left in the transition metals series, holes appear in the d band and it is found that this barrier disappears. For a more detailed discussion, the interested reader is referred to Harris.[41] The perspective plot (Figure 17.10) illustrates the topology of the H_2/Cu PES in the region near the activation barrier.

In conclusion it is probably fair to say that from the experimental results it is clear that barriers to H_2 dissociative chemisorption exist on Cu surfaces, the magnitude and location of which are debatable. Theoretical calculations have provided a detailed understanding of hydrogen adsorption and also predict the occurrence of an activation barrier to dissociative adsorption with $E_{act} \sim 1$ eV.

17.5. The Associative Desorption of H_2/Cu: A Classical Picture

In Section 17.3 it was shown that if the transition state is located at a bottleneck in the PES which restricts the number of recrossings, then the exact rate approaches that obtained from transition state theory. This situation is well met by the PES shown in Figure 17.9 for H_2/Cu.[43] The model we propose takes advantage of this feature and is tailored specifically to the associative desorption of light atoms that are assumed to obey classical mechanics. When the adsorbate is light, the excursions of the substrate atoms from their equilibrium positions are relatively small and can, as a first approximation, be ignored. The desorption is then governed by a six-dimensional PES, $V(\mathbf{R}_1, \mathbf{R}_2)$ with the substrate serving as a reservoir that establishes a thermal distribution for the reactants. Specifically, within the reactant region the probability of positions $\mathbf{R}_1$, $\mathbf{R}_2$ and velocities $\mathbf{V}_1$, $\mathbf{V}_2$ for a pair of adsorbed atoms is given by

$$P(\mathbf{R}_1, \mathbf{R}_2, \mathbf{V}_1, \mathbf{V}_2) \sim \exp\left[\frac{-(V\{\mathbf{R}_1, \mathbf{R}_2\} + \varepsilon\{\mathbf{V}_1, \mathbf{V}_2\})}{kT_s}\right] \qquad (17.5.1)$$

where ε is the total kinetic energy. It is assumed that the surface density is sufficiently low that multiatom correlations are unimportant. In spite of the restriction to classical mechanics, we have in mind H_2 desorption as prevailing in a typical permeation experiment. We will argue that some average quantities should be given reasonably by our approach even though, in view of the large vibrational quantum of H_2, the detailed distributions we obtain will not be correct in this case.†

---

† This assumption has in fact been substantiated quantitatively for the case of associative H_2 desorption.[44]

For a thermal distribution of reactants and products, equation (17.5.1) is of course exact. If we now imagine drawing a line in phase space between reactant and product regions such that the overwhelming number of trajectories that cross this line do so only once, we can then divide trajectories into four categories PP, RR, RP, and PR, where P and R stand for product and reactant respectively (see Figure 17.6). PP and RR trajectories correspond to scattering events and involve no crossing of the dividing line. RP trajectories give rise to associative desorption and PR trajectories to dissociative sticking. At the seam in the PES, only these two types of trajectories contribute to the distribution (17.5.1). The distribution of reactant trajectories at the seam that is relevant for desorption from a thermal ensemble *with the products absent* is therefore given by equation (17.5.1) with the proviso that the velocity vector normal to the seam points outward. Energy distributions for the products can then be obtained by starting trajectories at the seam with a weighting given by equation (17.5.1) and integrating out into the product region until the interaction potential attains its asymptotic form. We note that this relies crucially on the assumption that only very few multiple crossings of the seam occur, i.e., the seam, if it exists at all, is a special dividing line and cannot be placed arbitrarily. If a nonnegligible number of recrossings occurs, the division of trajectories into the four categories mentioned above is a complex affair and the distribution of RP trajectories is not given by equation (17.3.2).[15,45]

In order to restrict the number of trajectories to be run to a manageable value, we adopted a reduced dimensionality and used a PES which depends only on the coordinates Z and $D = X_2 - X_1$, with $Y_2 = Y_1$ and $Z_2 = Z_1$. Two "sleeping" coordinates corresponding to parallel translations and rotations were added to allow a rough determination of the angular and rotational distributions. The energy in these coordinates at the seam was preserved along the trajectory, i.e., conversion in and out of these coordinates in the exit channel was ignored. However, this energy was included in the weighting of the trajectories. Full account was taken of energy conversion between vibrational and normal-translational coordinates Z and D. The PES used in the calculations is depicted in Figure 17.9 and was constructed as follows. We assume V_{H_2} and V_{2H} denote the electronic energies of a proton pair in the "product" and "reactant" states. Quantity V_{H_2} is given by equation (17.4.6) with the assumption that the surface is uncorrugated. For V_{2H} we used a sum of Morse potentials. The ground-state adiabatic interaction potential $V(Z, D)$ was then constructed from V_{H_2} and V_{2H} via the usual Landau–Zener prescription,[11]

$$V(Z, d) = \frac{V_{H_2} + V_{2H} - \sqrt{(V_{H_2} - V_{2H})^2 + \xi^2}}{2} \qquad (17.5.2)$$

where ξ is an "interaction energy" that smooths the transition between the entrance channel, where $V_{2H} > V_{H_2}$, and the chemisorption region, where $V_{H_2} > V_{2H}$. ($\xi = 0$ would correspond to a sharp cusp or seam in the energy surface.) The value of ξ and the parameters of the Morse potentials were chosen so that the resulting energy contours resembled those calculated for Cu_2H_2 [41] with respect to the form of the seam and spatial location of the saddle point, and to give a saddle-point energy

E_{act} of about 1 eV. This figure was chosen as a compromise between the actual barrier heights calculated using a small cluster and jellium as a model for the substrate ($E_{act} \sim 1.3$ and 0.7 eV, respectively[41,46]). Each trajectory was integrated until sufficiently far within the product region that no further interchange of energy between product degrees of freedom occurred. The final translational, vibrational, and rotational energies of the trajectory were then calculated and binned with weighting

$$W_j^i = \mathbf{V}_j \cdot \mathbf{n}_s^i \exp\left[\frac{-H(\mathbf{V}_j, \mathbf{R}_i)}{kT_s}\right] \qquad (17.5.3)$$

to determine the distributions. Average values were calculated via

$$\langle E \rangle \equiv \sum_{i,j} E_j^i W_j^i \Big/ \sum_{i,j} W_j^i \qquad (17.5.4)$$

The weighting factor W_j^i is proportional to the number of trajectories crossing a seam element at s_i per unit time, per unit velocity element, and its sum over i and j to the number of molecules per second emitted by the surface. The distributions we give refer therefore to the total *number* of molecules emerging from the surface in a given time interval and *not* to the current of molecules across a plane above, and parallel to the surface. Trajectories that failed to emerge in the exit channel, but instead recrossed the seam into the chemisorption region, were abandoned. The number of such trajectories encountered was only a small fraction of the total. The results quoted below were obtained using 23 points evenly spaced along the seam about the saddle point and 25 values of each velocity coordinate. All trajectories were included whose total energies ε lay within the range $E_{act} \leq \varepsilon \leq (E_{act} + 7kT_s)$. Including "sleeping coordinates" the total number of trajectories involved was of order 10^6. This was sufficient to converge all average energies and yield reasonably smooth distributions.

Figure 17.11 shows the distributions of translational, vibrational, rotational, and total energy for H_2 desorption at a surface temperature of 850 K. The total energy distribution is zero below the barrier energy, $E_{act} \sim 1$ eV, and has average value 1.2 eV (corresponding approximately to $E_{act} + \frac{5}{2}kT_s$). Of the three contributions to this energy, the rotational component is thermal with an average value of 35 meV (about $\frac{1}{2}kT_s$) while the other two components show a dynamical enhancement due to the PES. This is most pronounced in the translational energy, with average value 0.99 eV, reflecting the major part of the barrier energy. The remainder appears in the vibrational energy, whose average value 0.18 eV exceeds by a factor of 5 the $\sim\frac{1}{2}kT_s$ thermal component. The origin of this enhancement, as mentioned earlier, is the distension of the molecular bond at the saddle point of the PES, where the proton separation is $D^* = 1.63$ au (Figure 17.9) compared with the molecular equilibrium separation of $D_{eq} = 1.41$ au. In the free molecule, this distension corresponds to a potential energy of 0.21 eV. At a surface temperature of 850 K, about 40% of this is converted by the PES into translational energy as the molecules exit through the surface potential. Since the parallel component of

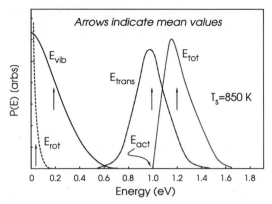

Figure 17.11. Distributions of the final-state energies of "classical H_2" formed by desorption from a thermal distribution of protons subject to the PES in Figure 17.9. The dashed, full, dot–dashed, and dotted lines refer respectively to rotational, vibrational, translational, and total energy. The average values are marked with arrows.

the kinetic energy is thermal ($\frac{1}{2}kT_s$) while the normal kinetic energy is enhanced by the barrier, the angular distribution is strongly peaked about the normal, with average exit angle of order $\sqrt{kT_s/2E_{act}} \sim 10°$. Our calculated value was 8.6°.

In Figure 17.12 we show the average vibrational and rotational energies, $\langle E_{vib} \rangle$ and $\langle E_{rot} \rangle$, of desorbing molecules as a function of the surface temperature. As mentioned above, the only source of rotational energy is the finite velocities possessed by the protons at the seam. In the absence of rot/vib or rot/trans couplings $\langle E_{rot} \rangle$ is the average energy of a Boltzmann distribution at the surface temperature, $\frac{1}{2}kT_s$. The same would be true for the vibrational energy if the influence of the PES were neglected. In this case, however, even after vib/trans couplings

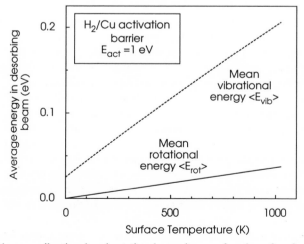

Figure 17.12. Average vibrational and rotational energies as a function of surface temperature.

in the exit channel have been accounted for, the PES gives rise to a strong enhancement, as is evident in Figure 17.12. The value of $\langle E_{vib} \rangle$ remains finite as T_s goes to zero (though, of course, the total current of molecules does go to zero with T_s) and increases markedly as T_s increases. This strong increase is due to the effect of the starting velocities on the efficiency of vib/trans conversion in the exit channel. At very low T_s the only source of proton pairs with substantial thermal weight is the region of the PES about the saddle point. The pairs start off with almost zero velocity and, in the presence of a strong friction which keeps this velocity low, would follow the potential out into the gas phase and emerge as a cold molecule with proton separation D_{eq}. In view of the adsorbate/substrate mass ratio, however, friction is extremely inefficient in H_2 desorption and the motion of the system should be close to mechanical. As the proton pair emerges over the col, therefore, it accelerates and the trajectory begins to depart from the ideal "classical path" and "overshoot."† The breakdown of the saddle-point energy into translational and vibrational components in the exit channel depends on the degree of "overshooting" during passage through the near-surface region. As Figure 17.12 shows, vib/trans conversion is very efficient at low T_s and only about 13% of the latent vibrational energy at the saddle point survives in the gas-phase product. As T_s increases, however, the enhanced starting velocity carries the proton pair away from the neighborhood of the saddle point, where conversion is strong, and the efficiency of the vib/trans conversion is reduced. In addition, the higher T_s, the larger the thermal weight for crossings high on the seam (Figure 17.9), where the proton separation is larger than D^*, the acceleration away from the seam is most rapid, and the time available for vib/trans conversion short.

A consequence of the decreasing efficiency of vib/trans conversion with increasing surface temperature is an extremely weak dependence of the average translational energy on T_s. This was found to increase by only 10 meV over the entire temperature range studied (50–1000 K). Since the parallel component is $\frac{1}{2}kT_s$ and increases by 40 meV over this range, it follows that the normal translational energy is actually a *decreasing* function of surface temperature.

In conclusion, the enhancement of the translational energy devolves from the entrance channel activation barrier for the inverse process of dissociative adsorption and has the same origin as in the Van Willigen model.[22] The enhancement of the vibrational energy, not given in the Van Willigen model, devolves from the distension of the interproton separation at and near the saddle point in the PES. The crucial distinction between our model and that of Van Willigen is a more detailed treatment of the manner in which the molecular bond forms. The desorption is viewed as arising from a thermal ensemble at the surface temperature, but this refers not to molecules that have already formed and are prevented from escaping by the barrier, but to *the prechemisorbed constituents from which the molecules are formed.* As the constituents merge and form the molecular bond, a considerable amount of energy is "stored" in the bond coordinate and part of this is retained in the gas phase, where it appears as vibrational excitation energy of the molecules.

---

† This effect is well known in bimolecular gas-phase scattering and is often referred to as the "classical bobsleigh effect." A discussion can be found elsewhere.[47]

17.6. The Dissociative Adsorption of H_2/Cu: A Quantum Wave-Packet Study

To follow the H_2/Cu story further, let us run the clock backward and study the dissociation probability of a beam of molecules approaching the surface. To make contact with the previous section, it makes sense to employ the same PES as used in Section 17.5, shown in Figure 17.9. Although the relatively large value of E_{act} (~1 eV) appears to conflict with the results of Balooch et al.[28] (Figure 17.8), the exercise is still useful in that we will compare the results of classical calculations with fully self-consistent quantum-mechanical scattering simulations.

To calculate the time evolution of the nonstationary wave function, the split operator method of Fleck, Morris, and Feit has been employed.[48] The application of this method to surface scattering problems has been extensively discussed elsewhere.[49,34] The PES was placed on a grid of 128 points in the Z direction and 64 points in the D direction and the asymptotic initial-state wave function was taken to be a product of a Morse oscillator in vibrational state v and a Gaussian function about a mean initial momentum p_Z,

$$\psi_{initial}(D, Z; t = 0) = \phi_v(D)g(Z - Z_i, p_Z) \tag{17.6.1}$$

In general, the incremental time development of the nonstationary wave function at position $\mathbf{r}$ and time $t + \Delta t$ is given by

$$\psi(\mathbf{R}, t + \Delta t) = \exp\{-i[\mathcal{T}_R + V(\mathbf{R})]\Delta t\}\psi(\mathbf{R}, t) \tag{17.6.2}$$

The time development over an interval Δt is achieved by symmetrically expanding the operator,

$$\psi(\mathbf{R}, t + \Delta t) = \exp(-i\mathcal{T}_R \Delta t/2)$$
$$\times \exp[-iV(\mathbf{R})\Delta t/2]\exp(-i\mathcal{T}_R\Delta t/2)\psi(\mathbf{R}, t) + O[(\Delta t)^3] \tag{17.6.3}$$

Thus one cycle involves a half time-step for a free particle, a full time-step phase change arising from the potential, and a further half-step free-space propagation. The free-particle time development is achieved with the aid of the band-limited Fourier-series representation of the wave function,

$$\psi(\mathbf{R}, t) = \sum_{l=-L/2}^{L/2-1} \sum_{m=-M/2}^{M/2-1} \sum_{n=-N/2}^{N/2-1} \psi_{lmn}(\mathbf{R}, t) \exp\left[2\pi i\left(\frac{l_x}{a_x} + \frac{m_y}{a_y} + \frac{n_z}{a_z}\right)\right] \tag{17.6.4}$$

The momentum-space time development is given by

$$\psi_{lmn}(t + \Delta t) = \psi_{lmn}(t) \exp\left\{-2\pi^2 i\Delta t\left[\left(\frac{l_x}{a_x}\right)^2 + \left(\frac{m_y}{a_y}\right)^2 + \left(\frac{n_z}{a_x}\right)^2\right]\right\} \tag{17.6.5}$$

It should be noted that since this operator is a constant for each k-space mesh point, the matrix need only be evaluated once and then stored, a procedure which results in a considerable saving of time. The free-space step is completed by back-transforming to give $\psi(\mathbf{R}, t + \Delta t)$. The potential-induced phase change is then obtained by multiplying each matrix element by the value $e^{-i\Delta t V(\mathbf{R})}$, for the appropriate mesh point. Again, since for our case $V(\mathbf{R})$ is time independent, this matrix need be evaluated only once and then stored. While in principle a series of half time-step propagations are necessary, in practice this is only the case for the first and last steps since in sequential application of equation (17.6.3), the half-steps combine, resulting in alternating free propagation with a potential-induced phase change. After a time t, when an asymptotic state has been reached, the reflection probability (and thereby S_0) is calculated by integrating the reflected portion of the packet $\psi_{\text{scatt}}(D, Z; t)$. Transition probabilities to excited vibrational states of the molecule may then be obtained by projecting $\psi_{\text{scatt}}(D, Z; t)$ out onto the complete set of $\phi_v(D)$. Figure 17.13 shows the time development during a typical scattering event for ground state H_2 with $E_\perp = 0.8$ eV.

The calculated dissociation probabilities for H_2 and D_2 molecules as a function of initial translational energy, $E_\perp$, for $v = 0$ and 1 are plotted in Figure 17.14a. For $E_{\text{tot}} (= E_\perp + E_v) < E_{\text{act}}$, the functional dependence on $E_\perp$ arises from tunneling

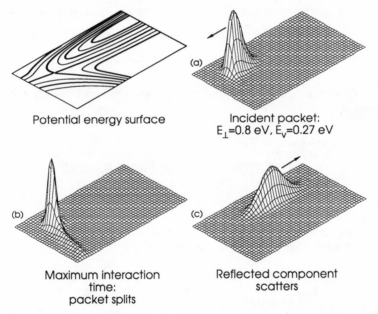

Potential energy surface

Incident packet:
$E_\perp$=0.8 eV, E_v=0.27 eV

Maximum interaction
time:
packet splits

Reflected component
scatters

Figure 17.13. Perspective sketch of the potential-energy surface shown in Figure 17.9 for H_2 approaching the bridge site of Cu(100). (a–c) The time evolution of an H_2 molecule in the vibrational ground state having 0.8 eV translational energy. The incident wave packet in (a), $\psi_{\text{initial}}(D, Z; t = 0)$, is turned around at (b) and it is here that tunneling into the dissociation channel occurs predominantly. In (c) the asymptotic final state evolves as a linear combination of displaced Morse oscillator states [equation (17.6.6)].

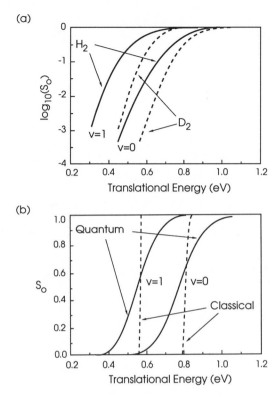

Figure 17.14. (a) The dissociation probability, S_0, plotted as a function of the initial translational energy, for H_2 and D_2 in initial vibrational states $v = 0$ and $v = 1$. (b) A comparison of quantum and classical results for the H_2 dissociation probability, S_0.

through the activation barrier which decreases in width as the molecule penetrates closer to the surface (see Figure 17.15). The significant enhancement of the dissociation probabilities of H_2 when compared to D_2 is due to the greater zero-point energy of H_2, and also because of the increased tunneling of H_2 particularly at low initial translational energies. For $E_{tot} > E_{act}$, there is a stronger antitunneling effect for the lighter isotope which, in Figure 17.14a, is manifested as a steeper energetic dependence for D_2 over H_2 which ultimately gives rise to a crossover in $S_0(E_\perp)$. The strong dependence of S_0 on v for a given species arises from the greater spatial extension of the $\phi_1(D)$ over $\phi_0(D)$ which in turn increases the tunneling probability (see Figure 17.15).[50] In passing, it is apparent from Figure 17.14a that vibrational energy is not as effective as translational energy in overcoming the barrier.

Table 17.1 illustrates the interesting effect of vibrational enhancement arising as a consequence of the reactive encounter. As a wave packet scatters from the surface it is distorted, since part of it remains trapped on the surface. Physically, this distortion implies that some fraction of those molecules which scatter are

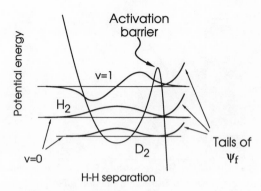

Figure 17.15. A section taken through the potential-energy surface along the vibrational coordinate for H_2 interacting with a metal displaying an activation barrier to dissociative chemisorption. Schematically shown are the initial-state (molecular vibrational) wave functions $\phi_v(D)$ for H_2 and D_2 in their ground states $(v = 0)$ and H_2 in its first excited state $(v = 1)$. The magnitude of the overlap of $\phi_v(D)$ with the dissociated state wave function $\psi_{\text{diss}}(D)$, which strongly affects the sticking coefficient S_0, is larger for H_2 $(v = 1)$ than H_2 $(v = 0)$ implying that $S_0(v = 1) > S_0(v = 0)$. Similar reasoning also explains why $S_0(H_2, v = 0) > S_0(D_2, v = 0)$.

vibrationally excited.[51] The asymptotic scattered wave function, centered on Z', is then

$$\psi_{\text{scatt}}(D, Z; t \to \infty) = \sum_v A_v \phi_v(D) g(Z - Z', p_Z), \qquad (17.6.6)$$

with energy being conserved in the overall process. in Table 17.1, the final-state vibrational distribution is presented for an initial beam of H_2 with $E_\perp = 1.05$ eV and a vibrational temperature $T_{\text{vib}} = 1250$ K. There is considerable vibrational excitation observed in the scattered beam and an approximate least-squares fit to the (non-Boltzmann) vibrational distribution yields an effective "vibrational temperature" of ~ 1700 K.

**Table 17.1. Vibrational State Distributions
Corresponding to an Incident Beam of Molecules
Having Vibrational Temperature of 1250 K and
Translational Energy 1.05 eV
(The effective "temperature" in the scattered
beam is ~1700 K)**

Vibrational state v	$\ln[P_v/P_0]$	
	Initial	Scattered
0	0.00	0.00
1	−4.58	−3.93
2	−8.89	−6.67

The dissociation probabilities for H_2 have also been calculated classically, integrating Hamilton's equations of motion and averaging over 200 initial phases for each $E_\perp$. The results are plotted in Figure 17.14b for the ground and first excited state, together with the quantum results. The classical probabilities, which bear little resemblance to the quantum results,[52] are essentially step functions switching from zero to unity over an energy range of only 0.05 eV. The reason for this sharp onset may be traced back to the topology of the crossing seam in the PES near to the saddle point (Figure 17.9). As E_{act} is exceeded, the phase space made available for reactive trajectories increases very quickly, a common feature of transition states which have a very low vibrational mode normal to the classical reaction path. In all but a very small transition region, phase variations are unimportant and this accounts for the abrupt "jump" from the classically forbidden to the classically allowed dissociation near the threshold.

17.7. Dissociative Adsorption: Multiple Activation Barriers and Diffraction

It was noted in Section 17.4 that Lapujoulade[31] had reported diffractive scattering of H_2 from a Cu surface for initial energies $76 \geq E_\perp \geq 218$ meV which, according to Balooch et al.,[28] should span E_{act}. If we return to Figure 17.3 then this implies is that as $E_\perp$ is steadily increased, the periodic repulsive potential V_{rep} (which is responsible for the diffraction although not explicitly shown) is sampled at increasingly larger values. Since $V_{rep}(R)$ is approximately proportional to the surface electron density $\rho(R)$, the effective corrugation felt by the incident beam h_{rep}, will increase with increasing $E_\perp$. What happens as the barrier is crossed, however, is unclear. It is at this point that the one-dimensional representation shown in Figure 17.3 ceases to provide useful guidance. To be more specific, we consider the two cases presented in Figure 17.16, where slices through the PES are shown for H_2 molecules incident at a top and a center site on a low-index surface.[34] In this figure neither the molecular orientation (θ, ϕ) nor bond length are included. The forms of V_{vw} and V_{rep} are given by equations (17.4.5)–(17.4.7) and it is seen that the top site is more repulsive than the center site for the incoming molecule. The V_{2H} PES is modeled by assuming that in the barrier region it is given approximately by

$$V_{2H}(R) = V_{diss} - V_1 \exp\{-\gamma[Z - g_{att}(R)]\} \tag{17.7.1}$$

where

$$g_{att}(R) = \frac{1}{2}\left[g_X\left(1 - \cos\frac{2\pi X}{a_X}\right) + g_Y\left(1 - \cos\frac{2\pi Y}{a_Y}\right)\right] \tag{17.7.2}$$

As discussed in Section 17.4, when a H_2 molecule approaches a surface, some

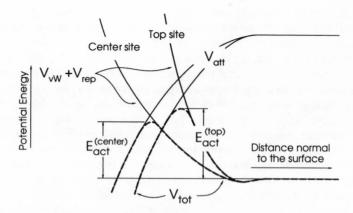

Figure 17.16. An extension of the "Lennard-Jones" potential-energy surface shown in Figure 17.3 to incorporate variations due to surface morphology. The diabatic potential describing the molecular interaction $[V_{vw}(Z) + V_{rep}(\mathbf{R})]$ shows a variation across the surface unit cell which gives rise to a corrugated potential similar to that experienced by a rare-gas atom. $V_{2H}(\mathbf{R})$ is the electronic state corresponding to atomic adsorption which also varies with surface position. As a consequence, the derived adiabatic potential which is experienced by a thermal H_2 beam has an activation barrier whose magnitude depends critically upon whether the incident molecule approaches a center or an on-top site.

charge will flow into the affinity level, which initiates a surface bond. Here we will be interested in essentially only what occurs on the *gas-phase* side of the activation barrier and the precise form of $V_{2H}(\mathbf{R})$ will not critically effect the quantitative results.[34] This functional form allows control over the variation in the surface unit cell of the energy range spanned by the activation barrier $E_{act}(\mathbf{R})$, as well as the exact spatial location of the minimum barrier energy $E_{act}^{\#}$ (an on-top or center site, for example, as shown in Figure 17.2). The "rounding off" of the barrier has been achieved by using the Landau–Zener expression given by equation (17.5.2).

Figure 17.17a shows a two-dimensional cut for an $H_2/Cu(110)$ PES where the minimum activation barrier $(E_{act}^{\#})$ is located at an on-top site, and Figure 17.17b is for the case when it is at a center site. Potential parameters have been carefully chosen to ensure that the reactive portions of the PES open at the same rate, in order to keep the PES as similar as possible. Figure 17.18 shows the real- and momentum-space time evolution of a Gaussian wave packet incident on a potential similar to those shown in Figure 17.17. The initial state (a, b) represents a plane wave incident normal to the surface. As the packet approaches the region of strongest interaction, it is seen (c) how part passes over the activation barrier while the majority scatters back into the gas phase. As time increases, it is seen (d) how the final diffraction pattern emerges (f) with the creation of parallel momentum states obeying the Bragg condition.

Results for H_2 diffraction from these two PES shown in Figure 17.17 as a function of increasing beam energy are shown in Figure 17.19. The gross features of the scattering distributions from the reactive surfaces are similar: they show a decrease of the specular beam with intensity being transferred into the diffracted beams as the primary energy is increased. Below $E_{act}^{\#}$ the behavior of the [00] and [10] beams is virtually identical with the [20] beam, being somewhat depressed for the center-site PES. As the barrier opens, flux begins to be consumed by the reactive channel. For the *elastic* channel, however, there is a sharp distinction between the two surfaces since the [00] beam for the top-site PES dives sharply by comparison to its center-site counterpart. Calculations for D_2 show equally large differences between the two PES.[34]

Since the calculations for diffracted intensities obey unitarity, by summing over the beams it is possible to obtain a value for the fraction of particles at a given energy which have incoherently scattered. *If* all of these particles were to accommodate with the surface, then this number would be the sticking coefficient. Since the calculations described here do not include surface degrees of freedom (and therefore inelastic processes) it is only possible to equate the incoherent fraction with an upper limit to the sticking fraction. Figure 17.20 shows the dependence of the absorbed H_2 fraction with primary energy for both PES shown in Figure 17.17. It is seen quite clearly that up to 20% tunneling is observed with an initial $E_{\perp}$ dependence that resembles an error function, approximately $(E_{\perp} - E_{\perp}^3/3)$, a *form* approximating well to results obtained for S_0/Cu[28] although differing in absolute magnitude. Also shown in this figure are the results of classical scattering from the two PES. When $E_{act}^{\#}$ is at the on-top site, the classical scattering shows an abrupt increase as the barrier minimum is attained but, interestingly, it does not saturate to unity at the barrier maximum but remains below even the quantum result! This effect arises from a steering of the classical molecules in their trajectory near the activation barrier. To clarify this, a third curve in Figure 17.20a shows the geometric fraction of the surface which is missing. This is obtained

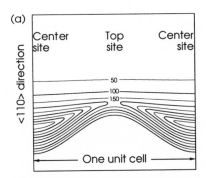
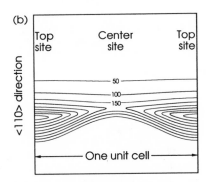

Figure 17.17. (a) A two-dimensional representation of the potential-energy surface for $H_2/\mathrm{Cu}(110)$. The equal energy contours are separated by 50 meV. The whole of one surface unit cell is shown and it is assumed that the surface corrugation is only appreciable in the $\langle 001 \rangle$ direction. The activation barrier has values $200 \le E_{act} \le 500$ meV, with the minimum being located above an on-top site. As the initial beam energy increases, (1) the effective corrugation will increase and (2) an increasing region of the surface will be exposed which is capable of dissociative adsorption. (b) The same as (a) except in this case the minimum value of the activation barrier is located at a center site in the surface. These potentials show quite clearly how both diffraction *and* dissociation can occur at a single primary energy.

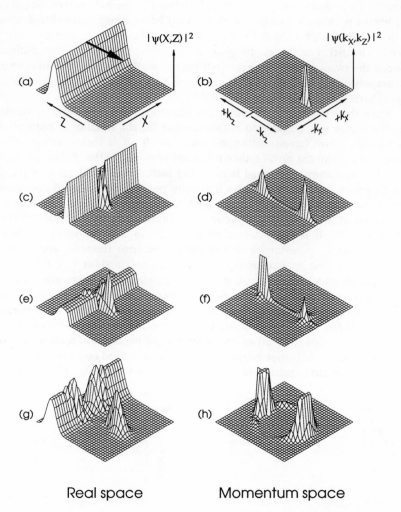

$|\psi(X,Z)|^2$

$|\psi(k_X,k_Z)|^2$

Real space   Momentum space

Figure 17.18. The time evolution of a wave packet interacting with a potential-energy surface based upon that shown in Figure 17.17. In the left panel (a, c, e, g), the real-space probabilities are shown and the corresponding momentum states are to the right (b, d, f, h). Initially a Gaussian packet is normally incident on a surface (a) with a mean energy such that, classically, it can overcome the activation barrier only within a restricted range of the surface unit cell. In (b) it is seen that the packet possesses only a zero parallel momentum component. As the packet encounters the strongly interacting region (c, e) (near the classical turning point) it begins to split and, although the majority is returned to the gas phase, a portion passes over the barrier into the dissociated state. In momentum space (d, f), positive states begin to populate. Finally, when the packet is in the region of zero potential, the wave function is composed of a coherent sum of diffracted states, as can be clearly seen in (h) where energy conservation dictates the emergent angles of the Bragg scattered beams.

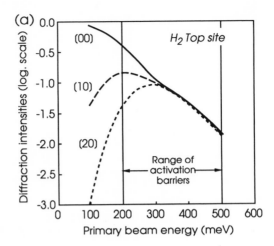

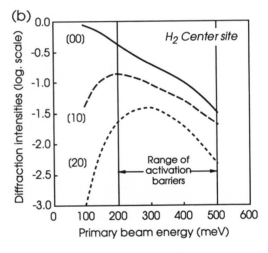

Figure 17.19. H_2 intensities for the first three diffraction states as a function of initial energy for a primary beam at normal incidence to a potential, (a) where the minimum value of the activation barrier is above a top site in the surface (Figure 17.17a) and (b) for scattering from the potential in Figure 17.17b where the minimum barrier is located above a center site. As the reaction zone in the surface opens, the specular beam intensity remains high and decays at a notably slower rate than its counterpart in (b). The reasons for this and other dissimilarities may be traced to topological differences in the two PES shown in Figure 17.17 and are discussed in detail elsewhere.[34]

simply from the contour plot shown in Figure 17.17. Essentially what is seen is that, for the scattered H_2 molecules, the *dynamic* reactive area is smaller than the geometric one due to potential gradients along the surface. Similar results, but for $E_{act}^{\#}$ at the center site, are shown in Figure 17.20b. This time the data obtained from classical scattering mirror the geometric reaction zone almost exactly and no prefocusing of the particles is observed. An examination of the two PES shown in Figure 17.17 and the classical scattering distributions in Figure 17.9 reveal why this is the case. The PES contours obtained with $E_{act}^{\#}$ at a center site are extremely flat and therefore there are almost no potential gradients (and hence forces) parallel to the surface.

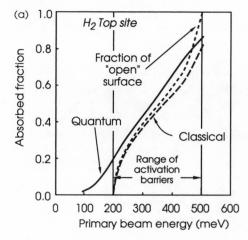

(a)

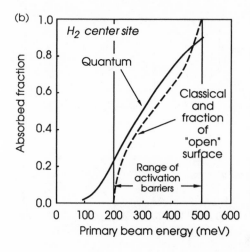

(b)

Figure 17.20. (a) The absorbed fraction as a function of primary energy for H_2 scattering from the potential in Figure 17.17a where the minimum activation barrier is located above a top site in the surface. The solid line is the result from the quantum-mechanical calculation described in Section 17.17. The short dashed line shows, for a given energy, the fraction of the unit cell which has a surmountable activation barrier. The long dashed line shows the results for a purely classical calculation which deviates considerably from the fraction of the surface which is effectively "open" for reaction. This arises from a dynamical funneling of the classical particles *away* from the reaction zone and toward the center sites which results in a reduced *normal* kinetic energy and makes the zone appear smaller than its simple geometric size. (b) As (a) but for the PES in Figure 17.17b where the minimum barrier is at a center site. In this case there is no observed steering of the classical particles and the reaction zone is the same size as that obtained from geometrical considerations.

Thus the apparent paradox of observing diffraction and dissociative adsorption simultaneously may be understood by a simple *multidimensional* PES. The activation barrier is corrugated and is modulated in size across the unit cell. This represents one of many possibilities why values of S_0 need not be either 1 or 0.

17.8. Summary

In this chapter, an attempt has been made to present a unified discussion of various theoretical methods currently employed to address problems in gas–surface dynamics. After reviewing the Born–Oppenheimer approximation, the origins of adiabatic and diabatic potential-energy surfaces were presented. The fundamental concepts behind transition state theory were then examined from the viewpoint of atomic desorption from surfaces. A classical calculation of the translational, rotational, and vibrational energy distribution functions for H_2 desorbing from a Cu surface was then presented. In this work, the potential-energy hypersurface employed was derived from an *ab initio* calculation for H_2 interacting with a small Cu cluster. This potential was then used as a basis for a quantum wave-packet study of H_2 dissociative adsorption. The split operator technique for wave-packet propagation was outlined and employed to study the translational and vibrational energy dependence of the H_2 sticking coefficient. Finally, results for the diffractive scattering of H_2 from Cu were presented. It was demonstrated that by extending the dimensionality of the familiar Lennard-Jones potential to include variations across the surface unit cell, the activation barrier to adsorption is no longer single-valued. Instead, its magnitude varies continuously, creating "holes" within the potential-energy surface where dissociation may occur. Thus molecules were able to coherently diffract *or* dissociatively adsorb on the surface. It is suggested that by inverting H_2 *elastic* scattering data it should be possible to obtain detailed information on the *reactive* potential-energy surface.

Acknowledgments

Thanks go to my collaborators David Halstead, Mick Hand, and John Harris without whom this would not appear. In addition I thank the organizers of the "Spring College" in Trieste of 1988 where the lectures on which this chapter is based were given.

References

1. C. T. Rettner, H. E. Pfnür, and D. J. Auerbach, *J. Chem. Phys.* **84**, 4163 (1985).
2. A. W. Kleyn, A. C. Luntz, and D. J. Auerbach, *Surf. Sci.* **117**, 33 (1982).
3. G. D. Kubiak, G. O. Sitz, and R. N. Zare, *J. Chem. Phys.* **81**, 6397 (1984).
4. J. A. Barker and D. J. Auerbach, *Surf. Sci. Rep.* **4**, 1 (1985); D. S. King and R. R. Cavanagh, in: *New Laser and Optical Investigations of Chemistry and Structure at Interfaces*, Verlag Chemie, Berlin (1985); M. C. Lin and G. Ertl, *Annu. Rev. Phys. Chem.* **37**, 587 (1986).
5. J. E. Lennard-Jones, *Trans. Faraday Soc.* **28**, 333 (1932).
6. J. C. Tully, in: *Dynamics of Molecular Collisions* (W. H. Miller, ed.), Plenum Press, New York (1976).
7. N. D. Lang, in: *Theory of the Inhomogeneous Electron Gas* (S. Lundqvist and N. H. March, eds.), Plenum Press, New York (1983).
8. T. F. O'Malley, *Adv. At. Mol. Phys.* **7**, 223 (1971).
9. J. W. Gadzuk, *Comments At. Mol. Phys.* **16**, 219 (1985).
10. J. von Neumann and E. P. Wigner, *Phys. Z.* **30**, 467 (1929).
11. E. E. Nikitin, *Chemische Elementarprozeße* (H. Hartman, ed.), Springer-Verlag, Berlin (1968).
12. F. T. Smith, *Phys. Rev.* **179**, 111 (1969).
13. B. Kasemo and L. Walldén *Surf. Sci.* **53**, 393 (1975); J. K. Nørskov, D. M. Newns, and B. I. Lundqvist, *Surf. Sci.* **80**, 179 (1979).
14. P. Pechukas, in: *Dynamics of Molecular Collisions* (W. H. Miller, ed.), Plenum Press, New York (1976).
15. E. K. Grimmelmann, J. C. Tully, and E. Helfand, *J. Chem. Phys.* **74**, 5300 (1981).
16. J. C. Keck, *Adv. Chem. Phys.* **13**, 85 (1967).

17. S. Holloway and J. L. Beeby, *J. Phys. C* **9**, 1907 (1976).
18. O. Beek, A. E. Smith, and A. Wheeler, *Proc. R. Soc. London, Ser. A* **177**, 62 (1940).
19. A. F. H. Ward, *Proc. R. Soc. London, Ser. A* **133**, 506 (1931).
20. D. D. Eley and D. R. Rossington, in: *Chemisorption* (W. Garner, ed.), Butterworths, London (1957).
21. C. S. Alexander and J. Prichard, *J. Chem. Soc., Faraday Trans.* **68**, 202 (1972).
22. W. Van Willigen, *Phys. Lett.* **28A**, 80 (1968).
23. T. L. Bradley and R. E. Stickney, *Surf. Sci.* **38**, 313 (1973).
24. M. Balooch and R. E. Stickney, *Surf. Sci.* **44**, 310 (1974).
25. G. Comsa and R. David, *Surf. Sci.* **117**, 77 (1982).
26. G. Comsa and R. David, *Surf. Sci. Rep.* **5**, 145 (1985).
27. G. Comsa and R. David, *Chem. Phys. Lett.* **49**, 512 (1977).
28. M. Balooch, M. J. Cardillo, D. R. Miller, and R. E. Stickney, *Surf. Sci.* **46**, 358 (1974).
29. T. Engel and G. Ertl, in: *The Chemical Physics of Solid Surfaces and Heterogeneous Catalysis* (D. A. King and D. P. Woodruff, eds.), Elsevier, Amsterdam (1983).
30. M. J. Cardillo, M. Balooch, and R. E. Stickney, *Surf. Sci.* **50**, 263 (1975).
31. J. Lapujoulade, Y. Le Cruer, M. Lefort, Y. Lejay, and E. Maurel, *Surf. Sci. Lett.* **103**, L85 (1981); J. Lapujoulade and J. Perreau, *Phys. Scr.* **T4**, 138 (1983).
32. J. Harris, S. Andersson, C. Holmberg, and P. Nordlander, *Phys. Scr.* **T13**, 155 (1986).
33. N. Esbjerg and J. K. Nørskov, *Phys. Rev. Lett.* **45**, 807 (1980).
34. D. Halstead and S. Holloway, *J. Chem. Phys.* **88**, 7197 (1988).
35. J. K. Nørskov and N. D. Lang, *Phys. Rev. B* **21**, 2136 (1980).
36. M. J. Stott and E. Zaremba, *Phys. Rev. B* **22**, 1564 (1980).
37. J. K. Nørskov, *Phys. Rev. B* **26**, 2875 (1982).
38. O. Gunnarsson, H. Hjelmberg, and B. I. Lundqvist, *Phys. Rev. Lett.* **37**, 292 (1976).
39. P. Nordlander, S. Holloway, and J. K. Nørskov, *Surf. Sci.* **136**, 59 (1984).
40. P. Nordlander and S. Holmström, *Surf. Sci.* **159**, 443 (1985).
41. J. Harris, *Z. Physik Appl. Phys. A* **47**, 63 1988.
42. J. K. Nørskov, A. Houmøller, P. Johansson, and B. I. Lundqvist, *Phys. Rev. Lett.* **76**, 257 (1981).
43. J. Harris, S. Holloway, T. Rahman, and K. Yang, *J. Chem. Phys.* **89**, 4427 (1988).
44. W. Brenig and H. Kasai, *Surf. Sci.* **213**, 170 (1989).
45. J. B. Anderson, *J. Chem. Phys.* **58**, 4684 (1973).
46. P. K. Johansson, *Surf. Sci.* **104**, 510 (1981).
47. R. D. Levine and R. B. Bernstein, *Molecular Reaction Dynamics and Chemical Reactivity*, Oxford University Press, Oxford (1987).
48. J. A. Fleck, Jr., J. R. Morris, and M. D. Feit, *Appl. Phys.* **10**, 129 (1976); M. D. Feit, J. A. Fleck, and A. Steiger, *J. Comput. Phys.* **47**, 412 (1982).
49. B. Jackson and H. Methiu, *J. Chem. Phys.* **86**, 1026 (1986).
50. S. Holloway, D. Halstead, and A. Hodgson, *J. Electron Spectrosc. Relat. Phenom.* **45**, 207 (1987); *Chem. Phys. Lett.* **147**, 425 (1988); M. R. Hand and S. Holloway, *J. Chem. Phys.* **91**, 7209 (1989).
51. S. Holloway, *J. Vac. Sci. Technol.* **A5**, 476 (1987).
52. Chao-Ming Chiang and Bret Jackson, *J. Chem. Phys.* **87**, 5497 (1987).

18

Elementary Chemical Processes on Surfaces and Heterogeneous Catalysis

A. Hamnett

18.1. Introduction

From the perspective of the chemist, the major thrust of research on surfaces has always been to further the understanding we have of the chemical processes that take place during heterogeneous catalysis. The last twenty years have seen an enormous growth in the detailed molecular information available at the solid–gas interface, and our ability both to comprehend the mode of action of traditional catalysts and to engage in the design of new catalysts has grown apace. Much of this information has emerged from a plethora of new techniques involving the elastic or inelastic scattering of electrons or atoms. These techniques are specific to the solid–gas interface; our level of understanding of the liquid–solid interface is at a much more primitive level, and this chapter will therefore not be concerned primarily with the latter.

Given that the main thrust is heterogeneous catalysis, we should first consider what catalysis is, and how it might be effected at the solid surface. The definition of a *catalyst* is that it is a "substance that increases the rate at which a chemical reaction approaches equilibrium without itself being consumed in the process." There are, in principle, two ways in which a catalyst may act: there may be a *concentration effect*, in which the local surface concentration of reactive species is substantially enhanced, or there may be an *energetic* effect, in which the activation energy for the desired reaction is reduced. In practice, the distinction is somewhat artificial, since the adsorption of molecules on the surface will only take place if the energetics are favorable, and the fundamental catalytic process is the lowering

A. Hamnett ● Department of Chemistry, Bedson Building, The University, Newcastle upon Tyne NE1 7RU, England.

Table 18.1. Preliminary Guide to the Types of Reaction and the Catalysis Commonly Used

Catalytic reaction	Example	Typical catalyst
Hydrogenation	$N_2 + 3H_2 \rightarrow 2NH_3$	Transition metals, e.g., Pd/C
Dehydrogenation	$C_2H_6 \rightarrow C_2H_4 + H_2$	Metal or metal oxides, e.g., $CuCr_2O_4$
Hydrogenolysis	$ROH + H_2 \rightarrow RH + H_2O$	Transition metals
Oxidation	$CH_3CH{=}CH_2 \rightarrow CH_2{=}CH \cdot CHO$	Mixed oxides, carbides
Hydrodesul-furization	(2,5 dialkyl)-thiophene $\rightarrow R{-}(CH_2)_4{-}R'$	Transition metal Sulfides, "nickel boride"
Condensation	$3C_2H_2 \rightarrow C_6H_6$	Transition metals
Hydration	$C_2H_2 + H_2O \rightarrow CH_3CHO$	HgO/Nafion
Dehydration	$HCOOH \rightarrow CO + H_2O$	Insulating oxides
Hydrohalogenation	$C_2H_2 + HCl \rightarrow CH_2{=}CHCl$	$HgCl_2/C$
Polymerization	$nC_2H_4 \rightarrow (CH_2)_{2n}$	Acids, SiO_2/Al_2O_3, zeolites
Isomerization	$CH_2{=}CH \cdot CH_2CH_3$ $\rightarrow CH_3CH{=}CHCH_3$	Transition metals and carbides of gp. VI
Cracking	$RC_2H_5 \rightarrow RH + C_2H_4$	Acidic oxides, e.g., alumino-silicates

of overall activation energy. Owing to the much smaller number of collisions on the surface as compared to the gas phase, the activation energy has to be lowered appreciably to achieve a reaction of the same intrinsic *rate* on the surface as in the gas phase, and estimates from collision theory suggest a lowering of at least 100 kJ mol^{-1} is required for worthwhile catalysis. As an example, the activation energy for the process $2N_2O \rightarrow 2N_2 + O_2$ is 245 kJ mol^{-1} in the gas phase, and 121 kJ mol^{-1} on an Au catalyst. Even taking into account the problems raised by collision theory, this represents an enhancement of *ca* 10^9 at room temperature.

It is interesting to inquire what forms of reaction have historically been effected by heterogeneous catalysis. Table 18.1 presents a preliminary guide to the types of reaction and the catalysts commonly used.

18.2. Fundamental Processes that Take Place at the Surface

18.2.1. Classical Models for Atomic and Molecular Scattering at Surfaces

We consider first a gas molecule or atom incident on a surface free of other adsorbed molecules. The processes that may take place are (1) an elastic collision, (2) an inelastic collision followed by escape of the incident molecule from the surface with higher or lower velocity, and (3) an inelastic collision in which sufficient energy is lost that the molecule can no longer escape and is trapped in a potential well at the surface. Initially, we shall be concerned with trapping processes that lead to *physisorption*; extension to *chemisorption* is not straightforward, and less is known about chemisorption than physisorption. However, it is now believed that, in a large number of cases, physisorption precedes chemisorption and an understanding of the former is essential to a treatment of the latter.

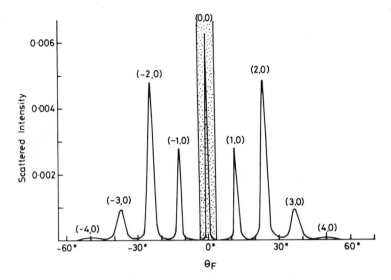

Figure 18.1. Diffraction of He from LiF(001) at $T_s = 10$ K, $\theta_i = 0°$. The detector scans along the $\langle 110 \rangle$ azimuth. The specular intensity is extrapolated from measurements at small θ_i. A nozzle beam with narrow velocity distribution was used $E_i \approx 58$ meV (after Boato et al.[3]).

Elastic collisions involving atomic beams have been used to probe surfaces and adsorbate structure, since diffraction effects can be observed in favorable circumstances, provided the beam is reasonably well collimated and monoenergetic.[1] The earliest experiments, carried out in the 1930s, were motivated by the desire to demonstrate the wave-like characteristics of light atoms such as He or of molecules such as H_2.[2] The surface used was that of a LiF crystal, in which alternating rows of cations and anions give a well-defined diffraction grating. In order to maximize elastic scattering, the temperature of the sample must be kept very low, so that phonon annihilation is negligible and phonon creation much reduced. Under these circumstances, diffraction patterns such as those seen in Figure 18.1 may be observed.

Atomic scattering from metals is much weaker, though diffraction of He from W(112) has been observed.[3] This is, in part, because the rows of metal atoms do not, in general, generate a sufficiently well-defined potential, though the (112) face of W does have a pronounced corrugated structure.

Atomic scattering is a difficult technique experimentally, though it does offer the possibility of surface structural determination for highly insulating materials which rapidly become charged if electron scattering techniques are employed. However, the technique suffers from the serious drawback that the probability of *inelastic* scattering is frequently very much greater than that of *elastic* scattering. We can estimate the probability of elastic specular reflection from the Debye–Waller factor $\exp(-2W)$, where

$$2W = [\Delta k_z(\theta_i, \theta_f)]^2 \langle u_z^2(T_s) \rangle$$

Δk_z being the total momentum transfer to the surface in the z-direction and $\langle u_z^2 \rangle$ the mean-square displacement of the surface atoms. In the high-temperature limit, we have

$$\langle u_z^2 \rangle \approx 3h^2 T_s / (4\pi^2 M_s k T_D^2)$$

with M_s the mass of the surface atom and T_D the surface Debye temperature. If the energy profile for an approaching atom can be approximated as in Figure 18.2 and we include the energy D in Δk_z,[5] then

$$(\Delta k_z)^2 = 32\pi^2 m_g E_i (\cos^2 \theta_i + D/E_i)/h^2$$

So finally, in the high-temperature limit,

$$2W = 24\mu E_i T_s (\cos^2 \theta_i + D/E_i)/kT_D^2$$

where $\mu = m_g/M_s$. As an example, we consider the case of Ar reflected from W: we can calculate W from $\mu \approx 0.2$, $D \approx 80$ meV, $T_D \approx T_D(\text{bulk})/\sqrt{2} \approx 270$ K ≈ 23.2 meV$/k$, and, if $E_i = 10$ meV, $T_s = 90$ K $\approx T_D/3$, and $\theta_i = 0$, then $2W \approx 6$ and $I_{sp}/I_0 \approx e^{-6} \approx 2 \times 10^{-3}$.

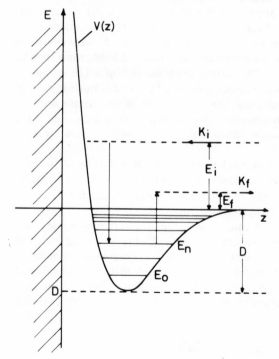

Figure 18.2. Energy diagram of the gas–solid interaction.

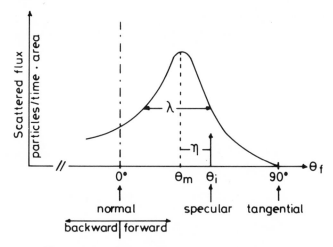

Figure 18.3. Lobular angular distribution (schematic).

It is clear that, unless conditions are favorable (small μ and T_s, large T_D, high angle of incidence, and small D), then the overwhelming probability is that the scattering will be *inelastic*. As indicated above, there are two possibilities: some energy may be lost to or gained from the lattice vibrations but the atom still bounces off the surface, or sufficient energy is lost to the lattice for a transition to take place to a *bound state* of energy E_n, i.e., the atom sticks.

The detailed distribution of atomic energies after impact will depend on the details of the interaction with the surface. In many gas–solid scattering systems, the interaction potential is of the van der Waals type. This leads to physisorption, and the well depth is small in comparison with the energies involved in chemisorption (such as He on LiF: $D = 7.6$ meV; Ar on graphite: $D \approx 100$ meV), and the shape of the potential well can usually be described adequately by a Lennard-Jones or Morse-type function.

Classically, scattering of a monoenergetic atomic beam by a surface will give rise to a Maxwellian distribution of reflected atomic velocities whose differential intensity takes the form

$$\frac{dI}{dv} = \frac{2I_0}{v_h}\left(\frac{v}{v_h}\right)^3 \exp\left[-\left(\frac{v}{v_h}\right)^2\right]$$

where $v_h = (2kT/m_g)^{1/2}$ and is the most probable escape velocity. If the *tangential* component of the velocity is conserved, then v is simple related to the angle θ_f for a given angle of incidence θ_i, and this is illustrated in Figure 18.3.

The relationship between v and θ_f gives rise to a characteristic lobular structure if intensity is plotted in polar coordinates as shown in Figure 18.4.

Typical features of these lobes, which are formed quite generally if $T_g < T_s$, are:

1. The maximum is shifted more or less against the specular direction.

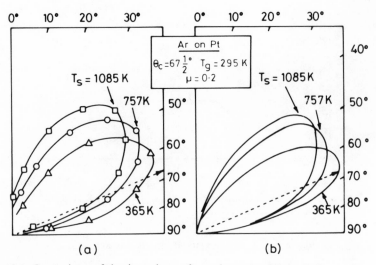

Figure 18.4. Comparisons of the dependence of experimental and theoretical scattering patterns on the surface temperature T_s: (a) experimental results for Ar on Pt; (b) corresponding theoretical results based on the hard-cube model (*vide infra*) (after Stickney[10]).

2. The lobes are broad.
3. The shift of the maximum, $\eta \equiv \theta_i - \theta_m$, depends in a characteristic way on θ_i, E_i, T_s, and μ.

Attempts to understand the energy exchange at a surface with classical theories have been surprisingly successful, at least at a semiquantitative level. Three approaches have been tried[6]:

1. Linear Harmonic Oscillator Model. This model is due to McCarroll and Ehrlich[7] and is a natural outgrowth of the Landau–Teller theory of energy transfer between molecules in the gas phase.[8] The solid is considered to be formed from chains of atoms which interact one with another through a force constant K. An incoming atom is conceived of as becoming (temporarily) attached to the outer atom of the chain with force constant K_0 as shown in Figure 18.5. Energy transfer then takes place from the translational energy of the incoming atom to the chain.

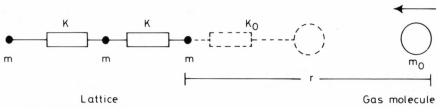

Figure 18.5. Linear harmonic oscillator model for energy exchange and trapping at a surface (after McCarroll and Ehrlich[7]).

The energy curve representing the interaction of the incident atom with the terminal atom of the lattice chain is a truncated parabola as shown in Figure 18.6 (this to allow the atom to escape). The analytical solution is complex, since a power series expansion must be used to take account of the strongly coupled vibrations withint he chain. However, even though the solution is complex, it takes on a simple *form* which can be expressed in terms of two parameters, β ($\equiv K_0/K$) and μ ($\equiv m_g/M_s$). We find

$$v(t)/v_i = 1 - \beta/\mu[f(t) + Z^{-1}f_1(t)]$$

where $v(t)$ is the velocity of the incident atom at time t, v_i its initial velocity, $f(t)$ and $f_1(t)$ are complex functions of time, and

$$Z = 2[\partial \log_e (r_0 - r)/\partial t]_{r=r_1}$$

where r, r_0, and r_1 are defined in Figure 18.6.

Detailed calculations reveal that:

1. At low values of β, the efficiency of energy transfer increases rapidly as β increases. Examination of Figure 18.6 reveals that β is a measure of the parabolicity of the potential energy curve; the steeper the repulsive part, the larger β, so we can say that the efficiency of energy exchange increases as the steepness of the repulsive part of U.

2. The efficiency of energy exchange increases with μ.

3. There will be a critical value of the kinetic energy (E_c) of the incident atom *below* which *trapping* will take place. If the binding energy of the terminal lattice atom to its neighbor is Q_L, then the ratio E_c/Q_L varies in a characteristic manner with β for various values of μ as shown in Figure 18.7. Clearly, as expected from (1) and (2) above, E_c increases with β, at least for small values of β, and with μ at constant β.

2. Hard Cube Model. This model was introduced by Logan and Stickney[9] in order to account more quantitatively for the results of the molecular-beam

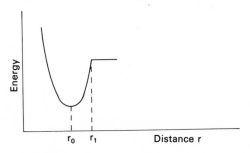

Figure 18.6. Truncated parabolic potential-energy curve for the linear harmonic oscillator model (after McCarroll and Ehrlich[7]).

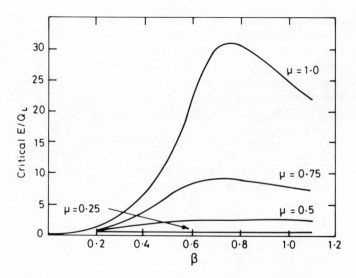

Figure 18.7. Critical kinetic energy for trapping of a gas molecule on a surface as a function of the force-constant ratio $\beta = K_0/K$ for various values of the mass ratio $\mu = (m_g/M_s)$ (after McCarroll and Ehrlich[7]).

experiments discussed above. The basic model can be summarized with reference to Figure 18.8.[10] The primary assumptions are:

1. The interaction potential $V(z) = \{0, z > z_0; \infty, z < z_0\}$, i.e., there are no attractive forces between gas and solid and the interaction is purely impulsive.
2. The surface is effectively flat, so that the tangential component of the gas–atom momentum is conserved.

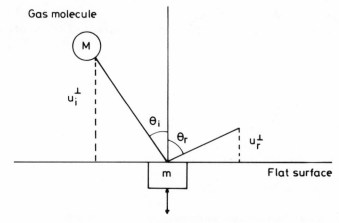

Figure 18.8. The hard-cube model (after Stickney[10]).

3. The solid is conceived as a collection of hard cubes confined within the volume of the solid by square-well potentials. The velocity distribution of these boxes is Maxwellian perpendicular to the surface.

The analysis of the model is straightforward, and the following results are found:

1. If $\theta_m = \theta_i$ (i.e., $\eta = 0$), then the temperatures of incident gas and surface are related by the expression

$$T_s/T_g = 9 \cos^2 \theta_i/8$$

In general, if T_s/T_g is *less* than this value, η ($\equiv \theta_i - \theta_m$) will be negative and *vice versa*.

2. The values of η are predicted qualitatively quite well, as shown in Figure 18.9, which is taken from the work of Yamamoto and Stickney[11] for the scattering of inert gas atoms from W(110).

There are, however, serious drawbacks to this model. Backward scattering is not included, the angular distribution does not fall off to zero as $\theta_f \to 90°$, and λ (see Figure 18.3) is not calculated correctly, especially for large values of μ. The theory is also misleading in that its qualitative success has been taken as evidence for the fundamental correctness of its axioms. However, even scattering from strongly periodic surfaces, such as those of the alkali halides, can show lobular structure[12] [e.g., Xe on LiF(001)]. Indeed, for the case of atomic O scattering

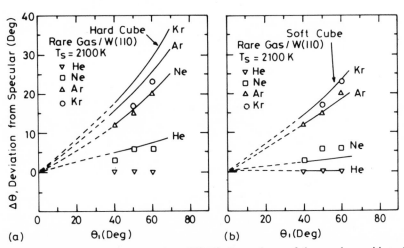

Figure 18.9. Scattering of inert gases from W(110). Dependence of the angular position of the lobe maximum on angle of incidence. The experimental data are compared witht he corresponding calculations based on (a) the hard-cube model and (b) the soft-cube model (after Yamamoto and Stickney[11]).

from LiF(001),[13] the agreement between the *direct* scattering lobe maximum and the behavior predicted by the hard-cube model was surprisingly good. The final difficulty is that, with so simple a potential $V(z)$, little can actually be deduced about the nature of the gas–surface interaction.

3. Soft-Cube Model. It is clear that a better description of atom–surface collisions would be afforded if a more realistic interaction potential could be incorporated into the theory, and this was done by Logan and Keck.[14] The interaction potential is shown schematically in Figure 18.10 and consists of a stationary attractive step and an exponential repulsive part that moves with the surface atom involved in the collision. This latter is assumed to be a linear harmonic oscillator in thermal equilibrium with the bulk.

The analysis is now more complex and three parameters are used in the theory: the depth D of the attractive step, a characteristic distance b over which the repulsive potential operates, and the frequency ω at which the surface atom vibrates. As might be expected, the fit to experimental data is now substantially improved, as illustrated in Figure 18.9. To some extent, this is to be expected given the larger number of parameters. To try to reduce the number of effective parameters to be fitted from experimental scattering data, D may be estimated from enthalpies of

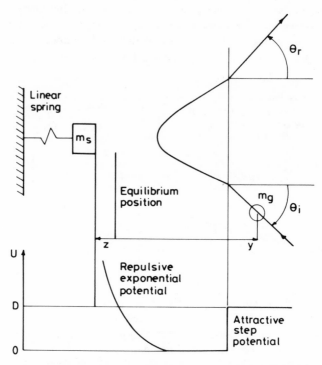

Figure 18.10. The soft-cube model (after Logan and Keck[14]).

adsorption and ω from the bulk Debye temperature. The result is that the fit in Figure 18.9 involves only one parameter, b, though the results are not entirely encouraging since the b values obtained are appreciably smaller than those derived from noble-gas pairwise interactions.

18.2.2. Quantum Effects in Atomic and Molecular Scattering

18.2.2.1. Bound State Resonances

The classical theories discussed above are sufficient to account for the general qualitative results of atomic scattering, but there is a number of effects that are clearly quantal in origin. The first is closely related to diffraction, and corresponds to the resonant transition into bound states, bound, that is to say, in the (x, y) plane. The diffraction experiments described above involve the scattering of atoms by the two-dimensional (2D) surface lattice. A concomitant of this is that, if the detector is maintained at the specular reflection angle and conditions are set to optimize elastic scattering, then rotation of the target crystal about the *azimuthal* angle will lead to minima appearing in the specular reflected intensity; an example is shown in Figure 18.11 for H˙ reflected from LiF(001).[15]

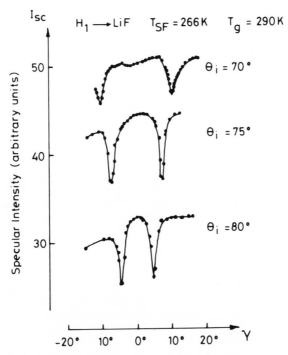

Figure 18.11. Resonant transition of H˙ into bound states on LiF(001), as indicated by minima in the angular distribution (after Hoinkes *et al.*[15]).

In order to understand the data of Figure 18.11 in more detail, we consider the Schroedinger equation for a particle on a surface:

$$[-(h^2/8\pi^2 m_g)\nabla^2 + V(\mathbf{r})]\Psi(\mathbf{r}) = (h^2 k^2/8\pi^2 m_g)\Psi(\mathbf{r})$$

This corresponds to a situation in which a gas atom of incident momentum $hk/2\pi$ and energy $h^2 k^2/8\pi^2 m_g$ is effectively trapped onto the surface following an elastic interaction. To solve this equation, we separate the variables into those parallel (x, y) and perpendicular (z) to the surface plane. Within the (x, y) plane, the potential is periodic but perpendicular to the surface, we assume a Morse potential. We can then write $V(\mathbf{r})$ as a Fourier series[16]:

$$V(\mathbf{r}) = \sum_G V_G(z) \exp(i\mathbf{G} \cdot \mathbf{R})$$

where $\mathbf{R}$ is the 2D vector (x, y) and $\mathbf{G}$ is a reciprocal lattice vector defined as

$$\mathbf{G} = n_1\mathbf{G}_1 + n_2\mathbf{G}_2$$

Here n_1 and n_2 are integers, while $\mathbf{G}_1$ and $\mathbf{G}_2$ are defined through the vector relations

$$\mathbf{G}_1 = \frac{2\pi\mathbf{L}_2 \wedge \mathbf{z}}{(\mathbf{L}_1 \wedge \mathbf{L}_2) \cdot \mathbf{z}} \quad \text{and} \quad \mathbf{G}_2 = \frac{2\pi\mathbf{L}_1 \wedge \mathbf{z}}{(\mathbf{L}_1 \wedge \mathbf{L}_2) \cdot \mathbf{z}}$$

where $\mathbf{L}_1$ and $\mathbf{L}_2$ are the fundamental translation vectors of the surface lattice. The momentum $\mathbf{k}$ on the surface can be written $(\mathbf{K}, \mathbf{k}_z)$, and the Fourier components V_G are given (uncorrected for dynamic effects) by

$$V_0(z)/D = \exp[-2\alpha(z - z_m)] - 2\exp[-\alpha(z - z_m)]$$

and

$$V_G(z)/D = \kappa_G \exp[-2\alpha(z - z_m)]$$

where z_m is the value of z at the Morse potential minimum. Clearly, if $\kappa_G \to 0$ we have a flat surface, and if $\kappa_G \gg 0$ we have a strongly periodic surface.

With these preliminaries, we can now solve the Schroedinger equation using Bloch's theorem to give

$$\Psi(\mathbf{r}) = \sum_G \Psi_G(z) \exp[i(\mathbf{G} + \mathbf{K}) \cdot \mathbf{R}]$$

where $\Psi_G(z)$ satisfies

$$\left[\frac{d^2}{dz^2} + k_{Gz}^2 - \frac{8\pi^2 m}{h^2} V_0(z)\right]\Psi_G(z) = \frac{8\pi^2 m}{h^2} \sum_{G' \neq G} V_{G-G'}(z)\Psi_{G'}(z)$$

and $k_{Gz}^2 = k^2 - (\mathbf{K} + \mathbf{G})^2$.

This equation will solve to give a finite sequence of N bound states, associated with quantum number n, in the z-direction, $0 \leq n \leq N - 1$, corresponding to vibrations perpendicular to the surface. In the event that $\kappa_G \to 0$, then motion parallel to the surface becomes essentially that of a free particle whose energy is purely kinetic. We then have

$$E_{K,n} = \varepsilon_n + h^2 K^2 / 8\pi^2 m_g$$

where

$$\varepsilon_n = -[2\pi(2m_g D)^{1/2}/\alpha h - n - \tfrac{1}{2}]^2 \cdot \alpha^2 h^2 / (8\pi^2 m_g)$$

If diffraction takes place, and if $\mathbf{K}_i$ is the projection of the incident wave vector onto the surface, then $\mathbf{K}_i$ is increased by a reciprocal lattice vector so that

$$\mathbf{K} = \mathbf{K}_i + \mathbf{G}$$

and if the equation

$$h^2 k^2 / (8\pi^2 m_g) = \varepsilon_n + h^2 (\mathbf{K}_i + \mathbf{G})^2 / (8\pi^2 m_g)$$

is satisfied, then a resonant transition may take place to a bound state of energy ε_n in the z-direction and kinetic energy $h^2(\mathbf{K}_i + \mathbf{G})^2/(8\pi^2 m_g)$ in the (x, y) plane. It should be emphasized that we have not introduced an *inelastic* transition; the total energy of the particle remains positive and only the energy in the perpendicular direction is negative.

Clearly, an important application of elastic bound-state resonance is as a probe for the energy spectrum of bound states on a surface. Thus, for ^{4}He on graphite, such experiments indicate a lowest bound state of -12.2 meV and -12.3 meV has been reported for H on LiF.[17] A more significant application, possibly, is to use these resonances as a probe of the atom–phonon interaction, and to extend the experiments to *sticking* or *trapping*. We consider first the kinematic conditions for nondiffractive inelastic scattering to a final polar angle θ_f in an event involving *one* phonon of energy $h\omega$ and wave vector Q along the surface

$$\pm h\omega = E_i - \frac{h^2}{2m_g}\left(\frac{\mathbf{K}_i \pm \mathbf{Q}}{\sin \theta_f}\right)^2$$

This may occur directly, but may also take place by a *resonant* process involving the particle diffracting into a bound state assisted by a phonon such that $\mathbf{K} = \mathbf{K}_i + \mathbf{G} \pm \mathbf{Q}$ and the bound state then diffracting into the final state $\mathbf{K}_f = \mathbf{K} - \mathbf{G}$ by the reverse of $\mathbf{G}$. The *resonance* condition is then

$$\pm h\omega = E_i - \varepsilon_n - (h^2/2m_g)(\mathbf{K}_i + \mathbf{G} \pm \mathbf{Q})^2$$

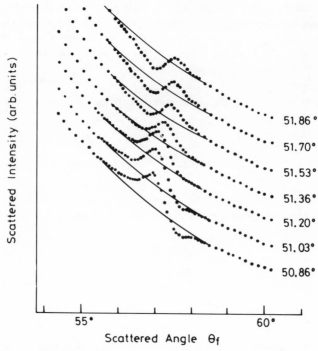

Figure 18.12. Scattered intensity showing interference pattern associated with phonon-assisted selective adsorption involving a bound state. The labels refer to the incident polar angles (after Cantini and Tatarek[18]).

Evidence for this resonant inelastic scattering is shown in Figure 18.12 for the scattering of He from graphite, where clear interference structure is seen on the otherwise smooth inelastic scattered intensity.[18] Further investigations on the He/LiF system[19] have led to the dispersion relations of phonons created or destroyed by inelastic resonant collisions, and have also been used to estimate the lifetimes of the bound resonant states. Thus, for the $n = 3$ state for He on LiF, an energy broadening of 0.08 meV was deduced, corresponding to an estimated lifetime of $5 \cdot 10^{-11}$ s. This implies that the He atom travels some 500 Å on the surface in the $n = 3$ state before re-emission. States of lower n have shorter lifetimes, since the greater proximity of the atom to the surface greatly enhances the possibility of scattering by phonons or impurities as well as diffractive effects.

18.2.2.2. Molecular Effects

The energy transfer from molecules to surfaces is more complex than that of atoms, since the energy of molecules may be distributed over internal degrees of freedom. The transfer of rotational energy is of particular significance and this has recently been studied using laser-induced fluorescence, which can probe the occupancy of the J levels. Scattering of NO from an NO-covered Pt(111) surface was found to give a rotational temperature identical to that of the surface, suggesting that the NO may become bonded to the surface.[20] This interpretation is supported by the cosine angular distribution of the scattered molecules. By contrast, CO scattered from LiF(100) shows incomplete accommodation.[21]

Rotational and translational energy may also be exchanged to permit resonant adsorption to a bound state to take place. Thus, if $\Delta E_J = E_i - [h^2 K^2/(8\pi^2 m_g) + \varepsilon_n]$, then a much enhanced sticking probability may result, as data for HD incident on Pt(111) has demonstrated.[22]

18.2.3. Trapping

If a transition to a truly bound state takes place, by emission of one or more phonons, the molecule or atom is said to be *trapped* by the surface; clearly, this process must take place as a precursor to real catalysis. Experimentally, the transfer of energy from substrate to incident gas, or *vice versa*, is defined in terms of an accommodation coefficient $\alpha(E_i, \Omega_i, T_s) = (\langle E_f \rangle - E_i)/(E_s - E_i)$, where $\langle E_f \rangle$ is the *mean* scattered energy for atoms incident in solid angle Ω_i with energy E_i, and E_s is the mean energy of a hypothetical molecule or atom that has been completely thermally equilibrated with the surface. Evidently, if trapping is to take place, $\langle E_f \rangle$ must be less than the well depth D at the surface, or

$$D/E_i \geq (1 - \alpha)/(1 - \alpha E_s/D)$$

The soft-cube model and its variants can be used to calculate the trapping probability, and the results are surprisingly good, as shown for He and Ne on W in Figure 18.13. It is clear that there is near quantitative agreement with the minimum in α as a function of T_s well predicted.[23] As expected, the trapping fraction f decreases monotonically with T_s.

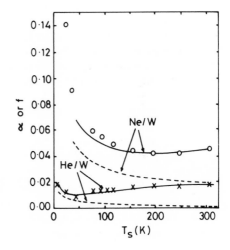

Figure 18.13. Theoretical trapping fractions (dashed curves) and equilibrium accommodation coefficients α (full curves) compared with experimental values of α for He (crosses) and Ne (circles) incident on W. Assumed well depths were 50 K and 350 K, respectively (after Logan[23]).

Table 18.2. Variation of f
with T_s and T_g

T_s	T_g	f
1240	3360	0
1190	1130	0.05
1200	420	0.22
610	2313	0.08
590	1120	0.16
450	3400	0.04
450	1100	0.14
430	290	0.63

Trapping of molecules has also been studied. A paradigmatic investigation of N_2 molecules on polycrystalline W showed that, for high beam and surface temperatures (T_g, $T_s > 1000$ K), the time-of-flight data were consistent with direct inelastic scattering. However, for temperatures below 1000 K, there were contributions to the scattering from both direct inelastic scattering and from trapping followed by phonon recapture and subsequent desorption. If the fraction of the latter is f, then the variation of f with T_g and T_s is given in Table 18.2.

18.3. Chemisorption

We may summarize the events that may take place when an atom or molecule is incident on a surface:

1. The particle may be scattered elastically. This process may take place with diffraction.
2. The particle may be scattered *inelastically* without, however, losing sufficient energy to become trapped. Certain essentially quantum events may take place associated with resonant transitions to a quasi-bound state.
3. The particle may lose sufficient energy to become trapped in an excited physisorbed state.
4. The particle may translate over the surface and then desorb by capture of a phonon or by changing its momentum through the gain of a reciprocal lattice vector.
5. The particle may form a real chemical bond to the surface, a process known as chemisorption.
6. Following chemisorption, the particle may translate, usually by an activated process involving vibrationally highly excited states.
7. Desorption of a chemisorbed state may take place either directly or by thermal excitation to an intermediate physisorbed state.

8. Reactions involving the chemisorbed state may lead to the formation of a
new molecule.

These processes are illustrated schematically in Figure 18.14 and the associated
energetics are shown in Figure 18.15.[6]

18.3.1. Kinetics of Chemisorption

The results of adsorption processes are usually expressed in terms of a
"sticking" probability, S, defined as the ratio of the rate of adsorption of the gas
to the rate of collision with the surface. In general, S is a strong function of the
overall coverage of the surface by adsorbed species, θ, and typical of the type of
variation found experimentally are the data of Figure 18.16.[24] The value of S as
$\theta \to 0$ is written S_0, and results are usually quoted as S/S_0.

We have seen that formation of a trapped physisorbed species is a common
precursor to the act of *chemisorption*. This idea was originally formulated by

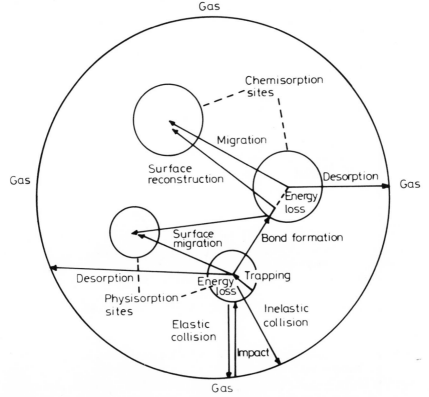

Figure 18.14. Schematic representation of surface processes. The circles represent the tops of the
potential-energy wells. The chemisorption well is much deeper than the physisorption well (after
Gasser[6]).

Kisliuk[25] and can be represented schematically as

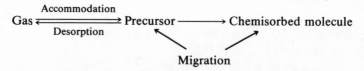

The energetics of one particular process are shown in Figure 18.17[6] for the case of O_2 adsorbing onto W. The precursor state is physisorbed O_2, and dissociative chemisorption takes place with small or zero activation energy to give O_{ads} on tungsten.

Another case in which the energetics are similar is that of H_2 adsorbed onto Ni. In this case, the physisorbed H_2 molecule is thought to lie some 3.2 Å above the surface, which compares to the chemisorbed $Ni—H_{ads}$ bond whose length can be estimated to be *ca* 1.6 Å. In this case, the dissociation energy of H_2 is 434 kJ mol^{-1}, which can be compared to overall enthalpy of adsorption $-\Delta H_c = 125$ kJ mol^{-1}.[26]

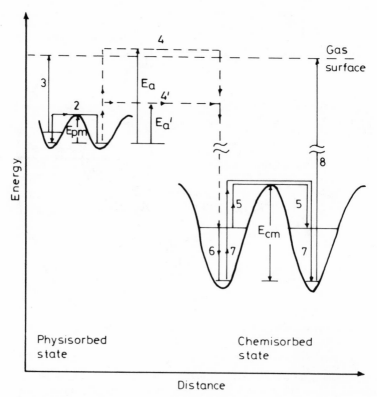

Figure 18.15. Potential-energy profiles for surface processes (not to scale) (after Gasser[6]).

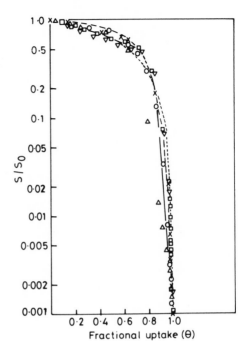

Figure 18.16. Sticking probability curve for hydrogen on rhodium at 160 K. Experimental results at pressures of 2.2×10^{-8} torr ($\times$), 2.8×10^{-8} torr ($\triangle$), 1.7×10^{-8} torr ($\square$), and 1.6×10^{-8} torr (∇) are compared with the theory outlined later in this section (after Edwards et al.[24]).

Returning to the basic model above, we can treat this with a simple kinetic model derived initially by Tamm and Schmidt.[27] If the gas is represented by A, then we have

$$A_{(gas)} \underset{k_d^*}{\overset{S^*f}{\rightleftarrows}} A^* \xrightarrow{k_a} A_{ads}$$

where S^* is the trapping probability to form the precursor, f is the flux of molecules to the surface, A^* is the precursor state, and A_{ads} is chemisorbed A. For a nonreversible chemisorption state, we may write

$$d[A^*]/dt = S^*f - k_d^*[A^*] - k_a[A^*]g(\theta)$$

where $g(\theta)$ is the coverage-dependent probability that an $[A^*]$ molecule finds a vacant state, and θ is the coverage by A_{ads}. If only *one* site is needed, then $g(\theta) \equiv 1 - \theta$, but if two adjacent sites are needed, then $g(\theta) \equiv (1 - \theta)^2$. If A^* is present only in very low stationary concentration, we may invoke the steady-state approximation:[28]

$$d[A^*]/dt \approx 0$$

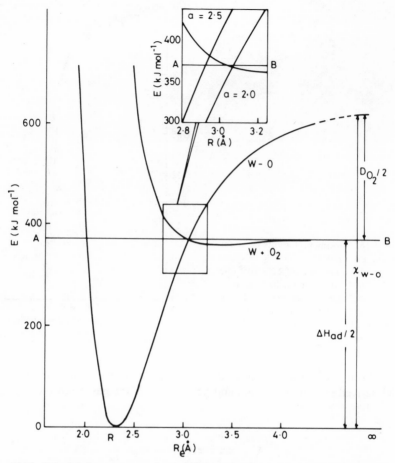

Figure 18.17. Calculated energy curves for the physisorption $(W + O_2)$ and chemisorption $(W - O)$ of oxygen on tungsten. The insert shows the effect of a change in the shape of the Morse potential, as reflected in the constant α. For $\alpha = 2.0$, the curves cross below the zero-energy line AB and chemisorption becomes nonactivated. When $\alpha = 2.5$, chemisorption becomes activated (after Gasser[6]).

In addition, we may use the definition of the sticking probability S as the ratio of the rate of adsorption $\{k_a[A^*]g(\theta)\}$ to the flux to the surface, f, to give

$$[A^*] = Sf/[k_a g(\theta)]$$

and from the steady-state approximation we have

$$S = \frac{S^*}{1 + k_d^*/[k_a g(\theta)]}$$

If we write $k_d^*/k_a \equiv K$, and note that $g(\theta) \to 1$ as $\theta \to 0$ whatever from $g(\theta)$ may take, then

$$S_0 = S^*/(1 + K) \qquad (18.3.1)$$

so, finally,

$$\frac{S}{S_0} = \frac{1 + K}{1 + K/g(\theta)} = \left(1 + \frac{\theta K'}{1 - \theta}\right)^{-1} \qquad (18.3.2)$$

where $K' = K/(1 + K)$ and we have assumed that $g(\theta) = 1 - \theta$. Typical values of S/S_0 for different K' values are shown in Figure 18.18. A comparison with the experimental results for hydrogen adsorption on Rh is shown in Figure 18.16, and in this case $g(\theta)$ was assumed to be $(1 - \theta)^2$.

If we assume for the moment that S^* does not vary with temperature, then, from equation (18.3.1) above, we have

$$\frac{1 - S_0/S^*}{S_0/S^*} = K = k_d^*/k_a \sim \exp\left[(E_a - E_d^*)/kT\right]$$

If $k_d^* \sim \exp(-E_d^*/RT)$ and $k_a \sim \exp(-E_a/RT)$, then, taking logarithms and differentiating, we find

$$\frac{\partial\{\ln\left[(1 - S_0/S^*)/(S_0/S^*)\right]\}}{\partial(1/RT)} = E_a - E_d^*$$

If chemisorption is nonactivated, i.e., $E_a < E_d^*$, then increasing the temperature will cause S_0 to *decrease*, while if chemisorption is activated, so $E_a > E_d^*$, then the reverse will take place.

The basic assumption that S^* is independent of temperature is, however, not always borne out experimentally. Thus, the trapping experiments for N_2 on polycrystalline W referred to earlier measure a fraction f of molecules that are trapped. The fraction of molecules that are trapped and then desorbed is proportional to $S^* - S_0$, so

$$f = (S^* - S_0)/(1 - S_0)$$

A comparison of the variation of f and S_0 with temperature reveals immediately that S^* must itself be a function of T.[23]

From the form of equation (18.3.2) we can also calculate the change in S/S_0 as T increases. If $E_a < E_d^*$, then K will increase with T and, since $g(\theta) < 1$, S/S_0 *decreases* with T. Physically, what occurs is that the weakly held precursor A^* increasingly tends to *desorb* as T increases, rather than becoming chemisorbed.

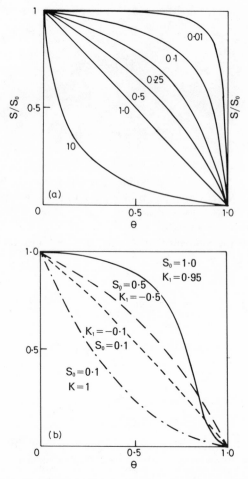

Figure 18.18. Theoretical sticking probability curves for (a) single-site adsorption according to the equation $S/S_0 = [1 + \theta K/(1 - \theta)]^{-1}$ for various K values and (b) two-site adsorption according to the equation $S/S_0 = \{(1 - \theta)^2/[1 - \theta(1 - K_1) + \theta^2 S_0]\}$ for various values of K_1 and S_0 (after Kisliuk[25]).

Our analysis has neglected any dependence of k_a on coverage, but this is unlikely in general to be valid. One well-known exception is that of N_2, which is dissociatively chemisorbed on W(100). In this case, the variation of S with coverage was modified by the existence of a repulsive potential W between adjacent N atoms and detailed statistical arguments[29] lead to the expression

$$g(\theta) = 1 - \theta - \left(\frac{2\theta(1 - \theta)}{[1 - 4\theta(1 - \theta)B]^{1/2} + 1}\right)$$

where $B = 1 - \exp(-W/kT)$. On $W(110)$, N_2 behaves quite differently. At ordinary temperatures, the surface shows very little chemisorption activity, but at lower temperatures (85–200 K), N_2 is weakly adsorbed to a state denoted γ-N_2 that is not dissociated and apparently sits upright on the surface.[30] The sticking probability for the γ-state actually rises with θ from $\theta = 0$ and goes through a maximum. This unusual behavior was ascribed to a cooperative transfer from precursor to γ-state when an adjacent site is already occupied.

18.3.2. Desorption

In general, the rate of desorption can be written in the form

$$-d\theta/dt = \nu\theta^\alpha \exp(-E_d/RT)$$

where θ is the coverage, α the kinetic order, and ν the attempt frequency. Experimentally, desorption kinetics are usually measured by heating the sample (often very rapidly) and measuring the amount of material released. If the temperature of the sample is increased at a rate linear with time, i.e., $T(t) = T_0 + \beta t$, then

$$-d\theta/dT = (\nu/\beta)\theta^\alpha \exp(-E_d/RT)$$

Under fast pumping, there will clearly be a peak in the pressure burst at some temperature T_ρ as the adsorbate is tripped off. At this point, $\partial^2\theta/\partial T^2 = 0$ and

$$(\nu\alpha\theta^{\alpha-1}/\beta) \exp(-E_d/RT_\rho) = E_d/RT_\rho^2$$

If desorption is first order, $E_d/RT_\rho^2 = (\omega/\beta) \exp(-E_d/RT)$, and T_ρ is *independent* of coverage. A typical result is shown in Figure 18.19 for desorption of Cl atoms from the $W(100)$ surface.[31]

More complex behavior is found for CO on polycrystalline tungsten, where several thermal desorption peaks are found. These peaks are shown in Figure 18.20, and T_ρ is again independent of coverage.[32]

This was taken as evidence of *molecular* adsorption, and it was assumed that both terminal $M-C\equiv O$ and bridging units were present. In fact, recent evidence suggests that strongly adsorbed CO may dissociate; if there are strong lateral repulsions between CO fragments, then the type of behavior shown in Figure 18.20 may be accounted for on this basis.[33]

Desorption has also been explored by molecular-beam methods. The scattering of CO from Pd(111) has been found to give rise to a cosine distribution indicative of extensive trapping and equilibration.[34] The detected intensity is

$$I_d(\theta, T) = Y\{I_0[1 - S(\theta, T)] + D(\theta, T)\}$$

where Y is a geometric factor, S the sticking coefficient, and D the desorption rate. We may define an "effective" sticking coefficient, S_{eff}, through the equation

$$I_d(\theta, T) = Y[1 - S_{eff}(\theta, T)]$$

so

$$D(\theta, T) = I_0[S(\theta, T) - S_{eff}(\theta, T)]$$

At 374 K, it is found that S and S_{eff} are identical. Also, there appears to be very little temperature dependence of S, so we have

$$D(\theta, T) \approx I_0[S(\theta, 374) - S(\theta, T)]$$

Analysis gives $\nu = 10^{14.4 \pm 0.8}$ and $E_d = 32 \text{ kcal mol}^{-1}$ at $\theta = 0$ and 26 kcal mol^{-1} at $\theta = 0.42$, consistent with direct isosteric heat measurements of 34 and 30 kcal mol^{-1}, respectively.

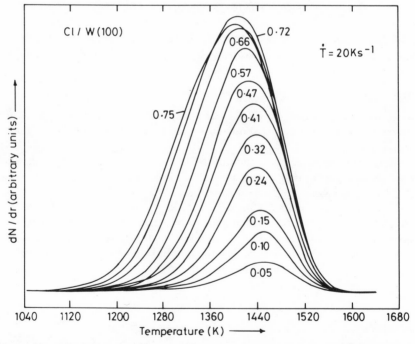

Figure 18.19. Desorption of chlorine atoms from tungsten (100). The mass spectrometer was in line of sight from the sample. The curves show the essentially temperature-independent maximum characteristic of first-order desorption. The curves can be analyzed to yield a coverage-independent energy of desorption of 82 kcal mol^{-1} (after Kramer and Bauer[31]).

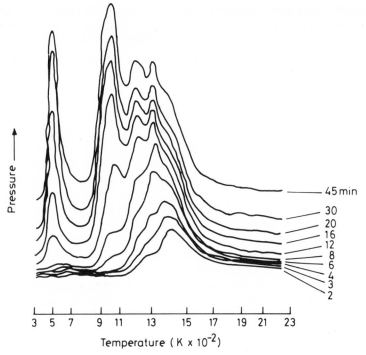

Figure 18.20. Desorption of CO from polycrystalline W after increasing exposure (after Redhead[32]).

18.3.3. Mobility on Surfaces

There are, in principle, two régimes that can describe mobility. In the first case, the ground state energy is sufficiently high to allow relatively uninhibited diffusion laterally, even at low temperature. This situation may arise either with very light gases (that have high zero-point energy) or when the surfaces show little potential variation (κ_G small). The diffusion coefficient, D, is then given by the standard expression

$$D = \langle [\mathbf{r}(t) - \mathbf{r}(0)]^2 \rangle / 4t$$

and only shows a small (power law) variation with T. An example is He on graphite.

The second case involves thermally activated hopping from site to site, which can be characterized as a random walk with step length l and time interval τ,

$$D = l^2/4\tau \equiv l^2 \nu \exp(-\Delta H_m/RT) \equiv D_0 \exp(-\Delta H_m/RT)$$

The value of ΔH_m is often *ca* 20% of the binding energy at a surface site and depends sensitively on the crystal face. A crude estimate of D_0 can be made by taking l as the lattice constant and multiplying by the Debye frequency; this leads to $D_0 \approx 10^{-2}$ cm^2 s^{-1}. Typical examples of activated diffusion are found for metal atoms on surfaces.

An interesting intermediate case is that of H$^{\cdot}$ on W(110);[35] at high temperatures the diffusion is activated, but at lower temperatures D varies more slowly with T, suggesting that quantum tunneling may play a major role in the low-temperature régime.

A variety of experimental methods have been used to probe diffusion on surfaces, including field ion and field emission microscopy (FIM and FEM), quasi-elastic neutron scattering (QUENS), NMR, and Mössbauer spectroscopy (MS). FIM has been used *inter alia* to study the migration of Si on W(110).[36,37] The silicon atom occupies three-coordinate sites on the W(110) surface, and data for isolated atom migration gave $\log_{10} D_0 = -3.5 \pm 1.3$ and $-\Delta H_m = 0.70 \pm 0.07$ eV, the latter being about 15% of the enthalpy of adsorption. This system has also yielded data on Si–Si interactions; analysis of the pair-correlation potential function suggests that V_{ij} is oscillatory, and so large as to lead to superlattice formation.

Understanding of surface diffusion still remains rather sketchy, particularly at high coverages, where the behavior may approximate that of a dense bidimensional fluid. Intermolecular forces in the 2D fluid are now dominant, and the effective *site* potential may become smeared out to form a slightly modulated broad energy well. This transformation, similar to that which is used to lead to the elctron-gas model of a metal, leads to a physical picture in which individual molecules spend a relatively long time within a very localized area until thermal motions open up a passage to allow translation to take place. Thus, the mobility becomes cooperative in nature, and the diffusion coefficient may, by analogy with the Cohen–Turnbull theory,[38] be written as

$$D \sim D_0 \exp\left[-\kappa/(T - T_0)\right]$$

where κ is a constant.[39] Physically, T_0 has a significance similar to that of the glass transition.

18.4. The Chemisorbed Layer

18.4.1. Thermodynamics of Chemisorbed Layers

We have seen that physisorption and chemisorption are clearly linked, and that the transfer from the one state to the other may take place more or less rapidly depending on the balance of rate constants as discussed in the previous section. We now turn to a consideration of the adsorbed layer on a variety of substrates. The behavior of the layer will differ qualitatively according to the energy of interaction between adsorbed species and we can treat this at the semiquantitative level using the Bragg–Williams approximation. We assume

1. That adsorbed atoms are localized on well-defined and equivalent sites of the substrate.
2. That interactions are restricted to nearest neighbors.
3. That there is a random distribution of adsorbed atoms on substrate sites (corresponding to the theory of *regular solutions*).

We let θ be the coverage, E_s the potential energy of a surface site, and w the energy of interaction between adsorbed nearest neighbors such that $w > 0$ corresponds to repulsion and $w < 0$ to attractive interactions. If z is the number of nearest neighbors in the layer, then

$$\mu_s = E_s + zw\theta + kT \ln [\theta/(1 - \theta)] - kT \ln [f_{vib}(T)]$$

where $f_{vib}(T)$ is the vibrational partition function (pf) of an adsorbed atom. If there is equilibrium between the surface and gas-phase atoms, and, in the gas phase,

$$\mu_g = \mu_g^0 + kT \ln (p)$$

where p is the gas pressure, then

$$p = [\theta/(1 - \theta)]\alpha(T) \exp [(E_s + zw\theta)/kT]$$

This result is termed the Fowler–Guggenheim isotherm[40] and it clearly reduces to the well-known Langmuir isotherm

$$p = \theta/[K(1 - \theta)]; \qquad K \sim \exp (-E_s/kT)$$

in the limit that $w \to 0$. If $p_{1/2}$ is the pressure of gas corresponding to $\theta = \frac{1}{2}$, then a plot of $\ln [p(\theta)/p_{1/2}]$ vs. θ will have the form shown in Figure 18.21 for different

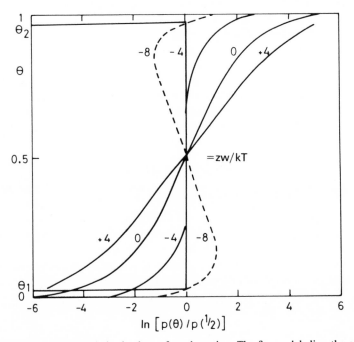

Figure 18.21. Fowler–Guggenheim isotherm for adsorption. The figures labeling the curves are values of zw/kT.

values of w. It will be seen that for $zw/kT < -4$, the isotherm shows a double loop characteristic of a *phase transition*. Physically, in this region, the layer consists of a dilute phase of local coverage θ_A which coexists with a dense phase of coverage θ_B. The relative proportions of these two phases are determined by the total coverage θ as calculated from the Fowler-Guggenheim isotherm.

Although this model does have qualitative appeal, quantitative agreement between experiment and theory is poor, even when single-crystal surfaces are used, since

1. The low coverage phase is extremely sensitive to the presence of impurities and defects on the surface.
2. Particularly in the case of chemisorption, thermodynamic equilibrium is difficult to achieve save at high temperatures and low pressures.
3. Adsorption on a particular site may profoundly modify the value of E_s not only for nearest-neighbor sites but for sites several interatomic distances away.
4. The underlying substrate may undergo surface reconstruction as is found for the adsorption of H_2 on Ni(110).
5. There may be an intrinsic variation in E_s on heterogeneous surfaces.

A paradigmatic example of a phase transition of the above type is found for the chemisorption of S on Ag(100). By using mixtures of H_2S and H_2 at temperatures in the region of 200 °C, two phases could be identified and characterized.[41] A high coverage-density phase has sulfur coordinated to two silver atoms and a low coverage-density phase has sulfur coordinated to four silver atoms. Both these phases are present between the limiting coverages θ_A and θ_B, and can be "frozen in" by rapid quenching under vacuum.

In general, the assumption of a constant coverage-independent value of E_s is a severe limitation as indicated above, even on single-crystal substrates. Even if we restrict ourselves to monolayer adsorption, the existence of a *range* of E_s values will lead to a modified form of the adsorption isotherm. The simplest case will be found if the heat of adsorption declines *linearly* with θ, such that

$$-\Delta H_{ads} = -\Delta H_0(1 - \beta\theta)$$

(since $\Delta H_{ads} < 0$). Then, approximately

$$\theta \approx -(RT/\beta\Delta H_0) \ln(\kappa p)$$

where κ is a constant related to the enthalpy of adsorption. This is termed the Temkin isotherm. If the heat of adsorption declines logarithmically with coverage over a range of (intermediate) θ values, such that

$$-\Delta H_{ads} = -\Delta H' \ln(\theta)$$

then, approximately, over the same intermediate range of θ values

$$\theta \sim kp^{1/n}$$

where $n > 1$ and is independent of θ. This is termed the Freundlich isotherm, and is frequently found for θ values between 0.2 and 0.8.

These two isotherms have been used and discussed extensively.[6,26,42] It must be emphasized that not only are data often insufficiently precise to distinguish the various isotherms, but the fact that one of them might appear to fit the data reasonably well cannot be taken as evidence for the veracity of the underlying assumptions.

18.4.2. Enthalpies of Chemisorption

The values of $-\Delta H_{ads}$ show wide variation for both gas and substrate. In general, the order of $-\Delta H_{ads}$ for gases over a wide range of substrates follows the order $O_2 > C_2H_2 > C_2H_4 > CO > H_2 > CO_2 > N_2$ with the notable exception of Au, which does not adsorb O_2. Metals may be classified, following Bond,[26] into seven classes according to their ability to chemisorb gases, and this classification is given in Table 18.3.

Examination of Table 18.3 shows that transition metals are strong adsorbers while main-group metals are not. It has been suggested that unpaired d electrons are necessary to stabilize the precursor state and therefore to permit transition to a strongly bonded chemisorbed state to take place without a high activation energy.

If we compare the enthalpies of *adsorption* and those of oxide *formation*, then the pattern is shown in Figure 18.22.

The first point to note is that there is a remarkable parallelism of the data for the first transition series. For the second and third rows, the bulk oxides are significantly *less* stable than the adsorbed oxygen layers, reflecting possibly the very high atomization energies found for the metals in this part of the periodic table. A similar situation is found in the nitrides, though data are less complete, but, for hydrogen, the data for bulk hydrides are too sparse for a meaningful comparison.

Table 18.3. Classification of Metals According to their Ability to Chemisorb Gases

Class	Metal or group in periodic table[a]	O_2	C_2H_2	C_2H_4	CO	H_2	CO_2	N_2
					Gases			
A	Groups IVA, VA, VIA, VIIIA$_1$	Y[b]	Y	Y	Y	Y	Y	Y
B$_1$	Ni, Co	Y	Y	Y	Y	Y	Y	N
B$_2$	Rh, Pd, Pt, Ir	Y	Y	Y	Y	Y	N	N
B$_3$	Mn, Cu	Y	Y	Y	Y	Y/N	N	N
C	Al, Au	Y	Y	Y	Y	N	N	N
D	Group IA	Y	Y[c]	N	N	N	N	N
E	Mg, Ag, Group IIB In, Groups IVB, VB	Y	N	N	N	N	N	N

[a] Group IVA: Ti, Zr, Hf; VA: V, Nb, Ta; VIA: Cr, Mo, W; IA: Li, Na, K; IIB: Zn, Cd; IVB: Si, Ge, Sn, Pb; VB: As, Sb, Bi.
[b] Y: strong chemisorption; Y/N: weak chemisorption; N: no observable chemisorption.
[c] Ethyne adsorbs on IA metals as $2M + C_2H_2 \rightarrow M^+C_2H^- + MH$.

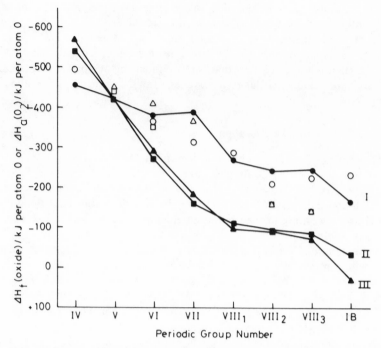

Figure 18.22. Enthalpies of formation of oxides (filled points) and of chemisorption on metals (open points) as a function of periodic group number. Circles: first transition series; squares: second transition series; triangles: third transition series (after Bond[26]).

For the adsorption of CO, the enthalpies of adsorption also show a semiquantitative similarity to the bond enthalpy data for the M—CO bond in metal carbonyls, at least in the later transition metals. However, the early transition metals show a *much* higher value of $-\Delta H_{ads}$ than bond enthalpy, a phenomenon associated with the probable *dissociative* chemisorption of CO on groups IV-VI.

A very similar set of data is available for the adsorption of CO_2 (see Fig. 18.23).

18.4.3. Chemisorption on Nonelemental Materials

The study of the adsorption of gases on metals has been central to the development of our theoretical understanding of catalysis, but it is a fact that an ever-increasing number of industrial catalysts are based on oxides, sulfides, and pnictides. Not only binary, but ternary and quaternary materials are known, and a major problem in modern catalytic studies is to establish the surface composition of often complex catalyst formulations. A second difficulty is that the surface composition may alter as a result of exposure to the reactant gas mixture, and a third problem is that even if single crystals of known surface composition can be synthesized, the much higher surface potentials associated with ionic or partly

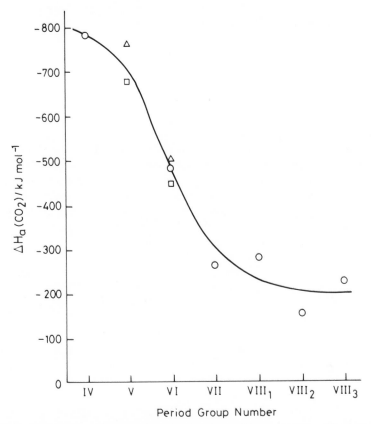

Figure 18.23. Enthalpies of chemisorption of carbon monoxide as a function of Periodic Group Number. Symbols as for Figure 18.22 (after Bond[26]).

ionic materials will lead to a much higher sensitivity of chemisorption to crystal face and to the presence of kink and step sites on the surface.

18.4.3.1. Surface Composition

If surface and bulk compositions differ in the intrinsic material, we have a phenomenon known as *segregation*. Simple segregation may be treated by an extension of the statistical treatment given at the beginning of this chapter. In dilute solution, the chemical potential of a dissolved atom will take the form

$$\mu_b = E_b + kT \ln (x) - kT \ln [f_b(T)]$$

where x is the mole fraction. E_b the potential energy of a dissolved atom, and $f_b(T)$ the associated change in vibrational partition function associated with the presence of the atom. Equating this to μ_s gives a relationship between x and θ very similar to the Fowler–Guggenheim isotherm:

$$x = [\theta/(1 - \theta)]\alpha'(T) \exp [(zw\theta - E)/kT]$$

where $E = E_b - E_s$.

Perhaps the simplest example of this kind is found for nickel metal containing a low (<1%) concentratoin of carbon in the *bulk* phase. At high temperature ($T > T_s$), the surface coverage of carbon on Ni(111) is low, and appears to approximate the bulk concentration, as shown in Figure 18.24.

At lower temperatures, the C–C interactions on the surface become dominant and a phase transition takes place to yield a monolayer phase. At a lower temperature still, bulk graphite forms at the surface. This remarkable behavior has been analyzed by Blakely and co-workers[43,44] who conclude that the binding energy of carbon in the monolayer phase is some 10% higher than E_b.

A similar result has been reported by Egdell and co-workers for antimony segregation onto $SnO_2(001)$.[45] It is well known that SnO_2 containing *ca* 3% Sb shows both metallic conductivity and transparency to visible light. This remarkable combination of properties has attracted the attention of many scientists, including catalyst chemists, and early results in the field of catalysis were interpreted in terms of the electronic theory of chemisorption. However, Egdell demonstrated clearly that, following equilibration at high temperatures, the surface coverage of Sb was far higher than the bulk composition value would suggest, being close to 50% in the (001) face.

In both the Ni(111)/C and $SnO_2(001)$/Sb cases, geometrical models showed that a stable chemical form could exist. For Ni(111)/C, an excellent fit of substrate and ordered hexagonal carbon layer was demonstrated, and the importance of this is pointed up by the fact that other Ni faces may exhibit quite different behavior, including that of showing no detectable segregation of carbon to the surface. In

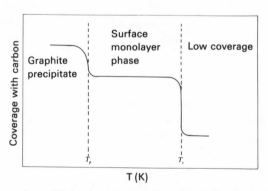

Figure 18.24. Schematic equilibrium temperature dependence of carbon coverage on the (111) surface of a carbon-doped nickel single crystal. A phase transition from a low coverage to a condensed state takes place at the segregation point T_s. Graphite precipitation starts at T_p (after Blakely and co-workers[43,44]).

the SnO_2 case, Sb is present as Sb(V) in the bulk but apparently as Sb(III) at the surface, and the stereochemical effect of the lone pair is believed to be of importance in stabilizing adsorption at low-coordination-number surface sites.

An interesting example of segregation is found for the mixed oxides CoO-Cr_2O_3 and NiO-Cr_2O_3. Measurements of the activation energy for surface diffusion of cations and XPS data both indicate a considerable segregation of Cr to the surface—even in samples containing less than 1% Cr_2O_3.[46] The surface diffusion measurements are, in fact, consistent with the formation of a spinel phase $CoCr_2O_4$ at the surface. Similar conclusions have also been drawn for the NiO-Cr_2O_3 case, and SIMS data reveal that the surface composition relaxes toward that of the bulk over an extremely narrow range of 10–20 monolayers.[47]

18.4.3.2. Dynamic Composition Effects

Many transition-metal oxides are effectively nonstoichiometric, and their composition will depend on the partial pressure of oxygen in the gas phase. Two types of behavior are found, depending on whether the metal ion has an easily accessible higher or lower oxidation state. In the first case, oxygen is *gained* on heating the crystal in O_2, usually by formation of metal vacancies such as

$$\tfrac{1}{2}O_2 \rightarrow V_M + O_O$$

$$M_M + V_M \rightarrow M_M^{\cdot} + V_M'$$

a situation found for NiO and CoO. In the second case, oxygen is *lost* either by formation of oxygen vacancies:

$$O_O \rightarrow V_O + \tfrac{1}{2}O_2$$

as in TiO_2 (at low vacancy concentration), or by formation of metal interstitials:

$$MO \rightarrow M_i + V_O + \tfrac{1}{2}O_2$$

as found for ZnO. In certain cases, at higher oxygen-vacancy concentrations, the vacancies may *order* to form shear planes, and examples include TiO_2 and V_2O_5.

The significance of these vacancies is two fold. First, they may confer electronic conductivity on the oxide and this permits redox reactions to take place on the surface. Second, they may also show a marked tendency to segregate to the surface under certain conditions. In addition, dynamic processes may result in the accommodation of defects by chemical reactions; thus, hydrogen is believed to adsorb heterolytically on reducible oxides such as

$$H_2 + M^{2+} + O^{2-} \rightarrow H-M^+ + O-H^-$$

Heating will cause the formation of V_O and reduced metal species. Similar effects are found for CO adsorption such as

$$CO + M^{2+} + O^{2-} \rightarrow M^0 + CO_2$$

In a similar way, O_2 may adsorb on oxidizable oxides such as

$$2Ni^{2+} + O_2 \rightarrow 2(O^- \cdots Ni^{3+})$$

and high surface coverages of O^- may result.

More subtle effects may involve the actual reconstruction of the surface. One example is that of $Cu_2^I Mo_3^{VI} O_{10}$. This is a selective oxidation catalyst that converts but-1-ene to butadiene but its activity and selectivity can be profoundly modified by exposing the surface to pulses of pure oxygen. It has been shown that this is associated with the formation of clusters of $Cu^{II} Mo^{VI} O_4$ which, on re-reduction to the parent compound, have a very high activity for the selective oxidation process. Of course, this process may result in the disintegration of the catalyst over a period of time, and this may make it quite unsuitable for technological applications.

The central point here is that the state of the surface during dynamic operation may be completely different from that expected on the basis of the known stoichiometry of the bulk catalyst, and that the state of the surface may have a profound influence on the course of a particular reaction.

18.4.3.3. Influence of Surface Geometry and Structure

Although speculation on the importance of surface geometry has a long history (see, e.g., Rienecker[48]), the first direct answer was given in an elegant series of papers by Samorjai and co-workers.[49] In these experiments, dehydrogenation and hydrogenolysis were studied on a wide variety of crystal planes of a single-crystal platinum substrate. High-index planes contain controlled concentrations of kink and step sites, and these can be monitored by LEED. The rate of hydrogenolysis was found to depend strongly on the concentration of steps and kinks, and it was clear that the reaction must involve low-coordinated platinum atoms present at these sites. By contrast, dehydrogenation does not appear to depend on such sites and apparently takes place on the terraces.

Within the selective oxidation field, equally interesting data have been provided by Gasiov and Machej[49] for the oxidation of o-xylene on V_2O_5. High conversion to phthalic anhydride was found for crystallites exposing the (001) face, which consists of V=O units arranged perpendicular to the surface plane. However, if the crystallites expose the (110) face, at which shear planes may nucleate and oxygen therefore removed more easily, complete oxidation to CO and CO_2 becomes the favored route. A similar phenomenon was observed on MoO_3 for the oxidation of methanol to HCHO or MeOMe.[50]

18.4.4. More Complex Adsorption Problems

18.4.4.1. Adsorption of CO on Metals

The adsorption of CO on transition metal substrates has been studied extensively and some degree of unanimity now exists in the literature. The commonest adsorbed form consists of CO *terminally* bonded to single surface atoms in the plane, and this form is generally prevalent at high coverages on those metals that do not dissociatively adsorb CO. At lower coverages, different ordered structures are possible, and evidence for these various structures has come from a variety of techniques.

LEED has now developed to the point where good agreement between experimental patterns and those calculated from the correct structural model may be anticipated, and Figure 18.25 shows both experimental data and the result of model calculations for Ni{001}c(2 × 2)–CO.[51]

The high oscillator strength associated with $\nu(C{\equiv}O)$ in the IR has led to the extensive exploitation of vibrational spectroscopy in the study of CO adsorption, and some results for the adsorption of CO on polycrystalline Rh particles on Al_2O_3 are shown in Figure 18.26.[52] The IR spectra show four peaks:

1. A broad peak at 1866 cm^{-1} that shifts to 1870 cm^{-1} at maximum uptake.
2. A weak band that shifts from 2050 to 2070 cm^{-1} as the coverage increases.
3. An intense doublet at 2101 and 2031 cm^{-1} whose frequencies are coverage-independent.

There are two observations that call for particular comment:

1. Why do bands (1) and (2) above shift to higher frequency with coverage?
2. What are the origins of the bands?

The assignment of the bands is straightforward. Band (1) is derived from CO adsorbed at *binary* sites

(2) is derived from CO adsorbed at *single* sites

and (3) is derived from *individual atoms* of Rh on the Al_2O_3 that can coordinate *two* carbonyls:

$$\underset{\displaystyle Rh}{\underset{\displaystyle \diagdown \qquad \diagup}{\overset{\displaystyle \underset{\displaystyle C}{\overset{\displaystyle O}{\|}} \qquad \underset{\displaystyle C}{\overset{\displaystyle O}{\|}}}{}}}$$

The shift of (1) and (2) was thought originally to be associated with the increased demands on the metal d_π donor orbitals. Back donation of these to CO reduces $\nu(C\equiv O)$, but as coverage increases, so does competition for the d_π orbitals and these therefore become less effective in terms of individual back-bonding. Plausible though this analysis is, it now appears not to be the whole story, and dipole–dipole interactions also play an important role. Very similar shifts in the frequency of terminal and bridge-bonded CO have been observed for CO on Pt(111) as shown in Figure 18.27.[53]

For Pd(100) at $\theta_{CO} < 0.5$, a single peak is observed that shifts from 1895 cm⁻¹ at low coverage to 1949 cm⁻¹ at $\theta_{CO} = 0.5$. This peak is associated with *bridge-bonded* CO and a structure proposed by Ertl[54] is shown in Figure 18.28.

For CO adsorption on Rh(111) the absorption frequencies vary with coverage as shown in Figure 18.29.[55] Clearly, the principal peak, at 1990 cm⁻¹, shifts to higher frequency with coverage, reaching a limit of 2070 cm⁻¹ at very high coverage (not shown). There is also a peak that has a frequency of 480 cm⁻¹ at low coverage and which shifts *down* in frequency with increasing coverage, eventually reaching 420 cm⁻¹. In addition, a shoulder appears on the higher-frequency band at 1870 cm⁻¹, and this does not shift with coverage. Once again, the 1990/2070 cm⁻¹ peak is derived from a linearly bonded CO and the 1870 cm⁻¹ peak is due to

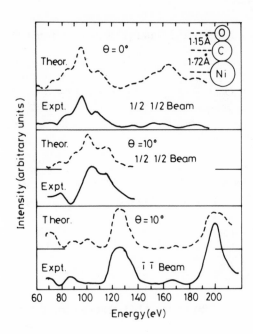

Figure 18.25. Comparison of the experimental and theoretical LEED spectra from Ni{001}c(2 × 2)-CO. The structural model is shown schematically at the top (after Passler *et al.*[51]).

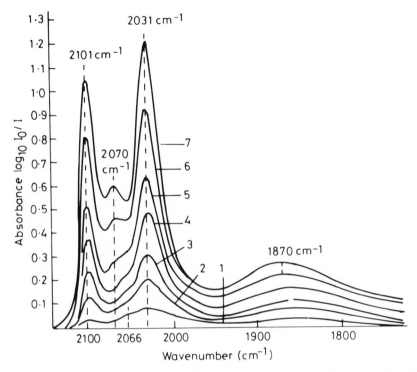

Figure 18.26. IR spectra for ^{12}CO adsorbed on rhodium for increasing CO coverage ($T = 295$ K): curve 1, $p_{CO} = 2.9 \times 10^{-3}$ torr; curve 2, $p_{CO} = 4.3 \times 10^{-3}$ torr; curve 3, $p_{CO} = 5.0 \times 10^{-3}$ torr; curve 4, $p_{CO} = 8.3 \times 10^{-3}$ torr; curve 5, $p_{CO} = 0.76$ torr; curve 6, $p_{CO} = 9.4$ torr; curve 7, $p_{CO} = 50$ torr (after Yates et al.[52]).

bridge-bonding CO. The 480/420 cm^{-1} band is associated with the Rh—C stretch, and it is interesting that the reduction of the d_π back-bonding postulated to account for in the $\nu(C\equiv O)$ increase would also predict a lowering of the bond order of the M—C bond and hence account for the decreasing frequency.

The adsorption of CO on Pd(111) shows how different *ordered* structures can evolve into one another. At $\theta = \frac{1}{3}$, the structure shown in Figure 18.30a is postulated in which CO is *triply* bonded, with a very low stretching frequency of 1823 cm^{-1}. If this structure is compressed, it evolves to that of Figure 18.30b, in which the CO are *doubly* bridging. The coverage is now 0.5 and the stretching frequency 1936 cm^{-1}. This coverage is the highest that can be attained at room temperature, but cooling results in further adsorption to give the highly compressed structure of Figure 18.30c with both triply and linearly bonded forms. The ease with which these films reorganize suggests that CO must be highly mobile on the surface of metals, and this is consistent with ^{13}C NMR results on carbonyl clusters.

The bonding of terminal CO has also been explored by UV-PES, and the major shift in electronic energy level on adsorption takes place in the 5σ (i.e., carbon lone pair) orbital. It can be seen from Figure 18.31 that this 5σ level is

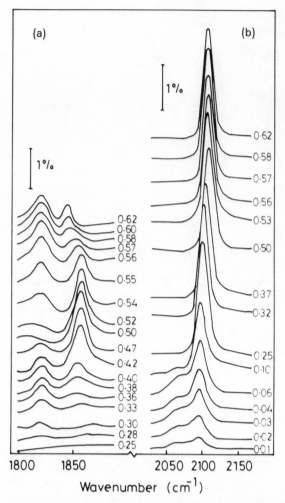

Figure 18.27. Reflection-absorption IR spectra for CO on Pt(111) at (a) 120 K and (b) 200 K. The band grows in intensity up to $\theta = \frac{1}{3}$ but is constant at higher coverages (after Hayden and Bradshaw[53]).

well separated from the 1π level in the free molecule, but on adsorption this orbital becomes accidentally degenerate with the 1π level,[57] a result similar to that found for CO on Pd(110).[58]

Dissociative adsorption of CO is of prime importance in understanding certain mechanisms, and has been explored using X-PE spectroscopy. An interesting example is iron: at low temperatures, on Fe(100), the UV-PE spectrum of adsorbed CO is typical of terminally bonded molecular CO with a pronounced differential shift of the 5σ orbital. Similar behavior is also encountered in polycrystalline Fe, and a study by X-PES shows a C 1s peak characteristic of molecularly adsorbed CO, provided the adsorption and subsequent measurement both take place well

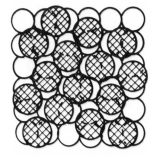 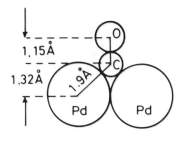

Figure 18.28. Structural model for CO adsorbed on palladium (100). The CO occupies bridging sites and gives a $(2\sqrt{2} \times \sqrt{2})R45°$ structure for which $\theta_{CO} = 0.5$ (after Ertl[54]).

below room temperature. However, if such an adsorbed layer is heated above 350 K, the C 1s signal of *molecular* CO vanishes and is replaced by one characteristic of adsorbed carbon *atoms*.[59] Further investigation along these lines has suggested that dissociative chemisorption is typically found on metals where $-\Delta H_{ads} >$ *ca* 250 kJ mol^{-1}. Iron, with $-\Delta H_{ads} \approx 180$ kJ mol^{-1}, is clearly borderline.

18.4.4.2. Adsorption of CO on oxides

On oxides, unlike metals, the frequency of the CO band may be shifted to *higher* frequency.[60,61] The absorption frequency of CO in the gas phase is

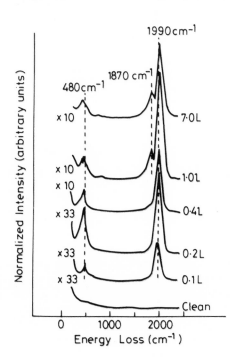

Figure 18.29. Vibrational spectra of CO chemisorbed on an initially clean rhodium (111) single-crystal surface at 300 K as a function of gas exposure (after Dubois and Samorjai[55]).

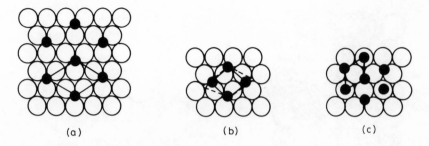

Figure 18.30. Models for the adsorption of CO on palladium (111): (a) a $(\sqrt{3} \times \sqrt{3})R30°$ structure at $\theta_{CO} = \frac{1}{3}$; (b) a $c(2 \times 4)$ structure at $\theta_{CO} = 0.5$; (c) a hexagonal structure at $\theta_{CO} = 0.66$ (after Conrad et al.[56]).

2143 cm^{-1} and reaches 2174 and 2187 cm^{-1} on V^{3+}/SiO_2, where the suggested geometry of the surface site is[62]

$$
\begin{array}{ccc}
\text{O} & & \text{O} \\
\text{C} & & \text{C} \\
\backslash & & \diagup \\
\text{O} & \text{—V}^{3+}\text{—} & \text{O} \\
& | & \\
& \text{O} &
\end{array}
$$

and 2185 cm^{-1} on TiO_2.[63] Interestingly, another peak, at 2115 cm^{-1}, can be seen on TiO_2, which has been ascribed to some form of CO acting as an intermediate

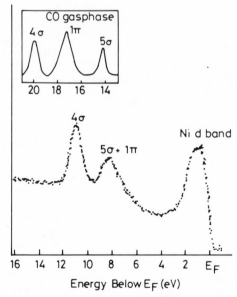

Figure 18.31. UV-PE spectrum for CO adsorbed on Ni(100) at $\theta = 0.6$ (after Bradshaw[57]).

Table 18.4. Species Formed under Various Conditions

Frequency (cm^{-1})	Species
2125, 2075	$Ni \overset{CO}{\underset{CO}{<}}$ and/or $Ni(CO)_3$
1970	$Ni-CO$
1910, 1880	$\overset{Ni}{\underset{Ni}{>}} CO$ and/or $Ni-CO$
1730, 1160	Organic-like carbonate
1655, 1533, 1280	Bulk carbonate
1630, 1318	Bidentate carbonate
1580, 1370	$Ni-C \overset{O}{\underset{O}{<}}$
1470	Monodentate carbonate
1215	Hydrogen carbonate

in the formation of surface adsorbed HCO_3^-. Adsorption of CO on NiO has been studied extensively,[64] and Table 18.4 shows those species formed under various conditions. Finally, CO adsorbed on ZnO has been reported to have a frequency of 2212 cm^{-1}.[65] As a rule of thumb, it has been suggested that $M^{2+}-CO$ will have a stretching frequency in excess of 2170 cm^{-1}, M^+-CO will lie between 2120 and 2160 cm^{-1}, and M^0-CO will lie below 2100 cm^{-1}.

18.5. Kinetics and Mechanism in Catalysis

18.5.1. Fundamentals

We shall first examine some of the ways in which the basic processes occurring at the surface can be described and the reactions formulated. The most straightforward models that have been suggested are the Langmuir–Hinshelwood mechanism, which envisages the reaction as taking place solely between surface adsorbed species, and the Eley–Rideal mechanism, which permits the involvement of both surface and gas-phase species in the overall reaction.

18.5.1.1. The Langmuir–Hinshelwood Mechanism

The main assumptions are:

1. The surface reaction is rate-limiting.

2. The Langmuir isotherm can be applied to describe the equilibrium between the gas phase and adsorbed reactants and, where necessary, products.
3. The adsorbed reactants *compete* for surface sites.

a. Unimolecular Decomposition Processes. The basic reaction is

$$A_{(gas)} \leftrightarrow A_{ads} \rightarrow B_{(gas)}$$

where $\leftrightarrow$ indicates a fast pre-equilibrium and $\rightarrow$ implies the rate-limiting step, while the rate of reaction is then given by

$$v = k'\theta_A = k'K_A p_A/(1 + K_A p_A)$$

Here k' is a heterogeneous rate constant and K_A is the Langmuir constant for gas A; we recall that $K_A \sim \exp(-\Delta H_{ads}/RT)$. There are two extreme cases:

1. If $K_A p_A \ll 1$, then $v \approx k'K_A p_A \sim k_{exp} p_A$ and is first-order in the gas pressure.
2. If $K p_A \gg 1$, then $v \approx k'$ and is zero-order in the gas pressure.

In case (1), the temperature dependence of the experimental rate constant is

$$\partial(\ln k_{exp})/\partial T = (E' + \Delta H_{ads})/RT^2 = E_{exp}/RT^2$$

where E' is the activation energy for the heterogeneous reaction on the surface. We note that $E_{exp} < E'$ since $\Delta H_{ads} < 0$. The overall activation energy is therefore *lowered* by adsorption of A.

If B is *adsorbed* and competes with A for sites on the surface, then

$$A_{(gas)} \leftrightarrow A_{ads} \rightarrow B_{ads} \leftrightarrow B_{(gas)}$$

and, from a simple extension of the Langmuir isotherm,

$$\theta_A = \frac{K_A p_A}{1 + K_A p_A + K_B p_B}$$

If $K_A p_A \ll 1 + k_B p_B$,

$$v \approx k'K_A p_A/(1 + k_B p_B)$$

$$\approx k'K_A p_A/k_B p_B \qquad \text{if } k_B p_B \gg 1$$

b. Bimolecular Surface Reactions. These have the general mechanistic form

$$
\begin{array}{cc}
A_{(gas)} & B_{(gas)} \\
\Updownarrow & \Updownarrow \\
A_{ads} + & B_{ads} \rightarrow C_{(gas)}
\end{array}
$$

and from the above we immediately see that

$$
v = \frac{k' K_A p_A K_B p_B}{(1 + K_A p_A + K_B p_B)^2}
$$

If K_A and K_B are *comparable*, then v will pass through a *maximum* as p_B is increased while p_A is fixed. If, by contrst, $K_A p_A$, $K_B p_B \ll 1$, then

$$
v \approx k_{exp} p_A p_B
$$

and is second order in gas-phase pressures. If A is weakly adsorbed so that $K_A p_A \ll K_B p_B + 1$, then

$$
v \approx \frac{k' K_A p_A K_B p_B}{(1 + K_B p_B)^2}
$$

which is first order in A. If B is *strongly* adsorbed in this case, so that $K_B p_B \gg 1$, then

$$
v \approx (k' K_A / K_B) \cdot p_A / p_B \sim k_{exp} p_A / p_B
$$

and the reaction is *inhibited* by B. The activation energy is given by

$$
E_{exp} = E' + \Delta H_{(ads)A} - \Delta H_{(ads)B} > E'
$$

since $-\Delta H_{(ads)A} < -\Delta H_{(ads)B}$. Thus, if one of the reactants is *strongly adsorbed* and the other weakly adsorbed, the reaction is slowed down.

If C (the product of A and B) is also adsorbed, then

$$
v = k' \theta_A \theta_B = \frac{k' K_A p_A K_B p_B}{(1 + K_A p_A + K_B p_B + K_C p_C)^2}
$$

and if C is *strongly* adsorbed such that $K_C p_C \gg 1 + K_A p_A + K_B p_B$, then

$$
v \approx (k' K_A K_B / K_C^2) \cdot p_A p_B / p_C^2
$$

and the rate is strongly inhibited by the product.

18.5.1.2. The Eley-Rideal Mechanism

The main distinction from the Langmuir-Hinshelwood mechanism is in assumption (1) above, which is replaced by the possibility of reaction between adsorbed and gas-phase species:

$$A_{ads} + B_{(gas)} \rightarrow \text{products}$$

whence the reaction velocity is

$$v = k'' \theta_A p_B = k'' K_A p_A p_B / (1 + K_A p_A)$$

and the major kinetic distinction lies in the behavior if p_A is kept constant and p_B is increased. In the corresponding Langmuir-Hinshelwood expression, it is seen that the rate will pass through a maximum, while the Eley-Rideal expression predicts an increase without limit.

Variation of Catalytic Rate with Temperature. We have seen that Arrhenius behavior is expected in certain of the cases quoted above, but in other cases the reaction rate may actually pass through a maximum as T increases. Physically, this is because there is a tendency for surface coverage to *decrease* at higher temperatures. This may be seen graphically by considering the Langmuir-Hinshelwood model for the bimolecular process above for which

$$v = \frac{k' K_A p_A K_B p_B}{(1 + K_A p_A + K_B p_B)^2}$$

Now if, at low temperatures, $K_B p_B \gg 1 + K_A p_A$, then $v \sim (k' K_A / K_B) \cdot p_A / p_B$ and the activation energy is

$$E_{exp} \sim E' + \Delta H_{(ads)A} - \Delta H_{(ads)B} > 0$$

so that, at low temperatures, the Arrhenius slope will be normal and the reaction will show an increase with temperature. At higher temperatures, we will assume that K_B has decreased to the point where $K_B p_B$, $K_A p_A \ll 1$ and

$$v \approx k' K_A p_A K_B p_B$$

The activation energy is now

$$E_{exp} = E' + \Delta H_{(ads)A} + \Delta H_{(ads)B}$$

and it is perfectly possible that E_{exp} will become *negative*, and the reaction rate *decreases* with T. From this analysis, it is clear that the reaction rate will actually pass through a maximum as the temperature is increased.

If, in a bimolecular reaction, the product C is strongly adsorbed at lower temperatures, then increasing the temperature will always increase the rate. However, the rate may remain very low until the temperature is reached at which flash desorption begins, in which case it will rise steeply. Above this point, it may remain relatively independent of T since the rate will now depend on the sticking fraction which, as we saw above, is not a strong function of temperature in many cases. One well-known example is the isotopic scrambling of CO, and results for Re are shown in Figure 18.32.[66]

18.5.2. Variation of Catalytic Rate with Substrate

The basic process of catalysis is shown in Figure 18.33 and it is clear that the steps that may be activated are: (1) chemisorption, (2) the heterogeneous reaction, and (3) desorption.

It can be seen that if the chemisorption is *weak*, surface coverage will be low and catalytic activity also low. As $-\Delta H_{ads}$ increases, the coverage will rise, but so may E'. At large enough values of $-\Delta H_{ads}$, the coverage will reach a limiting value, but E' may now be so large that the adsorbed species cannot be decomposed; catalytic activity will again be low. Thus there will, in general, be an optimal chemisorption strength as shown in Figure 18.34.

We may vary ΔH_{ads} by varying the substrate, and we therefore expect that a plot of catalytic activity vs. position of catalyst in the periodic table may also show a maximum. This is illustrated in Figure 18.35 for the hydrogenation of ethene, which is believed to be a Langmuir–Hinshelwood type of bimolecular process with C_2H_4 strongly chemisorbed.

If adsorption energy can be estimated, then the type of plot shown in Figure 18.35 may be converted to a true "volcano" plot by displaying rate as a function

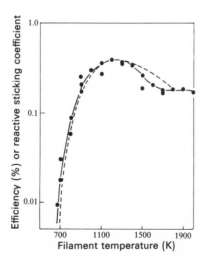

Figure 18.32. A comparison of two independent measurements of the efficiency of rhenium as a catalyst for the isotope reaction $^{12}C^{18}O + {}^{13}C^{17}O \rightarrow {}^{12}C^{16}O + {}^{13}C^{18}O$ (after Gasser and Holt[66]).

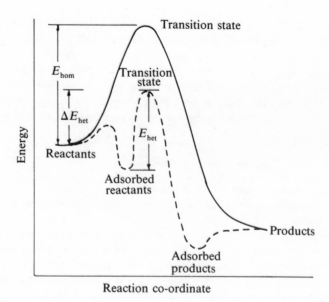

Figure 18.33. Schematic representation of the transition state theory of a homogeneous reaction (——) and a heterogeneous reaction (----) (after Gasser[6]).

of an enthalpy term. One elegant example is the rate of decomposition of formic acid. This chemisorbs very strongly on metals to form a surface formate species, so the enthalpy of formation of the metal formate may be taken as a measure of ΔH_{ads}. If the rate of the reaction is plotted as the temperature T_r at which the rate reaches a fixed value, then volcano plot of Figure 18.36 results.[67] We note that in Figure 18.36, T_r is plotted *down* the axis.

Experimental Distinction between Langmuir–Hinshelwood and Eley–Rideal Mechanisms. Immense efforts have been made in recent years to distinguish between these two mechanisms experimentally, and there are two broad categories of approach. In the first, a steady state is established by flowing gas mixtures over the catalyst, and then the temperature of the catalyst or pressures of the reacting gases are varied systematically. A second approach is to establish conditions remote from steady state by making abrupt changes in one or more variables and then to monitor the relaxation of the system.

To focus discussion, we consider the oxidation of CO on palladium. Two mechanisms have been proposed,[68] based on the Langmuir–Hinshelwood and Eley–Rideal models.

1. Langmuir–Hinshelwood (L–H) Mechanism

$$CO_{(gas)} \rightarrow CO_{ads}: k_1$$

$$CO_{ads} \rightarrow CO_{(gas)}: k_2$$

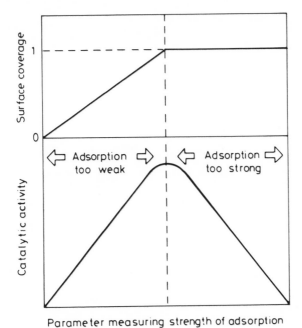

Figure 18.34. The "volcano" curve; dependence of catalytic activity on strength of reactant adsorption (lower part) and the corresponding variation in surface coverage (upper part) (after Bond[26]).

$$O_2(g) \rightarrow 2O^{\cdot}_{ads}: k_3$$

$$CO_{ads} + O^{\cdot}_{ads} \rightarrow CO_2(gas): k_4$$

2. Eley-Rideal (E-R) Mechanism

$$O_2(gas) \rightarrow 2O^{\cdot}_{ads}: k_3$$

$$O^{\cdot}_{ads} + CO_{(gas)} \rightarrow CO_2(g): k_5$$

Although a number of steady-state techniques has been deployed to distinguish these mechanisms, the most powerful tool has been the use of modulated molecular beams. In this experiment, one gas is flowed through at a steady pressure, and the other gas is in the form of a molecular beam that can be modulated by a shutter at a frequency of 100–1000 Hz. Thus, for oxidation of CO on Pd(111), a modulated beam of CO can be used, and the CO_2 formed detected mass spectrometrically. In addition to delays between opening the shutter and the detection of CO_2 that can be attributed to purely kinematic effects, there will be an additional delay occasioned by the finite rate of the surface reaction. If this additional delay is represented by a phase lag ϕ, then ϕ will be a strong function of the mechanism.

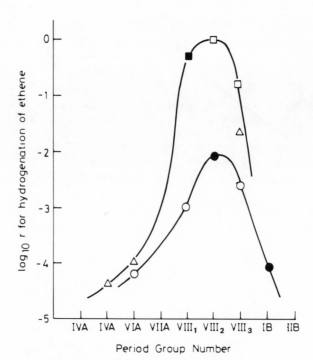

Figure 18.35. Logarithm of the rate of hydrogenation of ethene relative to that found on rhodium vs. periodic table group number. Open points: evaporated metal films; filled points: silica-supported metals; circles: first transition series; squares: second series; triangles: third series (after Bond[26]).

If the CO beam is modulated, and conditions fixed such that θ_{CO} is small (<0.03) and coverage by *atomic* oxygen large ($\theta_O > 0.1$), then we find[68]

$$\partial[CO_2]/\partial t = k_4 \theta_{CO} \theta_O \qquad \text{(L–H mechanism)}$$

$$\partial[CO_2]/\partial t = k_5 p_{CO} \theta_O \qquad \text{(E–R mechanism)}$$

where we have approximated the rather complex functional dependence of the experimental rate law on O coverage.

We now let the modulation of the beam be described in terms of the modulation of the pressure of CO, such that

$$p_{CO} = p_{CO}^0 + \alpha\, e^{i\omega t}$$

Then

$$\theta_{CO} = \theta_{CO}^0 + \beta\, e^{i\omega t}$$

For the L–H mechanism, by inserting these expressions for p_{CO} and θ_{CO} into the rate law for θ, namely

$$\partial\theta_{CO}/\partial t = k_1 p_{CO} - k_2 \theta_{CO} - k_4 \theta_{CO} \theta_O$$

we obtain, retaining just harmonic terms,

$$i\omega\beta = k_1 \alpha - k_2 \beta - k_4 \beta \theta_O$$

whence

$$\beta = k_1 \alpha\, e^{-i\phi}/[(k_2 + k_4\theta_O)^2 + \omega^2]^{1/2}$$

where $\tan\phi = \omega/(k_2 + k_4\theta_O)$.

The final expression for the rate of production of CO_2 in the L–H model is then given by

$$\partial[CO_2]/\partial t = k_4\theta_O\left(\theta_{CO}^0 + \frac{k_1 \alpha\, e^{i(\omega t - \phi)}}{[(k_2 + k_4\theta_O)^2 + \omega^2]^{1/2}}\right)$$

In a similar way, the E–R mechanism yields

$$\partial[CO_2]/\partial t = k_5\theta_O(p_{CO}^0 + \alpha\, e^{i\omega t})$$

and $\tan\phi = 0$.

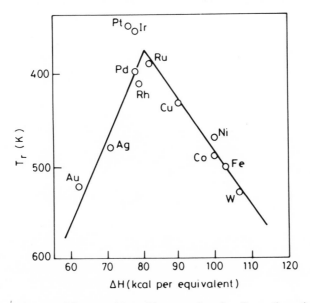

Figure 18.36. The rate of decomposition of formate, plotted as T_r, vs. the enthalpy of formation of the metal formate salt (after Rootsaert and Sachtler[67]).

Thus, for the L–H mechanism, tan ϕ is predicted to be temperature dependent through the rate constants k_2 and k_4, while for the E–R mechanism tan ϕ is independent of T. The results presented in Figure 18.37 for CO on Pd(111) show without any doubt that the L–H mechanism holds. In fact, under the conditions of the experiment $k_2 \gg k_4\theta_O$, so tan $\phi \approx \omega/k_2$ and a plot of $-\ln \tau \equiv -\ln (1/k_2)$ vs. $1/T$ gives the activation energy for desorption as 139 kJ mol^{-1}. This is very close to the isosteric heat of adsorption of CO on Pd(111) since adsorption of CO is not activated. The activation energy for k_4 may be obtained by plotting the CO_2 signal at a fixed ω vs. T since, with the approximations above, the phase-detected signal is given by const $\cdot k_4/(k_2^2 + \omega^2)^{1/2}$. This gives the activation energy for k_4 as 105 kJ mol^{-1}.

18.5.3. Mass Transport Limitations on Catalysis

It is clear that any heterogeneous catalytic process must involve five basic steps:

1. Transport of reactants to the surface.
2. Adsorption of the reactants on the catalyst.
3. Reaction on the catalyst involving one or more adsorbed species.
4. Desorption of the products.
5. Transport of products away from the catalyst.

Hitherto we have not considered the mass transport processes, and indeed, for many gas-phase reactions on simple catalysts, these processes are not rate limiting. However, at low pressures, or in the use of porous or reticulated catalytic supports,

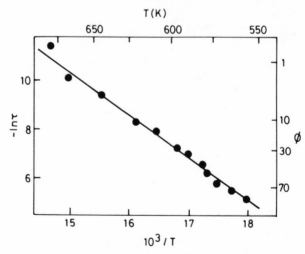

Figure 18.37. Arrhenius plot of $-\ln \tau \equiv -\ln (1/k_2)$ vs. $1/T$ for CO oxidation on Pd(111). The modulation frequency was 779 Hz (after Engel and Ertl[68]).

mass transport may become limiting. The problem of mass transport in liquid/solid systems is far more serious since diffusion coefficients are very much lower in the liquid than in the gas phase.

For porous catalyst supports or particles, the determination of whether the reaction is transport limited often presents a significant experimental problem, since stirring or agitation of the fluid will have only a marginal effect. The optimization of catalyst design is a complex multivariate problem, requiring advanced statistical techniques, and is beyond the scope of this chapter. However, one important effect may be that a transition can occur from a low-temperature régime, in which the rate is dominated by heterogeneous effects and has an associated activation energy that is quite large, to a high-temperature régime in which diffusion effects dominate and the activation energy falls considerably. There is a number of possible origins of this effect when it is found experimentally, but transport limitation is an important possibility.

18.5.4. Branching Effects

The simple situations envisaged at the beginning of Section 18.4 are rarely encountered in practice. Much more common are processes in which branching or sequential mechanisms are operative. Three typical examples are shown below.

$$
\begin{array}{ccc}
A + B \nearrow \!\!\!\! \begin{array}{c} C \\ \\ D \end{array} \!\!\!\! \searrow & A + B \to C \to D & A + B \nearrow \!\!\!\! \begin{array}{c} C \\ \downarrow \\ D \end{array} \!\!\!\! \searrow \\
\mathbf{I} & \mathbf{II} & \mathbf{III}
\end{array}
$$

The *selectivity* for a particular product i can be defined as

$$ S_i = \xi_i \Big/ \left(\sum_j \xi_j \right) $$

where the quantities ξ_i are the rates of formation of i. The selectivity can be contrasted with the *conversion*, which is defined as the fraction of reactant converted into all the products, and with the *yield* of product i, which can be defined as the amount of i present in the product stream at any given time. For the three mechanisms given above, plots of selectivity S_C for C and yield of the products C and D vs. conversion are shown in Figure 18.38.

18.5.5. The Fischer–Tropsch Process

We conclude this chapter with an examination of one of the most fascinating of the heterogeneously catalyzed reactions of CO, the Fischer–Tropsch reaction, which involves the conversion of mixtures of CO and H_2 into a wide range of hydrocarbon derivatives.

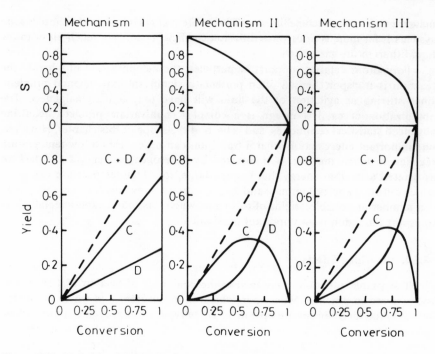

Figure 18.38. Variation of the yields of products C and D with conversion of reactants according to mechanisms I–III (lower part of each graph) and variation of selectivity for the intermediate product C (S_C) with conversion (upper curves) (after Bond[26]).

Historically, the first report of the hydrogenation of CO to give liquid hydrocarbons was in 1916 by the Badische Anilin u. Soda Fabrik, the catalyst being Co metal and the conditions severe (100 atm/300–400 °C). Later investigations at the Kaiser Wilhelm Institut für Kohleforschung at Mühlheim/Ruhr by Fischer and Tropsch established that much milder conditions (1 atm/250–350 °C) could be used with mixed Fe or Co catalysts, though these were rapidly poisoned by sulfur. Industrial scale-up, using Co/ThO_2 catalysts and CO from coal, was plagued by technological problems, though it was sufficiently successful in the end that, by 1941, nine Fischer–Tropsch plants were in operation. However, in the post-war period, the availability of cheap petroleum led to the abandonment of coal-based Fischer–Tropsch plants save in countries where special economic circumstances prevailed.

Catalysts for the general processes

$$nCO + 2nH_2 \rightarrow -(CH_2)_n- + nH_2O \qquad \text{favored by Co, Ni, Ru, } Cr_2O_3/ZnO$$

and

$$2nCO + nH_2 \rightarrow -(CH_2)_n- + nCO_2 \qquad \text{favored by Fe, } ThO_2/Al_2O_3$$

include Fe, Co, Ni, Ru, and ThO_2, with Fe being currently preferred. The active surface phase remains controversial, with metallic or metal-like components (such as nitrides or carbides) playing an important role for Ni, Co, and Ru. The case of Fe is more complex as dynamic modification to the surface occurs, with formation of mixed phases such as oxides and carbides. Each catalyst tends to produce different mixtures as shown in Figure 18.39, with some of the mixtures being very complex, particularly with Fe. The main point is that, by and large, straight-chain or methyl-branched isomers are produced, but more highly branched alkanes are not formed in significant quantities.

The basic mechanisms will now be examined. The essential processes that must take place are:

1. Scission of the original $C\equiv O$ bond.
2. Formation of new $C-C$, $C-H$, and $C-O$ bonds.
3. Formation of surface carbides and carbon itself.

It appears that the first step is the dissociative chemisorption of CO, which is known to take place on iron and believed to occur also on Co and Ni under the appropriate conditions. In the presence of hydrogen, however, these adsorbed carbon atoms will react, and a surface equilibrium mixture of CH_x will form, with x in the range 0–2. On nickel catalysts, the reaction of these CH_x species is believed to take place with each other through a chain growth mechanism and, from studies

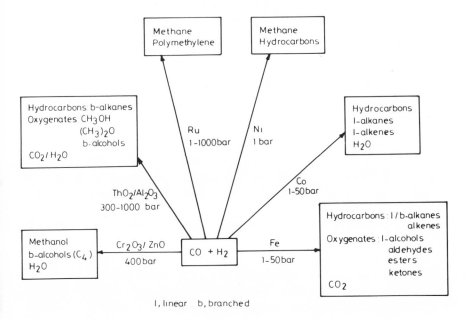

Figure 18.39. Products of hydrocondensation of CO on different catalysts (after Sneedon[69]).

with isotopically labeled CO, these oxygen-free species appear to act both as initiators and propagators. The weight of evidence is now that the dominant species on nickel is CH_2, and the basic mechanism is thought to be as follows:

$$RCH_2-M + CH_2M' \rightarrow RCH_2CH_2M' + M \qquad \text{propagation}$$

$$RCH_2-M + CH_3M' \rightarrow RCH_3 + M + CH_2M' \qquad \text{termination}$$

$$RCH_2M + CH_3M' \rightarrow M + R'CH=CH_2 + H_2 + CH_2M' \qquad \text{termination}$$

$$RCH_2-M + M'CO \rightarrow M + RCH_2COM'$$

$$\rightarrow M + M' + RCH_2CH_2OH \qquad \text{termination}$$

The methylene adsorbates may not be present on single sites, and one possibility is that they may occupy double Ni sites as

$$\begin{array}{c} CH_2 \\ \diagup \quad \diagdown \\ Ni \qquad Ni \end{array}$$

Some confirmation of the above scheme is provided by the observation that diazomethane, diluted by He or N_2, is converted on a Ni surface to ethene but, in the presence of H_2, the products are a mixture of linear C_1 to at least C_{18} alkanes and alkenes, similar to those obtained with CO/H_2 feedstock.

We have seen that CO may intervene in the chain propagation process to yield alcohols. In fact, aldehydes and carboxylates could also be produced as shown in equations (18.5.1) and (18.5.2),

$$\begin{array}{ccccc} R \ H \ H & RCH_2 & O & & \\ \diagdown | \diagup & \diagdown \quad \diagup\!\!/ & & H_2 & \\ C & \xrightarrow{\ CO\ } & C & \longrightarrow & RCH_2CHO \text{ or } RCH_2CH_2OH \qquad (18.5.1) \\ | & & | & & \\ M & & M & & \end{array}$$

$$\begin{array}{ccccc} R \qquad H & RCH=C\!\!\mp\!\!O & RCH_2-C\!\!\diagup\!\!/\!\!^O & & \\ \diagdown \ \diagup & & & & \\ C & \xrightarrow{\ CO\ } & \Big\downarrow \xrightarrow{\ M'OH\ } & O & \qquad (18.5.2) \\ \| & & & | & \\ M & M & M & \end{array}$$

and this process may be highly significant on Fe. In the case of Fe, the processes are very complex and are indicated schematically in Figure 18.40.

In this scheme, graphite represents a poison, inert to H_2 under the conditions of the catalysis, but the carbidic phase, which represents a complex mixture of iron carbides, does react with hydrogen to give CH_4 and other hydrocarbon products. Some support for the reactivity of C attached to iron is seen from the model reactions involving the iron carbonyl cluster $[(Et_4N)_2Fe_6C(CO)_{16}]^{2-}$, which can be cleaved oxidatively to yield a Fe_4 cluster as in equation (18.5.3).

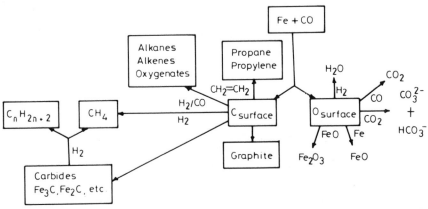

Figure 18.40. Possible evolution of $C_{surface}$ and $O_{surface}$ in Fischer-Tropsch synthesis (after Sneedon[69]).

$$[(Et_4N)_2Fe_6C(CO)_{16}]^{2-} \xrightarrow[CH_3OH]{oxidation} \begin{array}{c} Fe \\ \diagup \\ Fe \diagup \\ \diagdown \\ Fe \diagdown \\ Fe \end{array} C\!-\!C\!\!\begin{array}{c} O \\ \diagdown \\ OCH_3 \end{array} \xrightarrow{H_2} CH_3\!-\!C\!\!\begin{array}{c} O \\ \diagdown \\ OCH_3 \end{array}$$

$$(18.5.3)$$

In the case of Ru, conditions can be adjusted to give polymethylene with quite high molecular weight (5000–23,000). Again, the reaction is believed to proceed by methylene adsorbed on the surface. Growth may be by the mechanism indicated above, or by CH_2 insertion as in equation (18.5.4),

$$\begin{array}{cc} CH_2 & CH_2 \\ | & | \\ M & M \end{array} \longrightarrow \begin{array}{c} CH_3 \quad H \\ \diagdown \; C \; \diagup \\ \diagup \quad \diagdown \\ M \qquad M \end{array} \xrightarrow{CH_2} \begin{array}{c} CH_3CH_2 \quad H \\ \diagdown \; C \; \diagup \\ \diagup \quad \diagdown \\ M \qquad M \end{array} \qquad (18.5.4)$$

or CH_2 coupling as in equations (18.5.5) and (18.5.6).

$$\begin{array}{cc} CH_2 & CH_2 \\ | & | \\ M & M \end{array} \longrightarrow \begin{array}{c} CH_3 \quad H \\ \diagdown \; C \; \diagup \\ \diagup \quad \diagdown \\ M \qquad M \end{array} \xrightarrow{CH_2} \begin{array}{c} CH_3 \\ \diagdown \\ CH\!-\!CH_2 \\ | \qquad | \\ M \qquad M \end{array} \xrightarrow[transfer]{H} \begin{array}{c} CH_3CH_2 \quad H \\ \diagdown \; C \; \diagup \\ \diagup \quad \diagdown \\ M \qquad M \end{array}$$

$$(18.5.5)$$

$$\begin{array}{cc} CH_2 & CH_2 \\ | & | \\ M & M \end{array} \longrightarrow \begin{array}{c} CH_2\!-\!CH_2 \\ | \qquad | \\ M \qquad M \end{array} \xrightarrow{CH_2} \begin{array}{c} CH_2 \\ \diagup \; \diagdown \\ CH_2 \qquad CH_2 \\ | \qquad\qquad | \\ M \qquad\qquad M \end{array} \xrightarrow{CH_2} \begin{array}{c} CH_2\!-\!CH_2 \\ | \qquad | \\ CH_2 \qquad CH_2 \\ | \qquad | \\ M \qquad M \end{array} \quad (18.5.6)$$

References

1. H. Nahr, *Surface Science*, Vol. II, p. 9, Lectures presented at an International Course at Trieste from 16 January to 10 April 1974, International Atomic Energy Agency, Vienna (1975).
2. I. Eastermann, R. Frisch, and O. Stern, *Z. Phys.* **73**, 348 (1931); **84**, 430, 443 (1933).
3. G. Boato, P. Cantini, and L. Mattera, *Surf. Sci.* **55**, 141 (1976).
4. A. G. Stoll and R. P. Merill, *Surf. Sci.* **40**, 405 (1973).
5. J. L. Beeby, *J. Phys. C* **4**, 1359 (1971).
6. R. P. H. Gasser, *An Introduction to Chemisorption and Catalysis by Metals*, Clarendon Press, Oxford (1985).
7. B. McCarroll and G. Ehrlich, *J. Chem. Phys.* **38**, 523 (1968).
8. J. D. Lambert, *Vibrational and Rotational Relaxation in Gases*, Clarendon Press, Oxford (1975).
9. R. M. Logan and R. E. Stickney, *J. Chem. Phys.* **44**, 195 (1966).
10. R. E. Stickney, *Adv. At. Mol. Phys.* **3**, 143 (1967).
11. S. Yamamoto and R. E. Stickney, *J. Chem. Phys.* **53**, 1594 (1970).
12. D. R. O'Keefe, R. L. Palmer, and J. N. Smith, *J. Chem. Phys.* **55**, 4572 (1971).
13. H. Hoinkes, H. Nahr, and H. Wilsch, *J. Chem. Phys.* **58**, 3931 (1973).
14. R. M. Logan and J. C. Keck, *J. Chem. Phys.* **49**, 860 (1968).
15. H. Hoinkes, H. Nahr, and H. Wilsch, *Surf. Sci.* **30**, 363 (1972).
16. N. Cabrera, V. Celli, F. O. Goodman, and R. Manson, *Surf. Sci.* **19**, 67 (1970).
17. G. D. Derry, D. Wesner, W. E. Carlos, and D. R. Fackel, *Surf. Sci.* **87**, 629 (1979).
18. P. Cantini and R. Tatarek, *Phys. Rev. B* **23**, 3030 (1981).
19. G. Brusdeylins, R. B. Doak, and J. P. Toennies, *J. Chem. Phys.* **75**, 1784 (1981).
20. F. Frenkel, J. Hager, W. Krieger, H. Walther, C. T. Campbell, G. Ertl, H. Kuipers, and J. Segner, *Phys. Rev. Lett.* **46**, 152 (1981).
21. D. Ettinger, K. Honma, M. Keil, and J. C. Polanyi, *Phys. Rev. Lett.*
22. J. P. Cowin, C. F. Yu, S. J. Sibener, and J. E. Hurst, *J. Chem. Phys.* **75**, 1033 (1981).
23. R. M. Logan, *Solid State Surf. Sci.* **3**, 1 (1973).
24. S. M. Edwards, R. P. H. Gasser, D. P. Green, D. S. Hawkins, and A. J. Stevens, *Surf. Sci.* **72**, 213 (1978).
25. P. Kisliuk, *J. Phys. Chem. Solids* **3**, 95 (1957); **5**, 78 (1958).
26. G. C. Bond, *Heterogeneous Catalysis*, 2nd ed., Clarendon Press, Oxford (1986).
27. P. W. Tamm and L. D. Schmidt, *J. Chem. Phys.* **52**, 1150 (1970).
28. R. S. Berry, S. A. Rice, and J. Ross, *Physical Chemistry*, Wiley, New York (1980).
29. D. A. King and M. G. Wells, *Surf. Sci.* **29**, 454 (1972).
30. M. Bowker and D. A. King, *J. Chem. Soc., Faraday Trans. 1* **75**, 2100 (1979).
31. H. M. Kramer and E. Bauer, *Surf. Sci.* **107**, 1 (1981).
32. P. A. Redhead, *Trans. Faraday Soc.* **57**, 641 (1961).
33. C. G. Goymour and D. A. King, *J. Chem. Soc., Faraday Trans. 1* **69**, 736, 749 (1973).
34. J. Engel, *J. Chem. Phys.* **69**, 373 (1978).
35. R. Di Foggio and R. Gomer, *Phys. Rev. Lett.* **44**, 1258 (1980).
36. R. Casanova and T. T. Tsong, *Phys. Rev. B* **22**, 5590 (1980).
37. H. W. Fink, K. Faulian, and E. Bauer, *Phys. Rev. Lett.* **44**, 1008 (1980).
38. N. H. Cohen and D. Turnbull, *J. Chem. Phys.* **31**, 1164 (1959).
39. J. J. Fripiat and H. van Damme, *Bull. Cl. Sci., Acad. R. Belg.* **60**, 568 (1974).
40. R. H. Fowler and E. A. Guggenheim, *Statistical Thermodynamics*, p. 429, Cambridge University Press (1939).
41. P. Rousseau, P. Delescluze, F. Delamare, N. Barbouth, and J. Oudar, *Surf. Technol.* **7**, 91 (1978).
42. A. W. Adamson, *Physical Chemistry of Surfaces*, 2nd ed., Wiley, New York (1967).
43. M. Eizenberg and J. M. Blakely, *Surf. Sci.* **82**, 493 (1979).
44. J. M. Blakely, *Crit. Rev. Solid State Mater. Sci.* **7**, 333 (1978).
45. R. G. Egdell, W. R. Flavell, and P. Tavener, *J. Solid State Chem.* **51**, 345 (1984).

46. J. Nowotny, I. Sikora, and J. B. Wagner, *J. Am. Ceram. Soc.* **65**, 192 (1982).

47. J. Haber, in: *Surface Properties and Catalysis by Non-Metals* (J. P. Bonnelle, ed.), p. 1, D. Reidel, New York (1983).

48. G. Rienecker, *Angew. Chem.* **69**, 545 (1957).

49. G. Samorjai, *Catal. Rev.* **18**, 173 (1978) and references cited therein.

50. J. M. Tatibouet and J. P. Germain, *J. Catal.* **72**, 375 (1981).

51. M. Passler, A. Ignatiev, F. P. Jona, D. W. Jepson, and P. M. Marcus, *Phys. Rev. Lett.* **43**, 360 (1979).

52. J. T. Yates, T. M. Duncan, S. D. Worley, and R. Vaughan, *J. Chem. Phys.* **70**, 1219 (1979).

53. B. E. Hayden and A. M. Bradshaw, *Surf. Sci.* **125**, 787 (1983).

54. G. Ertl, *Pure Appl. Chem.* **52**, 2051 (1980).

55. L. H. Dubois and G. A. Samorjai, in: *Vibrational Spectroscopies for Adsorbed Species* (A. T. Bell and M. L. Hair, eds.), *ACS Symp. Ser.* **137**, 163 (1980).

56. H. Conrad, G. Ertl, and J. Küppers, *Surf. Sci.* **76**, 323 (1978).

57. A. M. Bradshaw, *Surf. Sci.* **80**, 215 (1979).

58. J. Küppers, H. Conrad, G. Ertl, and E. E. Latta, *Jap. J. Appl. Phys.*, *Suppl.* **2**(2), 225 (1974).

59. M. W. Roberts, *Chem. Soc. Rev.* **6**, 373 (1977).

60. J. B. Peri, *J. Catal.* **86**, 89 (1984).

61. N. S. Hush and M. L. Williams, *J. Mol. Spectrosc.* **50**, 349 (1974).

62. B. Rebenstorf, M. Berglund, and R. Lykvist, *Z. Phys. Chem. N.F.* **126**, 349 (1974).

63. K. Tanaka and J. W. White, *J. Phys. Chem.* **86**, 4708 (1982).

64. E. Guglielminotti, L. Ceruti, and E. Borello, *Gazz. Chim. Ital.* **107**, 503 (1977).

65. C. H. Amberg and D. A. Seanor, Proc. 3rd. Int. Congr. Catal., p. 450, Amsterdam (1965).

66. R. P. H. Gasser and D. E. Holt, *Surf. Sci.* **64**, 520 (1977).

67. W. J. M. Rootsaert and W. M. H. Sachtler, *Z. Phys. Chem. N.F.* **26**, 16 (1960).

68. T. Engel and G. Ertl, *J. Chem. Phys.* **69**, 1267 (1978).

69. R. P. A. Sneedon, in: *Comprehensive Organometallic Chemistry*, Vol. 8 (1982) (G. Wilkinson, F. G. A. Stone, and E. W. Abel, eds.), p. 19, Pergamon Press, Oxford (1985).

19

Growth Processes at Surfaces
Modeling and Simulations

M. Djafari Rouhani and D. Estève

19.1. Introduction

Thin films are of interest in industrial applications, and particularly in semiconductor devices technology.[1] From the physical point of view, they allow the study of two-dimensional systems, and their differences with three-dimensional states of matter. The confinement of particles in ultrathin layers places strong requirements on the quality of these layers: structural and chemical perfection, uniform and well-controlled thickness, uniformity of physical properties in the plane normal to the growth direction. Two major fabrication techniques have allowed the preparation of such high-quality thin films, namely, metal organic chemical vapor deposition (MOCVD) and molecular beam epitaxy (MBE). Major advantages of growth from a vapor phase are lower growth temperatures avoiding contamination and impurity diffusion, the control of the thickness, and the easy doping over a wide range of concentrations. However, the MBE technique is best suited for the observation of nucleation processes, since the reactant species are directly deposited at known rates on the substrate surface from the very beginning of the experiment. This is not the case in a MOCVD experiment where the stabilization time of the flow might be relatively long, and the arrival of species to the substrate is by diffusion through the gas phase.[2] Indeed, the MBE technique has been used to grow a variety of elemental or compound semiconductor structures.

While the early goals of theoretical studies were the physical understanding of basic growth mechanisms, the development of simulation programs using high-speed computers can now serve as a guide for further technological design.

M. Djafari Rouhani • Laboratoire Physique des Solides, Université P. Sabatier, 31062 Toulouse Cedex, France. **D. Estève** • LAAS–CNRS, 31077 Toulouse Cedex, France.

The main problems to be solved have been clearly defined in the early work of Burton *et al.*[3] and Bennema.[4] These are the rate of growth of either a surface or a step, the rate of nucleation, and the equilibrium structure of crystal surfaces. The solutions given were based on thermodynamic considerations and will be discussed in the next section, together with more recent developments. The advent of high-speed computers opened a new era in the theory of crystal growth by allowing the introduction of atomic-scale processes via the use of Monte Carlo methods. Here one is no longer concerned with thermodynamic averages, but rather with random motion of atomic species. These processes, and their implementation in Monte Carlo procedures, are developed in Section 19.3, and their application to physical problems are described in Section 19.4. In recent years, with the development of MBE techniques, atomic species have been replaced by molecules impinging on the surface. Therefore chemical reactions have to take place before the atoms are incorporated into the crystal. Section 19.5 is devoted to the interactions between the surface and the molecules present in the gas phase. Results of some simulations are reported in Section 19.6. In simulating epitaxial growth of lattice matched materials, the interaction between two atomic species is given by a single energy parameter. This is no more sufficient when treating heteroepitaxial growth with high lattice mismatch, due to strain effects; and the interaction has to be position-dependent. A review of several types of interaction potential is given in Section 19.7, while the theory of heteroepitaxial growth is developed in Section 19.8.

19.2. Thermodynamic and Kinetic Models

There are three modes of crystal growth.[5] In the island (Volmer–Weber) mode, three-dimensional nuclei are formed on the substrate, which grow and coalesce to cover the whole substrate. In the layer (Frank–Van der Merwe) mode, the adsorbed atoms are more strongly bound to the substrate than to each other. Therefore each layer will be completed before a new layer starts to grow. This mode is the most usual in semiconductor on semiconductor growth. An intermediate case (Stranski–Krastanov mode) has to be mentioned, where, after the growth of a few monolayers, islands are formed. These different modes are characterized by adsorption isotherms (coverage versus supersaturation) crossing the zero-supersaturation line at different levels of the coverage.[6,7] The zero supersaturation is defined with reference to the equilibrium pressure of the gas phase with its bulk solid phase. The negative supersaturation part of the isotherm corresponds to the layer growth mode, while the positive part favors the island growth regime.

The first theories[8,9] dealt with condensation of vapors into liquid droplets. Nuclei are formed by thermodynamic fluctuations of the density and have probability of growing or shrinking according to their size-dependent free energies. Since this energy contains a negative volume energy and a positive surface energy, a critical size can be defined for which the energy is maximum, i.e., the nucleus is in unstable equilibrium with its vapor. This assumption is only valid if the nuclei are of sufficiently large size that thermodynamic notions of volume and surface energies are valid. According to the kinetic theory of gases, the rate at which atoms strike the surface R_+, and the supersaturation $\Delta\mu$, are given respectively by[10]

$$R_+ = P/\sqrt{2\pi mkT} \quad \text{and} \quad \Delta\mu = kT/\ln(P/P_e)$$

where P and P_e are, respectively, the gas pressure and the equilibrium pressure of the gas phase with its bulk solid at temperature T, while m is the mass of the vapor atoms. According to the principle of microscopic reversibility under equilibrium, the rate of evaporation $R_- = R_+$ for $\Delta\mu = 0$. Therefore the net rate of growth is

$$R = R_+ - R_- = R_+[1 - \exp(-\Delta\mu/kT)]$$

The above relation is called the Wilson–Frenkel law[11] and gives the maximum limiting growth rate. At low supersaturations, the growth rate will be proportional to the supersaturation.

When dealing with crystals, the presence of small-size nuclei, and in particular single adsorbed atoms mobile on the surface, will introduce configuration-dependent energies; and the atomic-scale processes have to be taken into account. Figure 19.1 shows few specific configurations, according to Burton et al.[3] However, it has generally been assumed[3,12-14] that atoms are definitely attached to the crystal, or evaporated from it, via the kink sites. With this assumption, it can be shown that the Wilson–Frenkel law is still valid. Frenkel[15] and Burton and Cabrera[16] have shown that, owing to thermodynamic fluctuations, steps will contain a high concentration of kinks. The adsorbed species therefore migrate on the substrate surface to reach these kinks. This migration has been taken into account using macroscopic diffusion theory[3,17,18] to calculate the growth rate, which depends on the diffusion length of the adsorbed species. On the other hand, Burton and Cabrera[16] have shown that steps can only be created by thermodynamic fluctuations at temperatures near the melting point. Thus other sources for the creation

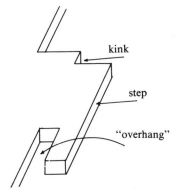

Figure 19.1. Schematic representation of different types of sites on a surface.

of steps had to be found. Following the idea introduced by Frank,[19] it has been shown[20,21] that screw dislocations present on the surface will act as these sources, and new steps are created as the growth proceeds.

If the starting surface is perfect, growth can only proceed by nucleation. Nucleation and growth theory has been described using kinetic rate equations,[22,23] which express the time dependence of surface densities n_j of clusters containing j atoms as

$$dn_1/dt = R_+ - n_1/\tau_1 - 2U_1 - \sum_{j=2}^{\infty} U_j$$

and

$$dn_j/dt = U_{j-1} - U_j \qquad \text{for } j \geq 2$$

where R_+ is the rate of impingement, τ_1 the mean lifetime of single adatoms, and U_j the net rate of capture of single adatoms by j-size clusters. In deriving these equations, it is assumed that only single adatoms are mobile on the surface. Further simplifications[8,9,24-28] assume that all subcritical clusters are in equilibrium with

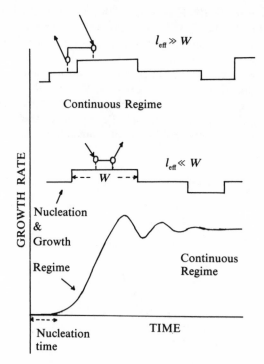

Figure 19.2. Physical conditions and the growth rate behavior for various growth regimes.

each other, such that $U_j = 0$. They divide clusters into subcritical and stable clusters, reducing the number of equations to three.

In conclusion, a macroscopic description using thermodynamic and kinetic rate equations can predict the two growth regimes, represented schematically in Figure 19.2, following Ghaisas and Madhukar.[29] Starting with a nearly perfect substrate surface, when the effective diffusion length $l_{eff} \ll W$, the average terrace width, one may observe the nucleation and growth regime. On the other hand, if the starting surface contains screw dislocations, or if steps appear on the substrate as a result of addition of subsequent monolayers, in such a way that $l_{eff} \gg W$, then the continuous growth regime will be favored.

19.3. Atomic-Scale Processes

Thermodynamic and kinetic models deal with average physical quantities, and therefore cannot take all possible configurations of growing clusters into account. In fact, this would require averaging over all possible paths connecting the clean substrate surface to the complete deposited layer. On the other hand, the Monte Carlo technique[30,31] chooses only one random path out of all possible ways, corresponding to an actual experiment. Starting with a given substrate, one follows individual and collective motions of atoms. The path is determined according to random numbers corresponding to transition probabilities between basic configurations and derived from physical considerations. With the high-speed computers available, an $(N \times N)$ lattice ($N = 20$ to 100) is used as the starting substrate, with periodic boundary conditions which prevent reproducing the exact stochastic nature of an actual experiment.

Basic elementary processes investigated in the literature are adsorption from, and reevaporation into the vapor phase,[32-35] intralayer and interlayer migration of species on the substrate,[12,36-39] condensation into kink sites, and formation of critical clusters.[40-42] According to their specific interests, various authors have chosen a set of basic processes from the above-mentioned mechanisms, but little attention has been paid to collective motions. Once the basic processes have been defined, a transition probability should be attributed to each of them. Obviously, these transition probabilities should depend on the local configuration of adsorbed species in order to conserve the stochastic nature of the experiment. They are generally assumed to have an Arrhenius from $K \exp(-E/kT)$, where K is a vibration frequency of the order of 10^{11}-10^{13} s^{-1} and E is the activation energy corresponding to the particular transition. To determine activation energies, a broken bond model,[36,12] a Lennard-Jones-type interaction potential,[40] or simply adjustable parameters[29,43] have been used. It is evident that the actual activation energy for the migration is much lower than that determined using a broken-bond model. However, this model has often been applied to semiconductor materials where the bonding is of covalent type, while the Lennard-Jones-type potential is mostly used in the case of metal–metal interaction or with adsorbed gases. An example of how an activation energy for migration is calculated is illustrated in Figure 19.3, where substrate atoms are denoted by "S" and adsorbed atoms by

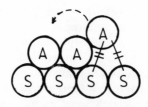

Figure 19.3. An example of migration activation-energy calculation, according to the broken-bond model.

"A". Given the particular type of migration and the local configuration represented in Figure 19.3, it is assumed that two "A—S" bonds have to be broken, while the "S—S" bond does not need to be broken.

Once the transition probabilities are known, a sequence of events can be determined by generating random numbers according to the above probabilities. Two methods have generally been used. Early calculations[12,32,36,41] have used random numbers to choose an event between a number of possible ones, according to their probabilities of occurrence, without being interested in their time of occurrence. Since the surface configuration is modified after each single event, transition probabilities are also modified after each event. In order to avoid the renormalization of the sum of the transition probabilities to one, or to any other constant, after each event, subintervals corresponding to no event have often been introduced.[12] The second technique[42,44,45] attributes a time t_i to each possible event, according to the Poisson probability law, $t_i = -(1/p_i) \ln r_i$, where p_i is the probability of occurrence of the event i per unit time, and r_i a random number uniformly distributed over the interval $(0, 1)$. The actual event is the one corresponding to the minimum time. In this way, a time of occurrence is given to each event in the sequence.

19.4. Application of Monte Carlo Simulations

The Monte Carlo technique has been widely applied to study the main crystal growth problems mentioned in Section 19.1. These are the morphology of the surface expressed as its roughness, and the rate of growth in different regimes, namely, the continuous or normal regime and the nucleation and growth regime. Thermodynamic considerations[3] show that surface roughening occurs at a critical temperature T_c given by $kT_c/u = 0.57$, where u is the interaction energy between nearest-neighbor atoms in a cubic lattice. The same results have been obtained using a Monte Carlo technique, which clearly shows[14,31,46,47] that few layers start growing at the same time above the critical temperature, especially in the case where surface migration is not taken into account. Surface migrations are seen to reduce the mean deviations of surface height from its average value.[37] Moreover, the Monte Carlo technique allows the study of surface morphology during growth, i.e., out of thermodynamic equilibrium. Simulations show that the actual roughness of grown layers might either be due to thermodynamic fluctuations or to the kinetics of the growth. In fact, experiments at low temperatures such that $kT/u = 0.25$, but in the presence of supersaturated vapor, show that roughening is observed after the deposition of a few layers.[14,46] However, this roughening can be recovered by interrupting the growth, as can be seen in RHEED measurements during MBE.[48-53] Interruption of growth has proved useful in obtaining flat surfaces during growth of multiple quantum wells.

Turning now to the calculation of growth rates,[12,31,54-57] Monte Carlo simulations show linear growth rates as a function of supersaturation, situated below the maximal Wilson–Frenkel law at temperatures $T \gtrsim T_R$, the roughening temperature. Moreover, it has been suggested[31] that Wilson–Frenkel law should be modified to

$$R = R_+ A_0 \frac{1 + x_s^2}{1 + A_0 x_s^2} [1 - \exp(-\Delta\mu/kT)]$$

where A_0 is the concentration of active sites (kink sites) and varies from 0.4 at T_R to 1 at $T \to \infty$; x_s represents the surface diffusion length of species and $\Delta\mu$ is the supersaturation. Simulations have also shown[41] that the edges tend to form facets and that the growth takes place at the concave parts of a step. This explains discontinuities in the time evolution of the spreading velocity of supercritical clusters. These results compare with another type of Monte Carlo calculation based on a polynuclear growth model.[42,58] This model consists of disk-shaped clusters that start with zero radius at random positions and expand with a constant radial velocity. A multilevel structure is formed by superposition of new clusters on the top of the old ones.

19.5. Gas–Surface Interactions

The Monte Carlo simulations described in Sections 19.3 and 19.4 dealt with an elemental solid grown from atomic vapor. However, a majority of vapor-phase crystal-growth experiments, such as molecular beam epitaxy (MBE) of III-V compounds, involve at least one species in molecular form. It is thus necessary to examine the role of surface molecular reaction kinetics on the growth mechanisms. Singh and Madhukar[44] and Ghaisas and Madhukar[29,38] investigated extensively these mechanisms and their influence on the MBE growth of III-V compounds, and especially of GaAs, leading to a more general growth process, which they referred to as the configuration-dependent reactive incorporation (CDRI) growth process.

In the MBE of GaAs, arsenic is supplied in the As_4 or As_2 molecular form. Surface molecular reactions have been investigated by Arthur,[59,60] Cho,[61] and Foxon and Joyce.[62,63] It is generally found that the sticking coefficients of As_4 and As_2 molecules are zero in the absence of Ga flux, and reach a maximum, depending on the nature (As_4 or As_2) of the molecules and the temperature, under high Ga flux. This behavior is explained by assuming that each Ga atom supplied consumes one arsenic atom. The excess As_4 or As_2 molecules are desorbed. The Ga atoms in excess of that producing the maximum As sticking coefficient is eventually consumed if the arsenic beam is maintained after the Ga beam is turned off.

Concerning As_4 molecules, the maximum sticking coefficient decreases from unity at 300 K to 0.5 at 450 K, and takes the constant value of 0.5 in the range between 450 K and 600 K.[59,63] Furthermore, As_4 desorption measurements show that the desorption rate changes from second order at low As_4 coverages to first order at high coverages.[63] The above facts imply that, in the 450–600 K range which is of interest in MBE experiments, chemisorption must be dissociative and can be illustrated by

$$As_4 \text{ (vapor)} \rightarrow As_4^* \text{ (physisorbed)}$$

$$As_4^* + As_4^* \rightarrow As_2^* \text{ (physisorbed)} + As_2^* + As_4 \text{ (vapor)}$$

In the As_2 case, the maximum sticking coefficient tends to unity at 600 K. Desorption measurements show the presence of As_4 molecules, explained by an association reaction, below 600 K. The basic reactions can therefore be represented by[63]

$$As_2 \text{ (vapor)} \rightleftarrows As_2^* \text{ (physisorbed)} \rightarrow As + As \text{ (chemisorbed)}$$

above 600 K, and can be completed by

$$As_2^* + As_2^* \text{ (physisorbed)} \rightarrow As_4^* \text{ (physisorbed)} \rightarrow As_4 \text{ (vapor)}$$

below 600 K.

These results show that, under MBE experimental conditions, incorporation of arsenic atoms is via the dissociative chemisorption of a precursor physisorbed As_2^* state, which requires two unfilled As sites on the surface. Ghaisas, Singh, and Madhukar[29,38,44] have considered the lowest-order Ga configurations capable of providing a pair of adjacent unfilled As sites. These are a single Ga, two adjacent Ga, and three consecutive Ga atoms along the bond orbital direction, and four Ga atoms in the surface nearest-neighbor square configuration (Figure 19.4). The atomic As is allowed to bind with single or two adjacent Ga atoms, thus creating a single or a double Ga–As bond. In addition to the intraplanar configurations, interplanar configurations involving two Ga atoms in two different planes, such as would occur at steps, can also provide sites for As_2 dissociative reactions.[64,65]

The above processes may become the rate-limiting step for arsenic incorporation under special growth conditions. Thus, in accordance with experimental results, even though the time-averaged growth rate of GaAs may be controlled by the arrival rate of the Ga under a sufficient population of As_2, the time-dependent growth rate can be controlled by the reactive incorporation kinetics of the As_2 molecules. This is the essence of the CDRI process described earlier in this section.[29,39,43] The two extremes of the CDRI process are called the reaction limited incorporation (RLI) and the configuration limited reactive incorporation (CLRI). In the first case, the limiting process is the dissociative incorporation of

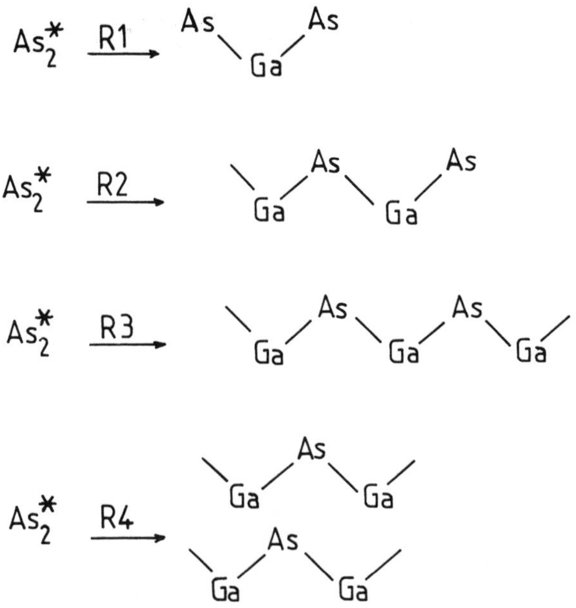

Figure 19.4. Lowest-order surface configurations used in the CDRI model (GaAs example).

As_2, while in the second case the probability of occurrence of lowest-order configurations of Ga atoms controls the growth rate behavior. It is generally assumed that the availability of physisorbed As_2 is not the rate-limiting step, i.e., the arsenic pressure is high enough.

19.6. Surface Reaction Simulations

It is found that the effective migration length l_{eff} is a function of three user-controlled growth parameters, namely, the substrate temperature, the Arsenic pressure, and the Ga flux. As mentioned in Section 19.2, the relative magnitudes of l_{eff} and the average terrace width W determine the growth regime (see Figure 19.2), which will therefore be known for each set of experimental conditions. When $W > l_{eff}$, a layer-by-layer mode of material addition occurs and gives rise to the oscillatory nature of the observed RHEED intensity. For $W < l_{eff}$, the group III atoms are predominantly able to migrate to the existing steps, and growth proceeds via the step propagation.[66] As_2 incorporation events, even small in number, can play a significant role in surface smoothness by enhancing the intralayer Ga migration.[65]

An example of the influence of As_2 incorporation rates on the growth rate is given by the case where only intraplanar configurations are considered and where R_3 and R_4 (see Figure 19.4) are sufficiently high.[29] A layer-by-layer mode of material addition is maintained, but for small R_1 and R_2 (respectively 10 and 20 per second) the As coverage lags behind the preceding Ga layer coverage until the near completion of the layer. For large R_1 and R_2 (respectively 1000 and 2000 per second) coverages of the As layers lead the coverage of the preceding GA layers. This is a consequence of rapid incorporation rates, leading to attachment of essentially two As atoms for each Ga deposited at the initiation of a given layer. Furthermore, while the Ga attachment rate is constant, the total growth rate exhibits oscillations.

For slow rates R_1 and R_2, the total growth rate is low in the initial stages of the growth since most of the Ga delivered are in a single configuration prohibiting As incorporation. Subsequently, two and higher Ga configurations occur, due to both migration and the increased coverage, increasing the incorporation rate of As_2. As a given Ga layer nears completion, the next higher Ga layer is initiated, reducing the As_2 incorporation rate, and giving rise to oscillations of the total growth rate. This is an example of the reaction limited incorporation (RLI) growth mechanism.

For fast rates R_1 and R_2, the total growth rate is higher since two As atoms are incorporated for each Ga atom. Subsequently, the As incorporation rate must slow down since rapid interlayer Ga migration inhibits the formation of the next higher Ga layer, until the given Ga layer is near completion. At this point, two As atoms can again be attached to a single Ga, and the total growth rate increases, giving rise to oscillations. In this case, the reactions are not the limiting step for the growth, but the unavailability of Ga in an exposed state slows down the process, and we are in the presence of configuration limited reactive incorporation (CLRI) growth.

19.7. Interaction Potential Models

Another feature which has to be accounted for in crystal growth is the lattice mismatch. When the epitaxial layer and the substrate are of different type materials, almost invariably their lattice constants are different, even when their crystal structures are identical. Under these conditions of strained layer and substrate, where the atoms are displaced from their equilibrium positions, the interaction energy between atoms can no longer be expressed as a constant. Rather, position-dependent interaction potentials should be used. The Born–Oppenheimer approximation[67] separates nuclear and electronic coordinates in the total Hamiltonian and defines a potential-energy function for the nuclei as the eigenvalues of the electronic Hamiltonian. In the absence of external fields, this potential which depends on atomic positions is expanded in series. Two types of expansion have generally been used in the literature.

The first is a many-body series expansion of the form[68-75]

$$E(x_1, \ldots, x_N) = (1/2!) \sum_{i \neq j} \sum V^{(2)}(r_{ij}) + (1/3!) \sum_{i \neq j \neq k} \sum \sum V^{(3)}(r_{ij}, r_{jk}, r_{ki})$$

$$+ \cdots + (1/N!) \sum_{i_1 \neq \cdots \neq i_N} \sum V^{(N)}(r_{i_1 i_2}, \ldots)$$

where $V^{(n)}$ is the n-body potential depending on the relative coordinates of the n atoms. Each quantity $V^{(n)}$ is given a basic functional form with few adjustable parameters to fit experimental values of cohesive energies and crystal structures. The two-body potential is usually assumed to take the Lennard-Jones form

$$V^{(2)}(r) = u[(r_0/r)^{12} - 2(r_0/r)^6]$$

with two adjustable parameters: the equilibrium distance r_0, and the interaction energy u between the two atoms. But two-body contributions alone cannot properly describe molecular configurations, even in the gas phase.[70] Indeed, the two-body term favors close-packed structures. To investigate open structures, one has to include three-body potentials in the calculations. Therefore, three-body interactions of Axilrod–Teller type[76]

$$V^{(3)}(r_1, r_2, r_3) = Z(1 + 3 \cos \theta_1 \cos \theta_2 \cos \theta_3)/(r_1 r_2 r_3)^3$$

are often taken into account. Here r_i and θ_i represent respectively the sides and angles of the triangle formed by the three atoms, and Z is the adjustable parameter. Large values of Z favor the linear cluster morphology. To our knowledge, higher-order terms in the expansion have always been neglected. It has also been shown[69] that using three-body potentials to calculate atomic relaxation on surfaces yields a contraction of the surface layer with respect to its crystal position, in agreement with experimental observations.

However, there is no direct justification for choosing these potentials, other than for purely pragmatic reasons.[69] Moreover, the number of adjustable parameters increases rapidly upon refinement,[71-73] from 3 to 18, for one-component materials. Turning to binary and ternary compounds, the minimum number of adjustable parameters reach 10 and 22, respectively. However, as a result of their simplicity, especially in the case where only pair potentials are used, they have been widely used in molecular dynamics calculations.[14,40,72,77-79]

The second type of expansion is a Taylor series expansion as a function of the coordinates of the atomic displacements, Δx_i, with respect to their equilibrium positions:

$$V = V_0 + \sum_i A_i^{(1)} \Delta x_i + (1/2!) \sum_{i,j} A_{ij}^{(2)} \Delta x_i \Delta x_j + (1/3!) \sum_{i,j,k} A_{ijk}^{(3)} \Delta x_i \Delta x_j \Delta x_k + \cdots$$

The first term V_0 is the equilibrium energy and can be accounted for as in homoepitaxial growth. The quantity $A^{(n)}$ is the nth order derivative of V at the equilibrium configuration. Therefore $A_i^{(1)} = 0$. Writing all quantities Δx_i as functions of bond length and bond angle variations, as in the valence force field approach,[80-82] the derivatives $A^{(i)}$ are expressed as functions of force constants. They have been adjusted, in the case of molecules, to obtain the best fit with physical properties such as geometries, conformational energies, heats of formation, etc.[83] In the case of crystals, second-order derivatives can easily be calculated using vibrational data or elastic constants.[82] Higher-order terms are related to anharmonic effects. This second type of expansion is usually referred to as molecular mechanics, because it has been widely used to determine the structural properties of molecules. But it has also been applied to crystals to explain the local order in ternary alloys.[84,85]

19.8. Heteroepitaxial Growth

A thermodynamic description of heteroepitaxial growth with lattice mismatch has been given by Frank and Van der Merwe[86,87] using continuous media within the elastic approximation. The interaction potential between atoms is replaced by an average periodic surface potential.[86,88] The equilibrium configuration is found by assuming that tensions at the interface should vanish. It has been shown[86,89] that, for a one-dimensional model, the limiting misfits at which a coherent monolayer loses stability and metastability, respectively, are approximately 9% and 14%. For a two-dimensional model, these numbers should be reduced by 25%, because of the Poisson ratio. For thick layers, the limiting misfits are further reduced. Inversely, with a given misfit, the growing layer is coherent with the substrate and stable, although stressed, until it reaches a critical thickness. At this point the layer becomes unstable and misfit dislocations are introduced at the free edges of the layer, while the homogeneous strain decreases.[89,90] These results are valid for continuous overlayers in thermodynamic equilibrium within the elastic approximation.

In recent years, there has been a growing interest in the study of heteroepitaxial films, since electronic quality epitaxial layers of materials with large lattice mismatches have been prepared. We should mention that it was long believed that good epitaxial growth was only possible if the lattice mismatch did not exceed 1%. An example is the growth of CdTe films of GaAs substrate, with 14.7% lattice mismatch, which has potential applications in optoelectronic devices and infrared imaging. This also opens a new field in three-dimensional heterogeneous integration where the basic device is made of different superimposed layers, each having a specific function according to its physical properties. However, several problems have still to be solved: the interdiffusion between constituent, the defect creation at the interface, and the direction of growth. In fact, experiments show that using a (100) GaAs substrate, the grown layer is either (100) or (111).[91-94]

In order to introduce atomistic processes into the growth, Cohen Solal et al.[95] used a model based on chemical-bonding and lattice-matching considerations.[96] Starting with a (100) As surface of GaAs with around half of the As atoms missing, Te atoms may bind to one As and two Ga, leaving one dangling bond. This configuration will lead to (111) growth with no defect in the CdTe layer. Let us now start with a stabilized (100) Ga surface. In a first step, Te atoms bind to Ga atoms, in the normal As positions. In a second step, two such Te are tied by a third Te not bound to any Ga atom, having therefore two dangling bonds and leading to (100) growth. With continued growth, stable clusters of atoms take place, bound by periodic lines of dislocations.

Recently, in order to study growth kintic effects as well as atomistic mechanisms, molecular mechanics has been associated with the Monte Carlo technique to simulate heteroepitaxial growth of materials exhibiting large lattice mismatches.[97] The model has been applied to the epitaxy of CdTe on GaAs with 14.7% lattice mismatch. A molecular-mechanics type of interaction potential within a harmonic approximation has been preferred to a many-body approach to avoid the multiplication of adjustable parameters, and in order to conserve the concept of chemical bond which has been present in all crystal-growth simulations up to now, and which is absent in many-body-type potentials. The role of the position-dependent interaction is twofold. First, stress energies are added to chemical bonding energies when calculating activation energies for transition probabilities. Second, after each event, the substrate and the layer arrange themselves to minimize the stress energy. In this way, transition probabilities become configuration-dependent, and *vice versa*, such that stressed configurations become unstable. Given the relative values of the Young modulus and shear modulus, it appears that bond-angle variations are 20 to 50 times larger than bond-length variations, in most common semiconductor materials. This creates an anistropy of deformations on a (100) surface along the bond direction. As a first approximation, it is assumed that displacements are limited toward the above direction.

Two types of defects are created after the deposition of the first layer: interstitials on the substrate with two dangling bonds, and vacancies in the deposited layer. Depending on local configuration, these defects might be stable or unstable, i.e., annihilated soon after their creation. In the case of rigid substrates, the first layer deposits coherently, showing that the critical thickness is beyond the first deposited layer, despite the lattice mismatch of 14.7% and the results obtained within a continuum model reported earlier in this section. Even when stable defects are created their density does not exceed one every 20 atoms, while the equilibrium density is one defect every seven atoms. This is due to the kinetics of the growth. In fact, two conditions must be fulfilled for the creation of a defect: high enough strain in the substrate and the existence of vacancies in the deposited layer. The first condition implies that the layer is near completion, while the second is best fulfilled at low coverages. This discrepancy limits the possibility of defect creation. The basic atomic-scale processes involved in heteroepitaxial growth certainly need further development.

19.9. Conclusion

After a brief review of thermodynamic treatments of crystal growth, we described the role of Monte Carlo techniques in the study of atomic-scale growth processes. Obviously, this technique has been of major importance in studying the kinetics of different processes, especially under nonsteady-state conditions. Furthermore, we showed how they take some account of the stochastic nature of a real experiment. It is clear that dealing with numerical methods, use of drastic approximations is not needed. Therefore one can approach experimental conditions, and even use simulations as a technological guide, to check the validity of analytical results.

However, it should be mentioned that the application of Monte Carlo techniques has only become possible with the advent of high-speed computers. Therefore, its further developments, especially toward the simulation of heteroepitaxial growths where collective motions of atoms have to be accounted for, would require the use of specialized computers or parallel computing techniques.

References

1. F. J. Grunthaner and A. Madhukar (eds.), Proc. First Int. Conf. on Metastable and Modulated Semiconductor Structures, Dec. 1982, Pasadena Calif., *J. Vac. Sci. Technol.* **B1** (1983).
2. B. A. Joyce, *Rep. Prog. Phys.* **37**, 363 (1974).
3. W. K. Burton, N. Cabrera, and F. C. Frank, *Phil. Trans. R. Soc. London, Ser. A* **243**, 299 (1951).
4. P. Bennema, *J. Cryst. Growth* **69**, 182 (1984).
5. E. Bauer, *Z. Kristallogr.* **110**, 372 (1958).
6. R. Kern, G. Le Lay, and J. J. Métois, in: *Current Topics in Materials Sciences* (E. Kaldis, ed.), Vol. 3, p. 139, North-Holland, Amsterdam (1979).
7. J. G. Dash, *Phys. Rev. B* **15**, 3136 (1977).
8. M. Volmer and A. Weber, *Z. Phys. Chem.* **119**, 227 (1926).
9. R. Becker and W. Doring, *Ann. Phys.* **24**, 719 (1935).
10. T. L. Hill, *Statistical Thermodynamics*, Addison Wesley, Mass. (1960).
11. H. A. Wilson, *Philos. Mag.* **50**, 238 (1900).
12. G. H. Gilmer and P. Bennema, *J. Appl. Phys.* **43**, 1347 (1971).
13. H. J. Leamy, G. H. Gilmer, and K. A. Jackson, in: *Surface Physics of Materials* (J. B. Blakely, ed.), Vol. I, Academic Press, New York (1975).
14. G. H. Gilmer and J. Q. Broughton, *J. Vac. Sci. Technol.* **B1**, 298 (1983).
15. J. Frenkel, *J. Phys. USSR* **9**, 392 (1945).
16. W. K. Burton and N. Cabrera, *Disc. Farad. Soc. No. 5*, **33**, 40 (1949).
17. G. H. Gilmer, R. Ghez, and N. Cabrera, *J. Cryst. Growth* **8**, 79 (1971).
18. G. H. Gilmer and H. H. Farrell, *J. Appl. Phys.* **47**, 3792 (1976); **47**, 4373 (1976).
19. F. C. Frank, *Disc. Faraday Soc. No. 5*, **48**, 67 (1949).
20. W. K. Burton, N. Cabrera, and F. C. Frank, *Nature* **163**, 398 (1949).
21. B. Van der Hoek, J. P. Van der Eerden, and P. Bennema, *J. Cryst. Growth* **56**, 108 (1982).
22. G. Zinsmeister, *Vacuum* **16**, 529 (1966).
23. G. Zinsmeister, *Thin Solid Films* **2**, 497 (1968).
24. D. Walton, *J. Chem. Phys.* **37**, 2182 (1962).
25. J. A. Venables, *Philos. Mag.* **27**, 693 (1973).
26. J. A. Venables and G. L. Price, in: *Epitaxial Growth* (J. W. Matthews, ed.), Academic, New York (1975).
27. J. A. Venables, G. D. T. Spiller, and M. Hanbücken, *Rep. Prog. Phys.* **47**, 399 (1984).
28. S. Stoyanov and D. Kashchiev, in: *Current Topics in Materials* (E. Kaldis, ed.), Vol. 7, p. 69 (1981).
29. S. V. Ghaisas and A. Madhukar, *Phys. Rev. Lett.* **56**, 1066 (1986).
30. N. Metropolis, A. W. Rosenbluth, M. N. Rosenbluth, A. Teller, and E. Teller, *J. Chem. Phys.* **21**, 1087 (1953).
31. J. P. Van der Eerden, P. Bennema, and T. A. Cherepanova, *Prog. Crystal Growth Charact.* **1**, 219 (1978).
32. J. D. Weeks, G. H. Gilmer, and K. A. Jackson, *J. Chem. Phys.* **65**, 712 (1976).
33. C. Van Leeuwen and F. H. Mischgofsky, *J. Phys. A*, **9**, 1827 (1976).
34. A. Trayanov and D. Kashchiev, *J. Cryst. Growth* **78**, 399 (1986).
35. D. Kashchiev, J. P. Van der Eerden, and C. Van Leeuwen, *J. Cryst. Growth* **40**, 47 (1977).
36. F. F. Abraham and G. M. White, *J. Appl. Phys.* **41**, 1841 (1970).
37. J. P. Chauvineau, *J. Cryst. Growth* **53**, 505 (1981).

38. A. Madhukar, *Surf. Sci.* **132**, 344 (1983).

39. A. Madhukar and S. V. Ghaisas, *Appl. Phys. Lett.* **47**, 247 (1985).

40. N. Tsai, F. F. Abraham, and G. M. Pound, *Surf. Sci.* **77**, 465 (1978).

41. C. Van Leeuwen and J. P. Van der Eerden, *Surf. Sci.* **64**, 237 (1977).

42. G. H. Gilmer, *J. Cryst. Growth* **49**, 465 (1980).

43. S. V. Ghaisas and A. Madhukar, *J. Vac. Sci. Technol.* **B3**, 540 (1985).

44. J. Singh and A. Madhukar, *J. Vac. Sci. Technol.* **20**, 716 (1982); **B1**, 305 (1983).

45. J. Singh and A. Madhukar, *Phys. Rev. Lett.* **51**, 794 (1983).

46. G. H. Gilmer, *Science* **208**, 355 (1980).

47. H. J. Lemy and G. H. Gilmer, *J. Cryst. Growth* **24/25**, 499 (1974).

48. A. Madhukar, T. C. Lee, M. Y. Yen, P. Chen, J. Y. Kim, S. V. Ghaisas, and P. G. Newman, *Appl. Phys. Lett.* **46**, 1148 (1985).

49. B. F. Lewis, F. J. Grunthaner, A. Madhukar, T. C. Lee, and R. Fernandez, *J. Vac. Sci. Technol.* **B3**, 1317 (1985).

50. M. Y. Yen, T. C. Lee, P. Chen, and A. Madhukar, *J. Vac. Sci. Technol.* **B4**, 590 (1986).

51. F. Voillot, A. Madhukar, J. Y. Kim, P. Chen, N. M. Cho, W. C. Tang, and P. G. Newman, *Appl. Phys. Lett.* **48**, 1009 (1986).

52. P. Chen, J. Y. Kim, A. Madhukar, and N. M. Cho, *J. Vac. Sci. Technol.* **B4**, 890 (1986).

53. A. Madhukar, P. Chen, F. Voillot, M. Thomsen, J. Y. Kim, W. C. Tang, and S. V. Ghaisas, *J. Cryst. Growth* **81**, 26 (1987).

54. R. H. Swendsen, P. J. Kortman, D. P. Landau, and H. Müller-Krumbhaar, *J. Cryst. Growth* **35**, 73 (1976).

55. G. H. Gilmer, *J. Cryst. Growth* **35**, 15 (1976).

56. A. E. Michaels, G. M. Pound, and F. F. Abraham, *J. Appl. Phys.* **45**, 9 (1974).

57. J. P. Van der Eerden, R. L. Kalf, and C. Van Leeuwen, *J. Cryst. Growth* **35**, 241 (1976).

58. U. Bertoci, *J. Electrochem. Soc.* **119**, 822 (1972).

59. J. R. Arthur, *J. Appl. Phys.* **37**, 3057 (1966); **39**, 4032 (1968).

60. J. R. Arthur, *Surf. Sci.* **43**, 449 (1974).

61. A. Y. Cho, *J. Appl. Phys.* **41**, 2780 (1970); **42**, 2074 (1971).

62. C. T. Foxon, M. R. Boudary, and B. A. Joyce, *Surf. Sci.* **44**, 69 (1974).

63. C. T. Foxon and B. A. Joyce, *Surf. Sci.* **50**, 434 (1975); **64**, 293 (1977).

64. J. Singh and K. K. Bajaj, *J. Vac. Sci. Technol.* **B2**, 276 (1984); **B2**, 576 (1984); **B3**, 520 (1985).

65. M. Thomsen and A. Madhukar, *J. Cryst. Growth* **80**, 275 (1987).

66. T. C. Lee, M. Y. Yen, P. Chen, and A. Madhukar, *J. Vac. Sci. Technol.* **A4** 884 (1986).

67. J. M. Ziman, *Principles of the Theory of Solids*, Cambridge University Press (1964).

68. T. Halicioglu, *Phys. Status Solidi B* **99**, 347 (1980).

69. E. Pearson, T. Takai, T. Halicioglu, and W. A. Tiller, *J. Cryst. Growth* **70**, 33 (1984).

70. W. A. Tiller, *J. Cryst. Growth* **70**, 13 (1984).

71. T. Takai, T. Halicioglu, and W. A. Tiller, *Surf. Sci.* **164**, 341 (1985).

72. F. Stillinger and T. Weber, *Phys. Rev. B* **31**, 5262 (1985).

73. R. Biswas and D. R. Hamann, *Phys. Rev. Lett.* **55**, 2001 (1985).

74. D. K. Choi, T. Takai, S. Erkoc, T. Halicioglu, and W. A. Teller, *J. Cryst. Growth* **85**, 9 (1987).

75. H. Balamane, T. Halicioglu, and W. A. Tiller, *J. Cryst. Growth* **85**, 16 (1987).

76. B. M. Axilrod and E. Teller, *J. Chem. Phys.* **11**, 299 (1943).

77. J. M. Chavazas, A. Bonissent, and B. Mutaftschiev, *J. Cryst. Growth* **76**, 9 (1986).

78. A. Kobayashi, S. M. Paik, K. E. Khor, and S. Das Sarma, *Surf. Sci.* **174**, 48 (1986).

79. A. Kobayashi, S. M. Paik, and S. Das Sarma, *J. Vac. Sci. Technol.* **B4**, 884 (1986).

80. M. J. P. Musgrave and J. A. Pople, *Proc. R. Soc. London. Ser. A* **268**, 474 (1962).

81. P. N. Keating, *Phys. Rev.* **145**, 637 (1966).

82. R. M. Martin, *Phys. Rev. B* **1**, 4005 (1970).

83. V. Burket and N. L. Allinger, *Molecular Mechanics*, ACS Monographs 177 (1982).

84. T. Fukui, *J. Appl. Phys.* **57**, 5188 (1985).

85. M. Ichimura and A. Sasaki, *Jap. J. Appl. Phys.* **25**, 976 (1986).

86. F. C. Frank and J. H. Van der Merwe, *Proc. R. Soc. London, Ser. A* **198**, 205, 216 (1949).

87. J. H. Van der Merwe, *J. Appl. Phys.* **34**, 117 (1963).
88. C. A. B. Ball, *Phys. Status Solidi* **42**, 357 (1970).
89. J. H. Van der Merwe, *Surf. Sci.* **31**, 198 (1972).
90. J. W. Matthews, *Epitaxial Growth*, Academic Press, New York (1975).
91. K. Nishitani, K. Okhata, and T. Murotani, *J. Electron. Mater.* **12**, 619 (1983).
92. P. P. Chow, D. K. Greenlaw, and D. Johnson, *J. Vac. Sci. Technol.* **A1**, 562 (1983).
93. H. A. Mar, N. Salansky, and K. T. Chee, *Appl. Phys. Lett.* **44**, 898 (1984).
94. C. J. Summers, E. L. Mecks, and N. W. Cox, *J. Vac. Sci. Technol.* **B2**, 224 (1984).
95. G. Cohen Solal, F. Bailly, and M. Barbe, *Appl. Phys. Lett.* **49**, 1519 (1986).
96. N. Otsuka, L. A. Kolodziejski, R. L. Gunshor, S. Datta, R. N. Bicknell, and J. F. Schetzina, *Appl. Phys. Lett.* **46**, 860 (1985).
97. D. Estève, M. Djafari Rouhani, V. V. Pham, A. Amrani, and J. J. Simonne, Proc. SPIE 88 Conf. Advances in Semiconductor Physics and Device Applications, New Port Beach, Calif., 13–18 March (1988).

Index